THIRD EDITION

FUNDAMENTALS OF ORGANIC CHEMISTRY

John McMurry
Cornell University

BROOKS/COLE PUBLISHING COMPANY
Pacific Grove, California

Brooks/Cole Publishing Company
A Division of Wadsworth, Inc.

Printed in the United States of America
10 9 8 7 6 5

LIBRARY OF CONGRESS CATALOGING-IN-PUBLICATION DATA

McMurry, John.
Fundamentals of organic chemistry / John McMurry. — Ed. 3.
p. cm.
Includes index.
ISBN 0-534-21210-7
1. Chemistry, Organic. I. Title.
QD251.2.M4 1994 93-25587
547—dc20 CIP

Sponsoring Editor: Lisa J. Moller
Editorial Assistant: Beth Wilbur
Production Editor: Phyllis Niklas
Production Coordinator: Joan Marsh
Manuscript Editor: Phyllis Niklas
Interior Design: Janet Bollow
Cover Design: Vernon T. Boes
Cover Photo: Shattil/Rozinski—Stock Imagery, Inc.
Photo Researcher: Stuart Kenter
Typesetting: Jonathan Peck Typographers
Cover Printing: Lehigh Press
Printing and Binding: R. R. Donnelley & Sons

COMPOUND TYPE	FUNCTIONAL GROUP	SIMPLE EXAMPLE	NAME ENDING
Ester	$-\overset{\overset{\displaystyle O}{\|}}{C}-O-\overset{/}{\underset{\backslash}{C}}-$	$CH_3\overset{\overset{\displaystyle O}{\|}}{C}-OCH_2CH_3$ Ethyl ethanoate (ethyl acetate)	*-oate*
Carboxylic acid	$-\overset{\overset{\displaystyle O}{\|}}{C}-O-H$	$CH_3CH_2CH_2\overset{\overset{\displaystyle O}{\|}}{C}-OH$ Butanoic acid	*-oic acid*
Carboxylic acid chloride	$-\overset{\overset{\displaystyle O}{\|}}{C}-Cl$	$CH_3CH_2\overset{\overset{\displaystyle O}{\|}}{C}-Cl$ Propanoyl chloride	*-yl chloride*
Amide	$-\overset{\overset{\displaystyle O}{\|}}{C}-N\langle$	$CH_3\overset{\overset{\displaystyle O}{\|}}{C}-NH_2$ Ethanamide	*-amide*
Amine	$-\overset{\backslash}{\underset{/}{C}}-N\langle$	$CH_3CH_2NH_2$ Ethylamine	*-amine*
Nitrile	$-C\equiv N$	$CH_3C\equiv N$ Ethanenitrile (acetonitrile)	*-nitrile*
Nitro	$-\overset{\backslash}{\underset{/}{C}}-\overset{+}{N}(=O)O^-$	$CH_3CH_2NO_2$ Nitroethane	None
Organometallic	$-\overset{\backslash}{\underset{/}{C}}-M$ M = metal	CH_3-Li Methyllithium	None

THIRD EDITION

FUNDAMENTALS OF ORGANIC CHEMISTRY

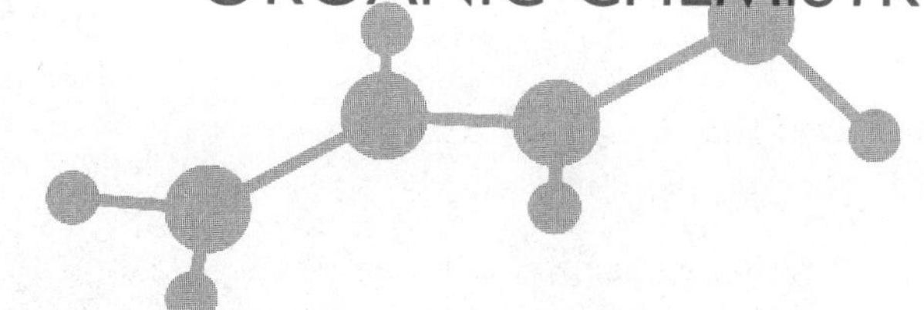

BRIEF CONTENTS

CONTENTS

7 ALKYL HALIDES 199

8 ALCOHOLS, ETHERS, AND PHENOLS 232

9 ALDEHYDES AND KETONES: NUCLEOPHILIC ADDITION REACTIONS 260

10 CARBOXYLIC ACIDS AND DERIVATIVES 285

11 CARBONYL ALPHA-SUBSTITUTION REACTIONS AND CONDENSATION REACTIONS 325

12 AMINES 350

13 STRUCTURE DETERMINATION 378

PREFACE

I wrote in the first edition of this text that my goal was to produce a readable and effective teaching text—one that presents only those subjects needed for a brief course in organic chemistry but that keeps the important pedagogical tools commonly found in larger books. Explanations are clear, the artwork is carefully done, important points are repeated, and varied end-of-chapter learning tools are used. The result, I believe, is a book that is easier to read and learn from than other short organic chemistry texts.

All the features that made the first two editions a success have been improved, and new ones have been added in this third edition. Among the changes:

- Full color has been added to Chapter 6 on stereochemistry and to Chapters 14–17 on biomolecules.
- A new chapter has been added. Titled "The Organic Chemistry of Metabolic Pathways," Chapter 17 ties together many common laboratory organic reactions with their biochemical counterparts.
- The writing, already clear and accessible, has been further refined at the sentence level on every page.
- Problem sets have been expanded, and many new drill problems have been added.
- Numerous reactions and reagents judged too complex for a course at this level have been removed from this edition.
- Review material on bonding, electronegativity, and acid–base chemistry has been added to Chapter 1.
- New "interludes" on toxicity and risk, ethanol, and magnetic resonance imaging have been added.

Organization

The primary organization of this book is by functional group, beginning in Chapter 2 with alkanes and going on to more complex compounds.

Within this primary organization, more emphasis is placed on explaining the fundamental similarities of organic reactions than is common in other short texts. Chapter 11, "Carbonyl Alpha-Substitution Reactions and Condensation Reactions," for example, helps to remove the artificial lines between ketones and esters by showing how all carbonyl compounds undergo similar reactions. Memorization is minimized and understanding maximized with this approach.

Spectroscopy

Spectroscopy is treated as a tool, not as a specialized field of study. Infrared, ultraviolet, ^{13}C NMR, and ^{1}H NMR spectroscopies are all covered by showing the kind of information that can be derived from each and how each can be used to answer specific structural questions.

Nomenclature

The IUPAC system of nomenclature is used throughout. For the most part, this involves the use of systematic names, although a few IUPAC-approved nonsystematic names such as acetic acid, acetone, ethylene, and phenol are also employed. Since it's unlikely that these few common names will disappear from everyday use in the near future, it's probably best for students to learn them.

Coverage

The coverage in this book is up-to-date, reflecting important advances of the past decade. For example, ^{13}C NMR is introduced as a routine spectroscopic tool, equal in importance to ^{1}H NMR. Similarly, the chemistry of nucleic acids is covered, including a section on DNA sequencing by the Maxam–Gilbert method.

Interludes

Brief "interludes" are included at the end of each chapter. Meant to serve as short breathers between chapters, these interludes show interesting applications of organic chemistry to industrial and biological systems. They can be covered by the instructor or left for student reading.

Practice Problems

Each chapter contains many worked-out examples that illustrate how problems can be solved. Each practice problem and solution is then followed by a similar problem for the reader to solve. These worked-out examples are valuable because of their appearance in the text, but are not meant to serve as a replacement for the accompanying *Study Guide and Solutions Manual.*

Pedagogy

In addition to the above features, every effort has been made to make this book as effective, clear, and readable as possible—in short, to make it easy to learn from:

- Paragraphs start with summary sentences.
- Transitions between paragraphs and between topics are smooth.
- Extensive use is made of computer-generated, three-dimensional art and carefully rendered stereochemical formulas.
- Extensive cross-referencing to earlier material ties ideas together.
- A second color is used to indicate the changes that occur during reactions, and full color is used for clarity in the chapters on stereochemistry and biomolecules.
- More than 900 problems are included, both within the text and at the end of every chapter. These include both drill and thought problems.
- Key terms are defined in the margin next to where they first appear in the text.
- An innovative vertical format is used to explain reaction mechanisms. The mechanisms are printed vertically, while explanations of the changes taking place in each step are printed next to the reaction arrow. This format allows the reader to see easily what is occurring at each step in a reaction without having to jump back and forth between the text and structures.

Study Guide and Solutions Manual

A carefully prepared *Study Guide and Solutions Manual* accompanies this text. Written by Susan McMurry, this companion volume answers all in-text and end-of-chapter problems and explains in detail how answers are obtained. In addition, many valuable supplemental materials are given, including a list of study goals for each chapter, a glossary, a summary of name reactions, a summary of organic reaction mechanisms, a summary of the uses of important reagents, tables of spectroscopic information, and a list of suggested readings.

ACKNOWLEDGMENTS

I sincerely thank the many people whose help and suggestions were so valuable in the creation of this book. Foremost is my wife Susan who read, criticized, and improved all aspects of the text, and who authored the accompanying *Study Guide and Solutions Manual*. Among the reviewers providing thoughtful comments were Claudia P. Cartaya, Appalachian State University—Rainking Science Center; Mildred V. Hall, Pennsylvania State University—Dubois Campus; John A. Miller, Western Washington University; David Minter, Texas Christian University; Roger K. Murray, University of Delaware; George V. Odell, Oklahoma State University; Stanley Raucher, University of Washington; David J. Rislove, Winona State University; Ronald Starkey, University of Wisconsin; and Kathleen M. Trahanovsky, Iowa State University.

Special thanks are due Harvey Pantzis, Lisa Moller, Joan Marsh, Phyllis Niklas, Kathy Lee, and all of the Brooks/Cole staff for their usual fine work.

A NOTE FOR STUDENTS

We have similar goals. Yours is to learn organic chemistry; mine is to do everything possible to help you learn. It's going to require work on your part, but the following suggestions should prove helpful:

Don't read the text immediately. As you begin each new chapter, look it over first. Read the introductory paragraphs, find out what topics will be covered, and then turn to the end of the chapter and read the summary. You'll be in a much better position to understand new material if you first have a general idea of where you're heading. Once you've begun a chapter, read it several times. First read the chapter rapidly, making checks or comments in the margin next to important or difficult points; then return for an in-depth study.

Keep up with the material. Who's likely to do a better job—the runner who trains five miles per day for weeks before a race, or the one who suddenly trains twenty miles the day before the race? Organic chemistry is a subject that builds on previous knowledge. You have to keep up with the material on a daily basis.

Work the problems. There are no shortcuts here. Working problems is the only way to learn organic chemistry. The practice problems show you how to approach the material, the in-text problems provide immediate practice, and the end-of-chapter problems provide additional drill and some real challenges. Answers and explanations for all problems are given in the accompanying *Study Guide and Solutions Manual.*

Ask questions. Faculty members and teaching assistants are there to help you. Most of them will turn out to be extremely helpful and genuinely interested in seeing you learn.

Use molecular models. Organic chemistry is a three-dimensional science. Although this book uses many careful drawings to help you visualize molecules, there's no substitute for building a molecular model, turning it in your hands, and looking at it from different views.

Use the study guide. The *Study Guide and Solutions Manual* that accompanies this text gives complete solutions to all problems and provides a wealth of supplementary material. Included are a list of study goals for each chapter, outlines of each chapter, a large glossary, a summary of name reactions, a summary of methods for preparing functional groups, a summary of the uses of important reagents, and tables of spectroscopic information. Find out ahead of time what's there so that you'll know where to go when you need help.

Good luck. I sincerely hope you enjoy learning organic chemistry and that you come to see the logic and beauty of its structure. I would be glad to receive comments and suggestions from any who have learned from this book.

THIRD EDITION

FUNDAMENTALS OF ORGANIC CHEMISTRY

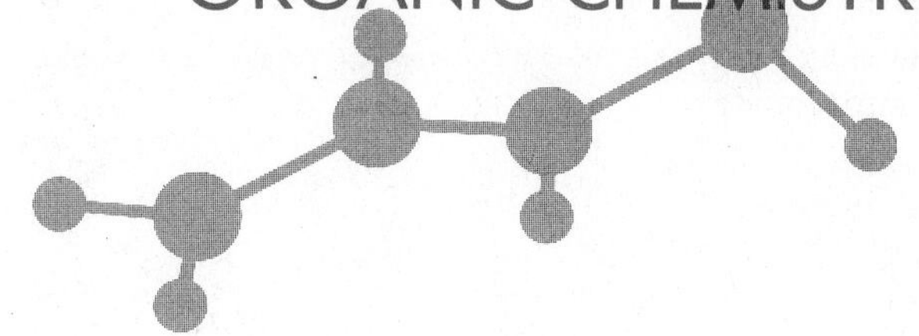

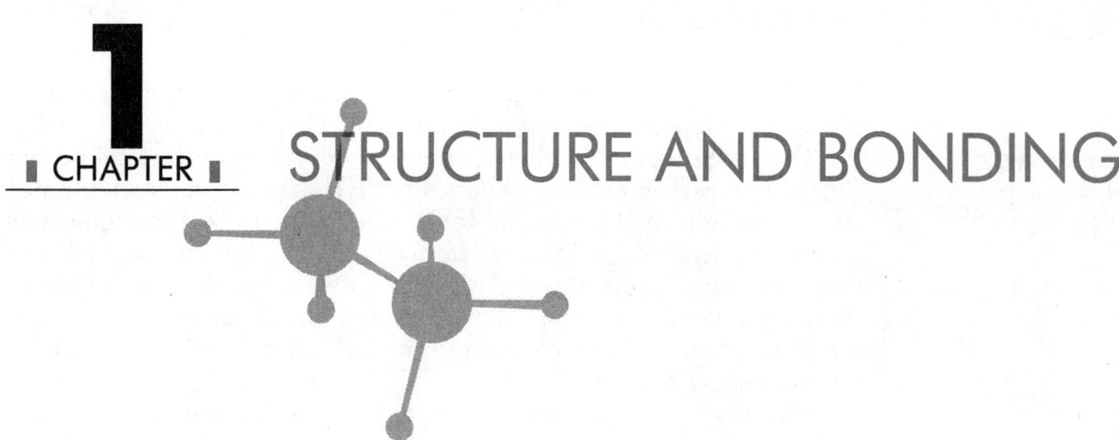

CHAPTER 1 STRUCTURE AND BONDING

What is organic chemistry, and why should you study it? The answers are everywhere. Every living organism is made of organic chemicals. The foods you eat; the medicines you take; the wood, paper, plastics, and fibers that make modern life possible, are all organic chemicals. Anyone with a curiosity about life and living things must have a fundamental understanding of organic chemistry.

The roots of organic chemistry can be traced to the mid-1700s when alchemists noticed unexplainable differences between compounds derived from living sources and those derived from minerals. Compounds from plants and animals were often difficult to isolate and purify. Even when pure, they were difficult to work with and tended to decompose more easily than compounds from minerals. The Swedish chemist Torbern Bergman was the first person to express this difference between "organic" and "inorganic" substances, and the term *organic chemistry* soon came to mean the chemistry of compounds from living organisms.

To many chemists of the time, the only explanation for the difference in behavior between organic and inorganic compounds was that organic compounds contained an undefinable "vital force" as a result of their origin in living sources. With time, however, it became clear that organic compounds could be manipulated in the laboratory just like inorganic compounds. Friedrich Wöhler discovered in 1828, for example, that it was possible to convert the "inorganic" salt ammonium cyanate into the known "organic" substance urea.

$$NH_4^+ \ ^-OCN \xrightarrow{\text{Heat}} H_2N-\overset{\overset{\large O}{\|}}{C}-NH_2$$

Ammonium cyanate → **Urea**

By the mid-1800s, the weight of evidence was against the vitalistic theory, and it had become clear that the same basic scientific principles are applicable to all compounds. The only distinguishing characteristic of organic compounds is that all contain the element carbon.

ORGANIC CHEMISTRY
Chemistry of the compounds of carbon

Organic chemistry, then, is the study of the compounds of carbon. But why is carbon special? What is it that sets carbon apart from all other elements in the periodic table? The answers to these questions derive from the unique ability of carbon atoms to bond together, forming rings and long chains. Carbon, alone of all elements, is able to form an immense diversity of compounds, from the simple to the staggeringly complex—from methane, containing one carbon atom, to DNA, which can contain tens of *billions*.

Not all organic compounds are derived from living organisms, of course. Modern chemists are enormously sophisticated in their ability to synthesize new organic compounds in the laboratory. Medicines, dyes, polymers, plastics, pesticides, and a host of other organic substances are all prepared in the laboratory. Organic chemistry is a subject that touches the lives of everyone. Its study can be a fascinating undertaking.

1.1 ATOMIC STRUCTURE

Before beginning a study of organic chemistry, let's review some general ideas about atoms and bonds. Atoms consist of a dense, positively charged *nucleus* surrounded at a relatively large distance by negatively charged electrons (Figure 1.1). The nucleus consists of subatomic particles called *neutrons*, which are electrically neutral, and *protons*, which are positively charged. Though extremely small—about 10^{-14} to 10^{-15} meter (m) in diameter—the nucleus nevertheless contains essentially all the mass of the atom. Electrons have negligible mass and orbit the nucleus at a distance of approximately 10^{-10} m. Thus, the diameter of a typical atom is about 2×10^{-10} m, or 2 *angstroms* (Å), where 1 Å = 10^{-10} m. To give you an idea of how small this is, a thin pencil line is about 3 *million* carbon atoms wide.

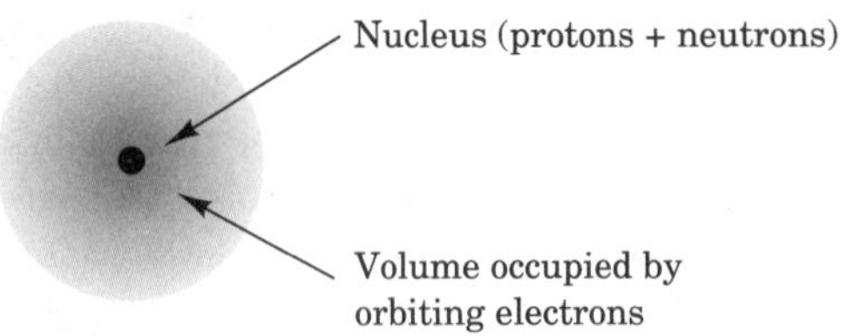

FIGURE 1.1 A schematic view of an atom. The dense, positively charged nucleus contains most of the atom's mass and is surrounded by negatively charged electrons.

ATOMIC NUMBER
Number of protons in an atom's nucleus

An atom is described by its **atomic number**, which is the number of protons in the atom's nucleus, and its **mass number**, which is the total of protons plus neutrons. All the atoms of a given element have the same

MASS NUMBER
Total number of protons and neutrons in an atom's nucleus

ATOMIC WEIGHT
Average mass of a large number of atoms of an element

ELECTRON SHELL
Imaginary layer around the nucleus occupied by electrons

atomic number (1 for hydrogen, 6 for carbon, 17 for chlorine, and so on), but they can have different mass numbers, depending on how many neutrons they contain. The *average* mass number of a great many atoms of an element is called the element's **atomic weight**. Because it's an average, the atomic weight often is not an integer: 1.008 for hydrogen, 12.011 for carbon, 35.453 for chlorine, and so on.

How are the electrons distributed in an atom? It turns out that electrons aren't completely free to move about, but are confined to different regions within the atom according to the amount of energy they have. Electrons may be thought of as belonging to different layers, or **shells**, around the nucleus. The larger the shell, the more electrons it can hold and the greater the energies of those electrons. Thus, an atom's lowest-energy electrons occupy the first shell, which is nearest the nucleus and can hold only 2 electrons. The second shell can hold 8 electrons, the third shell can hold 18 electrons, and so on, as shown in Table 1.1.

TABLE 1.1 Distribution of Electrons into Shells

NUMBER OF SHELL	ELECTRON CAPACITY OF SHELL	
		Higher energy
Fourth	32	↑
Third	18	
Second	8	
First	2	
		Lower energy

ORBITAL
Region of space occupied by a given electron or pair of electrons

Within each shell, electrons are further grouped into *subshells*, denoted *s*, *p*, *d*, and *f*, and within each subshell, electrons are grouped by pairs into **orbitals**. It's helpful to think of an orbital as a time-lapse photograph of an electron's movements around a nucleus. Such a photograph would show the orbital as a blurry cloud indicating the region of space surrounding the nucleus where the electron has been. This electron cloud doesn't have a sharp boundary, but for practical purposes we can set the limits by saying that an orbital represents the space where an electron spends most (90–95%) of its time.

What do orbitals look like? The shape of an orbital depends on what kind of subshell it's in, whether *s*, *p*, *d*, or *f*. Of the four, we'll be concerned only with *s* and *p* orbitals because most atoms found in living organisms use only these. The *s* orbitals have a spherical shape with the nucleus at the center, and *p* orbitals have a dumbbell shape, as shown in Figure 1.2. Note that a given shell contains three different *p* orbitals, oriented in space so that each is perpendicular to the other two. They are arbitrarily denoted p_x, p_y, and p_z.

Different shells have different numbers and kinds of orbitals. As indicated in Figure 1.3, the 2 electrons of the first shell occupy a single *s* orbital, designated 1*s*. The 8 electrons of the second shell occupy one *s* orbital (designated 2*s*) and three *p* orbitals (each designated 2*p*). The 18 electrons of the third shell occupy one *s* orbital (3*s*), three *p* orbitals (3*p*), and five *d* orbitals (3*d*).

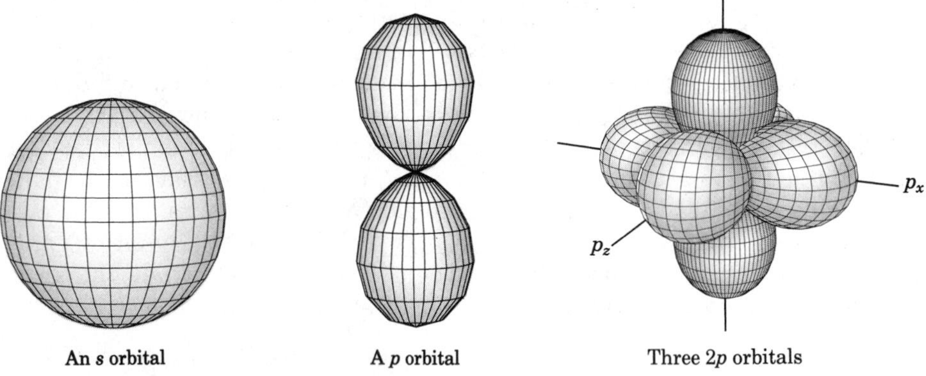

FIGURE 1.2 Computer-generated shapes of *s* and *p* orbitals. There are three *p* orbitals, denoted p_x, p_y, and p_z.

Energy

3rd shell (*capacity*—18 electrons)
3*d* ⇅ ⇅ ⇅ ⇅ ⇅
3*p* ⇅ ⇅ ⇅
3*s* ⇅

2nd shell (*capacity*—8 electrons)
2*p* ⇅ ⇅ ⇅
2*s* ⇅

1st shell (*capacity*—2 electrons)
1*s* ⇅

FIGURE 1.3 The distribution of electrons in an atom. The first shell holds a maximum of 2 electrons in one 1*s* orbital; the second shell holds a maximum of 8 electrons in one 2*s* and three 2*p* orbitals; the third shell holds a maximum of 18 electrons in one 3*s*, three 3*p*, and five 3*d* orbitals; and so on. The 2 electrons in each orbital are represented by up and down arrows ⇅. Although not shown, the 4*s* orbital has an energy level between 3*p* and 3*d*.

1.2 ELECTRONIC CONFIGURATION OF ATOMS

GROUND-STATE ELECTRONIC CONFIGURATION
Lowest-energy arrangement of electrons in an atom

The lowest-energy arrangement, or **ground-state electronic configuration**, of an atom is a description of the orbitals that the atom's electrons occupy. We can determine an atom's ground-state electronic configuration by following three rules:

RULE 1 The orbitals of lowest energy are filled first, according to the order 1*s* → 2*s* → 2*p* → 3*s* → 3*p* → 4*s* → 3*d*, as shown in Figure 1.3.

RULE 2 Only two electrons can occupy an orbital, and they must be of opposite spin.[1]

RULE 3 If two or more empty orbitals of equal energy are available, one electron is placed in each with spins parallel until all are half-full.

Some examples of how these rules are applied are shown in Table 1.2. Hydrogen, the lightest element, has only one electron, which must occupy the lowest-energy orbital. Thus, hydrogen has a 1*s* ground-state electronic configuration.[2] Carbon has six electrons and the ground-state electronic configuration $1s^2\ 2s^2\ 2p^2$.

TABLE 1.2 Ground-State Electronic Configuration of Some Elements

ELEMENT	ATOMIC NUMBER	CONFIGURATION	ELEMENT	ATOMIC NUMBER	CONFIGURATION
Hydrogen	1	1*s* ↑	Neon	10	2*p* ↑↓ ↑↓ ↑↓ 2*s* ↑↓ 1*s* ↑↓
Lithium	3	2*s* ↑ 1*s* ↑↓			
Carbon	6	2*p* ↑ ↑ — 2*s* ↑↓ 1*s* ↑↓	Chlorine	17	3*p* ↑↓ ↑↓ ↑ 3*s* ↑↓ 2*p* ↑↓ ↑↓ ↑↓ 2*s* ↑↓ 1*s* ↑↓

PRACTICE PROBLEM 1.1 Give the ground-state electronic configuration of nitrogen.

SOLUTION The periodic table on the rear inside cover shows that nitrogen has atomic number 7 and thus has seven electrons. Using Figure 1.3 to find the relative energy levels of orbitals, and applying the three rules to assign the seven electrons to orbitals, the first two electrons go into the lowest-energy orbital ($1s^2$), the next two electrons go into the second-lowest-energy orbital ($2s^2$), and the remaining three electrons go into the three next lowest-energy orbitals ($2p^3$). Thus, the configuration of nitrogen is $1s^2\ 2s^2\ 2p^3$.

PROBLEM 1.1 How many electrons does each of these elements have in its outermost electron shell?
(a) Potassium **(b)** Calcium **(c)** Aluminum

PROBLEM 1.2 Give the ground-state electronic configuration of these elements:
(a) Boron **(b)** Phosphorus **(c)** Oxygen **(d)** Argon

[1]Electrons can be thought of as spinning on an axis in much the same way that the earth spins. This spin can have two orientations, denoted as up ↑ and down ↓.

[2]A superscript is used to represent the number of electrons at a particular energy level. For example, $1s^2$ indicates that there are two electrons in the 1*s* orbital. No superscript is used when there is only one electron in an orbital.

1.3 DEVELOPMENT OF CHEMICAL BONDING THEORY

By the mid-1800s, the new science of chemistry was developing rapidly, and chemists had begun to probe the forces holding molecules together. In 1858, August Kekulé and Archibald Couper independently proposed that, in all organic compounds, carbon always has four "affinity units." That is, carbon is *tetravalent*: It always forms four bonds when it joins other elements to form stable compounds. Furthermore, said Kekulé, carbon atoms can bond to each other to form extended chains, and chains can double back on themselves to form rings.

Although Kekulé was correct in describing the tetravalent nature of carbon, chemistry was still viewed in a two-dimensional way until 1874. In that year, Jacobus van't Hoff and Joseph Le Bel added a third dimension to our ideas about chemistry. They proposed that the four bonds of carbon are not randomly oriented but have a specific spatial orientation. Van't Hoff went even further and proposed that the four atoms to which a carbon atom is bonded sit at the corners of a tetrahedron, with carbon in the center. A representation of a tetrahedral carbon atom is shown in Figure 1.4. Note the conventions used to show three-dimensionality: Solid lines represent bonds in the plane of the paper, heavy wedged lines represent bonds coming out of the plane of paper toward the viewer, and dashed lines represent bonds receding into the plane away from the viewer. Such representations will be used throughout this text.

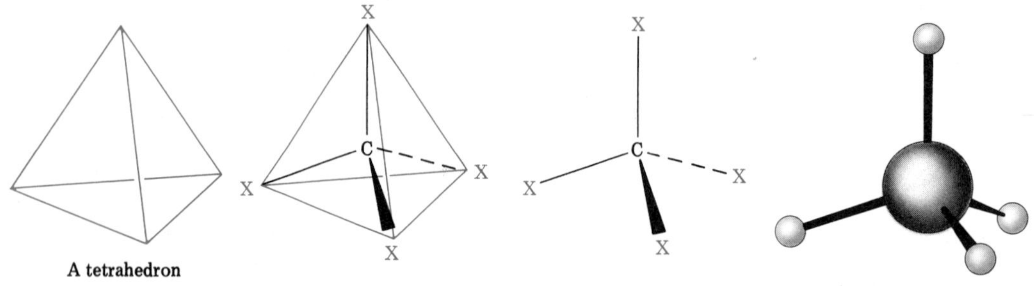

FIGURE 1.4 Van't Hoff's tetrahedral carbon atom. The solid lines are in the plane of the paper, the heavy wedged line comes out of the plane of the paper, and the dashed line goes back into the plane.

PROBLEM 1.3 Draw a molecule of chloromethane, CH_3Cl, using solid, wedged, and dashed lines to show its tetrahedral geometry.

1.4 THE NATURE OF CHEMICAL BONDS: IONIC BONDS

Why do atoms bond together, and how can bonds be described? The *why* question is relatively easy to answer: Atoms bond together because the product that results is more stable (has less energy) than the separate atoms themselves. Energy is always *released* and flows out of the system when chemical bonds are formed, just as water always flows downhill. The *how* question is more difficult. To answer it, we need to know more about the properties of atoms.

VALENCE SHELL
Outermost electron shell of an atom

We know through observation that eight electrons (an electron octet) in the outermost shell (the **valence shell**) impart special stability to the noble-gas elements in Group 8A: neon (2 + *8*), argon (2 + 8 + *8*), krypton (2 + 8 + 18 + *8*). We also know that the chemistry of many main-group elements is governed by a tendency for them to take on the stable noble-gas electronic makeup. The alkali metals in Group 1A, for example, have a single *s* electron in their outer shells. By losing this valence electron, they can achieve a noble-gas configuration. Such elements that tend to give up electrons easily are called **electropositive**.

ELECTROPOSITIVE
Element with a tendency to donate electrons

Just as electropositive metals on the left side of the periodic table tend to form positive ions by losing one or more electrons, the halogens (Group 7A elements) and other nonmetals on the right side of the periodic table tend to form *negative* ions by *gaining* one or more electrons. By so doing, these elements achieve noble-gas configurations. Such elements that tend to accept electrons are called **electronegative**.

ELECTRONEGATIVE
Element with a tendency to withdraw electrons

The simplest kind of chemical bonding is that between an electropositive element and an electronegative element. For example, when sodium metal (electropositive) reacts with chlorine gas (electronegative), sodium transfers an electron to chlorine to form Na^+ ions and Cl^- ions. When a vast number of sodium atoms transfer electrons to an equal number of chlorine atoms, a visible crystal of sodium chloride results. The NaCl product is said to have **ionic bonding**. That is, the Na^+ and Cl^- ions are held together by an electrical attraction between their unlike charges. Note, though, that there's no such thing as an individual NaCl "molecule," and we can't speak of specific ionic bonds between specific pairs of ions. Rather, there are many ionic bonds between an individual ion and its neighbors, and so we speak of the whole crystal as being an **ionic solid**.

IONIC BOND
Bond between ions due to the electrostatic attraction of unlike charges

IONIC SOLID
Crystal of positive and negative ions held together by ionic bonds

PROBLEM 1.4 How many outer-shell electrons do these elements have?
(a) Be **(b)** S **(c)** Br

PROBLEM 1.5 Judging from their positions in the periodic table, which is more electronegative?
(a) Potassium or oxygen **(b)** Calcium or bromine

1.5 THE NATURE OF CHEMICAL BONDS: COVALENT BONDS

We've just seen that elements at the left (sodium) and right (chlorine) of the periodic table form ionic bonds by gaining or losing electrons. How, though, do the elements in the middle of the periodic table form bonds? Look at methane, CH_4, the main constituent of natural gas, for example. The bonding in methane is not ionic because it would be very difficult for

COVALENT BOND
Bond formed by sharing electrons between two nuclei

MOLECULE
Group of atoms joined together by covalent bonds

LEWIS STRUCTURE
A way of representing a molecule using dots to indicate an atom's outer-shell electrons

carbon ($1s^2\ 2s^2\ 2p^2$) either to gain or lose *four* electrons to achieve a noble-gas configuration.[3] In fact, carbon bonds to other atoms, not by donating electrons, but by *sharing* them. Such shared-electron bonds, first proposed in 1916 by G. N. Lewis, are called **covalent bonds**. The neutral collection of atoms held together by covalent bonds is called a **molecule**.

A simple shorthand way of indicating covalent bonds in molecules is to use **Lewis structures**, or *electron-dot structures*, in which an atom's valence electrons are represented by dots. Thus, hydrogen has one dot ($1s$), carbon has four dots ($2s^2\ 2p^2$), oxygen has six dots ($2s^2\ 2p^4$), and so on. A stable molecule results when a valence octet of electrons has been achieved for all atoms in the molecule except hydrogen, as in the following examples:

$$4\,\text{H}\cdot + \cdot\dot{\underset{\cdot}{\text{C}}}\cdot \longrightarrow \text{H}:\underset{\text{H}}{\overset{\text{H}}{\ddot{\underset{\cdot\cdot}{\text{C}}}}}:\text{H}$$

Methane (CH_4)

$$3\,\text{H}\cdot + \cdot\dot{\underset{\cdot\cdot}{\text{N}}}\cdot \longrightarrow \text{H}:\overset{\text{H}}{\ddot{\underset{\cdot\cdot}{\text{N}}}}:\text{H}$$

Ammonia (NH_3)

$$2\,\text{H}\cdot + \cdot\ddot{\underset{\cdot}{\text{O}}}: \longrightarrow \text{H}:\underset{\text{H}}{\ddot{\underset{\cdot\cdot}{\text{O}}}}:$$

Water (H_2O)

$$3\,\text{H}\cdot + \cdot\dot{\underset{\cdot}{\text{C}}}\cdot + \cdot\ddot{\underset{\cdot}{\text{O}}}: + \text{H}\cdot \longrightarrow \text{H}:\underset{\text{H}}{\overset{\text{H}}{\ddot{\underset{\cdot\cdot}{\text{C}}}}}:\underset{\text{H}}{\ddot{\underset{\cdot\cdot}{\text{O}}}}:$$

Methanol (CH_3OH)

$$2\,\text{H}\cdot + \cdot\ddot{\underset{\cdot\cdot}{\text{O}}}\cdot + \text{H}^+ \longrightarrow \text{H}:\underset{\text{H}}{\overset{+}{\ddot{\underset{\cdot\cdot}{\text{O}}}}}:\text{H}$$

Hydronium ion (H_3O^+)

The number of covalent bonds an atom forms depends on how many valence electrons it has. Atoms with one, two, or three valence electrons form one, two, or three bonds, respectively, but atoms with four or more valence electrons form as many bonds as they need electrons to fill the s and p levels of their valence shells and thereby reach a stable octet. Thus, boron has three valence electrons ($2s^2\ 2p^1$) and forms three bonds, as in BH_3; carbon ($2s^2\ 2p^2$) fills its valence shell by forming four bonds, as in CH_4; nitrogen ($2s^2\ 2p^3$) forms three bonds, as in NH_3; and oxygen ($2s^2\ 2p^4$) forms two bonds, as in H_2O.

One bond	Two bonds	Three bonds	Four bonds
H— Cl— Br— F—	—O—	—N(—)— —B(—)—	—C(—)(—)—

[3]The electronic configuration of carbon can be written either as $1s^22s^22p^2$ or as $1s^22s^22p_x2p_y$. Both notations are correct, but the latter is more informative because it indicates that two of the three equivalent p orbitals are half filled.

NONBONDING ELECTRON Valence electron not used for bonding

LONE-PAIR ELECTRONS Pair of nonbonding electrons

Those valence electrons that are not used for bonding are called **nonbonding**, or **lone-pair, electrons**. The nitrogen atom in ammonia, for instance, shares six of its eight valence electrons in three covalent bonds with hydrogens, and has its remaining two valence electrons in a nonbonding lone pair.

Nonbonding, lone-pair electrons → $H:\ddot{N}:H$ (with H above and below N) or H—N(—H)—H

Ammonia

KEKULÉ STRUCTURE Representation of a molecule that indicates a covalent bond as a line between atoms

LINE-BOND STRUCTURE Alternative name for Kekulé structure

Lewis structures are valuable because they make electron bookkeeping possible and remind us of the number of valence electrons present. Simpler still is the use of **Kekulé structures**, or **line-bond structures**. In a line-bond structure, the two electrons in a covalent bond are indicated simply by a line. Lone pairs of nonbonding valence electrons are often ignored when drawing line-bond structures, but it's still necessary to keep them in mind. Some examples are shown in Table 1.3.

TABLE 1.3 Lewis and Kekulé Structures of Some Simple Molecules

NAME	LEWIS STRUCTURE	KEKULÉ STRUCTURE	NAME	LEWIS STRUCTURE	KEKULÉ STRUCTURE
Water (H_2O)	$H:\ddot{\underset{..}{O}}:H$	H—O—H	Methane (CH_4)	H:C:H (H above and below C)	H—C—H (H above and below C)
Ammonia (NH_3)	H:N:H (H above N, lone pair below)	H—N—H (H above N)	Methanol (CH_3OH)	H:C:Ö:H (H above and below C)	H—C—O—H (H above and below C)

PRACTICE PROBLEM 1.2 How many hydrogen atoms does phosphorus bond to in forming phosphine, $PH_?$?

SOLUTION Because phosphorus is in Group 5A of the periodic table, it has five valence electrons. It needs to share three more electrons to make an octet, and it therefore bonds to three hydrogen atoms, giving PH_3.

PRACTICE PROBLEM 1.3 Draw a Lewis structure for chloromethane, CH_3Cl.

SOLUTION Hydrogen has one valence electron, carbon has four valence electrons, and chlorine has seven valence electrons. Thus, chloromethane is represented as

H:C:C̈l: (H above and below C) **Chloromethane**

PROBLEM 1.6 What are the likely formulas of these molecules?
(a) $CCl_?$ (b) $AlH_?$ (c) $CH_?Cl_2$ (d) $SiF_?$

PROBLEM 1.7 Write both Lewis and line-bond structures for these molecules, showing all nonbonded electrons:
(a) $CHCl_3$, chloroform (b) H_2S, hydrogen sulfide
(c) CH_3NH_2, methylamine

PROBLEM 1.8 Which of these substances are likely to have covalent bonds and which ionic bonds?
(a) CH_4 (b) CH_2Cl_2 (c) LiI (d) KBr (e) $MgCl_2$ (f) Cl_2

PROBLEM 1.9 Write both a Lewis structure and a line-bond structure for ethane, C_2H_6.

1.6 FORMATION OF COVALENT BONDS

The simplest way to picture the formation of a covalent bond is to imagine an *overlapping* of two atomic orbitals, each of which contains one electron. For example, we can picture the hydrogen molecule (H–H) by imagining what might happen if two hydrogen atoms, each with one electron in its atomic 1*s* orbital, come together. As the two spherical atomic orbitals approach and combine, a new, egg-shaped H–H orbital results. The new orbital is filled by two electrons, one donated by each hydrogen:

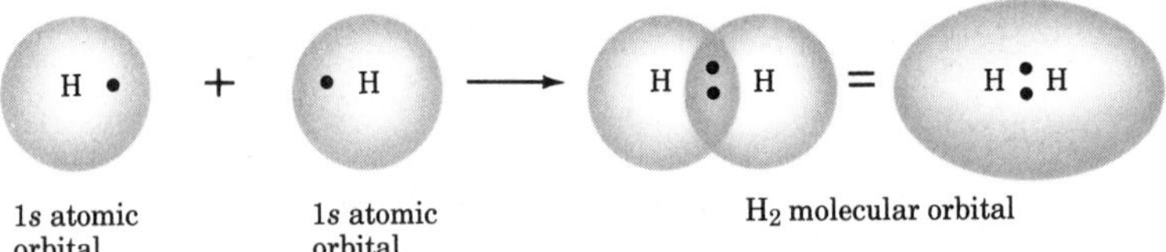

1*s* atomic orbital 1*s* atomic orbital H_2 molecular orbital

BOND STRENGTH Amount of energy needed to break a covalent bond

During the reaction 2 H· → H_2, 104 kcal/mol (435 kJ/mol) of energy is released.[4] Because the product H_2 molecule has 104 kcal/mol less energy than the starting 2 H·, we say that the product is more stable than the starting material and that the new H–H bond has a **bond strength** of 104 kcal/mol. In other words, we would have to put 104 kcal/mol of energy (heat) *into* the H–H bond to break the H_2 molecule into two H atoms.

104 kcal/mol released

H· + H· ⇌ H—H

104 kcal/mol absorbed

[4]Organic chemists still prefer to use kilocalories (kcal) as a measure of energy rather than kilojoules (kJ), the SI unit. This book will generally show values in both units. The conversion factor is: 1 kcal = 4.184 kJ.

BOND LENGTH
Distance between atoms in a covalent bond

How close are the two nuclei in the hydrogen molecule? If they're too close, they will repel each other since both are positively charged, yet if they're too far apart, they won't be able to share the bonding electrons. Thus, there is an optimum distance between nuclei that leads to maximum bond stability, a distance called the **bond length** (Figure 1.5). In the hydrogen molecule, the bond length is 0.74 Å. Every covalent bond has both a characteristic bond strength and bond length.

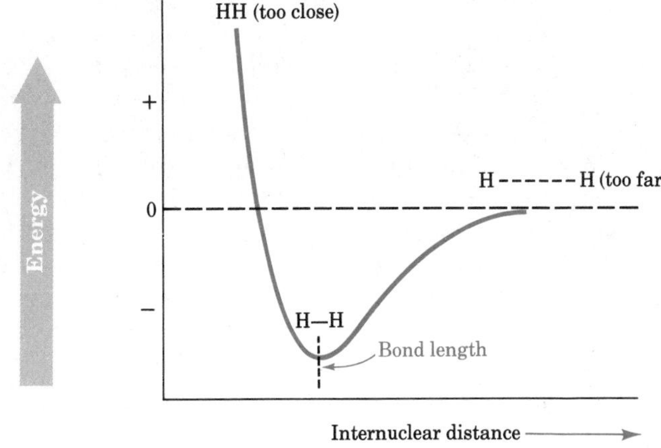

FIGURE 1.5 A plot of energy versus internuclear distance for two hydrogen atoms. The distance at the minimum energy point is called the bond length.

SIGMA (σ) BOND
Covalent bond formed by head-on overlap of atomic orbitals

The orbital in the hydrogen molecule has the elongated egg shape that we might get by pressing two spheres together, and the intersection of a plane cutting through the middle of the H–H bond looks like a circle. In other words, the H–H bond is *cylindrically symmetrical*, as shown in Figure 1.6. Such bonds, which have circular cross-sections and are formed by head-on overlap of two atomic orbitals, are called **sigma (σ) bonds**.

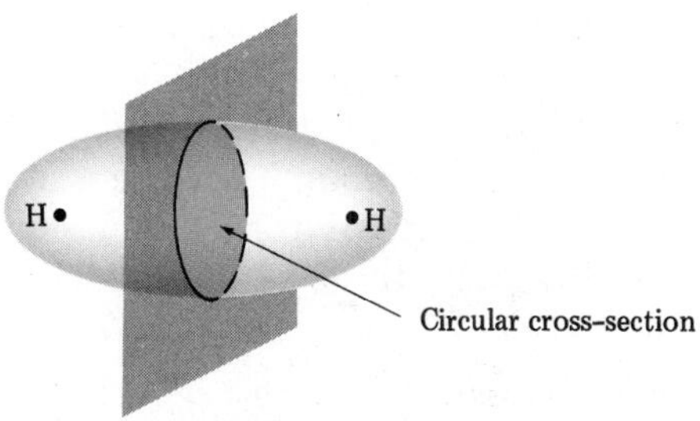

FIGURE 1.6 Cylindrical symmetry of the H–H sigma bond. The intersection of a plane cutting through the bond looks like a circle.

1.7 HYBRIDIZATION: THE FORMATION OF sp^3 ORBITALS

The bonding in the hydrogen molecule is fairly straightforward, but the situation becomes more complex when we turn to organic molecules with tetravalent carbon atoms. Let's start with the simplest case and consider methane, CH_4. Carbon has four electrons in its valence shell and can form four bonds to hydrogens. In Lewis structures:

$$\cdot\dot{\underset{\cdot}{C}}\cdot \xrightarrow{4\,H\cdot} H:\overset{H}{\underset{H}{\ddot{\underset{\cdot\cdot}{C}}}}:H$$

HYBRIDIZATION
Combination of atomic orbitals to form new orbitals with different spatial properties

***sp*³ HYBRID ORBITAL**
Hybrid orbital formed by combination of one *s* and three *p* atomic orbitals

What are the four C–H bonds in methane like? Because carbon uses *two* kinds of orbitals (2*s* and 2*p*) for bonding purposes, we might expect methane to have two kinds of C–H bonds. In fact, though, all four C–H bonds in methane are identical. How can we explain this?

The answer was provided in 1931 by Linus Pauling, who showed that an *s* orbital and three *p* orbitals can combine, or **hybridize**, to form four equivalent atomic orbitals that are spatially oriented toward the four corners of a tetrahedron. Shown in Figure 1.7, these tetrahedral orbitals are called ***sp*³ hybrids**[5] because they arise from a combination of one *s* orbital with three *p* orbitals.

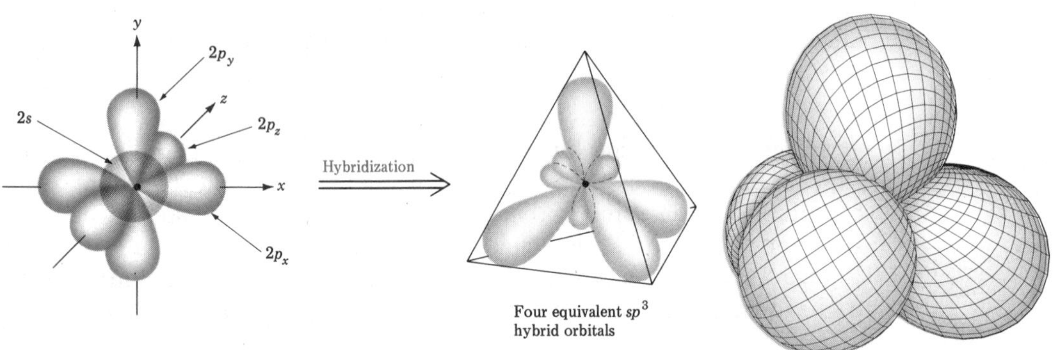

FIGURE 1.7 The formation of sp^3 hybrid orbitals by combination of one *s* orbital with three *p* orbitals.

The concept of hybridization explains *how* carbon forms four equivalent tetrahedral bonds but doesn't answer the question of *why* it does so. Viewing an sp^3 hybrid orbital from the side suggests the answer. When an *s* orbital hybridizes with three *p* orbitals, the resultant hybrid orbitals

[5]Note that the superscript used to identify an sp^3 hybrid orbital tells how many of each type of atomic orbital combine in the hybrid; it doesn't tell how many electrons occupy that orbital.

are unsymmetrical about the nucleus. One of the two lobes of an sp^3 orbital is much larger than the other (Figure 1.8) and can therefore overlap better with another orbital when it forms a bond. As a result, sp^3 hybrid orbitals form stronger bonds than do unhybridized *s* or *p* orbitals.

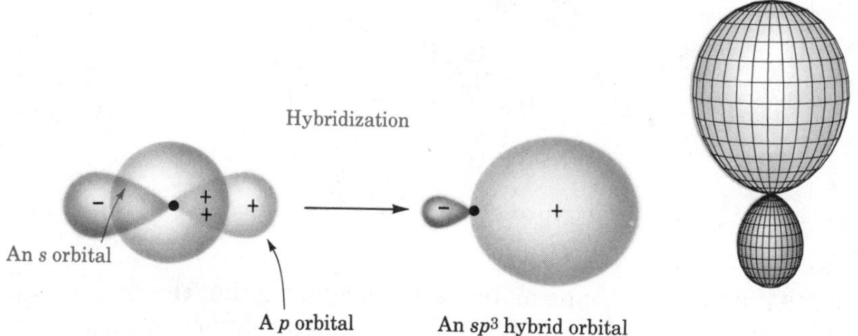

FIGURE 1.8 A side view of an sp^3 hybrid orbital, showing how it is strongly oriented in one direction.

1.8 THE STRUCTURE OF METHANE

BOND ANGLE
Angle formed by two adjacent bonds

When the four identical orbitals of an sp^3-hybridized carbon atom overlap with four hydrogen atoms, four identical C–H bonds are formed, and methane, CH_4, results. Each C–H bond of methane has a strength of 104 kcal/mol (435 kJ/mol) and a length of 1.10 Å. Because the four bonds have a specific geometry, we can also define a property called a **bond angle**. The angle formed by each H–C–H is 109.5°, the so-called tetrahedral angle. Methane thus has the structure shown in Figure 1.9.

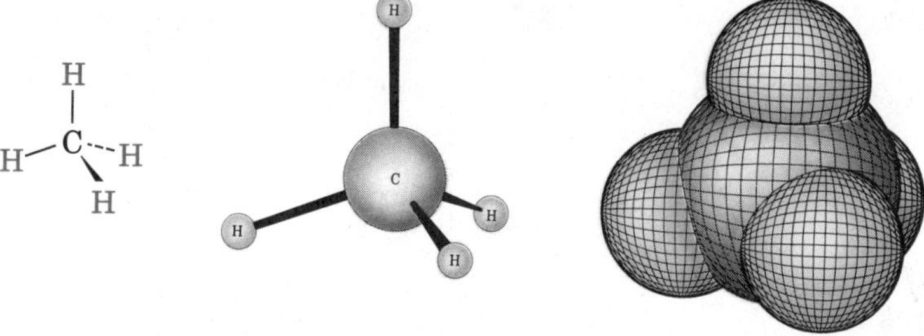

FIGURE 1.9 The structure of methane. The drawings are computer-generated for accuracy.

PROBLEM 1.10 Draw a tetrahedral representation of tetrachloromethane, CCl_4, using the standard convention of solid, dashed, and wedged lines.

PROBLEM 1.11 Why do you think a C–H bond (1.09 Å) is longer than an H–H bond (0.74 Å)?

1.9 THE STRUCTURE OF ETHANE

The same kind of hybridization that explains the methane structure also explains how one carbon atom can bond to another to form a chain. Ethane, C_2H_6, is the simplest molecule containing a carbon–carbon bond:

$$\mathrm{H{:}\overset{H}{\underset{H}{\ddot{\underset{\cdot\cdot}{C}}}}{:}\overset{H}{\underset{H}{\ddot{\underset{\cdot\cdot}{C}}}}{:}H} \qquad \mathrm{H{-}\overset{H}{\underset{H}{\overset{|}{\underset{|}{C}}}}{-}\overset{H}{\underset{H}{\overset{|}{\underset{|}{C}}}}{-}H} \qquad \mathrm{CH_3CH_3}$$

Some representations of ethane

We can picture the ethane molecule by imagining that the two carbon atoms bond to each other by sigma overlap of an sp^3 hybrid orbital from each. The remaining three sp^3 hybrid orbitals on each carbon form the six C–H bonds, as shown in Figure 1.10. The C–H bonds in ethane are similar to those in methane, though a bit weaker (98 kcal/mol for ethane versus 104 kcal/mol for methane). The C–C bond is 1.54 Å long and has a strength of 88 kcal/mol (368 kJ/mol). All the bond angles of ethane are near the tetrahedral value, 109.5°.

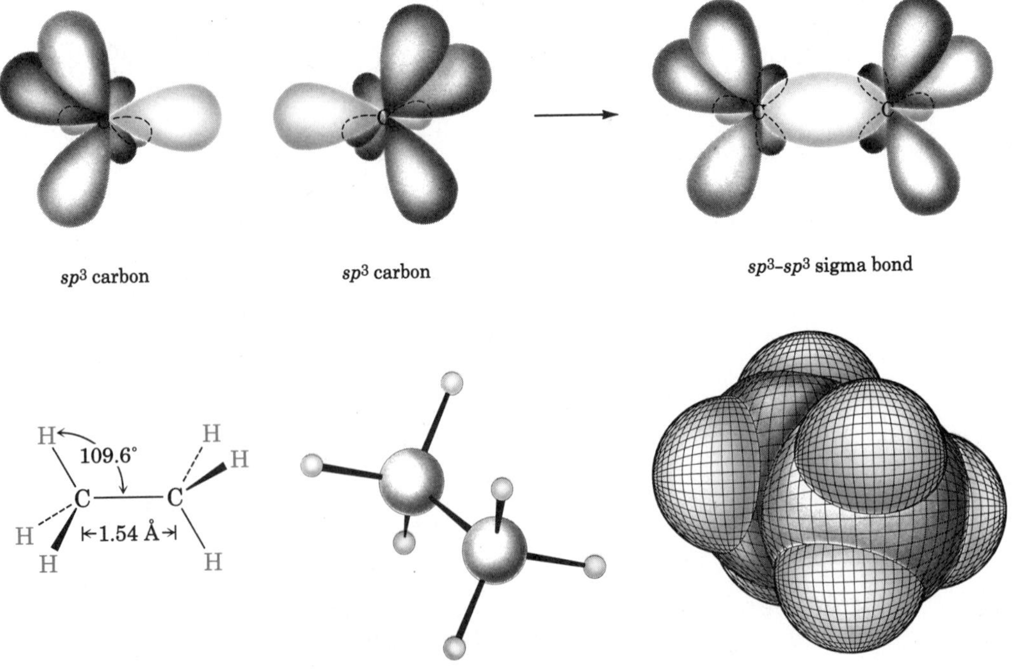

FIGURE 1.10 The structure of ethane. The carbon–carbon bond is formed by sigma overlap of two carbon sp^3 hybrid orbitals.

PROBLEM 1.12 Draw a line-bond structure for propane, $CH_3CH_2CH_3$. Predict the value of each bond angle and indicate the overall shape of the molecule.

PROBLEM 1.13 Why can't an organic molecule have the formula C_2H_7?

1.10 HYBRIDIZATION: sp^2 ORBITALS AND THE STRUCTURE OF ETHYLENE

Although sp^3 hybridization is the most common electronic state of carbon, it's not the only possibility. Look at ethylene, C_2H_4, for example. It was recognized over 100 years ago that ethylene carbon atoms can be tetravalent only if they share *four* electrons and are linked by a *double* bond.

Top view Side view

Ethylene

sp^2 HYBRID ORBITAL Hybrid orbital formed by combination of one *s* and two *p* atomic orbitals

When we imagined the formation of sp^3 hybrid orbitals to explain the bonding in methane, we combined all four of carbon's outer-shell atomic orbitals to construct four equivalent sp^3 hybrids. Imagine instead that we combine the carbon 2*s* orbital with only *two* of the three available 2*p* orbitals. Three hybrid orbitals called **sp^2 hybrids** result, and one unhybridized 2*p* orbital remains unchanged. The three sp^2 orbitals lie in a plane at angles of 120° to each other, with the remaining *p* orbital perpendicular to the sp^2 plane, as shown in Figure 1.11.

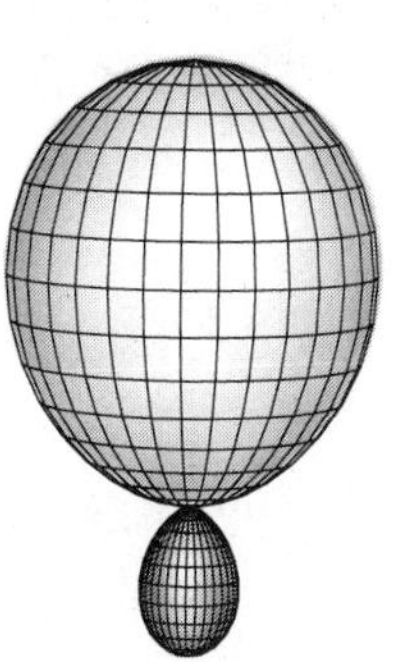

An sp^2 hybrid orbital

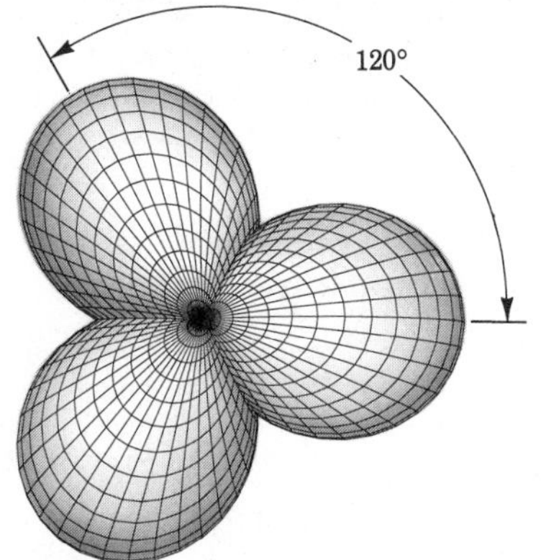

Top view of three sp^2 orbitals

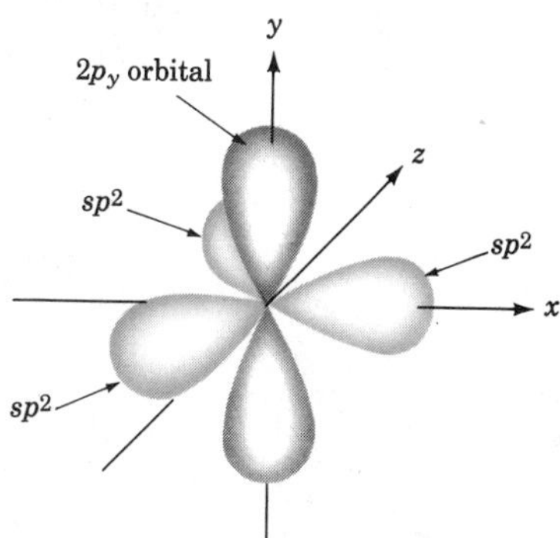

An sp^2-hybridized carbon

FIGURE 1.11 An sp^2-hybridized carbon atom.

When two sp^2-hybridized carbon atoms approach each other, they can form a strong sigma bond by sp^2–sp^2 overlap. At the same time, the unhybridized p orbitals on each carbon approach each other with the correct geometry for *sideways* rather than head-on overlap, leading to the formation of what is called a **pi (π) bond**. The combination of sp^2–sp^2 sigma overlap and $2p$–$2p$ pi overlap results in the net sharing of four electrons and the formation of a carbon–carbon double bond (Figure 1.12).

PI (π) BOND
Covalent bond formed by sideways overlap of two p orbitals

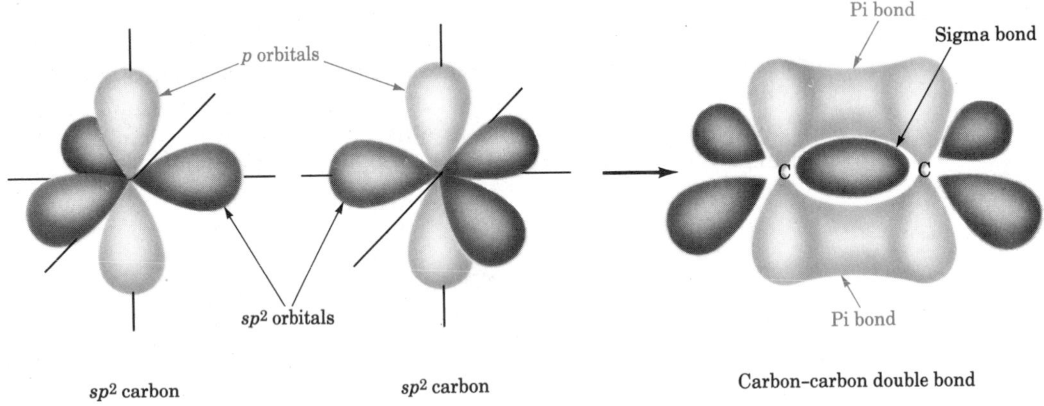

FIGURE 1.12 Orbital overlap of two sp^2-hybridized carbon atoms in a carbon–carbon double bond.

To complete the structure of ethylene, four hydrogen atoms form sigma bonds to the remaining four carbon sp^2 orbitals. The resultant ethylene molecule has a planar (flat) structure with H–C–H and H–C–C bond angles of approximately 120°. Each C–H bond has a length of 1.076 Å and a strength of 103 kcal/mol (431 kJ/mol).

As you might expect, the carbon–carbon double bond in ethylene is both shorter and stronger than the ethane single bond because it results from the sharing of four electrons rather than two. Ethylene has a C=C bond length of 1.33 Å and a bond strength of 152 kcal/mol (636 kJ/mol), whereas ethane has values of 1.54 Å and 88 kcal/mol, respectively. The structure of ethylene is shown in Figure 1.13.

PRACTICE PROBLEM 1.4 Formaldehyde, CH_2O, contains a carbon–*oxygen* double bond. Draw Lewis and line-bond structures of formaldehyde, and indicate the hybridization of the carbon atom.

SOLUTION There is only one way that two hydrogens, one carbon, and one oxygen can combine:

H:C::O: (with H) — **Lewis structure**

H₂C=O (H–C(–H)=O) — **Line-bond structure**

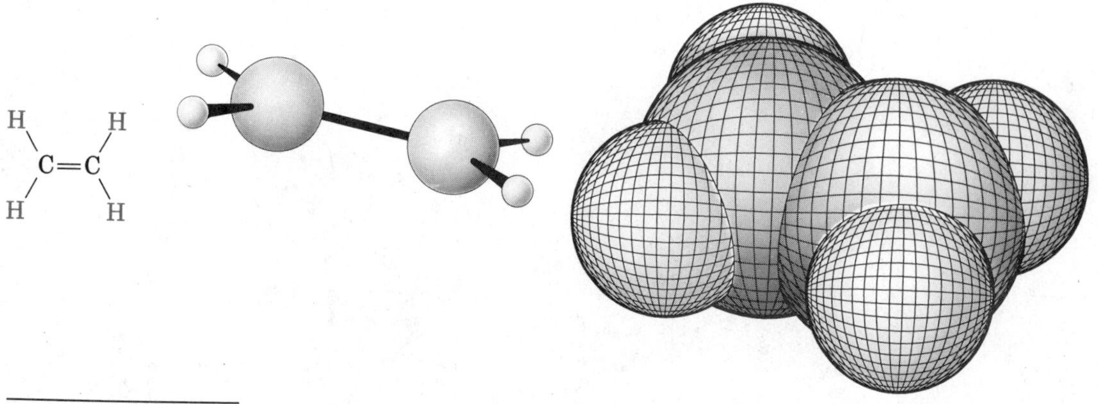

FIGURE 1.13 The structure of ethylene. Note that these computer-generated structures show only the connections between atoms and do not explicitly indicate the carbon–carbon double bond.

Like the carbon atoms in ethylene, the carbon atom in formaldehyde is sp^2-hybridized.

PROBLEM 1.14 Draw both a Lewis structure and a line-bond structure for acetaldehyde, CH_3CHO.

PROBLEM 1.15 Draw a line-bond structure for propene, $CH_3CH{=}CH_2$, indicate the hybridization of each carbon, and predict the value of each bond angle.

PROBLEM 1.16 Draw a line-bond structure for 1,3-butadiene, $H_2C{=}CH{-}CH{=}CH_2$, indicate the hybridization of each carbon, and predict a value for each bond angle.

1.11 HYBRIDIZATION: *sp* ORBITALS AND THE STRUCTURE OF ACETYLENE

***sp* HYBRID ORBITAL** Hybrid orbital formed by combination of one *s* and one *p* atomic orbital

In addition to being able to form single and double bonds by sharing two and four electrons, carbon can form *triple* bonds by sharing *six* electrons. To account for triple bonds like that in acetylene, C_2H_2, we need a third kind of hybrid orbital, an ***sp* hybrid**.

H:C:::C:H H—C≡C—H

Acetylene

Imagine that, instead of combining with two or three 2*p* orbitals, the carbon 2*s* orbital hybridizes with only a single 2*p* orbital. Two *sp* hybrid orbitals result, and two *p* orbitals remain unchanged. The two *sp* orbitals are linear, or 180° apart on the x-axis, while the remaining two p orbitals are perpendicular on the y-axis and the z-axis, as shown in Figure 1.14.

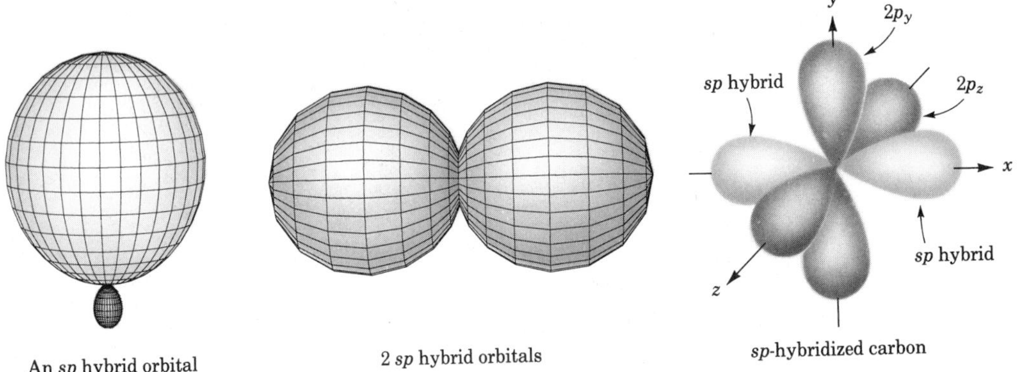

FIGURE 1.14 An *sp*-hybridized carbon atom. The two *sp* hybrid orbitals are oriented 180° apart.

When two *sp*-hybridized carbon atoms approach each other, *sp* hybrid orbitals from each carbon overlap head-on to form a strong *sp*–*sp* sigma bond. In addition, the p_z orbitals from each carbon form a p_z–p_z pi bond by sideways overlap, and the p_y orbitals from each carbon overlap similarly to form a p_y–p_y pi bond. The net effect is the formation of one sigma bond and two pi bonds—that is, a carbon–carbon triple bond. The remaining *sp* hybrid orbitals form sigma bonds to hydrogen to complete the acetylene molecule (Figure 1.15).

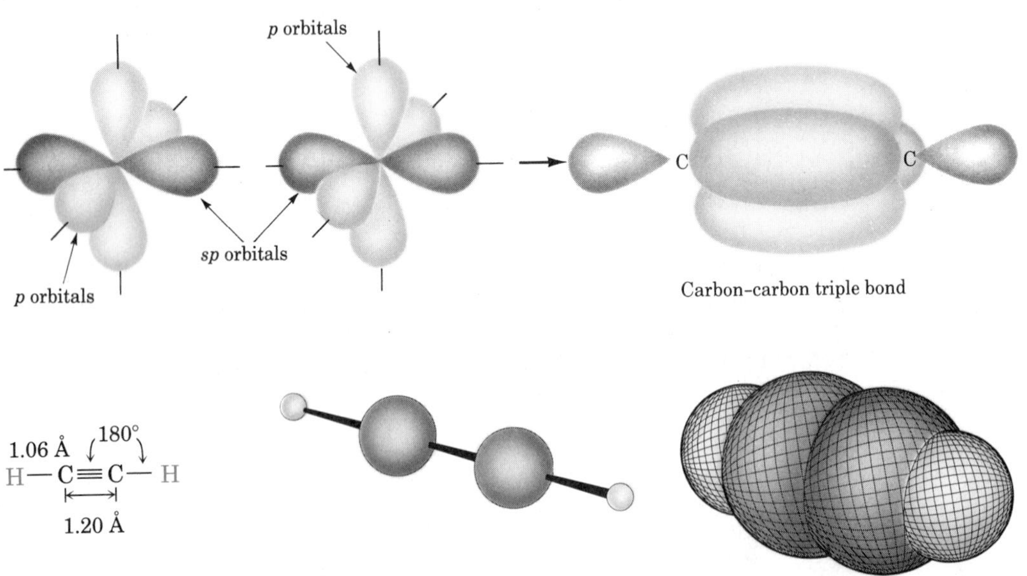

FIGURE 1.15 The carbon–carbon triple bond in acetylene.

Because of *sp* hybridization, acetylene is a linear molecule with H–C–C bond angles of 180°. The C–H bond has a length of 1.06 Å and a strength of 125 kcal/mol (523 kJ/mol). The C≡C bond length is 1.20 Å, and its strength is 200 kcal/mol (837 kJ/mol), making it the shortest and strongest of any carbon–carbon bond.

PROBLEM 1.17 Draw a line-bond structure for propyne, $CH_3C{\equiv}CH$, indicate the hybridization of each carbon, and predict a value for each bond angle.

1.12 BOND POLARITY AND ELECTRONEGATIVITY

Up to this point, we've viewed chemical bonding in an either/or manner: A given bond is either ionic or covalent. It's more accurate, though, to look at bonding as a continuum of possibilities, from a fully covalent bond with a symmetrical electron distribution on the one hand, to a fully ionic bond between anions and cations on the other (Figure 1.16).

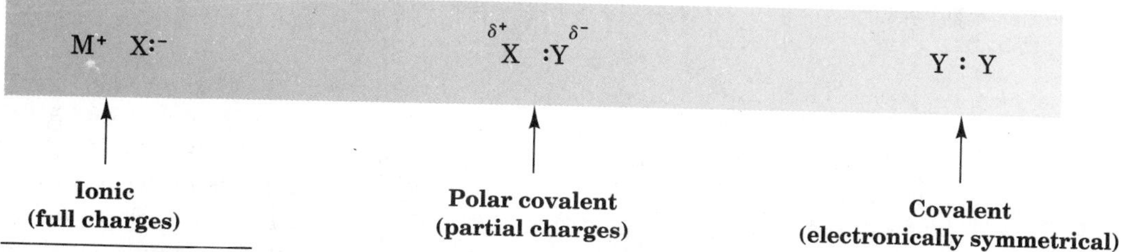

FIGURE 1.16 The bonding continuum from covalent to ionic is a result of unsymmetrical electron distribution. The symbol δ (Greek delta) means *partial* charge, either positive (δ^+) or negative (δ^-).

POLAR COVALENT BOND
Covalent bond in which the electrons are attracted more strongly by one atom than by the other

ELECTRONEGATIVITY
Ability of an atom to attract electrons in a covalent bond

The carbon–carbon bond in ethane, for example, is electronically symmetrical and therefore fully covalent; the two bonding electrons are equally shared by the two equivalent carbon atoms. The bond in sodium chloride, by contrast, is largely ionic;[6] an electron has been transferred from sodium to chlorine to give Na^+ and Cl^- ions. Between these two extremes lie the great majority of chemical bonds, in which the bonding electrons are attracted *somewhat* more strongly by one atom than by the other. We call such bonds **polar covalent bonds**.

Bond polarity is due to differences in **electronegativity**, the intrinsic ability of an atom to attract electrons in a covalent bond. As shown in Figure 1.17, metallic elements on the left side of the periodic table attract electrons weakly, whereas the halogens and other elements on the right side of the periodic table attract electrons strongly.

[6]Even in NaCl, the bond is only about 80% ionic rather than 100%.

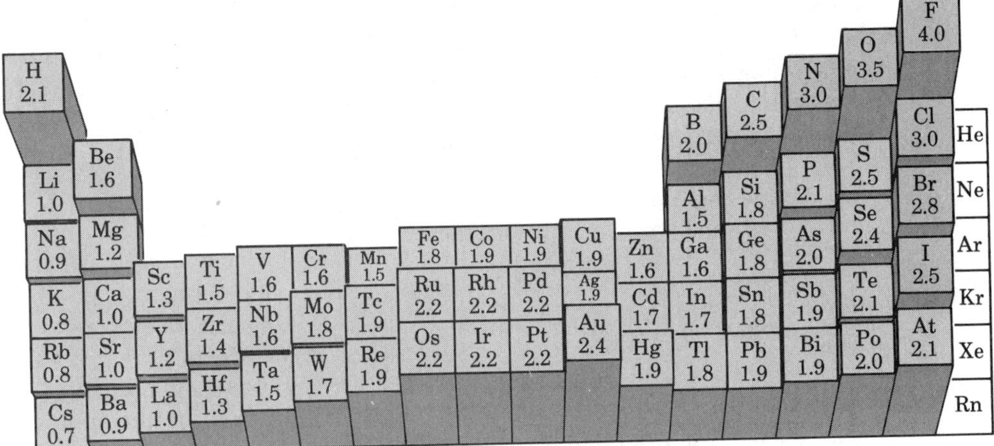

FIGURE 1.17 Electronegativity trends in the periodic table. Elements on the right side of the table are more electronegative than elements on the left side. The values shown are based on an arbitrary scale, with H = 2.1 and F = 4.0. Carbon has an electronegativity value of 2.5. Any element more electronegative than carbon has a value greater than 2.5, and any element less electronegative than carbon has a value less than 2.5.

As a general rule, bonds between atoms with similar electronegativities are covalent, bonds between atoms whose electronegativities (EN) differ by less than 2 units are polar covalent, and bonds between atoms whose electronegativities differ by more than 2 units are largely ionic. The bonds between carbon and more electronegative elements such as oxygen, fluorine, and chlorine, for example, are polar covalent. The electrons in such bonds are drawn away from carbon toward the electronegative atom, leaving the carbon with a partial positive charge (denoted by δ^+; δ is the Greek letter delta) and the electronegative atom with a partial negative charge (δ^-). For example,

$Cl^{\delta-}$ — $C^{\delta+}H_3$ **Chloromethane**

Chlorine: EN = 3.0
Carbon: EN = 2.5
ΔEN = 0.5

An arrow $\leftrightarrow$ is used to indicate the direction of polarity. By convention, *electrons move in the direction of the arrow*. The tail of the arrow is electron-poor (δ^+), and the head of the arrow is electron-rich (δ^-).

Bonds between carbon and less electronegative elements are polarized so that carbon bears a partial negative charge and the other atom bears a partial positive charge. Organometallic compounds such as tetraethyllead, the "lead" in gasoline, provide good examples of this kind of polar covalent bond.

$$\begin{array}{c} \delta^- \\ CH_2CH_3 \\ \delta^- \quad \delta^+ | \quad \delta^- \\ CH_3CH_2-Pb-CH_2CH_3 \\ \delta^- | \\ CH_2CH_3 \end{array}$$

Tetraethyllead

Carbon: EN = 2.5
Lead: EN = 1.9
ΔEN = 0.6

INDUCTIVE EFFECT Ability of an atom to polarize a covalent bond by donating or withdrawing the bonding electrons

When speaking of an atom's ability to polarize a bond, we often use the term **inductive effect**. An inductive effect is simply the shifting of electrons in a bond in response to the electronegativity of nearby atoms. Electropositive elements such as lithium and magnesium inductively donate electrons, whereas electronegative elements such as oxygen and chlorine inductively withdraw electrons. Inductive effects play a major role in understanding chemical reactivity, and we'll use them many times throughout this text to explain a wide variety of chemical phenomena.

PRACTICE PROBLEM 1.5 Predict the amount and direction of polarization of an O–H bond in water.

SOLUTION Oxygen (EN = 3.5) is more electronegative than hydrogen (EN = 2.1), according to Figure 1.17, and therefore attracts electrons more strongly. The difference in electronegativities (ΔEN = 3.5 − 2.1 = 1.4) implies that an O–H bond is strongly polarized.

$$\overset{\delta^+}{H}-\overset{\delta^-}{O}-\overset{\delta^+}{H}$$

PROBLEM 1.18 Which element in each of these pairs is more electronegative?
(a) Li or H **(b)** Be or Br **(c)** Cl or I

PROBLEM 1.19 Use the δ^-/δ^+ convention to indicate the direction of expected polarity for each of the bonds shown:
(a) Br–CH_3 **(b)** H_2N–CH_3 **(c)** Li–CH_3 **(d)** H–NH_2
(e) HO–CH_3 **(f)** BrMg–CH_3 **(g)** F–CH_3

PROBLEM 1.20 Order the bonds in the following compounds according to their increasing ionic character: CCl_4, $MgCl_2$, $TiCl_3$, Cl_2O.

1.13 ACIDS AND BASES

Acidity and *basicity* are related to the concepts of electronegativity and bond polarity just discussed, and it's a good idea at this point to review some fundamental ideas about these topics. There are two main ways of defining acidity: the *Brønsted–Lowry definition* and the *Lewis definition.*

BRØNSTED–LOWRY ACID Substance that donates a hydrogen ion, H^+

BRØNSTED–LOWRY BASE Substance that accepts a hydrogen ion, H^+

Brønsted–Lowry Acids and Bases

According to the **Brønsted–Lowry** definition, an acid is a substance that loses a proton (hydrogen ion, H^+), and a base is a substance that gains a proton. When hydrogen chloride dissolves in water, for example, HCl

CONJUGATE BASE
Product that results when an acid loses H+

CONJUGATE ACID
Product that results when a base accepts H+

loses a proton and H_2O gains it, yielding Cl^- and H_3O^+. Chloride ion (Cl^-), the product that results when the acid HCl loses a proton, is called the **conjugate base** of the acid, and H_3O^+, the product that results when the base H_2O gains a proton, is called the **conjugate acid** of the base. In a general sense:

$$\underset{\textbf{Acid}}{\text{H—A}} + \underset{\textbf{Base}}{\text{:B}} \rightleftharpoons \underset{\textbf{Conjugate base}}{\text{A:}^-} + \underset{\textbf{Conjugate acid}}{\text{H—B}^+}$$

For example:

$$\underset{\textbf{Acid}}{\text{H—}\ddot{\text{C}}\text{l:}} + \underset{\textbf{Base}}{\text{:}\ddot{\text{O}}\text{—H (—H)}} \rightleftharpoons \underset{\textbf{Conjugate base}}{\text{:}\ddot{\text{C}}\text{l:}^-} + \underset{\textbf{Conjugate acid}}{\text{H—}\ddot{\text{O}}^+\text{—H (—H)}}$$

ACIDITY CONSTANT, K_a
Value that expresses the strength of an acid in water solution

Acids differ in the ease with which they lose a proton. Stronger acids such as HCl react almost completely with water, whereas weaker acids such as acetic acid (CH_3COOH) react only slightly. The exact strength of a given acid in water solution is expressed by its **acidity constant, K_a**.[7]

For the reaction

$$HA + H_2O \rightleftharpoons H_3O^+ + A^-$$

the *equilibrium constant*, K_{eq}, and acidity constant, K_a, are calculated as

$$K_{eq} = \frac{[H_3O^+][A^-]}{[HA][H_2O]} \quad \text{and} \quad K_a = \frac{[H_3O^+][A^-]}{[HA]}$$

Stronger acids have their equilibria toward the right and thus have larger acidity constants, whereas weaker acids have their equilibria toward the left and have smaller acidity constants. The range of K_a values for different acids is enormous, running from about 10^{15} for the strongest acids to about 10^{-60} for the weakest. The common inorganic acids such as H_2SO_4, HNO_3, and HCl have K_a's in the range 10^2–10^9, while most organic acids have K_a's in the range 10^{-5}–10^{-15}. As you gain more experience, you'll develop a rough feeling for which acids are "strong" and which are "weak" (always remembering that the terms are relative).

Acid strengths are normally expressed using pK_a values, where the pK_a is equal to the negative logarithm of the acidity constant:

$$pK_a = -\log K_a$$

[7]Remember that brackets [ ] refer to the molar concentration of the species indicated. Note also that the concentration of water [H_2O] is left out of the expression for K_a, since it remains effectively constant.

A stronger acid (larger K_a) has a *lower* pK_a; a weaker acid (smaller K_a) has a *higher* pK_a. Table 1.4 lists the pK_a's of some common acids in order of strength.

TABLE 1.4 Relative Strengths of Some Common Acids

	ACID	NAME	pK_a	CONJUGATE BASE	NAME	
Weaker acid	CH_3CH_2OH	Ethanol	16.00	$CH_3CH_2O^-$	Ethoxide ion	Stronger base
	H_2O	Water	15.74	HO^-	Hydroxide ion	
↓	HCN	Hydrocyanic acid	9.2	CN^-	Cyanide ion	↑
	CH_3COOH	Acetic acid	4.72	CH_3COO^-	Acetate ion	
	HF	Hydrofluoric acid	3.2	F^-	Fluoride ion	
	HNO_3	Nitric acid	−1.3	NO_3^-	Nitrate ion	
Stronger acid	HCl	Hydrochloric acid	−7.0	Cl^-	Chloride ion	Weaker base

Notice in Table 1.4 that there is an inverse relationship between the acid strength of an acid and the base strength of its conjugate base. To understand this inverse relationship, think about what is happening to the acidic proton: A *strong acid* loses its proton easily, meaning that its conjugate base does not hold the proton tightly and is therefore a *weak base*. A *weak acid* loses its proton with difficulty, meaning that its conjugate base holds the proton tightly and is therefore a *strong base*. The fact that HCl is a strong acid, for example, means that Cl^- does not hold the proton tightly and is thus a weak base. Water, on the other hand, is a weak acid, meaning that OH^- holds the proton tightly and is a strong base.

In general, an acid will lose a proton to the conjugate base of any acid with a higher pK_a. Conversely, the conjugate base of an acid will remove a proton from any acid with a lower pK_a. For example, the data in Table 1.4 indicate that OH^- will react with acetic acid, CH_3COOH, to yield acetate ion, $CH_3CO_2^-$, and H_2O. Since water (pK_a = 15.74) is a weaker acid than acetic acid (pK_a = 4.72), hydroxide ion has a greater affinity for a proton than acetate ion has.

$$CH_3COOH + {}^-{:}\ddot{O}{-}H \rightleftharpoons CH_3CO\ddot{O}{:}^- + H{-}\ddot{O}{-}H$$

Acetic acid (pK_a = 4.72) **Hydroxide ion** **Acetate ion** **Water (pK_a = 15.74)**

Another way of predicting acid–base reactivity is to remember that the products must be more stable than the reactants in order for reaction to occur. In other words, the product acid and base must be less reactive (weaker) than the starting acid and base. In the reaction of acetic acid with hydroxide ion, for example, the product conjugate base (acetate ion) is weaker than the starting base (hydroxide ion), and the product conjugate acid (water) is weaker than the starting acid (acetic acid).

$$\underset{\text{Stronger acid}}{CH_3\overset{O}{\overset{\|}{C}}OH} + \underset{\text{Stronger base}}{HO^-} \rightleftharpoons \underset{\text{Weaker acid}}{H_2O} + \underset{\text{Weaker base}}{CH_3\overset{O}{\overset{\|}{C}}O^-}$$

PRACTICE PROBLEM 1.6 Water has $pK_a = 15.74$ and acetylene has $pK_a = 25$. Which of the two is more acidic? Will hydroxide ion react with acetylene?

$$H{-}C{\equiv}C{-}H + H{-}O^- \xrightarrow{?} H{-}C{\equiv}C{:}^- + H{-}O{-}H$$

SOLUTION In comparing two acids, the one with the lower pK_a is stronger. Thus, water is a stronger acid than acetylene. Since water loses a proton more easily than acetylene, the $H{-}O^-$ ion has less affinity for a proton than the $H{-}C{\equiv}C{:}^-$ ion. In other words, the anion of acetylene is a stronger base than hydroxide ion, and the reaction will not proceed as written.

PROBLEM 1.21 Formic acid, HCO_2H, has $pK_a = 3.7$, and picric acid, $C_6H_3N_3O_7$, has $pK_a = 0.3$. Which is the stronger acid?

PROBLEM 1.22 Amide ion, H_2N^-, is a stronger base than hydroxide ion, HO^-. Which is the stronger acid, $H_2N{-}H$ (ammonia) or $HO{-}H$ (water)? Explain.

PROBLEM 1.23 Is either of these reactions likely to take place, according to the pK_a data in Table 1.4?

(a) $H{-}CN + CH_3COO^-\ Na^+ \longrightarrow Na^+\ {}^-CN + CH_3COO{-}H$

(b) $CH_3CH_2O{-}H + Na^+\ {}^-CN \longrightarrow CH_3CH_2O^-\ Na^+ + H{-}CN$

Lewis Acids and Bases

LEWIS ACID Substance that accepts an electron pair

LEWIS BASE Substance that donates an electron pair

The Lewis definition of acids and bases differs from the Brønsted–Lowry definition in that it isn't limited to substances that gain or lose protons. A **Lewis acid** is a substance that accepts an electron pair, and a **Lewis base** is a substance that donates an electron pair. The donated pair of electrons is shared between Lewis acid and base in a newly formed covalent bond.

Lewis acids include not only proton donors but many other species as well. Thus, a proton (H^+) is a Lewis acid because it accepts a pair of electrons to fill its vacant 1*s* orbital when it reacts with a base. Compounds such as BF_3 and $AlCl_3$ are Lewis acids because they too accept electron pairs from Lewis bases to fill vacant valence orbitals.

$$H^+ + :\ddot{O}H_2 \rightleftharpoons H_3\ddot{O}^+$$

Hydrogen ion (a Lewis acid) **Water (a Lewis base)** **Hydronium ion**

$$BF_3 + :\ddot{O}(CH_3)_2 \rightleftharpoons F_3\bar{B}-\ddot{O}^+(CH_3)_2$$

Boron trifluoride (a Lewis acid) **Dimethyl ether (a Lewis base)**

$$AlCl_3 + :N(CH_3)_3 \rightleftharpoons Cl_3\bar{Al}-N^+(CH_3)_3$$

Aluminum trichloride (a Lewis acid) **Trimethylamine (a Lewis base)**

The Lewis definition of basicity—a compound with a pair of nonbonding electrons that it can use in forming a bond to a Lewis acid—is similar to the Brønsted–Lowry definition. Thus, H_2O, with its two pairs of nonbonding electrons on oxygen, acts as a Lewis base by donating an electron pair to a proton in forming the hydronium ion, H_3O^+:

$$:\ddot{Cl}-H + :\ddot{O}H_2 \rightleftharpoons H_3\ddot{O}^+ + :\ddot{Cl}:^-$$

Acid **Lewis base** **Hydronium ion**

In a more general sense, most oxygen- and nitrogen-containing organic compounds are Lewis bases because they have lone pairs of electrons. Divalent oxygen compounds each have two lone pairs of electrons, and trivalent nitrogen compounds have one lone pair. Note in the following examples that some compounds can act as both acids and bases, just as water can. Alcohols and carboxylic acids, for instance, act as *acids* when they lose their –OH proton but as *bases* when their oxygen atom accepts a proton.

Some Lewis bases

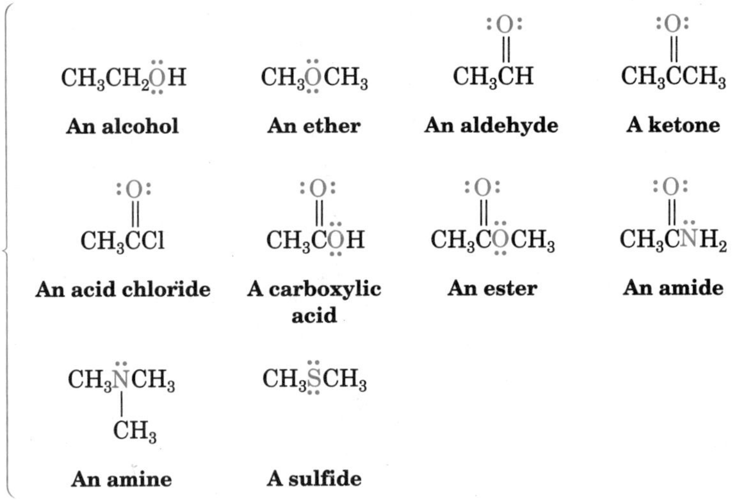

$CH_3CH_2\ddot{O}H$	$CH_3\ddot{O}CH_3$	$CH_3\overset{:O:}{\overset{\|\|}{C}}H$	$CH_3\overset{:O:}{\overset{\|\|}{C}}CH_3$
An alcohol	**An ether**	**An aldehyde**	**A ketone**
$CH_3\overset{:O:}{\overset{\|\|}{C}}Cl$	$CH_3\overset{:O:}{\overset{\|\|}{C}}\ddot{O}H$	$CH_3\overset{:O:}{\overset{\|\|}{C}}\ddot{O}CH_3$	$CH_3\overset{:O:}{\overset{\|\|}{C}}\ddot{N}H_2$
An acid chloride	**A carboxylic acid**	**An ester**	**An amide**
$CH_3\ddot{N}(CH_3)CH_3$	$CH_3\ddot{S}CH_3$		
An amine	**A sulfide**		

PRACTICE PROBLEM 1.7 Show how acetone can act as a Lewis base.

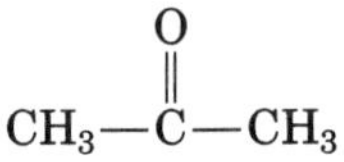

$$CH_3-\overset{O}{\overset{\|}{C}}-CH_3 \quad \textbf{Acetone}$$

SOLUTION The oxygen atom of acetone has two lone pairs of electrons that it can donate to a Lewis acid such as H^+.

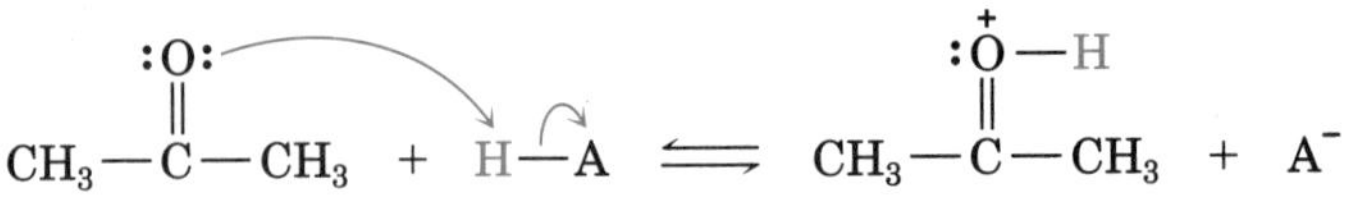

$$CH_3-\overset{:O:}{\overset{\|}{C}}-CH_3 + H-A \rightleftharpoons CH_3-\overset{:\overset{+}{O}-H}{\overset{\|}{C}}-CH_3 + A^-$$

PROBLEM 1.24 Which of the following are Lewis acids and which are Lewis bases?

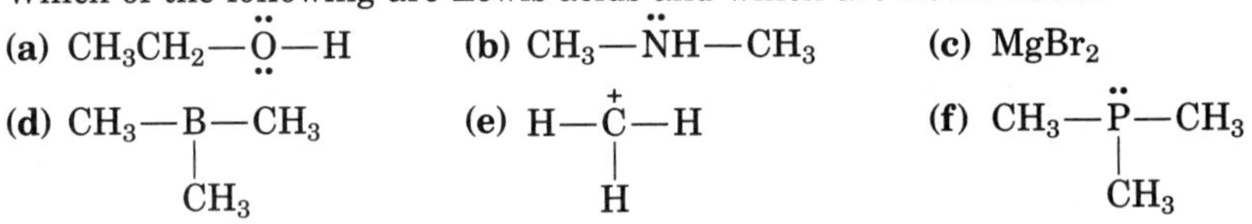

(a) $CH_3CH_2-\ddot{O}-H$ (b) $CH_3-\ddot{N}H-CH_3$ (c) $MgBr_2$

(d) $CH_3-\underset{CH_3}{\underset{|}{B}}-CH_3$ (e) $H-\underset{H}{\underset{|}{\overset{+}{C}}}-H$ (f) $CH_3-\underset{CH_3}{\underset{|}{\ddot{P}}}-CH_3$

INTERLUDE

CHEMICALS, TOXICITY, AND RISK

We hear and read a lot these days about the dangers of chemicals—about pesticide residues on food, dangerous food additives, unsafe medicines, and so forth. What's a person to believe?

Life is not risk-free—we all take many risks each day. We decide to ride a bike rather than drive, even though there is a ten times greater likelihood per mile of dying in a bicycling accident than in a car. We decide to walk down stairs rather than take an elevator, even though 7000 people die from falls each year in the United States. We decide to smoke cigarettes, even though it increases our chance of getting cancer by 50%. Making judgments that affect our health is something we do every day without thinking about it.

What about risks from chemicals? Risk evaluation of chemicals is usually carried out by exposing test animals (usually rats) to a chemical, and then monitoring them for signs of harm. To limit the expense and time needed for tests, the amounts administered are hundreds or thousands of times greater than those a human might normally encounter. Once the animal data are available, the interpretation of the data involves many assumptions. If a substance is harmful to animals, is it necessarily harmful to humans? How can a large dose for a small animal be translated into a small dose for a large human? The standard method for evaluating acute chemical toxicity, as opposed to long-term toxicity, is to report an *LD_{50} value*, the amount of a substance per kilogram body weight that is lethal to 50% of the test animals. The LD_{50} values of various substances are shown in Table 1.5.

TABLE 1.5 Some LD_{50} values

SUBSTANCE	LD_{50} (g/kg)
Aflatoxin B_1	4×10^{-4}
Aspirin	1.7
Chloroform	3.2
Ethyl alcohol	10.6
Formaldehyde	2.4
Sodium cyclamate	17

How we evaluate risk is strongly influenced by familiarity. The presence of chloroform in municipal water supplies at a barely detectable level of 0.000 000 01% has caused an outcry in many cities, yet chloroform has a lower acute toxicity than aspirin. Many foods contain natural ingredients far more toxic than synthetic food additives or pesticide residues, but the ingredients are ignored because the foods are familiar. Peanut butter, for example, contains tiny amounts of aflatoxin, a far more potent cancer threat than sodium cyclamate, an artificial sweetener that has been banned because of its "risk."

All decisions involve tradeoffs. Does the benefit of a pesticide that will increase the availability of food outweigh the health risk to 1 person in 1 million who are exposed? Do the beneficial effects of a new drug outweigh a potentially dangerous side effect to 0.001% of users? The answers aren't always obvious, but it's the responsibility of legislators and well-informed citizens to keep their responses on a factual level rather than an emotional one.

SUMMARY AND KEY WORDS

Organic chemistry is the study of carbon compounds. Although a division into inorganic and organic chemistry occurred historically, there is no scientific reason for the division.

Atoms are composed of a positively charged nucleus surrounded by negatively charged electrons that occupy specific regions of space called **orbitals**. Different orbitals have different energy levels and shapes. For example, *s* orbitals are spherical, and *p* orbitals are dumbbell shaped.

There are two fundamental kinds of chemical bonds: **ionic bonds** and **covalent bonds**. The ionic bonds commonly found in inorganic salts result from the electrical attraction of unlike charges. The covalent bonds found in organic molecules are formed by the sharing of an electron pair between atoms. Electron sharing occurs when two atoms approach each other and their atomic orbitals overlap. Bonds that have a circular cross-section and are formed by head-on overlap of atomic orbitals are called **sigma (σ) bonds**. Bonds formed by sideways overlap of two *p* orbitals are called **pi (π) bonds**.

In order to form bonds in organic compounds, carbon first **hybridizes** to an excited-state configuration. When forming only single bonds, carbon is sp^3-hybridized and has four equivalent **sp^3 hybrid orbitals** with tetrahedral geometry. When forming double bonds, carbon is sp^2-hybridized, has three equivalent **sp^2 orbitals** with planar geometry, and has one unhybridized *p* orbital. When forming triple bonds, carbon is *sp*-hybridized, has two equivalent ***sp* orbitals** with linear geometry, and has two unhybridized *p* orbitals.

Most covalent bonds are **polar** because of unsymmetrical electron sharing. For example, a carbon–chlorine bond is polar because chlorine is more **electronegative** than carbon and therefore attracts the bonding electrons more strongly. Carbon–metal bonds, however, are usually polarized in the opposite sense because carbon attracts electrons more strongly than most metals. Carbon–hydrogen bonds are relatively nonpolar.

A **Brønsted–Lowry acid** is a substance that can lose a proton (hydrogen ion, H^+); a **Brønsted–Lowry base** is a substance that can accept a proton. The strength of an acid or base is expressed by the **acidity constant, K_a**. A **Lewis acid** is a substance that has a low-energy unfilled orbital and can accept an electron pair. A **Lewis base** is a substance that donates an unshared electron pair. Many organic molecules that contain oxygen and nitrogen are weak Lewis bases.

Working Problems

Learning organic chemistry means knowing a large number of facts. Although careful reading and rereading of the text is important, reading alone isn't enough. You must also be able to apply the information you read and be able to use your knowledge in new situations. Working problems gives you the opportunity to do this. There's no better way to learn organic chemistry than by working problems.

Each chapter in this book provides many problems of different sorts. The in-chapter problems are placed for immediate reinforcement of ideas just learned. The end-of-chapter problems provide additional practice, and are of two types: Early problems tend to be the drill type, which provide an opportunity for you to practice your command of the fundamentals; later problems tend to be more challenging and thought provoking.

As you study organic chemistry, take the time to work the problems. Work those you can and ask for help with those you can't do. If stumped by a particular problem, check the accompanying *Study Guide and Solutions Manual* for an explanation that should help clarify the difficulty. Working problems takes effort, but the payoff in knowledge and understanding is immense.

ADDITIONAL PROBLEMS

1.25 How many valence (outer-shell) electrons does each of these atoms have?
(a) Oxygen **(b)** Magnesium **(c)** Fluorine

1.26 Give the ground-state electronic configuration of the following elements. For example, carbon is $1s^2\ 2s^2\ 2p^2$.
(a) Lithium **(b)** Sodium **(c)** Aluminum **(d)** Sulfur

1.27 What are the likely formulas of these molecules?
(a) $AlCl_?$ **(b)** $CF_2Cl_?$ **(c)** $NI_?$

1.28 Identify the bonds in these molecules as covalent, polar covalent, or ionic:
(a) BeF_2 **(b)** SiH_4 **(c)** CBr_4

1.29 Write Lewis (electron-dot) structures for these molecules:
(a) H–C≡C–H **(b)** AlH_3 **(c)** CH_3OH **(d)** $H_2C{=}CHCl$

1.30 Write a Lewis (electron-dot) structure for acetonitrile, $CH_3C{\equiv}N$. How many electrons does the nitrogen atom have in its valence shell? How many are used for bonding and how many are not used for bonding?

1.31 What is the hybridization of each carbon atom in acetonitrile, $CH_3C{\equiv}N$?

1.32 Fill in any unshared electrons that are missing from the following line-bond structures:
(a) $CH_3{-}O{-}CH_3$ **(b)** CH_3NH_2 **(c)** CH_2Cl_2 **(d)** $H_3C{-}C({=}O){-}CH_3$

1.33 There are two structures that correspond to the formula C_4H_{10}. Draw them.

1.34 Convert these molecular formulas into line-bond structures:
(a) C_3H_8 **(b)** C_3H_7Br (2 possibilities) **(c)** C_3H_6 (2 possibilities)
(d) C_2H_6O (2 possibilities)

1.35 Indicate the kind of hybridization (*sp*, *sp*^{2}, or *sp*^{3}) you would expect for each carbon atom in these molecules:

(a) Butane, $CH_3CH_2CH_2CH_3$ (b) 1-Butene, $CH_3CH_2CH{=}CH_2$

(c) Cyclobutene, (structure: four-membered ring C=C–C–C with H on each sp² carbon and two H on each sp³ carbon) (d) 1-Buten-3-yne, $H_2C{=}CH{-}C{\equiv}CH$

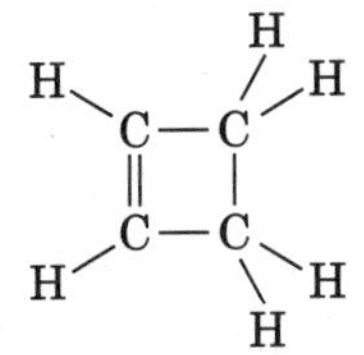

1.36 What is the hybridization of each carbon atom in benzene? What shape would you expect benzene to have?

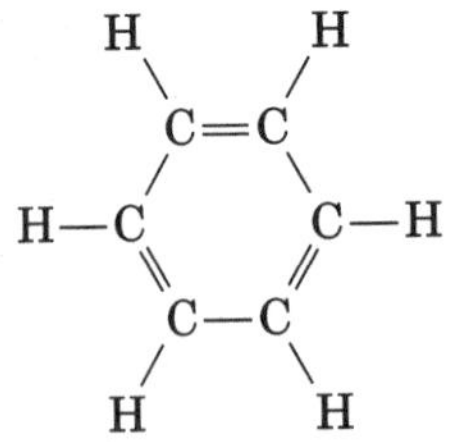

(structure: six-membered ring of C atoms with alternating double bonds, one H on each carbon) **Benzene**

1.37 Write Lewis (electron-dot) structures for these molecules:
(a) $CH_3{-}Be{-}CH_3$ (b) $CH_3{-}\underset{\displaystyle CH_3}{\underset{|}{P}}{-}CH_3$ (c) $TiCl_4$

1.38 Draw line-bond structures for these covalent molecules:
(a) Br_2 (b) CH_3Cl (c) HF (d) CH_3CH_2OH

1.39 Indicate which of the bonds in the structures you drew for Problem 1.38 are polar covalent. Indicate bond polarity by using the symbols δ^+ and δ^-.

1.40 Identify all the bonds in these molecules as either ionic or covalent:
(a) NaOH (b) HOH (c) CH_3OH (d) CH_3OCH_3 (e) FF

1.41 Sodium methoxide, $NaOCH_3$, contains both ionic and covalent bonds. Indicate which is which.

1.42 Identify the most electronegative element in each of these molecules:
(a) CH_2FCl (b) $FCH_2CH_2CH_2Br$ (c) $HOCH_2CH_2NH_2$ (d) CH_3OCH_2Li

1.43 Use the electronegativity table (Figure 1.17) to predict which bond in each of the following sets is more polar.
(a) $Cl{-}CH_3$ or $Cl{-}Cl$ (b) $H{-}CH_3$ or $H{-}Cl$ (c) $HO{-}CH_3$ or $(CH_3)_3Si{-}CH_3$

1.44 Indicate the direction of polarity for each bond in Problem 1.43.

1.45 Which atoms in these structures have unshared valence electrons? Draw in these unshared electrons.

(a) CH_3SH (b) $CH_3{-}\underset{\displaystyle CH_3}{\underset{|}{N}}{-}CH_3$ (c) CH_3CH_2Br (d) $CH_3\overset{\displaystyle O}{\overset{\|}{C}}{-}OH$ (e) $CH_3\overset{\displaystyle O}{\overset{\|}{C}}{-}Cl$

1.46 Draw a three-dimensional representation of chloroform, $CHCl_3$, using the standard convention of solid, wedged, and dashed lines. Do the same for the oxygen-bearing carbon atom in ethanol, CH_3CH_2OH.

1.47 Ammonia, $H_2N{-}H$, has $pK_a \approx 36$ and acetone has $pK_a \approx 20$. Will the following reaction take place? Explain.

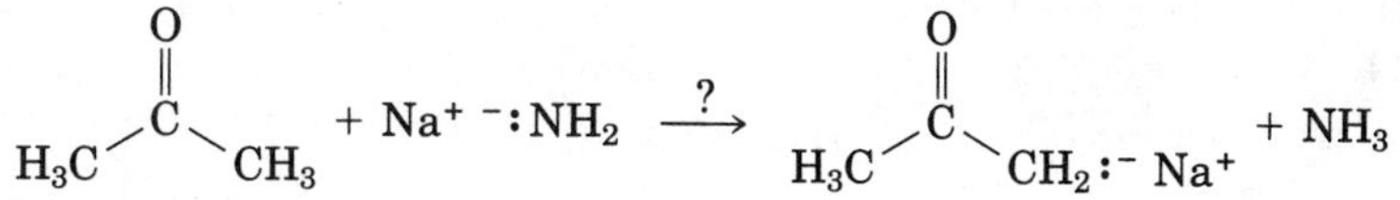

$$H_3C{-}C({=}O){-}CH_3 + Na^+ \ {}^-{:}NH_2 \xrightarrow{?} H_3C{-}C({=}O){-}CH_2{:}^- \ Na^+ + NH_3$$

Acetone

1.48 Classify these reagents as either Lewis acids or Lewis bases:
(a) $AlBr_3$ **(b)** $CH_3CH_2NH_2$ **(c)** HF **(d)** CH_3SCH_3

1.49 Is the bicarbonate anion (HCO_3^-) a strong enough base to react with methanol (CH_3OH)? In other words, does the following reaction take place as written? (The pK_a of methanol is 15.5; the pK_a of H_2CO_3 is 6.4.)

$$CH_3OH + HCO_3^- \longrightarrow CH_3O^- + H_2CO_3$$

1.50 Identify the acids and bases in these reactions:

(a) $CH_3OH + H^+ \longrightarrow CH_3\overset{+}{O}H_2$ (b) $CH_3OH + {}^-NH_2 \longrightarrow CH_3O^- + NH_3$

(c) 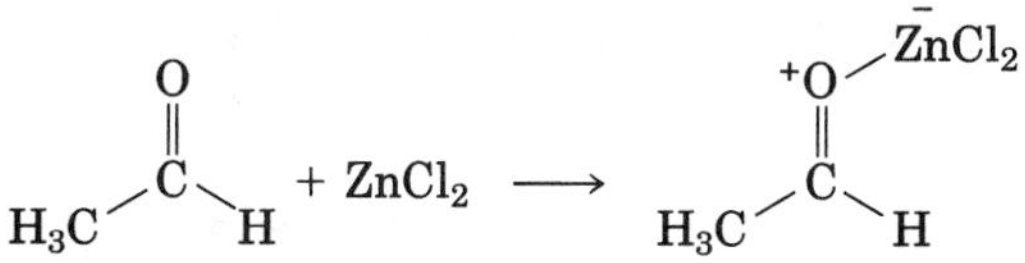 $H_3C{-}C({=}O){-}H + ZnCl_2 \longrightarrow H_3C{-}C({=}\overset{+}{O}{-}\overset{-}{Z}nCl_2){-}H$

1.51 The ammonium ion, NH_4^+, has a geometry identical to that of methane, CH_4. What kind of hybridization do you think the nitrogen atom has? Explain.

1.52 Draw a three-dimensional representation of ethane, CH_3CH_3, using normal, dashed, and wedged lines for both carbons.

1.53 Indicate the kind of hybridization you would expect for each carbon atom in these molecules:

(a) 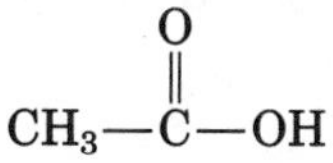 $CH_3{-}C({=}O){-}OH$ (b) 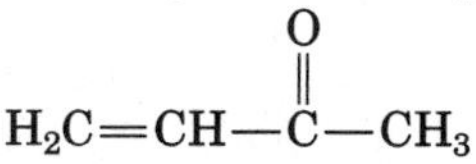 $H_2C{=}CH{-}C({=}O){-}CH_3$ (c) $H_2C{=}CH{-}C{\equiv}N$

Acetic acid **3-Buten-2-one** **Acrylonitrile**

1.54 Use Figure 1.17 to order the following molecules according to increasing positive character of the carbon atom: CH_3F, CH_3OH, CH_3Li, CH_3I, CH_3CH_3, CH_3NH_2.

1.55 Draw an orbital picture of allene, $H_2C{=}C{=}CH_2$. What hybridization must the central carbon atom have in order to form two double bonds? What shape do you predict for allene?

1.56 Draw a Lewis structure and an orbital picture for carbon dioxide, CO_2. What kind of hybridization do you think the carbon atom has? What is the relationship between CO_2 and allene (see Problem 1.55)?

1.57 Although most stable organic compounds have tetravalent carbon atoms, high-energy species with trivalent carbon atoms also exist. *Carbocations* are one such class of compounds. If the positively charged carbon atom has planar geometry, what hybridization do you think it has? How many valence electrons does the carbon have?

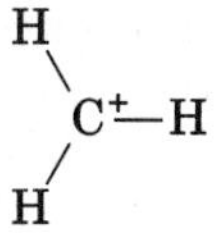

$$H_2\overset{+}{C}{-}H$$ **A carbocation**

2 CHAPTER

THE NATURE OF ORGANIC COMPOUNDS: ALKANES

There are more than *10 million* known organic compounds, each of which has its own unique physical and chemical properties. Fortunately, chemists have learned through years of experience that these compounds can be classified into families according to their structural features and that the chemical reactivity of the members of a given family is similar. Instead of 10 million compounds with random reactivity, there are a few dozen families of compounds whose chemistry is reasonably predictable. We'll study the chemistry of specific families of organic molecules throughout this book, beginning in this chapter with a look at the simplest family, the *alkanes*.

2.1 FUNCTIONAL GROUPS

FUNCTIONAL GROUP
Atom or group of atoms within a molecule that has a characteristic chemical behavior

The structural features that make it possible to classify compounds by reactivity are called functional groups. A **functional group** is a part of a larger molecule; it is composed of an atom or group of atoms that have a characteristic chemical behavior. Chemically, a given functional group behaves in nearly the same way in every molecule it's a part of. For example, one of the simplest functional groups is the carbon–carbon double bond. Because the electronic structure of the carbon–carbon double bond remains essentially the same in all molecules where it occurs, its chemical reactivity also remains the same. Ethylene, the simplest compound with a carbon–carbon double bond, undergoes reactions that are remarkably similar to those of α-pinene, a seemingly more complicated molecule (and major component of turpentine). Both, for example, react with bromine to give products in which a bromine atom has added to each of the double-bond carbons (Figure 2.1).

The example shown in Figure 2.1 is typical: *The chemistry of every organic molecule, regardless of size and complexity, is determined by the functional groups it contains.* Table 2.1 lists many of the common func-

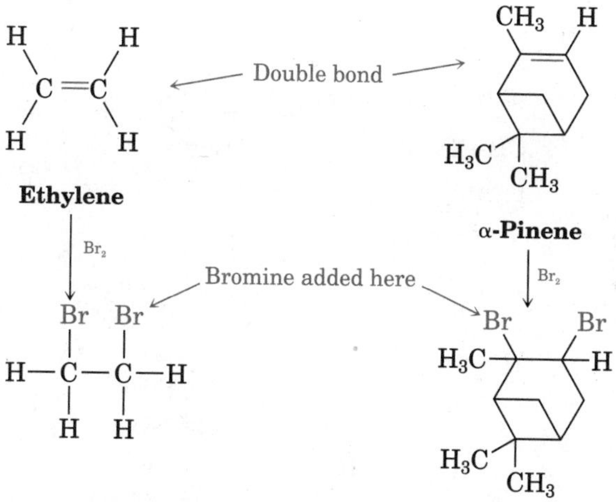

Double bond

Ethylene

α-Pinene

Br_2

Bromine added here

FIGURE 2.1 The reactions of ethylene and α-pinene with bromine. In both cases, bromine reacts with the C=C double-bond functional group in exactly the same way. The size and nature of the remainder of the molecule doesn't matter.

tional groups and gives simple examples of their occurrence. Look carefully at this table to see the many types of functional groups found in organic compounds. Some functional groups, such as alkenes, alkynes, and aromatic rings, have only carbon–carbon double or triple bonds; others have halogen, and still others have oxygen, nitrogen, or sulfur. Much of the chemistry you'll be studying in the remainder of this book is the chemistry of these functional groups.

It's a good idea at this point to familiarize yourself with the structures of the functional groups shown in Table 2.1 so that they'll be familiar when you see them again. They can be grouped into several categories:

Functional Groups with Carbon–Carbon Multiple Bonds

Alkenes, alkynes, and arenes (aromatic compounds) all contain carbon–carbon multiple bonds. Alkenes have a double bond, alkynes have a triple bond, and arenes have three alternating double and single bonds in a six-membered ring of carbon atoms. Because of their structural similarities, these compounds also have some chemical similarities.

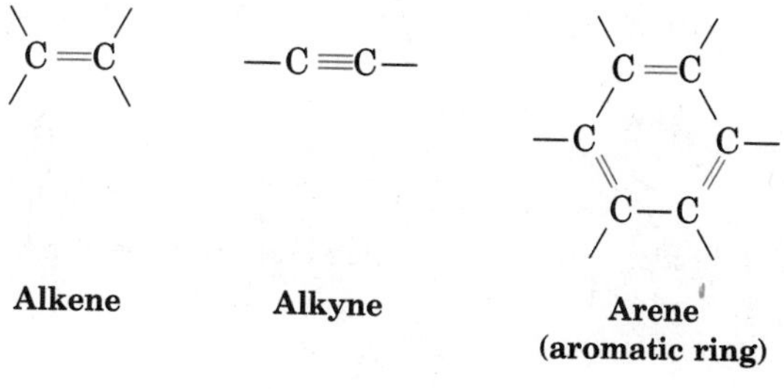

Alkene **Alkyne** **Arene (aromatic ring)**

TABLE 2.1 Structure of Some Important Functional Groups

FAMILY NAME	FUNCTIONAL GROUP STRUCTURE[a]	SIMPLE EXAMPLE	NAME ENDING
Alkane	(Contains only C–H and C–C single bonds)	CH_3CH_3	*-ane* Ethane
Alkene	$>C=C<$	$H_2C=CH_2$	*-ene* Ethene (Ethylene)
Alkyne	$-C\equiv C-$	$H-C\equiv C-H$	*-yne* Ethyne (Acetylene)
Arene	Six-membered ring of C atoms with alternating single and double bonds, each C with one open bond	C_6H_6 (each ring C bearing one H)	None Benzene
Halide	$-\overset{\vert}{\underset{\vert}{C}}-\ddot{\underset{..}{X}}:$ (X = F, Cl, Br, I)	H_3C-Cl	None Chloromethane
Alcohol	$-\overset{\vert}{\underset{\vert}{C}}-\ddot{\underset{..}{O}}-H$	H_3C-O-H	*-ol* Methanol
Ether	$-\overset{\vert}{\underset{\vert}{C}}-\ddot{\underset{..}{O}}-\overset{\vert}{\underset{\vert}{C}}-$	$H_3C-O-CH_3$	*ether* Dimethyl ether
Amine	$-\overset{\vert}{\underset{\vert}{C}}-\ddot{N}H_2$, $-\overset{\vert}{\underset{\vert}{C}}-\underset{\vert}{\ddot{N}}-H$, $-\overset{\vert}{\underset{\vert}{C}}-\underset{\vert}{\ddot{N}}-$	H_3C-NH_2	*-amine* Methylamine
Nitrile	$-\overset{\vert}{\underset{\vert}{C}}-C\equiv N:$	$H_3C-C\equiv N$	*-nitrile* Ethanenitrile (Acetonitrile)
Sulfide	$-\overset{\vert}{\underset{\vert}{C}}-\ddot{\underset{..}{S}}-\overset{\vert}{\underset{\vert}{C}}-$	$H_3C-S-CH_3$	*-sulfide* Dimethyl sulfide
Thiol	$-\overset{\vert}{\underset{\vert}{C}}-\ddot{\underset{..}{S}}-H$	H_3C-SH	*-thiol* Methanethiol

[a]The bonds whose connections aren't specified are assumed to be attached to carbon or hydrogen atoms in the rest of the molecule.

TABLE 2.1 (Continued)

FAMILY NAME	FUNCTIONAL GROUP STRUCTURE[a]	SIMPLE EXAMPLE	NAME ENDING
CARBONYL,	$-\overset{\overset{\large :O:}{\|\|}}{C}-$		
Aldehyde	$-\overset{\vert}{\underset{\vert}{C}}-\overset{\overset{:O:}{\|\|}}{C}-H$	$H_3C-\overset{\overset{O}{\|\|}}{C}-H$	*-al* Ethanal (Acetaldehyde)
Ketone	$-\overset{\vert}{\underset{\vert}{C}}-\overset{\overset{:O:}{\|\|}}{C}-\overset{\vert}{\underset{\vert}{C}}-$	$H_3C-\overset{\overset{O}{\|\|}}{C}-CH_3$	*-one* Propanone (Acetone)
Carboxylic acid	$-\overset{\vert}{\underset{\vert}{C}}-\overset{\overset{:O:}{\|\|}}{C}-\ddot{\underset{\cdot\cdot}{O}}H$	$H_3C-\overset{\overset{O}{\|\|}}{C}-OH$	*-oic acid* Ethanoic acid (Acetic acid)
Ester	$-\overset{\vert}{\underset{\vert}{C}}-\overset{\overset{:O:}{\|\|}}{C}-\ddot{\underset{\cdot\cdot}{O}}-\overset{\vert}{\underset{\vert}{C}}-$	$H_3C-\overset{\overset{O}{\|\|}}{C}-O-CH_3$	*-oate* Methyl ethanoate (Methyl acetate)
Amide	$-\overset{\vert}{\underset{\vert}{C}}-\overset{\overset{:O:}{\|\|}}{C}-\ddot{N}H_2$, $-\overset{\vert}{\underset{\vert}{C}}-\overset{\overset{:O:}{\|\|}}{C}-\underset{\vert}{\ddot{N}}-H$, $-\overset{\vert}{\underset{\vert}{C}}-\overset{\overset{:O:}{\|\|}}{C}-\underset{\vert}{\ddot{N}}-$	$H_3C-\overset{\overset{O}{\|\|}}{C}-NH_2$	*-amide* Ethanamide (Acetamide)

Functional Groups with Carbon Singly Bonded to an Electronegative Atom

Alkyl halides, alcohols, ethers, amines, thiols, and sulfides all have a carbon atom singly bonded to an electronegative atom—a halogen, an oxygen, a nitrogen, or a sulfur. Alkyl halides have a carbon atom bonded to a halogen (–X), alcohols have a carbon atom bonded to a hydroxyl (–OH) group, ethers have two carbon atoms bonded to the same oxygen, amines have a carbon atom bonded to a nitrogen, thiols have a carbon atom bonded to an –SH group, and sulfides have two carbon atoms bonded to the same sulfur. In all cases, the bonds are polar, with the carbon atom bearing a slight positive charge (δ^+) and the electronegative atom bearing a slight negative charge (δ^-).

Alkyl halide Alcohol Ether Amine Thiol Sulfide

Functional Groups with a Carbon–Oxygen Double Bond (Carbonyl Groups)

CARBONYL GROUP
The carbon–oxygen double bond, C=O

Note particularly in Table 2.1 the different families of compounds that contain the **carbonyl group, C=O** (pronounced car-bo-**neel**). Carbon–oxygen double bonds are present in some of the most important compounds in organic chemistry. These compounds are similar in many respects but differ depending on the identity of the atoms bonded to the carbonyl-group carbon. Aldehydes have one carbon and one hydrogen bonded to the C=O, ketones have two carbons bonded to the C=O, carboxylic acids have one carbon and one –OH group bonded to the C=O, esters have one carbon and one ether-like oxygen bonded to the C=O, amides have one carbon and one nitrogen bonded to the C=O, acid chlorides have one carbon and one chlorine bonded to the C=O, and so on.

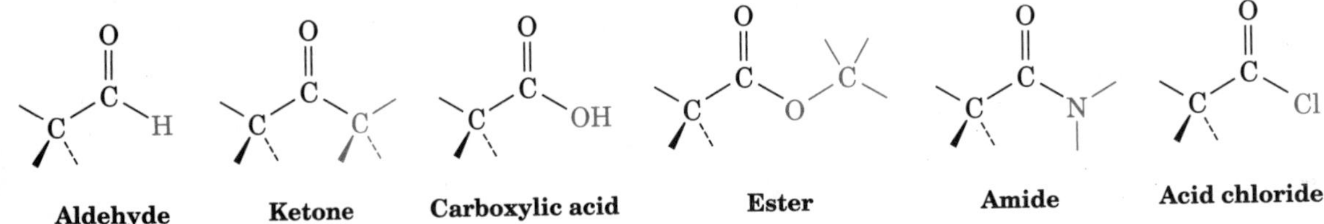

PROBLEM 2.1 Circle and identify the functional groups in these molecules:

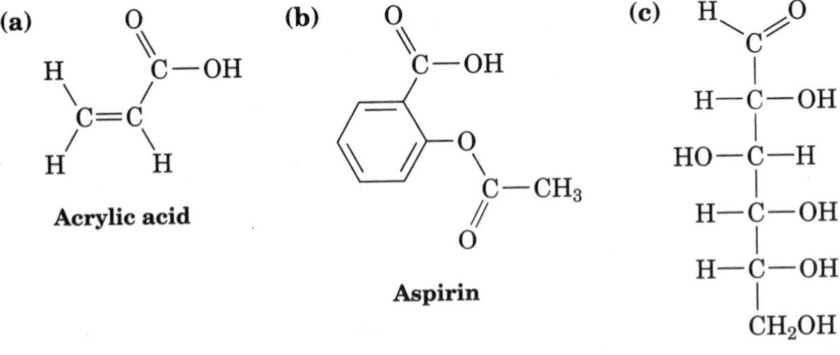

Glucose

PROBLEM 2.2 Propose structures for simple molecules that contain these functional groups:

(a) Alcohol **(b)** Aromatic ring **(c)** Carboxylic acid
(d) Amine **(e)** Both ketone and amine **(f)** Two double bonds

2.2 ALKANES AND ALKYL GROUPS: ISOMERS

ALKANE
Compound of carbon and hydrogen that has only single bonds

We saw in Section 1.9 that the carbon–carbon single bond in ethane results from sigma (head-on) overlap of carbon sp^3 orbitals. If we imagine joining three, four, five, or even more carbon atoms by carbon–carbon single bonds, we can generate the large family of molecules called **alkanes.**

```
   H           H  H            H  H  H             H  H  H  H
   |           |  |            |  |  |             |  |  |  |
H—C—H      H—C—C—H      H—C—C—C—H      H—C—C—C—C—H     . . . and so on
   |           |  |            |  |  |             |  |  |  |
   H           H  H            H  H  H             H  H  H  H
```

Methane **Ethane** **Propane** **Butane**

HYDROCARBON
Compound that has only carbon and hydrogen

SATURATED
Compound that has only single bonds

ALIPHATIC
Alternative word describing alkanes

Alkanes are often described as **saturated hydrocarbons**: **hydrocarbons** because they contain only carbon and hydrogen; saturated because they have only C–C and C–H single bonds and thus contain the maximum possible number of hydrogens per carbon. They have the general formula C_nH_{2n+2}, where n is any integer. Alkanes are also occasionally referred to as **aliphatic** compounds, a name derived from the Greek *aleiphas*, meaning "fat." We'll see later that animal fats contain long carbon chains similar to alkanes.

Think about the ways that carbon and hydrogen can combine to make alkanes. With one carbon and four hydrogens, only one structure is possible: methane, CH_4. Similarly, there is only one possible combination of two carbons with six hydrogens (ethane, CH_3CH_3) and only one possible combination of three carbons with eight hydrogens (propane, $CH_3CH_2CH_3$). If larger numbers of carbons and hydrogens combine, however, more than one kind of molecule can form. For example, there are *two* ways that molecules with the formula C_4H_{10} can form: The four carbons can be in a row (butane), or they can branch (isobutane). Similarly, there are three ways in which C_5H_{12} molecules can form, and so on for larger alkanes:

CH_4 C_2H_6 C_3H_8

```
   H            H  H             H  H  H
   |            |  |             |  |  |
H—C—H       H—C—C—H       H—C—C—C—H
   |            |  |             |  |  |
   H            H  H             H  H  H
```

Methane **Ethane** **Propane**

C_4H_{10}

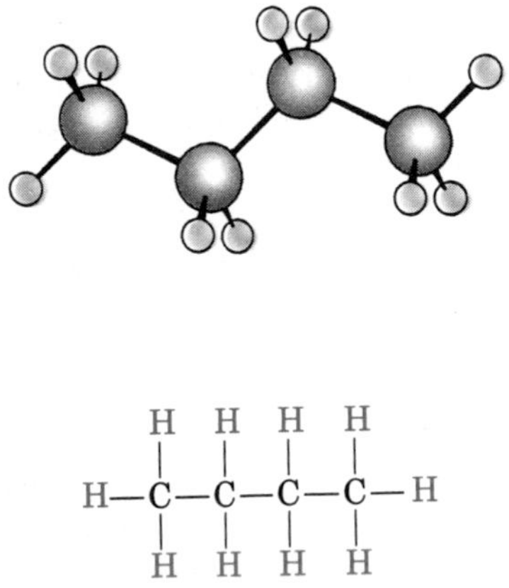

Butane

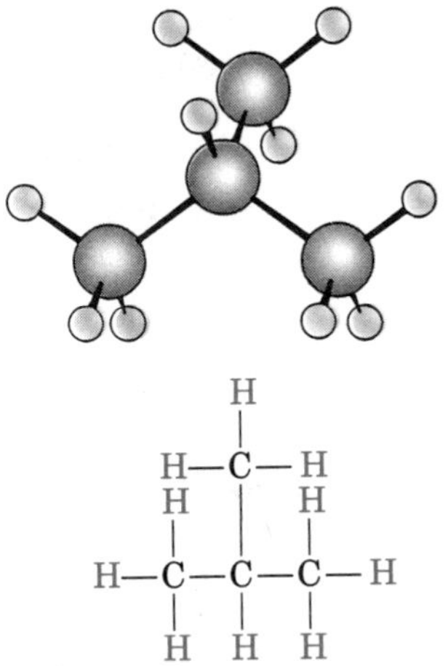

Isobutane (2-methylpropane)

C_5H_{12}

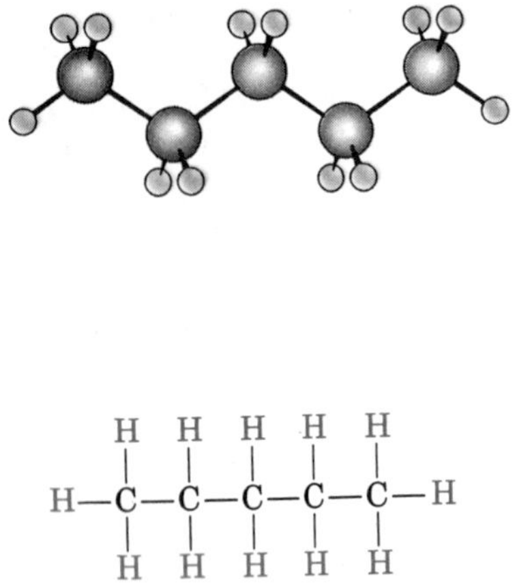

Pentane

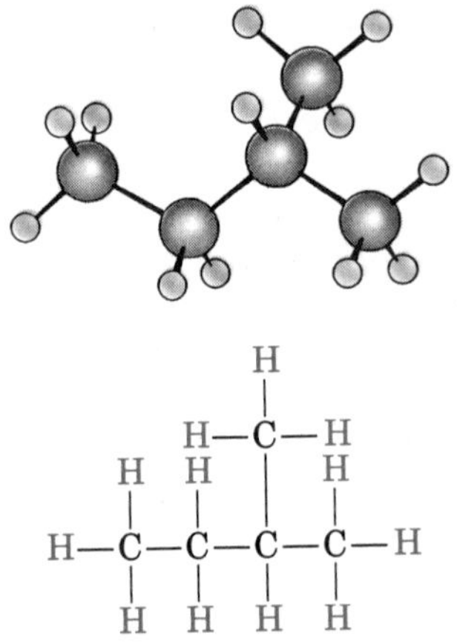

2-Methylbutane

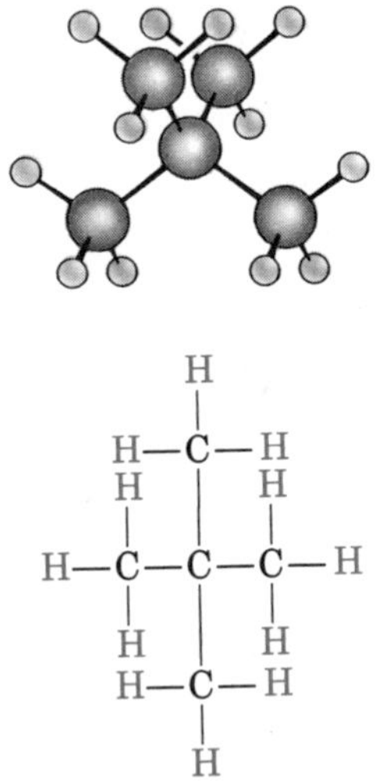

2,2-Dimethylpropane

STRAIGHT-CHAIN ALKANE Alkane with carbon atoms connected in a row

NORMAL ALKANE Straight-chain alkane

Compounds like butane, with carbons connected in a row, are called **straight-chain alkanes**, or **normal alkanes**, whereas compounds with branched carbon chains, such as isobutane (2-methylpropane), are called **branched-chain alkanes**. The difference between the two is that you can draw a line connecting all the carbons of a straight-chain alkane without retracing your path or lifting your pencil from the paper. For a

BRANCHED-CHAIN ALKANE
Alkane that contains a branching arrangement of carbon atoms in its chain

ISOMERS
Compounds that have the same formula but different chemical structures

CONSTITUTIONAL ISOMERS
Isomers that have their atoms connected in a different order

branched-chain alkane, however, you either have to retrace your path or lift your pencil from the paper in order to draw a line connecting all the carbons.

Compounds like the two C_4H_{10} molecules that have the same formula but different structures are called **isomers**, from the Greek *isos* + *meros* meaning "made of the same parts." Isomers have the same numbers and kinds of atoms but differ in the way these atoms are arranged. Compounds like butane and isobutane, whose atoms are connected differently, are called **constitutional isomers.** We'll see shortly that other kinds of isomerism are also possible, even among compounds whose atoms are connected in the same order. As Table 2.2 shows, the number of possible alkane isomers increases dramatically as the number of carbon atoms increases.

TABLE 2.2 Number of Alkane Isomers

FORMULA	NUMBER OF ISOMERS	FORMULA	NUMBER OF ISOMERS
C_6H_{14}	5	$C_{10}H_{22}$	75
C_7H_{16}	9	$C_{15}H_{32}$	4,347
C_8H_{18}	18	$C_{20}H_{42}$	366,319
C_9H_{20}	35	$C_{30}H_{62}$	4,111,846,763

CONDENSED STRUCTURE
Shorthand way of writing structures in which bonds are understood rather than shown

A given alkane can be arbitrarily drawn in many ways. For example, the straight-chain, four-carbon alkane called butane can be represented by any of the structures shown in Figure 2.2. These structures don't imply any particular three-dimensional geometry for butane; they only indicate the connections among its atoms. In practice, we soon tire of drawing all the bonds in a molecule and usually refer to butane by the shorthand **condensed structure**, $CH_3CH_2CH_2CH_3$, or even more simply as n-C_4H_{10}, where n signifies *normal*, straight-chain butane. In condensed structures like $CH_3CH_2CH_2CH_3$, the C–C and C–H bonds aren't usually shown but are "understood." If a carbon has three hydrogens bonded to it, we write CH_3; if a carbon has two hydrogens bonded to it, we write CH_2; and so on.

$CH_3—CH_2—CH_2—CH_3$

$CH_3—CH_2—CH_2$ (with CH_3 attached above the last CH_2)

$CH_3—CH_2$ (with $CH_2—CH_3$ attached above the CH_2)

$CH_2—CH_2—CH_3$ (with CH_3 attached below the first CH_2)

$CH_3(CH_2)_2CH_3$

$CH_3CH_2CH_2CH_3$

FIGURE 2.2 Some representations of butane, n-C_4H_{10}. The molecule is the same regardless of how it's drawn. These structures imply only that butane has a continuous chain of four carbon atoms.

Straight-chain alkanes are named according to the number of carbon atoms in the chain, as shown in Table 2.3. With the exception of the first four compounds—methane, ethane, propane, and butane—whose names have historical origins, the alkanes are named based on Greek numbers, according to the number of carbons in the molecule. The suffix *-ane* is added to the end of each name to identify the molecule as an alkane.

TABLE 2.3 Names of Straight-Chain Alkanes

NUMBER OF CARBONS (n)	NAME	FORMULA (C_nH_{2n+2})	NUMBER OF CARBONS (n)	NAME	FORMULA (C_nH_{2n+2})
1	Methane	CH_4	8	Octane	C_8H_{18}
2	Ethane	C_2H_6	9	Nonane	C_9H_{20}
3	Propane	C_3H_8	10	Decane	$C_{10}H_{22}$
4	Butane	C_4H_{10}	11	Undecane	$C_{11}H_{24}$
5	Pentane	C_5H_{12}	12	Dodecane	$C_{12}H_{26}$
6	Hexane	C_6H_{14}	20	Icosane	$C_{20}H_{42}$
7	Heptane	C_7H_{16}	30	Triacontane	$C_{30}H_{62}$

ALKYL GROUP
Partial structure formed by removing a hydrogen from an alkane

If one hydrogen atom is removed from an alkane, the partial structure that remains is called an **alkyl group**. Alkyl groups are named by replacing the *-ane* ending of the parent alkane with an *-yl* ending. For example, removal of a hydrogen atom from methane, CH_4, generates a *methyl group*, $-CH_3$, and removal of a hydrogen atom from ethane, CH_3CH_3, generates an *ethyl group*, $-CH_2CH_3$. Similarly, removal of a hydrogen atom from the end carbon of any *n*-alkane gives the series of *n*-alkyl groups shown in Table 2.4.

TABLE 2.4 Some Straight-Chain Alkyl Groups

ALKANE	NAME	ALKYL GROUP	NAME
CH_4	Methane	$-CH_3$	Methyl (Me)
CH_3CH_3	Ethane	$-CH_2CH_3$	Ethyl (Et)
$CH_3CH_2CH_3$	Propane	$-CH_2CH_2CH_3$	Propyl (Pr)
$CH_3CH_2CH_2CH_3$	Butane	$-CH_2CH_2CH_2CH_3$	Butyl (Bu)
$CH_3CH_2CH_2CH_2CH_3$	Pentane	$-CH_2CH_2CH_2CH_2CH_3$	Pentyl

Just as *n*-alkyl groups are generated by removing a hydrogen from an *end* carbon, branched alkyl groups are generated by removing a hydrogen atom from an *internal* carbon. Two three-carbon alkyl groups and four four-carbon alkyl groups are possible (Figure 2.3).

One further word of explanation about naming alkyl groups: The prefixes used for the C_4 alkyl groups in Figure 2.3—*sec* (for secondary) and *tert* (for tertiary)—refer to the degree of alkyl substitution at a carbon

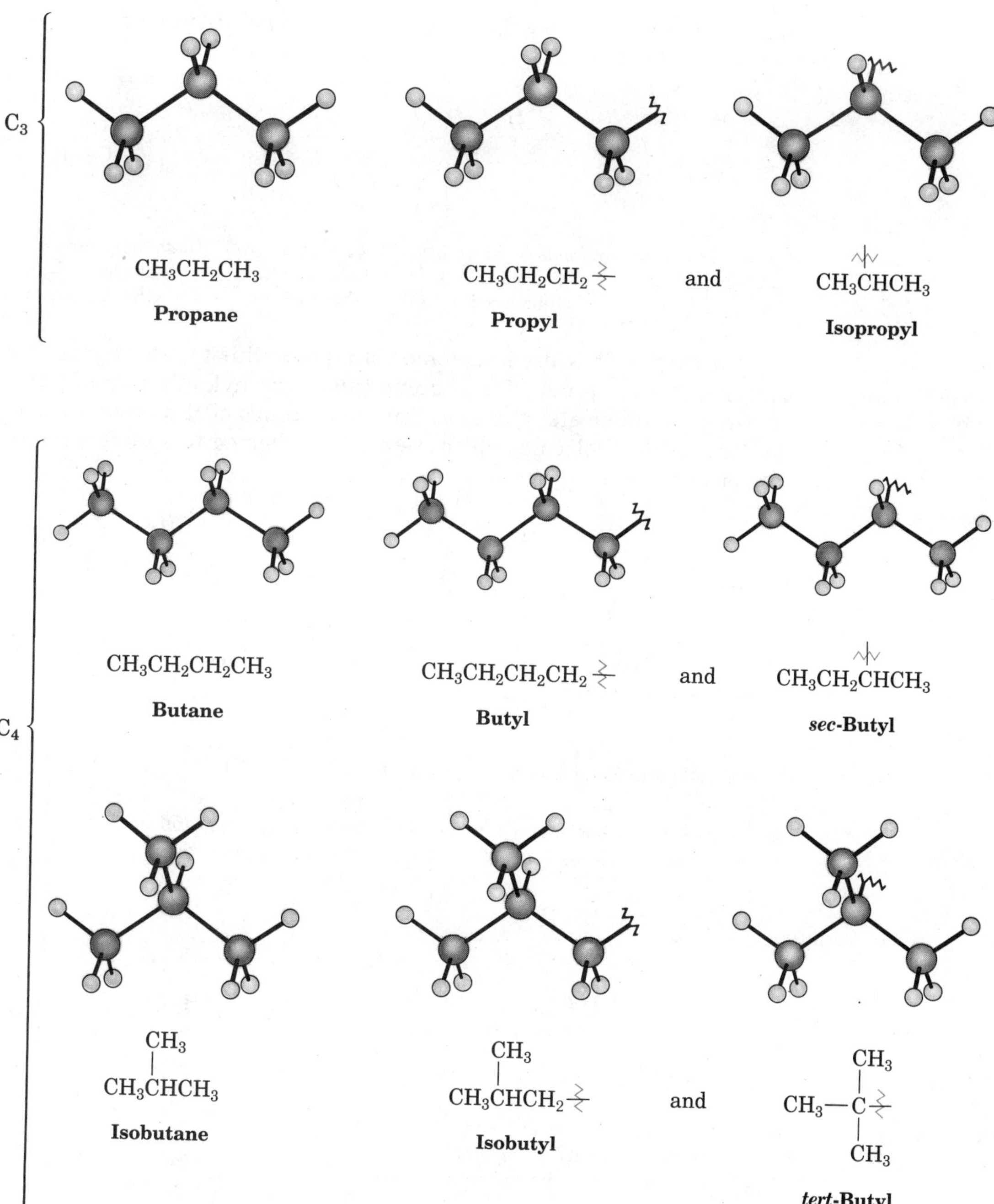

FIGURE 2.3 Generation of straight-chain and branched-chain alkyl groups from *n*-alkanes.

atom. There are four possible degrees of alkyl substitution for carbon, denoted 1°, 2°, 3°, and 4°:

```
     H              H              R              R
     |              |              |              |
  R—C—H          R—C—H          R—C—H          R—C—R
     |              |              |              |
     H              R              R              R
```

Primary carbon (1°) is bonded to one other carbon	*Secondary* carbon (2°) is bonded to two other carbons	*Tertiary* carbon (3°) is bonded to three other carbons	*Quaternary* carbon (4°) is bonded to four other carbons

R GROUP General symbol used for an organic partial structure

The symbol **R** is used here and throughout this text to represent a generalized alkyl group. The R group can be methyl, ethyl, or any of a multitude of other alkyl groups. You might think of R as representing the **R**est of the molecule, which we aren't bothering to specify because it's not important.

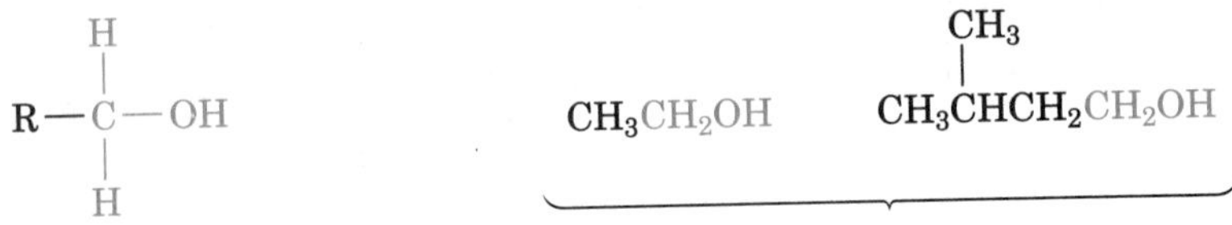

General class of primary alcohols, RCH_2OH

Specific examples of primary alcohols, RCH_2OH

PRACTICE PROBLEM 2.1 Propose structures for two isomers with the formula C_2H_6O.

SOLUTION We know that carbon forms four bonds, oxygen forms two, and hydrogen forms one. Putting the pieces together yields two isomeric structures:

```
   H  H                   H       H
   |  |                   |       |
H—C—C—O—H     and     H—C—O—C—H
   |  |                   |       |
   H  H                   H       H
```

PROBLEM 2.3 Draw structures of the five isomers of C_6H_{14}.

PROBLEM 2.4 Draw structures that meet these descriptions:
(a) Three isomers with the formula C_8H_{18}
(b) Two isomers with the formula $C_4H_8O_2$

PROBLEM 2.5 Draw the eight possible five-carbon alkyl groups (pentyl isomers).

PROBLEM 2.6 Draw alkanes that meet these descriptions:
(a) An alkane with two tertiary carbons
(b) An alkane that contains an isopropyl group
(c) An alkane that has one quaternary and one secondary carbon

PROBLEM 2.7 Identify the carbon atoms in these molecules as primary, secondary, tertiary, or quaternary:

(a) $CH_3CH(CH_3)CH_2CH_2CH_3$

$$\begin{array}{l} \quad\;\; CH_3 \\ \quad\;\;\; | \\ CH_3CHCH_2CH_2CH_3 \end{array}$$

(b) $CH_3CH_2CH(CH(CH_3)_2)CH_2CH_3$

$$\begin{array}{l} \qquad\qquad CH_3CHCH_3 \\ \qquad\qquad\quad\;\; | \\ CH_3CH_2CHCH_2CH_3 \end{array}$$

(c) $CH_3CH(CH_3)CH_2C(CH_3)_2CH_3$

$$\begin{array}{l} \quad\;\; CH_3 \qquad\; CH_3 \\ \quad\;\;\; | \qquad\qquad | \\ CH_3CHCH_2CCH_3 \\ \qquad\qquad\quad\; | \\ \qquad\qquad\;\; CH_3 \end{array}$$

2.3 NAMING BRANCHED-CHAIN ALKANES

In earlier times when few pure organic chemicals were known, new compounds were named at the whim of their discoverer. Thus, urea (CH_4N_2O) is a crystalline substance isolated from urine, and barbituric acid is a tranquilizing agent named by its discoverer in honor of his friend Barbara. As the science of organic chemistry grew in the nineteenth century, however, so too did the number of known compounds and the need for a rational and systematic method of naming them. The system of nomenclature we'll use in this book is that devised by the International Union of Pure and Applied Chemistry (IUPAC, usually spoken as **eye**-you-pac).

A chemical name has three parts in the IUPAC system: prefix, parent, and suffix. The parent name specifies the overall size of the molecule by identifying the number of carbon atoms in its main chain; the suffix identifies the family that the molecule belongs to; and the prefix specifies the location of various substituents on the main chain:

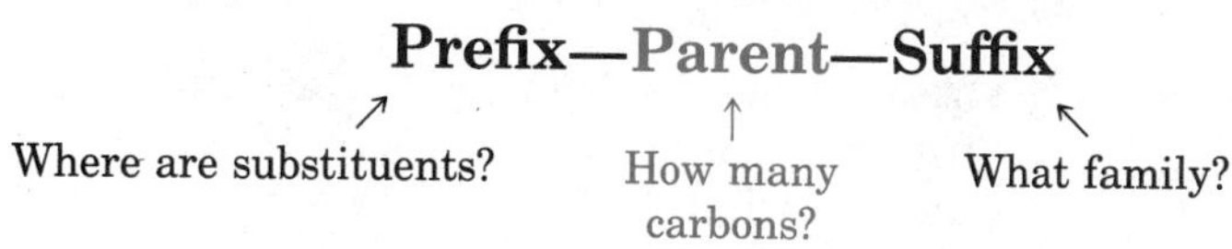

As we cover new functional groups in later chapters, the applicable IUPAC rules of nomenclature will be given. In addition, the Appendix gives an overall view of organic nomenclature and shows how compounds that contain more than one functional group are named. For the present, though, let's see how branched-chain alkanes are named. All but the most complex branched-chain alkanes can be named by following four steps:

STEP 1 Find the parent hydrocarbon.

a. Find the *longest continuous carbon chain* in the molecule and use the name of that chain as the parent name. The longest chain may not always be obvious; you may have to "turn corners":

$$\begin{array}{l} \qquad\qquad\quad\;\; CH_2CH_3 \\ \qquad\qquad\qquad | \\ CH_3CH_2CH_2CH—CH_3 \end{array}$$

Named as a substituted *hexane*

b. If two chains of equal length are present, choose the one with the larger number of branch points as the parent:

```
      CH3
      |
CH3CHCHCH2CH2CH3          Named as a hexane with two substituents
        |
        CH2CH3
                          NOT
      CH3
      |
CH3CH—CHCH2CH2CH3         as a hexane with one substituent
         |
         CH2CH3
```

STEP 2 Beginning at the end *nearer the first branch point*, number each carbon atom in the parent chain:

```
      2    1                             6    7
      CH2CH3                             CH2CH3
      |                                  |
CH3—CHCH—CH2CH3         NOT       CH3—CHCH—CH2CH3
      3  4|                              5  4|
         CH2CH2CH3                          CH2CH2CH3
         5    6   7                         3    2   1
```

The first branch occurs at C3 in the proper system of numbering, *not* at C4 as shown in the improper system.

STEP 3 Assign a number to each substituent according to its point of attachment on the parent chain. If there are two substituents on the same carbon, assign them both the same number. There must always be as many numbers in the name as there are substituents.

```
10  9  8                                     Substituents:
CH3CH2CH2                CH2CH3              On C3, –CH2CH3    (3-ethyl)
       |                 |                   On C4, –CH3       (4-methyl)
CH3—CHCH2CH2CHCHCH2CH3                       On C7, –CH3       (7-methyl)
       7   6    5  |4  3   2   1
                   CH3

       CH3     CH3                           Substituents:
       |       |                             On C2, –CH3       (2-methyl)
CH3CH2—C—CH2CHCH3                            On C4, –CH3       (4-methyl)
 6   5  4|  3  2  1                          On C4, –CH2CH3    (4-ethyl)
       CH2CH3
```

STEP 4 Write the name as a single word, using hyphens to separate the various prefixes and commas to separate numbers. If two or more different side chains are present, cite them in alphabetical order. If two or more identical side chains are present, use one of the prefixes *di-*, *tri-*, *tetra-*, and so forth. Don't use these prefixes for alphabetizing, though.

```
           2   1
           CH2CH3
6  5   4  3|
CH3CH2CH2CH—CH3
```

3-Methylhexane

```
H3C CH2CH3
  |  |
CH3CHCHCH2CH2CH3
1   2  3  4   5  6
```

3-Ethyl-2-methylhexane

```
          CH3      CH3
           |        |
CH3CH2—C—CH2CHCH3
6    5    4|   3    2   1
          CH2CH3
```

4-Ethyl-2,4-dimethylhexane

```
       9   8
       CH3CH2              CH2CH3
           |                 |
CH3—CHCH2CH2CHCHCH2CH3
         7   6    5    |4  3   2   1
                       CH3
```

3-Ethyl-4,7-dimethylnonane

PRACTICE PROBLEM 2.2 What is the IUPAC name of this alkane?

```
   CH2CH3        CH3
   |             |
CH3CHCH2CH2CH2CHCH3
```

SOLUTION The molecule has a chain of eight carbons (octane) with two methyl substituents. Numbering from the end nearer the first methyl substituent indicates that the methyls are at C2 and C6, giving the name 2,6-dimethyloctane.

```
   7   8
   CH2CH3        CH3
   |             |
CH3CHCH2CH2CH2CHCH3
   6  5   4   3   2  1
```

PRACTICE PROBLEM 2.3 Draw the structure of 3-isopropyl-2-methylhexane.

SOLUTION First, look at the parent name (hexane) and draw its carbon structure:

C—C—C—C—C—C **Hexane**

Next, find the substituents (3-isopropyl and 2-methyl), and place them on the proper carbons:

```
      CH3CHCH3   ←—   An isopropyl group at C3
          |
C—C—C—C—C—C
1  2|  3   4   5   6
    CH3   ←———————   A methyl group at C2
```

Finally, add hydrogens to complete the structure:

```
         CH3CHCH3
            |
      CH3CHCHCH2CH2CH3        3-Isopropyl-2-methylhexane
          |
         CH3
```

PROBLEM 2.8 Give IUPAC names for these alkanes:

(a) The three isomers of C_5H_{12}

(b)
```
                CH3
                |
     CH3CH2CHCHCH3
                  |
                 CH2CH3
```

(c)
```
      CH3    CH3
       |      |
   CH3CHCH2CHCH3
```

(d)
```
            CH3          CH2CH3
             |             |
     CH3—C—CH2CH2CHCH3
             |
            CH3
```

PROBLEM 2.9 Draw structures corresponding to these IUPAC names:

(a) 3,4-Dimethylnonane
(b) 3-Ethyl-4,4-dimethylheptane
(c) 2,2-Dimethyl-4-propyloctane
(d) 2,2,4-Trimethylpentane

2.4 PROPERTIES OF ALKANES

PARAFFIN Alternative name for alkanes

Alkanes are sometimes referred to as **paraffins**, a word derived from the Latin *parum affinis* meaning "slight affinity." This term aptly describes their behavior, since alkanes show little chemical affinity for other molecules and are inert to most laboratory reagents. They do, however, react with oxygen, chlorine, and a few other substances under appropriate conditions.

The reaction of alkanes with oxygen occurs during combustion in an engine or furnace when the alkane is used as a fuel. Carbon dioxide and water are formed as products, and a large amount of heat is released. For example, methane (natural gas) reacts with oxygen according to the equation

$$CH_4 + 2\ O_2 \longrightarrow CO_2 + 2\ H_2O + 213\ \text{kcal}\ (890\ \text{kJ})$$

The reaction of an alkane with chlorine occurs when a mixture of the two is irradiated with ultraviolet light (denoted $h\nu$, where ν is the Greek letter nu). Depending on the relative amounts of the two reactants and on the time allowed for reaction, a sequential replacement of the alkane hydrogen atoms by chlorine occurs, leading to a mixture of chlorinated products. Methane, for instance, reacts with chlorine to yield a mixture of chloromethane (CH_3Cl), dichloromethane (CH_2Cl_2), trichloromethane ($CHCl_3$), and tetrachloromethane (CCl_4). We'll see how this chlorination reaction occurs when we take up the chemistry of alkyl halides in Chapter 7.

$$CH_4 + Cl_2 \longrightarrow CH_3Cl + HCl \xrightarrow{Cl_2} CH_2Cl_2 + HCl \xrightarrow{Cl_2} CHCl_3 + HCl \xrightarrow{Cl_2} CCl_4 + HCl$$

Alkanes show regular increases in both boiling point and melting point as molecular weight increases (Figure 2.4). Average carbon–carbon bond parameters are nearly the same in all alkanes, with bond lengths of 1.54 ± 0.01 Å and bond strengths of 85 ± 3 kcal/mol (355 ± 10 kJ/mol). Carbon–hydrogen bond parameters are also nearly constant at 1.09 ± 0.01 Å and 95 ± 3 kcal/mol (400 ± 10 kJ/mol).

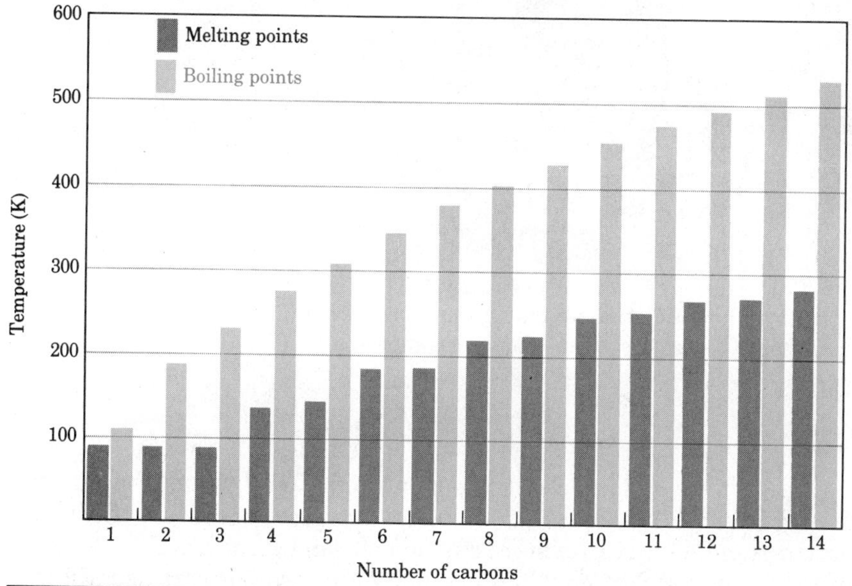

FIGURE 2.4 A plot of the number of carbons versus melting and boiling points for the C_1–C_{14} alkanes. There is a regular increase with molecular size.

2.5 CONFORMATIONS OF ETHANE

We saw earlier that carbon–carbon bonds in alkanes result from sigma overlap of two tetrahedral carbon sp^3 orbitals. Let's now look into the three-dimensional consequences of such bonding. What are the spatial relationships between the hydrogens on one carbon and the hydrogens on the other?

We know that sigma bonds result from the head-on overlap of two atomic orbitals and that a cross section through a sigma bond is circular. As a consequence of this circular symmetry, *rotation* is possible around carbon–carbon single bonds. Orbital overlap in the C–C bond is exactly the same for all geometric arrangements of the hydrogens (Figure 2.5).

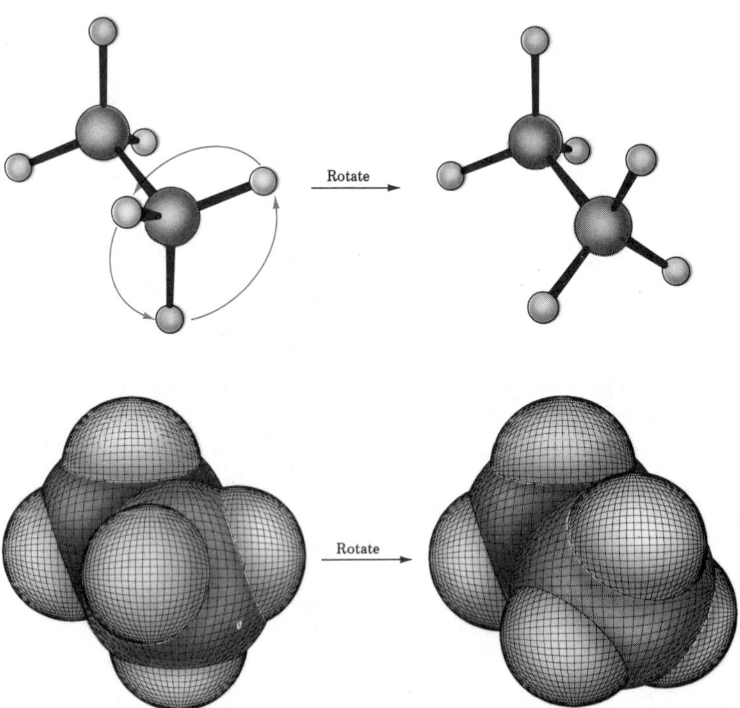

FIGURE 2.5 Two conformations of ethane. Rotation around the carbon–carbon single bond interconverts the forms. The drawings are computer-generated.

CONFORMATION
Specific three-dimensional arrangement of atoms in a molecule

CONFORMERS
Conformational isomers that differ only in rotation around a single bond

SAWHORSE REPRESENTATION
A way of viewing a molecule's spatial arrangement by looking at a carbon–carbon bond from an oblique angle

NEWMAN PROJECTION
A way of viewing a molecule's spatial arrangement by looking end-on at a carbon–carbon bond

The different spatial arrangements of atoms caused by rotation around a single bond are called **conformations**, and a specific conformation is called a **conformer** (**confor**mational iso**mer**). Unlike constitutional isomers (Section 2.2), though, different conformers usually can't be isolated because they interconvert too rapidly.

Chemists represent conformational isomers in two ways, as shown in Figure 2.6. **Sawhorse representations** view the carbon–carbon bond from an oblique angle and indicate spatial relationships by showing all the C–H bonds. **Newman projections** view the carbon–carbon bond end-on and represent the two carbon atoms by a circle. Bonds attached to the front carbon are represented by lines to the center of the circle, and bonds attached to the rear carbon are represented by lines to the edge of the circle.

In spite of what we've just said about sigma-bond symmetry, we don't observe *completely* free rotation in ethane. Experiments show that there is a slight (2.9 kcal/mol; 12 kJ/mol) barrier to rotation and that some conformations are more stable than others. The lowest-energy, most stable conformation is the one in which all six carbon–hydrogen bonds are as

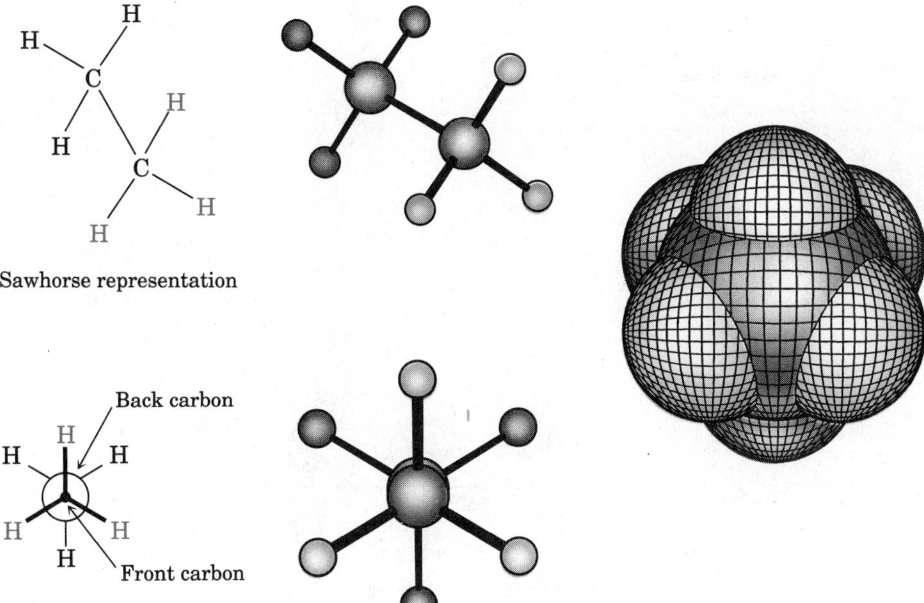

FIGURE 2.6 A sawhorse representation and a Newman projection of ethane.

STAGGERED The conformation around a carbon–carbon single bond that places all attached atoms as far away from each other as possible

ECLIPSED The conformation around a carbon–carbon single bond that places attached atoms as close as possible

far away from each other as possible (**staggered** when viewed end-on in a Newman projection). The highest-energy, least stable conformation is the one in which the six carbon–hydrogen bonds are as close as possible (**eclipsed** in a Newman projection). At any given instant, about 99% of ethane molecules have an approximately staggered conformation, and only about 1% are near the eclipsed conformation.

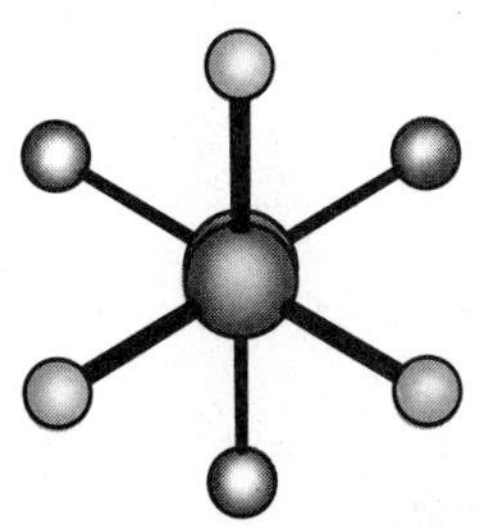

H H H H H H

Staggered conformation of ethane

Rotate rear carbon 60°

H 1 kcal/mol H H H H H 1 kcal/mol

Eclipsed conformation of ethane (C – H bonds on front and back carbon atoms are parallel)

The barrier to bond rotation in ethane is easier to see if you draw a graph of potential energy versus amount of bond rotation, as shown in Figure 2.7. The minimum energy geometries correspond to staggered

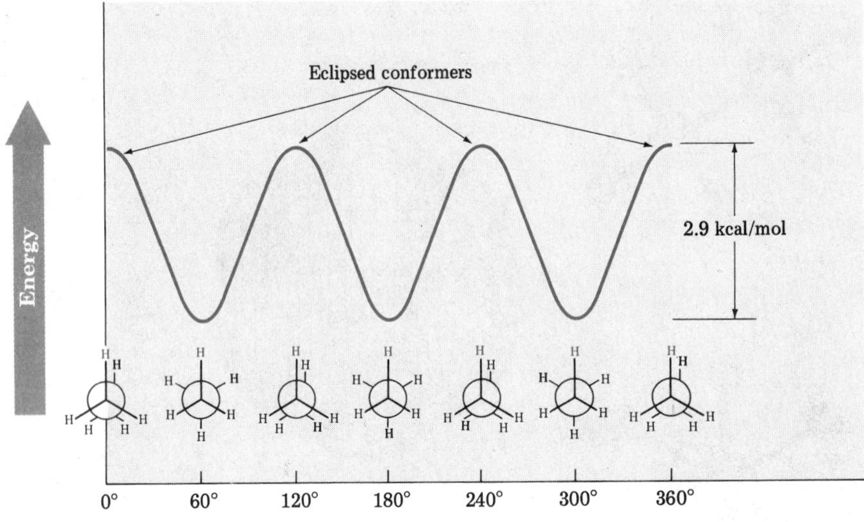

FIGURE 2.7 A graph of potential energy versus bond rotation in ethane. The staggered conformers are 2.9 kcal/mol lower in energy than the eclipsed conformers.

conformations, and the maximum energy geometries correspond to eclipsed conformations. The barrier is caused by the slight repulsion between electron clouds in the carbon–hydrogen bonds as they pass by each other at close quarters in the eclipsed conformer.

What is true for ethane is also true for propane, butane, and all higher alkanes. The most favored conformation for any alkane is the one in which all bonds have staggered arrangements (Figure 2.8).

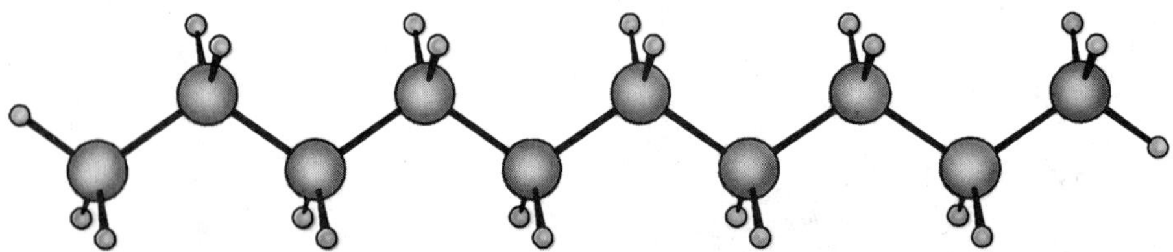

FIGURE 2.8 The most stable conformation of any alkane is the one in which the bonds on adjacent carbons are staggered and the carbon chain is fully extended, as in this computer-generated structure of decane.

PROBLEM 2.10 Sight along a carbon–carbon bond of propane and draw a Newman projection of the most stable conformation. Draw a Newman projection of the least stable conformation.

PROBLEM 2.11 Draw a graph, similar to Figure 2.7, of energy versus degree of bond rotation for propane.

PROBLEM 2.12 Looking along the C2–C3 bond of butane, there are two different staggered conformations and two different eclipsed conformations. Draw them.

PROBLEM 2.13 Which of the butane conformations you drew in Problem 2.12 do you think is the most stable? Why?

2.6 DRAWING CHEMICAL STRUCTURES

SKELETAL STRUCTURE Shorthand way of drawing structures that shows only bonds, not atoms

In the structures we've been using, a line between atoms represents the two electrons in a covalent bond. Most chemists find themselves drawing many structures each day, however, and it would soon become awkward if every bond and every atom had to be indicated. Chemists have therefore devised a shorthand way of drawing **skeletal structures** that greatly simplifies matters, particularly for the cyclic compounds that we'll see shortly.

The rules for drawing skeletal structures are simple:

RULE 1 Carbon atoms usually aren't shown. Instead, a carbon atom is assumed to be at the intersection of two lines (bonds) and at the end of each line. Occasionally, a carbon atom might be indicated for emphasis or for clarity.

RULE 2 Hydrogen atoms bonded to carbon aren't shown. Because carbon always has a valence of four, we mentally supply the correct number of hydrogen atoms for each carbon.

RULE 3 All atoms other than carbon and hydrogen *are* shown.

The following examples indicate how these rules are applied in specific cases:

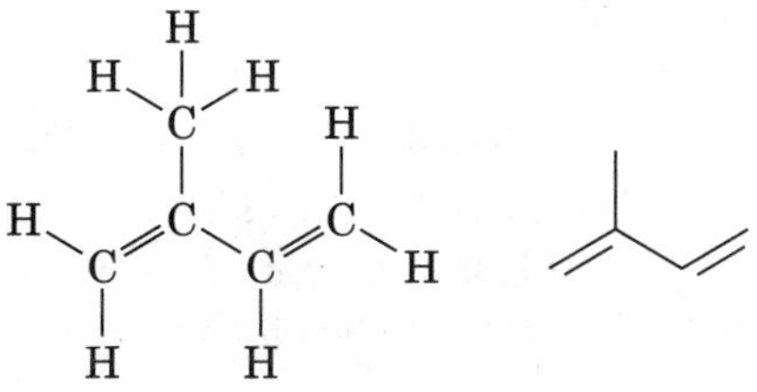

Isoprene, C_5H_8

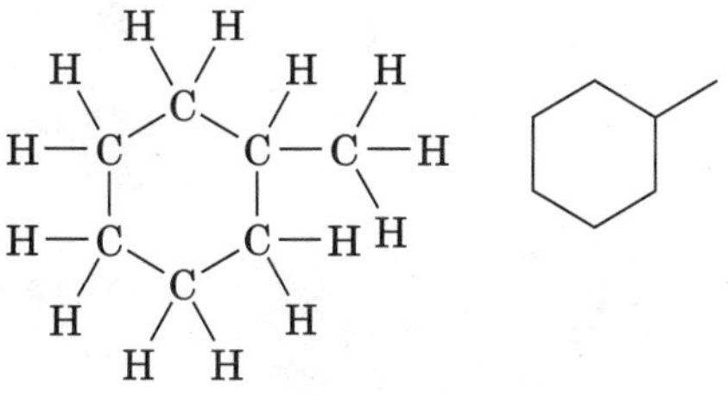

Methylcyclohexane, C_7H_{14}

PRACTICE PROBLEM 2.4 Convert this skeletal structure of adrenaline into a molecular formula.

HO, OH, HO—, $CHCH_2NHCH_3$ **Adrenaline**

SOLUTION Remember that each intersection of lines is a carbon atom: $C_9H_{13}NO_3$

PROBLEM 2.14 Convert these skeletal structures into molecular formulas:

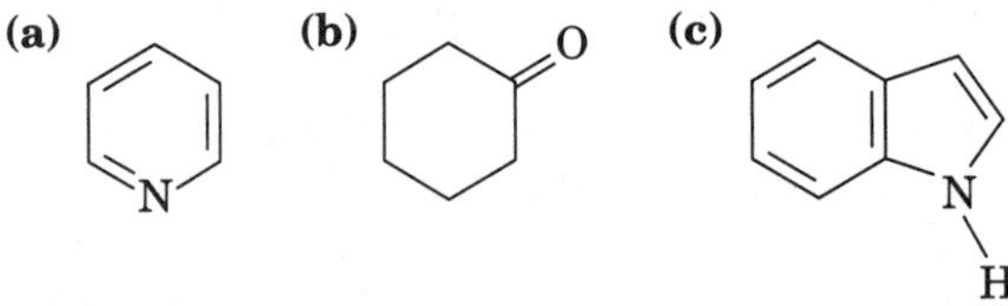

Pyridine **Cyclohexanone** **Indole**

PROBLEM 2.15 Propose skeletal structures for these molecular formulas:
(a) C_4H_8 (b) C_3H_6O (c) C_4H_9Cl

2.7 CYCLOALKANES

CYCLOALKANE Alkane with a ring of carbons

ALICYCLIC Alternative term for cycloalkane

Though we've only discussed open-chain alkanes up to now, chemists have known for over 100 years that compounds with *rings* of carbon atoms also exist. Such compounds are called **cycloalkanes**, or **alicyclic** (**ali**phatic **cyclic**) compounds. Since cycloalkanes consist of rings of $-CH_2-$ units, they have the general formula $(CH_2)_n$, or C_nH_{2n}, and are represented by polygons in skeletal drawings:

Cyclopropane **Cyclobutane** **Cyclopentane** **Cyclohexane** **Cycloheptane**

Substituted cycloalkanes are named by rules similar to those used for open-chain alkanes. For most compounds, there are only two steps:

STEP 1 Count the number of carbon atoms in the ring and add the prefix *cyclo-* to the name of the corresponding alkane. If a substituent is present on the ring, the compound should be named as an alkyl-substituted cycloalkane rather than as a cycloalkyl-substituted alkane.

$-CH_3$ **Methylcyclopentane**

STEP 2 Start at a point of attachment and number the substituents on the ring so as to arrive at the lowest sum. If two or more different substituents are present, number them in order of their alphabetical priority.

CH_3 CH_3 1 2 3 4 5 6 *NOT* 1 2 3 4 5 6 CH_3 CH_3

1,3-Dimethylcyclohexane **1,5-Dimethylcyclohexane**

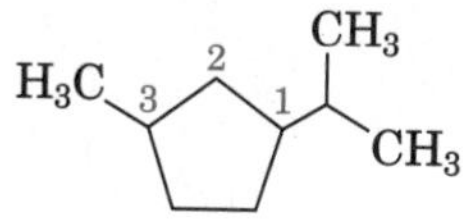

H_3C 3 2 1 CH_3 CH_3

1-Isopropyl-3-methylcyclopentane
(*NOT* 1-methyl-3-isopropylcyclopentane)

PROBLEM 2.16 Give IUPAC names for these cycloalkanes:

(a) 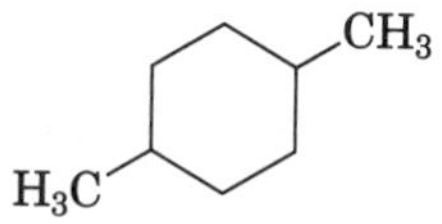 CH_3 H_3C **(b)** 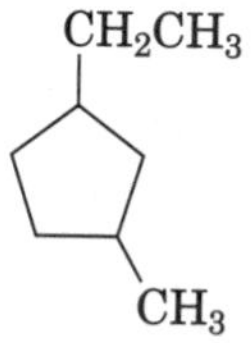 CH_2CH_3 CH_3 **(c)** 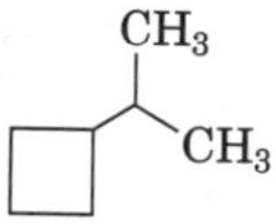 CH_3 CH_3

PROBLEM 2.17 Draw structures corresponding to these IUPAC names:
(a) 1-*tert*-Butyl-2-methylcyclopentane **(b)** 1,1-Dimethylcyclobutane
(c) 1-Ethyl-4-isopropylcyclohexane

2.8 CIS–TRANS ISOMERISM IN CYCLOALKANES

In most respects, the behavior of cycloalkanes mimics that of open-chain, acyclic alkanes. Both classes of compounds are nonpolar and are chemically inert to most reagents. There are, however, some important differences. One difference is that cycloalkanes have less conformational freedom than their open-chain counterparts. Although open-chain alkanes have nearly free rotation around their carbon–carbon single bonds, cycloalkanes have less freedom of rotation around bonds and are more constrained in their geometry. For example, cyclopropane is constrained by geometry to be a flat, planar molecule with a rigid structure. No rotation around a carbon–carbon bond is possible in cyclopropane without breaking open the ring.

Because of their cyclic structure, cycloalkanes have two sides, a "top" side and a "bottom" side. As a result, isomerism is possible in substituted cycloalkanes. For example, there are two 1,2-dimethylcyclopropane isomers, one with the two methyl groups on the same side of the ring and one with them on opposite sides. Both isomers are stable compounds and can't be interconverted without breaking chemical bonds.

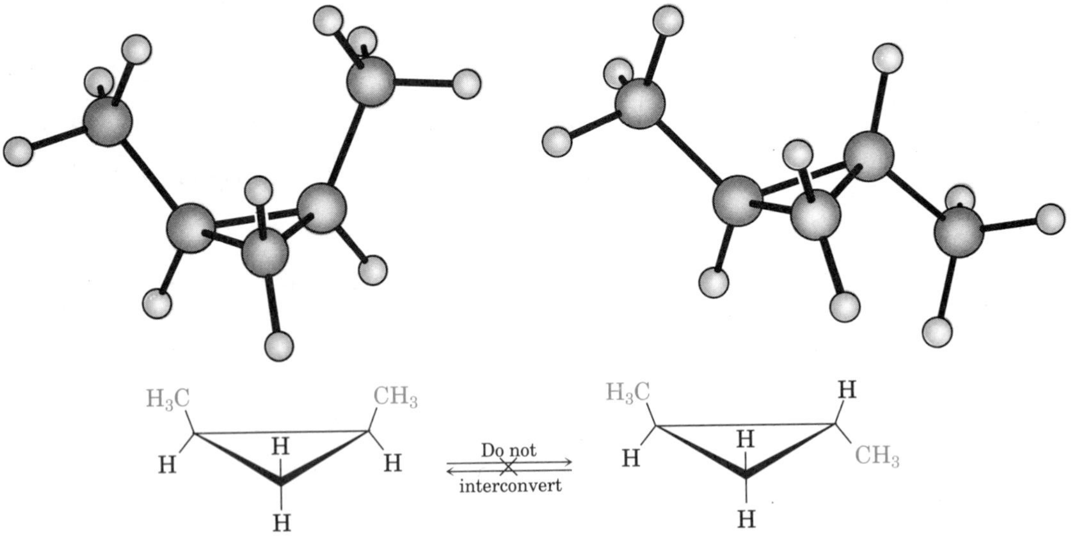

H_3C H H H H CH_3 — Do not interconvert — H_3C H H H H CH_3

***cis*-1,2-Dimethylcyclopropane**
(methyl groups on same side of ring)

***trans*-1,2-Dimethylcyclopropane**
(methyl groups on opposite sides of the ring)

STEREOISOMERS
Isomers that have their atoms connected in the same order but with a different three-dimensional arrangement

CIS–TRANS ISOMERS
Stereoisomers that differ in having substituent groups attached to the same side (cis) or opposite sides (trans) of the molecule

Unlike the constitutional isomers butane and isobutane (Section 2.2), which have different connections among atoms, the two 1,2-dimethylcyclopropanes have the *same* connections but differ in the spatial orientation of their atoms. Such compounds that have their atoms connected in the same way but that differ in three-dimensional orientation are called **stereoisomers**. The 1,2-dimethylcyclopropanes are special kinds of stereoisomers called **cis-trans isomers**; the prefixes *cis-* (Latin, "on the same side") and *trans-* (Latin, "across") are used to distinguish between them.

PRACTICE PROBLEM 2.5 Draw *cis*-1,4-dimethylcyclohexane.

SOLUTION *cis*-1,4-Dimethylcyclohexane contains a ring of six carbon atoms with methyl substituents on the same side of the ring at carbons 1 and 4.

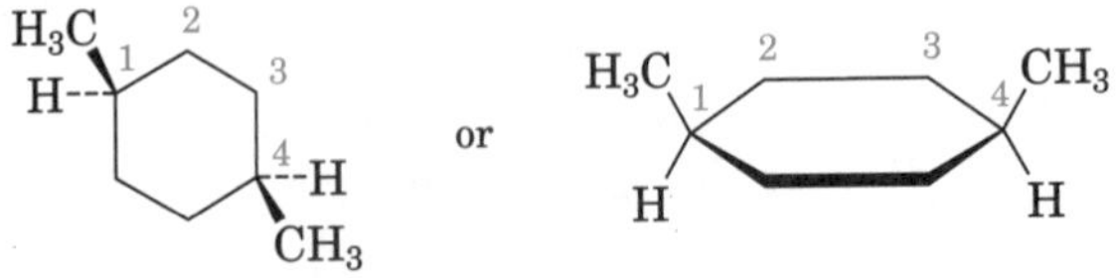

H_3C H 1 2 3 4 H CH_3 or H_3C H 1 2 3 4 CH_3 H

***cis*-1,4-Dimethylcyclohexane**

PROBLEM 2.18 Draw *cis*-1-chloro-3-methylcyclopentane.

PROBLEM 2.19 Draw both cis and trans isomers of 1,2-dibromocyclobutane.

2.9 CONFORMATIONS OF SOME COMMON CYCLOALKANES

In the early days of organic chemistry, cycloalkanes provoked a good deal of consternation among chemists. The problem was that if carbon prefers to have bond angles of 109.5°, how is it possible for cyclopropane and cyclobutane to exist? After all, cyclopropane must have a triangular shape with bond angles near 60°, and cyclobutane must have a square or rectangular shape with bond angles near 90°. Nonetheless, these compounds *do* exist and are stable.

Let's look at the most common cycloalkanes.

Cyclopropane

Cyclopropane is a symmetrical molecule with C–C–C bond angles of 60°, as indicated in Figure 2.9. The three carbons form an equilateral triangle, with three hydrogens protruding above and three below the plane of the carbons. All six of the C–H bonds have an eclipsed, rather than staggered, arrangement with their neighbors.

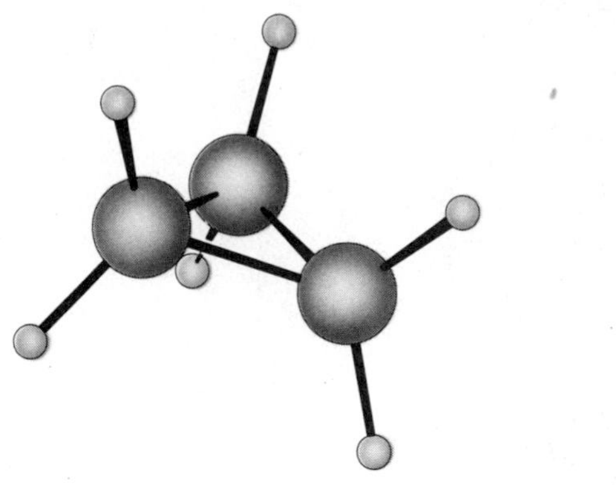

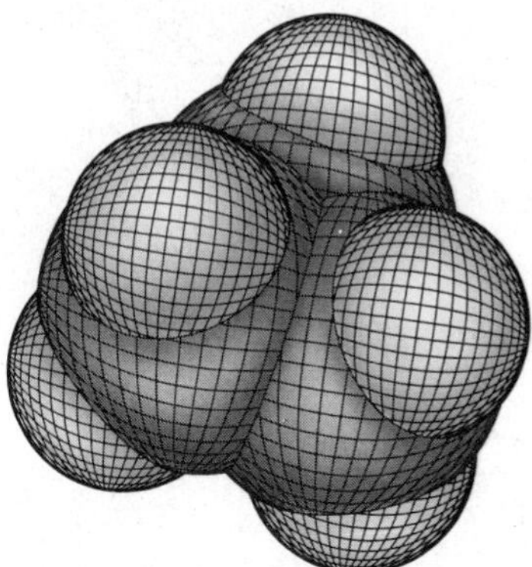

FIGURE 2.9 Computer-generated models of cyclopropane.

The simplest way to account for the distortion of the C–C–C bond angles from their ideal value of 109.5° to a value of 60° in cyclopropane is to think of cyclopropane as having *bent bonds* (Figure 2.10). In an open-chain alkane, maximum bonding efficiency is achieved when two atoms are located so that their overlapping orbitals point directly toward each other. In cyclopropane, however, the orbitals can't point directly toward each other but must instead overlap at a slight angle. The results of this

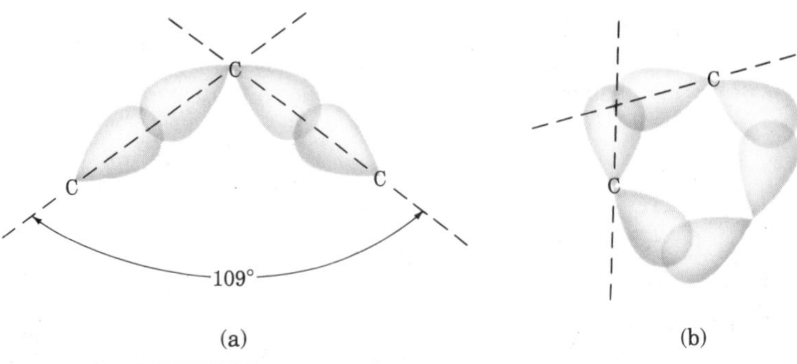

FIGURE 2.10 An orbital view of cyclopropane. (a) Normal C–C bonds have good overlap of orbitals, but (b) cyclopropane bent bonds have poor overlap of orbitals.

ANGLE STRAIN Strain in a molecule caused by expanding or compressing a bond angle away from its normal value

poor overlap are that cyclopropane carbon–carbon bonds are weaker than normal alkane bonds because of what is called **angle strain** and that cyclopropane is therefore more reactive than normal alkanes.

Cyclobutane and Cyclopentane

Cyclobutane and cyclopentane are slightly puckered rather than flat, as indicated in Figure 2.11. This puckering makes the C–C–C bond angles

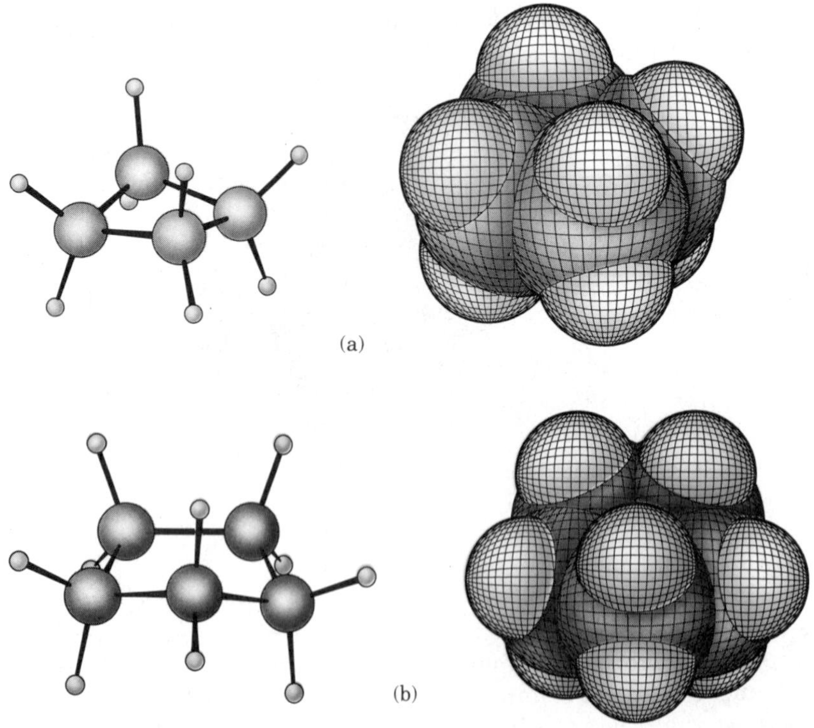

FIGURE 2.11 Computer-generated models of (a) cyclobutane and (b) cyclopentane.

a bit smaller than they would otherwise be and increases the angle strain. At the same time, though, the puckering relieves the eclipsing interactions of adjacent C–H bonds that would occur if the ring were flat.

Cyclohexane

Cyclohexane rings are the most important of all cycloalkanes because of their wide occurrence in nature. A large number of compounds, including steroids and many pharmaceutical agents, have cyclohexane rings.

CHAIR CONFORMATION The energetically favored conformation of cyclohexane, in which all bond angles are near 109° and all bonds have a staggered arrangement

Cyclohexane is not flat. Rather, it is puckered into a three-dimensional arrangement called a **chair conformation**, in which the C–C–C bond angles have close to the ideal 109.5° tetrahedral value (Figure 2.12). In addition to having ideal bond angles and being free of angle strain, chair cyclohexane is also free of all C–H eclipsing interactions because neighboring C–H bonds are staggered.

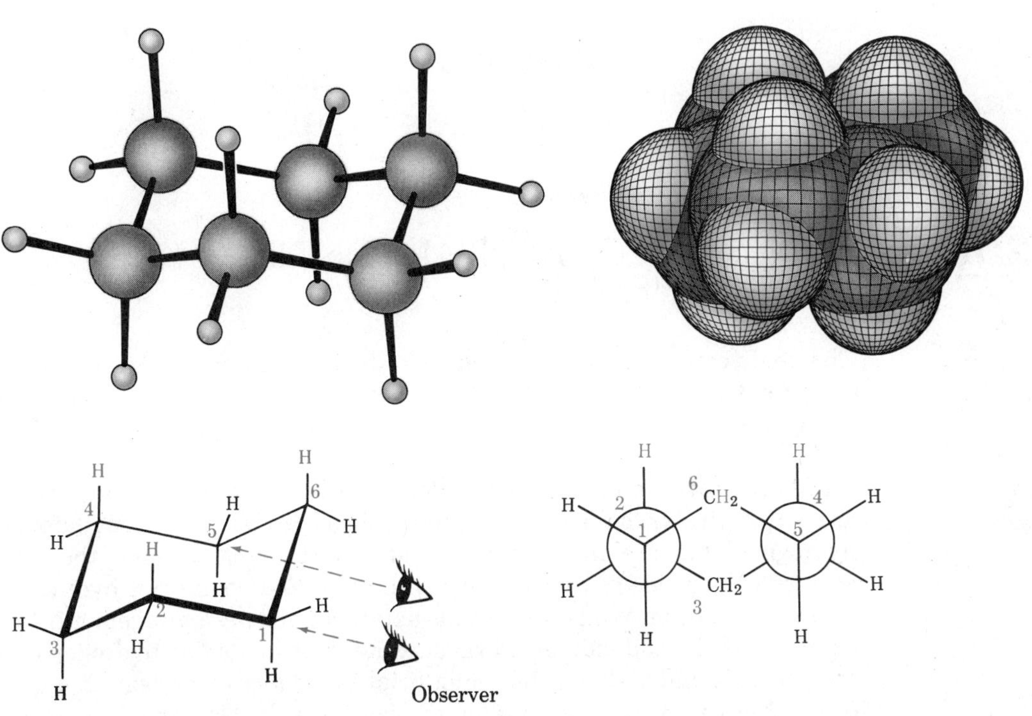

FIGURE 2.12 The chair conformation of cyclohexane. This conformation has no eclipsing between neighboring C–H bonds, and all bond angles are close to 109°.

Chair conformations are drawn by following three steps:

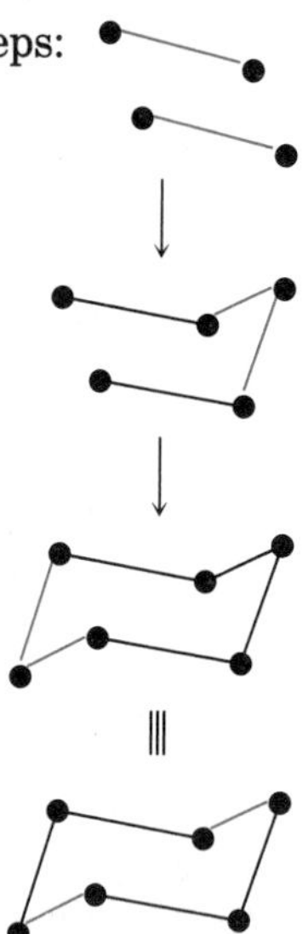

STEP 1 Draw two parallel lines, slanted downward and slightly offset from each other. These lines show that four of the cyclohexane carbon atoms lie in a plane.

STEP 2 Place the topmost carbon atom above and to the right of the plane of the other four and connect the bonds.

STEP 3 Place the bottommost carbon atom below and to the left of the plane of the middle four and connect the bonds. Note that the bonds to the bottommost carbon atom are parallel to the bonds to the topmost carbon.

It's important to remember when viewing cyclohexane that the lower bond is in front, and the upper bond is in back. If this convention is not defined, an optical illusion can make it appear that the reverse is true.

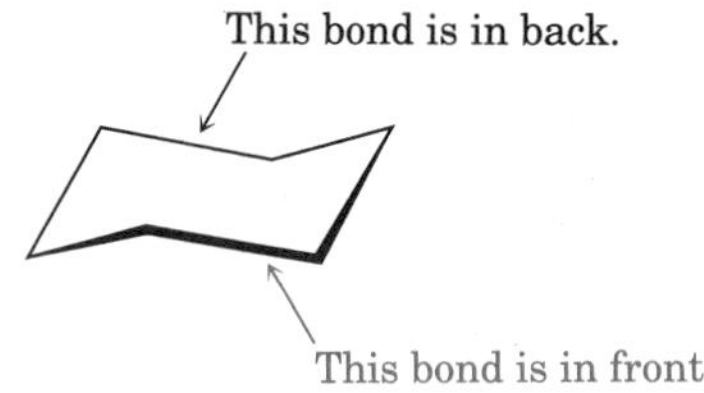
This bond is in back.
This bond is in front.

2.10 AXIAL AND EQUATORIAL BONDS IN CYCLOHEXANE

The chair conformation of cyclohexane leads to many important consequences. One such consequence is that there are two kinds of positions for hydrogens on the ring—*axial* positions and *equatorial* positions (Figure 2.13).

As indicated in Figure 2.13, cyclohexane has six axial hydrogens that are perpendicular to the ring (parallel to the ring *axis*), and six equatorial hydrogens that are in the rough plane of the ring (around the ring *equator*). Each side of the ring has both axial and equatorial hydrogens in an alternating axial–equatorial–axial–equatorial arrangement. For example, if the top side of a cyclohexane ring has axial hydrogens on carbons 1, 3, and 5, then it has equatorial hydrogens on carbons 2, 4, and 6. Exactly the reverse is true for the bottom side: Carbons 1, 3, and 5 have equatorial hydrogens, but carbons 2, 4, and 6 have axial hydrogens.

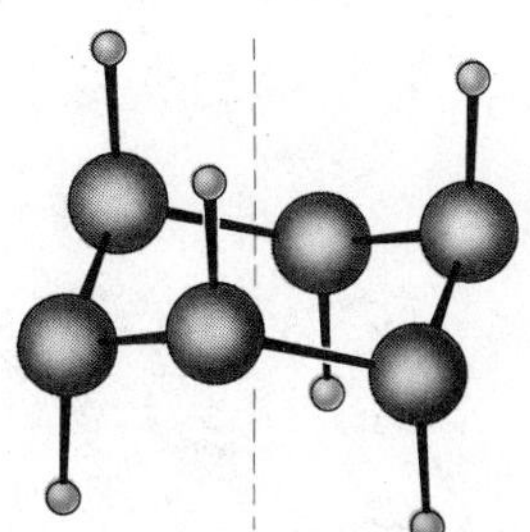

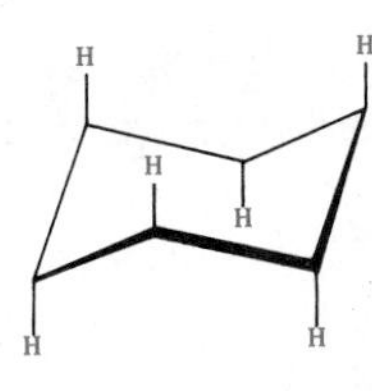

(a) Six *axial* hydrogens (parallel to axis of ring)

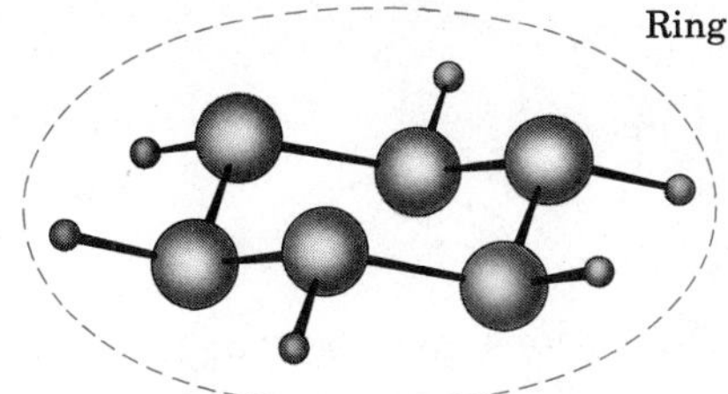

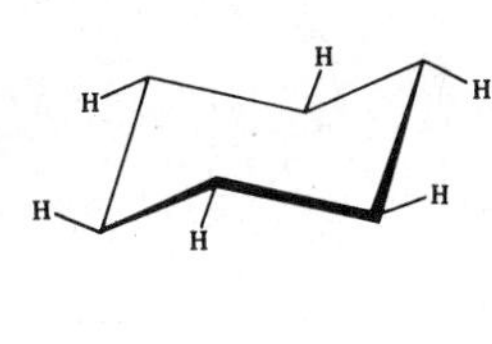

(b) Six *equatorial* hydrogens (in a band around the equator of the ring)

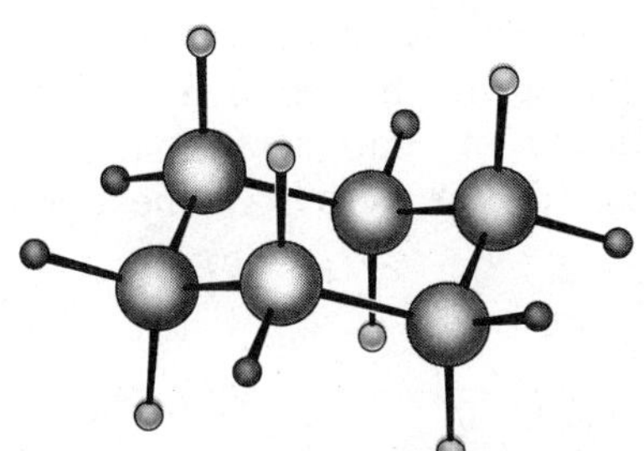

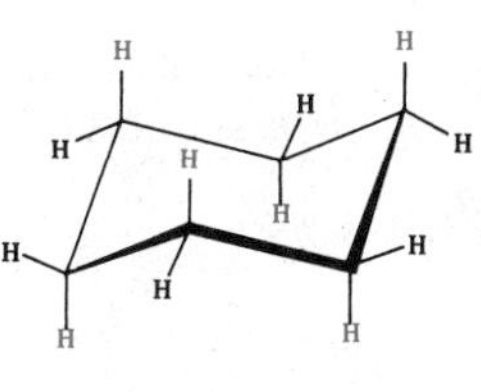

(c) Chair cyclohexane with all its hydrogen atoms

FIGURE 2.13 Axial and equatorial hydrogen atoms in cyclohexane.

Note that we haven't used the words *cis* and *trans* in this discussion of cyclohexane geometry. Two hydrogens on the same side of a ring are always cis, regardless of whether they're axial or equatorial and regardless of whether they're adjacent. Similarly, two hydrogens on opposite sides of the ring are always trans, regardless of whether they're axial or equatorial.

Axial and equatorial bonds can be drawn in the following way:

AXIAL BONDS
Bonds on chair cyclohexane that are roughly perpendicular to the plane of the ring

Axial bonds: The six axial bonds, one on each carbon, are parallel and have an alternating up–down relationship.

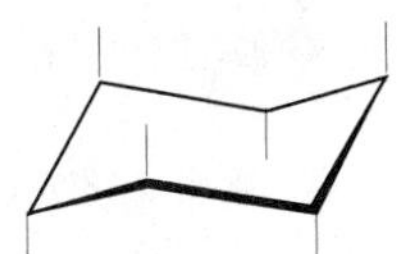

EQUATORIAL BONDS Bonds on chair cyclohexane that are roughly parallel to the plane of the ring

Equatorial bonds: The six equatorial bonds, one on each carbon, come in three sets of two parallel lines. Each set is also parallel to two ring bonds. Equatorial bonds alternate between top and bottom sides around the ring.

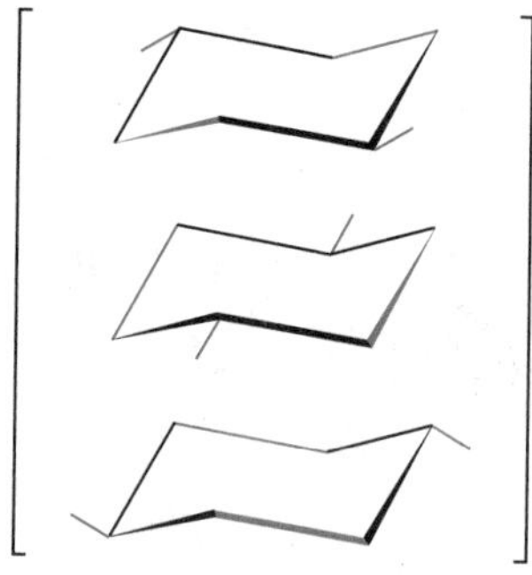

PROBLEM 2.20 Draw two structures for methylcyclohexane, one with the methyl group axial and one with the methyl group equatorial.

2.11 CONFORMATIONAL MOBILITY OF CYCLOHEXANE

Since chair cyclohexane has two kinds of positions, axial and equatorial, we might expect to find two isomeric forms of a monosubstituted cyclohexane. In fact, this expectation is wrong. At room temperature, there is only one methylcyclohexane, one bromocyclohexane, and so forth, because cyclohexane rings are *conformationally mobile*. The two chair cyclohexane conformations readily interconvert, resulting in the interchange of axial and equatorial positions. This interconversion of chair conformations, referred to as a *ring-flip*, is shown in Figure 2.14 for methylcyclohexane.

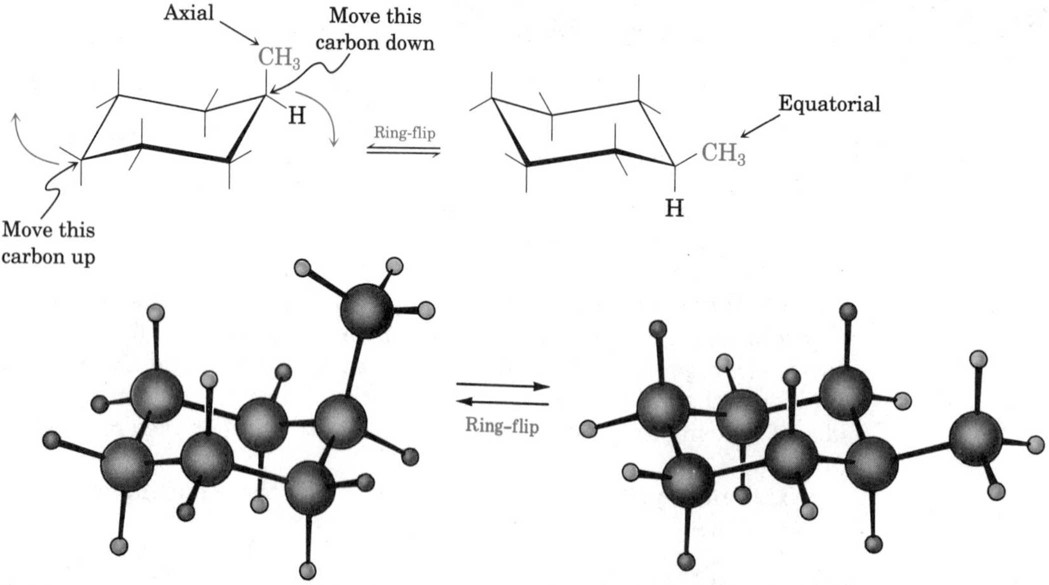

FIGURE 2.14 A ring-flip in chair cyclohexane interconverts axial and equatorial positions.

We can imagine the ring-flip of a chair cyclohexane by mentally keeping the middle four carbon atoms in place while folding the two ends in opposite directions. The net result of a ring-flip is that axial and equatorial positions interconvert. An axial substituent in one chair form becomes an equatorial substituent in the ring-flipped chair form, and vice versa. For example, axial methylcyclohexane becomes equatorial methylcyclohexane after ring-flip. Since this interconversion occurs rapidly at room temperature with an energy barrier of only 10.8 kcal/mol (45 kJ/mol), we can isolate only an interconverting mixture rather than distinct axial and equatorial isomers.

1,3-DIAXIAL INTERACTION Spatial interaction between an axial substituent and a hydrogen atom three carbons away in a substituted chair cyclohexane

STERIC STRAIN Strain in a molecule caused by having atoms too close together

Even though axial and equatorial methylcyclohexanes interconvert rapidly, they aren't equally stable. At any given instant about 95% of methylcyclohexane molecules have their methyl group equatorial and only 5% have their methyl group axial. Equatorial methylcyclohexane is relatively unstrained, but axial methylcyclohexane has an unfavorable steric (spatial) interaction between the methyl group on carbon 1 and the axial hydrogen atoms on carbons 3 and 5 (Figure 2.15). This so-called **1,3-diaxial interaction** introduces 1.8 kcal/mol (7.5 kJ/mol) of **steric strain** into the molecule because the axial methyl group and the nearby axial hydrogen are too close together and are trying to occupy the same space.

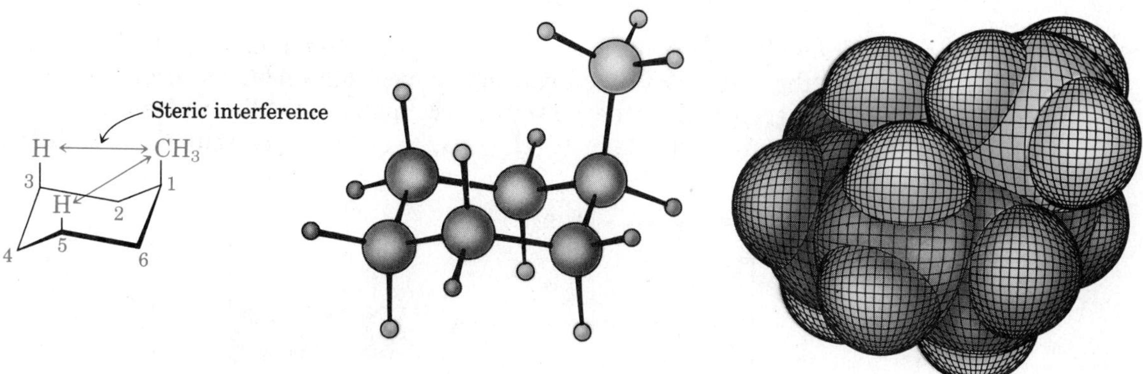

FIGURE 2.15 1,3-Diaxial steric interactions in axial methylcyclohexane.

What is true for methylcyclohexane is also true for all other monosubstituted cyclohexanes: A substituent is always more stable in an equatorial position than in an axial position. As you might expect, the amount of steric strain increases as the size of the axial substituent group increases.

PRACTICE PROBLEM 2.6 Draw 1,1-dimethylcyclohexane, indicating whether each methyl group is axial or equatorial.

SOLUTION First draw a chair cyclohexane ring and then put two methyl groups on the same carbon. The methyl group in the rough plane of the ring is equatorial and the other (above or below the ring) is axial.

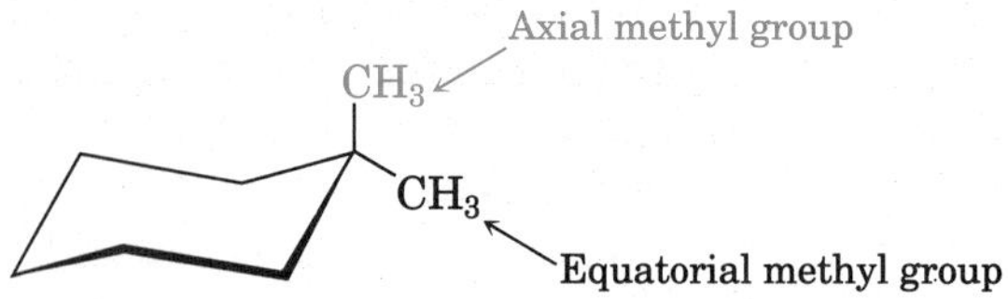

Axial methyl group
CH_3
CH_3
Equatorial methyl group

PROBLEM 2.21 Draw two different chair conformations of bromocyclohexane showing all hydrogen atoms. Label all positions as axial or equatorial. Which of the two conformations do you think is more stable?

PROBLEM 2.22 Explain why a *cis*-1,2-disubstituted cyclohexane such as *cis*-1,2-dichlorocyclohexane must have one group axial and one equatorial.

PROBLEM 2.23 Explain why a *trans*-1,2-disubstituted cyclohexane must have either both groups axial or both equatorial.

INTERLUDE

PETROLEUM

Many alkanes occur naturally in the plant and animal world. For example, the waxy coating on cabbage leaves contains nonacosane ($C_{29}H_{60}$), and the wood oil of the Jeffrey pine common to the Sierra Nevada mountains contains heptane (C_7H_{16}). By far the major sources of alkanes,

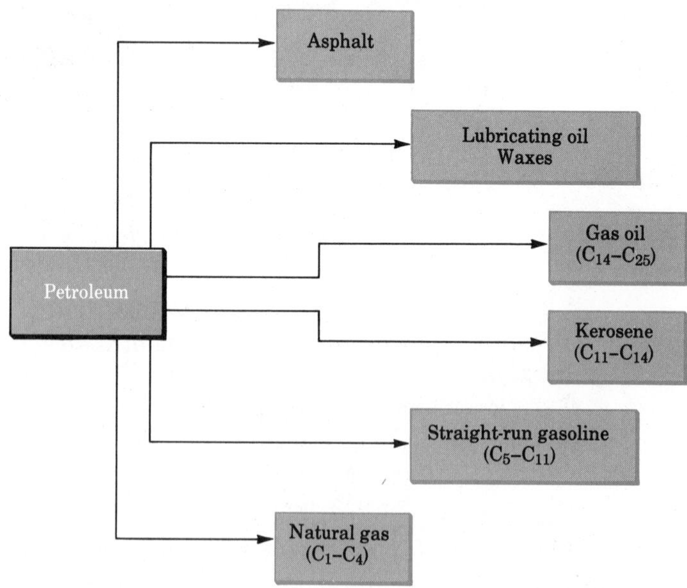

however, are the world's natural gas and petroleum deposits. Laid down eons ago, these natural deposits are derived from the decomposition of organic matter, primarily of marine origin.

Natural gas consists chiefly of methane but also contains ethane, propane, and butane. *Petroleum* is a complex mixture of hydrocarbons that must be refined into different fractions before it can be used. Refining begins by fractional distillation of crude petroleum into three principal cuts: straight-run gasoline (boiling point, bp, 30–200°C), kerosene (bp 175–300°C), and gas oil (bp 275-400°C). Finally, distillation under reduced pressure yields lubricating oils and waxes and leaves an undistillable tarry residue of asphalt.

It turns out that straight-run gasoline is a rather poor fuel because of the phenomenon of *engine knock*. Thus, the simple distillation of petroleum into fractions is just the beginning of the process by which automobile fuel is made. In the ordinary automobile engine, a piston draws a mixture of fuel and air into a cylinder on its downward stroke and compresses the mixture on its upward stroke. Just before the end of the compression, a spark plug ignites the fuel and smooth combustion occurs, pushing the piston downward. Not all fuels burn equally well, though. When poor fuels are used, combustion can be initiated in an uncontrolled manner by a hot surface in the cylinder before the spark plug fires. This *preignition*, detected as an engine knock, can destroy the engine by putting irregular forces on the crankshaft.

The *octane number* of a fuel is the measure by which its antiknock properties are judged. It was recognized long ago that straight-chain alkanes are far more prone to induce engine knock than are branched-chain compounds. Heptane, a particularly bad fuel, is assigned a base value octane number = 0; 2,2,4-trimethylpentane (commonly known as isooctane) has a rating of 100.

$$CH_3CH_2CH_2CH_2CH_2CH_2CH_3$$

Heptane
Octane number = 0

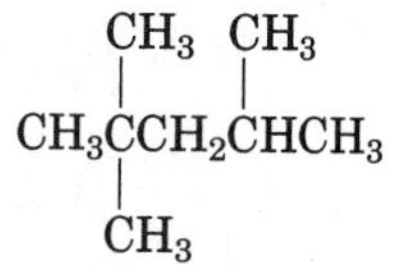

$$\begin{array}{c} \quad CH_3 \;\; CH_3 \\ \quad | \qquad | \\ CH_3CCH_2CHCH_3 \\ | \qquad\quad \\ CH_3 \qquad\quad \end{array}$$

2,2,4-Trimethylpentane (isooctane)
Octane number = 100

Because straight-run gasoline has a high percentage of unbranched alkanes and is therefore a poor fuel, petroleum chemists have devised sophisticated methods for producing better fuels. One of these methods, *catalytic cracking*, involves taking the kerosene cut (C_{11}–C_{14}) and "cracking" it into smaller molecules at high temperature on a silica-alumina catalyst. The major products of cracking are light hydrocarbons in the C_3-C_5 range. These small hydrocarbons are then catalytically recombined to yield C_7-C_{10} branched-chain molecules that are perfectly suited for use as high-octane fuels.

SUMMARY AND KEY WORDS

A **functional group** is an atom or group of atoms within a larger molecule that has a characteristic chemical reactivity. Because functional groups behave approximately the same way in all molecules where they occur, the chemical reactions of an organic molecule are largely determined by its functional groups.

Alkanes are a class of **hydrocarbons** having the general formula C_nH_{2n+2}. They contain no functional groups, are chemically rather inert, and can be either straight-chain or branched. Alkanes can be named by a series of IUPAC rules of nomenclature. Isomerism is possible for all but the simplest alkanes. Compounds that have the same chemical formula but different structures are called **isomers**. Compounds such as butane and isobutane, which have the same formula but differ in the way their atoms are connected, are called **constitutional isomers**.

As a result of their symmetry, rotation is possible about carbon–carbon single bonds. Alkanes can therefore adopt any of a large number of rapidly interconverting **conformations**. **Newman projections** allow us to visualize the spatial consequences of bond rotation by sighting directly along a carbon–carbon bond axis. **Staggered conformations** are more stable than **eclipsed conformations**.

Staggered conformation　　Eclipsed conformation

Cycloalkanes contain rings of carbon atoms and have the general formula C_nH_{2n}. Because conformational mobility (rotation) is reduced in cycloalkanes, complete rotation around carbon–carbon bonds is not possible. Disubstituted cycloalkanes can therefore exist as **cis-trans stereoisomers**. In the cis isomer, both substituents are on the same side of the ring, whereas in the trans isomer, the substituents are on opposite sides of the ring. Cyclohexanes are the most important of all rings because of their wide occurrence in nature. Cyclohexane exists in a puckered, strain-free **chair conformation** in which all bond angles are near 109°

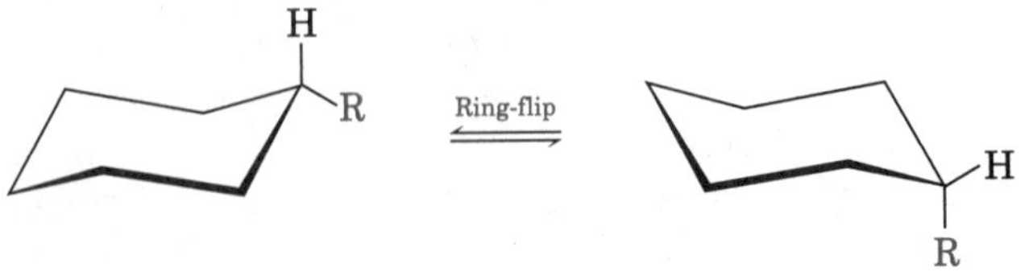

Ring-flip

R equatorial, more stable　　R axial, less stable

and all neighboring C–H bonds are staggered. Chair cyclohexane has two kinds of bonds, axial and equatorial. **Axial bonds** are directed up and down, parallel to the ring axis; **equatorial bonds** lie in a belt around the ring equator. Chair cyclohexanes can undergo a **ring-flip** that interconverts axial and equatorial positions. Substituents on the ring are more stable in the equatorial than in the axial position.

ADDITIONAL PROBLEMS

2.24 Locate and identify the functional groups in these molecules:

(a) OH

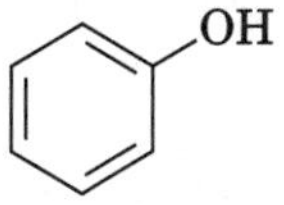

Phenol

(b) O

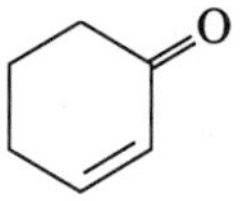

2-Cyclohexenone

(c) O ‖ CH_3CHCOH | NH_2

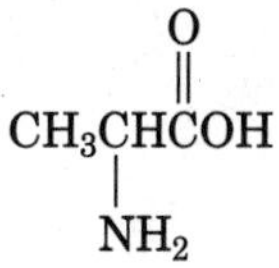

Alanine

(d) O

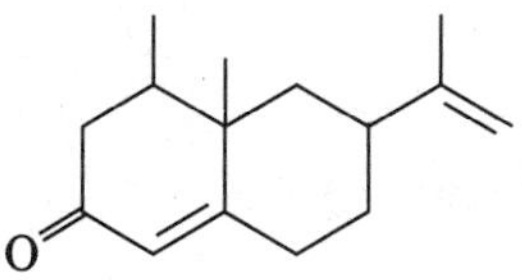

Nootkatone (from grapefruit)

(e) O HO

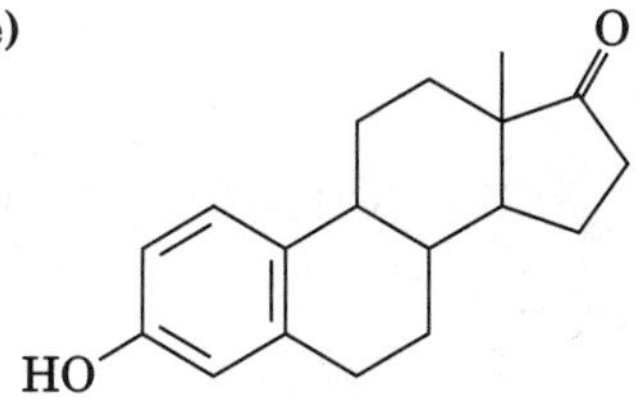

Estrone

2.25 Propose structures for molecules that fit these descriptions:
(a) An alkene with six carbons (b) A cycloalkene with five carbons
(c) A ketone with five carbons (d) An amide with four carbons
(e) A five-carbon ester (f) An aromatic alcohol

2.26 Propose suitable structures for the following:
(a) An alkene, C_7H_{14} (b) A cycloalkene, C_3H_4 (c) A ketone, C_4H_8O
(d) A nitrile, C_5H_9N (e) A dialkene, C_5H_8 (f) A dialdehyde, $C_4H_6O_2$

2.27 Write as many structures as you can that fit these descriptions:
(a) Alcohols with formula $C_4H_{10}O$ (b) Amines with formula $C_5H_{13}N$
(c) Ketones with formula $C_5H_{10}O$ (d) Aldehydes with formula $C_5H_{10}O$
(e) Ethers with formula $C_4H_{10}O$ (f) Esters with formula $C_4H_8O_2$

2.28 Draw all monobromo derivatives of *n*-pentane, $C_5H_{11}Br$.

2.29 Draw all monochloro derivatives of 2,5-dimethylhexane.

2.30 How many constitutional isomers are there with the formula C_3H_8O? Draw them.

2.31 Propose structures for compounds that contain the following:
(a) A quaternary carbon **(b)** Four methyl groups **(c)** An isopropyl group
(d) Two tertiary carbons **(e)** An amino group ($-NH_2$) bonded to a secondary carbon

2.32 What hybridization do you expect for the carbon atom in these functional groups?
(a) Ketone **(b)** Nitrile **(c)** Ether **(d)** Alcohol

2.33 Which of these structures represent the same compound and which represent different compounds?

```
                                          H
                                          |
                                        H—C—H
        H  H  H  H                 H  H   |                H  H  H
        |  |  |  |                 |  |   |                |  |  |
(a) H—C—C—C—C—H               H—C—C—C—H              H—C—C—C—H
        |  |  |  |                 |  |   |                |  |  |
        H  H  H  H                 H  H   H                H  |  H
                                                             H—C—H
                                                               |
                                                               H
```

```
                                                           H
                                                           |
                                                         H—C—H
        H  H  Br H                 H  Br H  H              |  H  H
        |  |  |  |                 |  |  |  |              |  |  |
(b) H—C—C—C—C—H               H—C—C—C—C—H           H—C—C—C—H
        |  |  |  |                 |  |  |  |              |  |  |
        H  H  H  H                 H  H  H  H              H  Br H
```

```
               CH3             CH3
               |               |
(c) CH3CHBrCHCH3         CH3CHCHBrCH3        (CH3)2CHCHBrCH2CH3
```

(d) [Three dihydroxybenzene structures: OH/OH on adjacent ring positions; OH and OH in 1,3 positions; HO and HO on adjacent ring positions]

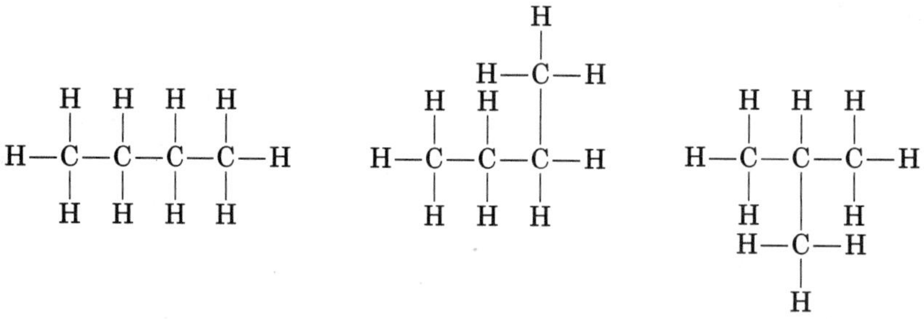

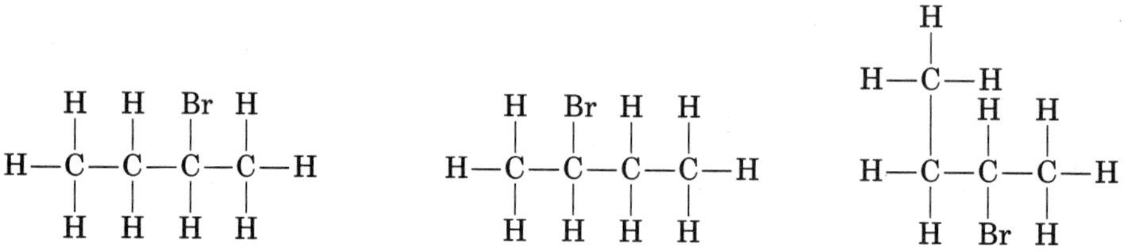

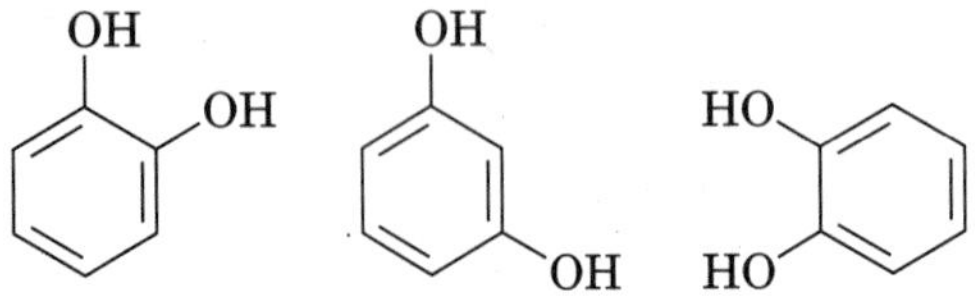

2.34 Draw structural formulas for the following:
(a) 2-Methylheptane **(b)** 4-Ethyl-2-methylhexane **(c)** 4-Ethyl-3,4-dimethyloctane
(d) 2,4,4-Trimethylheptane **(e)** 1,1-Dimethylcyclopentane **(f)** 4-Isopropyl-3-methylheptane

2.35 Give IUPAC names for these alkanes:

```
                      CH3                              CH3
                      |                                |
(a) CH3CH2CH2CHCHCH3               (b) CH3CH2CH2CHCHCH3
                   |                                |
                   CH3                              CH2CH2CH2CH3

       CH3   CH2CH3                                CH2CH3
       |     |                                     |
(c) CH3CHCH2CCH3                   (d) CH3CH2CCH2CH3
             |                                     |
             CH2CH3                                CH2CH3
```

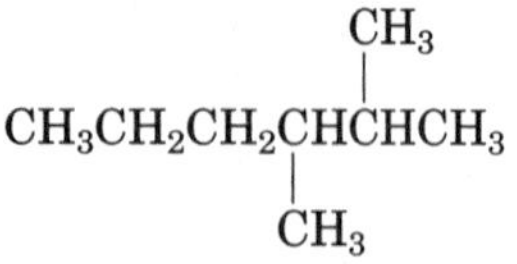

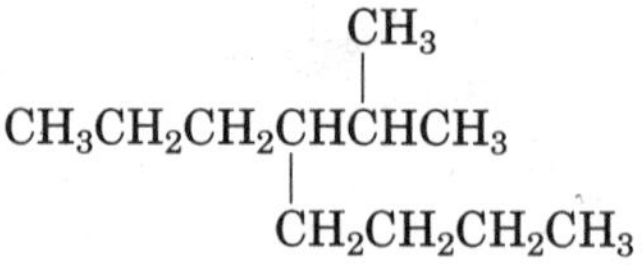

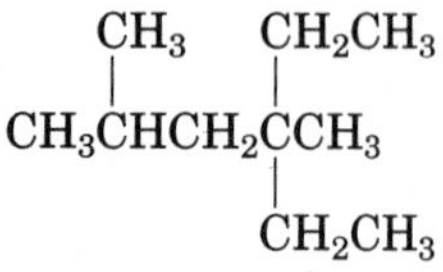

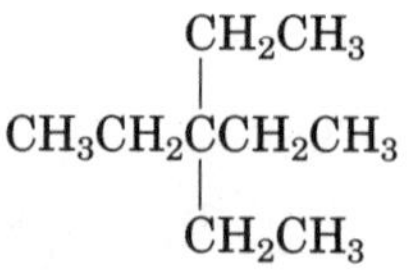

2.36 Convert the following line-bond structures into skeletal drawings:

(a) 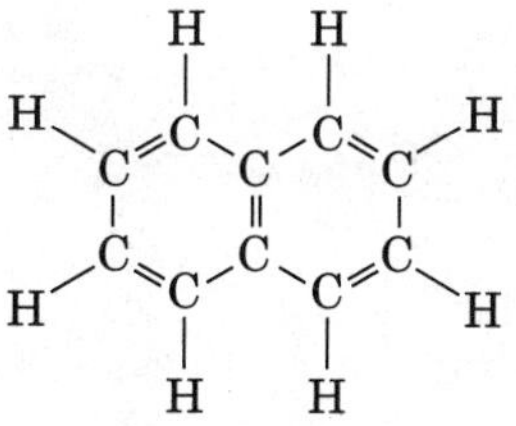 **Naphthalene**

(b) 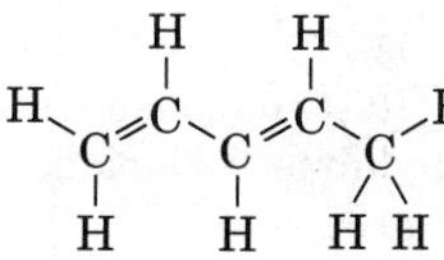 **1,3-Pentadiene**

2.37 For each of these compounds, draw a constitutional isomer having the same functional groups:

(a) 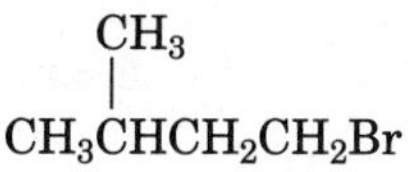 $CH_3CH(CH_3)CH_2CH_2Br$ (b) 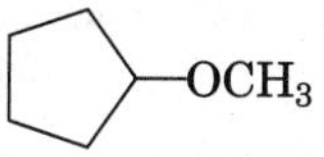 cyclopentyl–OCH_3 (c) $CH_3CH_2CH_2C{\equiv}N$

(d) 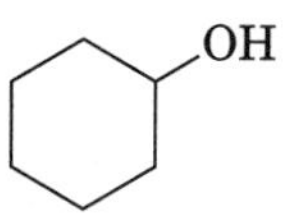 cyclohexyl–OH (e) 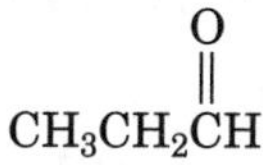 CH_3CH_2CHO (f) 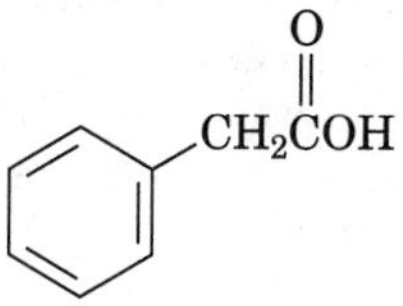 phenyl–CH_2COOH

2.38 Sighting along the C2–C3 bond of 2-methylbutane, there are two different staggered conformations. Draw them both in Newman projections, and tell which you think is more stable. Explain.

2.39 Sighting along the C2–C3 bond of 2-methylbutane (see Problem 2.38), there are also two possible eclipsed conformations. Draw them both in Newman projections, and tell which you think is lower in energy. Explain.

2.40 *cis*-1-*tert*-Butyl-4-methylcyclohexane exists almost exclusively in the conformation shown. What does this tell you about the relative size of a *tert*-butyl substituent and a methyl substituent?

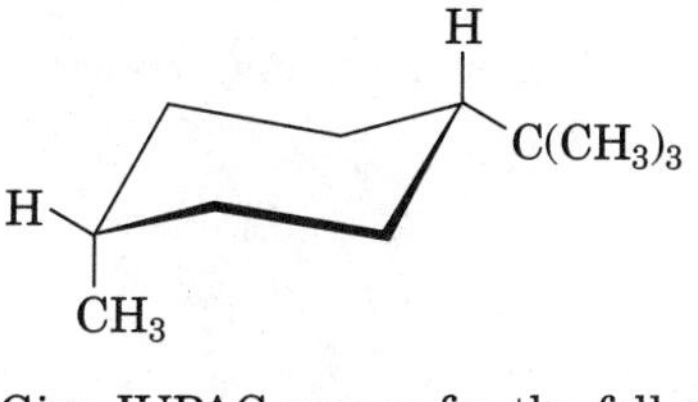

H, $C(CH_3)_3$, H, CH_3

***cis*-1-*tert*-Butyl-4-methylcyclohexane**

2.41 Give IUPAC names for the following compounds:

(a) 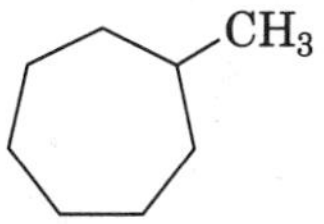 cycloheptyl–CH_3 (b) 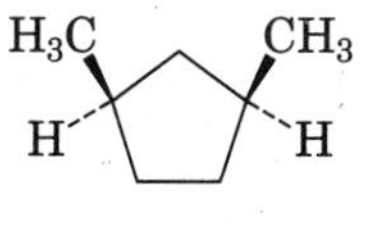 H_3C, CH_3, H, H (cyclopentane) (c) 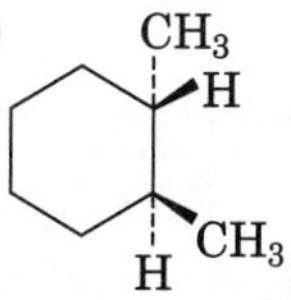 CH_3, H, CH_3, H (cyclohexane)

(d) 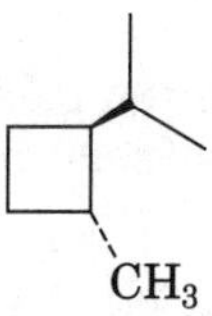 cyclobutane with isopropyl and CH_3 (e) 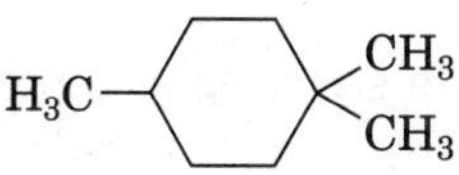 H_3C–cyclohexane–$(CH_3)_2$

2.42 Give IUPAC names for the five isomers of C_6H_{14}.

2.43 Draw structures for the nine isomers of C_7H_{16}.

2.44 Propose structures and give correct IUPAC names for the following:
(a) A dimethyloctane (b) A diethyldimethylhexane
(c) A cycloalkane with three methyl groups

2.45 The following names are *incorrect*. Give the proper IUPAC names.
(a) 2,2-Dimethyl-6-ethylheptane (b) 4-Ethyl-5,5-dimethylpentane
(c) 3-Ethyl-4,4-dimethylhexane (d) 5,5,6-Trimethyloctane

2.46 The barrier to rotation about the C–C bond in bromoethane is 3.6 kcal/mol. If each hydrogen–hydrogen interaction in the eclipsed conformation is responsible for 0.9 kcal/mol, how much is the H–Br eclipsing interaction responsible for?

2.47 Make a graph of energy versus degree of bond rotation around the C–C bond in bromoethane (see Problem 2.46).

2.48 Malic acid, $C_4H_6O_5$, has been isolated from apples. Because malic acid reacts with 2 equivalents of base, it can be formulated as a dicarboxylic acid.
(a) Draw at least five possible structures for malic acid.
(b) If malic acid is also a secondary alcohol (has an –OH group attached to a secondary carbon), what is its structure?

2.49 Cyclopropane was first prepared by reaction of 1,3-dibromopropane with sodium.
(a) Formulate the reaction.
(b) What product might the following reaction give? What geometry would you expect for the product?

$$C(CH_2Br)_4 \xrightarrow{4\ Na} ?$$

2.50 Tell whether the following pairs of compounds are identical, constitutional isomers, or stereoisomers.
(a) *cis*-1,3-Dibromocyclohexane and *trans*-1,4-dibromocyclohexane
(b) 2,3-Dimethylhexane and 2,5,5-trimethylpentane
(c)

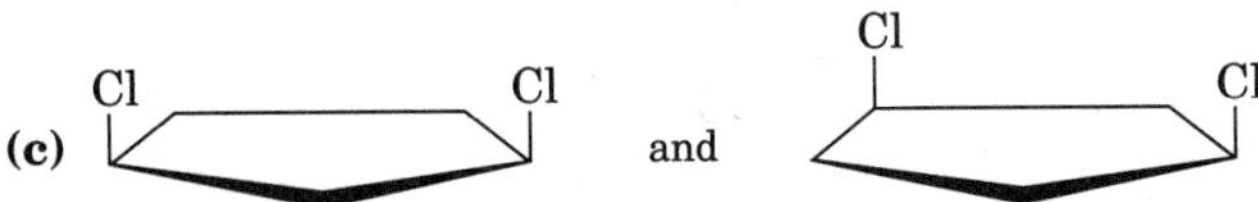

2.51 Draw two constitutional isomers of *cis*-1,2-dibromocyclopentane.

2.52 Draw a stereoisomer of *trans*-1,3-dimethylcyclobutane.

2.53 Draw *trans*-1,2-dimethylcyclohexane in its more stable chair conformation. Are the methyl groups axial or equatorial?

2.54 Draw *cis*-1,2-dimethylcyclohexane in its more stable chair conformation. Are the methyl groups axial or equatorial? Which do you think is more stable, *cis*-1,2-dimethylcyclohexane or *trans*-1,2-dimethylcyclohexane (Problem 2.53)?

2.55 Which do you think is more stable, *cis*-1,3-dimethylcyclohexane or *trans*-1,3-dimethylcyclohexane? Draw chair conformations of both and explain your answer.

2.56 *N*-Methylpiperidine has the conformation shown. What does this tell you about the relative steric requirements of a methyl group versus an electron lone pair?

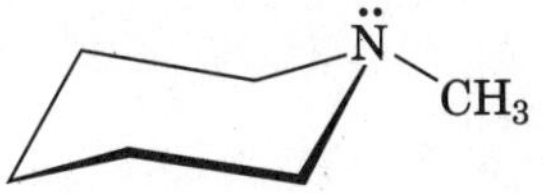

$\ddot{N}$ CH_3 ***N*-Methylpiperidine**

2.57 Glucose contains a six-membered ring in which all the substituents are equatorial. Draw glucose in its more stable chair conformation.

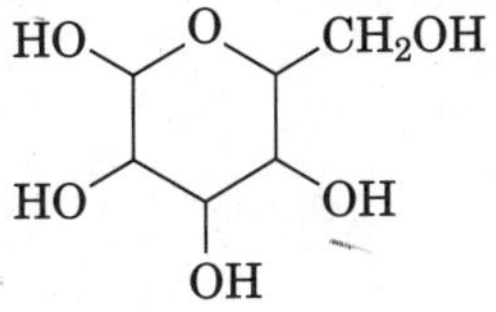

HO O CH_2OH

Glucose

HO OH

OH

2.58 Draw 1,3,5-trimethylcyclohexane using a hexagon to represent the ring. How many cis–trans stereoisomers are possible?

2.59 Here's a tough one. There are two different substances named *trans*-1,2-dimethylcyclopentane. What do you think is the relationship between them? We'll explore this kind of isomerism in Chapter 6.

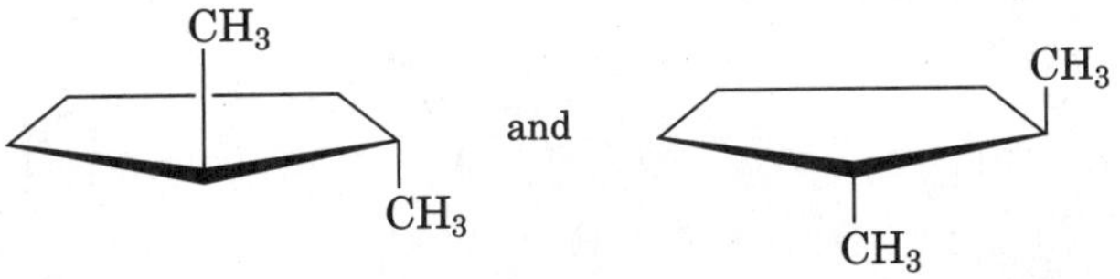

CH_3 CH_3

and

CH_3 CH_3

3 CHAPTER ALKENES: THE NATURE OF ORGANIC REACTIONS

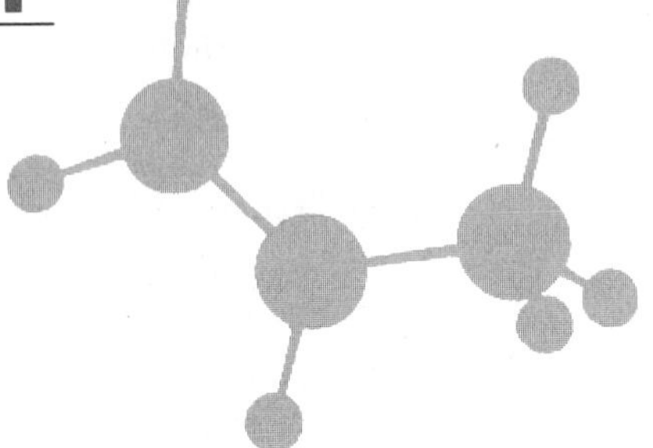

ALKENE
Hydrocarbon with one or more carbon–carbon double bonds

Alkenes are hydrocarbons that contain a carbon–carbon double bond functional group. They occur abundantly in nature, and many have important biological roles. For example, ethylene is a plant hormone that induces ripening in fruit, and α-pinene is the major constituent of turpentine.

Ethylene **α-Pinene**

We'll see in this chapter how and why alkenes behave the way they do, and we'll develop some general ideas about organic chemical reactivity that can be applied to all molecules.

3.1 NAMING ALKENES

UNSATURATED
Containing one or more double or triple bonds

Because of their double bond, alkenes have fewer hydrogens per carbon than related alkanes and are therefore referred to as **unsaturated**. Ethylene, for example, has the formula C_2H_4 whereas ethane has the formula C_2H_6.

Ethylene, C_2H_4
(fewer hydrogens: *unsaturated*)

Ethane, C_2H_6
(more hydrogens: *saturated*)

Alkenes are named according to a series of rules similar to those used for naming alkanes, with the suffix *-ene* used in place of *-ane* to identify the family. There are three steps:

STEP 1 Name the parent hydrocarbon. Find the longest carbon chain that contains the double bond, and name the compound using the suffix *-ene*.

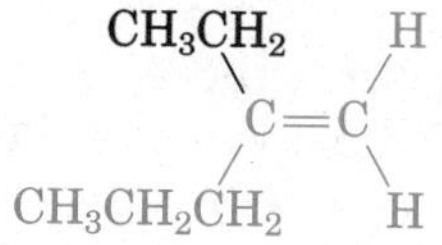

$$\begin{array}{c} CH_3CH_2 \quad\quad H \\ \diagdown \quad \diagup \\ C{=}C \\ \diagup \quad \diagdown \\ CH_3CH_2CH_2 \quad\quad H \end{array} \qquad\qquad \begin{array}{c} CH_3CH_2 \quad\quad H \\ \diagdown \quad \diagup \\ C{=}C \\ \diagup \quad \diagdown \\ CH_3CH_2CH_2 \quad\quad H \end{array}$$

Named as a *pentene* *NOT* as a hexene, since the double bond is not contained in the six-carbon chain

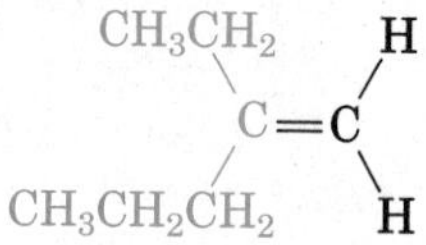

STEP 2 Number the carbon atoms in the chain, beginning at the end nearer the double bond. If the double bond is equidistant from the two ends, begin numbering at the end nearer the first branch point:

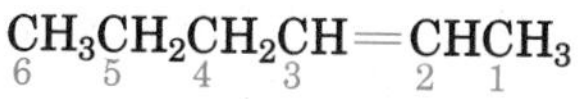

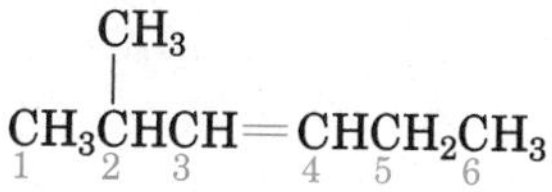

$$\underset{6}{CH_3}\underset{5}{CH_2}\underset{4}{CH_2}\underset{3}{CH}{=}\underset{2}{CH}\underset{1}{CH_3} \qquad \underset{1}{CH_3}\underset{2}{\overset{\overset{\displaystyle CH_3}{|}}{C}H}\underset{3}{CH}{=}\underset{4}{CH}\underset{5}{CH_2}\underset{6}{CH_3}$$

STEP 3 Write the full name, numbering the substituents according to their position in the chain and listing them alphabetically. Indicate the position of the double bond by giving the number of the *first* alkene carbon. If more than one double bond is present, give the position of each and use one of the suffixes *-diene*, *-triene*, and so on.

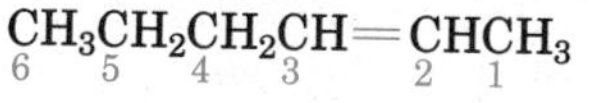

$$\underset{6}{CH_3}\underset{5}{CH_2}\underset{4}{CH_2}\underset{3}{CH}{=}\underset{2}{CH}\underset{1}{CH_3} \qquad \underset{1}{CH_3}\underset{2}{\overset{\overset{\displaystyle CH_3}{|}}{C}H}\underset{3}{CH}{=}\underset{4}{CH}\underset{5}{CH_2}\underset{6}{CH_3}$$

2-Hexene 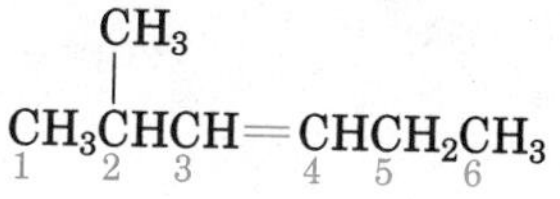 **2-Methyl-3-hexene**

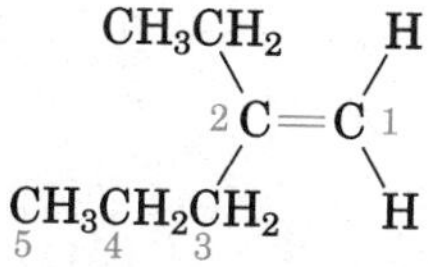

$$\begin{array}{c} CH_3CH_2 \quad\quad H \\ \diagdown \quad \diagup \\ {}_2C{=}C_1 \\ \diagup \quad \diagdown \\ \underset{5}{CH_3}\underset{4}{CH_2}\underset{3}{CH_2} \quad\quad H \end{array} \qquad \underset{1}{H_2C}{=}\underset{2}{\overset{\overset{\displaystyle CH_3}{|}}{C}}{-}\underset{3}{CH}{=}\underset{4}{CH_2}$$

2-Ethyl-1-pentene 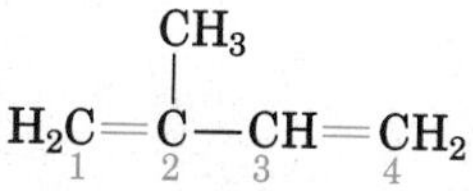 **2-Methyl-1,3-butadiene**

Cycloalkenes are named in a similar way, but because there is no chain end to begin from, we number the cycloalkene so that the double bond is between C1 and C2 and so that the first substituent has as low a number as possible:

1-Methylcyclohexene 1,4-Cyclohexadiene 1,5-Dimethylcyclopentene

For historical reasons, there are a few alkenes whose names don't conform to the rules. For example, the alkene corresponding to ethane should be called *ethene*, but the name *ethylene* has been used for so long that it is accepted by IUPAC. Table 3.1 lists some other common names.

TABLE 3.1 Common Names of Some Alkenes

COMPOUND	SYSTEMATIC NAME	COMMON NAME
$H_2C{=}CH_2$	Ethene	Ethylene
$CH_3CH{=}CH_2$	Propene	Propylene
$CH_3C(CH_3){=}CH_2$	2-Methylpropene	Isobutylene
$H_2C{=}C(CH_3){-}CH{=}CH_2$	2-Methyl-1,3-butadiene	Isoprene

PRACTICE PROBLEM 3.1 What is the IUPAC name of this alkene?

$$CH_3C(CH_3)_2CH_2CH_2CH{=}C(CH_3)CH_3$$

SOLUTION First, find the longest chain containing the double bond. In this case, it's a heptene. Next, number the chain beginning at the end nearer the double bond, and identify the substituents at each position. In this case, there are methyl groups at C2 and C6 (two):

$$\overset{7}{C}H_3\overset{6}{C}(CH_3)_2\overset{5}{C}H_2\overset{4}{C}H_2\overset{3}{C}H{=}\overset{2}{C}(CH_3)\overset{1}{C}H_3$$

The full name is 2,6,6-trimethyl-2-heptene.

PROBLEM 3.1 Give IUPAC names for these compounds:

(a) $\overset{\displaystyle CH_3}{\underset{}{}}$ $H_2C{=}CHCH_2\overset{|}{C}HCH_3$ (with CH_3 on the fourth carbon)

(b) $CH_3CH_2CH{=}CHCH_2CH_2CH_3$

(c) $H_2C{=}CHCH_2CH_2CH{=}CHCH_3$

(d) $CH_3CH_2CH{=}CHCH(CH_3)_2$

PROBLEM 3.2 Name these cycloalkenes:

(a) CH_3, CH_3 (b) 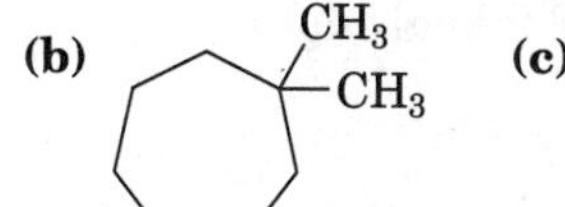

(c) 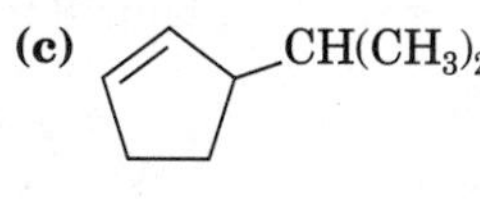

PROBLEM 3.3 Draw structures corresponding to these IUPAC names:

(a) 2-Methyl-1-hexene (b) 4,4-Dimethyl-2-pentene
(c) 2-Methyl-1,5-hexadiene (d) 3-Ethyl-2,2-dimethyl-3-heptene

3.2 ELECTRONIC STRUCTURE OF ALKENES

We saw in Section 1.10 that the carbon atoms in a double bond are sp^2-hybridized and have three equivalent orbitals that lie in a plane at angles of 120° to one another. The fourth carbon orbital is an unhybridized p orbital, which is perpendicular to the sp^2 plane. When two such carbon atoms approach each other, they form two kinds of bonds: a sigma bond, formed by head-on overlap of sp^2 orbitals, and a pi bond, formed by sideways overlap of p orbitals. The doubly bonded carbons and the four attached atoms lie in a plane, with bond angles of approximately 120° (Figure 3.1).

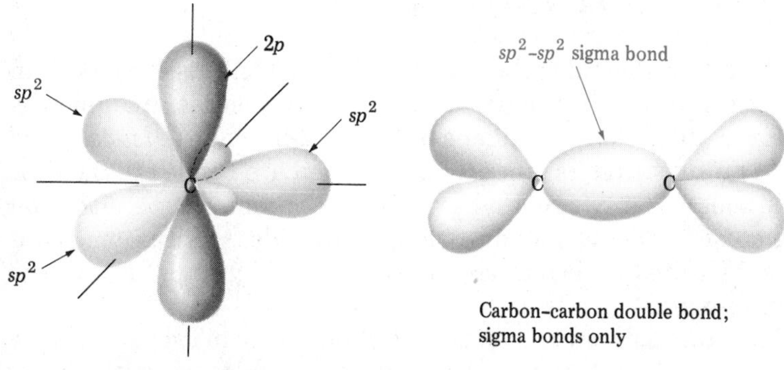

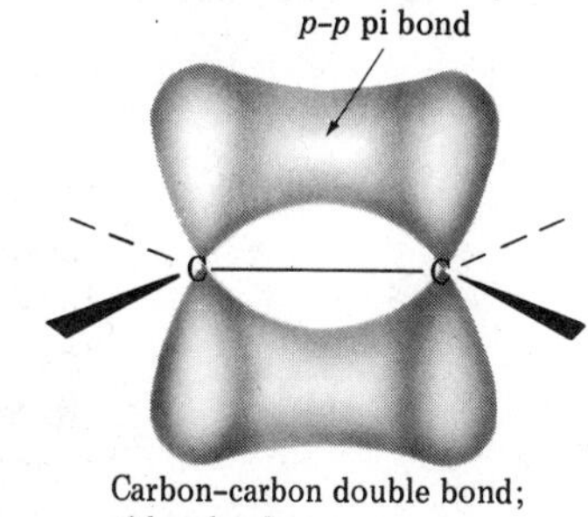

FIGURE 3.1 An orbital picture of the carbon–carbon double bond.

We know from Section 2.5 that free rotation is possible around single bonds, and that open-chain alkanes like ethane and propane therefore have many rapidly interconverting conformations. The same is not true for double bonds, however. No rotation can take place around carbon–carbon double bonds, because this would break the pi part of the bond (Figure 3.2). In fact, the energy barrier to rotation around a double bond is as large as the strength of the pi bond itself, an estimated 64 kcal/mol (268 kJ/mol). (Recall that the rotation barrier for a single bond is only about 2.9 kcal/mol.)

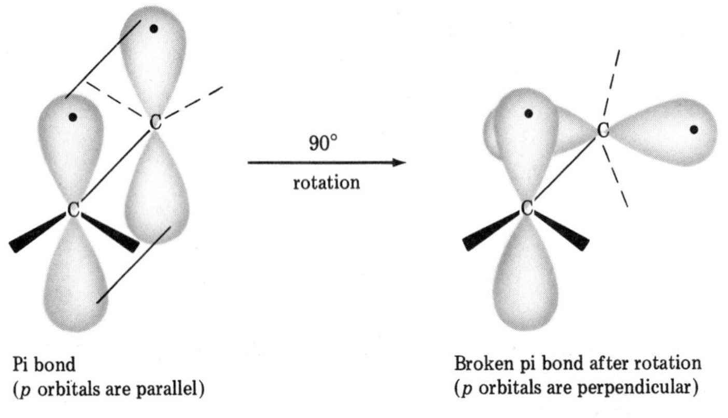

FIGURE 3.2 Breaking the pi bond is necessary for rotation around a carbon–carbon double bond to take place.

3.3 CIS–TRANS ISOMERS

The lack of rotation around carbon–carbon double bonds is of more than just theoretical interest; it also has chemical consequences. Imagine the situation for a disubstituted alkene such as 2-butene. (*Disubstituted* means that two substituents other than hydrogen are bonded to the double-bond carbons.) The two methyl groups in 2-butene can be either on the same side of the double bond or on opposite sides, a situation reminiscent of substituted cycloalkanes (Section 2.8). Figure 3.3 shows the two 2-butene isomers.

Since bond rotation can't occur, the two 2-butenes can't spontaneously interconvert; they are different chemical compounds. As with disubstituted cycloalkanes (Section 2.8), we call such compounds *cis–trans isomers*. The isomer with both substituents on the same side is called *cis*-2-butene, and the isomer with substituents on opposite sides is *trans*-2-butene.

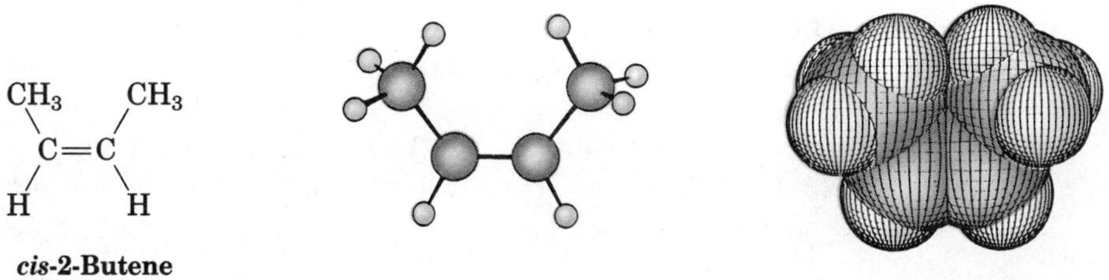

CH_3 CH_3 C=C H H

cis-2-Butene

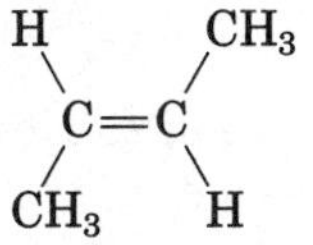

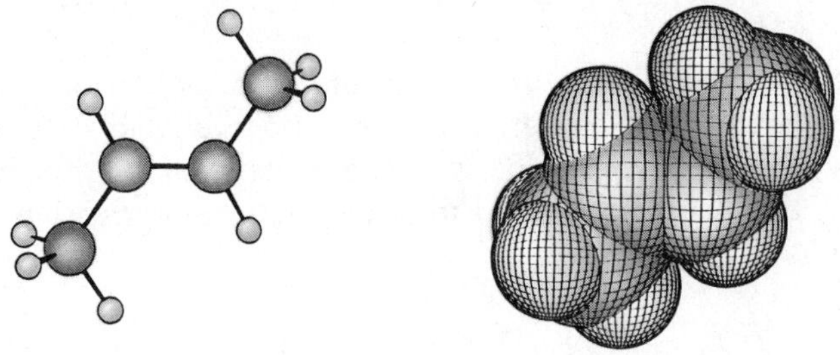

H CH_3 C=C CH_3 H

trans-2-Butene

FIGURE 3.3 Cis and trans isomers of 2-butene. The cis isomer has the two methyl groups on the same side of the double bond, and the trans isomer has the methyl groups on opposite sides.

Cis–trans isomerism is not limited to disubstituted alkenes. It can occur whenever both of the double-bond carbons are attached to two different groups. If one of the double-bond carbons is attached to two identical groups, however, then cis–trans isomerism is not possible (Figure 3.4).

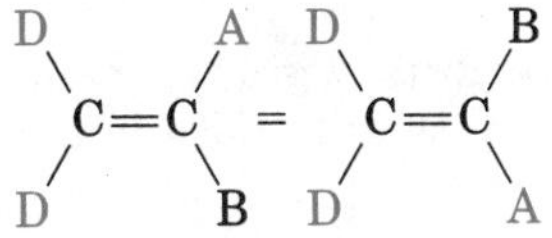

D A C=C D B = D B C=C D A

These two compounds are identical; they aren't cis–trans isomers.

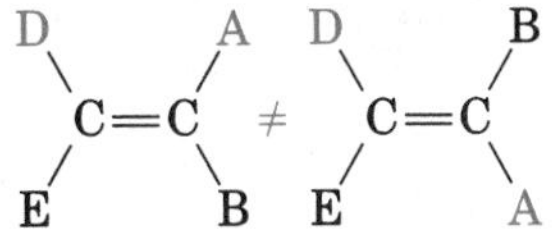

D A C=C E B ≠ D B C=C E A

These two compounds are *not* identical; they are cis–trans isomers.

FIGURE 3.4 The requirement for cis–trans isomerism in alkenes. Both double-bond carbons must be attached to two different groups.

Although the cis–trans interconversion of alkene isomers doesn't occur spontaneously, it can be made to happen by treating the alkene with a strong acid catalyst. If we interconvert *cis*-2-butene with *trans*-2-butene and allow them to reach equilibrium, we find that they aren't of equal stability. The trans isomer is more favored than the cis isomer by a ratio of 76 (trans) to 24 (cis) at equilibrium.

$$\underset{\textbf{\textit{cis}-2-Butene (24\%)}}{(H_3C)(H)C{=}C(CH_3)(H)} \xrightleftharpoons[\text{catalyst}]{H^+} \underset{\textbf{\textit{trans}-2-Butene (76\%)}}{(H)(H_3C)C{=}C(CH_3)(H)}$$

Cis alkenes are less stable than their trans isomers because of steric (spatial) interference between the bulky substituents on the same side of the double bond. This is the same kind of interference, or *steric strain*, that we saw in axial methylcyclohexane (Section 2.11).

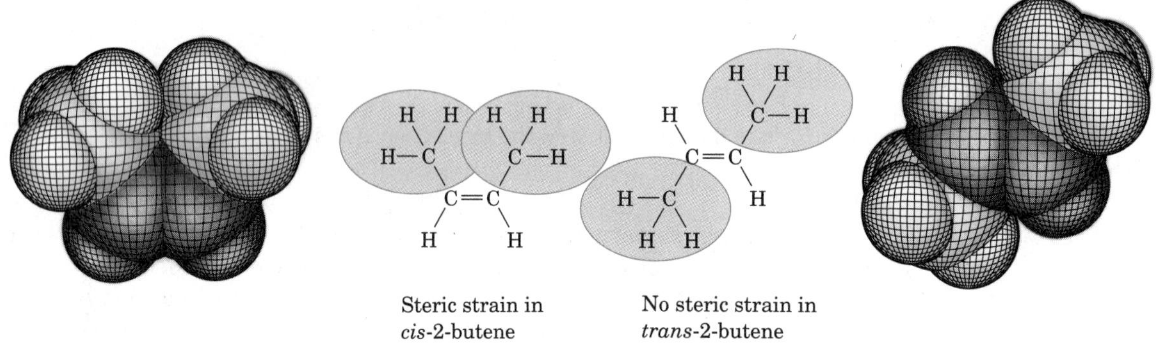

Steric strain in *cis*-2-butene

No steric strain in *trans*-2-butene

PRACTICE PROBLEM 3.2 Draw the cis and trans isomers of 5-chloro-2-pentene.

SOLUTION 5-Chloro-2-pentene is $ClCH_2CH_2CH{=}CHCH_3$. The two substituent groups are on the same side of the double bond in the cis isomer and on opposite sides in the trans isomer.

$$\underset{\textbf{\textit{cis}-5-Chloro-2-pentene}}{(ClCH_2CH_2)(H)C{=}C(CH_3)(H)} \qquad \underset{\textbf{\textit{trans}-5-Chloro-2-pentene}}{(H)(ClCH_2CH_2)C{=}C(CH_3)(H)}$$

PROBLEM 3.4 Which of these compounds can exist as pairs of cis–trans isomers? Draw each cis–trans pair.

(a) $CH_3CH{=}CH_2$ **(b)** $(CH_3)_2C{=}CHCH_3$
(c) $ClCH{=}CHCl$ **(d)** $CH_3CH_2CH{=}CHCH_3$
(e) $CH_3CH_2CH{=}CBrCH_3$ **(f)** 3-Methyl-3-heptene

PROBLEM 3.5 Which is more stable, *cis*-2-methyl-3-hexene or *trans*-2-methyl-3-hexene?

PROBLEM 3.6 How can you account for the observation that cyclohexene does not show cis–trans isomerism?

3.4 SEQUENCE RULES: THE *E*,*Z* DESIGNATION

In the previous discussion of isomerism in the 2-butenes, we used the terms *cis* and *trans* to specify alkenes whose two substituents were on the same side and opposite sides of a double bond, respectively. This cis–trans naming system is unambiguous for disubstituted alkenes, but how do we denote the geometry of *tri*substituted and *tetra*substituted double bonds? (*Trisubstituted* means three substituents other than hydrogen on the double bond; *tetrasubstituted* means four substituents other than hydrogen.)

SEQUENCE RULES
Series of rules for assigning priorities to substituent groups

The answer is provided by the *E*,*Z* system of nomenclature, which uses a system of **sequence rules** to assign priorities to the substituent groups on the double-bond carbons. Considering each of the double-bond carbons separately, we use the sequence rules to decide which of the two groups on each carbon is higher in priority. If the higher-priority groups are on the same side of the double bond, the alkene is designated *Z* (for the German *zusammen*, "together"). If the higher-priority groups are on opposite sides, the alkene is designated *E* (for the German *entgegen*, "opposite"). A simple way to remember which is which is that in the *Z* isomer, the groups are on "ze zame zide." These assignments are shown in Figure 3.5.

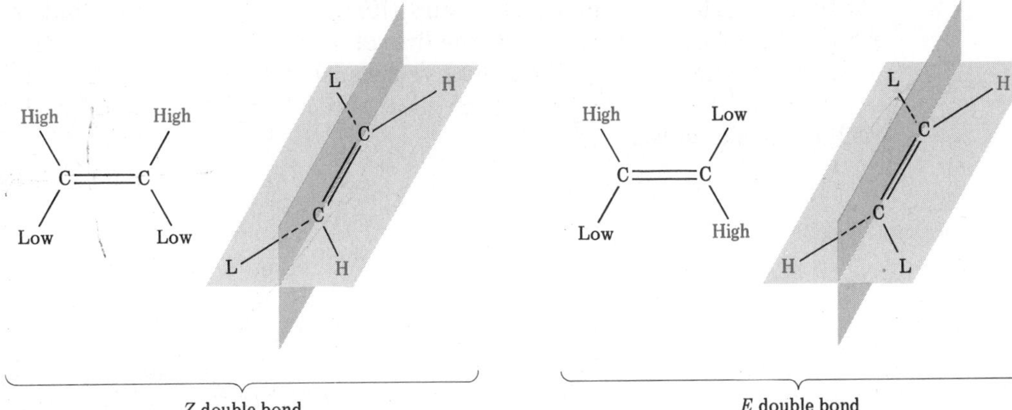

FIGURE 3.5 The *E*,*Z* system of nomenclature for substituted alkenes.

The sequence rules used in assigning priorities are as follows:

SEQUENCE RULE 1 Look at the atoms directly attached to each of the double-bond carbons and rank them in order of decreasing atomic number. That is, an atom with a higher atomic number like Cl (17) is higher in priority than an atom with a lower atomic number like H (1). Thus, the atoms that we commonly find attached to a double-bond carbon are assigned priorities as follows:

$$\overset{35}{\text{Br}} > \overset{17}{\text{Cl}} > \overset{8}{\text{O}} > \overset{7}{\text{N}} > \overset{6}{\text{C}} > \overset{1}{\text{H}}$$

For example:

Low priority H, High priority CH_3 | Cl High priority, CH_3 Low priority

(a) (*E*)-2-Chloro-2-butene

Low priority H, High priority CH_3 | CH_3 Low priority, Cl High priority

(b) (*Z*)-2-Chloro-2-butene

Because chlorine has a higher atomic number than carbon, it receives higher priority than a methyl (CH_3) group. Methyl receives higher priority than hydrogen, however, and isomer (a) is therefore assigned *E* geometry (high-priority groups on opposite sides of the double bond). Isomer (b) has *Z* geometry (high-priority groups on ze zame zide of the double bond).

SEQUENCE RULE 2 If a decision can't be reached by ranking the first atoms in the substituents, look at the second, third, or fourth atoms away from the double-bond carbons until a difference is found. Thus, an ethyl substituent, $-CH_2CH_3$, and a methyl substituent, $-CH_3$, are equivalent by rule 1 since both have carbon as the first atom. By rule 2, however, ethyl receives higher priority than methyl because the next atoms are one carbon and two hydrogens rather than three hydrogens. Look at the following examples:

$-CH_3$ (Lower) vs. $-CH_2CH_3$ (Higher)

$-O-H$ (Lower) vs. $-O-CH_3$ (Higher)

$-CH(CH_3)CH_3$ (Higher) vs. $-CH_2CH_3$ (Lower)

$-CH(CH_3)NH_2$ (Lower) vs. $-CH_2Cl$ (Higher)

SEQUENCE RULE 3 Multiple-bonded atoms are equivalent to the same number of single-bonded atoms. For example, an aldehyde substituent (–CH=O), which has a carbon atom *doubly* bonded to *one* oxygen, is equivalent to a substituent with a carbon atom *singly* bonded to *two* oxygens.

This carbon is bonded to H, O, O	This oxygen is bonded to C, C	This carbon is bonded to H, O, O	This oxygen is bonded to C, C

(–CH=O is equivalent to –CH(O)–O–C)

In the following examples, the indicated pairs are equivalent:

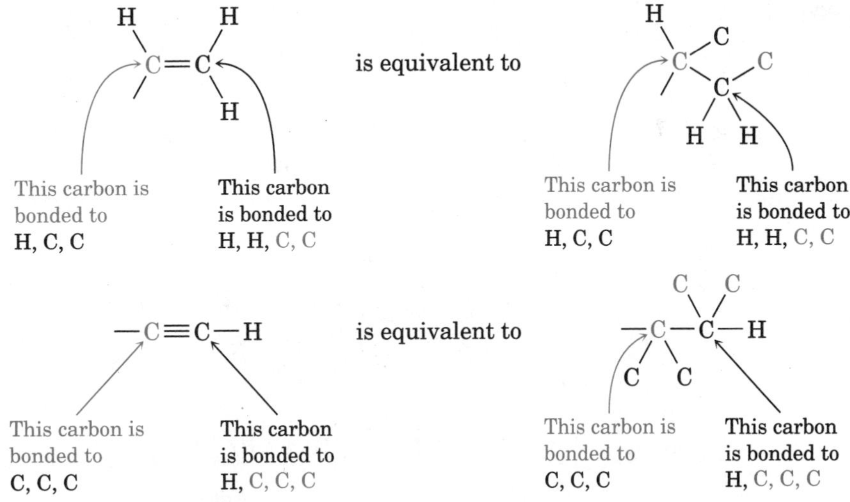

PRACTICE PROBLEM 3.3 Assign *E* or *Z* configuration to the double bond in this compound:

(H and H_3C on the left carbon; $CH(CH_3)_2$ and CH_2OH on the right carbon of C=C)

SOLUTION Look at the two double-bond carbons individually. The left-hand carbon has two substituents, –H and $-CH_3$, of which $-CH_3$ receives higher priority by rule 1. The right-hand carbon also has two substituents, $-CH(CH_3)_2$ and $-CH_2OH$, which are equivalent by rule 1. By rule 2, however, $-CH_2OH$ receives higher priority than $-CH(CH_3)_2$ because $-CH_2OH$ has an *oxygen* and two hydrogens as the next atoms, whereas $-CH(CH_3)_2$ has two *carbons* and one hydrogen. Thus, the two high-priority groups are on the same side of the double bond, and we assign *Z* configuration.

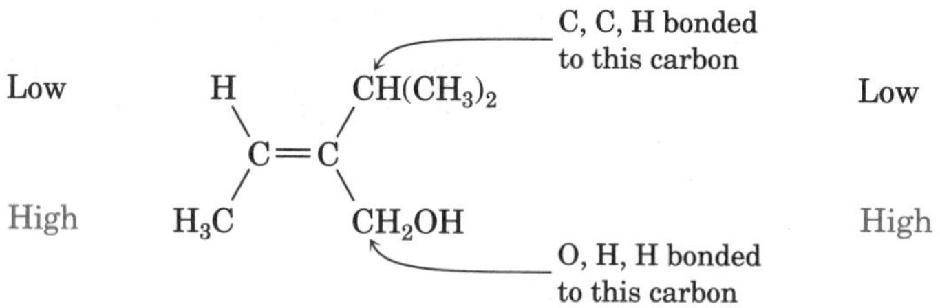

Low H $CH(CH_3)_2$ ← C, C, H bonded to this carbon; Low

$C{=}C$

High H_3C CH_2OH ← O, H, H bonded to this carbon; High

***Z* configuration**

PROBLEM 3.7 Which member in each set is higher in priority?
(a) –H or –Br **(b)** –Cl or –Br **(c)** $-CH_3$ or $-CH_2CH_3$ **(d)** $-NH_2$ or –OH
(e) $-CH_2OH$ or $-CH_3$ **(f)** $-CH_2OH$ or $-CH{=}O$

PROBLEM 3.8 Which is higher in priority, $-\overset{O}{\overset{\|}{C}}-OH$ or $-\overset{O}{\overset{\|}{C}}-OCH_3$? Explain.

PROBLEM 3.9 Which is higher in priority, isopropyl or *n*-octyl? Explain.

PROBLEM 3.10 Assign *E* or *Z* configuration to these alkenes:

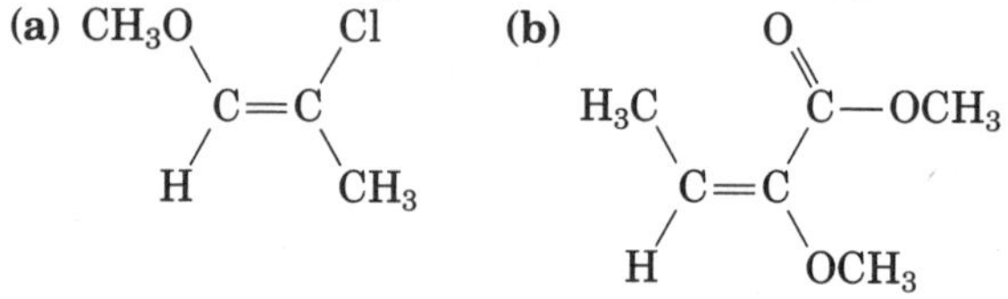

(a) $(CH_3O)(H)C{=}C(Cl)(CH_3)$ — CH_3O and Cl on top, H and CH_3 on bottom

(b) $(H_3C)(H)C{=}C(\overset{O}{\overset{\|}{C}}-OCH_3)(OCH_3)$ — H_3C and $C(=O)OCH_3$ on top, H and OCH_3 on bottom

3.5 KINDS OF ORGANIC REACTIONS

Now that we know something about alkenes and the double-bond functional group, it's time to learn about their chemical reactivity. As an introduction, we'll first look at some of the basic principles that underlie all organic reactions. In particular, we'll develop some general notions about why compounds react the way they do, and we'll see some methods that have been developed to help understand how reactions take place.

Organic chemical reactions can be organized either by *what kinds* of reactions occur or by *how* reactions occur. We'll begin by looking at the kinds of reactions that take place. There are four particularly important kinds of organic reactions: additions, eliminations, substitutions, and rearrangements.

ADDITION REACTION Reaction that occurs when two reagents combine to form a single new product with no atoms left over

Addition reactions occur when two reactants add together to form a single new product with no atoms "left over." We can generalize the process as

These reactants add together . . . $A + B \longrightarrow C$ . . . to give this single product.

As an example of an important addition reaction that we'll be studying soon, alkenes react with HCl to yield alkyl chlorides:

These two reactants . . . H—Cl + $H_2C{=}CH_2$ ⟶ $H_3C{-}CH_2Cl$ . . . add to give this product.

Ethylene (an alkene) **Chloroethane (an alkyl halide)**

ELIMINATION REACTION
Reaction that occurs when a single reactant splits apart into two products

Elimination reactions are, in a sense, the opposite of addition reactions. Eliminations occur when a single reactant splits into two products, a process we can generalize as

This one reactant . . . $A \longrightarrow B + C$ . . . splits apart to give these two products.

As an example of an important elimination reaction, alkyl halides split apart into an acid and an alkene when treated with base:

This one reactant . . . $H_3C{-}CH_2Cl \xrightarrow{NaOH} H_2C{=}CH_2 + H{-}Cl$ . . . gives these two products.

Chloroethane (an alkyl halide) **Ethylene (an alkene)**

SUBSTITUTION REACTION
Reaction that occurs when two reactants exchange parts to give two new products

Substitution reactions occur when two reactants exchange parts to give two new products, a process we can generalize as

These two reactants exchange parts . . . $A{-}B + C{-}D \longrightarrow A{-}C + B{-}D$ . . . to give these two new products.

As an example of a substitution reaction, we saw in Section 2.4 that alkanes react with chlorine gas in the presence of ultraviolet light to yield alkyl chlorides. A –Cl group from chlorine substitutes for the –H group of the alkane, and two new products result:

These two reactants . . . $CH_4 + Cl{-}Cl \xrightarrow{Light} CH_3Cl + H{-}Cl$ . . . give these two products.

Methane (an alkane) **Chloromethane (an alkyl halide)**

REARRANGEMENT REACTION
Reaction that occurs when a single reactant undergoes a reorganization of bonds and atoms to give an isomeric product

Rearrangement reactions occur when a single reactant undergoes a reorganization of bonds and atoms to yield a single isomeric product, a process we can generalize as

This single reactant . . . A $\longrightarrow$ B . . . gives this isomeric product.

As an example of a rearrangement reaction, we saw in Section 3.3 that *cis*-2-butene can be converted into its isomer *trans*-2-butene by treatment with an acid catalyst:

$$(H_3C)(H)C{=}C(CH_3)(H) \underset{\text{catalyst}}{\overset{H^+}{\rightleftharpoons}} (H)(H_3C)C{=}C(CH_3)(H)$$

***cis*-2-Butene (24%)** ***trans*-2-Butene (76%)**

PROBLEM 3.11

Classify these reactions as additions, eliminations, substitutions, or rearrangements:
(a) $CH_3Br + KOH \longrightarrow CH_3OH + KBr$
(b) $CH_3CH_2OH \longrightarrow H_2C{=}CH_2 + H_2O$
(c) $H_2C{=}CH_2 + H_2 \longrightarrow CH_3CH_3$

3.6 HOW REACTIONS OCCUR: MECHANISMS

REACTION MECHANISM
Description of the details by which a reaction occurs

Having looked at the kinds of reactions that take place, let's now see how reactions occur. An overall description of how a reaction occurs is called a **reaction mechanism**. A mechanism describes in detail exactly what takes place at each stage of a chemical transformation—which bonds are broken and in what order, which bonds are formed and in what order, and what the relative rate of each step is.

HOMOLYTIC
Electronically symmetrical breaking of a covalent bond to yield two radicals

HETEROLYTIC
Electronically unsymmetrical breaking of a covalent bond to yield an anion and a cation

All chemical reactions involve bond breaking and bond making. When two reactants come together, react, and yield products, specific chemical bonds in the reactants are broken, and specific bonds in the products are formed. Fundamentally, there are two ways in which a covalent two-electron bond can break: in an electronically *symmetrical* way so that one electron remains with each product fragment, or in an electronically *unsymmetrical* way so that both electrons remain with one product fragment, leaving the other fragment with an empty orbital. The symmetrical cleavage is called a **homolytic** process, and the unsymmetrical cleavage is called a **heterolytic** process.

$$A{:}B \longrightarrow A\cdot + \cdot B$$ **Homolytic bond breaking (radical)**

$$A{:}B \longrightarrow A^+ + {:}B^-$$ **Heterolytic bond breaking (polar)**

HOMOGENIC
Electronically symmetrical formation of a covalent bond by combination of two radicals

HETEROGENIC
Electronically unsymmetrical formation of a covalent bond by combination of an anion and a cation

RADICAL REACTION
Reaction that involves electronically symmetrical bond making and bond breaking

RADICAL
A species that contains an odd number of electrons

POLAR REACTION
Reaction that involves electronically unsymmetrical bond making and bond breaking

ORGANOMETALLIC
Containing a carbon–metal bond

Conversely, there are two ways in which a covalent two-electron bond can form: in an electronically symmetrical (**homogenic**) way when one electron is donated to the new bond by each reactant, or in an electronically unsymmetrical (**heterogenic**) way when both bonding electrons are donated to the new bond by one reactant.

$$\text{A}\cdot + \cdot\text{B} \longrightarrow \text{A}:\text{B} \qquad \textbf{Homogenic bond making (radical)}$$

$$\text{A}^+ + :\text{B}^- \longrightarrow \text{A}:\text{B} \qquad \textbf{Heterogenic bond making (polar)}$$

Processes that involve symmetrical bond breaking and making are called **radical reactions**. A **radical** is a chemical species that contains an *odd* number of valence electrons and thus has an orbital with only one electron. Processes that involve unsymmetrical bond breaking and making are called **polar reactions**. Polar reactions always involve species that contain an *even* number of valence electrons and have only electron pairs in their orbitals. Polar processes are the more common reaction type in organic chemistry, and a large part of this book is devoted to their description.

To see how polar reactions occur, we need to recall the discussion of polar covalent bonds in Section 1.12 and to look more deeply into the effects of bond polarity on organic molecules. We've seen that certain bonds in a molecule, particularly the bonds in functional groups, are often polar. When a carbon atom bonds to an electronegative atom such as chlorine or oxygen, the bond is polarized so that the carbon bears a partial positive charge (δ^+) and the electronegative atom bears a partial negative charge (δ^-). Conversely, when carbon bonds to an atom that is less electronegative than itself, the opposite polarity results. Such is the case with most carbon–metal (**organometallic**) bonds.

Where Y = O, N, Cl, Br, I   Where M = a metal such as Mg or Li

What does bond polarity mean with respect to chemical reactions? *Because unlike charges attract each other, the fundamental characteristic of all polar reactions is that the electron-rich sites in one molecule react with the electron-poor sites in another molecule.* Bonds are made in a polar reaction when the electron-rich reactant donates a *pair* of electrons to the electron-poor reactant, and bonds are broken in a polar reaction when one of the two product fragments leaves with the electron *pair*.

Chemists usually indicate the electron movement that occurs during a polar reaction by curved arrows. A curved arrow shows where electrons move during the reaction. It means that an electron pair moves *from* the atom at the tail of the arrow *to* the atom at the head of the arrow.

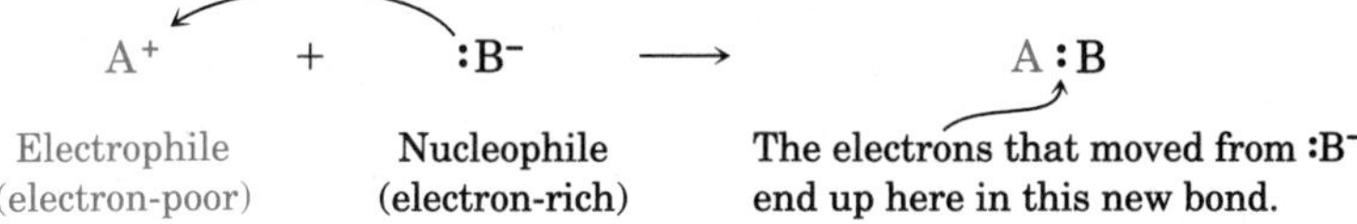
This curved arrow shows that electrons are moving from :B⁻ (electron-rich) to A⁺ (electron-poor)
A⁺ + :B⁻ ⟶ A:B
Electrophile (electron-poor)
Nucleophile (electron-rich)
The electrons that moved from :B⁻ end up here in this new bond.

NUCLEOPHILE
Electron-rich reagent that can donate an electron pair to an electrophile in a polar reaction

ELECTROPHILE
Electron-poor reagent that can accept an electron pair from a nucleophile in a polar reaction

In referring to polar reactions, chemists have coined the words *nucleophile* and *electrophile*. A **nucleophile** is a reagent that is "nucleus-loving"; a nucleophile has an electron-rich site and forms a bond by donating an electron pair to an electron-poor site. Nucleophiles are often, though not always, negatively charged. An **electrophile**, by contrast, is "electron-loving"; an electrophile has an electron-poor site and forms a bond by accepting an electron pair from a nucleophile. Electrophiles are often, though not always, positively charged.

PRACTICE PROBLEM 3.4 What is the direction of bond polarity in the amine functional group, $C{-}NH_2$?

SOLUTION According to the electronegativity table (Figure 1.17), nitrogen is more electronegative than carbon. Thus, an amine is polarized with carbon as δ^+ and nitrogen as δ^-.

PROBLEM 3.12 What is the direction of bond polarity in these functional groups?
(a) Ketone **(b)** Alkyl chloride **(c)** Alcohol **(d)** Alkyllithium

PROBLEM 3.13 Identify the functional groups and show the direction of bond polarity in each of these molecules:

(a) Acetone, $CH_3\overset{\overset{\displaystyle O}{\|}}{C}CH_3$ **(b)** Chloroethane, CH_3CH_2Cl
(c) Methanethiol, CH_3SH
(d) Tetraethyllead, $(CH_3CH_2)_4Pb$ (the "lead" in gasoline)

PROBLEM 3.14 Which of the following would you expect to behave as electrophiles, and which as nucleophiles? Explain.
(a) H^+ **(b)** HO^- **(c)** Br^+ **(d)** NH_3 **(e)** $H{-}C{\equiv}C{-}H$ **(f)** CO_2

3.7 AN EXAMPLE OF A POLAR REACTION: ADDITION OF HCl TO ETHYLENE

Let's look in detail at a typical polar reaction, the reaction of ethylene with HCl. When ethylene is treated with hydrogen chloride at room temperature, chloroethane is produced. Overall, the reaction can be formulated as follows:

$$H_2C{=}CH_2 + H{-}Cl \longrightarrow CH_3{-}CH_2Cl$$

Ethylene (nucleophile) **Hydrogen chloride** (electrophile) **Chloroethane**

ELECTROPHILIC ADDITION Addition of an electrophile to an unsaturated acceptor, usually an alkene

This reaction, an example of a general polar reaction type known as an **electrophilic addition**, can be understood in terms of the general concepts just discussed. We'll begin by looking at the nature of the two reactants.

What do we know about ethylene? We know from Sections 1.10 and 3.2 that a carbon–carbon double bond results from orbital overlap of two sp^2-hybridized carbon atoms. The sigma part of the double bond results from sp^2–sp^2 overlap, and the pi part results from p–p overlap.

What kind of chemical reactivity might we expect of carbon–carbon double bonds? We know that *alkanes* are rather inert because all their valence electrons are tied up in strong, nonpolar, carbon–carbon and carbon–hydrogen bonds. Furthermore, alkane bonding electrons are inaccessible to external reagents because they are sheltered in sigma orbitals between atoms. The situation for ethylene and other alkenes is quite different. For one thing, double bonds have greater electron density than single bonds—four electrons in a double bond versus only two electrons in a single bond. Equally important, the electrons in the pi bond are accessible to external reagents because they are located above and below the plane of the double bond rather than between the nuclei (Figure 3.6).

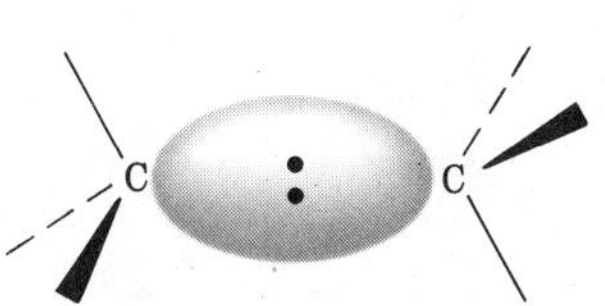

Carbon–carbon sigma bond:
stronger; less accessible bonding electrons

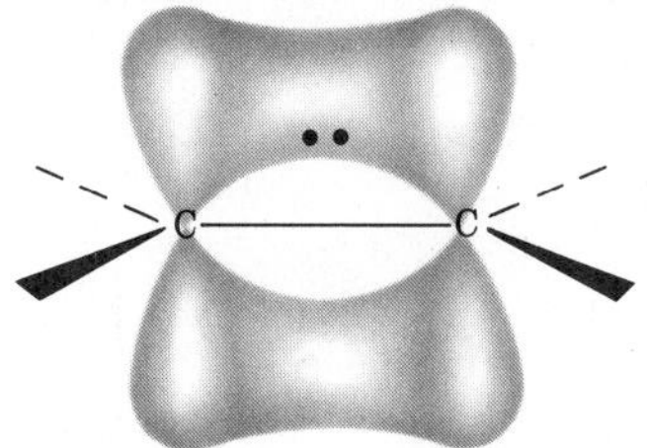

Carbon–carbon pi bond:
weaker; more accessible electrons

FIGURE 3.6 A comparison of carbon–carbon single and double bonds. A double bond is both more electron-rich (more nucleophilic) and more accessible to reaction with external reagents than a single bond.

Both electron richness and electron accessibility lead to a prediction of high reactivity for carbon–carbon double bonds. In the terminology of polar reactions, we might predict that carbon–carbon double bonds should behave as *nucleophiles*. That is, the chemistry of alkenes should involve reaction of the electron-rich double bond with electron-poor reagents. This is exactly what we find: The most important reaction of alkenes is their reaction with electrophiles.

What about HCl? As a strong acid, HCl is a powerful proton (H^+) donor. Because a proton is positively charged and electron-poor, it is a good electrophile. Thus, the reaction of H^+ with ethylene is a typical electrophile-nucleophile combination, characteristic of all polar reactions. (Although chemists often talk about "H^+" when referring to acids, there is really no such free species. Protons are always associated with another molecule for stability—for example, with water in H_3O^+.)

3.8 THE MECHANISM OF AN ORGANIC REACTION: ADDITION OF HCl TO ETHYLENE

We can view the electrophilic addition reaction between ethylene and HCl as proceeding by the mechanism shown in Figure 3.7. The reaction takes place in two steps, beginning with an attack on the electrophile H^+ by the electron pair from the nucleophilic ethylene pi bond. Two electrons from the pi bond form a new sigma bond between the entering hydrogen and one of the ethylene carbons, as shown by tracing the path of the curved arrow in Figure 3.7. (Remember: A curved arrow is used to indicate how electrons move in a polar reaction. In this case, the electrons move away from the carbon–carbon pi bond to form a new bond with the incoming H^+.)

The electrophile H^+ is attacked by the pi electrons of the double bond, and a new C–H sigma bond is formed. This leaves the other carbon atom with a + charge and a vacant *p* orbital.

Carbocation intermediate

Cl^- donates an electron pair to the positively charged carbon atom, forming a C–Cl sigma bond and yielding the neutral addition product.

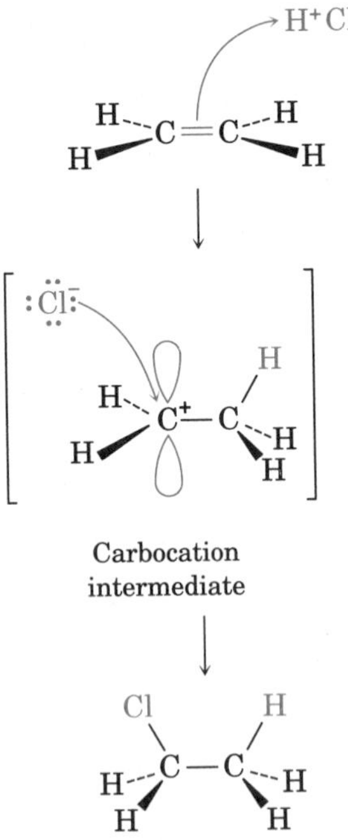

FIGURE 3.7 The mechanism of the electrophilic addition of HCl to ethylene. The reaction takes place in two steps and involves an intermediate carbocation.

The other ethylene carbon atom, having lost its share of the pi electrons, is now trivalent and is left with a vacant *p* orbital. Because the double-bond pi electrons were used in the formation of the new C–H bond, the trivalent carbon has only six valence electrons and therefore carries a positive charge. In the second step, this positively charged species, a carbon-cation or **carbocation**, is itself an electrophile that can accept an electron pair from the nucleophilic chloride anion to form a C–Cl bond, yielding the neutral addition product.

CARBOCATION
A species that has a positively charged, trivalent carbon atom

PRACTICE PROBLEM 3.5 What product would you expect from reaction of HCl with cyclohexene?

SOLUTION HCl should add to the double-bond functional group in cyclohexene in exactly the same way it adds to ethylene, yielding an addition product.

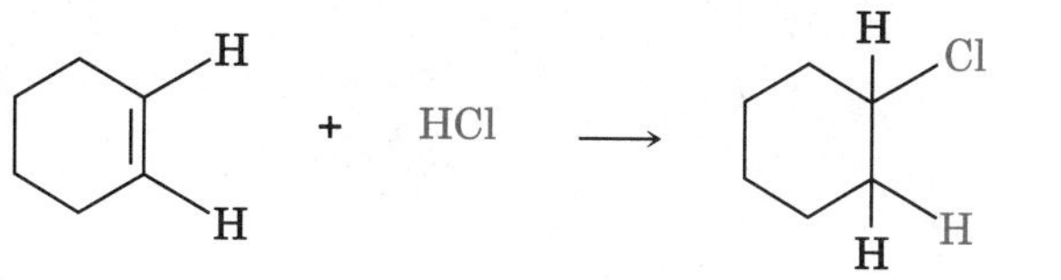

Cyclohexene + HCl ⟶ Chlorocyclohexane

PROBLEM 3.15 Reaction of HCl with 2-methylpropene yields 2-chloro-2-methylpropane. Formulate the mechanism of the reaction. What is the structure of the carbocation formed during the reaction?

$$(CH_3)_2C{=}CH_2 + HCl \longrightarrow (CH_3)_3C{-}Cl$$

PROBLEM 3.16 Reaction of HCl with 2-pentene yields a mixture of two addition products. Write the reaction, and show the two products.

3.9 DESCRIBING A REACTION: RATES AND EQUILIBRIA

Every chemical reaction can go in two directions. Reactants can give products, and products can revert to reactants. The position of the resultant chemical equilibrium is expressed by an equation in which K_{eq}, the **equilibrium constant**, is equal to the concentration of products, divided by the concentration of reactants. For the generalized reaction

EQUILIBRIUM CONSTANT (K_{eq})
Value that expresses the extent to which a given reaction takes place at equilibrium

$$a\text{A} + b\text{B} \longrightarrow c\text{C} + d\text{D}$$

we have

$$K_{eq} = \frac{[\text{Products}]}{[\text{Reactants}]} = \frac{[\text{C}]^c[\text{D}]^d}{[\text{A}]^a[\text{B}]^b}$$

The magnitude of the equilibrium constant tells which side of the reaction arrow is energetically favored. If K_{eq} is larger than 1, then the product concentration term $[C]^c[D]^d$ is larger than the reactant concentration term $[A]^a[B]^b$, and the reaction proceeds as written from left to right. If K_{eq} is smaller than 1, the reaction does not take place as written but instead goes from right to left.

What the equilibrium equation does *not* tell is the *rate* of the reaction, or how fast the equilibrium is established. Some reactions are extremely slow even though they have favorable equilibrium constants. Gasoline is stable at room temperature, for example, because the rate of its reaction with oxygen is slow under these conditions. Under other conditions, however (contact with a lighted match, for example), gasoline reacts rapidly with oxygen and undergoes complete conversion to water and carbon dioxide. Rates (*how fast* a reaction occurs) and equilibria (*how much* a reaction occurs) are entirely different.

Rate ⟶ Is reaction fast or slow?

Equilibrium ⟶ In what direction does reaction proceed?

EXOTHERMIC
Reaction that gives off energy (heat)

ENDOTHERMIC
Reaction that absorbs energy (heat)

HEAT OF REACTION (ΔH)
Amount of heat released or absorbed in a reaction

What determines whether a reaction takes place? For a reaction to have a favorable equilibrium constant, the energy level of the products must be lower than the energy level of the reactants. In other words, energy must be given off.[1] Such reactions are said to be **exothermic** (from the Greek *exo*, "outside," and *therme*, "heat"). If the energy level of the products is higher than the energy level of the reactants, then the equilibrium constant for the reaction is unfavorable, and energy must be added to make the reaction take place. Such reactions are said to be **endothermic** (Greek *endon*, "within").

A good analogy for the relationship between energy and chemical reactivity (stability) is that of a rock poised near the top of a hill. The rock in its unstable position has stored the energy that was required to get it up there. When it rolls downhill, it releases its energy until it reaches a stable, low-energy position at the bottom of the hill. In the same way, the energy level in a chemical reaction goes downhill as the energy stored in the chemical bonds of a reactant is released and a more stable product is formed (Figure 3.8).

The exact amount of heat released in an exothermic reaction or absorbed in an endothermic reaction is called the **heat of reaction, ΔH** (spoken as delta-H). By convention, ΔH has a negative value in an exothermic reaction since heat is released, and a positive value in an endothermic reaction since heat is absorbed. Since ΔH is a measure of the

[1]Strictly speaking, the "energy" released in a favorable reaction is the Gibbs free energy (ΔG), which is the sum of two contributions, an *enthalpy* contribution (ΔH) that measures the heat change in the reaction and a temperature-dependent *entropy* contribution ($T\Delta S$) that measures the change in molecular disorder during the reaction ($\Delta G = \Delta H - T\Delta S$). The entropy contribution is normally small and is often ignored when making qualitative arguments.

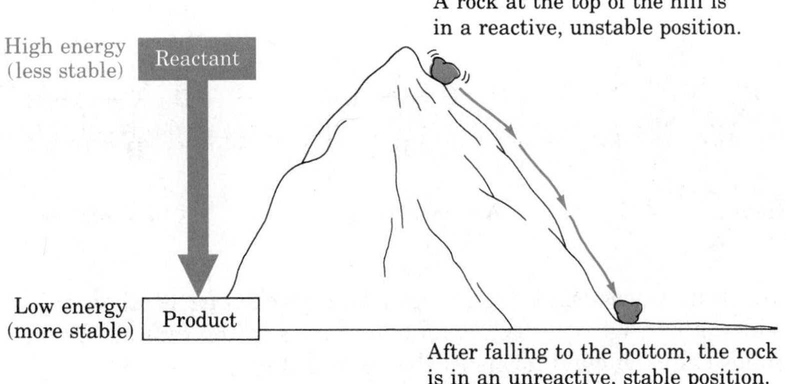
A rock at the top of the hill is in a reactive, unstable position.
High energy (less stable)
Reactant
Low energy (more stable)
Product
After falling to the bottom, the rock is in an unreactive, stable position.

FIGURE 3.8 The relationship between energy and stability. Like a rock near the top of a hill, high-energy substances are unstable. They release their energy when they drop downhill to form low-energy, stable products.

difference in energy between reactants and products, the size of ΔH determines the size of the equilibrium constant, K_{eq}. Favorable reactions with large K_{eq}'s are highly exothermic and have negative heats of reaction, whereas unfavorable reactions with small K_{eq}'s are endothermic and have positive heats of reaction.

$$A + B \rightleftharpoons C + D$$

$$K_{eq} = \frac{[C][D]}{[A][B]} \quad \begin{array}{l} \text{Exothermic if } K_{eq} > 1\text{; negative value of } \Delta H \\ \text{Endothermic if } K_{eq} < 1\text{; positive value of } \Delta H \end{array}$$

PRACTICE PROBLEM 3.6 Which reaction is more favorable, one with $\Delta H = -15$ kcal/mol or one with $\Delta H = +15$ kcal/mol?

SOLUTION Reactions with negative ΔH are exothermic and thus are favorable, but reactions with positive ΔH are endothermic and unfavorable.

PROBLEM 3.17 Which reaction is more exothermic, one with $\Delta H = -10$ kcal/mol or one with $\Delta H = +10$ kcal/mol?

PROBLEM 3.18 Which reaction is more exothermic, one with $K_{eq} = 1000$ or one with $K_{eq} = 0.001$?

3.10 DESCRIBING A REACTION: REACTION ENERGY DIAGRAMS AND TRANSITION STATES

For a reaction to take place, reactant molecules must collide, and reorganization of atoms and bonds must occur. Let's look again at the addition reaction between ethylene and HCl:

$$H_2C{=}CH_2 + H^+ \longrightarrow \left[H{-}CH_2{-}\overset{+}{C}H_2 \right] \xrightarrow{:\ddot{Cl}:^-} H{-}CH_2{-}CH_2Cl$$

Ethylene (nucleophile) **A carbocation** **Chloroethane**

As the reaction proceeds, ethylene and HCl approach each other, the pi bond breaks, a new carbon–hydrogen bond forms in the first step, and a new carbon–chlorine bond forms in the second step.

REACTION ENERGY DIAGRAM Method for depicting the energy changes that occur during a reaction

Over the years, chemists have developed a method for depicting the energy changes that occur during a reaction using **reaction energy diagrams** of the sort shown in Figure 3.9. The vertical axis of the diagram represents the total energy of all reactants, and the horizontal axis represents the progress of the reaction from beginning (left) to end (right). Let's take a careful look at the reaction, one step at a time, and see how the addition of HCl to ethylene can be described in a reaction energy diagram.

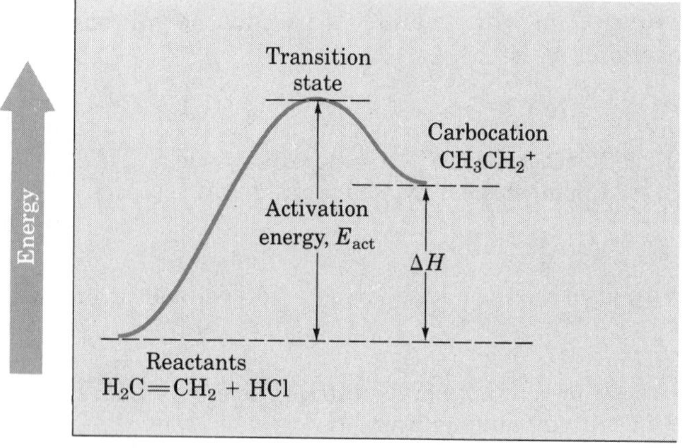

FIGURE 3.9 A reaction energy diagram for the first step in the reaction of ethylene with HCl. The energy difference between reactants and transition state, E_{act}, determines the reaction rate. The energy difference between reactants and carbocation product, ΔH, determines the position of the equilibrium.

At the beginning of the reaction, ethylene and HCl have the total amount of energy indicated by the reactant level on the left side of the diagram. As the two molecules crowd together, their electron clouds repel each other, causing the energy level to rise. If the collision has occurred with sufficient force and proper orientation, the reactants continue to approach each other despite the repulsion until the new carbon–hydrogen bond starts to form. At some point, a structure of maximum energy is reached, a structure we call the **transition state**.

TRANSITION STATE
Hypothetical structure of maximum energy formed during the course of a reaction

The transition state represents the highest-energy structure involved in this step of the reaction and can't be isolated. Nevertheless, we can imagine the transition state to be a kind of activated complex of the two reactants in which the carbon–carbon pi bond is partially broken and the new carbon–hydrogen bond is partially formed (Figure 3.10).

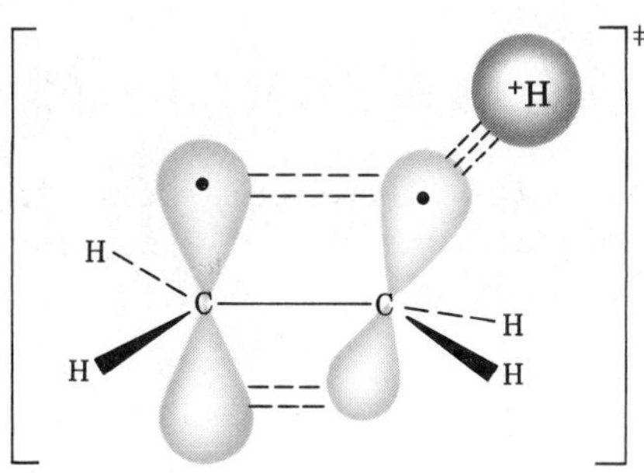

FIGURE 3.10 A hypothetical transition-state structure for the first step of the reaction of ethylene with HCl. The C–C pi bond is just beginning to break, and the C–H bond is just beginning to form.

ACTIVATION ENERGY (E_{act})
The energy difference between starting material and transition state

The energy difference between reactants and transition state, called the **activation energy, E_{act}**, measures how rapidly the reaction occurs. A large activation energy corresponds to a large energy difference between reactants and transition state, and results in a slow reaction because few of the reacting molecules collide with enough energy to climb the high barrier. A small activation energy results in a rapid reaction because almost all reacting molecules are energetic enough to climb to the transition state.

The situation of reactants needing enough energy to climb the barrier from reactant to transition state is like the situation of hikers who need enough energy to climb over a mountain pass. If the pass is a high one, the hikers need a lot of energy and surmount the barrier slowly. If the pass is low, however, the hikers need less energy and reach the top quickly.

Most organic reactions have activation energies in the range 10–35 kcal/mol (40–150 kJ/mol). Reactions with activation energies less than 20 kcal/mol take place spontaneously at room temperature or below, whereas reactions with higher activation energies normally require heating. Heat provides the energy necessary for the reactants to climb the activation barrier.

Once the high-energy transition state has been reached, the reaction proceeds to the carbocation product. Energy is released as the new C–H bond forms fully, and the curve in the reaction energy diagram in Figure 3.9 therefore turns downward until it reaches a minimum. This minimum point represents the energy level of the carbocation product of the first step. The energy change, ΔH, between reactants and carbocation product is simply the difference between the two levels in the diagram.[2] Since the carbocation is less stable than the alkene reactant, the first step is endothermic, and energy is absorbed.

PROBLEM 3.19 Which reaction is faster, one with $E_{act} = 15$ kcal/mol or one with $E_{act} = 20$ kcal/mol? Is it possible to predict which of the two has the larger K_{eq}?

3.11 DESCRIBING A REACTION: INTERMEDIATES

How can we describe the carbocation structure formed in the first step of the reaction of ethylene with HCl? The carbocation is clearly different from the reactants, yet it isn't a transition state and it isn't a final product.

$$H_2C{=}CH_2 + H^+ \longrightarrow \left[H{-}\underset{H}{\overset{H}{C}}{-}\underset{H}{\overset{+}{C}}{-}H \right] \xrightarrow{Cl^-} H{-}\underset{H}{\overset{H}{C}}{-}\underset{H}{\overset{Cl}{C}}{-}H$$

Ethylene Reaction intermediate **Chloroethane**

INTERMEDIATE Species formed during the course of a multistep reaction that lies at a minimum in a reaction energy diagram

We call the carbocation, which is formed fleetingly during the course of the multistep reaction, a reaction **intermediate**. As soon as the intermediate is formed in the first step by reaction of ethylene with H^+, it reacts with Cl^- in a second step to give the final product, chloroethane. This second step has its own activation energy, E_{act}; its own transition state; and its own energy change, ΔH. We can view the second transition state as an activated complex between the electrophilic carbocation intermediate and nucleophilic chloride anion, a complex in which the new C–Cl bond is just starting to form.

A complete energy diagram for the overall reaction of ethylene with HCl can be constructed as shown in Figure 3.11. In essence, we draw diagrams for each of the two individual steps and join them in the middle so that the product of step 1 (the carbocation) is the starting material for step 2. As indicated in Figure 3.11, the reaction intermediate lies at an energy minimum between steps 1 and 2. Since the energy level of this intermediate is higher than the level of either the reactants (ethylene + HCl) or the product (chloroethane), the intermediate is highly reactive and can't be isolated. It is, however, more stable than either of the two transition states that surround it.

[2]As noted previously, the energy difference between starting materials and products is actually the Gibbs free energy, $\Delta G = \Delta H - T\Delta S$. Since the entropy contribution is normally small, we make the simplifying assumption that ΔG and ΔH are approximately equal.

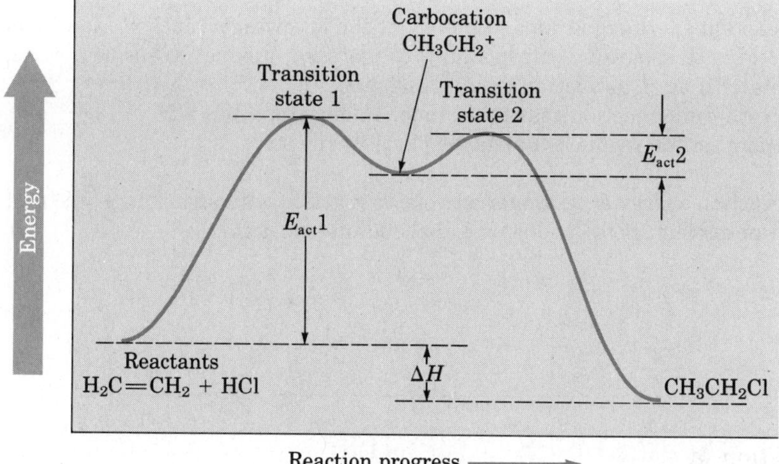

FIGURE 3.11 An overall reaction energy diagram for the reaction of ethylene with HCl. Two separate steps are involved, each with its own transition state. The energy minimum between the two steps represents the reaction intermediate.

Each step in a multistep process can always be considered separately. Each step has its own E_{act} (rate) and its own ΔH (energy change). The overall ΔH of the reaction, however, is the energy difference between initial reactants (far left) and final products (far right). This is always true regardless of the shape of the reaction energy curve. Note, for example, that the energy diagram for the reaction of HCl with ethylene in Figure 3.11 shows the energy level of the final product to be lower than the energy level of the reactants. Thus, the overall reaction is exothermic.

PRACTICE PROBLEM 3.7 Sketch a reaction energy diagram for a one-step reaction that is very fast and highly exothermic.

SOLUTION A very fast reaction has a low E_{act}, and a highly exothermic reaction has a large negative ΔH. Thus, the diagram will look like this:

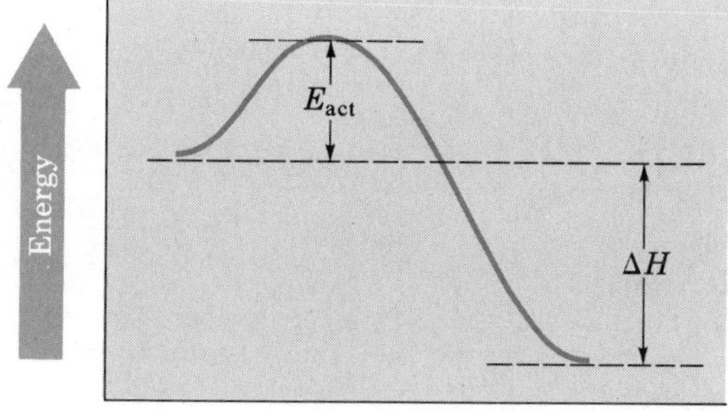

PROBLEM 3.20 Sketch reaction energy diagrams to represent the following situations, and label the parts of each diagram corresponding to reactant, product, transition state, intermediate (if any), activation energy, and ΔH.

(a) An exothermic reaction that takes place in one step.

(b) An endothermic reaction that takes place in one step.

PROBLEM 3.21 Draw a reaction energy diagram for a two-step reaction with an endothermic first step and an exothermic second step. Label the intermediate.

INTERLUDE

CARROTS, ALKENES, AND THE CHEMISTRY OF VISION

Your mother may have told you when you were growing up that eating carrots is good for your eyes. Although that's probably not true for healthy adults on a proper diet (your mother lied), there's no question that the chemistry of carrots and the chemistry of vision are related. Alkenes play a role in both.

Carrots are rich in β-carotene, a purple-orange alkene that is an excellent dietary source of vitamin A. β-Carotene is converted to vitamin A in the liver, where enzymes first cut the molecule in half and then change the geometry of the C11–C12 double bond to produce 11-*cis*-retinal, the light-sensitive pigment on which the visual systems of all living things are based.

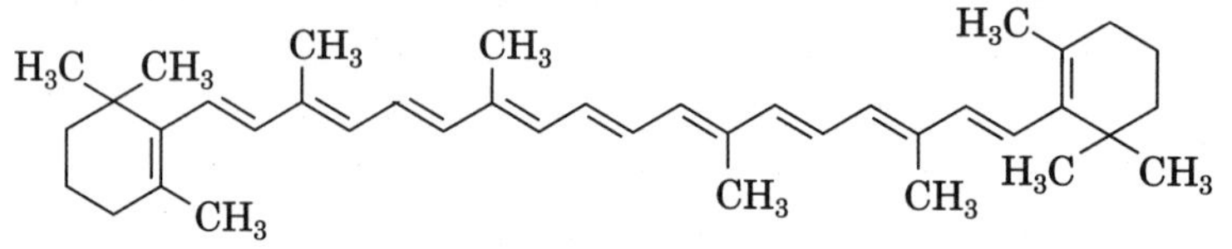

β-Carotene

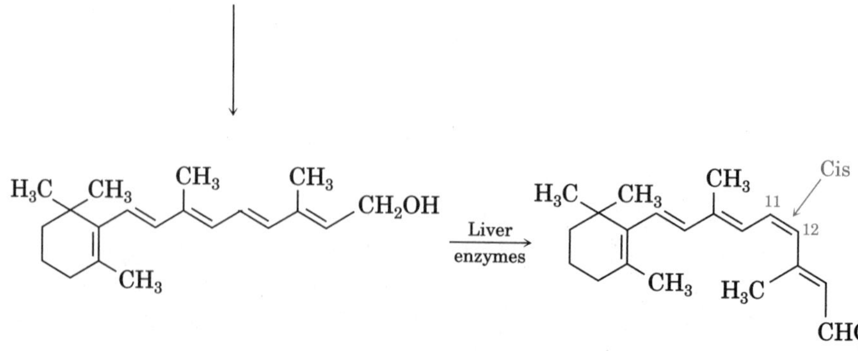

Liver enzymes

Vitamin A

11-*cis*-Retinal

The retina of the eye contains two types of light-sensitive receptor cells, *rod cells* and *cone cells*. Rod cells are primarily responsible for seeing in dim light, whereas cone cells are responsible for seeing in bright light and for the perception of colors. In the rod cells of the eye, 11-*cis*-retinal is converted into *rhodopsin*, a light-sensitive substance formed from the protein *opsin* and 11-*cis*-retinal. When light strikes the rod cell, isomerization of the C11–C12 double bond occurs, and 11-*trans*-rhodopsin, also called metarhodopsin II, is produced. This cis–trans isomerization of rhodopsin is accompanied by a change in molecular geometry, which in turn causes a nerve impulse to be sent to the brain where it is perceived as vision.

Rhodopsin

Metarhodopsin II

Metarhodopsin II is then recycled into rhodopsin by a multistep sequence involving cleavage into all-*trans*-retinal, conversion to vitamin A, cis–trans isomerization to 11-*cis*-vitamin A, and conversion back to 11-*cis*-retinal.

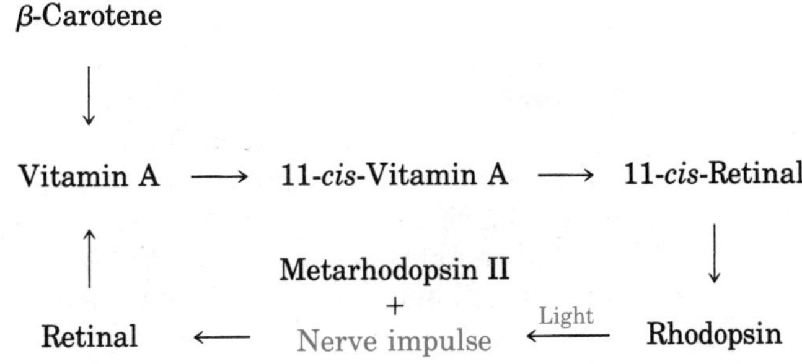

SUMMARY AND KEY WORDS

Alkenes are hydrocarbons that contain carbon–carbon double bonds. A double bond consists of two parts: a **sigma bond** formed by head-on overlap of two sp^2 orbitals and a **pi bond** formed by sideways overlap of two *p* orbitals. The bond strength of an alkene double bond is greater than that of a carbon–carbon single bond, with the strength of the pi part estimated to be 64 kcal/mol. Rotation around the double bond is restricted, and substituted alkenes can therefore exist as **cis–trans isomers**. The geometry of a double bond can be described as either ***Z*** (*zusammen*) or ***E*** (*entgegen*) by application of a series of **sequence rules**.

All organic reactions involve bond making and bond breaking. Covalent two-electron bonds can break or form in two ways. Bonds can break in an electronically symmetrical (**homolytic**) way so that each product retains one electron or in an electronically unsymmetrical (**heterolytic**) way so that one product retains both electrons, leaving the other product with a vacant valence orbital. Conversely, bonds can form in an electronically symmetrical (**homogenic**) way if each of two reactants donates one electron or in an electronically unsymmetrical (**heterogenic**) way if one reactant donates two electrons. Electronically symmetrical bond making and bond breaking occur in **radical reactions** whereas electronically unsymmetrical bond making and bond breaking occur in **polar reactions**:

$$A\cdot + \cdot B \longrightarrow A{:}B \qquad \textbf{Radical reaction}$$

$$A^+ + {:}B^- \longrightarrow A{:}B \qquad \textbf{Polar reaction}$$

The energy changes that take place during a reaction can be described by *rates* (how fast a reaction occurs) and **equilibria** (to what extent the reaction occurs). The equilibrium position of a reaction is determined by ΔH, the energy change that takes place during the reaction. If the reaction is **exothermic**, energy is given off, and the reaction has a favorable equilibrium constant. If the reaction is **endothermic**, however, energy is absorbed and the reaction has an unfavorable equilibrium constant. Reactions can be described pictorially by using **reaction energy diagrams**, which follow the course of a reaction from reactant to product.

Every reaction proceeds through a **transition state**, which represents the highest energy point reached. Transition-state structures can't be isolated because they are unstable, but we can imagine them to be activated complexes between reactants in which old bonds are beginning to break and new bonds are beginning to form. The amount of energy needed by reactants to reach the transition state is the **activation energy, E_{act}**. The larger the activation energy, the slower the reaction.

Many reactions, such as the addition of HCl to ethylene, take place in more than one step and involve the formation of an **intermediate.** A reaction intermediate is a structure that is formed during the course of a multistep reaction and that lies in an energy minimum between two transition states. Intermediates are more stable than transition states but are often too reactive to be isolated.

ADDITIONAL PROBLEMS

3.22 Identify the functional groups in these molecules:

(a) $CH_3CH_2C{\equiv}N$ (b) 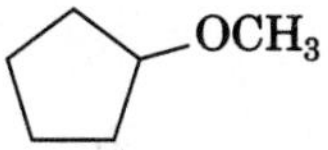 [cyclopentane ring]–OCH_3 (c) $CH_3C(=O)CH_2C(=O)OCH_3$

(d) 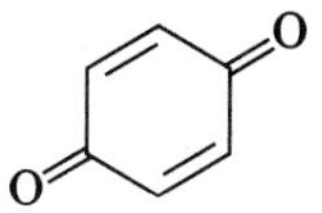 [cyclohexadiene ring with two =O] (e) 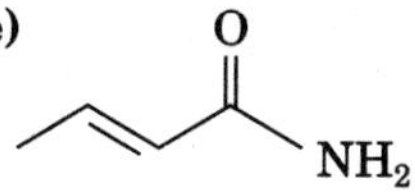 [$CH_3CH{=}CHC(=O)NH_2$ skeletal] (f) 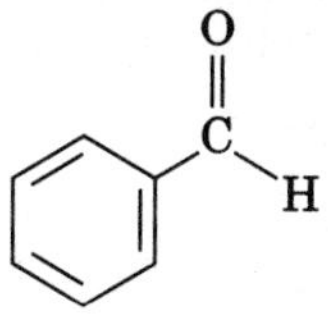 [benzene ring]–C(=O)H

3.23 Predict the direction of polarization of the functional groups you identified in Problem 3.22.

3.24 Identify the functional groups in these molecules:

(a) 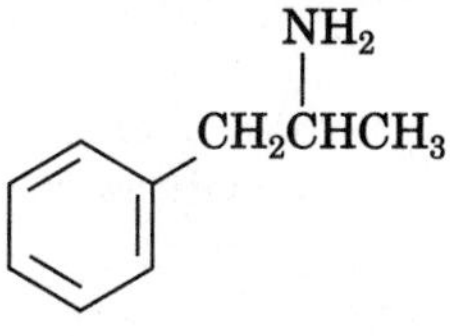 [benzene ring]–$CH_2CH(NH_2)CH_3$

Amphetamine

(b) 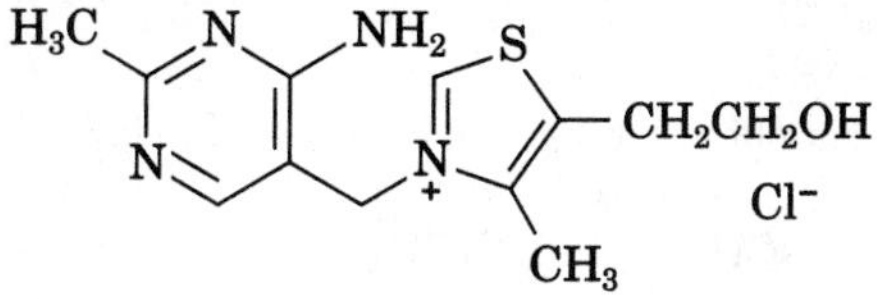 H_3C, N, NH_2, S, CH_2CH_2OH, N, N^+, Cl^-, CH_3

Thiamine

3.25 Explain the differences between addition, elimination, substitution, and rearrangement reactions.

3.26 Classify the following as either nucleophiles or electrophiles:
(a) Cl^- (b) CH_3NH_2 (c) Mg^{2+} (d) CN^- (e) $CH_3{}^+$

3.27 Give IUPAC names for these alkenes:

(a) $CH_3CH{=}CHCH(CH_3)CH_2CH_3$ (b) $CH_3CH{=}CHCH(CH_2CH_2CH_3)CH_2CH_2CH_3$ (c) $H_2C{=}C(CH_2CH_3)CH_2CH_3$

(d) $H_2C{=}C{=}CHCH_3$

3.28 Name these cycloalkenes by IUPAC rules:

(a) 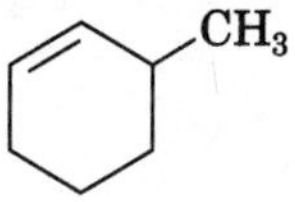 [cyclohexene]–CH_3 (b) 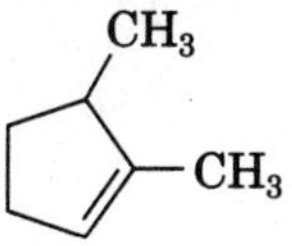 [cyclopentene]–CH_3, CH_3 (c) 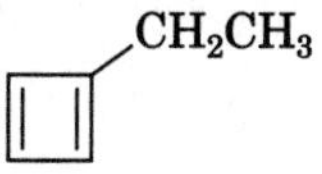 [cyclobutene]–CH_2CH_3 (d) 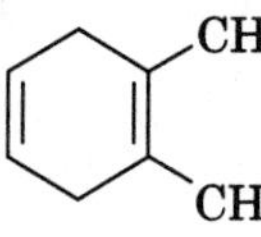 [cyclohexene]–CH_3, CH_3

3.29 Draw structures corresponding to these IUPAC names:
(a) 3-Propyl-2-heptene **(b)** 2,4-Dimethyl-2-hexene **(c)** 1,5-Octadiene
(d) 4-Methyl-1,3-pentadiene **(e)** *cis*-4,4-Dimethyl-2-hexene **(f)** (*E*)-3-Methyl-3-heptene

3.30 Draw the structures of these cycloalkenes:
(a) *cis*-4,5-Dimethylcyclohexene **(b)** 3,3,4,4-Tetramethylcyclobutene

3.31 The following names are incorrect. Draw each molecule and give its correct name.
(a) 1-Methyl-2-cyclopentene **(b)** 1-Methyl-1-pentene **(c)** 6-Ethylcycloheptene
(d) 3-Methyl-2-ethylcyclohexene

3.32 Which of the following molecules show cis–trans isomerism?

(a) 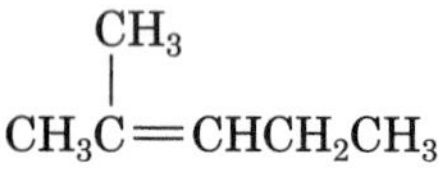
(b) 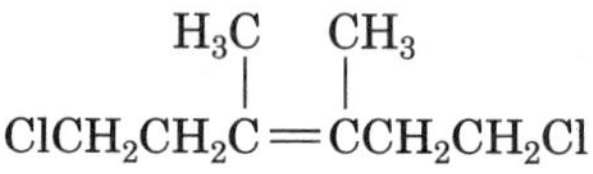
(c) 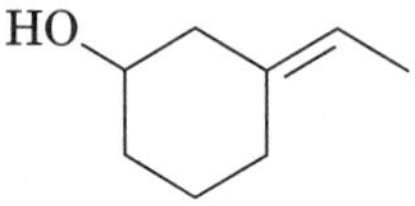

3.33 Draw and name molecules that meet these descriptions:
(a) An alkene, C_6H_{12}, that does not show cis–trans isomerism
(b) The *E* isomer of a trisubstituted alkene, C_6H_{12}
(c) A cycloalkene, C_7H_{12}, with a tetrasubstituted double bond

3.34 Neglecting cis–trans isomers, there are five possible isomers with the formula C_4H_8. Draw and name them.

3.35 Which of the molecules you drew in Problem 3.34 show cis–trans isomerism? Draw and name their cis–trans isomers.

3.36 Draw four possible structures for each of these formulas:
(a) C_6H_{10} **(b)** C_8H_8O **(c)** $C_7H_{10}Cl_2$

3.37 How can you explain the fact that cyclohexene does not show cis–trans isomerism but cyclodecene does?

3.38 Rank the following sets of substituents in order of priority according to the sequence rules:
(a) $-CH_3$, $-Br$, $-H$, $-I$ **(b)** $-OH$, $-OCH_3$, $-H$, $-COOH$ **(c)** $-CH_3$, $-COOH$, $-CH_2OH$, $-CHO$
(d) $-CH_3$, $-CH{=}CH_2$, $-CH_2CH_3$, $-CH(CH_3)_2$

3.39 Assign *E* or *Z* configuration to these alkenes:

(a) 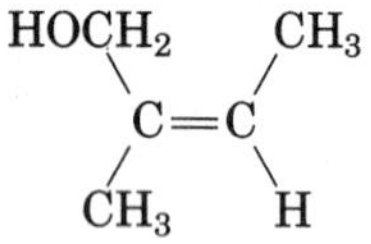
(b) 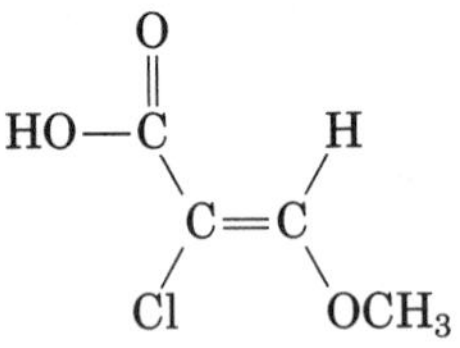

3.40 Draw and name the five possible C_5H_{10} alkene isomers. Ignore cis–trans isomers.

3.41 Menthene, a hydrocarbon found in mint plants, has the IUPAC name 1-isopropyl-4-methylcyclohexene. What is the structure of menthene?

3.42 Name these cycloalkenes by IUPAC rules:

(a) 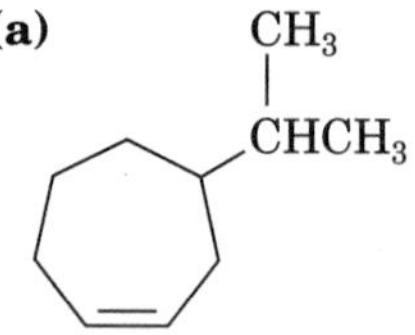
(b) 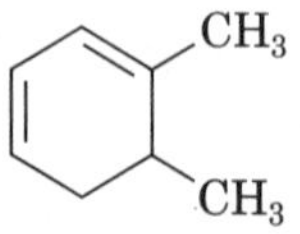
(c) 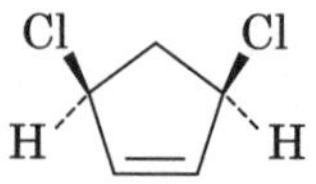

3.43 Classify these reagents as either electrophiles or nucleophiles:

(a) Zn^{2+} (b) $CH_3\ddot{N}H_2$ (c) $CH_3-\overset{\overset{\Large :O:}{\|}}{C}-\ddot{\underset{..}{O}}:^-$ (d) $H\ddot{\underset{..}{S}}:^-$

3.44 α-Farnesene is a constituent of the natural waxy coating found on apples. What is its IUPAC name?

α-Farnesene

3.45 Indicate E or Z configuration for each of the double bonds in α-farnesene (see Problem 3.44).

3.46 Define these terms:
(a) Polar reaction **(b)** Radical reaction **(c)** Functional group
(d) Reaction intermediate

3.47 Give an example of each of the following:
(a) An electrophile **(b)** A nucleophile **(c)** An oxygen-containing functional group

3.48 If a reaction has $K_{eq} = 0.001$, is it likely to be exothermic or endothermic? Explain.

3.49 If a reaction has $E_{act} = 5$ kcal/mol, is it likely to be fast or slow at room temperature? Explain.

3.50 If a reaction has $\Delta H = 12$ kcal/mol, is it exothermic or endothermic? Is it likely to be fast or slow at room temperature? Explain.

3.51 Draw a reaction energy diagram for a two-step exothermic reaction whose first step is faster than its second step. Label the parts of the diagram corresponding to reactants, products, transition state, activation energies, and overall ΔH.

3.52 Draw a reaction energy diagram for a two-step reaction whose second step is faster than its first step. Label the parts of the diagram corresponding to reactants, products, transition state, activation energies, and overall ΔH.

3.53 Draw a reaction energy diagram for a reaction with $K_{eq} = 1$.

3.54 Describe the difference between a transition state and a reaction intermediate.

3.55 Consider the reaction energy diagram shown here and answer the following questions:

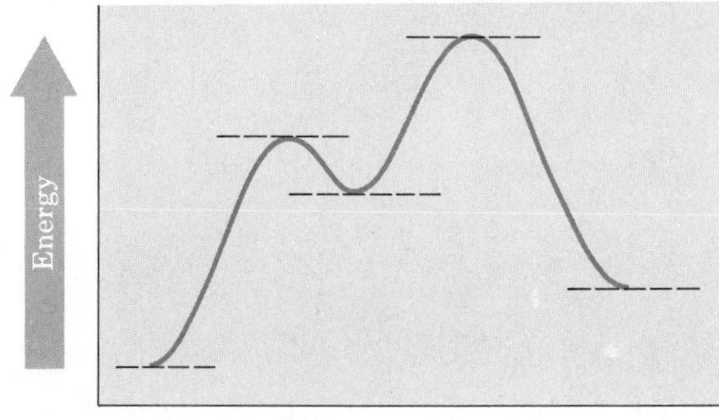

(a) Indicate the overall ΔH for the reaction. Is it positive or negative?
(b) How many steps are involved in the reaction?
(c) Which step is faster (has the lower activation energy)?
(d) How many transition states are there? Label them.

3.56 When isopropylidenecyclohexane is treated with strong acid at room temperature, isomerization occurs by the mechanism shown below to yield 1-isopropylcyclohexene:

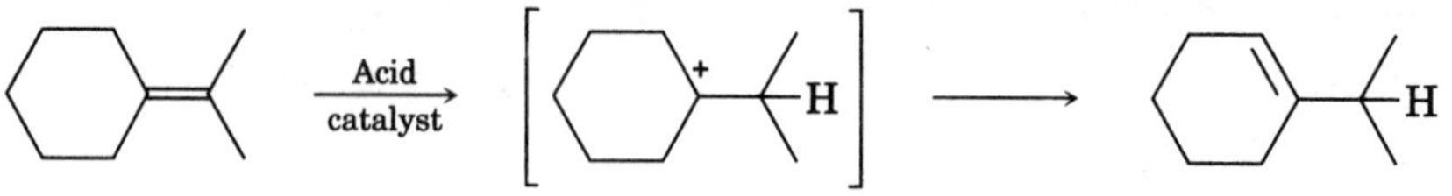

Acid catalyst

H

H

Isopropylidenecyclohexane **1-Isopropylcyclohexene**

At equilibrium, the product mixture contains about 30% isopropylidenecyclohexane and about 70% 1-isopropylcyclohexene.

(a) What kind of reaction is occurring? Is the mechanism polar or radical?

(b) Draw curved arrows to indicate electron flow in each step.

(c) Calculate K_{eq} for the reaction.

CHAPTER

4 ALKENES AND ALKYNES

We saw in the previous chapter how organic reactions can be classified, and we developed some general ideas about how reactions can be described. In this chapter, we'll apply those ideas to a systematic study of the alkene and alkyne families of compounds. In particular, we'll see that the most important reaction of these two functional groups is the addition to the multiple bond of various reagents X–Y to yield saturated products:

$$\mathrm{>C{=}C<} + \mathrm{X{-}Y} \longrightarrow \mathrm{-\overset{X}{\underset{|}{\overset{|}{C}}}-\overset{Y}{\underset{|}{\overset{|}{C}}}-}$$

An alkene — An addition product

4.1 ADDITION OF HX TO ALKENES

We saw in Section 3.8 that alkenes react with HCl to yield alkyl chloride addition products. For example, ethylene reacts with HCl to give chloroethane. The reaction takes place in two steps and involves a carbocation intermediate.

$$H_2C{=}CH_2 + H^+ \longrightarrow \left[H{-}\overset{H}{\underset{H}{C}}{-}\overset{+}{\underset{H}{C}}{-}H \right] \xrightarrow{Cl^-} H{-}\overset{H}{\underset{H}{C}}{-}\overset{Cl}{\underset{H}{C}}{-}H$$

Ethylene — Carbocation intermediate — **Chloroethane**

The addition of halogen acids, HX, to alkenes is a general reaction that allows chemists to prepare a variety of halo-substituted alkane products. Thus, HCl, HBr, and HI all add to alkenes:[1]

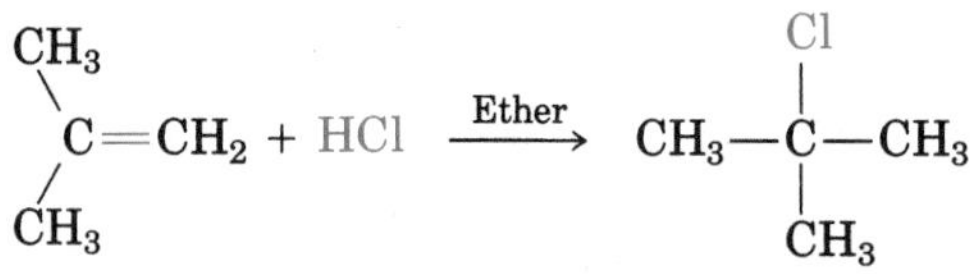

$$(CH_3)_2C{=}CH_2 + HCl \xrightarrow{\text{Ether}} CH_3{-}C(Cl)(CH_3){-}CH_3$$

2-Methylpropene → **2-Chloro-2-methylpropane (94%)**

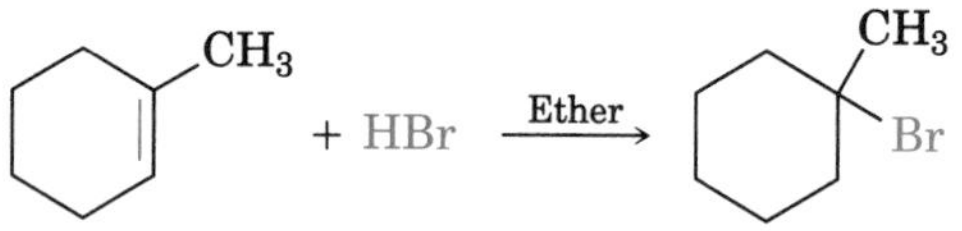

$$\text{1-Methylcyclohexene} + HBr \xrightarrow{\text{Ether}} \text{1-Bromo-1-methylcyclohexane}$$

1-Methylcyclohexene → **1-Bromo-1-methylcyclohexene (91%)**

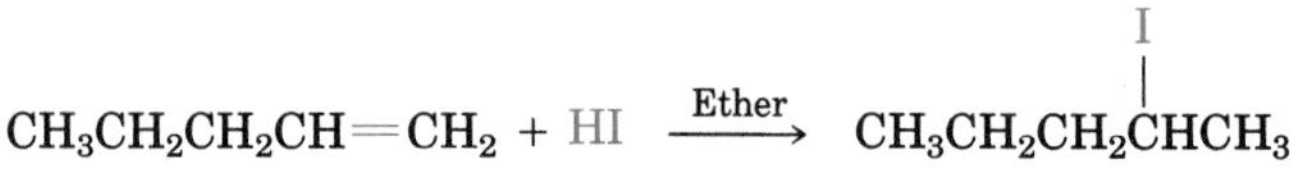

$$CH_3CH_2CH_2CH{=}CH_2 + HI \xrightarrow{\text{Ether}} CH_3CH_2CH_2CH(I)CH_3$$

1-Pentene → **2-Iodopentane**

---

[1]Organic reaction equations can be written in different ways to emphasize different points. For example, the reaction of ethylene with HCl might be written in the format A + B → C to emphasize that both reaction partners are equally important for the purposes of the discussion. The reaction solvent and notes about other reaction conditions such as temperature are usually written either above or below the reaction arrow:

$$H_2C{=}CH_2 + HCl \xrightarrow[25°C]{\text{Ether}} CH_3CH_2Cl$$

(Solvent → Ether)

Alternatively, we might choose to write the same reaction in the format

$$A \xrightarrow{B} C$$

to emphasize that reagent A is the organic starting material whose chemistry is of greater interest. Reagent B is then placed above the reaction arrow, together with notes about solvent and reaction conditions. For example:

$$H_2C{=}CH_2 \xrightarrow[\text{Ether, }25°C]{HCl} CH_3CH_2Cl$$

(Reagent → HCl; Solvent → Ether)

Both reaction formats are frequently used in chemistry, and you sometimes have to look carefully at the overall transformation to identify the different roles of the substances shown next to the reaction arrows.

4.2 ORIENTATION OF ALKENE ADDITION REACTIONS: MARKOVNIKOV'S RULE

REGIOSELECTIVE Describing the orientation of an addition reaction that occurs on an unsymmetrical substrate and leads to a predominance of one product

Look carefully at the reactions shown in the previous section. In each case, an unsymmetrically substituted alkene has given a *single* addition product rather than the mixture that might have been expected. For example, 2-methylpropene might have added HCl to give 1-chloro-2-methylpropane, but it didn't; it gave only 2-chloro-2-methylpropane. We say that reactions are **regioselective** (**ree**-jee-oh-selective) when only one of the two possible directions of addition predominates.

$$(CH_3)_2C{=}CH_2 + HCl \longrightarrow CH_3{-}\overset{Cl}{\underset{CH_3}{C}}{-}CH_3 \qquad \left[CH_3\overset{CH_3}{C}HCH_2Cl\right]$$

2-Methylpropene — **2-Chloro-2-methylpropane** (*sole product*) — **1-Chloro-2-methylpropane** (*not formed*)

From an examination of many such reactions, the Russian chemist Vladimir Markovnikov proposed in 1869 what has come to be known as **Markovnikov's rule:**

MARKOVNIKOV'S RULE Rule for predicting the orientation of alkene electrophilic addition reactions

MARKOVNIKOV'S RULE *In the addition of HX to an alkene, the H attaches to the carbon that has fewer alkyl substituents, and the X attaches to the carbon that has more alkyl substituents.*

2 alkyl groups on this carbon — no alkyl groups on this carbon

$$(CH_3)_2C{=}CH_2 + HCl \xrightarrow{Ether} CH_3{-}\overset{Cl}{\underset{CH_3}{C}}{-}CH_3$$

2-Methylpropene — **2-Chloro-2-methylpropane**

2 alkyl groups on this carbon

1-Methylcyclohexene + HBr $\xrightarrow{Ether}$ 1-Bromo-1-methylcyclohexane

1-Methylcyclohexene — **1-Bromo-1-methylcyclohexane**

1 alkyl groups on this carbon

When both ends of the double bond have the same degree of substitution, a mixture of addition products results:

$$\underset{\text{2-Pentene}}{CH_3CH_2CH{=}CHCH_3} + HBr \xrightarrow{\text{Ether}} \underset{\text{2-Bromopentane}}{CH_3CH_2CH_2CH(Br)CH_3} + \underset{\text{3-Bromopentane}}{CH_3CH_2CH(Br)CH_2CH_3}$$

(1 alkyl group on this carbon; 1 alkyl group on this carbon)

Since carbocations are involved as intermediates in these reactions (Section 3.11), another way to express Markovnikov's rule is to say that, in the addition of HX to alkenes, the more highly substituted carbocation intermediate is formed in preference to the less highly substituted one. For example, addition of H^+ to 2-methylpropene yields the intermediate *tertiary* carbocation rather than the alternative primary carbocation. Why should this be?

$$\underset{\text{2-Methylpropene}}{(CH_3)_2C{=}CH_2} + HCl \longrightarrow \left[CH_3-\overset{+}{C}(CH_3)-CH_2H\right] \xrightarrow{Cl^-} CH_3-C(Cl)(CH_3)-CH_3$$

***tert*-Butyl carbocation** (tertiary; 3°) → **2-Chloro-2-methylpropane**

$$(CH_3)_2C{=}CH_2 + HCl \not\longrightarrow \left[CH_3-CH(CH_3)-\overset{+}{C}H_2\right] \xrightarrow{Cl^-} CH_3-CH(CH_3)-CH_2Cl$$

Isobutyl carbocation (primary; 1°) → **1-Chloro-2-methylpropane** (not formed)

PRACTICE PROBLEM 4.1 What product would you expect from the reaction of HCl with 1-ethylcyclopentene?

$$\text{1-ethylcyclopentene } (\text{cyclopentene ring with } CH_2CH_3) + HCl \longrightarrow ?$$

SOLUTION Markovnikov's rule predicts that the hydrogen will add to the double-bond carbon that has one alkyl group (C2 on the ring), and the chlorine will add to the double-bond carbon that has two alkyl groups (C1 on the ring). The expected product is 1-chloro-1-ethylcyclopentane.

2 alkyl groups on this carbon

CH_2CH_3 (cyclopentene, C1–C2) + HCl ⟶ CH_2CH_3, Cl (cyclopentane)

1 alkyl groups on this carbon

1-Chloro-1-ethylcyclopentane

PROBLEM 4.1 Predict the products of these reactions:

(a) $CH_3CH_2CH{=}CH_2$ + HCl ⟶ ?

(b) $CH_3C(CH_3){=}CHCH_2CH_3$ + HI ⟶ ?

(c) cyclohexene + HCl ⟶ ?

PROBLEM 4.2 What alkenes would you start with to prepare these alkyl halides?

(a) Bromocyclopentane

(b) $CH_3CH_2CH(Br)CH_2CH_2CH_3$

(c) 1-Iodo-1-isopropylcyclohexane

(d) cyclohexyl–CH(CH_3)–Br

4.3 CARBOCATION STRUCTURE AND STABILITY

To understand why Markovnikov's rule works, we need to learn more about the structure and stability of substituted carbocations. Regarding structure, evidence shows that carbocations are *planar*. The positively charged carbon atom is sp^2-hybridized, and the three substituents are oriented to the corners of an equilateral triangle (Figure 4.1). Since there are only six electrons in the carbon valence shell and all six are used in the three sigma bonds, the p orbital extending above and below the plane is vacant.

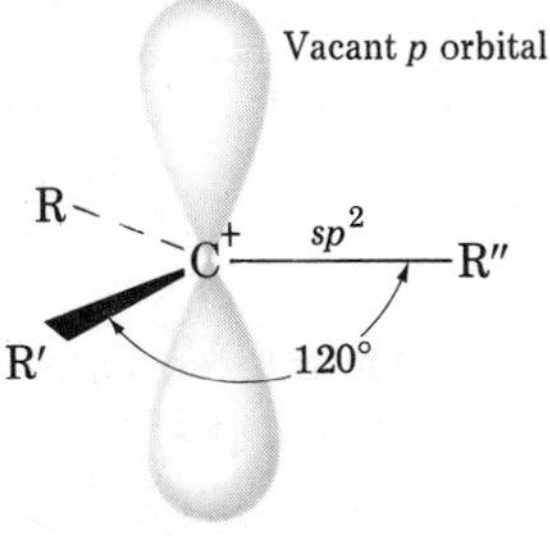

FIGURE 4.1 Carbocation structure. The carbon is sp^2-hybridized and has a vacant p orbital.

Regarding stability, measurements show that carbocation stability increases with increasing alkyl substitution: More highly substituted carbocations are more stable than less highly substituted ones because alkyl groups tend to donate electrons to the positively charged carbon atom. The more alkyl groups there are, the more electron donation there is, and the more stable the carbocation.

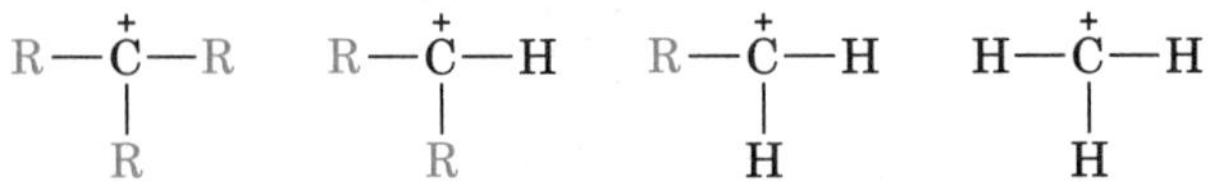

$$\mathrm{R{-}\overset{+}{C}(R){-}R} \qquad \mathrm{R{-}\overset{+}{C}(R){-}H} \qquad \mathrm{R{-}\overset{+}{C}(H){-}H} \qquad \mathrm{H{-}\overset{+}{C}(H){-}H}$$

Tertiary (3°) > Secondary (2°) > Primary (1°) > Methyl

More stable ← Less stable

With the above information, we can now explain Markovnikov's rule. In the reaction of 2-methylpropene with HCl, for example, the intermediate carbocation might have either *three* alkyl substituents (a tertiary cation, 3°) or *one* alkyl substituent (a primary cation, 1°). Since the tertiary cation is more stable than the primary one, it's the tertiary cation that forms as the reaction intermediate, thus leading to the observed tertiary alkyl chloride product.

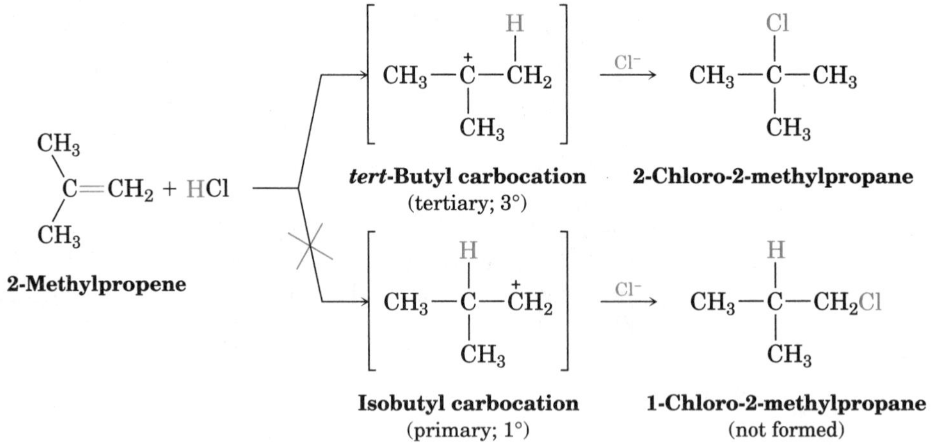

$$(CH_3)_2C{=}CH_2 + HCl \longrightarrow \left[CH_3{-}\overset{+}{C}(CH_3){-}CH_2H\right] \xrightarrow{Cl^-} CH_3{-}C(Cl)(CH_3){-}CH_3$$

2-Methylpropene → *tert*-Butyl carbocation (tertiary; 3°) → 2-Chloro-2-methylpropane

$$(CH_3)_2C{=}CH_2 + HCl \not\longrightarrow \left[CH_3{-}C(H)(CH_3){-}\overset{+}{C}H_2\right] \xrightarrow{Cl^-} CH_3{-}C(H)(CH_3){-}CH_2Cl$$

2-Methylpropene → Isobutyl carbocation (primary; 1°) → 1-Chloro-2-methylpropane (not formed)

PROBLEM 4.3 Show the structures of the carbocation intermediates you would expect in these reactions:

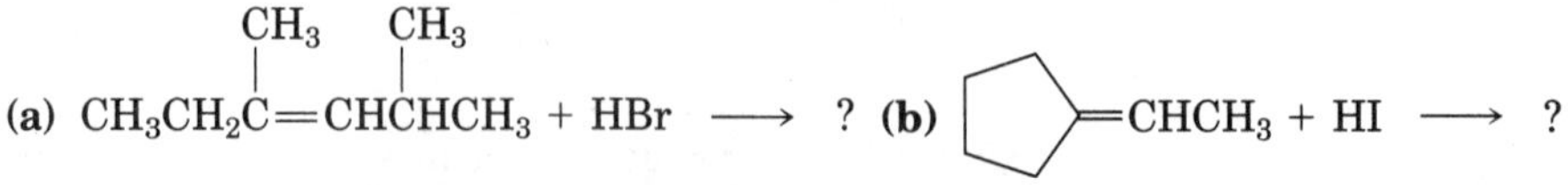

(a) $CH_3CH_2C(CH_3){=}CHCH(CH_3)CH_3 + HBr \longrightarrow ?$

(b) (cyclopentylidene)$=CHCH_3 + HI \longrightarrow ?$

4.4 HYDRATION OF ALKENES

HYDRATION
Addition of water to a substrate, usually an alkene

Water can be added to simple alkenes like ethylene and 2-methylpropene in a **hydration** reaction to yield alcohols, ROH. Industrially, more than 300,000 tons of ethanol are produced each year in the United States by this method:

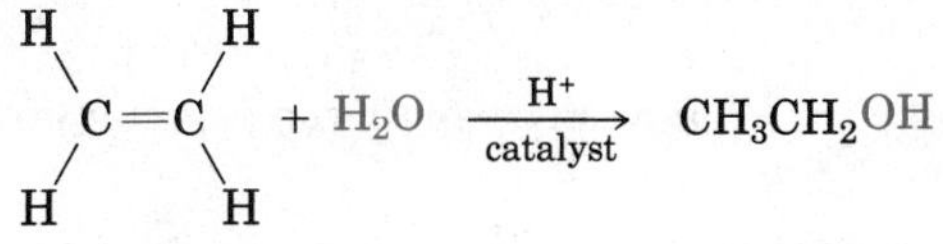

$$H_2C{=}CH_2 + H_2O \xrightarrow[\text{catalyst}]{H^+} CH_3CH_2OH$$

Ethylene **Ethanol**

The hydration of an alkene takes place on reaction with aqueous acid by a mechanism similar to that of HX addition. Reaction of the alkene double bond with H^+ yields a carbocation intermediate, which then reacts with water as nucleophile to yield a protonated alcohol (ROH_2^+) product. Loss of H^+ from the protonated alcohol gives the neutral alcohol and regenerates the acid catalyst (Figure 4.2). The addition of water to an

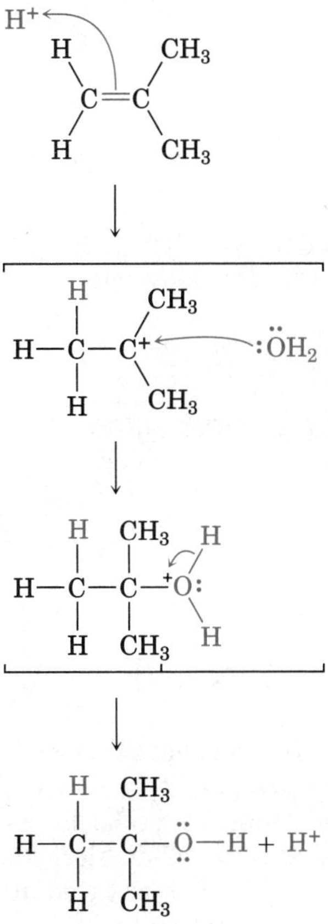

$$H^+ + H_2C{=}C(CH_3)_2$$

The alkene double bond reacts with H^+ to yield a carbocation intermediate.

$$H_3C{-}C^+(CH_3)_2 \;\leftarrow\; :\ddot{O}H_2$$

Water acts as a nucleophile to donate a pair of electrons to form a carbon–oxygen bond and produce a protonated alcohol intermediate.

$$H_3C{-}C(CH_3)_2{-}\overset{+}{O}H_2$$

Loss of H^+ from the protonated alcohol intermediate then gives the neutral alcohol product and regenerates the acid catalyst.

$$H_3C{-}C(CH_3)_2{-}\ddot{O}{-}H + H^+$$

FIGURE 4.2 Mechanism of the acid-catalyzed hydration of an alkene.

unsymmetrical alkene follows Markovnikov's rule just as addition of HX does, giving the more highly substituted alcohol as product.

Unfortunately, the reaction conditions required for hydration are so severe that molecules are sometimes destroyed by the high temperatures and strongly acidic conditions. For example, the hydration of ethylene to produce ethanol requires a sulfuric acid catalyst and reaction temperatures of up to 250°C.

PRACTICE PROBLEM 4.2 What product would you expect from addition of water to methylenecyclopentane?

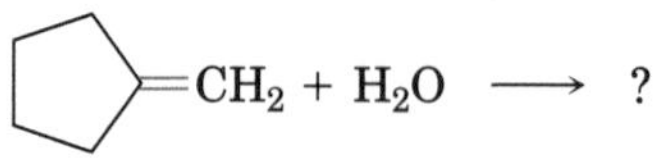

$$\text{C}_5\text{H}_8{=}CH_2 + H_2O \longrightarrow ?$$

Methylenecyclopentane

SOLUTION According to Markovnikov's rule, H^+ will add to the carbon that already has more hydrogens (the CH_2 carbon), and OH will add to the carbon that has fewer hydrogens (the ring carbon). Thus, the product will be a tertiary alcohol.

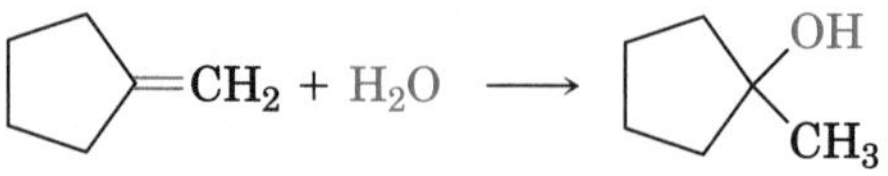

$$\text{C}_5\text{H}_8{=}CH_2 + H_2O \longrightarrow \text{1-methylcyclopentanol (ring C bearing OH and } CH_3)$$

PROBLEM 4.4 What product would you expect to obtain from addition of water to these alkenes?

(a) $CH_3CH_2C(CH_3){=}CHCH_2CH_3$ (b) 1-Methylcyclopentene
(c) 2,5-Dimethyl-2-heptene

PROBLEM 4.5 What alkenes do you suppose these alcohols were made from?

(a) $CH_3CH_2CH(OH)CH_3$ (b) $CH_3CH_2{-}C(OH)(CH_3){-}CH_2CH_3$ (c) cyclohexane ring bearing OH and CH_3 on one carbon and CH_3 on the adjacent carbon

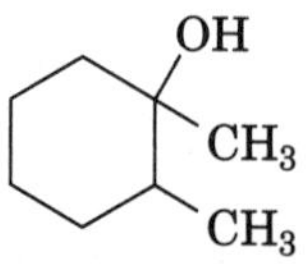

4.5 ADDITION OF HALOGENS TO ALKENES

Many other reagents besides HX and H_2O add to alkenes. Bromine and chlorine are particularly effective, and their reaction with alkenes provides a general method for preparing 1,2-dihaloalkanes. More than 5 million tons of 1,2-dichloroethane (also called ethylene dichloride) are synthesized each year by addition of Cl_2 to ethylene. The product is used both as a solvent and as starting material for the synthesis of poly(vinyl chloride), PVC.

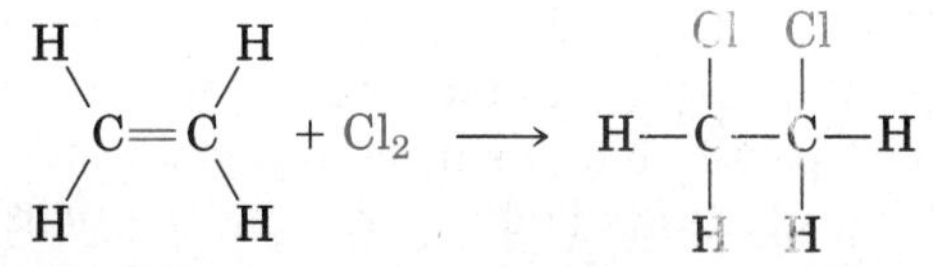

$$\mathrm{H_2C{=}CH_2} + \mathrm{Cl_2} \longrightarrow \mathrm{H{-}CHCl{-}CHCl{-}H}$$

Ethylene **1,2-Dichloroethane (ethylene dichloride)**

Addition of bromine also serves as a simple and rapid laboratory test for the presence of a carbon–carbon double bond in a molecule of unknown structure. A sample of unknown structure is dissolved in dichloromethane, CH_2Cl_2, and several drops of bromine are added. Immediate disappearance of the reddish bromine color signals a positive test, indicating that the sample is an alkene.

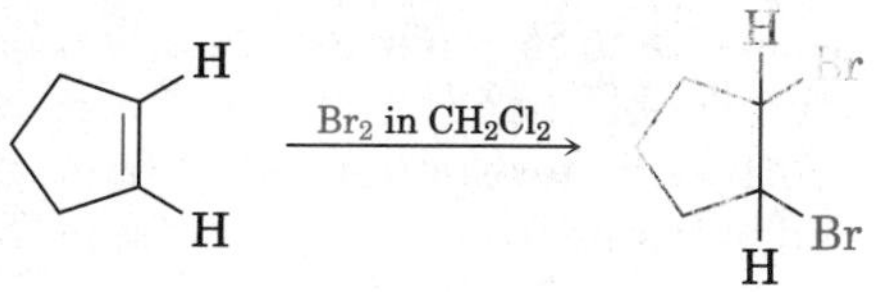

$\xrightarrow{Br_2 \text{ in } CH_2Cl_2}$

Cyclopentene (95%) **1,2-Dibromocyclopentane (95%)**

Bromine and chlorine react with alkenes by the pathway shown in Figure 4.3. The pi electrons of the alkene attack the Br_2 molecule, displacing Br^-. The net result is that electrophilic Br^+ adds to the alkene in much the same way that H^+ does, yielding an intermediate carbocation that immediately reacts further with Br^- to give the dibromo addition product.

δ^+ δ^- Br—Br

The electron pair from the double bond attacks the polarized bromine, forming a C–Br bond and causing the Br–Br bond to break. Bromide ion departs with both electrons from the former Br–Br bond.

:Br:$^-$

Bromide ion uses an electron pair to attack the carbocation intermediate, forming a C–Br bond and giving the neutral addition product.

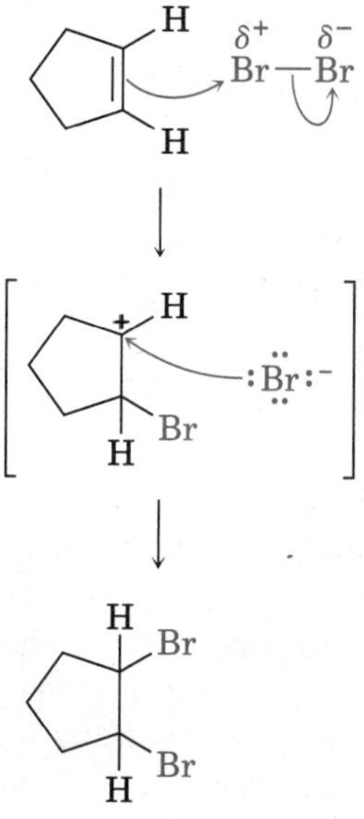

FIGURE 4.3 A possible mechanism for the addition of bromine to cyclopentene.

The mechanism of halogen addition to alkenes shown in Figure 4.3 looks reasonable, but it's not completely consistent with known facts. In particular, the mechanism doesn't explain the *stereochemistry* of halogen addition. That is, the mechanism doesn't explain what product *stereoisomers* (Section 2.8) are formed in the reaction.

Let's look again at the reaction of Br_2 with cyclopentene and assume that Br^+ adds from the bottom side to form the cation intermediate shown in Figure 4.4. (The addition could just as well occur from the top side, but we'll consider only one possibility for simplicity.) Because the positively charged carbon is planar and sp^2-hybridized, it could be attacked by bromide ion in the second step of the reaction from either the top or the bottom side. Thus, a mixture of products might result, in which the two bromine atoms are either on the same side of the ring (cis) or on opposite sides (trans). We find, however, that only *trans*-1,2-dibromocyclopentane is produced. The two bromine atoms add to opposite sides of the double bond, a result described by saying that the reaction occurs with **anti stereochemistry**. (*Anti* means that the two bromines that have added came from opposite sides of the molecule approximately 180° apart.)

ANTI STEREOCHEMISTRY
Referring to a reaction in which both sides of a reactant are involved

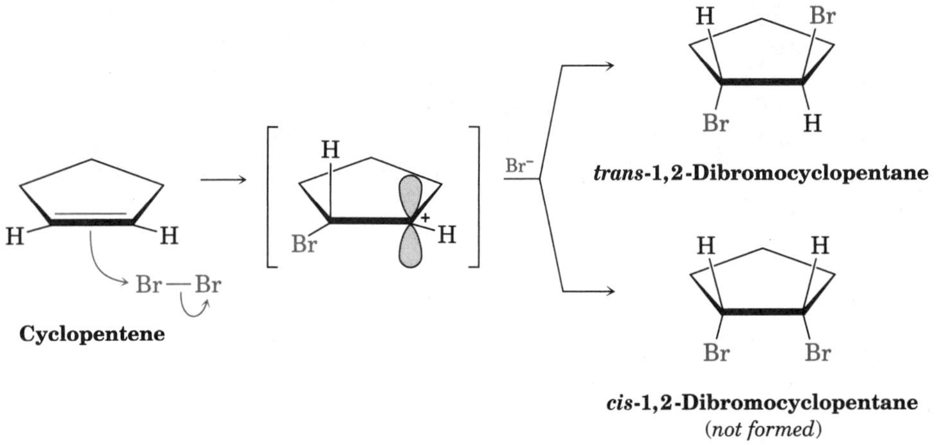

H H Br—Br
Cyclopentene

H Br H Br
Br^-

H Br Br H
***trans*-1,2-Dibromocyclopentane**

H H Br Br
***cis*-1,2-Dibromocyclopentane**
(*not formed*)

FIGURE 4.4 Stereochemistry of the addition of bromine to cyclopentene. Only the trans product is formed.

The stereochemistry of bromine addition is best explained by imagining that the reaction intermediate is not a true carbocation but is instead a **bromonium ion**, R_2Br^+. The bromonium ion in this case is a three-membered ring, formed by the overlap of the vacant carbocation *p* orbital with a lone pair of electrons on the neighboring bromine atom (Figure 4.5). Since the bromine atom shields one side of the molecule, reaction with bromide ion in the second step occurs from the opposite, more accessible side to give the anti product.

BROMONIUM ION
Species with a positively charged, divalent bromine atom

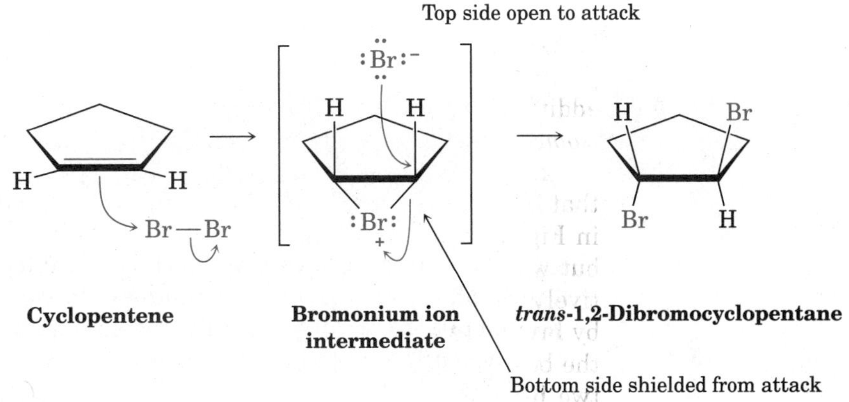

Top side open to attack

Cyclopentene → [Bromonium ion intermediate] → *trans*-1,2-Dibromocyclopentane

Bottom side shielded from attack

FIGURE 4.5 Formation of a bromonium ion intermediate by addition of Br^+ to an alkene.

PROBLEM 4.6 What product would you expect to obtain from addition of Br_2 to 1,2-dimethylcyclohexene? Show the stereochemistry of the product.

PROBLEM 4.7 Show the structure of the intermediate bromonium ion formed in Problem 4.6.

4.6 HYDROGENATION OF ALKENES

HYDROGENATION
Addition of H_2 to a molecule, usually an alkene

REDUCTION
Addition of hydrogen to a molecule or the removal of oxygen from it

Addition of hydrogen to the double bond occurs when alkenes are exposed to an atmosphere of hydrogen gas in the presence of a catalyst. We describe the result by saying that the double bond is **hydrogenated**, or **reduced**. (The word *reduction* in organic chemistry refers to the addition of hydrogen or removal of oxygen from a molecule.) For most alkene hydrogenations, either palladium metal or platinum (as PtO_2) is used as the catalyst.

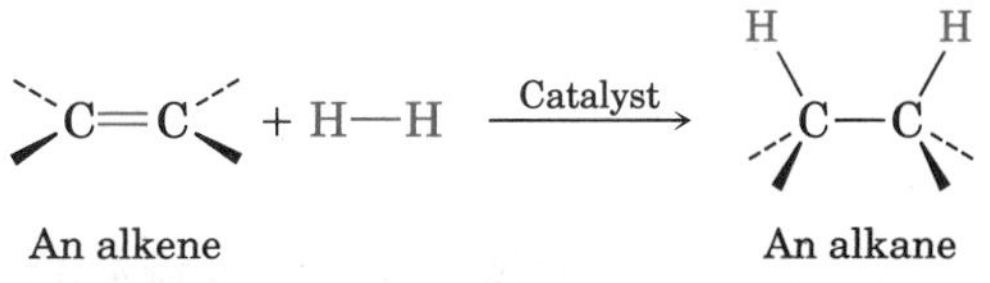

$$\text{C=C} + \text{H—H} \xrightarrow{\text{Catalyst}} \text{H—C—C—H}$$

An alkene An alkane

SYN STEREOCHEMISTRY
Referring to a reaction in which only one side of a reactant is involved

Catalytic hydrogenation of alkenes is unlike most other organic reactions in that it is a heterogeneous process, rather than a homogeneous one. That is, the hydrogenation reaction occurs on the surface of solid catalyst particles rather than in solution. The reaction occurs with **syn stereochemistry** (the opposite of *anti*), meaning that both hydrogens add to the double bond from the same side.

1,2-Dimethylcyclohexene $\xrightarrow[CH_3CO_2H]{H_2,\ PtO_2}$ ***cis*-1,2-Dimethylcyclohexane (82%)**

In addition to its usefulness in the laboratory, alkene hydrogenation is a reaction of great commercial value. In the food industry, unsaturated vegetable oils are catalytically hydrogenated on a vast scale to produce the saturated fats used in margarine.

PROBLEM 4.8

What product would you expect to obtain from catalytic hydrogenation of these alkenes?

(a) $(CH_3)_2C{=}CHCH_2CH_3$ **(b)** 3,3-Dimethylcyclopentene

4.7 OXIDATION OF ALKENES

HYDROXYLATION Addition of one or more –OH groups to a molecule

OXIDATION Addition of oxygen to a molecule or the removal of hydrogen from it

DIOL Dialcohol

Hydroxylation of an alkene—the addition of a hydroxyl group to each of the alkene carbons—can be carried out by treatment of the alkene with potassium permanganate, $KMnO_4$, in basic solution. Since oxygen is added to the alkene during the reaction, we call this an **oxidation**. The reaction occurs with syn stereochemistry and yields a cis 1,2-dialcohol (**diol**) product. For example, cyclohexene gives *cis*-1,2-cyclohexanediol in 37% yield.

Cyclohexene + $KMnO_4$ $\xrightarrow[NaOH]{H_2O}$ ***cis*-1,2-Cyclohexanediol (37%)**

When the reaction of the alkene with $KMnO_4$ is carried out in either neutral or acidic solution, cleavage of the double bond occurs, giving carbonyl-containing products in moderate yield. If the double bond is tetrasubstituted, the two carbonyl-containing products are ketones; if a hydrogen is present on the double bond, one of the carbonyl-containing products is a carboxylic acid; and if two hydrogens are present on one carbon, CO_2 is formed:

$$\text{Isopropylidenecyclohexane} + KMnO_4 \xrightarrow{H_2O} \text{Cyclohexanone} + O{=}C(CH_3)_2 \text{ (Acetone)}$$

Isopropylidenecyclohexane **Cyclohexanone** **Acetone** (two ketones)

$$CH_3CH_2CH(CH_3)CH{=}CH_2 + KMnO_4 \xrightarrow{H_2O} CH_3CH_2CH(CH_3)COOH + CO_2$$

3-Methyl-1-pentene **2-Methylbutanoic acid (45%)**

An alternative method for oxidatively cleaving carbon–carbon double bonds is to treat an alkene with ozone, O_3. Conveniently prepared by passing a stream of oxygen through a high-voltage electrical discharge, ozone adds rapidly to alkenes at low temperature to yield **ozonides**.

OZONIDE
Addition product of ozone and an alkene

$$>C{=}C< + O_3 \xrightarrow[-78°C]{CH_2Cl_2} \text{ozonide} \xrightarrow[CH_3COOH]{Zn} >C{=}O + O{=}C<$$

An ozonide

Since they are sometimes explosive, ozonides aren't usually isolated. Instead, they are treated with a reducing agent such as zinc metal in acetic acid to convert them to carbonyl compounds. The net result of the ozonolysis/zinc reduction sequence is that the carbon–carbon double bond is cleaved, and oxygen becomes doubly bonded to each of the original alkene carbons. If a tetrasubstituted double bond is ozonized, two ketones result; if a trisubstituted double bond is ozonized, one ketone and one aldehyde result; and so on.

$$CH_3CH_2CH_2CH{=}CHCH_2CH_2CH_3 \xrightarrow[2.\ Zn,\ H_3O^+]{1.\ O_3} 2\ CH_3CH_2CH_2CHO$$

4-Octene **Butanal** (two aldehydes)

$$\beta\text{-Pinene} \xrightarrow[2.\ Zn,\ H_3O^+]{1.\ O_3} \text{Nopinone} + O{=}CH_2$$

β-Pinene (disubstituted) **Nopinone** **Formaldehyde**

75%; one ketone, one aldehyde

PRACTICE PROBLEM 4.3 Predict the product of reaction of 2-pentene with aqueous acidic $KMnO_4$.

SOLUTION Reaction of acidic $KMnO_4$ with an alkene yields carbonyl-containing products in which the double bond is broken and the two fragments have C=O in place of the original alkene C=C. If a hydrogen is present on the double bond, a carboxylic acid is produced. Thus, 2-pentene gives the following reaction:

$$\underset{\textbf{2-Pentene}}{CH_3CH_2CH{=}CHCH_3} + KMnO_4 \xrightarrow{H_2O} \underset{\textbf{Propanoic acid}}{CH_3CH_2\overset{\overset{O}{\parallel}}{C}OH} + \underset{\textbf{Acetic acid}}{HO\overset{\overset{O}{\parallel}}{C}CH_3}$$

PRACTICE PROBLEM 4.4 What alkene gives a mixture of acetone and propanal on ozonolysis followed by reduction with zinc?

$$? \xrightarrow[2.\ Zn,\ H_3O^+]{1.\ O_3} CH_3\overset{\overset{O}{\parallel}}{C}CH_3 + CH_3CH_2\overset{\overset{O}{\parallel}}{C}H$$

SOLUTION To find out what starting alkene gives the ozonolysis products shown, simply remove the oxygen atoms from the two products and rejoin the carbon fragments with a double bond:

$$\underset{\textbf{2-Methyl-2-pentene}}{CH_3\overset{\overset{CH_3}{|}}{C}{=}CHCH_2CH_3} \xrightarrow[2.\ Zn,\ H_3O^+]{1.\ O_3} \underset{\textbf{Acetone}}{CH_3\overset{\overset{O}{\parallel}}{C}CH_3} + \underset{\textbf{Propanal}}{CH_3CH_2\overset{\overset{O}{\parallel}}{C}H}$$

PROBLEM 4.9 Predict the product of the reaction of 1,2-dimethylcyclohexene with the following:
(a) Aqueous acidic $KMnO_4$ **(b)** Ozone, followed by zinc

PROBLEM 4.10 Propose structures for alkenes that yield these products on ozonolysis/reduction:
(a) $(CH_3)_2C{=}O + H_2C{=}O$ **(b)** 2 equiv $CH_3CH_2CH{=}O$
(c) 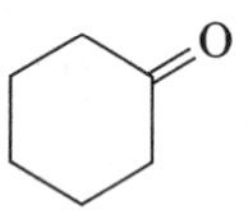 Cyclohexanone + $CH_3\overset{\overset{O}{\parallel}}{C}{-}H$

4.8 ALKENE POLYMERS

POLYMER
Large molecule built up by repetitive bonding together of smaller units

No other group of synthetic organic compounds has had as great an impact on our day-to-day living as the synthetic polymers. A **polymer** is a large molecule built up by the bonding together of many smaller units, called **monomers**. As we'll see in later chapters, nature makes wide use of biological polymers. Cellulose, for example, is a polymer built of repeating

MONOMER
Small building block from which polymers are made

sugar units; proteins are polymers built of repeating amino acid units; and nucleic acids are polymers built of repeating nucleotide units. Although synthetic polymers are chemically much simpler than biopolymers, there is an immense diversity to the structures and properties of synthetic polymers, depending on the nature of the monomers and on the reaction conditions used for polymerization.

Radical Polymerization of Alkenes

Many simple alkenes undergo rapid polymerization when treated with a small amount of a radical catalyst. Ethylene, for example, yields polyethylene. Ethylene polymerization is usually carried out at high pressure (1000–3000 atm) and high temperature (100–250°C) with a radical catalyst such as benzoyl peroxide. The resultant polymer may have anywhere from a few hundred to a few thousand monomer units incorporated into the chain.

$$H_2C{=}CH_2 \xrightarrow[\text{peroxide}]{\text{Benzoyl}} \sim CH_2CH_2{-}CH_2CH_2{-}CH_2CH_2{-}CH_2CH_2{-}CH_2CH_2 \sim$$

Ethylene **A segment of polyethylene**

Radical polymerizations of alkenes involve three kinds of steps: *initiation*, *propagation*, and *termination*. *Initiation* occurs when small amounts of radicals are generated by the catalyst (step 1, below). For example, when benzoyl peroxide is used as initiator, the oxygen–oxygen bond is broken on heating to yield benzoyloxy radicals. One of these radicals adds to the double bond of an ethylene molecule to generate a new carbon radical (step 2), and the polymerization is off and running. Note that this radical addition step results in formation of a bond between the initiator and the ethylene molecule in which one electron has been contributed by each partner. The remaining electron from the ethylene pi bond remains on carbon as the new radical site.

Initiation

STEP 1 $C_6H_5C(=O){-}O{-}O{-}C(=O)C_6H_5 \xrightarrow{\Delta} 2\ C_6H_5C(=O)O\cdot = In\cdot$

Benzoyl peroxide **Benzoyloxy radical**

STEP 2 $In\cdot + H_2C{=}CH_2 \longrightarrow In{-}CH_2CH_2\cdot$

Propagation of the reaction occurs when the carbon radical formed in step 2 adds to another ethylene molecule (step 3). Repetition of step 3 for hundreds or thousands of times builds the polymer chain:

Propagation

STEP 3 $In—CH_2CH_2\cdot + H_2C=CH_2 \longrightarrow In—CH_2CH_2CH_2CH_2\cdot$

$$\xrightarrow[\text{many times}]{\text{Repeat}} In—(CH_2CH_2)_nCH_2CH_2\cdot$$

Eventually, the polymer chain is *terminated* by reactions that consume the radical. For example, combination of two chains by chance meeting (step 4) is a possible chain-terminating reaction:

Termination

STEP 4 $2\,R—CH_2CH_2\cdot \longrightarrow R—CH_2CH_2—CH_2CH_2—R$

Polymerization of Substituted Ethylenes

VINYL MONOMER Simple substituted ethylene used to make polymers

Many substituted ethylenes (**vinyl monomers**) undergo radical-initiated polymerization to yield polymers with substituent groups (denoted by a circled S) regularly spaced along the polymer backbone.

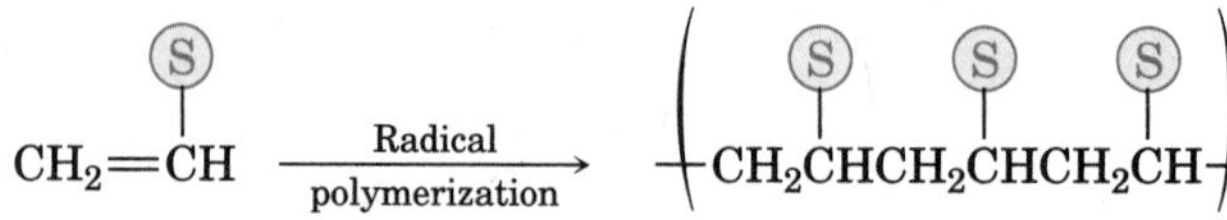

$$\underset{\text{Monomer}}{CH_2=\overset{\text{(S)}}{CH}} \xrightarrow[\text{polymerization}]{\text{Radical}} \underset{\text{Polymer}}{\left(CH_2\overset{\text{(S)}}{CH}CH_2\overset{\text{(S)}}{CH}CH_2\overset{\text{(S)}}{CH}\right)}$$

Table 4.1 shows some of the more important vinyl monomers and lists the industrial uses of the different polymers that result.

TABLE 4.1 Some Alkene Polymers and Their Uses

MONOMER NAME	FORMULA	TRADE OR COMMON NAME OF POLYMER	USES
Ethylene	$H_2C=CH_2$	Polyethylene	Packaging, bottles, cable insulation, films and sheets
Propene (propylene)	$H_2C=CHCH_3$	Polypropylene	Automotive moldings, rope, carpet fibers
Chloroethylene (vinyl chloride)	$H_2C=CHCl$	Poly(vinyl chloride), Tedlar	Insulation, films, pipes
Styrene	$H_2C=CHC_6H_5$	Polystyrene, Styron	Foam and molded articles
Tetrafluoroethylene	$F_2C=CF_2$	Teflon	Valves and gaskets, coatings
Acrylonitrile	$H_2C=CHCN$	Orlon, Acrilan	Fibers
Methyl methacrylate	$H_2C=\overset{CH_3}{\overset{\vert}{C}}CO_2CH_3$	Plexiglas, Lucite	Molded articles, paints
Vinyl acetate	$H_2C=CHOCOCH_3$	Poly(vinyl acetate)	Paints, adhesives

PRACTICE PROBLEM 4.5 Show the structure of poly(vinyl chloride) by drawing several repeating units. Vinyl chloride is $H_2C{=}CHCl$.

SOLUTION The general structure of poly(vinyl chloride) is

$$\underset{\textbf{Vinyl chloride}}{CH_2{=}\overset{\overset{\large Cl}{|}}{C}H} \longrightarrow \underset{\textbf{Poly(vinyl chloride)}}{\leftarrow CH_2\overset{\overset{\large Cl}{|}}{C}HCH_2\overset{\overset{\large Cl}{|}}{C}HCH_2\overset{\overset{\large Cl}{|}}{C}H \rightarrow}$$

PROBLEM 4.11 Show the structure of polypropylene by drawing several repeating units. Propylene is $CH_3CH{=}CH_2$.

4.9 PREPARATION OF ALKENES: ELIMINATION REACTIONS

Just as addition reactions account for most of the chemistry that alkenes undergo, *elimination reactions* account for most of the ways used to prepare alkenes. Additions and eliminations are, in many respects, two sides of the same coin:

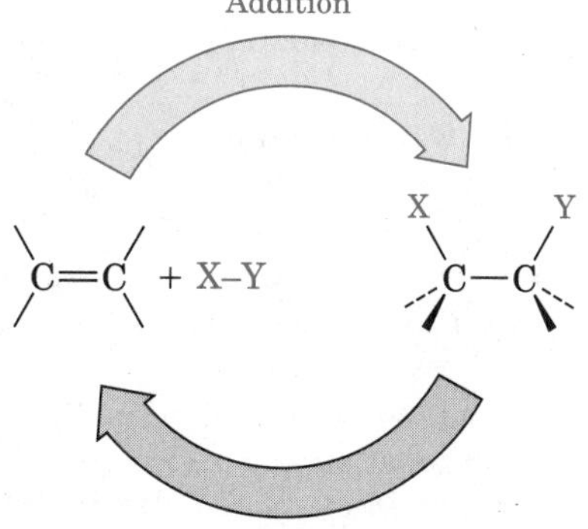

DEHYDROHALOGENATION Elimination of HX from an alkyl halide to yield an alkene

DEHYDRATION Loss of water from an alcohol to yield an alkene

Let's look briefly at two elimination reactions: the **dehydrohalogenation** of an alkyl halide (elimination of HX) and the **dehydration** of an alcohol (elimination of water, H_2O). We'll return for a closer look at how these reactions take place in Chapter 7.

Elimination of HX from Alkyl Halides: Dehydrohalogenation

Alkyl halides can be synthesized by addition of HX to alkenes. Conversely, alkenes can be synthesized by elimination of HX from alkyl halides. Dehydrohalogenation is usually effected by treating the alkyl halide with a strong base. Thus, bromocyclohexane yields cyclohexene when treated with potassium hydroxide in alcohol solution:

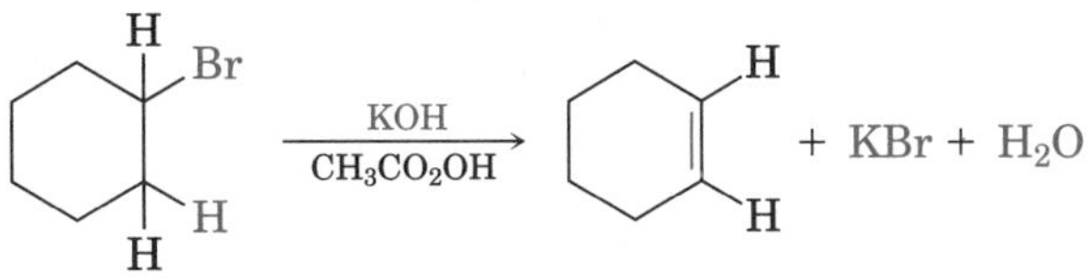

Bromocyclohexane $\xrightarrow[CH_3CO_2OH]{KOH}$ Cyclohexene (81%) + KBr + H_2O

Bromocyclohexane **Cyclohexene (81%)**

Elimination reactions are somewhat more complex than addition reactions because of the regioselectivity problem: What product results from the dehydrohalogenation of an unsymmetrical halide? In fact, elimination reactions almost always give mixtures of alkene products. The best we can usually do is to predict which product will be major.

ZAITSEV'S RULE
Rule for predicting the major product of an alkene-forming elimination reaction

According to **Zaitsev's rule**, a predictive rule formulated by the Russian chemist Alexander Zaitsev, base-induced elimination reactions generally give the more highly substituted alkene product. For example, if 2-bromobutane is treated with KOH in ethanol, Zaitsev's rule predicts that 2-butene (disubstituted; two alkyl-group substituents on the double-bond carbons) should predominate over 1-butene (monosubstituted; one alkyl-group substituent on the double-bond carbons). This is exactly what is found.

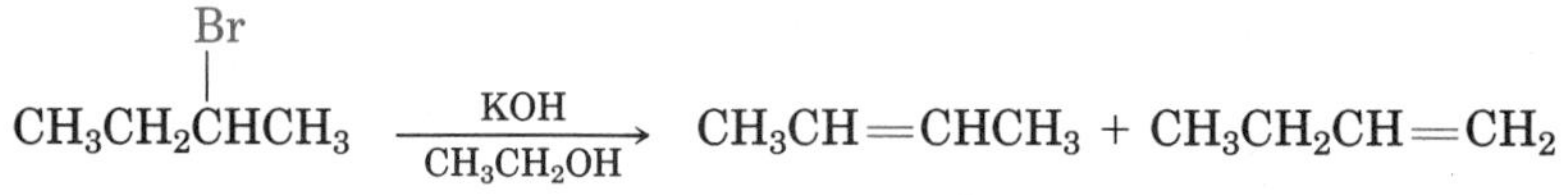

$$CH_3CH_2CH(Br)CH_3 \xrightarrow[CH_3CH_2OH]{KOH} CH_3CH{=}CHCH_3 + CH_3CH_2CH{=}CH_2$$

2-Bromobutane **2-Butene (81%)** **1-Butene (19%)**

PRACTICE PROBLEM 4.6 What product would you expect from reaction of 1-chloro-1-methylcyclohexane with KOH in ethanol?

SOLUTION Treatment of an alkyl halide with a strong base such as KOH causes dehydrohalogenation and yields an alkene. To find the products in a specific case, draw the structure of the starting material and locate the hydrogen atoms on each neighboring carbon. Then generate the potential alkene products by removing HX in as many ways as possible. The major product will be the one that has the most highly substituted double bond—in this case, 1-methylcyclohexene.

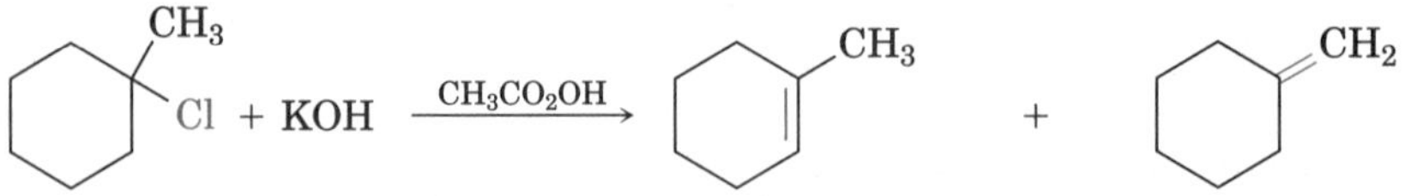

1-Chloro-1-methylcyclohexane + KOH $\xrightarrow{CH_3CO_2OH}$ 1-Methylcyclohexene + Methylenecyclohexane

1-Chloro-1-methylcyclohexane **1-Methylcyclohexene** (major) **Methylenecyclohexane** (minor)

PROBLEM 4.12 What products would you expect from the reaction of 2-bromo-2-methylbutane with KOH in ethanol? Which will be major?

PROBLEM 4.13 What alkyl halide starting materials might these alkenes have come from?

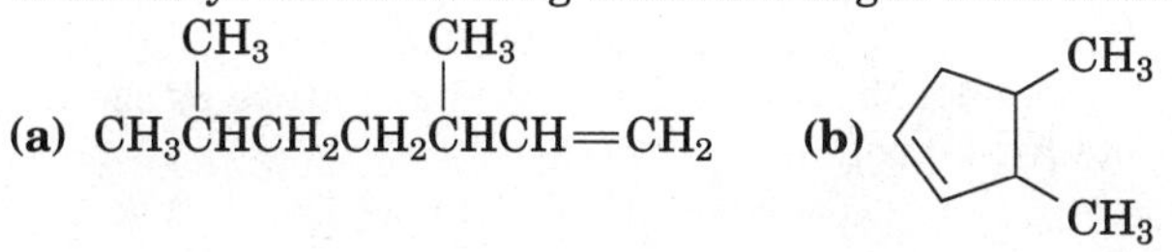

(a) $CH_3CH(CH_3)CH_2CH_2CH(CH_3)CH{=}CH_2$ (b) CH_3 CH_3

Elimination of H_2O from Alcohols: Dehydration

Dehydration of an alcohol is one of the most useful methods of alkene synthesis, and many ways of carrying out the reaction have been devised. A method that works particularly well for tertiary alcohols is acid-catalyzed dehydration. For example, when 1-methylcyclohexanol is treated with aqueous sulfuric acid, dehydration occurs to yield 1-methylcyclohexene:

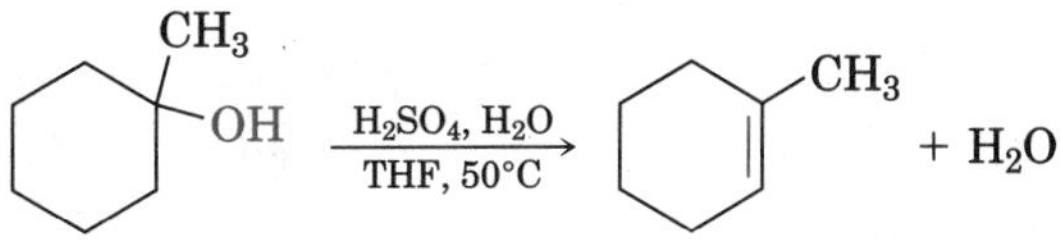

$$\text{1-Methylcyclohexanol} \xrightarrow[\text{THF, 50°C}]{H_2SO_4,\ H_2O} \text{1-Methylcyclohexene (91\%)} + H_2O$$

1-Methylcyclohexanol **1-Methylcyclohexene (91%)**

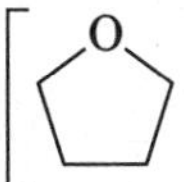

[**Tetrahydrofuran (THF)—a common solvent**]

Acid-catalyzed dehydrations usually follow Zaitsev's rule and yield the more highly substituted alkene as major product. Thus, 2-methyl-2-butanol gives primarily 2-methyl-2-butene (trisubstituted) rather than 2-methyl-1-butene (disubstituted):

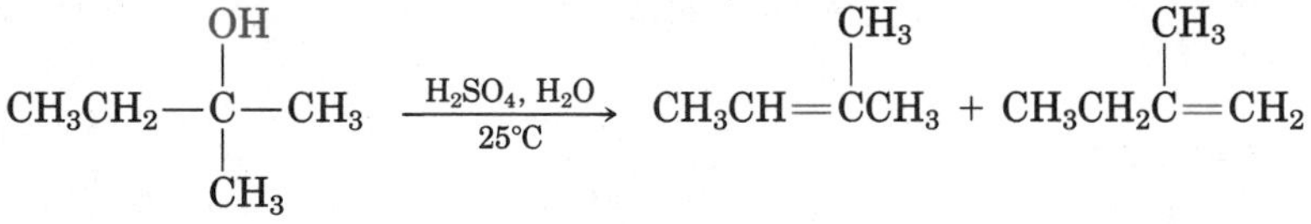

$$CH_3CH_2{-}C(OH)(CH_3){-}CH_3 \xrightarrow[25°C]{H_2SO_4,\ H_2O} CH_3CH{=}C(CH_3)CH_3 + CH_3CH_2C(CH_3){=}CH_2$$

2-Methyl-2-butanol **2-Methyl-2-butene** (major) **2-Methyl-1-butene** (minor)

PRACTICE PROBLEM 4.7 Predict the major product of this reaction:

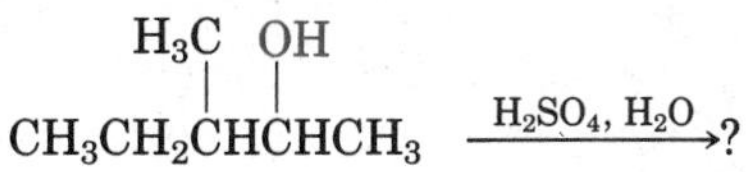

$$CH_3CH_2CH(CH_3)CH(OH)CH_3 \xrightarrow{H_2SO_4,\ H_2O} ?$$

SOLUTION Treatment of an alcohol with acid leads to dehydration and formation of the more highly substituted alkene product (Zaitsev's rule). Thus, dehydration of 3-methyl-2-pentanol should yield 3-methyl-2-pentene as the major product rather than 3-methyl-1-pentene.

$$\underset{\textbf{3-Methyl-2-pentanol}}{CH_3CH_2\overset{H_3C}{\overset{|}{C}}H\overset{OH}{\overset{|}{C}}HCH_3} \xrightarrow{H_2SO_4,\ H_2O} \underset{\substack{\textbf{3-Methyl-2-pentene}\\ \text{(major)}}}{CH_3CH_2\overset{CH_3}{\overset{|}{C}}{=}CHCH_3} + \underset{\substack{\textbf{3-Methyl-1-pentene}\\ \text{(minor)}}}{CH_3CH_2\overset{CH_3}{\overset{|}{C}}HCH{=}CH_2}$$

PROBLEM 4.14 Predict the products you would expect from these reactions. Indicate the major product in each case.

(a) 2-Bromo-2-methylpentane + KOH $\longrightarrow$?

(b) $CH_3\overset{H_3C}{\overset{|}{C}}H{-}\underset{CH_3}{\underset{|}{\overset{OH}{\overset{|}{C}}}}{-}CH_2CH_3 \xrightarrow{H_2SO_4}$?

PROBLEM 4.15 What alcohols might these alkenes have come from?

(a) 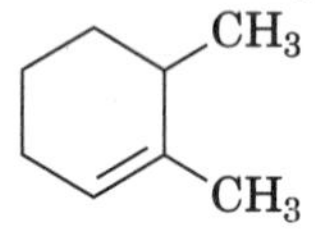 1,6-Dimethylcyclohexene (cyclohexene ring with CH_3 on each of two adjacent carbons, one at the double bond) (b) $CH_3CH_2CH{=}CHCH_2CH_2CH_3$

4.10 CONJUGATED DIENES

CONJUGATION
Two or more multiple bonds alternating with single bonds in a molecule

CONJUGATED DIENE
Diene whose two double bonds are separated by a single bond

Multiple bonds that alternate with single bonds are said to be **conjugated**. Thus, 1,3-butadiene is a **conjugated diene**, whereas 1,4-pentadiene is a nonconjugated diene with isolated double bonds.

$H_2C{=}CH{-}CH{=}CH_2$	$H_2C{=}CH{-}CH_2{-}CH{=}CH_2$
1,3-Butadiene A conjugated diene with alternating single and double bonds	**1,4-Pentadiene** A nonconjugated diene with nonalternating single and double bonds

What's so special about conjugated dienes that we need to look at them separately? The orbital view of 1,3-butadiene shown in Figure 4.6 provides a clue to the answer: *There is an electronic interaction between the two double bonds of a conjugated diene* because of *p* orbital overlap across the central single bond. This interaction of *p* orbitals across a single bond gives conjugated dienes some unusual properties.

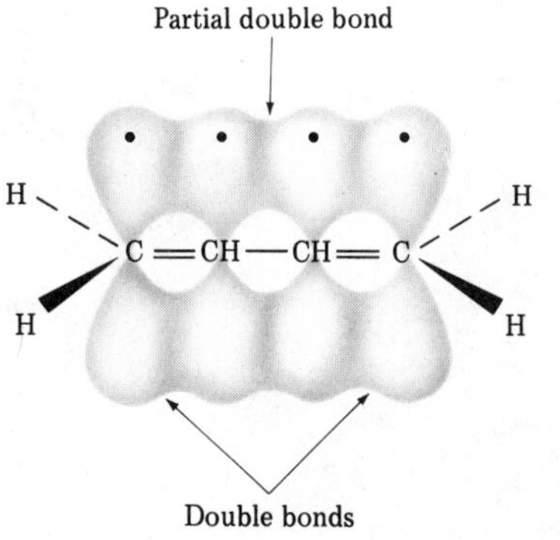

Partial double bond

H H
C=CH—CH=C
H H

Double bonds

FIGURE 4.6 An orbital view of 1,3-butadiene. Each of the four carbon atoms has a *p* orbital, allowing for an electronic interaction across the C2–C3 single bond.

Although much of the chemistry of conjugated dienes and isolated alkenes is similar, there is a striking difference in their addition reactions with electrophiles like HX and X_2. When HX adds to an isolated alkene, Markovnikov's rule usually predicts the formation of a single product. When HX adds to a conjugated diene, though, mixtures of products are usually obtained. For example, reaction of HBr with 1,3-butadiene yields two products:

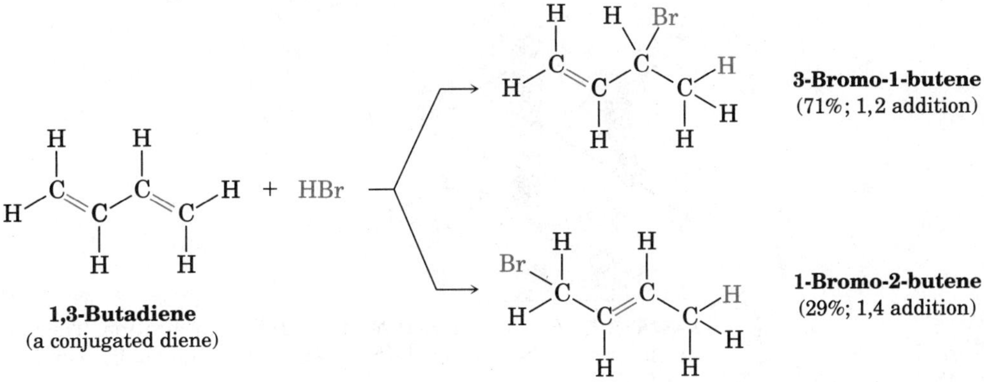

$$CH_2{=}CH{-}CH{=}CH_2 + HBr \longrightarrow CH_2{=}CH{-}CHBr{-}CH_3 + BrCH_2{-}CH{=}CH{-}CH_3$$

1,3-Butadiene (a conjugated diene)

3-Bromo-1-butene (71%; 1,2 addition)

1-Bromo-2-butene (29%; 1,4 addition)

1,4 ADDITION Addition of an electrophile to carbons 1 and 4 of a conjugated diene

ALLYLIC Next to a double bond

3-Bromo-1-butene (a secondary bromide) is the normal product of Markovnikov addition, but 1-bromo-2-butene (a primary bromide) is unexpected. The double bond in this product has moved to a position between carbons 2 and 3, and HBr has added to carbons 1 and 4. How can we account for the formation of this **1,4-addition** product?

The answer is that an *allylic carbocation* is involved as an intermediate in the reaction (**allylic** means next to a double bond). When H^+ adds to an electron-rich pi bond of 1,3-butadiene, two carbocation intermediates are possible—a primary nonallylic carbocation and a secondary allylic carbocation. Allylic carbocations are very stable and therefore form in preference to less stable, nonallylic carbocations.

$H_2C{=}CH{-}CH{=}CH_2$ + HBr → $H_2C{=}CH{-}\overset{+}{C}H{-}CH_3$ + Br^-

1,3-Butadiene

Secondary, allylic carbocation

$H_2C{=}CH{-}CH_2{-}\overset{+}{C}H_2$ + Br^-

Primary carbocation (*not formed*)

4.11 STABILITY OF ALLYLIC CARBOCATIONS: RESONANCE

Why are allylic carbocations stable? To get an idea of the reason, look at the orbital picture of an allylic carbocation in Figure 4.7. The positively charged carbon atom has a vacant *p* orbital that can overlap the *p* orbitals of the neighboring double bond.

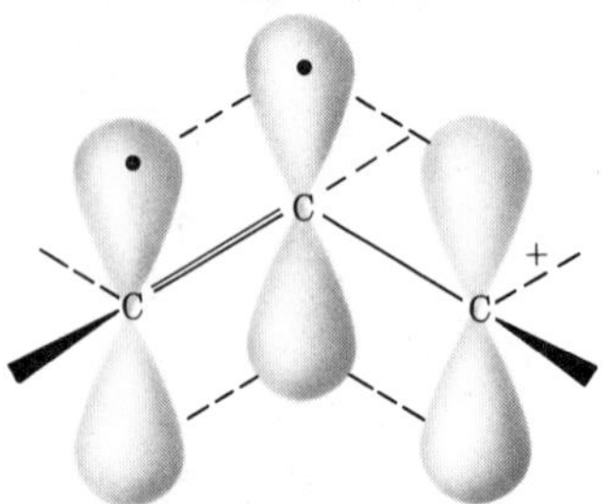

FIGURE 4.7 An orbital picture of an allylic carbocation. The vacant *p* orbital on the positively charged carbon can overlap the double-bond *p* orbitals.

From a *p* orbital point of view, an allylic carbocation is symmetrical. All three carbon atoms are sp^2-hybridized, and each has a *p* orbital. Thus, the *p* orbital on the central carbon can overlap equally well with *p* orbitals on *either* of the two neighboring carbons. The two electrons are free to move about and spread out over the entire three-orbital array, as indicated in Figure 4.7.

One consequence of this orbital picture is that there are two ways to draw an allylic carbocation. We can draw it with the vacant orbital on the left and the double bond on the right, or we can draw it with the vacant orbital on the right and the double bond on the left. *Neither structure is completely correct: The true structure of the allylic carbocation is somewhere in between the two.*

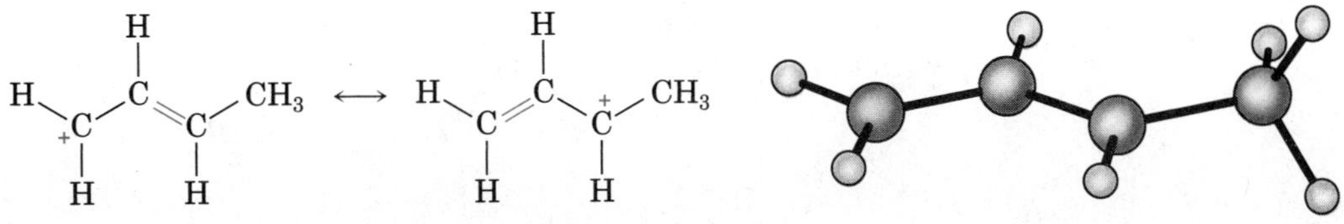

Two resonance forms of an allylic carbocation

RESONANCE FORMS
Representations of a molecule that differ only in where the bonding electrons are placed

The two individual structures are called **resonance forms**, and their special relationship is indicated by the double-headed arrow between them. The only difference between the resonance forms is the position of the bonding electrons. The nuclei don't move but occupy exactly the same places in both resonance forms.

The best way to think about resonance is to realize that a species like an allylic carbocation is no different from any other organic substance. An allylic carbocation doesn't jump back and forth between two resonance forms, spending part of its time looking like one and the rest of its time looking like the other; rather, it has a single, unchanging structure that we call a **resonance hybrid**. (A useful analogy is to think of a resonance hybrid as being like a mutt, or mixed-breed dog. Just as a dog that's a mixture of dachshund and german shepherd doesn't change back and forth from one to the other, a resonance hybrid doesn't change back and forth.)

RESONANCE HYBRID
Composite structure of a molecule described by different resonance forms

The difficulty in understanding resonance hybrids is visual, because we can't draw an accurate single picture of a resonance hybrid by using familiar kinds of structures. The line-bond structures that work so well to represent most organic molecules just don't work well for resonance hybrids like allylic carbocations. We might try to represent the allylic carbocation by using a dashed line to indicate that the two C–C bonds are equivalent and that each is approximately $1\frac{1}{2}$ bonds, as shown below, but such a drawing is ambiguous and won't be used again in this book.

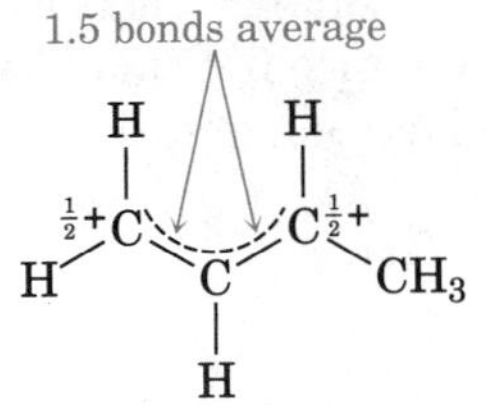

1.5 bonds average

An allylic carbocation

One of the most important postulates of resonance theory is that *the greater the number of possible resonance forms, the greater the stability of the substance.* Because an allylic carbocation is a resonance hybrid of two line-bond structures, it's therefore more stable than a normal carbocation. This stability is due to the fact that the pi electrons can be spread out, or *delocalized,* over an extended *p* orbital network rather than centered on only one site.

In addition to affecting stability, the resonance picture of an allylic carbocation also has chemical consequences. When the allylic carbocation produced by protonation of 1,3-butadiene reacts with bromide ion to complete the addition reaction, attack can occur at either C1 or C3, because both share the positive charge. The result is a mixture of 1,2- and 1,4-addition products.

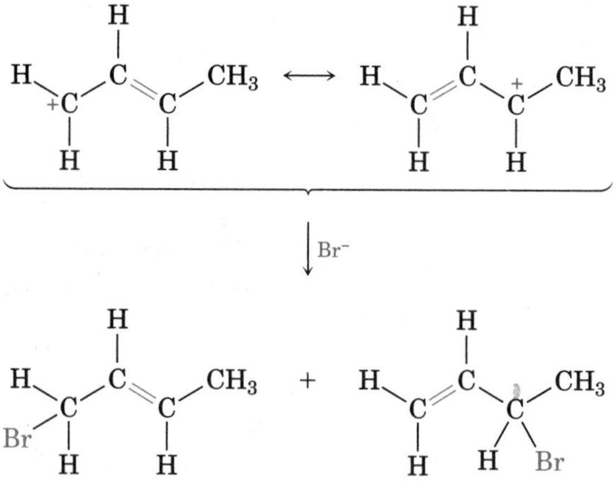

1,4 addition (29%)

1,2 addition (71%)

PROBLEM 4.16 1,3-Butadiene reacts with Br_2 to yield a mixture of 1,2- and 1,4-addition products. Show the structure of each.

4.12 DRAWING AND INTERPRETING RESONANCE FORMS

Resonance is an extremely useful concept for explaining a variety of phenomena. In inorganic chemistry, for example, the carbonate ion (CO_3^{2-}) has identical bond lengths for its three C–O bonds. Although there is no single line-bond structure that can account for this equality of C–O bonds, resonance theory accounts for it nicely. The carbonate ion is simply a resonance hybrid of three resonance forms. The three oxygens share the pi electrons and the negative charges equally:

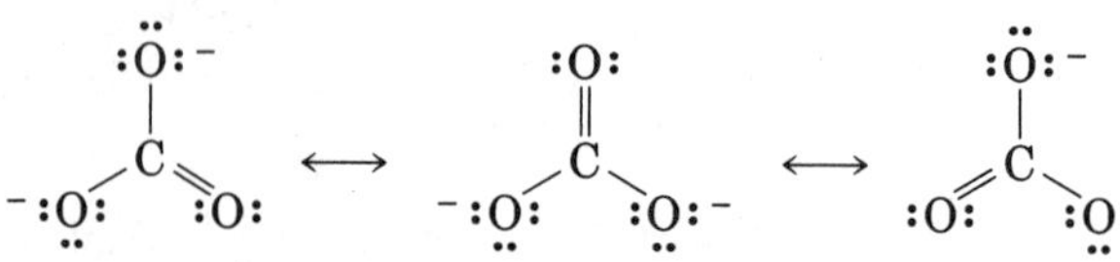

As an example from organic chemistry, we'll see in the next chapter that the six C–C bonds in aromatic compounds like benzene are equivalent because benzene is a resonance hybrid of two forms. Each form has alternating single and double bonds, and neither form is correct by itself. The true benzene structure is a hybrid of the two forms.

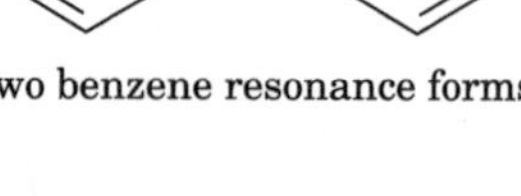

Two benzene resonance forms

When first dealing with resonance theory, it's often useful to have a set of guidelines that describe how to draw and interpret resonance forms. The following five rules should prove helpful:

RULE 1 *Resonance forms are imaginary, not real.* The real structure is a composite, or hybrid, of the different forms. Substances like the allylic carbocation, the carbonate ion, and benzene are no different from any other substance in having single, unchanging structures. The only difference is in the way they must be represented on paper.

RULE 2 *Resonance forms differ from each other only in the placement of their pi or nonbonding electrons.* Neither the position nor the hybridization of atoms changes from one resonance form to another. In benzene, for example, the pi electrons in the double bonds move, but the six carbon atoms remain in place:

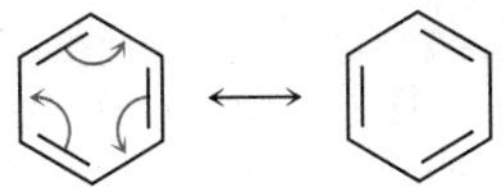

By contrast, two structures like 1,3-cyclohexadiene and 1,4-cyclohexadiene are *not* resonance structures, because their hydrogen atoms don't occupy the same positions. Instead, the two dienes are constitutional isomers.

Constitutional isomers, *not* resonance forms

1,3-Cyclohexadiene **1,4-Cyclohexadiene**

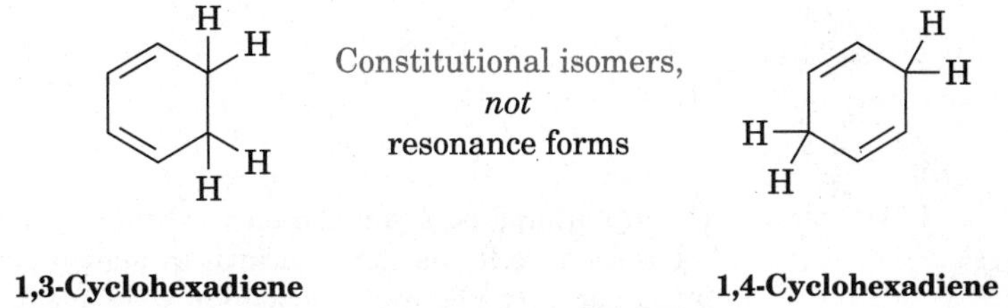

RULE 3 *Different resonance forms of a substance don't have to be equivalent.* For example, the allylic carbocation obtained by reaction of 1,3-butadiene with H^+ is unsymmetrical. One end of the delocalized pi electron system has a methyl substituent, and the other end is unsubstituted. Even though the two resonance forms aren't equivalent, they both contribute to the overall resonance hybrid.

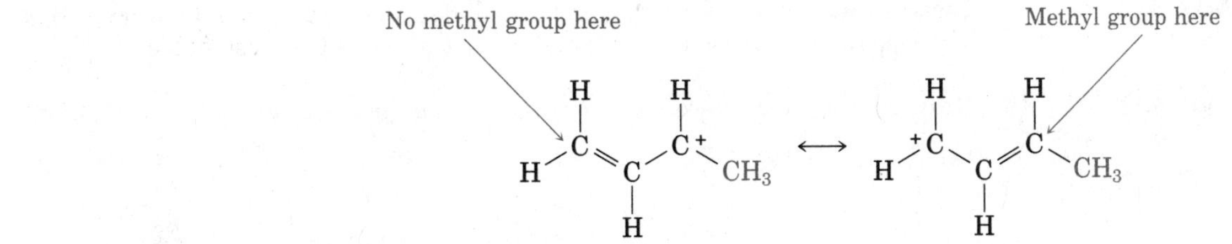

No methyl group here

Methyl group here

In general, when two resonance forms are not equivalent, the actual structure of the resonance hybrid is closer to the more stable form than to the less stable form. Thus, we might expect the butenyl carbocation to look a bit more like a secondary carbocation than like a primary one.

RULE 4 *All resonance forms must obey normal rules of valency.* Resonance forms are like any other structure: The octet rule still holds. For example, one of the following structures for the carbonate ion is not a valid resonance form because the carbon atom has five bonds and ten electrons:

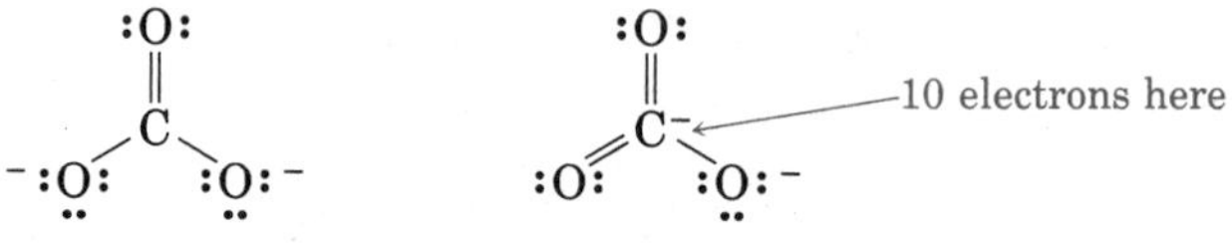

10 electrons here

Carbonate ion *Not* a resonance form

RULE 5 *The resonance hybrid is more stable than any single resonance form.* In other words, resonance leads to stability. The greater the number of resonance forms possible, the more stable the substance. We've already seen, for example, that an allylic carbocation is more stable than a normal carbocation. In a similar manner, we'll see in the next chapter that a benzene ring is more stable than a cyclic alkene.

PRACTICE PROBLEM 4.8 Use resonance structures to explain why the two C–O bonds of sodium formate are equivalent.

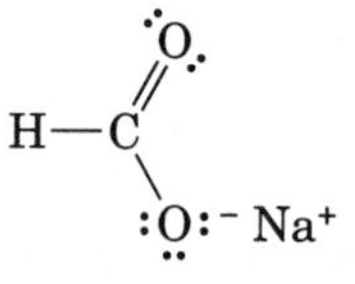

Sodium formate

SOLUTION The formate anion is a resonance hybrid of two equivalent resonance forms. The two resonance forms can be drawn by showing the double bond either to the top oxygen or to the bottom oxygen. Only the positions of the electrons are different in the two structures.

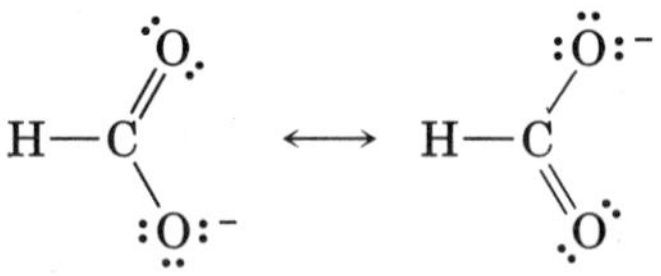

PROBLEM 4.17 Give the structure of all possible monoadducts of HCl and 1,3-pentadiene.

PROBLEM 4.18 Look at the possible carbocation intermediates produced during addition of HCl to 1,3-pentadiene (Problem 4.17) and predict which is the most stable.

PROBLEM 4.19 Draw as many resonance structures as you can for these species:

(a) 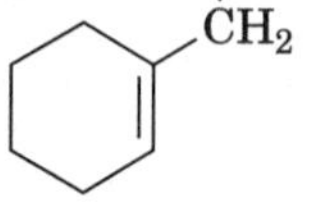 (b) 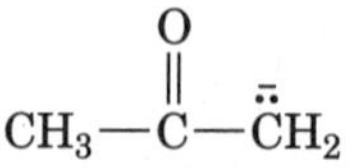 $CH_3-\overset{O}{\overset{\|}{C}}-\ddot{C}H_2$ (c)

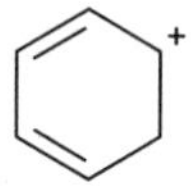

4.13 ALKYNES

ALKYNE
Hydrocarbon that has a carbon–carbon triple bond

Alkynes are hydrocarbons that contain a carbon–carbon triple bond. Since four hydrogens must be removed from an alkane, C_nH_{2n+2}, to generate a triple bond, the general formula for an alkyne is C_nH_{2n-2}.

A carbon–carbon triple bond results from the overlap of two *sp*-hybridized carbon atoms, as we saw in Section 1.11. The two *sp* hybrid orbitals of carbon lie at an angle of 180° to each other along an axis that is perpendicular to the axes of the unhybridized $2p_y$ and $2p_z$ orbitals. When two such *sp*-hybridized carbons approach each other for bonding, the geometry is perfect for the formation of one *sp–sp* sigma bond and two *p–p* pi bonds—a net triple bond (Figure 4.8). The two remaining *sp* orbitals form bonds to other atoms at an angle of 180° from the carbon–carbon sigma bond. Acetylene, H–C≡C–H, is thus a linear molecule with H–C–C bond angles of 180°.

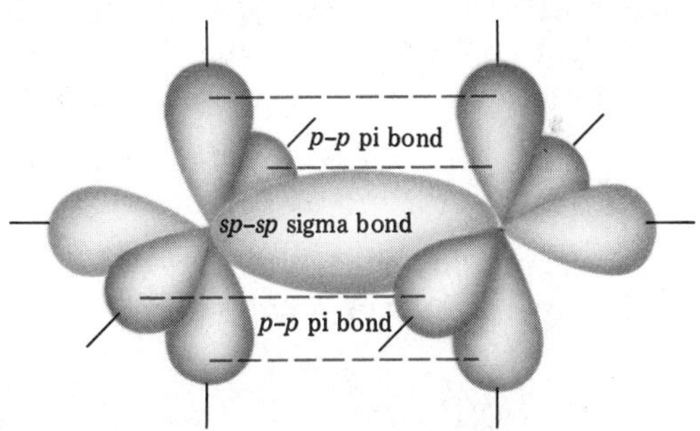

FIGURE 4.8 The electronic structure of a carbon–carbon triple bond.

Alkynes follow closely the general rules of hydrocarbon nomenclature already discussed for alkanes (Section 2.3) and alkenes (Section 3.1). The suffix *-yne* is used in the base hydrocarbon name to denote an alkyne, and the position of the triple bond is indicated by its number in the chain. Numbering always begins at the chain end nearer the triple bond so that the triple bond receives as low a number as possible.

$$\underset{7\ \ 6}{CH_3CH_2}\underset{5}{\overset{\overset{\displaystyle CH_3}{|}}{C}}\underset{5\ \ 4}{HCH_2}\underset{3}{C}{\equiv}\underset{2\,1}{CCH_3}$$

Begin numbering carbons at the end nearer the triple bond

5-Methyl-2-heptyne

Compounds containing both double and triple bonds are called *enynes*, not *ynenes*. Numbering of the hydrocarbon chain always starts from the end nearer the first multiple bond, but if there's a choice in numbering, double bonds receive lower numbers than triple bonds. For example,

$$\underset{1}{CH_3}\underset{2}{CH}{=}\underset{3}{CH}\underset{4}{CH_2}\underset{5}{CH_2}\underset{6}{C}{\equiv}\underset{7}{C}\underset{8}{CH_3}$$

2-Octen-6-yne (*not* 6-octen-2-yne)

PROBLEM 4.20 Give IUPAC names for these compounds:

(a) $CH_3CH_2C{\equiv}CCH_2CH(CH_3)_2$ (b) $HC{\equiv}CC(CH_3)_3$

(c) $CH_3CH(CH_3)CH_2C{\equiv}CCH_3$ (d) $CH_3CH{=}CHCH_2C{\equiv}CCH_3$

4.14 REACTIONS OF ALKYNES: ADDITION OF H_2, HX, AND X_2

Based on their structural similarity, we might expect alkynes and alkenes to show chemical similarities also. As a general rule, this prediction is true: Alkynes react in much the same way that alkenes do.

Addition of H_2 to Alkynes

Alkynes are easily converted into alkanes by reduction with 2 molar equivalents of hydrogen over a palladium catalyst:

$$\underset{\textbf{4-Octyne}}{CH_3CH_2CH_2C{\equiv}CCH_2CH_2CH_3} + 2\ H_2 \xrightarrow[\text{catalyst}]{\text{Pd}} \underset{\textbf{Octane (95\%)}}{CH_3CH_2CH_2CH_2CH_2CH_2CH_2CH_3}$$

The catalytic hydrogenation of an alkyne to yield an alkane proceeds through an intermediate alkene, and the reaction can be stopped at the alkene stage if the right catalyst is used. The catalyst most often used for this purpose is the Lindlar catalyst, a specially prepared form of palladium metal. Because hydrogenation occurs with syn stereochemistry, alkynes are catalytically reduced to give cis alkenes. For example,

$$\underset{\textbf{4-Octyne}}{CH_3CH_2CH_2C{\equiv}CCH_2CH_2CH_3} + H_2 \xrightarrow[\text{catalyst}]{\text{Lindlar}} \underset{\textbf{\textit{cis}-4-Octene (92\%)}}{(H)(CH_3CH_2CH_2)C{=}C(H)(CH_2CH_2CH_3)}$$

Addition of HX to Alkynes

Alkynes give the expected addition products with HCl, HBr, and HI. Although the reactions usually can be stopped after addition of 1 molar

VINYLIC
Attached to a double bond

equivalent of HX to yield a **vinylic** halide (*vinylic* means on the double bond), an excess of reagent leads to formation of a dihalide product. As the following example indicates, the regioselectivity of addition to monosubstituted alkynes follows Markovnikov's rule. The H atom adds to the terminal carbon of the triple bond, and the X atom adds to the internal, more highly substituted, carbon:

$$CH_3CH_2CH_2CH_2C{\equiv}CH + HBr \longrightarrow CH_3CH_2CH_2CH_2C(Br){=}CH_2$$

1-Hexyne → **2-Bromo-1-hexene**

Addition of X_2 to Alkynes

Bromine and chlorine add to alkynes to give addition products with trans stereochemistry:

$$CH_3CH_2CH_2CH_2C{\equiv}CH + Br_2 \xrightarrow{CCl_4} (CH_3CH_2CH_2CH_2)(Br)C{=}C(Br)(H)$$

1-Hexyne → **(*E*)-1,2-Dibromo-1-hexene**

PROBLEM 4.21

What products would you expect from these reactions?

(a) $CH_3CH_2CH_2C{\equiv}CH + 1\ \text{equiv}\ Cl_2 \longrightarrow\ ?$

(b) $CH_3CH_2CH_2C{\equiv}CCH_2CH_3 + 1\ \text{equiv HBr} \longrightarrow\ ?$

(c) $CH_3CH(CH_3)CH_2C{\equiv}CCH_2CH_3 + H_2 \xrightarrow[\text{catalyst}]{\text{Lindlar}}\ ?$

4.15 ADDITION OF WATER TO ALKYNES

Addition of water takes place when an alkyne is treated with aqueous sulfuric acid in the presence of mercuric sulfate catalyst:

$$CH_3CH_2CH_2C{\equiv}CH + H_2O \xrightarrow[HgSO_4]{H_2SO_4} \left[CH_3CH_2CH_2C(OH){=}CH_2\right] \longrightarrow CH_3CH_2CH_2C({=}O)CH_3$$

1-Pentyne → An enol → **2-Pentanone (78%)**

Markovnikov regioselectivity is found for the hydration reaction, with the H attaching to the less substituted carbon and the OH attaching to the more substituted carbon. Interestingly, though, the expected alkenyl alcohol, or *enol* (*ene* = alkene; *ol* = alcohol), is not isolated.

TAUTOMERISM
Word used to describe two rapidly interconverting constitutional isomers

Instead, this intermediate enol rearranges to a more stable isomer, a ketone ($R_2C{=}O$). It turns out that enols and ketones rapidly interconvert—a process called **tautomerism**, which we'll discuss in more detail in Section 11.1. With few exceptions, the tautomeric equilibrium heavily favors the ketone; enols are almost never isolated.

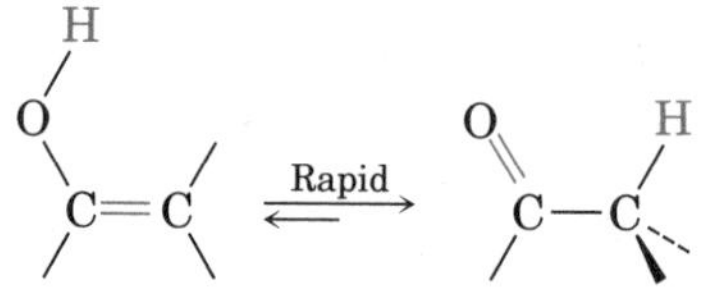

$$\text{(H–O)C=C} \xrightleftharpoons[]{\text{Rapid}} \text{(O=)C–C–H}$$

Enol tautomer (less favored) **Keto tautomer** (more favored)

A mixture of both possible ketones results when an internal alkyne (R–C≡C–R′) is hydrated, but only a single product is formed from reaction of a terminal alkyne (R–C≡C–H).

$$CH_3CH_2C{\equiv}CCH_3 + H_2O \xrightarrow[HgSO_4]{H_2SO_4} CH_3CH_2\overset{O}{\overset{\|}{C}}CH_2CH_3 + CH_3CH_2CH_2\overset{O}{\overset{\|}{C}}CH_3$$

2-Pentyne (an internal alkyne) **3-Pentanone** **2-Pentanone**

$$CH_3CH_2CH_2C{\equiv}CH + H_2O \xrightarrow[HgSO_4]{H_2SO_4} CH_3CH_2CH_2\overset{O}{\overset{\|}{C}}CH_3$$

1-Pentyne (a terminal alkyne) **2-Pentanone**

PRACTICE PROBLEM 4.9 What product would you obtain by hydration of 4-methyl-1-hexyne?

SOLUTION Addition of water to 4-methyl-1-hexyne according to Markovnikov's rule should yield a product with the OH group attached to C2 rather than to C1. This enol then isomerizes to yield a ketone:

$$CH_3CH_2\overset{CH_3}{\overset{|}{C}}HCH_2C{\equiv}CH + H_2O \xrightarrow[HgSO_4]{H_2SO_4} \left[CH_3CH_2\overset{CH_3}{\overset{|}{C}}HCH_2\overset{OH}{\overset{|}{C}}{=}CH_2\right] \longrightarrow CH_3CH_2\overset{CH_3}{\overset{|}{C}}HCH_2\overset{O}{\overset{\|}{C}}CH_3$$

4-Methyl-1-hexyne **4-Methyl-2-hexanone**

PROBLEM 4.22 What product would you obtain by hydration of 4-octyne?

PROBLEM 4.23 What alkynes would you start with to prepare these ketones by a hydration reaction?

(a) $CH_3CH_2CH_2\overset{O}{\overset{\|}{C}}CH_3$ **(b)** $CH_3CH_2CH_2\overset{O}{\overset{\|}{C}}CH_2CH_3$

4.16 ALKYNE ACIDITY: FORMATION OF ACETYLIDE ANIONS

The most striking difference between the chemistry of alkenes and alkynes is that terminal alkynes (R–C≡C–H) are weakly acidic, with $pK_a \approx 25$ (Section 1.13). Alkenes, by contrast, have $pK_a \approx 44$. When a terminal alkyne is treated with a strong base such as sodium amide, $NaNH_2$, the terminal hydrogen is removed, and an acetylide anion is formed:

$$R{-}C{\equiv}C{-}H + :\ddot{N}H_2Na^+ \longrightarrow R{-}C{\equiv}C:^-\ Na^+ + :NH_3$$

A terminal alkyne An acetylide anion

The presence of an unshared electron pair on carbon makes acetylide anions both basic and nucleophilic. As a result, acetylide anions react with alkyl halides such as bromomethane to substitute for the halogen and yield a new alkyne product. We won't study the mechanism of this substitution reaction until Chapter 7, but will note that it is a very useful method for preparing substituted alkynes from simpler precursors. Terminal alkynes can be prepared by reaction of acetylene itself, and internal alkynes can be prepared by further reaction of a terminal alkyne:

$$HC{\equiv}CH \xrightarrow{NaNH_2} HC{\equiv}C^-\ Na^+ \xrightarrow{RCH_2Br} HC{\equiv}CCH_2R$$

Acetylene A terminal alkyne

$$RC{\equiv}CH \xrightarrow{NaNH_2} RC{\equiv}C^-\ Na^+ \xrightarrow{R'CH_2Br} RC{\equiv}CCH_2R'$$

A terminal alkyne An internal alkyne

As an example, conversion of 1-hexyne into its anion, followed by reaction with 1-bromobutane, yields 5-decyne:

$$CH_3CH_2CH_2CH_2C{\equiv}CH \xrightarrow[\text{2. } CH_3CH_2CH_2CH_2Br,\ THF]{\text{1. } NaNH_2,\ NH_3} CH_3CH_2CH_2CH_2C{\equiv}CCH_2CH_2CH_2CH_3$$

1-Hexyne **5-Decyne (76%)**

The only real limitation to the reaction of an acetylide anion with an alkyl halide is that only primary alkyl halides, RCH_2X, can be used (for reasons to be discussed in Chapter 7). If a secondary or tertiary alkyl halide is used instead, the acetylide anion causes dehydrohalogenation of the alkyl halide, giving an alkene (Section 4.9).

PRACTICE PROBLEM 4.10 What alkyne and what alkyl halide would you use to prepare 1-pentyne?

SOLUTION Draw the structure of the target molecule, and identify the alkyl group(s) attached to the triple-bonded carbons. In the present case, one of the triple-bonded carbons has a propyl group attached to it, and the other triple-bonded carbon has a hydrogen attached to it. Thus, 1-pentyne could be prepared by treatment of acetylene with sodium amide to yield sodium acetylide, followed by reaction with 1-bromopropane:

$$\text{H—C}\equiv\text{C—H} + :\ddot{\text{N}}\text{H}_2^-\ \text{Na}^+ \longrightarrow \text{H—C}\equiv\text{C:}^-\ \text{Na}^+ + :\text{NH}_3$$

Acetylene **Sodium acetylide**

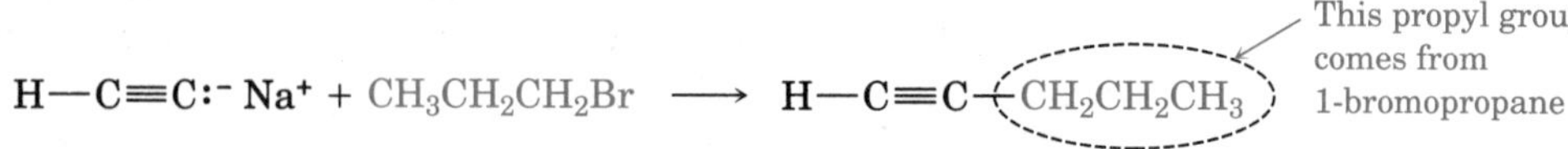

1-Bromopropane **1-Pentyne**

PROBLEM 4.24 Show the terminal alkyne and alkyl halide starting materials from which the following products can be obtained. Where two routes look feasible, list both choices.

(a) 5-Methyl-1-hexyne **(b)** 2-Hexyne **(c)** 4-Methyl-2-pentyne

INTERLUDE

NATURAL RUBBER

Rubber—an unusual name for a most unusual substance—is a naturally occurring alkene polymer produced by more than 400 different plants. The major source, however, is the so-called rubber tree, *Hevea brasiliensis*, from which the crude material is harvested as it drips from a slice made through the bark. The name *rubber* was coined by Joseph Priestley, the discoverer of oxygen and early researcher of rubber chemistry, for the simple reason that one of rubber's early uses was to rub out pencil marks on paper.

Unlike polyethylene and other simple alkene polymers, natural rubber is a polymer of a conjugated diene, *isoprene*, or 2-methyl-1,3-butadiene. The polymerization takes place by 1,4 addition (Section 4.10) of each isoprene monomer unit to the growing chain, leading to formation of a polymer that still contains double bonds spaced regularly at four-carbon intervals. As the following structure shows, these double bonds have *Z* stereochemistry:

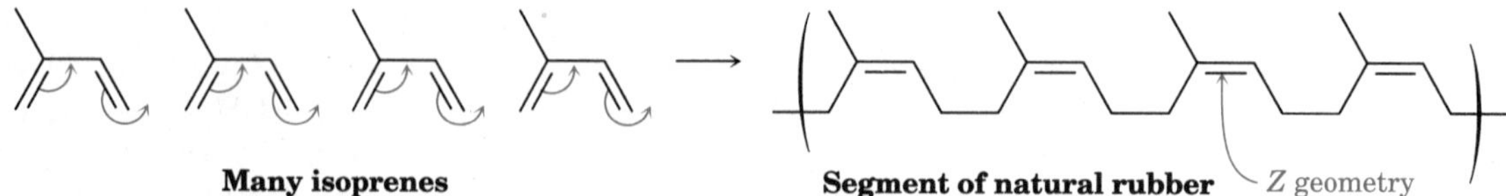

Many isoprenes (1,3-butadiene) **Segment of natural rubber** *Z* geometry

Crude rubber (latex) is collected from the tree as an aqueous dispersion that is washed, dried, and coagulated by warming in air to give a polymer with chains that average about 5000 monomer units in length and have molecular weights of 200,000–500,000. This crude coagulate is too soft and tacky to be useful until it is hardened by heating with elemental sulfur, a process called *vulcanization*. By mechanisms that are still not fully understood, vulcanization cross-links the rubber chains by forming carbon-sulfur bonds between them, thereby hardening and stiffening the polymer. The exact degree of hardening can be varied, yielding material soft enough for automobile tires or hard enough for bowling balls (*ebonite*).

The remarkable ability of rubber to stretch and then contract to its original shape is due to the irregular shapes of the polymer chains caused by the double bonds. These double bonds introduce bends and kinks into the polymer chains, thereby preventing neighboring chains from nestling together into tightly packed, semicrystalline regions. When stretched, the randomly coiled chains straighten out and orient along the direction of the pull but are kept from sliding over each other by the cross-links. When the stretch is released, the polymer reverts to its original random state (Figure 4.9).

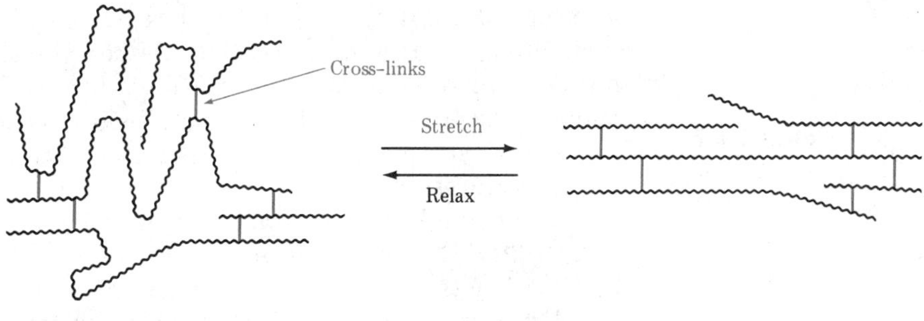

FIGURE 4.9 Unstretched and stretched sections of cross-linked rubber chains.

SUMMARY AND KEY WORDS

The chemistry of alkenes is dominated by **addition reactions** of electrophiles. When HX reacts with an alkene, **Markovnikov's rule** predicts that the hydrogen will add to the carbon that has fewer alkyl substituents and the X group will add to the carbon that has more alkyl substituents. For example,

$$(H_3C)_2C{=}CH_2 + HCl \longrightarrow H_3C{-}\underset{CH_3}{\overset{Cl}{C}}{-}CH_3$$

Many other electrophiles besides HX add to alkenes. Thus, bromine and chlorine add to give 1,2-dihalide addition products having **anti stereochemistry**. Addition of water (**hydration**) takes place on reaction of the alkene with aqueous acid, and hydrogen can be added to alkenes (**hydrogenation**) by reaction in the presence of a metal catalyst such as platinum or palladium.

Oxidation of alkenes is carried out using potassium permanganate, $KMnO_4$. Under basic conditions, $KMnO_4$ reacts with alkenes to yield cis 1,2-diols. Under neutral or acidic conditions, $KMnO_4$ cleaves double bonds to yield carbonyl-containing products. Double-bond cleavage can also be effected by reaction of the alkene with ozone, followed by treatment with zinc in acetic acid.

Alkenes are prepared from alkyl halides and alcohols by **elimination reactions**. Treatment of an alkyl halide with a strong base effects **dehydrohalogenation**, and treatment of an alcohol with acid effects **dehydration**. These elimination reactions usually give a mixture of alkene products in which the more highly substituted alkene predominates (**Zaitsev's rule**).

Conjugated dienes like 1,3-butadiene contain alternating single and double bonds. Conjugated dienes undergo **1,4 addition** of electrophiles through the formation of a resonance-stabilized allylic carbocation intermediate. No single line-bond representation can depict the true structure of an allylic carbocation. Rather, the true structure is a **resonance hybrid** somewhere intermediate between two contributing resonance forms. The only difference between two **resonance forms** is in the location of bonding electrons: The nuclei remain in the same places in both structures.

Many simple alkenes undergo **polymerization** when treated with a radical catalyst. **Polymers** are large molecules built up by the repetitive bonding together of many small **monomer** units.

Alkynes are hydrocarbons that contain carbon–carbon triple bonds. Much of the chemistry of alkynes is similar to that of alkenes. For example, alkynes react with 1 equiv of HBr and HCl to yield **vinylic** halides, and with 1 equiv of Br_2 and Cl_2 to yield 1,2-dihalides. Alkynes can also be hydrated by reaction with aqueous sulfuric acid in the presence of mercuric sulfate catalyst. The reaction leads to an intermediate enol that immediately isomerizes (**tautomerizes**) to a ketone. Alkynes can also be hydrogenated with the Lindlar catalyst to yield a cis alkene. Terminal alkynes are weakly acidic and can be converted into **acetylide anions** by treatment with a strong base. Reaction of the acetylide anion with a primary alkyl halide then gives an internal alkyne.

SUMMARY OF REACTIONS

Note: No stereochemistry is implied unless specifically stated or indicated with wedged, solid, and dashed lines.

1. Reactions of alkenes
   (a) Addition of HX, where X = Cl, Br, or I (Sections 4.1–4.2)

$$>C{=}C< \xrightarrow[\text{Ether}]{\text{HX}} >\overset{H}{C}-\overset{X}{C}<$$

Markovnikov's rule: H adds to the less highly substituted carbon, and X adds to the more highly substituted one.

(b) Addition of H_2O (Section 4.4)

$$\mathrm{C{=}C} + H_2O \xrightarrow[\text{catalyst}]{H^+} \mathrm{H{-}C{-}C{-}OH}$$

Markovnikov's rule: H adds to the less highly substituted carbon, and OH adds to the more highly substituted one.

(c) Addition of X_2, where X = Cl or Br (Section 4.5)

$$\mathrm{C{=}C} \xrightarrow[CCl_4]{X_2} \mathrm{X{-}C{-}C{-}X}$$

Anti addition

(d) Addition of H_2 (hydrogenation; Section 4.6)

$$\mathrm{C{=}C} \xrightarrow{H_2,\ \text{catalyst}} \mathrm{H{-}C{-}C{-}H}$$

Syn addition

(e) Hydroxylation with $KMnO_4$ (Section 4.7)

$$\mathrm{C{=}C} \xrightarrow[\text{NaOH},\ H_2O]{KMnO_4} \mathrm{HO{-}C{-}C{-}OH}$$

Syn addition

(f) Oxidative cleavage of alkenes with ozone (Section 4.7)

$$\mathrm{C{=}C} \xrightarrow[\text{2. Zn, } H_3O^+]{\text{1. } O_3} \mathrm{C{=}O} + \mathrm{O{=}C}$$

(g) Radical-induced polymerization of alkenes (Section 4.8)

$$n\ H_2C{=}CH_2 \xrightarrow[\text{initiator}]{\text{Radical}} {+}CH_2CH_2{+}_n$$

2. Synthesis of alkenes by elimination reactions

(a) Dehydrohalogenation of alkyl halides (Section 4.9)

$$\mathrm{H{-}C{-}C{-}X} \xrightarrow{\text{Base}} \mathrm{C{=}C}$$

Zaitsev's rule: Major product formed is the alkene with the more highly substituted double bond.

(b) Dehydration of alcohols (Section 4.9)

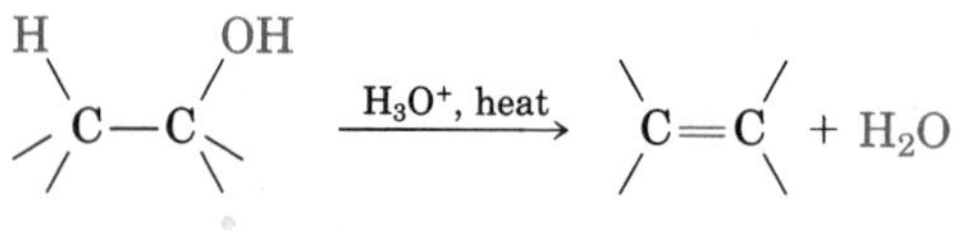

$$\text{H(C)}-\text{C(OH)} \xrightarrow{H_3O^+,\ \text{heat}} \text{C}=\text{C} + H_2O$$

Zaitsev's rule: Major product formed is the alkene with the more highly substituted double bond.

3. Reactions of alkynes

(a) Addition of H_2 (hydrogenation; Section 4.14)

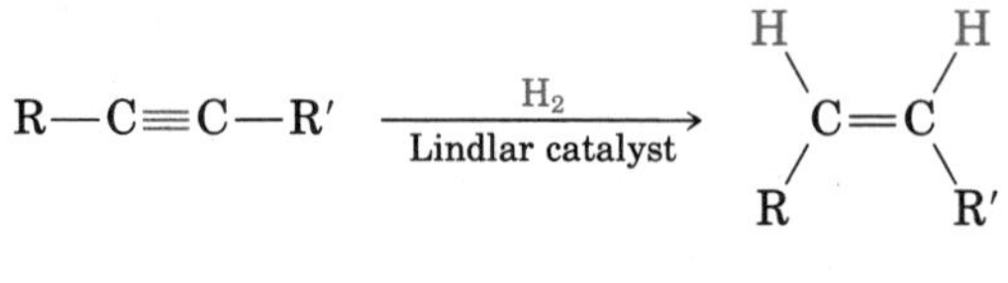

$$R-C\equiv C-R' \xrightarrow[\text{Lindlar catalyst}]{H_2} \text{(H)(R)C}=\text{C(H)(R')}$$

A cis alkene

Syn addition

(b) Addition of HX, where X = Cl, Br, or I (Section 4.14)

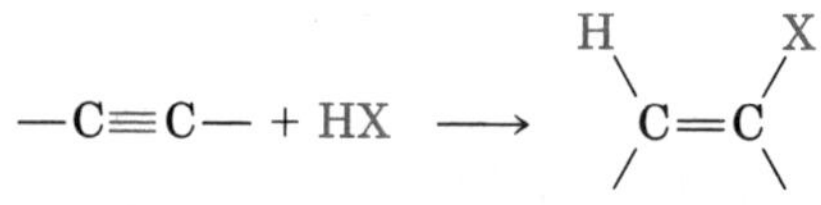

$$-C\equiv C- + HX \longrightarrow \text{(H)C}=\text{C(X)}$$

Markovnikov's rule: H adds to the less highly substituted carbon, and X adds to the more highly substituted one.

(c) Addition of X_2, where X = Cl or Br (Section 4.14)

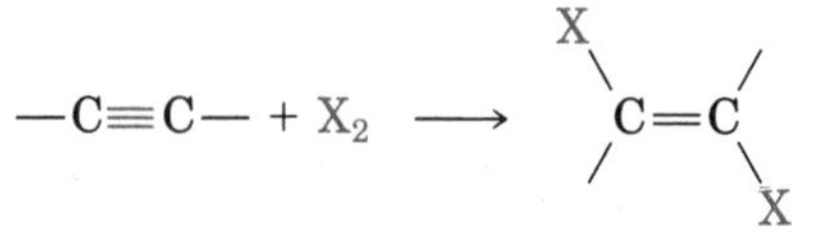

$$-C\equiv C- + X_2 \longrightarrow \text{(X)C}=\text{C(X)}$$

Trans addition

(d) Addition of H_2O to yield ketones (Section 4.15)

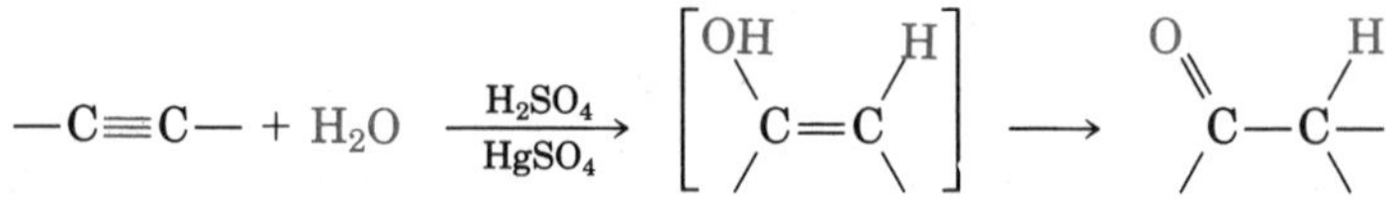

$$-C\equiv C- + H_2O \xrightarrow[HgSO_4]{H_2SO_4} \left[\text{(OH)C}=\text{C(H)}\right] \longrightarrow \text{(O=)C}-\text{C(H)}-$$

(e) Acidity: conversion into acetylide anions (Section 4.16)

$$R-C\equiv C-H \xrightarrow{NaNH_2} R-C\equiv C:^-\ Na^+ + NH_3$$

(f) Reaction of acetylide ions with alkyl halides (Section 4.16)

$$R-C\equiv C:^-\ Na^+ + R'CH_2X \longrightarrow R-C\equiv C-CH_2R' + NaX$$

ADDITIONAL PROBLEMS

4.25 Give IUPAC names for these compounds:

(a) $CH_3CH{=}CH\overset{CH_3}{\overset{|}{C}}{=}CHCH_3$ (b) $CH_3CH{=}CH\overset{CH_2CH_2CH_3}{\overset{|}{C}H}CH_2C{\equiv}CH$ (c) $CH_2{=}C{=}\overset{CH_3}{\overset{|}{C}}CH_3$

(d) $HC{\equiv}CCH_2C{\equiv}C\overset{CH_3}{\overset{|}{C}H}CH_3$

4.26 Draw structures corresponding to these IUPAC names:
(a) 3-Ethyl-1-heptyne (b) 3,5-Dimethyl-4-hexen-1-yne (c) 1,5-Heptadiyne
(d) 1-Methyl-1,3-cyclopentadiene

4.27 Draw three possible structures for each of these formulas:
(a) C_6H_8 (b) C_6H_8O

4.28 Name these alkynes according to IUPAC rules:
(a) $CH_3CH_2C{\equiv}CCH_2CH_2CH_3$ (b) $CH_3CH_2C{\equiv}CC(CH_3)_3$ (c) $CH_3C{\equiv}CCH_2C{\equiv}CCH_2CH_3$
(d) $H_2C{=}CHCH{=}CHC{\equiv}CH$

4.29 Draw structures corresponding to these IUPAC names:
(a) 3-Heptyne (b) 3,3-Dimethyl-4-octyne (c) 3,4-Dimethylcyclodecyne
(d) 2,2,5,5-Tetramethyl-3-hexyne

4.30 Draw and name all the possible pentyne isomers, C_5H_8.

4.31 Draw and name the six possible diene isomers of formula C_5H_8. Which of the six are conjugated dienes?

4.32 The following two hydrocarbons have been isolated from plants in the sunflower family. Name them according to IUPAC rules.
(a) $CH_3CH{=}CHC{\equiv}CC{\equiv}CCH{=}CHCH{=}CHCH{=}CH_2$ (all trans)
(b) $CH_3C{\equiv}CC{\equiv}CC{\equiv}CC{\equiv}CC{\equiv}CCH{=}CH_2$

4.33 Predict the products of the following reactions. Indicate regioselectivity where relevant. (The aromatic ring is inert to all the indicated reagents.)

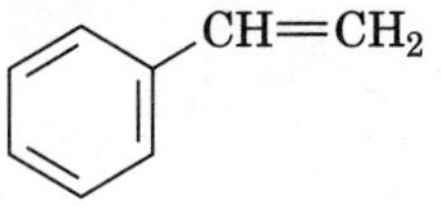

$CH{=}CH_2$

Styrene

(a) Styrene + H_2 $\xrightarrow{Pd}$? (b) Styrene + Br_2 $\longrightarrow$?

(c) Styrene + HBr $\longrightarrow$? (d) Styrene + $KMnO_4$ $\xrightarrow{NaOH,\ H_2O}$?

4.34 Suggest structures for alkenes that give the following reaction products. There may be more than one answer for some cases.

(a) ? $\xrightarrow{H_2/Pd\ catalyst}$ 2-Methylhexane

(b) ? $\xrightarrow{Br_2\ in\ CCl_4}$ 2,3-Dibromo-5-methylhexane

(c) ? $\xrightarrow{HBr}$ 2-Bromo-3-methylheptane

(d) ? $\xrightarrow{KMnO_4,\ OH^-}$ $CH_3\overset{CH_3}{\overset{|}{C}H}CH_2\overset{HO}{\overset{|}{C}H}\overset{OH}{\overset{|}{C}H}CH_2CH_3$

4.35 Using an oxidative cleavage reaction, explain how you would distinguish between these two isomeric cyclohexadienes:

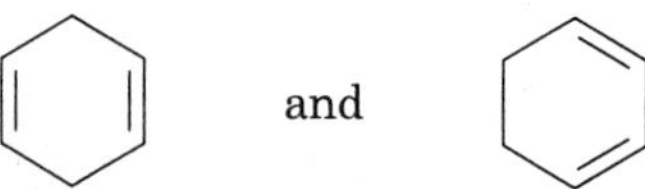

4.36 Formulate the reaction of cyclohexene with Br_2, showing the reaction intermediate and the final product with correct stereochemistry.

4.37 What products would you expect to obtain from reaction of 1,3-cyclohexadiene with each of the following?
(a) 1 mol Br_2 in CCl_4 **(b)** O_3, followed by Zn **(c)** 1 mol HCl **(d)** 1 mol DCl (D = deuterium)
(e) H_2 over a Pd catalyst

4.38 Draw the structure of a hydrocarbon that reacts with only 1 mol equiv of hydrogen on catalytic hydrogenation and gives only pentanal, $CH_3CH_2CH_2CH_2CHO$, on treatment with ozone. Write the reactions involved.

4.39 Give the structure of an alkene that yields the following keto aldehyde on reaction with ozone, followed by treatment with Zn.

$$? \xrightarrow[\text{2. Zn, } H_3O^+]{\text{1. } O_3} \overset{O}{\overset{\|}{HC}}CH_2CH_2CH_2CH_2\overset{O}{\overset{\|}{C}}CH_3$$

4.40 What alkenes would you hydrate to obtain these alcohols?

(a) $CH_3CH_2\overset{OH}{\overset{|}{C}}HCH_3$ **(b)** OH **(c)** OH CH$_3$

4.41 What alkynes would you hydrate to obtain these ketones?

(a) $CH_3\overset{CH_3}{\overset{|}{C}}HCH_2\overset{O}{\overset{\|}{C}}CH_3$ **(b)** O CH$_3$

4.42 Draw the structure of a hydrocarbon that reacts with 2 mol equiv of hydrogen on catalytic hydrogenation and gives only butanedial on reaction with ozone.

$$\overset{O}{\overset{\|}{HC}}CH_2CH_2\overset{O}{\overset{\|}{C}}H$$ **Butanedial**

4.43 Predict the products of the following reactions on 1-hexyne:

$CH_3CH_2CH_2CH_2C{\equiv}CH$ **1-Hexyne**

(a) $\xrightarrow{\text{1 equiv HBr}}$? **(b)** $\xrightarrow{\text{1 equiv } Cl_2}$? **(c)** $\xrightarrow{H_2\text{, Lindlar catalyst}}$?

4.44 Predict the products of the following reactions on 5-decyne:

$$CH_3CH_2CH_2CH_2C{\equiv}CCH_2CH_2CH_2CH_3 \quad \textbf{5-Decyne}$$

(a) $\xrightarrow{H_2,\ \text{Lindlar catalyst}}$? **(b)** $\xrightarrow{2\ \text{equiv}\ Br_2}$? **(c)** $\xrightarrow{H_2O,\ H_2SO_4,\ HgSO_4}$?

4.45 In planning the synthesis of a compound, it's as important to know what *not* to do as to know what to do. What is wrong with each of the following reactions?

(a) 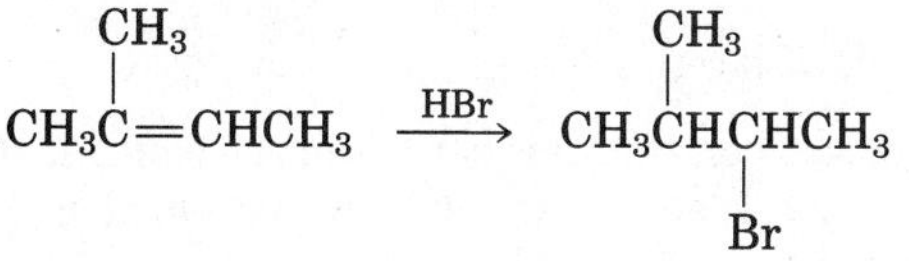 $CH_3C(CH_3){=}CHCH_3 \xrightarrow{HBr} CH_3CH(CH_3)CH(Br)CH_3$

(b) 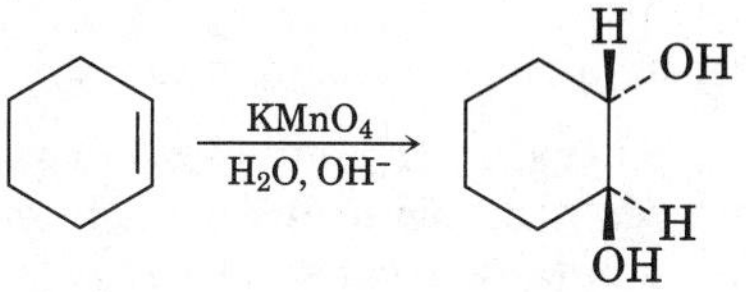 cyclohexene $\xrightarrow[H_2O,\ OH^-]{KMnO_4}$ trans-1,2-cyclohexanediol (H, OH / H, OH)

(c) 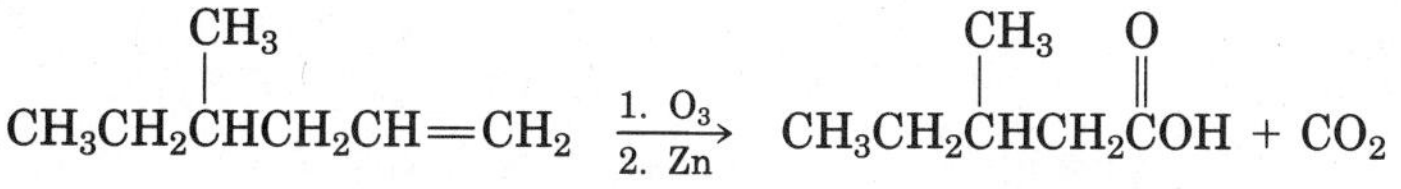 $CH_3CH_2CH(CH_3)CH_2CH{=}CH_2 \xrightarrow[2.\ Zn]{1.\ O_3} CH_3CH_2CH(CH_3)CH_2COOH + CO_2$

4.46 Acrylonitrile, $H_2C{=}CHC{\equiv}N$, contains a carbon–carbon double bond and a carbon-nitrogen triple bond. Sketch the orbitals involved in the multiple bonding in acrylonitrile, and indicate the hybridization of the carbons. Is acrylonitrile conjugated?

4.47 How would you prepare *cis*-2-butene starting from 1-propyne, an alkyl halide, and any other reagents needed? (This problem can't be worked in a single step. You'll have to carry out more than one chemical reaction.)

4.48 Using 1-butyne as the only organic starting material, along with any inorganic reagents needed, how would you synthesize the following compounds? More than one step may be needed.
(a) Butane **(b)** 1,1,2,2-Tetrachlorobutane **(c)** 2-Bromobutane
(d) 2-Butanone ($CH_3CH_2COCH_3$)

4.49 Give the structure of an alkene that provides only acetone, $(CH_3)_2C{=}O$, on reaction with ozone.

4.50 Compound A has the formula C_8H_8. It reacts rapidly with acidic $KMnO_4$ but reacts with only 1 equiv of H_2 over a palladium catalyst. On hydrogenation under conditions that reduce aromatic rings, A reacts with 4 equiv of H_2, and hydrocarbon B, C_8H_{16}, is produced. The reaction of A with $KMnO_4$ gives CO_2 and a carboxylic acid C, $C_7H_6O_2$. What are the structures of A, B, and C? Write all the reactions.

4.51 The sex attractant of the common housefly is a hydrocarbon named *muscalure*, $C_{23}H_{46}$. On treatment of muscalure with aqueous acidic $KMnO_4$, two products are obtained, $CH_3(CH_2)_{12}COOH$ and $CH_3(CH_2)_7COOH$. Propose a structure for muscalure.

4.52 How would you synthesize muscalure (see Problem 4.51) starting from acetylene and any alkyl halides needed? (The double bond in muscalure is cis.)

4.53 Draw a reaction energy diagram for the addition of HBr to 1-pentene. Let one curve on your diagram show the formation of 1-bromopentane product and another curve on the same diagram show the formation of 2-bromopentane product. Label the positions for all reactants, intermediates, and products.

4.54 Make sketches of what you imagine the transition-state structures to look like in the reaction of HBr with 1-pentene (see Problem 4.53).

4.55 Methylenecyclohexane, on treatment with strong acid, isomerizes to yield 1-methylcyclohexene:

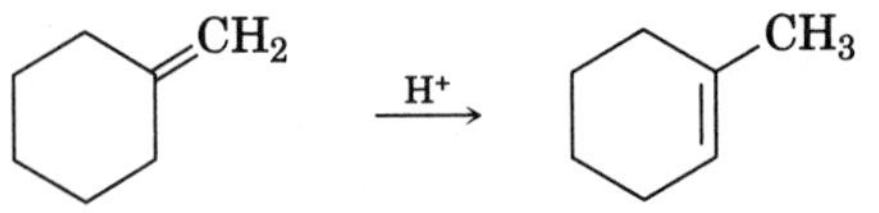

$$\text{Methylenecyclohexane} \xrightarrow{H^+} \text{1-Methylcyclohexene}$$

Methylenecyclohexane **1-Methylcyclohexene**

Propose a mechanism by which this reaction might occur.

4.56 α-Terpinene, $C_{10}H_{16}$, is a pleasant-smelling hydrocarbon that has been isolated from oil of marjoram. On hydrogenation over a palladium catalyst, α-terpinene reacts with 2 mol equiv of hydrogen to yield a new hydrocarbon, $C_{10}H_{20}$. On ozonolysis, followed by reduction with zinc and acetic acid, α-terpinene yields glyoxal and 6-methyl-2,5-heptanedione. Propose a structure for α-terpinene.

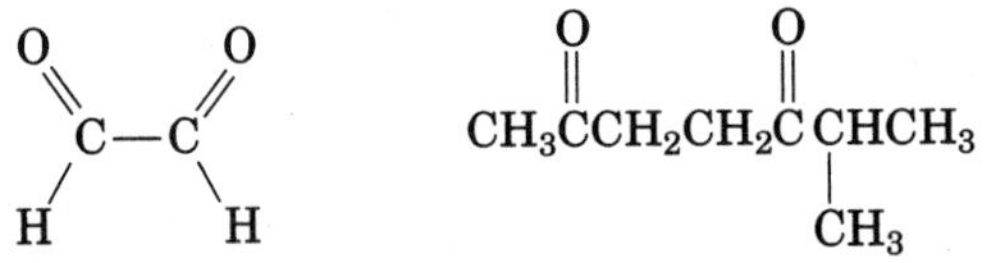

OHC—CHO $\qquad$ $CH_3\overset{O}{\overset{\|}{C}}CH_2CH_2\overset{O}{\overset{\|}{C}}\underset{CH_3}{\underset{|}{C}H}CH_3$

Glyoxal **6-Methyl-2,5-heptanedione**

5 CHAPTER

AROMATIC COMPOUNDS

In the early days of organic chemistry, the word *aromatic* was used to describe fragrant substances such as benzaldehyde (from cherries, peaches, and almonds), toluene (from tolu balsam), and benzene (from coal distillate). It was soon realized, however, that the substances grouped as aromatic differ from most other compounds in their chemical behavior.

AROMATIC
Referring to the class of compounds that contain a benzene-like six-membered ring with three double bonds

Today, we use the word **aromatic** to refer to the class of compounds that contain benzene-like, six-membered rings with three double bonds. Many important compounds are aromatic in part, including the steroidal hormone estrone and the analgesic ibuprofen. Benzene itself causes a depressed white blood cell count (leukopenia) on prolonged exposure and should not be used as a laboratory solvent. We'll see in this chapter how aromatic substances behave and why they're different from the alkanes, alkenes, and alkynes we've studied up to this point.

CH_3 O H H H HO CH_3 CH_3CHCH_2 O $CHCOH$ CH_3

Benzene **Estrone** **Ibuprofen**

5.1 STRUCTURE OF BENZENE: THE KEKULÉ PROPOSAL

By the mid-1800s, benzene was known to have the molecular formula C_6H_6, and its chemistry was being actively explored. It was known that, although benzene is relatively unreactive toward most reagents that attack alkenes, it reacts with Br_2 in the presence of iron to give the *substitution* product C_6H_5Br rather than the possible *addition* product $C_6H_6Br_2$. Furthermore, only one monobromo substitution product was known; no isomers had been prepared.

$$C_6H_6 + Br_2 \xrightarrow{Fe} C_6H_5Br + HBr \quad \left[C_6H_6Br_2 \right]$$

Benzene — **Bromobenzene** (substitution product) — (addition product; *not formed*)

On the basis of these and other results, August Kekulé proposed in 1865 that benzene consists of a ring of carbon atoms and can be formulated as 1,3,5-cyclohexatriene. Kekulé reasoned that this structure would readily account for the isolation of only a single monobromo substitution product, since all six carbon atoms and all six hydrogens in 1,3,5-cyclohexatriene are equivalent.

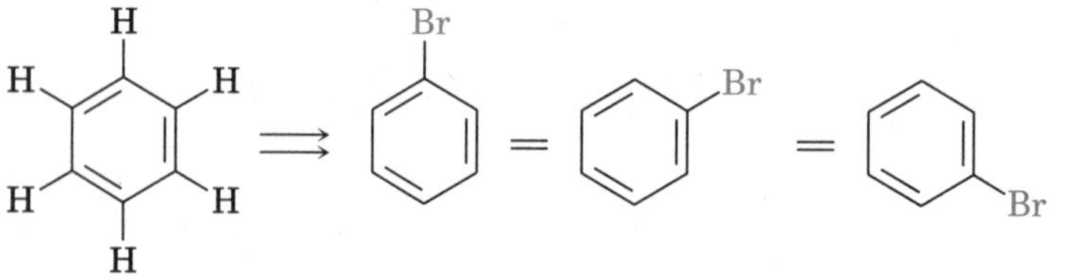

All six hydrogens are equivalent

Only one possible monobromo substitution product

Kekulé's proposal was widely criticized by other chemists of the day. Although it satisfactorily accounts for the correct number of monosubstituted benzene isomers, the proposal doesn't answer the critical questions of why benzene is unreactive compared with other alkenes and why benzene gives a substitution product rather than an addition product on reaction with Br_2.

PROBLEM 5.1 How many dibromobenzene derivatives are possible according to Kekulé's theory? Draw them.

5.2 STABILITY OF BENZENE

The unusual stability of benzene was a great puzzle to early chemists. Although its formula, C_6H_6, indicates that several double or triple bonds must be present, benzene shows none of the behavior characteristic of alkenes. For example, alkenes readily react with $KMnO_4$ to give 1,2-diols; they react with aqueous acid to give alcohols; and they react with gaseous HCl to give saturated chloroalkanes. Benzene does none of these things. *Benzene does not undergo electrophilic addition reactions.*

$$\text{Benzene} \xrightarrow{H_3O^+} \text{No reaction}$$
$$\text{Benzene} \xrightarrow{KMnO_4} \text{No reaction}$$
$$\text{Benzene} \xrightarrow[\text{Ether}]{HCl} \text{No reaction}$$

Benzene

Further evidence for the unusual nature of benzene is that all carbon–carbon bonds in benzene have the same length, intermediate between a normal single and a normal double bond. Most carbon–carbon single bonds have lengths near 1.54 Å, and most carbon–carbon double bonds are about 1.34 Å long, but the carbon–carbon bonds in benzene are 1.39 Å long (Figure 5.1).

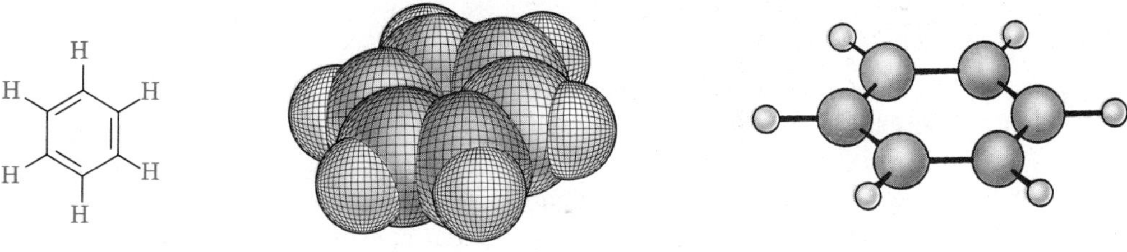

FIGURE 5.1 The structure of benzene. All six carbon–carbon bonds are identical.

5.3 STRUCTURE OF BENZENE: THE RESONANCE PROPOSAL

How can we account for benzene's properties, and how can we best represent its structure? To answer these questions, we need to look again at the concept of resonance. We saw in Sections 4.11–4.12 that an allylic carbocation is best described as a resonance hybrid of two contributing resonance forms. Neither form is correct by itself; the true structure of an allylic carbocation can't be drawn using a single line-bond structure, because it is intermediate between the two resonance forms:

$$H_2C{=}CH{-}\overset{+}{C}H_2 \longleftrightarrow H_2\overset{+}{C}{-}CH{=}CH_2$$

In the same way, resonance theory says that benzene can't be described satisfactorily by a single line-bond structure. Rather, benzene is a resonance hybrid of two equivalent structures. Benzene doesn't oscillate back and forth between two extremes; its true structure is somewhere between the two and is impossible to draw with our usual conventions. Each carbon–carbon connection is an average of 1.5 bonds, midway between a single and a double bond.

1.5 bonds on average

An orbital view of benzene shows the situation more clearly, emphasizing the cyclic conjugation of the benzene molecule and the equivalence of the six carbon–carbon bonds. Benzene is a flat, symmetrical molecule in the shape of a regular hexagon. All C–C–C bond angles are 120°, each carbon atom is sp^2-hybridized, and each carbon has a *p* orbital perpendicular to the plane of the six-membered ring. Since all six *p* orbitals are equivalent, it's impossible to define three localized alkene pi bonds in which a given *p* orbital overlaps only one neighboring *p* orbital. Rather, each *p* orbital overlaps equally well with both neighboring *p* orbitals, leading to a structure for benzene in which the pi electrons are delocalized around the ring in two doughnut-shaped clouds (Figure 5.2).

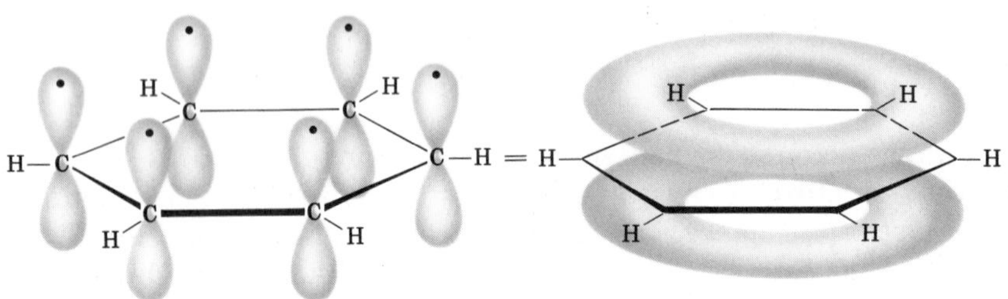

FIGURE 5.2 An orbital picture of benzene.

We can now see why benzene is unusually stable. According to resonance theory, the more resonance forms a substance has, the more stable it is. Benzene, with two resonance forms of equal energy, is thus more stable than a normal alkene.

PROBLEM 5.2 How does resonance theory account for the fact that there is only one *o*-dibromobenzene rather than the two isomers that Kekulé's theory would suggest?

5.4 NAMING AROMATIC COMPOUNDS

Aromatic substances, more than any other class of organic compounds, have acquired a large number of nonsystematic names. Although the use of such names is discouraged, IUPAC rules allow for the more common ones shown in Table 5.1 to be retained. Thus, methylbenzene is commonly known as toluene, hydroxybenzene as phenol, aminobenzene as aniline, and so on.

TABLE 5.1 Common Names of Some Aromatic Compounds

STRUCTURE	NAME	STRUCTURE	NAME
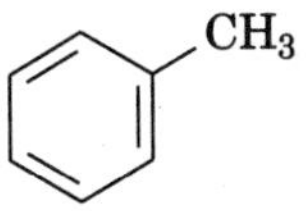 C_6H_5–CH_3	Toluene (bp 110°C)	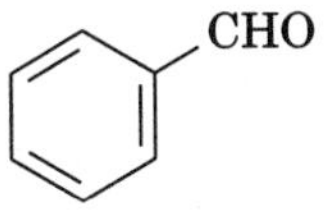 C_6H_5–CHO	Benzaldehyde (bp 178°C)
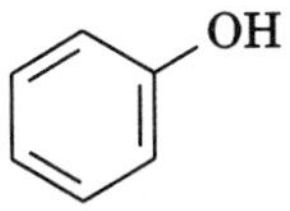 C_6H_5–OH	Phenol (mp 43°C)	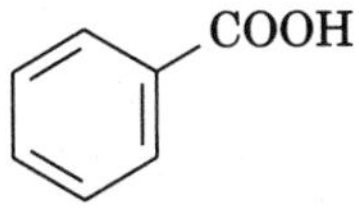 C_6H_5–COOH	Benzoic acid (mp 122°C)
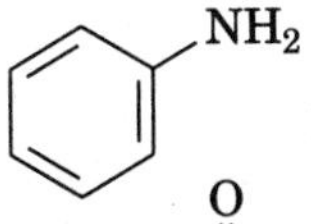 C_6H_5–NH_2	Aniline (bp 184°C)	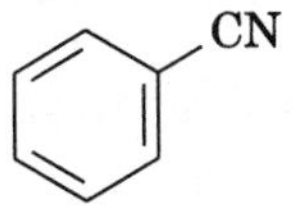 C_6H_5–CN	Benzonitrile (bp 191°C)
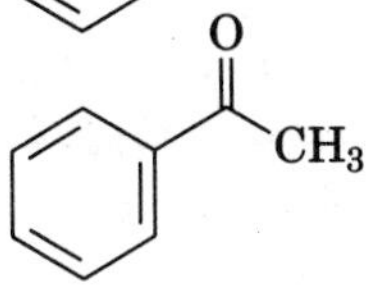 C_6H_5–C(=O)–CH_3	Acetophenone (mp 21°C)	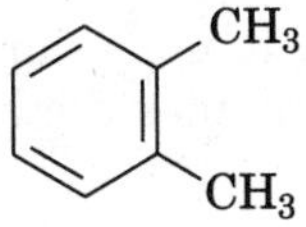 $C_6H_4(CH_3)_2$	*ortho*-Xylene (bp 144°C)

PHENYL
The C_6H_5– group

BENZYL
The $C_6H_5CH_2$– group

Monosubstituted benzene derivatives are systematically named in the same manner as other hydrocarbons, with *-benzene* as the parent name. Thus, C_6H_5Br is bromobenzene, and $C_6H_5CH_2CH_3$ is ethylbenzene. The name **phenyl** (**fen**-nil) is used for the $-C_6H_5$ unit when the benzene ring is considered a substituent group, and the name **benzyl** is used for the $C_6H_5CH_2$– alkyl group.

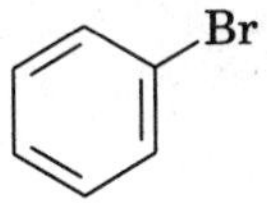

Br — **Bromobenzene**; CH_2CH_3 — **Ethylbenzene**;  **A phenyl group**; CH_2 — **A benzyl group**

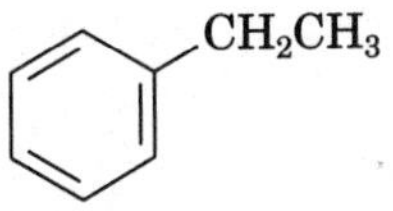

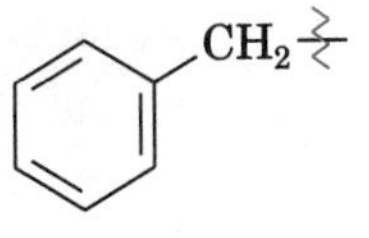

Disubstituted benzenes are named using one of the prefixes *ortho-* (*o*), *meta-* (*m*), or *para-* (*p*). An ortho-disubstituted benzene has its two substituents in a 1,2 relationship on the ring; a meta-disubstituted benzene has its two substituents in a 1,3 relationship; and a para-disubstituted benzene has its substituents in a 1,4 relationship:

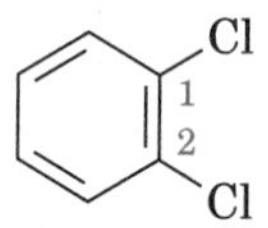

***ortho*-Dichlorobenzene** (1,2-disubstituted)

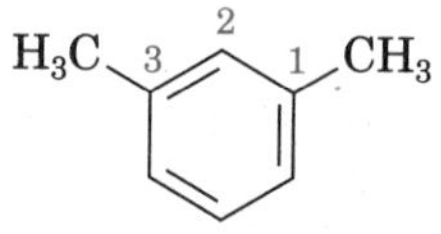

***meta*-Xylene** (1,3-disubstituted)

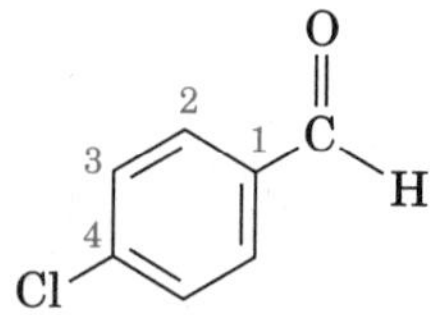

***para*-Chlorobenzaldehyde** (1,4-disubstituted)

Benzenes with more than two substituents are named by numbering the position of each substituent on the ring so that the lowest possible numbers are used. The substituents are listed alphabetically when writing the name.

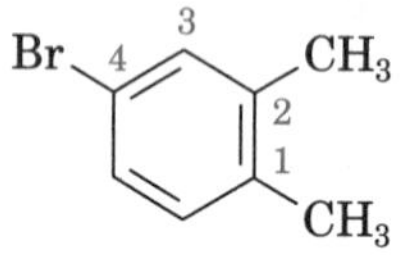

4-Bromo-1,2-dimethylbenzene

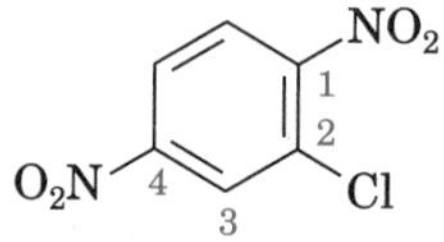

2-Chloro-1,4-dinitrobenzene

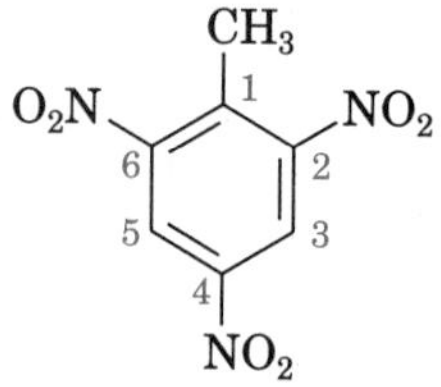

2,4,6-Trinitrotoluene (TNT)

In the third example shown, note that *-toluene* is used as the parent name rather than *-benzene.* Any of the monosubstituted aromatic compounds shown in Table 5.1 can serve as a parent name, with the principal substituent ($-CH_3$ in toluene, for example) assumed to be on carbon 1. The following two examples further illustrate this practice:

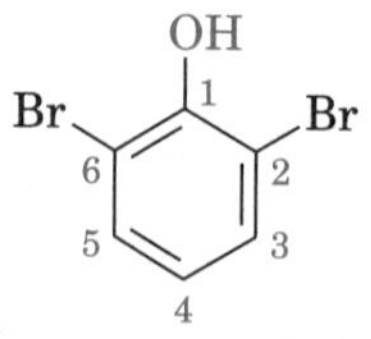

2,6-Dibromo*phenol*

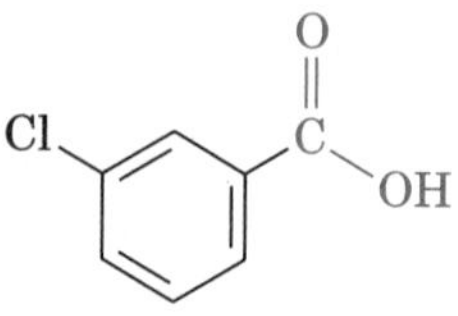

m*-Chloro*benzoic acid

PRACTICE PROBLEM 5.1 What is the IUPAC name of this compound?

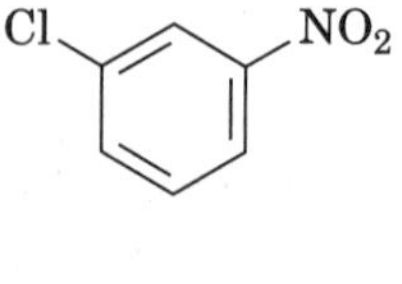

SOLUTION Because the nitro group ($-NO_2$) and chloro group are on carbons 1 and 3, they have a meta relationship. Citing the two substituents in alphabetical order gives the IUPAC name *m*-chloronitrobenzene.

PROBLEM 5.3 Tell whether these compounds are ortho, meta, or para disubstituted:

(a) 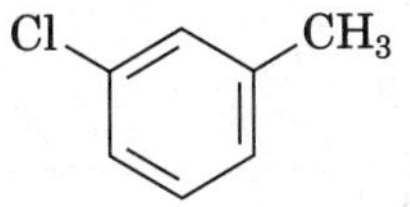 Cl ... CH_3 (b) 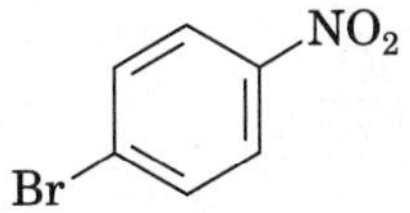 NO_2 ... Br (c) 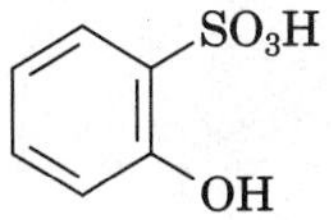 SO_3H ... OH

PROBLEM 5.4 Give IUPAC names for these compounds:

(a) 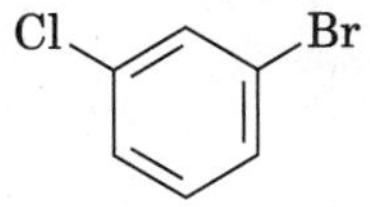 Cl ... Br (b) 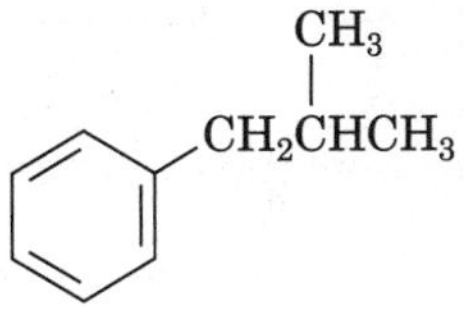 CH_2CHCH_3 with CH_3 (c) 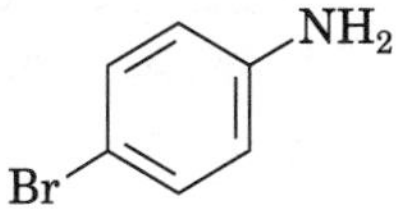 NH_2 ... Br

PROBLEM 5.5 Draw structures corresponding to these IUPAC names:
(a) *p*-Bromochlorobenzene **(b)** *p*-Bromotoluene **(c)** *m*-Chloroaniline
(d) 1-Chloro-3,5-dimethylbenzene

5.5 CHEMISTRY OF BENZENE: ELECTROPHILIC AROMATIC SUBSTITUTION

ELECTROPHILIC AROMATIC SUBSTITUTION
Substitution of an electrophile for a hydrogen atom on a benzene ring

The most important reaction of aromatic compounds is **electrophilic aromatic substitution**. That is, an electron-poor reagent (an electrophile, E^+) reacts with an aromatic ring and substitutes for one of the ring hydrogens:

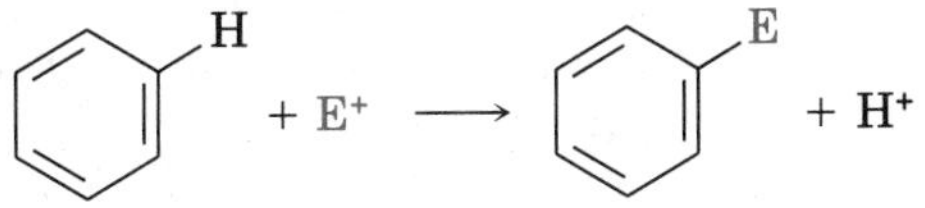

$$C_6H_5-H + E^+ \longrightarrow C_6H_5-E + H^+$$

Many different substituents can be introduced onto the aromatic ring by electrophilic substitution reactions. By choosing the proper reagents, it's possible to halogenate the aromatic ring (substitute a halogen: –F, –Cl, –Br, or –I), nitrate it (substitute a nitro group: $-NO_2$), sulfonate it (substitute a sulfonic acid group: $-SO_3H$), or alkylate it (substitute an alkyl group: –R). Starting with only a few simple materials, we can prepare many thousands of substituted aromatic compounds (Figure 5.3).

All these reactions (and many more as well) take place by a similar mechanism. Let's begin a study of this fundamental reaction type by looking at one reaction in detail, the bromination of benzene.

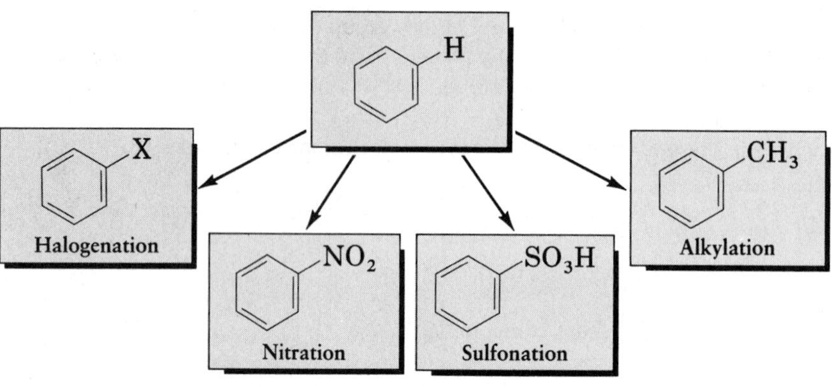

FIGURE 5.3 Some electrophilic aromatic substitution reactions.

PROBLEM 5.6 There are three products that might form on bromination of toluene. Draw and name them.

5.6 BROMINATION OF BENZENE

Benzene reacts with Br_2 in the presence of $FeBr_3$ as catalyst to yield the substitution product bromobenzene:

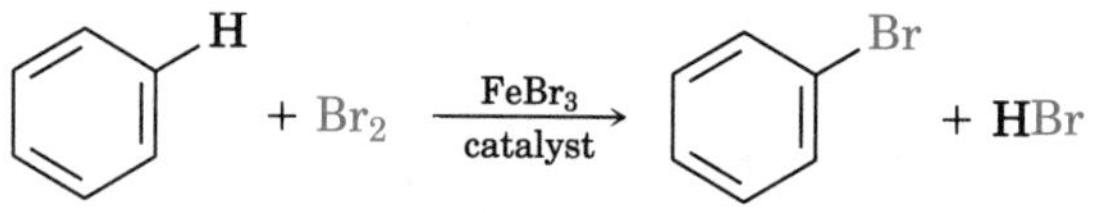

$$\text{Benzene} + Br_2 \xrightarrow[\text{catalyst}]{FeBr_3} \text{Bromobenzene (80\%)} + HBr$$

Benzene **Bromobenzene (80%)**

Before seeing how this electrophilic *substitution* reaction occurs, let's briefly recall what was said about electrophilic *additions* to alkenes in Sections 3.8–3.11. When an electrophile such as H^+ adds to an alkene, it approaches the *p* orbitals of the double bond and forms a bond to one carbon, leaving a positive charge on the other carbon. The carbocation intermediate is then attacked by a nucleophile such as chloride ion to yield the addition product (Figure 5.4).

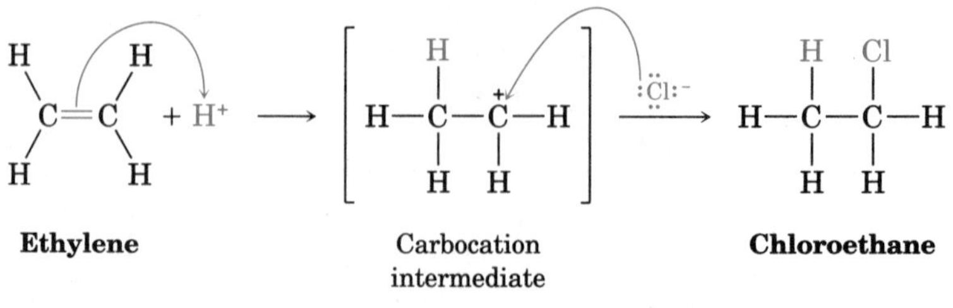

$$H_2C{=}CH_2 + H^+ \longrightarrow \left[H{-}CH_2{-}\overset{+}{C}H_2 \right] \xrightarrow{:\ddot{Cl}:^-} H{-}CH_2{-}CHCl{-}H$$

Ethylene Carbocation intermediate **Chloroethane**

FIGURE 5.4 The mechanism of an electrophilic addition to an alkene.

An electrophilic aromatic substitution reaction begins in a similar way, but there are a number of differences. One difference is that aromatic rings are much less reactive toward electrophiles than alkenes are. For example, Br_2 in CCl_4 solution reacts instantly with most alkenes but does not react with benzene. For bromination of benzene to take place, a catalyst such as $FeBr_3$ is needed. The catalyst acts by reacting with Br_2 to form $FeBr_4^-$ and Br^+, a highly reactive electrophile:

$$FeBr_3 + Br_2 \longrightarrow FeBr_4^- + Br^+$$

The electrophilic Br^+ then reacts with the electron-rich (nucleophilic) benzene ring to yield a nonaromatic carbocation intermediate. This carbocation is allylic (Section 4.11) and can be drawn in three resonance forms:

Although stable by comparison with nonallylic carbocations, the intermediate in electrophilic aromatic substitution is nevertheless much less stable than the starting aromatic reactant. Thus, reaction of an electrophile with a benzene ring has a relatively high activation energy and is rather slow. Figure 5.5 gives reaction energy diagrams that compare the reaction of an electrophile E^+ with an alkene and with benzene. The benzene reaction is slower (has a higher E_{act}) because the starting material is so stable.

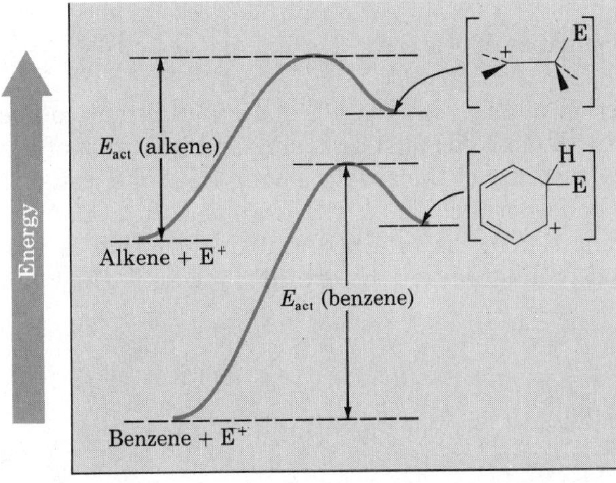

FIGURE 5.5 A comparison of the reactions of an electrophile with an alkene and with benzene: E_{act} (alkene) < E_{act} (benzene).

A second difference between alkene addition and aromatic substitution reactions occurs after the electrophile has added to the benzene ring to give the carbocation intermediate. Although it would presumably be possible for a nucleophile such as Br^- to react with the carbocation intermediate to yield the addition product dibromocyclohexadiene, this is not observed. Instead, a base removes H^+, yielding the neutral aromatic substitution product plus HBr. The net effect is the substitution of Br^+ for H^+ by the overall mechanism shown in Figure 5.6.

An electron pair from the benzene ring attacks Br^+, forming a new C–Br bond and leaving a carbocation intermediate.

The carbocation intermediate loses a proton, and the neutral substitution product froms as two electrons from the C–H bond move into the new aromatic ring.

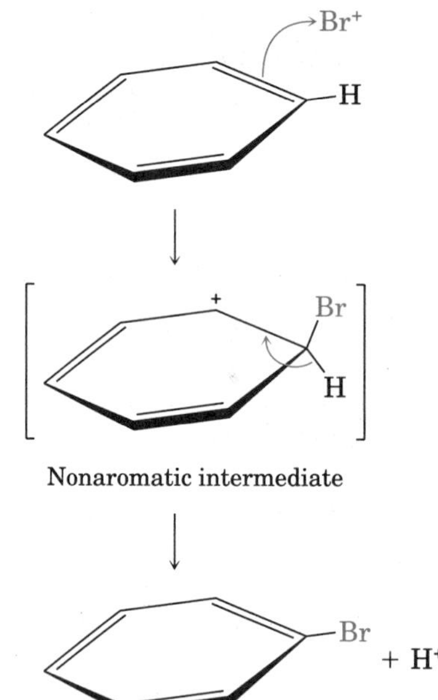

FIGURE 5.6 The mechanism of the electrophilic bromination of benzene.

Why does the reaction of Br_2 with benzene take a different course than its reaction with an alkene? The answer is simple: If *addition* of Br_2 to benzene occurred, the stability of the aromatic ring would be lost, and the reaction would be endothermic. When *substitution* occurs, though, the stability of the aromatic ring is retained and the reaction is exothermic. A reaction energy diagram for the overall process is shown in Figure 5.7.

5.7 OTHER ELECTROPHILIC AROMATIC SUBSTITUTION REACTIONS

Many electrophilic aromatic substitutions besides bromination occur by the same general mechanism. Let's look briefly at some of these other reactions.

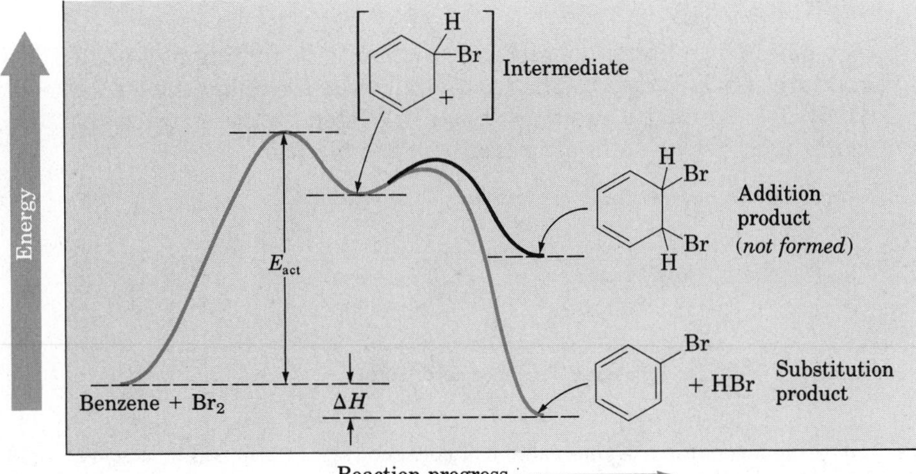

FIGURE 5.7 A reaction energy diagram for the electrophilic bromination of benzene.

Chlorination

Aromatic rings react with Cl_2 in the presence of $FeCl_3$ catalyst to yield chlorobenzenes. This kind of reaction is used in the synthesis of numerous pharmaceutical agents, including the tranquilizer diazepam (Valium).

$$\text{Benzene (C}_6\text{H}_5\text{H)} + Cl_2 \xrightarrow[\text{catalyst}]{FeCl_3} \text{Chlorobenzene (C}_6\text{H}_5\text{Cl) (86\%)} + HCl$$

Benzene **Chlorobenzene (86%)**

Diazepam

Nitration

Aromatic rings are nitrated by reaction with a mixture of concentrated nitric and sulfuric acids. The electrophile in this reaction is the nitronium ion, NO_2^+, which reacts with benzene in much the same way as does Br^+. Nitration of aromatic rings is a key step in the synthesis of explosives such as TNT (2,4,6-trinitrotoluene), dyes, and many pharmaceutical agents.

$$HNO_3 + H_2SO_4 \longrightarrow NO_2^+ + HSO_4^- + H_2O$$

$$\text{Benzene (C}_6\text{H}_5\text{H)} + HNO_3 \xrightarrow[\text{catalyst}]{H_2SO_4} \text{Nitrobenzene (C}_6\text{H}_5\text{NO}_2\text{) (85\%)} + H_2O$$

Benzene **Nitrobenzene (85%)**

TNT

Sulfonation

Aromatic rings are sulfonated by reaction with fuming sulfuric acid, a mixture of $SO_3 + H_2SO_4$. The reactive electrophile under these conditions is HSO_3^+. Aromatic sulfonation is a key step in the synthesis of such compounds as the sulfa drug family of antibiotics.

$$SO_3 + H_2SO_4 \longrightarrow SO_3H^+ + HSO_4^-$$

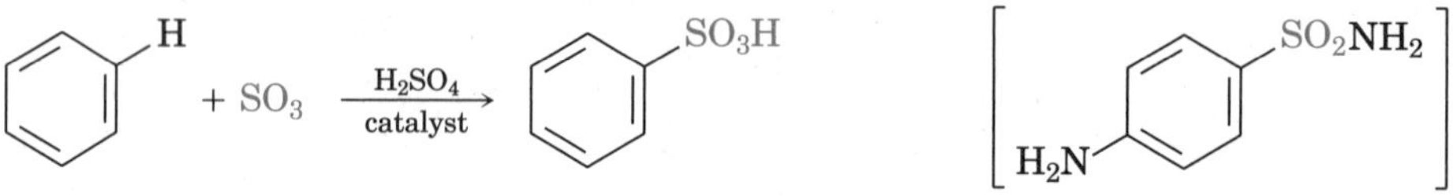

Benzene + SO_3 $\xrightarrow[\text{catalyst}]{H_2SO_4}$ Benzenesulfonic acid (95%)

Benzene **Benzenesulfonic acid (95%)** **Sulfanilimide** (a sulfa drug)

PRACTICE PROBLEM 5.2 Show the mechanism of the reaction of benzene with fuming sulfuric acid to yield benzenesulfonic acid.

SOLUTION The electrophile in sulfonation reactions is HSO_3^+, and the reaction occurs by the same two-step process common to all electrophilic aromatic substitutions.

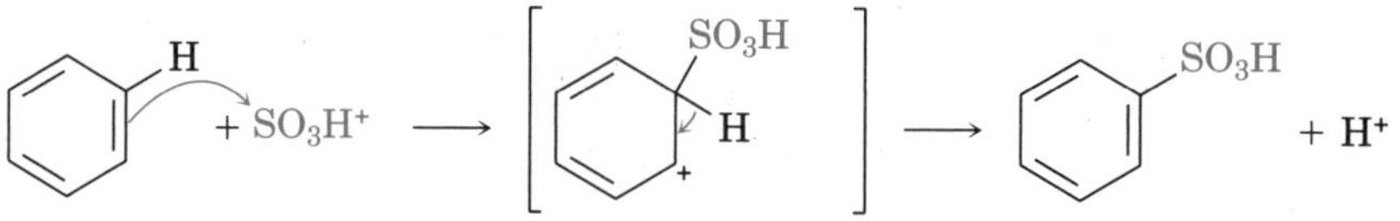

Benzene + SO_3H^+ ⟶ [Carbocation intermediate] ⟶ Benzenesulfonic acid + H^+

Carbocation intermediate

PROBLEM 5.7 Show the mechanism of the reaction of benzene with nitric acid and sulfuric acid to yield nitrobenzene.

PROBLEM 5.8 Chlorination of *o*-xylene (dimethylbenzene) yields a mixture of two products, but chlorination of *p*-xylene yields a single product. Explain.

PROBLEM 5.9 How many products might be formed on chlorination of *m*-xylene?

PROBLEM 5.10 How can you account for the fact that deuterium slowly replaces hydrogen in the aromatic ring when benzene is treated with D_2SO_4?

5.8 THE FRIEDEL–CRAFTS ALKYLATION AND ACYLATION REACTIONS

FRIEDEL–CRAFTS ALKYLATION REACTION
Introduction of an alkyl group onto a benzene ring by electrophilic substitution

An alkyl group is attached to an aromatic ring on reaction with an alkyl chloride in the presence of $AlCl_3$ catalyst, a process called the **Friedel–Crafts alkylation reaction**. For example, benzene reacts with 2-chloropropane in the presence of $AlCl_3$ to yield isopropylbenzene (also called cumene):

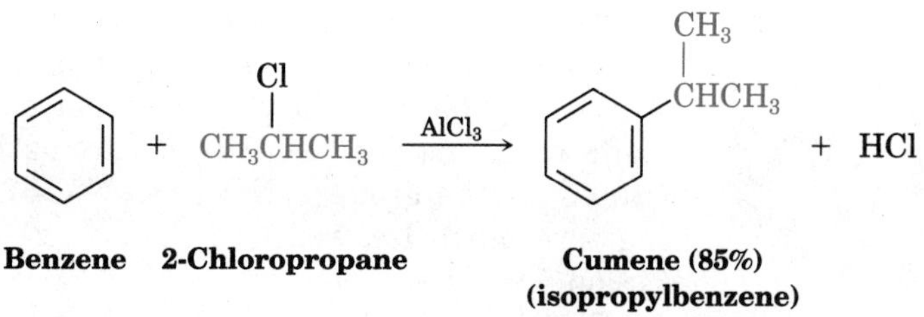

$$\text{Benzene} + CH_3CH(Cl)CH_3 \xrightarrow{AlCl_3} C_6H_5CH(CH_3)CH_3 + HCl$$

Benzene **2-Chloropropane** **Cumene (85%) (isopropylbenzene)**

The Friedel–Crafts alkylation reaction is an aromatic substitution in which the aromatic ring attacks a carbocation electrophile. Loss of a proton from the intermediate then yields the alkylated aromatic ring. The carbocation is usually generated by reaction of an alkyl chloride with $AlCl_3$ catalyst, and it's thought that the catalyst acts by helping the alkyl chloride to ionize, in much the same way that $FeBr_3$ helps Br_2 to ionize (Section 5.6). The overall Friedel–Crafts mechanism for the synthesis of isopropylbenzene is shown in Figure 5.8.

$$CH_3CH(Cl)CH_3 \quad AlCl_3 \longrightarrow CH_3\overset{+}{C}HCH_3 \quad AlCl_4^-$$

$$C_6H_6 \quad CH_3\overset{+}{C}HCH_3 \quad AlCl_4^-$$

An electron pair from the aromatic ring attacks the carbocation, forming a C–C bond and yielding a new carbocation intermediate.

$$\left[\text{carbocation intermediate bearing } CH(CH_3)CH_3 \text{ and } H \quad :\ddot{Cl}:^-\right]$$

Loss of a proton then gives the neutral alkylated substitution product.

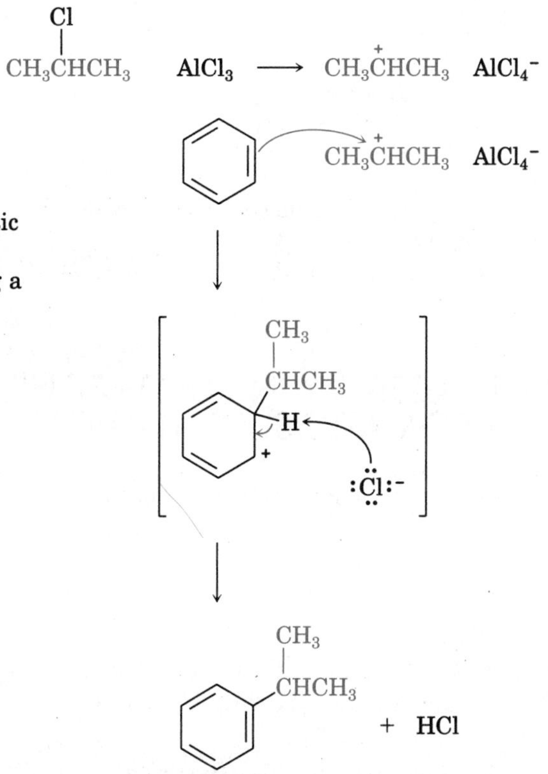

$$C_6H_5CH(CH_3)CH_3 + HCl$$

FIGURE 5.8 Mechanism of the Friedel–Crafts alkylation reaction in the synthesis of isopropylbenzene.

Although extremely useful, the Friedel–Crafts alkylation reaction has several important limitations. For example, only *alkyl* halides can be used; aryl halides like chlorobenzene don't react. In addition, Friedel–Crafts reactions don't succeed on aromatic rings that are already substituted by the groups $-NO_2$, $-C{\equiv}N$, $-SO_3H$, or $-COR$. Such aromatic rings are much less reactive than benzene for reasons we'll discuss in Section 5.9.

FRIEDEL–CRAFTS ACYLATION REACTION Introduction of an acyl group onto a benzene ring by electrophilic substitution

ACYL GROUP The $-\overset{O}{\overset{\|}{C}}R$ group

Closely related to the Friedel–Crafts alkylation reaction, is the **Friedel–Crafts acylation reaction**. When an aromatic compound is allowed to react with a carboxylic acid chloride, RCOCl, in the presence of $AlCl_3$, an **acyl** (**a**-sil) **group**, $-COR$, is introduced onto the ring. For example, reaction of benzene with acetyl chloride yields the ketone, acetophenone:

$$C_6H_6 + CH_3\overset{O}{\overset{\|}{C}}Cl \xrightarrow[80°C]{AlCl_3} C_6H_5\overset{O}{\overset{\|}{C}}CH_3 + HCl$$

Benzene **Acetyl chloride** **Acetophenone (95%)**

PROBLEM 5.11 What products would you expect to obtain from the reaction of these compounds with chloroethane and $AlCl_3$?
(a) Benzene (b) *p*-Xylene

PROBLEM 5.12 What products would you expect to obtain from the reaction of benzene with these reagents?
(a) $(CH_3)_3CCl$ and $AlCl_3$ (b) CH_3CH_2COCl and $AlCl_3$

5.9 REACTIVITY IN ELECTROPHILIC AROMATIC SUBSTITUTION

Only one monosubstitution product is possible when electrophilic substitution occurs on benzene, but what would happen if we were to carry out an electrophilic substitution reaction on a ring that is already substituted? Substituents already present on an aromatic ring have two effects:

1. Substituents affect the *reactivity* of the aromatic ring. Some substituents make the ring more reactive than benzene, and some make it less reactive.
2. Substituents affect the *orientation* of the reaction. Three disubstituted products are possible: ortho, meta, and para. These three products aren't formed in random ratios, though; instead, the nature of the substituent already present on the benzene ring determines the position of the second substitution.

Let's look first at how substituents can affect reactivity. Substituents can be classified into two groups: those that activate the aromatic ring for further electrophilic substitution and those that deactivate it. Rings that contain an activating substituent are more reactive than benzene, whereas those that contain a deactivating substituent are less reactive than benzene. Figure 5.9 shows some groups in both categories.

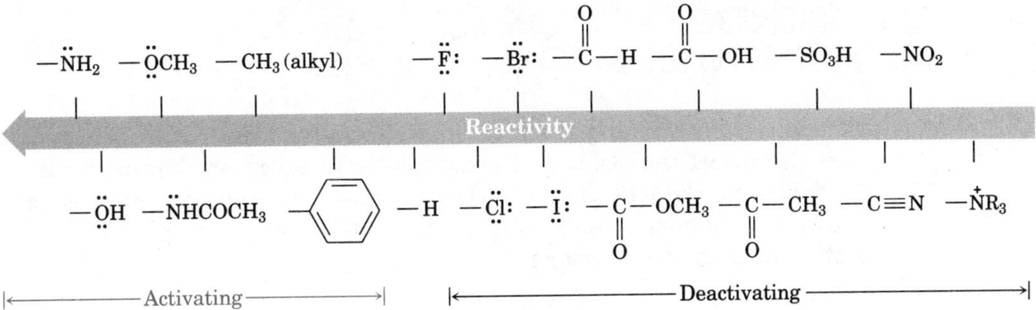

FIGURE 5.9 Activating and deactivating substituents for electrophilic aromatic substitution.

The common feature of all substituents within a category is that all activating groups are able to donate electrons to the ring, thereby stabilizing the carbocation intermediate and lowering the activation energy for its formation. All deactivating groups are able to withdraw electrons from the ring, thereby destabilizing the carbocation intermediate and raising the activation energy for its formation.

Y E+ E H Y +

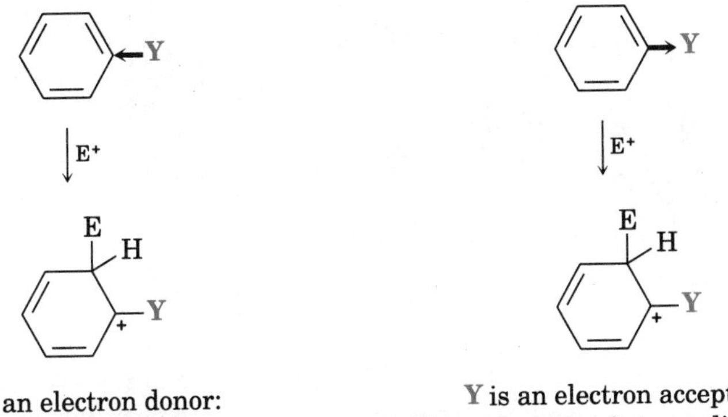

Y is an electron donor:
The carbocation intermediate is more stabilized, and the ring is activated.

Y is an electron acceptor:
The carbocation intermediate is less stabilized, and the ring is deactivated.

PRACTICE PROBLEM 5.3 Which would you expect to react faster in an electrophilic substitution reaction, chlorobenzene or ethylbenzene? Explain.

SOLUTION According to Figure 5.9, a chloro substituent is electron-withdrawing and deactivating, whereas an alkyl group is electron-donating and activating. Thus, ethylbenzene is more reactive than chlorobenzene.

PROBLEM 5.13 Use Figure 5.9 to rank the compounds in each of the following groups in order of their reactivity to electrophilic substitution:
(a) Nitrobenzene, phenol (hydroxybenzene), toluene
(b) Phenol, benzene, chlorobenzene, benzoic acid
(c) Benzene, bromobenzene, benzaldehyde, aniline (aminobenzene)

5.10 ORIENTATION OF REACTIONS ON SUBSTITUTED AROMATIC RINGS

In addition to affecting the reactivity of an aromatic ring, substituents can also direct the position of electrophilic substitution. For example, a hydroxyl substituent shows a strong ortho- and para-directing effect. Nitration of phenol yields *o*-nitrophenol (50%) and *p*-nitrophenol (50%), but none of the meta isomer:

HNO_3, H_2SO_4, 25°C

Phenol → **o-Nitrophenol (50%)** + **m-Nitrophenol (0%)** + **p-Nitrophenol (50%)**

On the other hand, a cyano substituent shows a strong meta-directing effect. Nitration of benzonitrile (cyanobenzene) yields 81% *m*-nitrobenzonitrile, along with only 17% of the ortho isomer and 2% of the para isomer:

HNO_3, H_2SO_4, 25°C

Benzonitrile → **o-Nitrobenzonitrile (17%)** + **m-Nitrobenzonitrile (81%)** + **p-Nitrobenzonitrile (2%)**

Figure 5.10 shows some of the groups in ortho, para-directing and in meta-directing categories. Note that ortho, para directors can be either activating or deactivating, whereas all meta directors are deactivating.

Ortho, Para Directors

Let's look at the nitration of phenol as an example of how ortho, para-directing substituents work. In the first step, attack on the nitronium ion electrophile (NO_2^+) can occur either ortho, meta, or para to the hydroxyl group, giving the carbocation intermediates shown in Figure 5.11. The

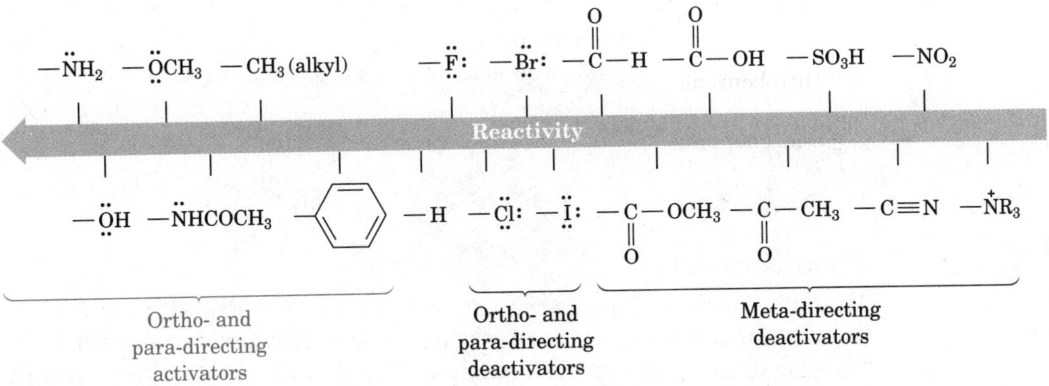

$-\ddot{N}H_2$ $-\ddot{O}CH_3$ $-CH_3$ (alkyl) $-\ddot{F}:$ $-\ddot{Br}:$ $-C(=O)-H$ $-C(=O)-OH$ $-SO_3H$ $-NO_2$

Reactivity

$-\ddot{O}H$ $-\ddot{N}HCOCH_3$ $-C_6H_5$ $-H$ $-\ddot{Cl}:$ $-\ddot{I}:$ $-C(=O)-OCH_3$ $-C(=O)-CH_3$ $-C\equiv N$ $-\overset{+}{N}R_3$

Ortho- and para-directing activators

Ortho- and para-directing deactivators

Meta-directing deactivators

FIGURE 5.10 Classification of directing effects for various substituents.

ortho and para intermediates are more stable than the meta intermediate because they have resonance forms that allow the positive charge to be stabilized by the substituent oxygen atom. Since the ortho and para intermediates are more stable than the meta intermediate, they are formed faster, and we say that the hydroxyl group is an ortho, para director.

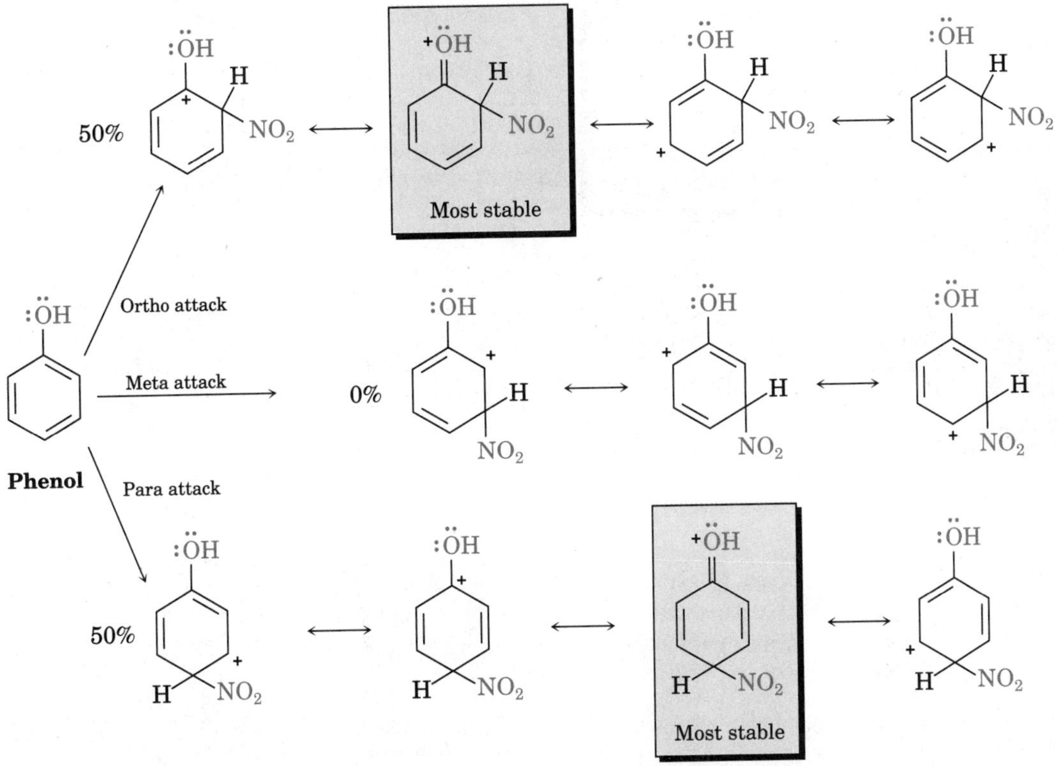

:ÖH H NO2 50% ⟷ +ÖH H NO2 Most stable ⟷ :ÖH H NO2 ⟷ :ÖH H NO2

:ÖH Phenol Ortho attack Meta attack Para attack

0% :ÖH H NO2 ⟷ :ÖH H NO2 ⟷ :ÖH H NO2

50% :ÖH H NO2 ⟷ :ÖH H NO2 ⟷ +ÖH H NO2 Most stable ⟷ :ÖH H NO2

FIGURE 5.11 Intermediates in the nitration of phenol. The ortho and para intermediates are more stable than the meta intermediate because of additional resonance forms involving the oxygen atom.

In general, any substituent that has a lone pair of electrons on the atom directly bonded to the aromatic ring allows a stabilizing resonance interaction to occur, and thus acts as an ortho, para director:

Ortho, para directors: ÖH, N̈H_2, C̈l:, B̈r:

Meta Directors

The influence of meta-directing substituents can be explained using the same kinds of arguments used for ortho, para directors. Look at the chlorination of benzaldehyde, for example (Figure 5.12). Of the three possible carbocation intermediates, those produced by reaction at ortho and para positions are least stable. In both cases, the unfavorable resonance forms indicated in Figure 5.12 place the positive charge directly on the carbon that bears the aldehyde group, where it is disfavored by a repulsive interaction with the positively polarized carbon atom of the C=O group. Hence, the meta intermediate is most favored and is formed faster than the ortho and para intermediates.

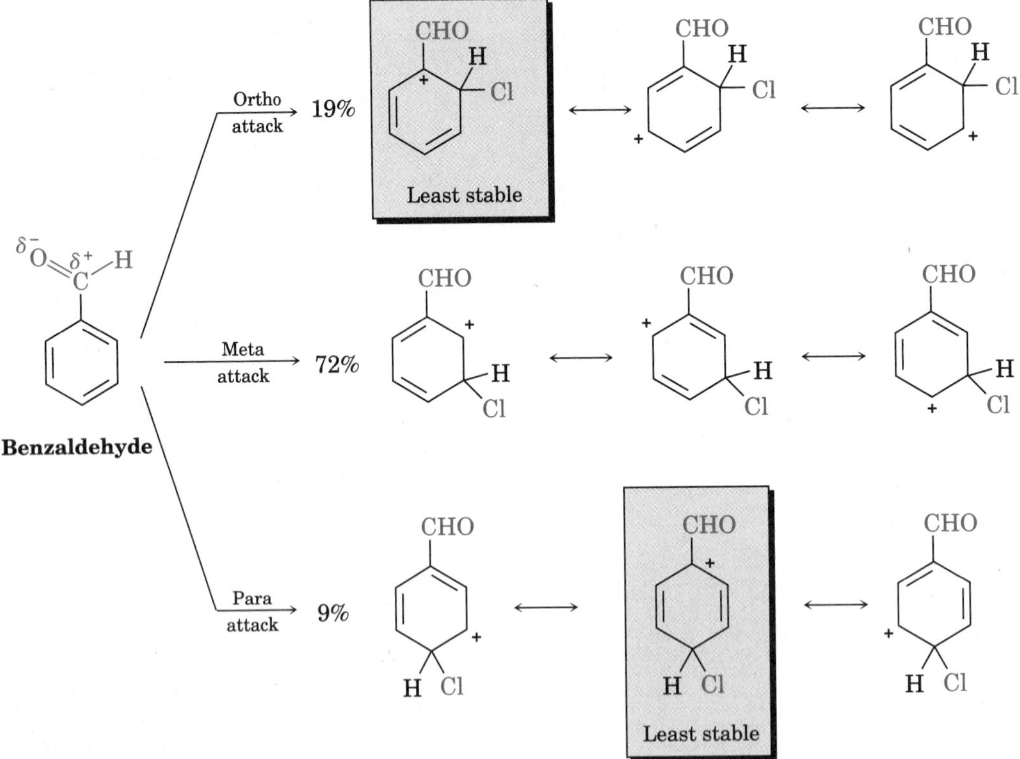

FIGURE 5.12 Intermediates in the chlorination of benzaldehyde. The ortho and para intermediates are less stable than the meta intermediate because they have unfavorable resonance forms.

In general, any substituent that has directly attached to the ring an atom that is polarized δ^+ allows a destabilizing resonance interaction to occur at ortho and para positions, and thus acts as a meta director:

Meta directors

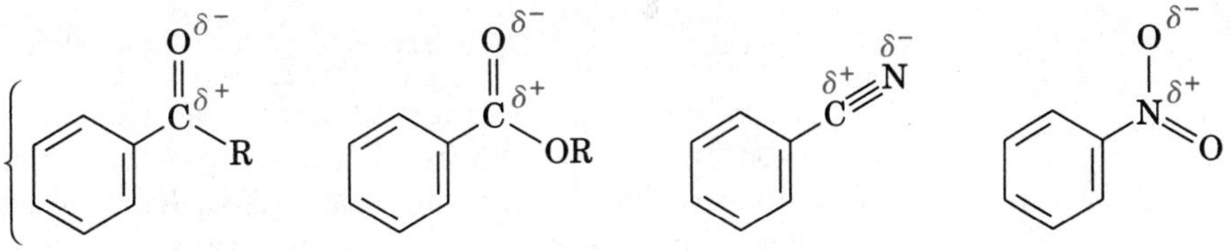

PRACTICE PROBLEM 5.4 What product(s) would you expect from monobromination of aniline, $C_6H_5NH_2$?

SOLUTION Figure 5.10 indicates that an amino group, $-NH_2$, is ortho- and para-directing. We therefore expect to obtain a mixture of *o*-bromoaniline and *p*-bromoaniline.

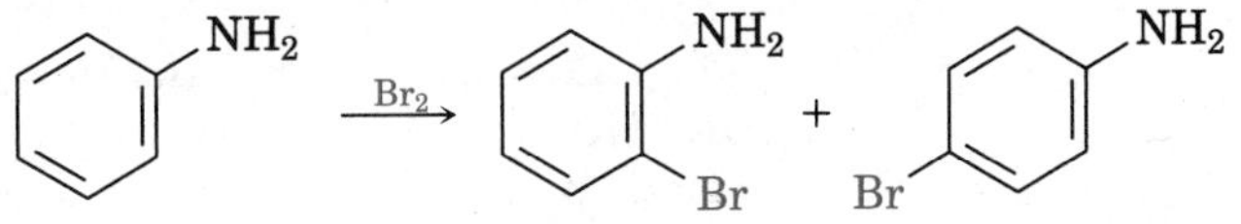

Aniline **o-Bromoaniline** **p-Bromoaniline**

PROBLEM 5.14 What product(s) would you expect from mononitration of these compounds?
(a) Nitrobenzene **(b)** Bromobenzene **(c)** Toluene **(d)** Benzoic acid
(e) *p*-Xylene

PROBLEM 5.15 Draw resonance structures of the potential carbocation intermediates to show how a methoxyl group ($-OCH_3$) directs bromination toward ortho and para positions.

PROBLEM 5.16 Draw resonance structures of the potential carbocation intermediates to show how an acetyl group, $CH_3C(=O)-$, directs bromination toward the meta position.

5.11 OXIDATION AND REDUCTION OF AROMATIC COMPOUNDS

The benzene ring, despite its unsaturation, is normally inert to strong oxidizing agents such as potassium permanganate. (Recall from Section 4.7 that this reagent cleaves alkene carbon–carbon double bonds.) Alkyl groups attached to the aromatic ring are readily attacked by these reagents, however, and are converted into carboxyl groups (–COOH). For example, butylbenzene is oxidized by $KMnO_4$ in high yield to give benzoic acid. The mechanism of this side-chain oxidation reaction isn't fully understood but probably involves attack on the side-chain C–H bonds at the position next to the aromatic ring (the **benzylic position**) to give radical intermediates.

BENZYLIC POSITION
Position next to an aromatic ring

$$\text{Butylbenzene } (C_6H_5CH_2CH_2CH_2CH_3) \xrightarrow[H_2O]{KMnO_4} \text{Benzoic acid } (C_6H_5COOH)$$

Butylbenzene **Benzoic acid (85%)**

As well as being inert to oxidation, aromatic rings are also inert to reduction under standard alkene hydrogenation conditions. Only if high temperatures and pressures are used does reduction of aromatic rings occur. For example, *o*-dimethylbenzene (*o*-xylene) gives 1,2-dimethylcyclohexane if reduced at high pressure:

$$\text{o-Xylene} \xrightarrow[\text{2000 psi, 25°C}]{H_2\text{, Pt; ethanol}} \text{1,2-Dimethylcyclohexane}$$

***o*-Xylene** **1,2-Dimethylcyclohexane (100%)**

PROBLEM 5.17 What aromatic products do you expect to obtain from the $KMnO_4$ oxidation of these substances?

(a) *m*-Chloroethylbenzene **(b)** 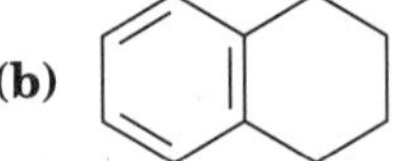 **Tetralin**

5.12 POLYCYCLIC AROMATIC HYDROCARBONS

POLYCYCLIC AROMATIC
Molecule that has two or more benzene rings fused together

The idea of aromaticity—the unusual chemical stability that arises in cyclic conjugated molecules like benzene—can be extended beyond simple monocyclic compounds to include **polycyclic aromatic compounds**. Naphthalene, with two benzene-like rings fused together, and anthracene, with three fused rings, are two of the simplest polycyclic aromatic molecules.

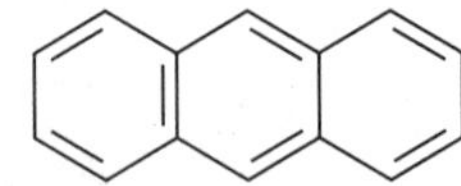

Naphthalene **Anthracene**

Naphthalene and other polycyclic aromatic hydrocarbons show many of the chemical properties we associate with aromaticity. Both, for example, react with electrophilic reagents such as Br_2 to give substitution products rather than double-bond addition products.

$$\text{Naphthalene} \xrightarrow[\Delta]{Br_2\text{, Fe}} \text{1-Bromonaphthalene} + HBr$$

Naphthalene **1-Bromonaphthalene (75%)**

We'll see in Chapter 12 that nitrogen-containing compounds like pyridine and pyrrole are also aromatic, even though they don't contain benzene rings.

Pyridine **Pyrrole**

PROBLEM 5.18 There are three resonance structures of naphthalene, of which only one is shown. Draw the other two.

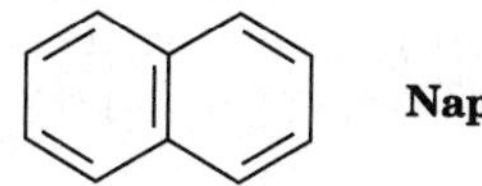

Naphthalene

5.13 ORGANIC SYNTHESIS

The laboratory synthesis of organic molecules from simple precursors is carried out for many reasons. In the pharmaceutical industry, new organic molecules are often designed and synthesized for evaluation as medicines. In the chemical industry, syntheses are often undertaken to devise more economical routes to known compounds. In this book, too, we'll sometimes devise syntheses of complex molecules from simpler precursors, but our purpose is simply to learn organic chemistry. Attempts at syntheses make us approach chemical problems in a logical way, drawing on our knowledge and organizing that knowledge into workable plans.

The only trick to planning an organic synthesis is to *work backward*. Look at the final product and ask yourself, "What is the immediate precursor of that product?" Having found an immediate precursor, proceed backward again, one step at a time, until a suitable starting material is found. Let's try some examples.

PRACTICE PROBLEM 5.5 Synthesize *m*-chloronitrobenzene starting with benzene.

SOLUTION Ask, "What is an immediate precursor of *m*-chloronitrobenzene?"

? ⟶

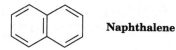

***m*-Chloronitrobenzene**

There are two substituents on the ring, a chloro group, which is ortho, para-directing, and a nitro group, which is meta-directing. We can't nitrate chlorobenzene, because the wrong isomers (*o*- and *p*-chloronitrobenzenes) would result, but chlorination of nitrobenzene should give the desired product.

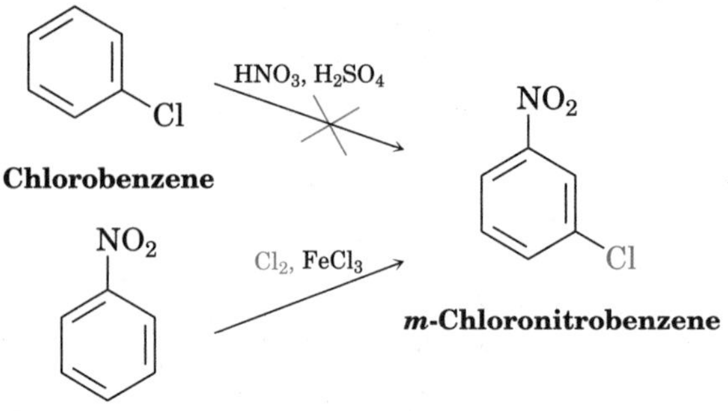

Chlorobenzene —(HNO_3, H_2SO_4; ✕)→ *m*-Chloronitrobenzene

Nitrobenzene —(Cl_2, $FeCl_3$)→ *m*-Chloronitrobenzene

"What is an immediate precursor of nitrobenzene?" Benzene, which can be nitrated. Thus, in two steps, we've solved the problem.

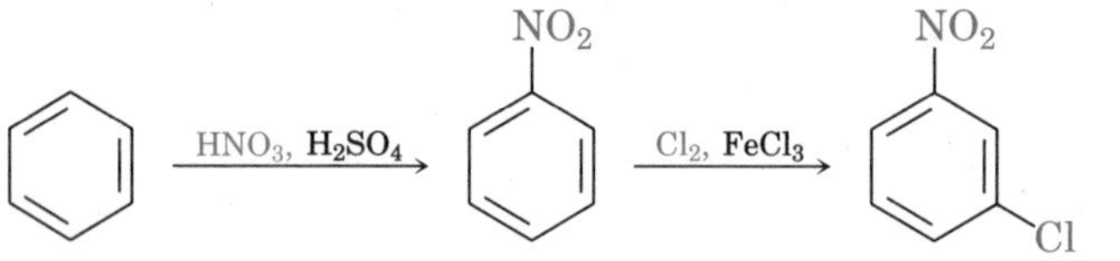

Benzene —(HNO_3, H_2SO_4)→ Nitrobenzene —(Cl_2, $FeCl_3$)→ *m*-Chloronitrobenzene

PRACTICE PROBLEM 5.6 Synthesize *p*-bromobenzoic acid starting from benzene.

SOLUTION Ask, "What is an immediate precursor of *p*-bromobenzoic acid?"

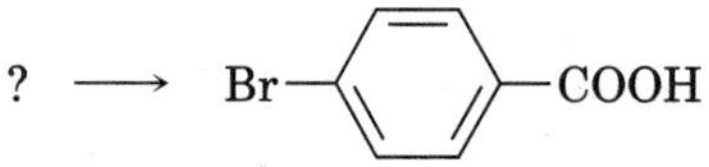

? ⟶ Br–C$_6$H$_4$–COOH

***p*-Bromobenzoic acid**

There are two substituents on the ring, a carboxyl group (–COOH), which is meta-directing, and a bromine, which is ortho, para-directing. We can't brominate benzoic acid, because the wrong isomer (*m*-bromobenzoic acid) would be formed. We know, however, that oxidation of alkylbenzene side chains yields benzoic acids. An immediate precursor of our target molecule might therefore be *p*-bromotoluene.

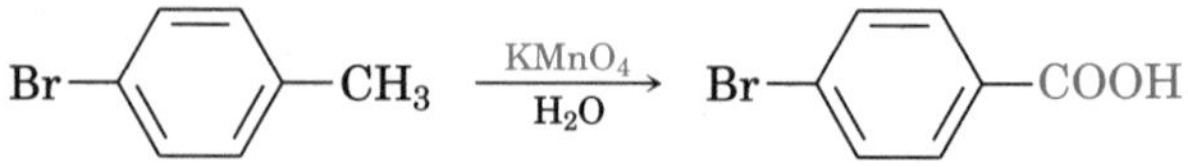

Br–C$_6$H$_4$–CH_3 —($KMnO_4$, H_2O)→ Br–C$_6$H$_4$–COOH

***p*-Bromotoluene** ***p*-Bromobenzoic acid**

"What is an immediate precursor of *p*-bromotoluene?" Perhaps toluene, because the methyl group would direct bromination to the ortho and para positions, and we could then separate isomers. Alternatively, bromobenzene might be an immediate precursor, because we could carry out a Friedel–Crafts alkylation and obtain the para product. Both methods are satisfactory.

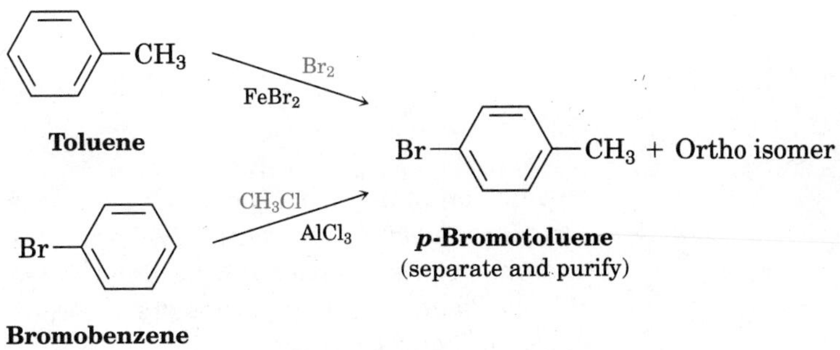

Toluene $\xrightarrow[\text{FeBr}_2]{\text{Br}_2}$ p-Bromotoluene + Ortho isomer

Bromobenzene $\xrightarrow[\text{AlCl}_3]{\text{CH}_3\text{Cl}}$ **p-Bromotoluene** (separate and purify)

"What is an immediate precursor of toluene?" Benzene, which can be methylated in a Friedel–Crafts reaction:

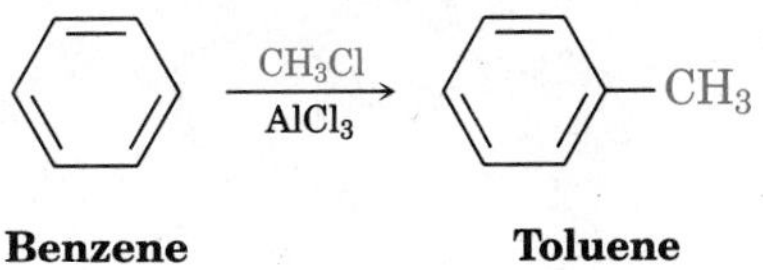

Benzene $\xrightarrow[\text{AlCl}_3]{\text{CH}_3\text{Cl}}$ **Toluene**

"Alternatively, what is an immediate precursor of bromobenzene?" Benzene, which can be brominated:

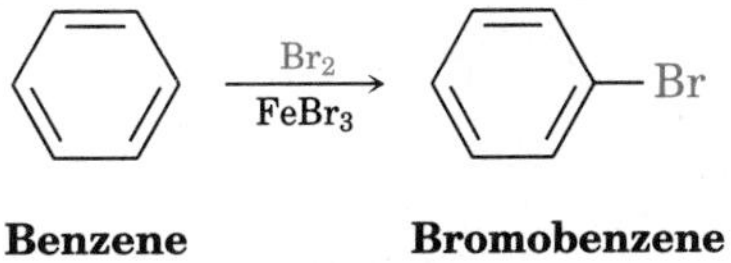

Benzene $\xrightarrow[\text{FeBr}_3]{\text{Br}_2}$ **Bromobenzene**

Our backward synthetic (*retrosynthetic*) analysis has provided two workable routes from benzene to *p*-bromobenzoic acid.

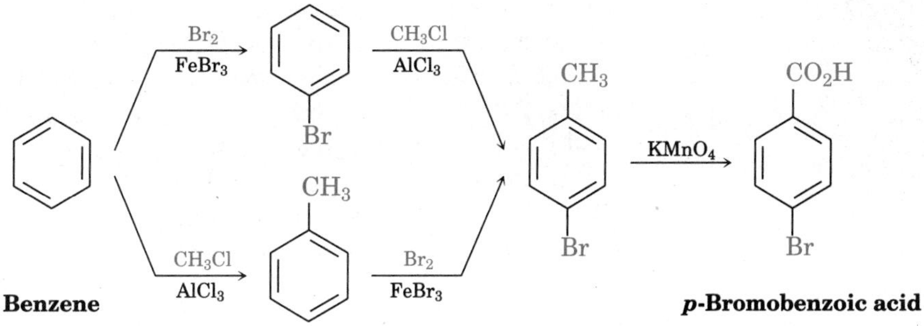

Benzene $\xrightarrow[\text{FeBr}_3]{\text{Br}_2}$ bromobenzene $\xrightarrow[\text{AlCl}_3]{\text{CH}_3\text{Cl}}$ p-bromotoluene

Benzene $\xrightarrow[\text{AlCl}_3]{\text{CH}_3\text{Cl}}$ toluene $\xrightarrow[\text{FeBr}_3]{\text{Br}_2}$ p-bromotoluene

p-bromotoluene $\xrightarrow{\text{KMnO}_4}$ **p-Bromobenzoic acid**

PROBLEM 5.19 Propose a synthesis of each of these substances, starting with benzene:
(a) *p*-Methylacetophenone **(b)** *p*-Chloronitrobenzene

PROBLEM 5.20 Synthesize these substances starting from benzene:
(a) *o*-Bromotoluene **(b)** 2-Bromo-1,4-dimethylbenzene

PROBLEM 5.21 How would you prepare *m*-chlorobenzoic acid from benzene?

INTERLUDE

POLYCYCLIC AROMATIC HYDROCARBONS AND CANCER

In addition to naphthalene and anthracene, there are a great many more complex *polycyclic aromatic hydrocarbons (PAH's)*. Benzo[a]pyrene, for example, contains five benzene rings joined together, and ordinary graphite (the "lead" in pencils) consists of essentially infinite, two-dimensional sheets of benzene rings.

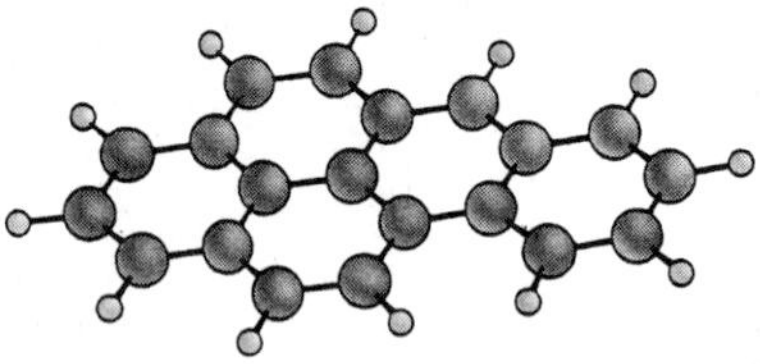

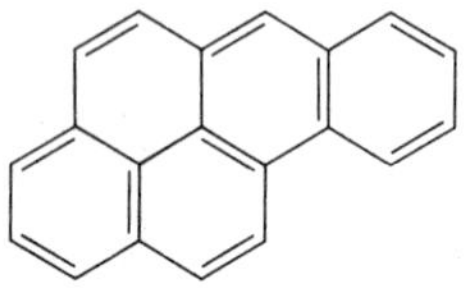

Benzo[a]pyrene

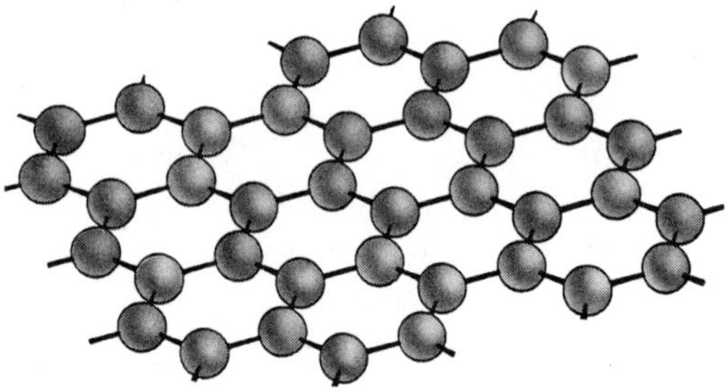

A graphite segment

Benzo[a]pyrene is a particularly important and well-studied PAH because it is one of the cancer-causing (*carcinogenic*) substances found in chimney soot and cigarette smoke. Exposure to a tiny amount is sufficient to induce a skin tumor in susceptible mice.

Recent studies have given a clear picture of how these PAH's cause tumors. After a PAH is absorbed by eating or inhaling, the body attempts to rid itself of the foreign substance by converting it into a water-soluble metabolite that can be excreted. In the case of benzo[a]pyrene, oxidation in the liver converts it into an oxygenated product called a *diol epoxide.* Unfortunately, this metabolite is able to bind to cellular DNA, causing mutations and cancers.

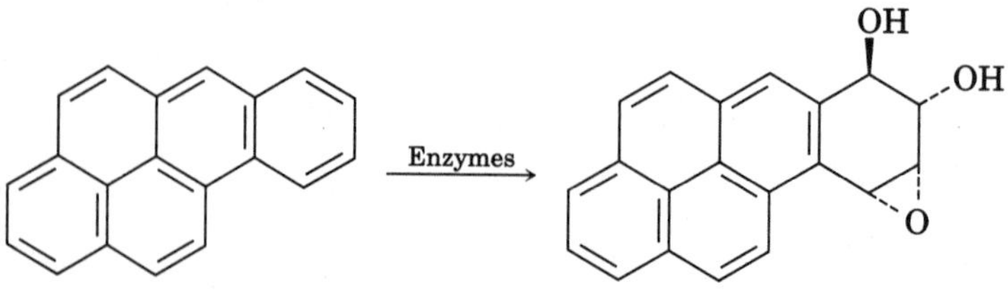

Enzymes →

OH OH O

Benzo[a]pyrene

A diol epoxide

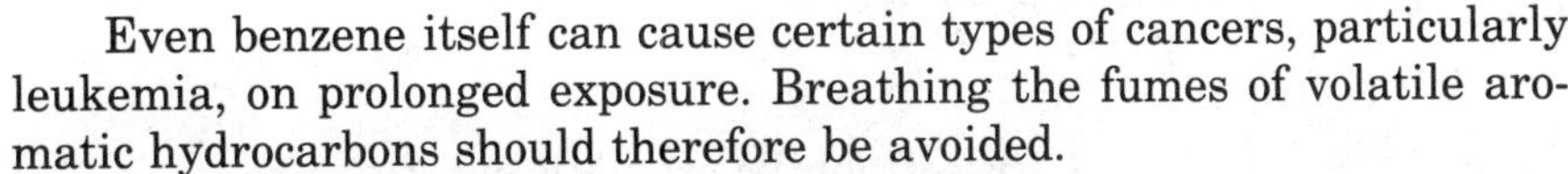

Even benzene itself can cause certain types of cancers, particularly leukemia, on prolonged exposure. Breathing the fumes of volatile aromatic hydrocarbons should therefore be avoided.

SUMMARY AND KEY WORDS

The word **aromatic** refers to the class of compounds structurally related to benzene. Aromatic compounds are named according to IUPAC rules, with disubstituted benzenes referred to as either **ortho** (1,2-disubstituted), **meta** (1,3-disubstituted), or **para** (1,4-disubstituted). Benzene is a resonance hybrid of two equivalent line-bond structures. Neither structure is correct by itself; the true structure of benzene is intermediate between the two:

The single most important reaction of aromatic compounds is **electrophilic aromatic substitution**. In this two-step polar reaction, the pi electrons of the aromatic ring first attack the electrophile to yield a resonance-stabilized carbocation intermediate. Loss of H^+ from this intermediate regenerates the stable aromatic ring and gives a substituted product. Bromination, chlorination, iodination, nitration, sulfonation, Friedel–Crafts alkylation, and Friedel–Crafts acylation can all be carried out with the proper choice of reagent. **Friedel–Crafts alkylation** is a particularly useful reaction for preparing a variety of alkylbenzenes but is limited because only alkyl halides can be used and strongly deactivated rings don't react.

Substituents on the benzene ring affect both the reactivity of the ring toward further substitution and the orientation of further substitution. We can classify substituents either as *activators* or *deactivators*, and either as *ortho, para directors* or as *meta directors*.

The side chains of alkylbenzenes have unique reactivity because of the neighboring aromatic ring. Thus, an alkyl group attached to the aromatic ring can be degraded to a carboxyl group (–COOH) by oxidation with aqueous $KMnO_4$. Also, aromatic rings can be reduced to cyclohexanes on catalytic hydrogenation at high pressure.

SUMMARY OF REACTIONS

1. Electrophilic aromatic substitution
   (a) Bromination (Section 5.6)

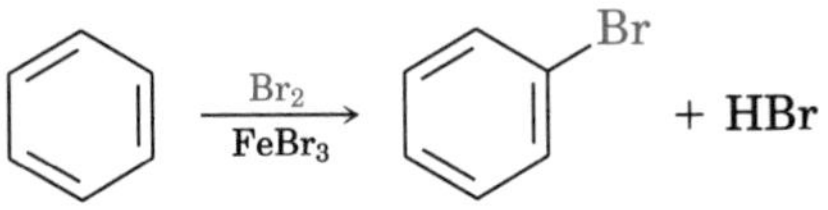

$$C_6H_6 \xrightarrow[FeBr_3]{Br_2} C_6H_5Br + HBr$$

   (b) Chlorination (Section 5.7)

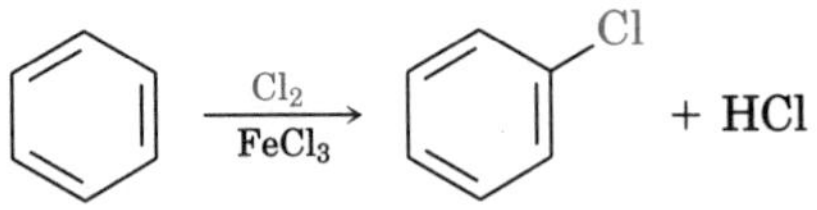

$$C_6H_6 \xrightarrow[FeCl_3]{Cl_2} C_6H_5Cl + HCl$$

   (c) Nitration (Section 5.7)

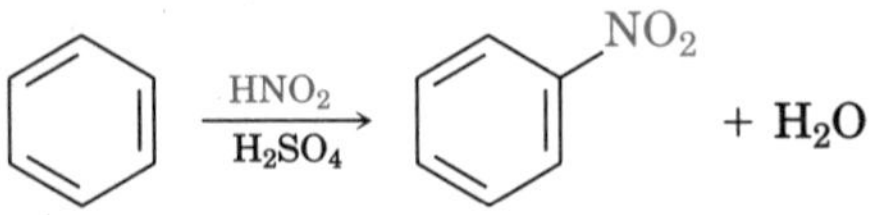

$$C_6H_6 \xrightarrow[H_2SO_4]{HNO_2} C_6H_5NO_2 + H_2O$$

   (d) Sulfonation (Section 5.7)

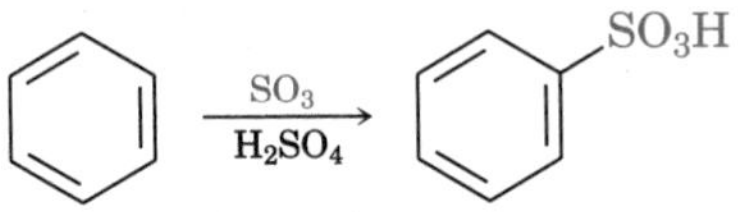

$$C_6H_6 \xrightarrow[H_2SO_4]{SO_3} C_6H_5SO_3H$$

   (e) Friedel–Crafts alkylation (Section 5.8)

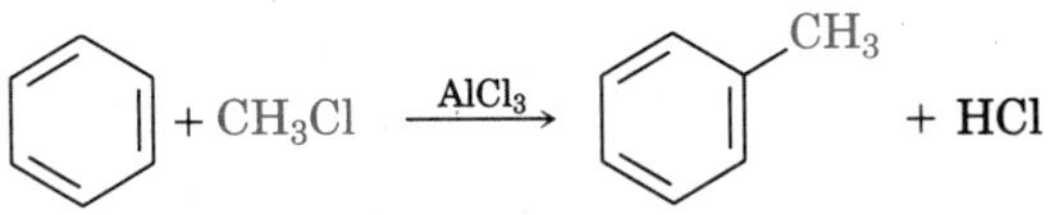

$$C_6H_6 + CH_3Cl \xrightarrow{AlCl_3} C_6H_5CH_3 + HCl$$

   (f) Friedel–Crafts acylation (Section 5.8)

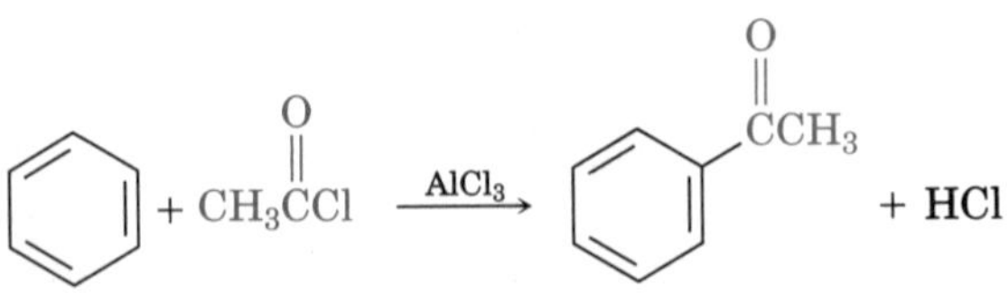

$$C_6H_6 + CH_3\overset{O}{\overset{\|}{C}}Cl \xrightarrow{AlCl_3} C_6H_5\overset{O}{\overset{\|}{C}}CH_3 + HCl$$

2. Oxidation of aromatic side chains (Section 5.11)

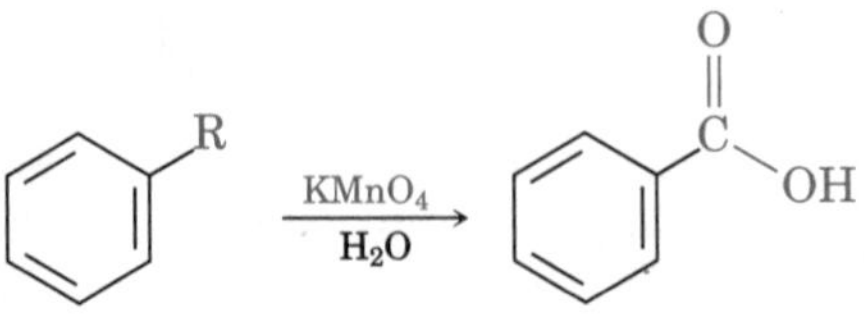

$$C_6H_5R \xrightarrow[H_2O]{KMnO_4} C_6H_5\overset{O}{\overset{\|}{C}}OH$$

3. Hydrogenation of aromatic rings (Section 5.11)

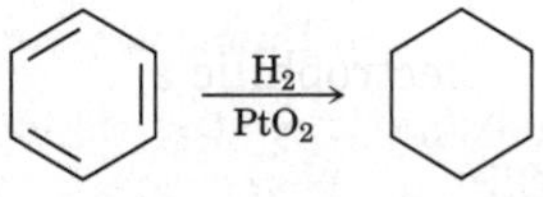

Benzene $\xrightarrow[PtO_2]{H_2}$ Cyclohexane

ADDITIONAL PROBLEMS

5.22 Give IUPAC names for these compounds:

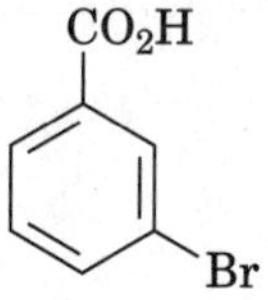

(a) $C_6H_5CH_2CH_2CH_2CH(CH_3)CH_3$

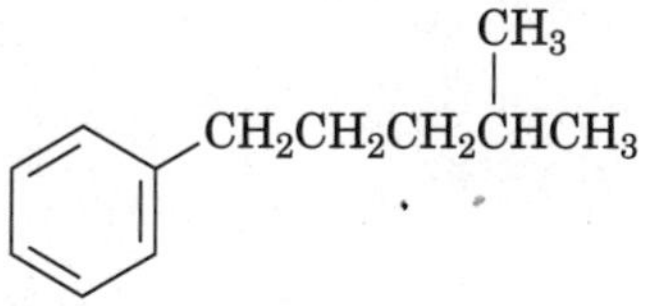

(b) Benzene ring with CO_2H and Br (meta)

(c) Benzene ring with Br, H_3C, CH_3 (1,3,5)

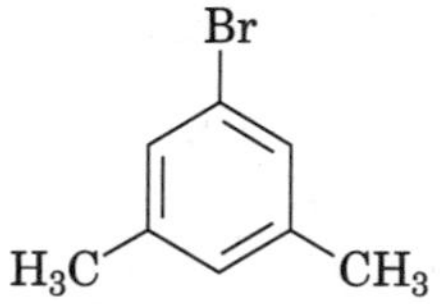

(d) Benzene ring with Br and $CH_2CH_2CH_3$ (ortho)

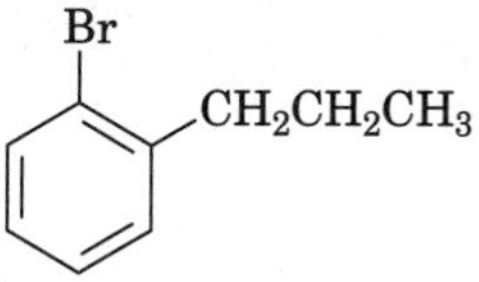

5.23 Draw structures corresponding to these names:
(a) *m*-Bromophenol (b) 1,3,5-Benzenetriol (c) *p*-Iodonitrobenzene
(d) 2,4,6-Trinitrotoluene (TNT) (e) *o*-Aminobenzoic acid (f) 3-Methyl-2-phenylhexane

5.24 Draw and name all aromatic compounds with the formula C_7H_7Cl.

5.25 Draw and name all isomeric:
(a) Dinitrobenzenes (b) Bromodimethylbenzenes

5.26 Propose structures for aromatic hydrocarbons meeting these descriptions:
(a) C_9H_{12}; can give only one product on aromatic bromination
(b) C_8H_{10}; can give three products on aromatic chlorination
(c) $C_{10}H_{14}$; can give two products on aromatic nitration

5.27 Formulate the reaction of benzene with 2-chloro-2-methylpropane in the presence of $AlCl_3$ catalyst to give *tert*-butylbenzene.

5.28 Predict the major product(s) of mononitration of these substances:
(a) Bromobenzene (b) Benzonitrile (cyanobenzene) (c) Benzoic acid
(d) Nitrobenzene (e) Phenol (f) Benzaldehyde

5.29 Which of the substances listed in Problem 5.28 react faster than benzene and which react slower?

5.30 Rank the compounds in each group according to their reactivity toward electrophilic substitution:
(a) Chlorobenzene, *o*-dichlorobenzene, benzene (b) *p*-Bromonitrobenzene, nitrobenzene, phenol
(c) Fluorobenzene, benzaldehyde, *o*-dimethylbenzene

5.31 Show the steps involved in the Friedel–Crafts reaction of benzene with CH_3Cl.

5.32 The orientation of electrophilic aromatic substitution on a disubstituted benzene ring is usually controlled by whichever of the two groups already on the ring is the more powerful activator. Name and draw the structure(s) of the major product(s) of electrophilic chlorination of these substances:
(a) *m*-Nitrophenol (b) *o*-Methylphenol (c) *p*-Chloronitrobenzene

5.33 Predict the major product(s) you would expect to obtain from sulfonation of these substances (see Problem 5.32):
(a) Bromobenzene **(b)** *m*-Bromophenol **(c)** *p*-Nitrotoluene

5.34 Rank the following aromatic compounds in the expected order of their reactivity toward Friedel–Crafts acylation. Which compounds are unreactive?
(a) Bromobenzene **(b)** Toluene **(c)** Anisole ($C_6H_5OCH_3$) **(d)** Nitrobenzene
(e) *p*-Bromotoluene

5.35 What is the structure of the compound with formula C_8H_9Br that gives *p*-bromobenzoic acid on oxidation with $KMnO_4$?

5.36 Draw the four resonance structures of anthracene.

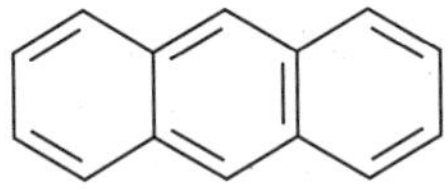

Anthracene

5.37 Draw the five resonance structures of phenanthrene.

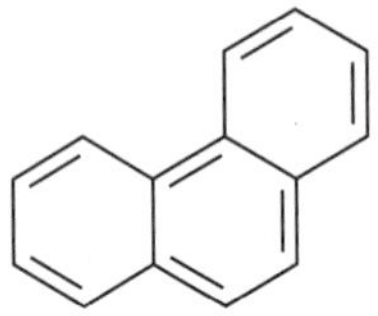

Phenanthrene

5.38 Suggest a reason for the observation that bromination of biphenyl occurs at ortho and para positions rather than at meta. Use resonance structures of the carbocation intermediates to explain your answer.

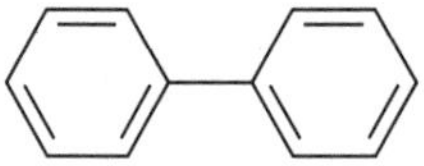

Biphenyl

5.39 In light of your answer to Problem 5.38, at what position and on which ring would you expect nitration of 4-bromobiphenyl to occur?

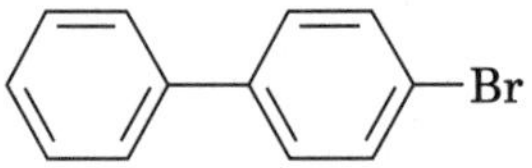

Br **4-Bromobiphenyl**

5.40 Starting with benzene, how would you synthesize the following substances? Assume that you can separate ortho and para isomers if necessary.
(a) *m*-Bromobenzenesulfonic acid **(b)** *o*-Chlorobenzenesulfonic acid **(c)** *p*-Chlorotoluene

5.41 Starting from any aromatic hydrocarbon necessary, how would you synthesize the following substances? Ortho and para isomers can be separated if necessary.
(a) *o*-Nitrobenzoic acid **(b)** *p-tert*-Butylbenzoic acid

5.42 Explain by drawing resonance structures of the intermediate carbocations why naphthalene undergoes electrophilic aromatic substitution at C1 rather than at C2.

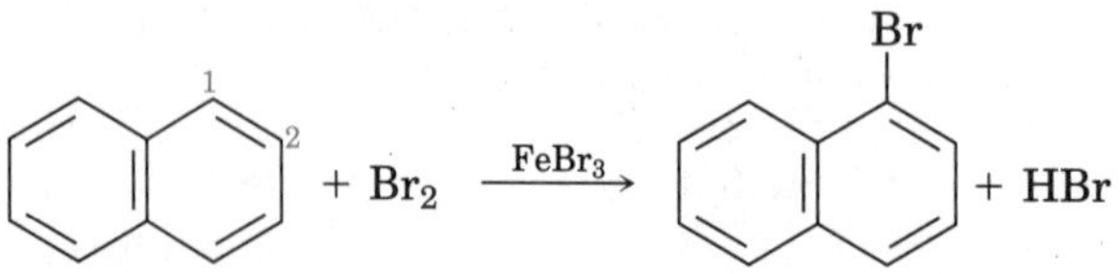

1 2 + Br_2 $\xrightarrow{FeBr_3}$ Br + HBr

5.43 We said in Section 4.11 that allylic carbocations are stabilized by resonance. Draw resonance structures to account for a similar stabilization of benzylic carbocations.

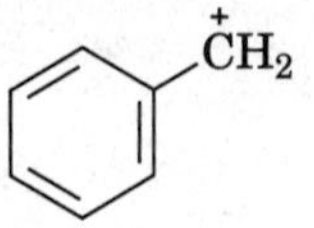

$\overset{+}{C}H_2$

A benzylic carbocation

5.44 Addition of HBr to 1-phenylpropene yields (1-bromopropyl)benzene as the exclusive product. Propose a mechanism for the reaction, and explain why none of the other regioisomer is produced (see Problem 5.43).

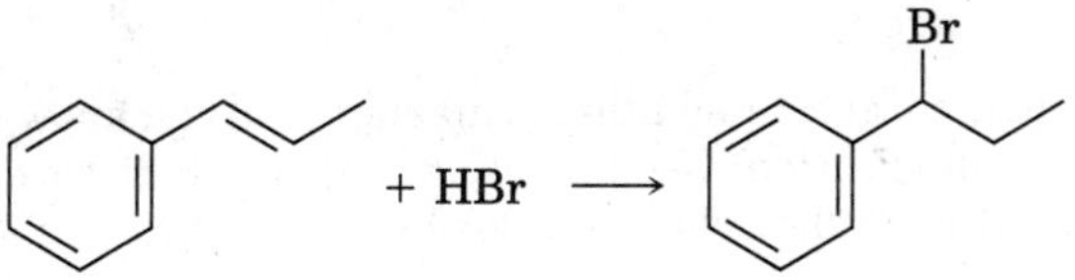

+ HBr ⟶ Br

5.45 The following syntheses have flaws in them. What is wrong with each?

(a) CH_3 $\xrightarrow[\text{2. KMnO}_4]{\text{1. Cl}_2\text{, FeCl}_3}$ COOH, Cl

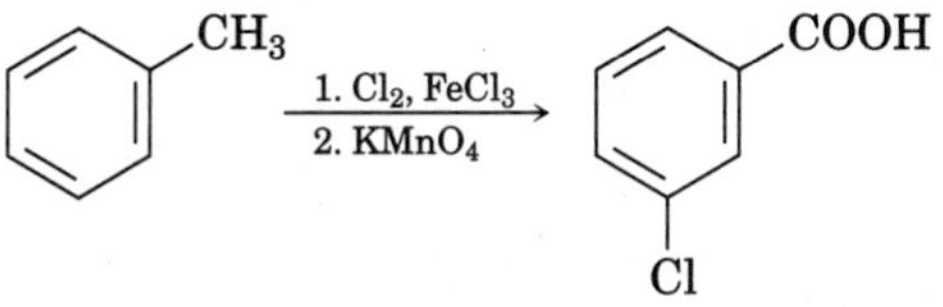

(b) Cl $\xrightarrow[\text{2. KMnO}_4\text{, H}_2\text{O}]{\text{1. (CH}_3\text{)}_3\text{CCl, AlCl}_3}$ Cl, COOH

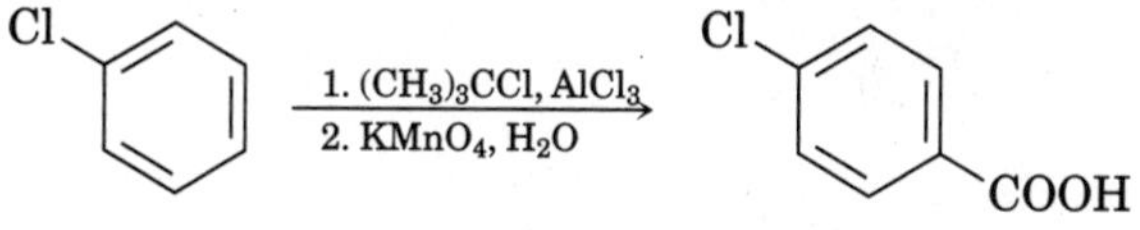

5.46 Pyridine is a cyclic nitrogen-containing organic compound that shows many of the properties associated with aromaticity. For example, pyridine undergoes electrophilic substitution reactions. Draw an orbital picture of pyridine and account for its aromatic properties.

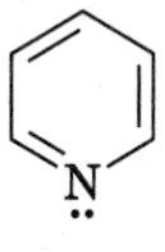

N

Pyridine

5.47 Would you expect the trimethylammonium group to be an activating or deactivating substituent? Explain.

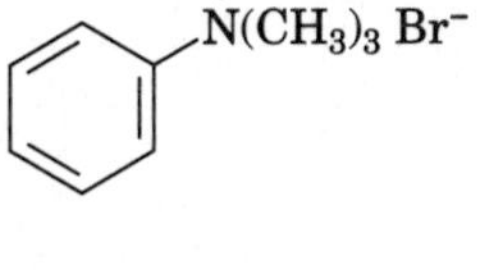

$\overset{+}{N}(CH_3)_3\ Br^-$

Phenyltrimethylammonium bromide

5.48 Starting with toluene, how would you synthesize the three nitrobenzoic acids?

5.49 Carbocations generated by reaction of an alkene with a strong acid catalyst can react with aromatic rings in a Friedel–Crafts reaction. Propose a mechanism to account for the industrial synthesis of the food preservative BHT from *p*-cresol and 2-methylpropene:

$$p\text{-Cresol (4-}CH_3C_6H_4OH) + CH_3C(CH_3){=}CH_2 \xrightarrow[\text{catalyst}]{H^+} \text{BHT (2,6-}(CH_3)_3C\text{)}_2\text{-4-}CH_3C_6H_2OH$$

p-Cresol **BHT**

5.50 You know the mechanism of HBr addition to alkenes, and you know the effects of various substituent groups on aromatic substitution. Use this knowledge to predict which of the following two alkenes reacts faster with HBr. Explain your answer by drawing resonance structures of the carbocation intermediates.

p-CH_3O–C_6H_4–$CH{=}CH_2$ and p-O_2N–C_6H_4–$CH{=}CH_2$

6 CHAPTER

STEREOCHEMISTRY

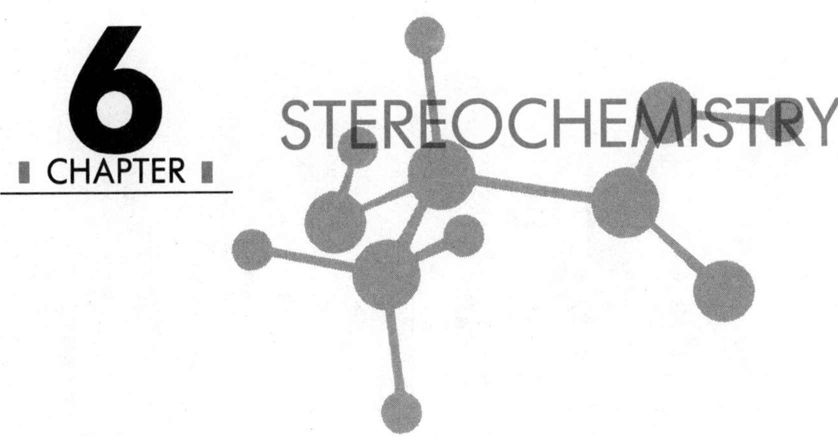

Up to this point, we've been concerned only with the general nature of chemical reactions and with the specific chemistry of hydrocarbon functional groups. Although we took a brief look at constitutional isomers of alkanes in Section 2.2 and cis–trans stereoisomers of cycloalkanes in Section 2.8, we've given little thought to any chemical consequences arising from the spatial arrangement of atoms in molecules. It's now time to look more deeply into these consequences. **Stereochemistry** is the branch of chemistry concerned with the three-dimensional nature of molecules.

STEREOCHEMISTRY Branch of chemistry concerned with three-dimensional consequences of structure

6.1 STEREOCHEMISTRY AND THE TETRAHEDRAL CARBON

Are you right-handed or left-handed? Though most of us don't often think about it, handedness plays a surprisingly large role in our daily activities. Musical instruments like oboes and clarinets have a handedness to them, the last available softball glove always fits the wrong hand, and left-handed people write in a "funny" way. The fundamental reason for these difficulties is that our hands aren't identical—they're *mirror images*. When you hold your *right* hand up to a mirror, the reflection looks like a *left* hand. Try it.

Handedness is also important in organic chemistry and is crucial in biochemistry, where carbohydrates, amino acids, nucleic acids, and many other naturally occurring molecules are handed. To see how molecular handedness arises, look at the molecules shown in Figure 6.1. On the left of Figure 6.1 are three molecules, and on the right are their images reflected in a mirror. The CH_3X and CH_2XY molecules are identical to their mirror images and thus are not handed. If you make molecular models of each molecule and of its mirror image, you find that they are identical and that you can superimpose one on the other. Unlike the CH_3X

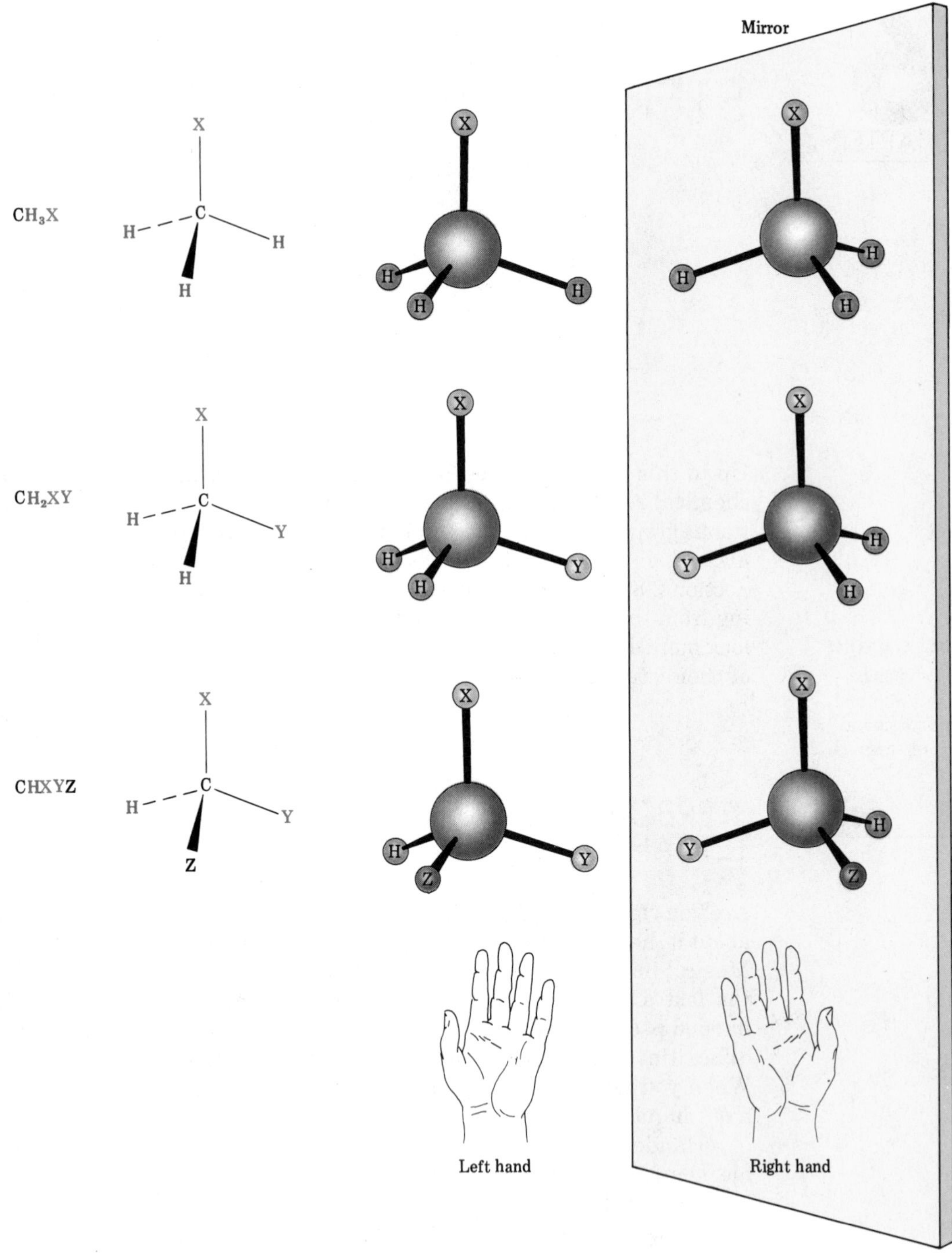

FIGURE 6.1 Three tetrahedral carbon atoms and their mirror images. Molecules of the type CH_3X and CH_2XY are identical to their mirror images, but a molecule of the type CHXYZ is not. A CHXYZ molecule is related to its mirror image in the same way that a right hand is related to a left hand.

and CH_2XY molecules, though, the CHXYZ molecule is *not* identical to its mirror image. You can't superimpose a model of the molecule on a model of its mirror image for the same reason that you can't superimpose a left hand on a right hand: They simply aren't the same. You might superimpose *two* of the substituents, X and Y for example, but H and Z would be reversed. If the H and Z substituents were superimposed, X and Y would be reversed.

ENANTIOMERS
Stereoisomers that have a mirror-image relationship

A molecule that is not identical to its mirror image is a special kind of stereoisomer called an **enantiomer** (e-**nan**-tee-o-mer; Greek *enantio*, "opposite"). Enantiomers are related to each other as a right hand is related to a left hand and result whenever a tetrahedral carbon is bonded to four different substituents (one need not be H). For example, lactic acid (2-hydroxypropanoic acid) exists as a pair of enantiomers because there are four different groups ($-H, -OH, -CH_3, -COOH$) attached to the central carbon atom:

H
CH₃—C—COOH
OH

H
X—C—Z
Y

Lactic acid: a molecule of general formula CHXYZ

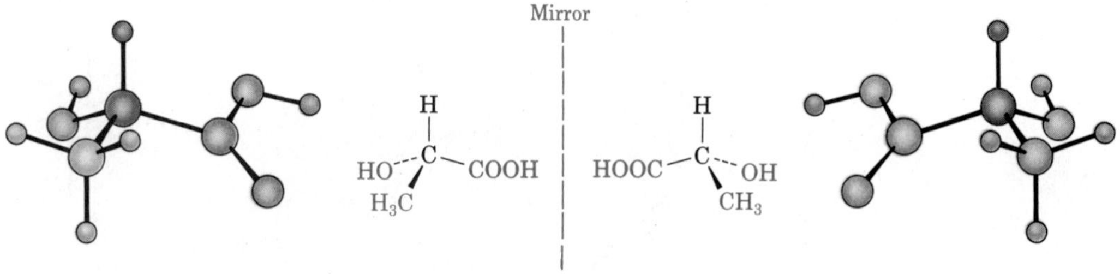

Mirror

H
HO—C—COOH
H₃C

H
HOOC—C—OH
CH₃

No matter how hard you try, you can't superimpose a molecule of "right-handed" lactic acid on top of a molecule of "left-handed" lactic acid; the two molecules aren't identical, as shown in Figure 6.2.

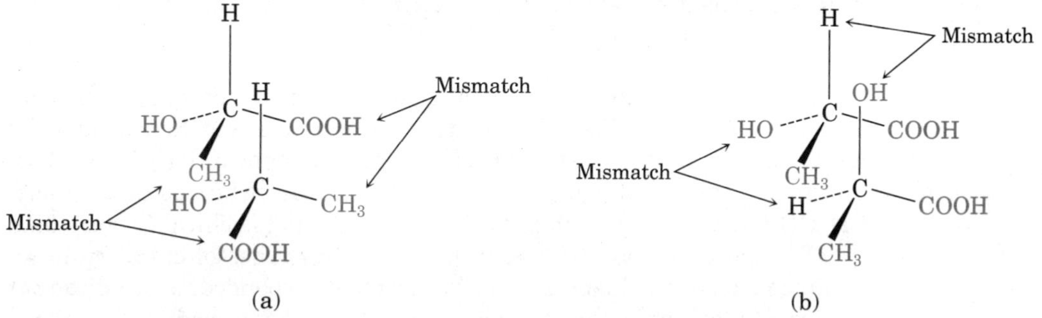

H, H, HO, C, COOH, CH₃, HO, C, CH₃, COOH, Mismatch, Mismatch

(a)

H, Mismatch, OH, HO, C, COOH, Mismatch, CH₃, H, C, COOH, CH₃

(b)

FIGURE 6.2 Attempts at superimposing the mirror-image forms of lactic acid: (a) When the $-H$ and $-OH$ substituents match up, the $-COOH$ and $-CH_3$ substituents don't; (b) when $-COOH$ and $-CH_3$ match up, $-H$ and $-OH$ don't. Regardless of how the molecules are oriented, they aren't identical.

6.2 THE REASON FOR HANDEDNESS IN MOLECULES: CHIRALITY

CHIRAL
Having handedness

Compounds that are not identical to their mirror images and thus exist in two enantiomeric forms are said to be **chiral** (**ky**-ral; Greek *cheir*, "hand"). You can't take a chiral molecule and its mirror image (enantiomer) and place one on top of the other so that all atoms coincide.

PLANE OF SYMMETRY
Imaginary plane that bisects an object or molecule so that one half is a mirror image of the other half

How can you predict whether a given molecule is or is not chiral? *A compound is not chiral if it contains a plane of symmetry.* A **plane of symmetry** is an imaginary plane that cuts through an object (or molecule) so that one half of the object is an exact mirror image of the other half. A flask, for example, has a plane of symmetry. If you were to cut the flask in half, one half would be an exact mirror image of the other half. A hand, however, does not have a plane of symmetry. One "half" of a hand is not a mirror image of the other "half" (Figure 6.3).

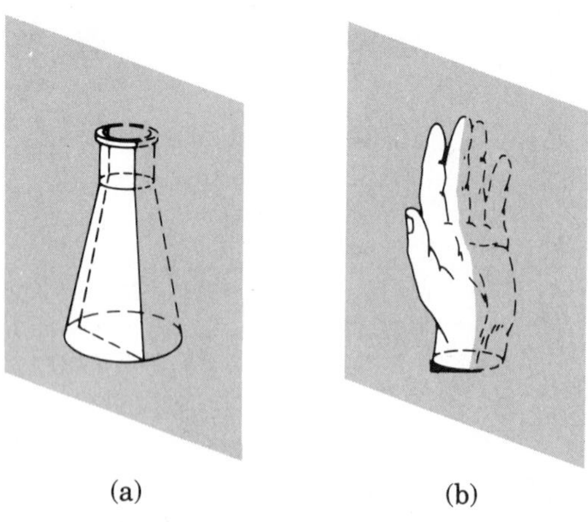

(a) (b)

FIGURE 6.3 The meaning of *symmetry plane*. An object like the flask (a) has a plane of symmetry passing through it, making the right and left halves mirror images. An object like a hand (b) has no symmetry plane; the right "half" of a hand is not a mirror image of the left "half."

ACHIRAL
Not having handedness

A molecule that has a plane of symmetry in any of its possible conformations *must* be identical to its mirror image and hence must be nonchiral, or **achiral** (a-**ky**-ral). Thus, propanoic acid, CH_3CH_2COOH, contains a plane of symmetry and is achiral, whereas lactic acid, $CH_3CH(OH)COOH$, has no plane of symmetry and is chiral (Figure 6.4).

STEREOGENIC CENTER
Atom in a molecule that is a local center of chirality

The most common (although not the only) cause of chirality in an organic molecule is the presence of a carbon atom bonded to four different groups—for example, the central carbon atom in lactic acid. Such carbons are referred to as *chiral centers*, or **stereogenic centers**. Note that chirality itself is a property of the entire molecule, whereas the cause of chirality is the presence of a certain kind of atom within the molecule.

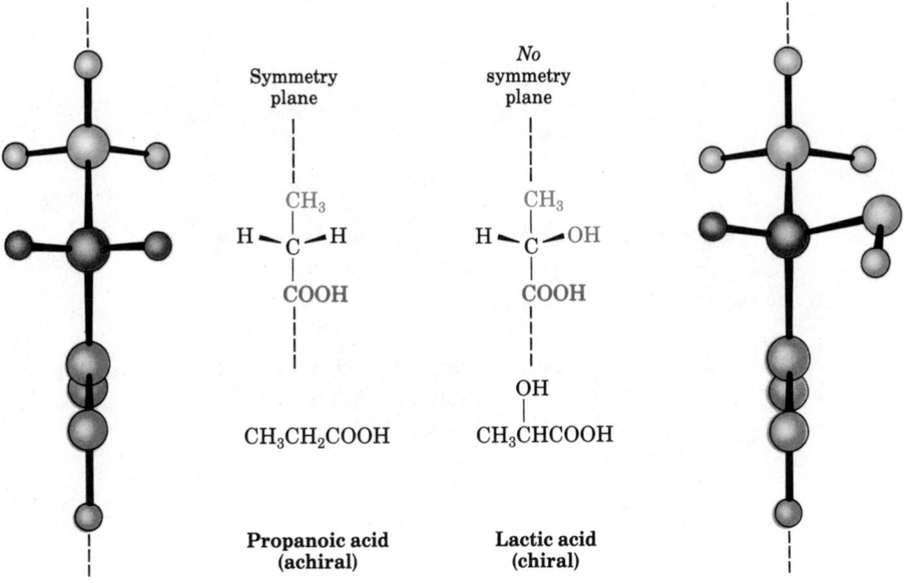

FIGURE 6.4 The achiral propanoic acid molecule versus the chiral lactic acid molecule. Propanoic acid has a plane of symmetry that makes one side of the molecule a mirror image of the other side. Lactic acid, however, has no such symmetry plane.

Detecting stereogenic centers in a complex molecule takes practice, because it's not always immediately apparent that four different groups are bonded to a given carbon. The differences don't necessarily appear right next to the stereogenic center. For example, 5-bromodecane is a chiral molecule because four different groups are bonded to C5, the stereogenic center (marked by an asterisk):

Br
|
$CH_3CH_2CH_2CH_2CH_2\overset{*}{C}CH_2CH_2CH_2CH_3$
|
H

5-Bromodecane (chiral)

SUBSTITUENTS ON CARBON 5

—H
—Br
—$CH_2CH_2CH_2CH_3$ (butyl)
—$CH_2CH_2CH_2CH_2CH_3$ (pentyl)

A butyl substituent is very *similar* to a pentyl substituent, but it isn't identical. The difference isn't apparent until four carbons away from the stereogenic center, but there's still a difference.

In the examples of chiral molecules that follow, check for yourself that the labeled centers are indeed chiral. (When checking for stereogenic centers, it's helpful to realize that $-CH_2-$, $-CH_3$, C=C, −C≡C−, and C=O carbons *can't* be chiral, because they have at least two identical bonds.)

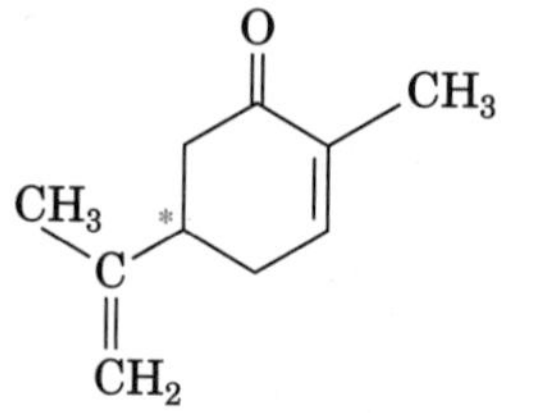

Carvone (spearmint oil) 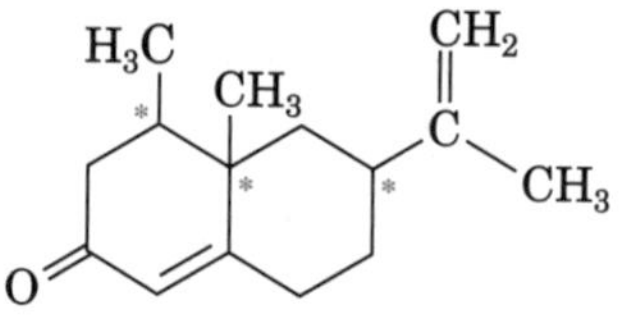 **Nootkatone (grapefruit oil)**

PRACTICE PROBLEM 6.1 Draw the structure of a chiral alcohol.

SOLUTION An alcohol is a compound that contains the –OH functional group. To make an alcohol chiral, we need to have four different groups bonded to a single carbon atom, say –H, –OH, $-CH_3$, and $-CH_2CH_3$:

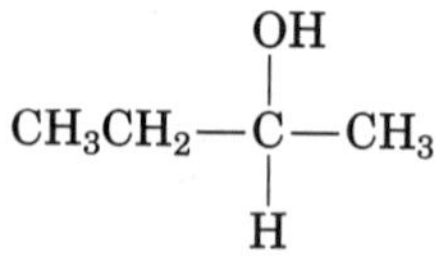

$$CH_3CH_2-\underset{H}{\overset{OH}{C}}-CH_3$$ **2-Butanol**

PRACTICE PROBLEM 6.2 Is 3-methylhexane chiral?

SOLUTION Draw the structure of 3-methylhexane and cross out all the $-CH_2-$ and $-CH_3$ carbons because they can't be chiral. Then look closely at any carbon that remains to see if it's bonded to four different groups. Since C3 is bonded to –H, $-CH_3$, $-CH_2CH_3$, and $-CH_2CH_2CH_3$, the molecule is chiral.

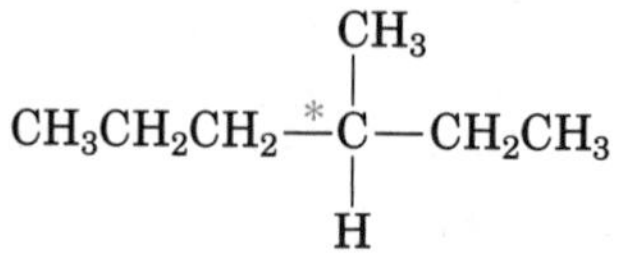

$$CH_3CH_2CH_2-\underset{H}{\overset{CH_3}{C^*}}-CH_2CH_3$$ **3-Methylhexane (chiral)**

PRACTICE PROBLEM 6.3 Is 2-methylcyclohexanone chiral?

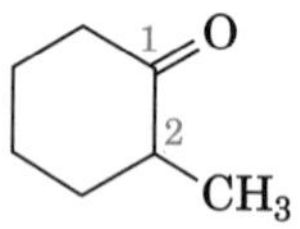

2-Methylcyclohexanone

SOLUTION Ignoring the $-CH_3$ carbon, the four $-CH_2-$ carbons in the ring, and the C=O carbon, look carefully at C2. Carbon 2 is bonded to four different groups: a $-CH_3$ group, an –H atom, a –C=O carbon in the ring, and a $-CH_2-$ ring carbon. Thus, 2-methylcyclohexanone is chiral.

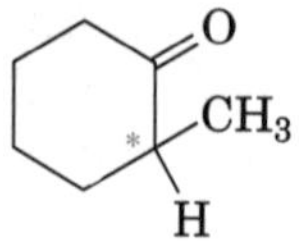

2-Methylcyclohexanone (chiral)

PROBLEM 6.1 Which of these objects are chiral (handed)?
(a) Bean stalk (b) Screwdriver (c) Screw (d) Shoe

PROBLEM 6.2 Which of these compounds are chiral?
(a) 3-Bromopentane (b) 1,3-Dibromopentane (c) 3-Methyl-1-hexene
(d) *cis*-1,4-Dimethylcyclohexane

PROBLEM 6.3 Which of the following molecules are chiral? Identify the stereogenic center(s) in each.

(a)

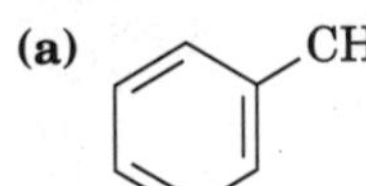

Toluene

(b) $CH_2CH_2CH_3$ N H

Coniine
(from poison hemlock)

(c)

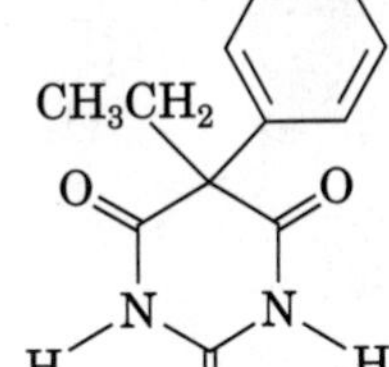

Phenobarbital
(tranquilizer)

PROBLEM 6.4 Place asterisks at all the stereogenic centers in these molecules:

(a)

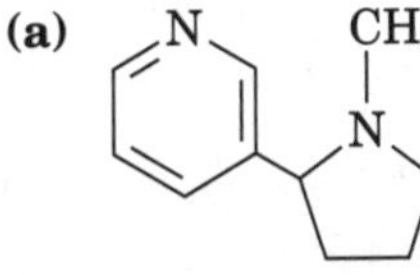

Nicotine

(b) O CH_3

Muscone
(from musk oil)

(c)

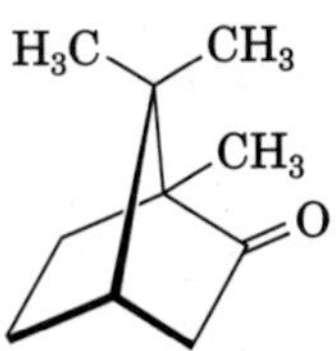

Camphor

PROBLEM 6.5 Alanine, an amino acid found in proteins, is a chiral molecule. Use the standard convention of wedged, solid, and dashed lines to draw the two enantiomers of alanine.

Alanine

6.3 OPTICAL ACTIVITY

The study of stereochemistry has its origins in the work of the French scientist, Jean Baptiste Biot, in the early nineteenth century. Biot, a physicist, was investigating the nature of *plane-polarized light*. A beam of ordinary light consists of electromagnetic waves that oscillate in an infinite number of planes at right angles to the direction of light travel.

When a beam of ordinary light passes through a device called a *polarizer*, though, only the light waves oscillating in a single plane get through, hence the name plane-polarized light. Light waves in all other planes are blocked out. The polarization process is represented in Figure 6.5.

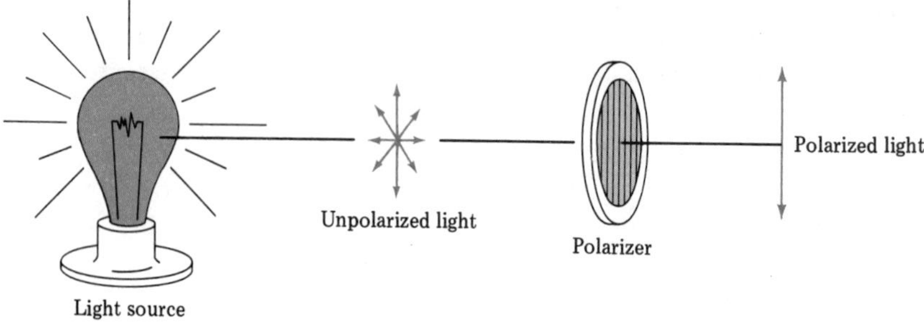

FIGURE 6.5 The nature of plane-polarized light. Only electromagnetic waves that oscillate in a single plane pass through the polarizer.

OPTICAL ACTIVITY
Ability of a molecule to rotate plane-polarized light

Biot made the remarkable observation that, when a beam of plane-polarized light passes through solutions of certain organic molecules such as sugar or camphor, the plane of polarization is *rotated*. Not all organic molecules exhibit this property, but those that do are said to be **optically active**.

The amount of rotation can be measured with an instrument called a *polarimeter*, represented schematically in Figure 6.6. Optically active organic molecules are placed in a sample tube, plane-polarized light is passed through the tube, and rotation of the plane occurs. The light then goes through a second polarizer known as the *analyzer*. By rotating the analyzer until light passes through it, we can find the new plane of polarization and can tell to what extent rotation has occurred. The amount of rotation observed is denoted by α (Greek alpha) and is expressed in degrees.

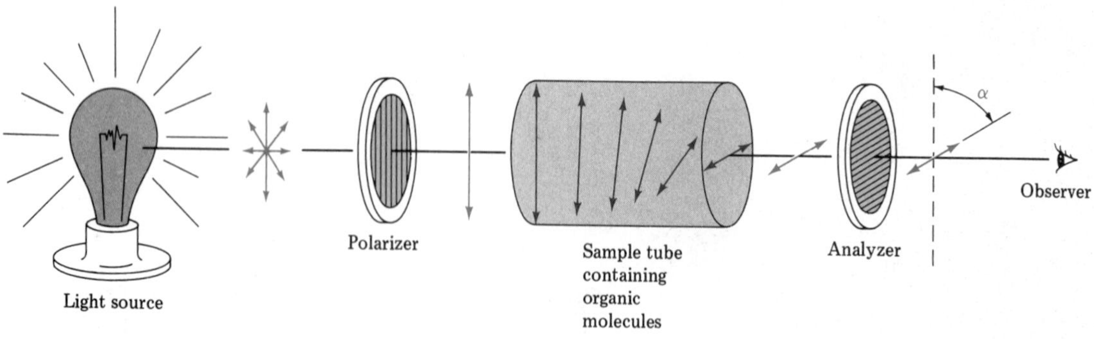

FIGURE 6.6 Schematic representation of a polarimeter.

LEVOROTATORY
Rotating plane-polarized light in a counterclockwise (left-handed) direction

DEXTROROTATORY
Rotating plane-polarized light in a clockwise (right-handed) direction

In addition to determining the extent of rotation, we can also find the direction. From the vantage point of the observer looking at the analyzer, some optically active molecules rotate plane-polarized light to the left (counterclockwise) and are said to be **levorotatory**, whereas other molecules rotate light to the right (clockwise) and are said to be **dextrorotatory**. By convention, rotation to the left is given a minus sign (−), and rotation to the right is given a plus sign (+). For example, (−)-morphine is levorotatory and (+)-sucrose is dextrorotatory.

6.4 SPECIFIC ROTATION

The amount of rotation observed in a polarimetry experiment depends on the structure of the specific molecule and on the number of molecules encountered by the light beam. The number of molecules encountered depends, in turn, on sample concentration and sample path length. If the concentration of the sample in a tube is doubled, the observed rotation is doubled. If the concentration is kept constant but the length of the sample tube is doubled, the observed rotation is doubled.

SPECIFIC ROTATION, $[\alpha]_D$
Amount that an optically active compound rotates plane-polarized light under standard conditions

To express optical rotation data in a meaningful way so that comparisons can be made, we have to choose standard conditions. The **specific rotation, $[\alpha]_D$**, of a compound is defined as the observed rotation when light of 5896 Å wavelength is used with a sample path length l of 1 decimeter (1 dm = 10 cm) and a sample concentration C of 1 g/mL. (Light of 5896 Å, the so-called sodium D line, is the yellow light emitted from common sodium lamps.)

$$[\alpha]_D = \frac{\text{Observed rotation (degrees)}}{\text{Path length (dm)} \times \text{Concentration (g/mL)}} = \frac{\alpha}{l \times C}$$

When optical rotation data are expressed in this standard way, the specific rotation $[\alpha]_D$ is a physical constant that is characteristic of each optically active compound. Some examples are listed in Table 6.1.

TABLE 6.1 Specific Rotations of Some Organic Molecules

COMPOUND	$[\alpha]_D$ (DEGREES)	COMPOUND	$[\alpha]_D$ (DEGREES)
Camphor	+44.26	Penicillin V	+223
Morphine	−132	Monosodium glutamate	+25.5
Sucrose	+66.47	Benzene	0
Cholesterol	−31.5	Hexane	0

PRACTICE PROBLEM 6.4 A 1.20 g sample of cocaine, $[\alpha]_D = -16°$, was dissolved in 7.50 mL chloroform and placed in a sample tube having a path length of 5.00 cm. What was the observed rotation?

SOLUTION Observed rotation, α, is equal to specific rotation $[\alpha]_D$ times sample concentration C times path length l:

$$\alpha = [\alpha]_D \times C \times l$$

where $[\alpha]_D = -16°$; $l = 5.00 \text{ cm} = 0.500 \text{ dm}$; $C = 1.20 \text{ g}/7.50 \text{ mL} = 0.160 \text{ g/mL}$.

Thus,

$$\alpha = -16° \times 0.500 \times 0.160 = -1.3°$$

PROBLEM 6.6 Is cocaine (see Practice Problem 6.4) dextrorotatory or levorotatory?

PROBLEM 6.7 A 1.50 g sample of coniine, the toxic extract of poison hemlock, was dissolved in 10.0 mL ethanol and placed in a sample tube with a path length of 5.00 cm. The observed rotation at the sodium D line was +1.21°. Calculate the specific rotation $[\alpha]_D$ for coniine.

6.5 PASTEUR'S DISCOVERY OF ENANTIOMERS

Little was done after Biot's discovery of optical activity until Louis Pasteur began work in 1848. Pasteur was working on crystalline salts of tartaric acid derived from wine when he made a surprising observation. Upon recrystallizing a concentrated solution of sodium ammonium tartrate below 28°C, two distinct kinds of crystals precipitated. Furthermore, the two kinds of crystals were *mirror images* of each other and were related to each other in the same way that a right hand is related to a left hand.

Working carefully with a pair of tweezers, Pasteur was able to separate the crystals into two piles, one of "right-handed" crystals and one of "left-handed" crystals like those shown in Figure 6.7. Although the original sample (a 50:50 mixture of right and left) was optically inactive, *solutions of crystals from each of the sorted piles were optically active*, and their specific rotations were equal in amount but opposite in sign.

Pasteur was far ahead of his time. Although the structural theory of Kekulé had not yet been proposed, Pasteur explained his results by speaking of the molecules themselves, saying "There is no doubt that [in the *dextro* tartaric acid] there exists an asymmetric arrangement having a nonsuperimposable image. It is no less certain that the atoms of the *levo* acid possess precisely the inverse asymmetric arrangement." Pasteur's vision was extraordinary, for it was not until 25 years later that his theories regarding the asymmetry of chiral molecules were confirmed.

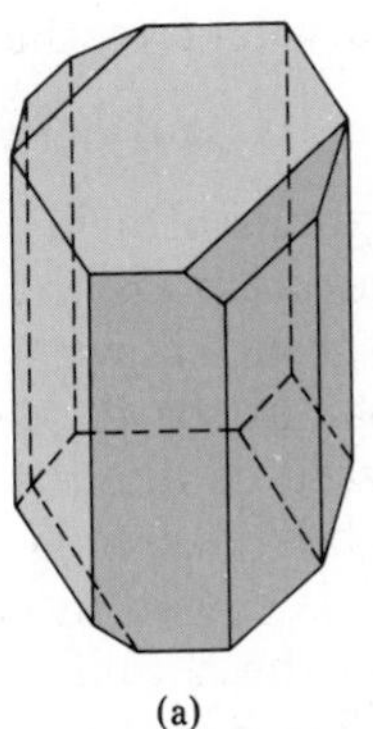
(a)

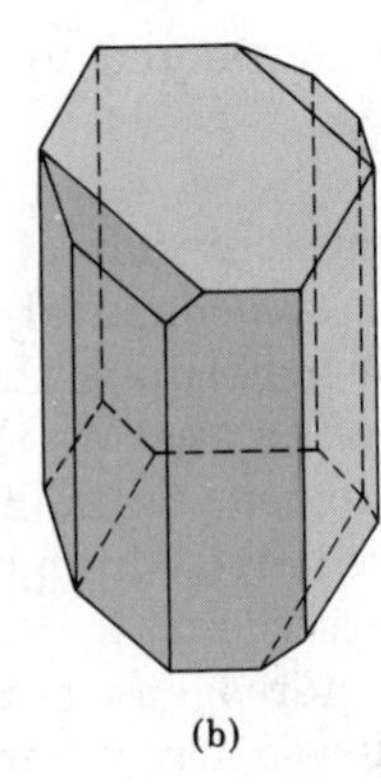
(b)

$$\begin{array}{c} COO^- \; Na^+ \\ | \\ H-C-OH \\ | \\ HO-C-H \\ | \\ COO^- \; NH_4^+ \end{array}$$

Sodium ammonium tartrate

FIGURE 6.7 Crystals of sodium ammonium tartrate. One of the crystals is dextrorotatory in solution, and the other is levorotatory. The drawings are taken from Pasteur's original sketches.

Today, we would describe Pasteur's work by saying that he had discovered the phenomenon of enantiomerism. Enantiomers (also called *optical isomers*) have identical physical properties, such as melting points and boiling points, but differ in the direction in which they rotate plane-polarized light.

6.6 SEQUENCE RULES FOR SPECIFICATION OF CONFIGURATION

CONFIGURATION Three-dimensional arrangement of atoms in space

Although drawings provide pictorial representations of stereochemistry, they are difficult to translate into words. Thus, a verbal method for specifying the three-dimensional arrangement (the **configuration**) of substituents around a stereogenic center is also necessary. The method used employs the same sequence rules that we used for the specification of alkene stereochemistry (Z versus E) in Section 3.4. Let's briefly review these sequence rules and see how they're used to specify the configuration of a stereogenic center. For a more thorough review, you should refer to Section 3.4.

RULE 1 Look at the four atoms directly attached to the stereogenic center and assign priorities in order of decreasing atomic number. The atom with highest atomic number is ranked first; the atom with lowest atomic number is ranked fourth.

RULE 2 If a decision about priority can't be reached by applying rule 1, compare atomic numbers of the second atoms in each substituent, continuing on as necessary through the third or fourth atoms outward until the first point of difference is reached.

RULE 3 Multiple-bonded atoms are considered to be equivalent to the same number of single-bonded atoms.

Having assigned priorities to the four substituent groups attached to a chiral carbon, we describe the stereochemical configuration around that carbon by mentally orienting the molecule so that the group of lowest priority (4) is pointing directly away from us. We then look at the three remaining substituents, which now appear to radiate toward us like the spokes on a steering wheel. If a curved arrow from the highest to second-highest to third-highest priority substituent (1 → 2 → 3) is drawn in a clockwise direction, we say that the stereogenic center has the *R* configuration (Latin *rectus*, "right"). If a curved arrow from 1 → 2 → 3 is drawn in a counterclockwise direction, the stereogenic center has the *S* configuration (Latin *sinister*, "left"). To remember these assignments, think of a car's steering wheel when making a **R**ight (clockwise) turn.

Look at (−)-lactic acid to see how configurations can be assigned:

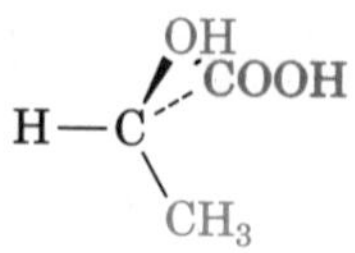

OH, COOH, H—C, CH_3

(−)-Lactic acid

PRIORITIES		
4	−H	(Low)
3	$-CH_3$	
2	−C(=O)—OH	
1	−OH	(High)

Sequence rule 1 says that −OH has first priority (1) and −H has fourth priority (4), but it doesn't distinguish between $-CH_3$ and −COOH since both groups have carbon as their first atom. Sequence rule 2, however, says that −COOH is higher in priority than $-CH_3$, because oxygen outranks hydrogen (the second atom in each group).

We next orient the molecule so that the fourth-priority group (−H) points away from us. Since the direction of the arrow 1 → 2 → 3 is clockwise (right turn of the steering wheel), we assign the *R* configuration to (−)-lactic acid (Figure 6.8). Applying the same procedure to (+)-lactic acid should (and does) lead to the opposite assignment. Try it for yourself.

As another example, look at (−)-glyceraldehyde, which has the *S* configuration shown in Figure 6.9. Note that the sign of optical rotation is not related to the *R*,*S* designation. (*R*)-Lactic acid and (*S*)-glyceraldehyde both happen to be levorotatory (−), but there's no simple correlation between the direction of rotation and *R*,*S* configuration.

PRACTICE PROBLEM 6.5 Draw a tetrahedral representation of (*R*)-2-chlorobutane.

SOLUTION The four substituents bonded to the chiral carbon of (*R*)-2-chlorobutane can be assigned the following priorities: (1) −Cl, (2) $-CH_2CH_3$, (3) $-CH_3$, (4) −H. To draw a tetrahedral representation of the molecule, first orient the low-priority −H group toward the rear and imagine that the other three groups are coming out of the

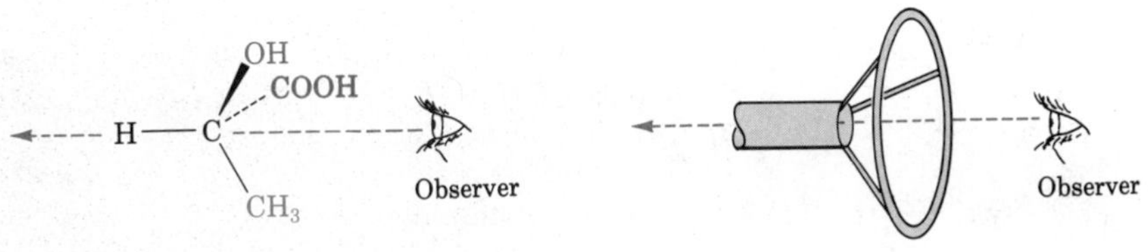

Observer

Observer

(−)-Lactic acid

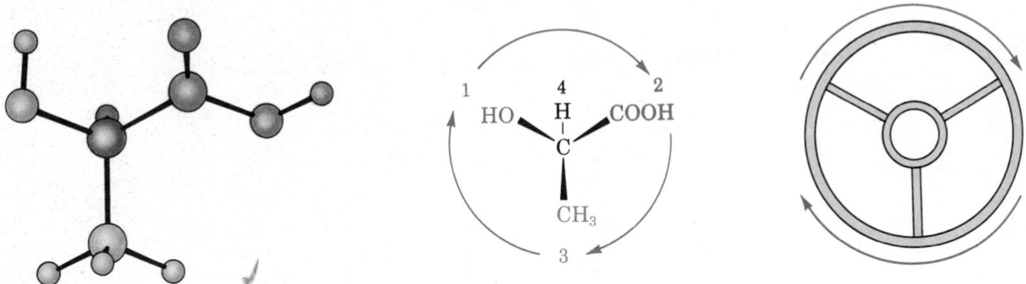

R configuration (right turn of steering wheel)

FIGURE 6.8 Assignment of configuration to (*R*)-(−)-lactic acid.

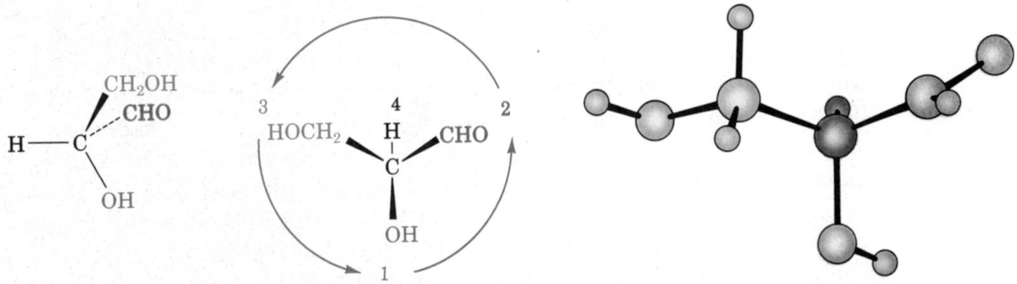

(*S*)-Glyceraldehyde
[(*S*)-(−)-2,3-Dihydroxypropanal]
$[\alpha]_D = -8.7°$

FIGURE 6.9 Configuration of (*S*)-(−)-glyceraldehyde.

page toward you. Place the remaining three substituents in order so that the direction of travel from 1 → 2 → 3 is clockwise (right turn), and then tilt the molecule to bring the rear hydrogen into view.

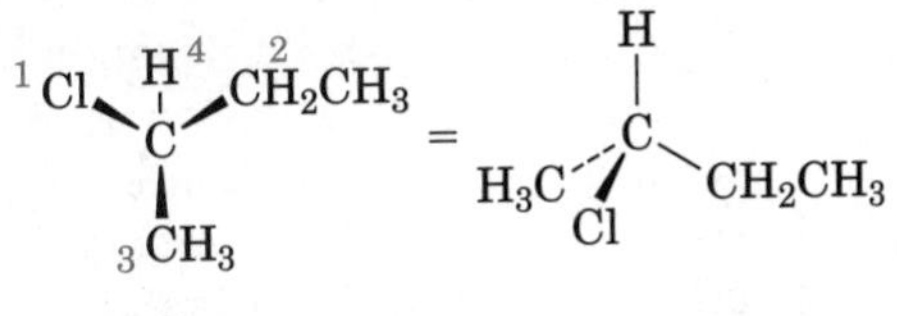

Using molecular models is a great help in working problems of this sort.

PROBLEM 6.8 Assign priorities to the substituents in each set:
(a) $-H, -Br, -CH_2CH_3, -CH_2CH_2OH$ **(b)** $-COOH, -COOCH_3, -CH_2OH, -OH$
(c) $-Br, -CH_2Br, -Cl, -CH_2Cl$

PROBLEM 6.9 Assign *R*,*S* configurations to these molecules:

(a) Br above C; H_3C, H, COOH on C
(b) H above C; HO, CH_3, COOH on C
(c) NH_2 above C; NC, H, CH_3 on C

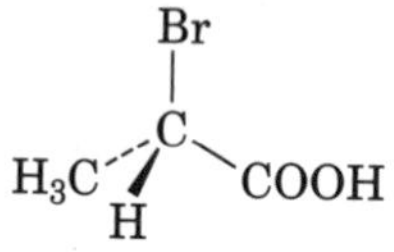

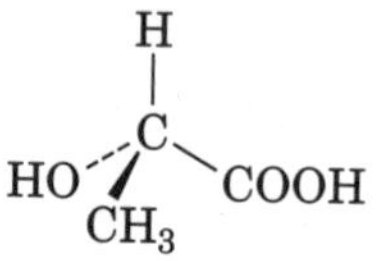

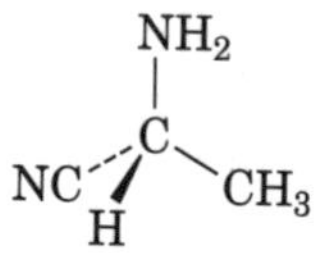

PROBLEM 6.10 Draw a tetrahedral representation of (*S*)-2-hydroxypentane (2-pentanol).

6.7 DIASTEREOMERS

Molecules like lactic acid and glyceraldehyde are relatively simple to deal with because each has only one stereogenic center and only two enantiomeric forms. The situation becomes more complex, however, for molecules that have more than one stereogenic center.

Take the amino acid threonine (2-amino-3-hydroxybutanoic acid) as an example. Since threonine has two stereogenic centers (C2 and C3), there are four possible stereoisomers, which are shown in Figure 6.10. (Check for yourself that the *R*,*S* configurations are correct as shown).

The four stereoisomers of 2-amino-3-hydroxybutanoic acid can be classified into two pairs of enantiomers (mirror images). The 2*R*,3*R* stereoisomer is the mirror image of 2*S*,3*S*, and the 2*R*,3*S* stereoisomer is the mirror image of 2*S*,3*R*. But what is the relationship between any two configurations that are not mirror images? What, for example, is the relationship between the 2*R*,3*R* compound and the 2*R*,3*S* compound? These two compounds are stereoisomers, yet they aren't enantiomers. To describe such a relationship, we need a new term.

DIASTEREOMER
Stereoisomers that aren't mirror images

Diastereomers are stereoisomers that are not mirror images of each other. Since we used the right-hand/left-hand analogy to describe the relationship between two enantiomers, we might extend the analogy by saying that the relationship between diastereomers is that of hands from two different people. Your hand and your friend's hand look *similar*, but they aren't identical, and they aren't mirror images. The same is true of diastereomers; they're similar, but not identical and not mirror images. Note carefully the difference between enantiomers and diastereomers: Enantiomers must have opposite (mirror-image) configurations at *all* stereogenic centers; diastereomers must have opposite configurations at *some* (one or more) stereogenic centers, but the same configuration at others. A full description of the four threonine stereoisomers is given in Table 6.2.

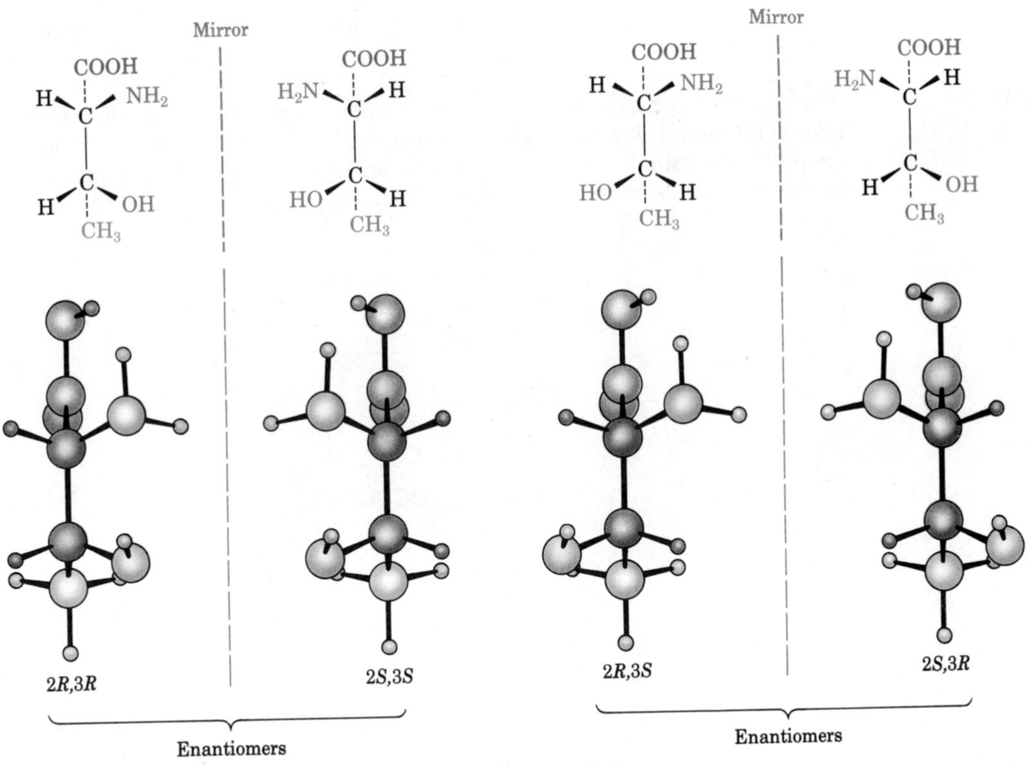

FIGURE 6.10 Four stereoisomers of 2-amino-3-hydroxybutanoic acid.

TABLE 6.2 Relationships Among the Four Threonine Stereoisomers

STEREOISOMER	ENANTIOMERIC WITH	DIASTEREOMERIC WITH
2*R*,3*R*	2*S*,3*S*	2*R*,3*S* and 2*S*,3*R*
2*S*,3*S*	2*R*,3*R*	2*R*,3*S* and 2*S*,3*R*
2*R*,3*S*	2*S*,3*R*	2*R*,3*R* and 2*S*,3*S*
2*S*,3*R*	2*R*,3*S*	2*R*,3*R* and 2*S*,3*S*

Of the four possible threonine stereoisomers, only the 2*S*,3*R* isomer, $[\alpha]_D = -28.3°$, occurs naturally and is an essential human nutrient. Most biologically important organic molecules are chiral, and usually only one stereoisomer is found in nature.

PROBLEM 6.11 Assign *R*,*S* configurations to each stereogenic center in these molecules:

(a) Br, H, CH_3 on C; C bearing H, OH, CH_3

(b) CH_3, H, Br on C; C bearing H_3C, H, OH

(c) CH_3, Br, H on C; C bearing H, CH_3, OH

PROBLEM 6.12 Which of the compounds in Problem 6.11 are enantiomers and which are diastereomers?

PROBLEM 6.13 Chloramphenicol is a powerful antibiotic isolated from the *Streptomyces venezuelae* bacterium. It is active against a broad spectrum of bacterial infections and is particularly valuable against typhoid fever. Assign *R*,*S* configurations to the stereogenic centers in chloramphenicol.

NO_2 — HO, H — C — C — H, $NHCOCHCl_2$ — CH_2OH

Chloramphenicol, $[\alpha]_D = +18.6°$

6.8 MESO COMPOUNDS

Let's look at one more example of a compound with two stereogenic centers: tartaric acid. We're already acquainted with tartaric acid because of its role in Pasteur's discovery of optical activity, and we can now draw the four stereoisomers:

Mirror Mirror

2*R*,3*R* **2*S*,3*S*** **2*R*,3*S*** **2*S*,3*R***

The mirror-image 2*R*,3*R* and 2*S*,3*S* structures are not identical and therefore represent an enantiomeric pair. A careful look, however, shows that the 2*R*,3*S* and 2*S*,3*R* structures *are* identical, as can be seen by rotating one structure 180°:

COOH, H, OH, C, C, H, OH, COOH — Rotate 180° → COOH, HO, H, C, C, HO, H, COOH

2R,3S **2S,3R**

Identical

The 2R,3S and 2S,3R structures are identical because the molecule is achiral and has a plane of symmetry. The symmetry plane cuts through the C2–C3 bond, making one half of the molecule a mirror image of the other half (Figure 6.11).

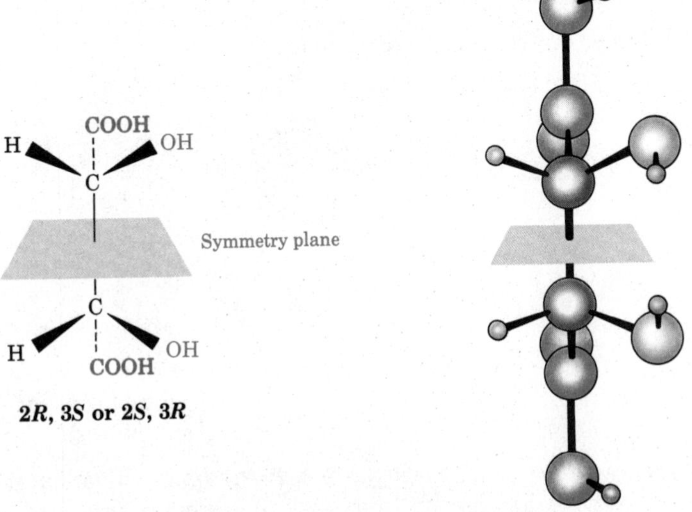

FIGURE 6.11 A symmetry plane through the C2–C3 bond of *meso*-tartaric acid.

MESO COMPOUNDS
Compounds that have chiral centers but are achiral overall because of a symmetry plane

Because of the plane of symmetry, the tartaric acid stereoisomer shown in Figure 6.11 is achiral, despite the fact that it has two stereogenic centers. Such compounds that are achiral, yet contain stereogenic centers, are called **meso compounds** (**me**-zo). Thus, tartaric acid exists in only three stereoisomeric configurations: two enantiomers and one meso form.

PRACTICE PROBLEM 6.6 Does *cis*-1,2-dimethylcyclobutane have any stereogenic centers? Is it a chiral molecule?

SOLUTION Looking at the structure of *cis*-1,2-dimethylcyclobutane, we see that both of the methyl-bearing ring carbons (C1 and C2) are chiral. Overall, though, the compound is achiral because there's a symmetry plane bisecting the ring between C1 and C2. Thus, the molecule is a meso compound.

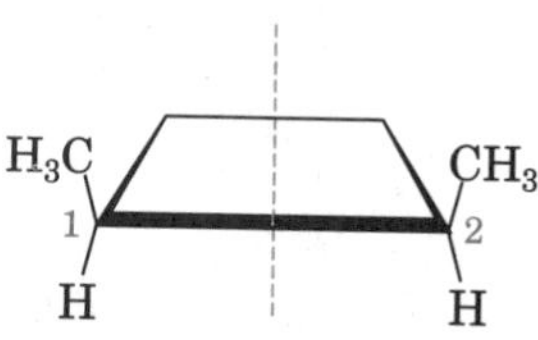

PROBLEM 6.14 Which of these substances have a meso form?

(a) 2,3-Dibromobutane (b) 2,3-Dibromopentane
(c) 2,4-Dibromopentane

PROBLEM 6.15 Which of these structures represent meso compounds?

(a) (b) (c)

6.9 MOLECULES WITH MORE THAN TWO STEREOGENIC CENTERS

One stereogenic center gives rise to two stereoisomers (one pair of enantiomers), and two stereogenic centers give rise to a maximum of four stereoisomers (two pairs of enantiomers). In general, a molecule with n stereogenic centers has a maximum of 2^n stereoisomers (2^{n-1} pairs of enantiomers). For example, cholesterol has eight stereogenic centers. Thus, $2^8 = 256$ stereoisomers of cholesterol, or 128 pairs of enantiomers, are possible in principle, though many would be too strained to exist. Only one, however, is produced in nature.

Cholesterol (eight stereogenic centers)

PROBLEM 6.16

Nandrolone is an anabolic steroid used illegally by some athletes to build muscle mass. How many stereogenic centers does nandrolone have? How many stereoisomers of nandrolone are possible in principle?

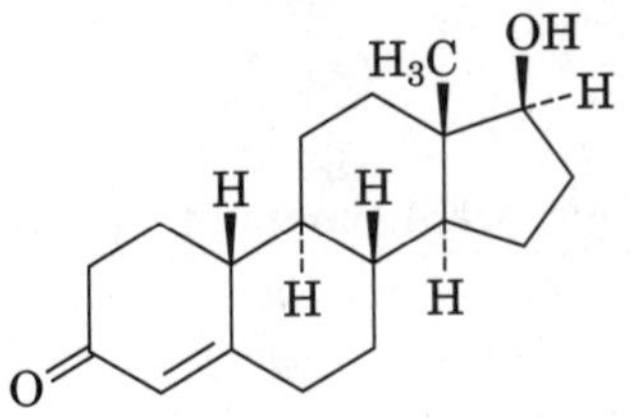

H_3C OH H H H H H O

Nandrolone

6.10 RACEMIC MIXTURES

To conclude this discussion of stereoisomerism, let's return for a last look at Pasteur's pioneering work. Pasteur took an optically inactive tartaric acid salt and found that he could crystallize from it two optically active forms having the 2*R*,3*R* and 2*S*,3*S* configurations. But what was the optically inactive form he started with? It couldn't have been *meso*-tartaric acid, because *meso*-tartaric acid is a different chemical compound and can't interconvert with the two chiral enantiomers without breaking and re-forming chemical bonds.

RACEMIC MIXTURE A 50:50 mixture of enantiomers

RESOLUTION Separation of a racemic mixture into its pure component enantiomers

The answer is that Pasteur started with a 50:50 mixture of the two chiral tartaric acid enantiomers. Such a mixture is called a **racemic** (ray-see-mic) **mixture**, or **racemate**. Racemic mixtures, often denoted by the symbol (±), show no optical activity because they contain equal amounts of (+) and (−) forms. The (+) rotation from one enantiomer exactly cancels the (−) rotation from the other. Through good fortune, Pasteur was able to separate, or **resolve**, (±)-tartaric acid into its (+) and (−) enantiomers. Unfortunately, the fractional crystallization technique he used doesn't work for most racemic mixtures, and other methods are required. We'll discuss a better method in Section 12.4.

6.11 PHYSICAL PROPERTIES OF STEREOISOMERS

If such seemingly simple compounds as tartaric acid can exist in different stereoisomeric configurations, the question arises whether the different stereoisomers have different physical properties. The answer is yes, they do.

Some physical properties of the three stereoisomers of tartaric acid and of the racemic mixture are shown in Table 6.3. As indicated, the (+) and (−) enantiomers have identical melting points, solubilities, and densities. They differ only in the sign of their rotation of plane-polarized light. The meso isomer, by contrast, is diastereomeric with the (+) and (−) forms, is a different compound altogether, and has different physical properties. The racemic mixture is different still. For reasons beyond our present scope, racemates act as though they were pure compounds, different from either enantiomer. Thus, the physical properties of racemic tartaric acid differ from those of the two enantiomers and from those of the meso form.

TABLE 6.3 Some Properties of the Stereoisomers of Tartaric Acid

STEREOISOMER	MELTING POINT (°C)	$[\alpha]_D$ (DEGREES)	DENSITY (g/cm^3)	SOLUBILITY AT 20°C (g/100 mL H_2O)
(+)	168–170	+12	1.7598	139.0
(−)	168–170	−12	1.7598	139.0
Meso	146–148	0	1.6660	125.0
(±)	206	0	1.7880	20.6

6.12 A BRIEF REVIEW OF ISOMERISM

We've seen several kinds of isomers in the past few chapters, and it's a good idea at this point to see how they relate to one another. As noted earlier, isomers are compounds that have the same chemical formula but different structures. There are two fundamental types of isomerism, both of which we've now encountered: constitutional isomerism and stereoisomerism.

Constitutional isomers (Section 2.2) are compounds whose atoms are connected differently. Among the kinds of constitutional isomers we've seen are skeletal, functional, and positional isomers.

Constitutional isomers—different connections among atoms:

Different carbon skeletons: $CH_3CH(CH_3)CH_3$ and $CH_3CH_2CH_2CH_3$

Isobutane **Butane**

Different functional groups: CH_3CH_2OH and CH_3OCH_3

Ethyl alcohol **Dimethyl ether**

Different position of functional groups: $CH_3CH(NH_2)CH_3$ and $CH_3CH_2CH_2NH_2$

Isopropylamine **Propylamine**

Stereoisomers (Section 2.8) are compounds whose atoms are connected in the same way but with a different geometry. Among the kinds of stereoisomers we've seen are enantiomers, diastereomers, and cis–trans isomers (both in alkenes and in cycloalkanes). To be accurate, though, cis–trans isomers are really just another kind of diastereomers since they meet the definition of non-mirror-image stereoisomers:

Stereoisomers—same connections among atoms but different geometry:

Enantiomers (mirror-image stereoisomers)

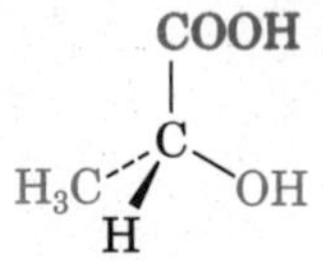

(R)-Lactic acid 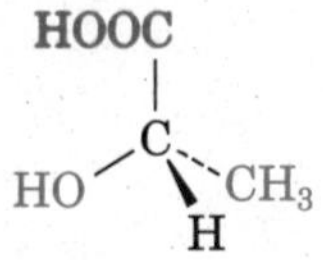 **(S)-Lactic acid**

Diastereomers (non-mirror-image stereoisomers)

Configurational diastereomers

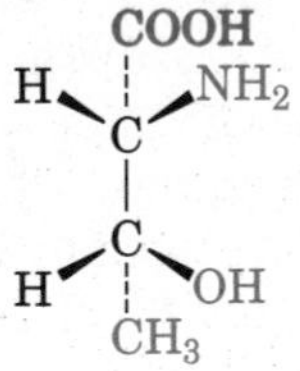

(2R,3R)-2-Amino-3-hydroxybutanoic acid 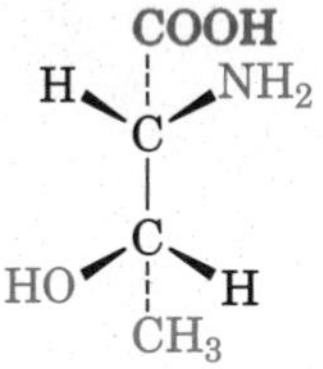 **(2R,3S)-2-Amino-3-hydroxybutanoic acid**

Cis–trans diastereomers

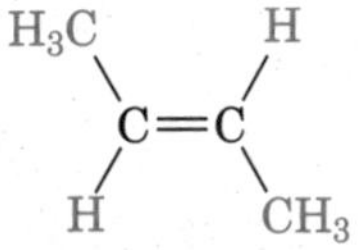

trans-2-Butene and **cis-2-Butene**

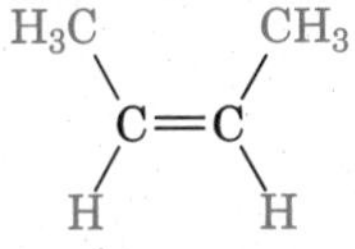

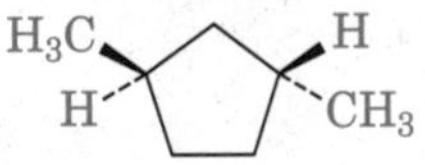

trans-1,3-Dimethylcyclopentane and **cis-1,3-Dimethylcyclopentane**

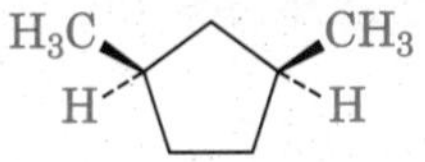

PROBLEM 6.17 What kinds of isomers are the following pairs?

(a) (*S*)-5-Chloro-2-hexene and chlorocyclohexane

(b) (2*R*,3*R*)-Dibromopentane and (2*S*,3*R*)-dibromopentane

6.13 STEREOCHEMISTRY OF REACTIONS: ADDITION OF HBr TO ALKENES

Many organic reactions, including some that we've studied, yield products with stereogenic centers. For example, addition of HBr to 1-butene yields 2-bromobutane, a chiral molecule. What predictions can we make about the stereochemistry of this chiral product? If a single enantiomer is formed, is it *R* or *S*? If a mixture of enantiomers is formed, how much of each? In fact, the 2-bromobutane produced is a racemic mixture of *R* and *S* enantiomers.

$$CH_3CH_2CH{=}CH_2 \xrightarrow[\text{Ether}]{\text{HBr}} CH_3CH_2\overset{\text{Br}}{\overset{|}{C}}HCH_3$$

1-Butene (achiral) — **(±)-2-Bromobutane** (chiral)

To understand *why* a racemic product results, think about what happens during the reaction. 1-Butene is first protonated to yield an intermediate secondary carbocation. Since the trivalent carbon is sp^2-hybridized, it has a plane of symmetry and is achiral. As a result, it can be attacked by bromide ion equally well from either the top or the bottom (Figure 6.12). Attack from the top leads to (*S*)-2-bromobutane, and attack from the bottom leads to (*R*)-2-bromobutane. Since both pathways occur with equal probability, a racemic product mixture results.

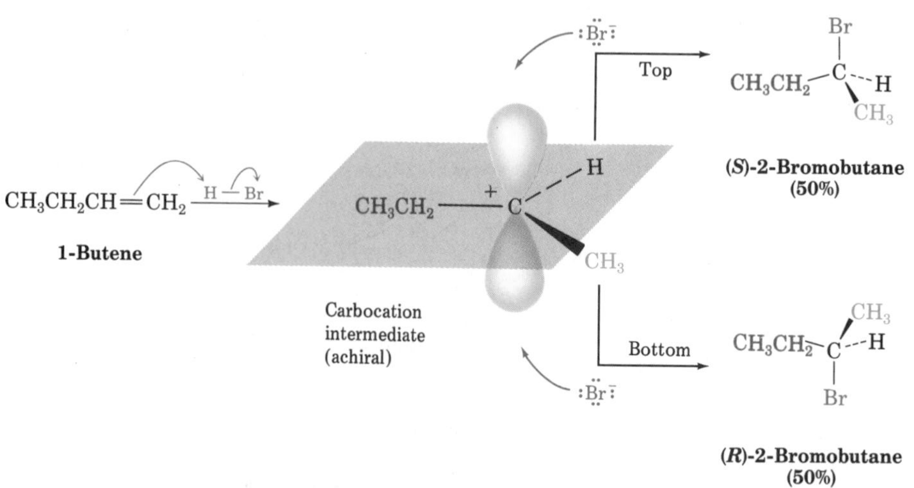

FIGURE 6.12 Stereochemistry of the addition of HBr to 1-butene. The intermediate carbocation is attacked equally well from both top and bottom, giving a racemic mixture of products that is 50% *R* and 50% *S*.

What's true for the reaction of 1-butene with HBr is also true for all other reactions: *Reaction of achiral starting materials always leads to optically inactive products.* Optically active products can't be produced from optically inactive intermediates or starting materials.

INTERLUDE

CHIRALITY IN NATURE

Just as the different stereoisomers of a substance have different physical properties, they also have different chemical and biological properties. For example, (+)-lactic acid in the body is rapidly converted into pyruvic acid by the enzyme lactic acid dehydrogenase; (−)-lactic acid, however, is unaffected by the enzyme.

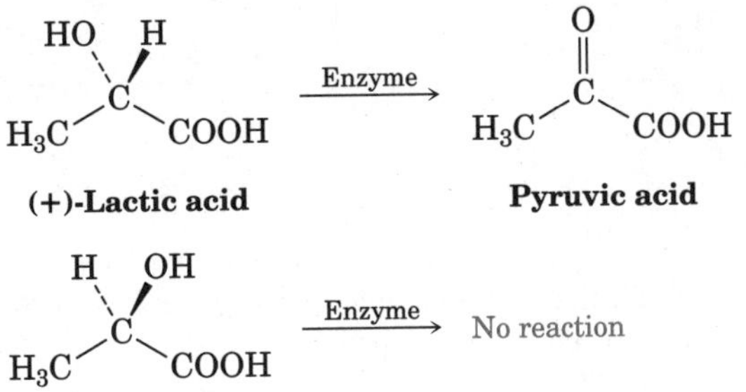

HO H / H_3C–C–COOH —Enzyme→ O / H_3C–C–COOH

(+)-Lactic acid **Pyruvic acid**

H OH / H_3C–C–COOH —Enzyme→ No reaction

(−)-Lactic acid

A remarkable example of how a simple change in chirality can affect the biological properties of a molecule is found in the amino acid, dopa. More properly named 2-amino-3-(3,4-dihydroxyphenyl)propanoic acid, dopa has a single stereogenic center and thus exists in two stereoisomeric forms. Although the dextrorotatory enantiomer, called D-dopa, has no physiological effect on humans, the levorotatory enantiomer, called L-dopa, has dramatic activity against Parkinsonism, a chronic disease of the central nervous system.

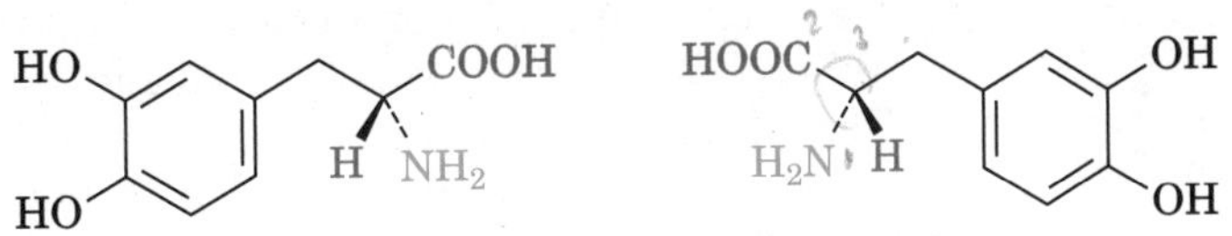

HO, HO, COOH, H NH_2 HOOC, H_2N H, OH, OH

D-Dopa
(no biological effect)

L-Dopa
(anti-Parkinsonism agent)

Why do different stereoisomers have such different biological properties? To exert its biological effect, a chiral molecule must fit into a chiral receptor at the target site, much as a hand fits into a glove. Just as a right hand can fit only into a right-hand glove, so a particular stereoisomer can fit only into a receptor with the proper complementary shape. Any other stereoisomer is a misfit like a right hand in a left-hand glove (Figure 6.13).

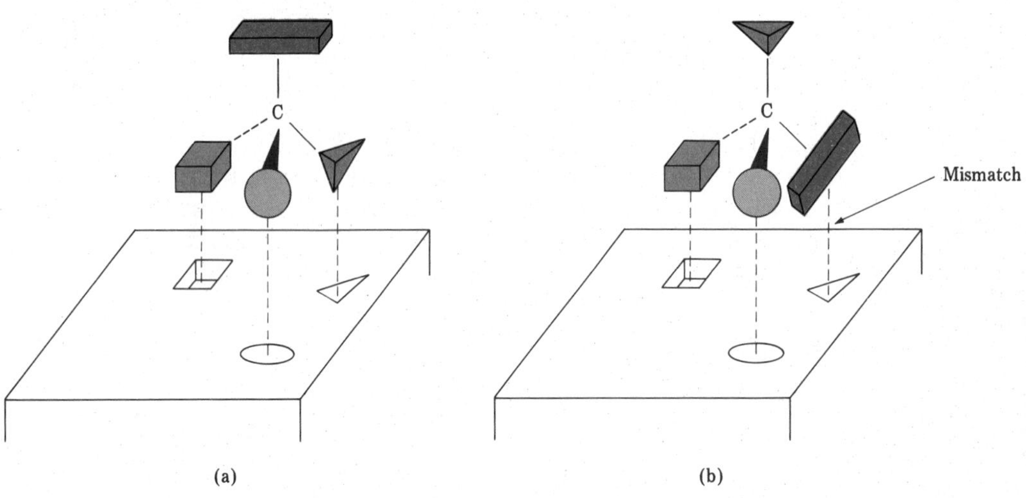

FIGURE 6.13 (a) L-Dopa fits into a complementary receptor site, but (b) D-dopa can't fit into the same receptor and therefore has no physiological effect.

SUMMARY AND KEY WORDS

When a beam of **plane-polarized light** is passed through a solution of certain organic molecules, the plane of polarization is rotated. Compounds that exhibit this behavior are called **optically active**. Optical activity is due to the handedness of the molecules.

An object that is not identical to its mirror image is said to be **chiral**, meaning "handed." For example, a glove is chiral, but a coffee cup is **achiral**. A chiral object is one that does not contain a **plane of symmetry**. The usual cause of chirality in organic molecules is the presence of a tetrahedral carbon atom bonded to four different groups. Chiral compounds can exist as a pair of mirror-image stereoisomers called **enantiomers**. Enantiomers are identical in their physical properties except for the direction in which they rotate plane-polarized light, and are related to each other as a right hand is related to a left hand.

The stereochemical **configuration** of chiral carbon centers is specified as either ***R*** (*rectus*) or ***S*** (*sinister*). ***Sequence rules*** are used to assign priorities to the four substituents on the chiral carbon, and the molecule is then oriented so that the lowest-priority group points directly away from the viewer. If a curved arrow from the highest to second-highest to third-highest priority groups is drawn in a clockwise direction, the configuration is labeled R. If the direction of the arrow is counterclockwise, the configuration is labeled S.

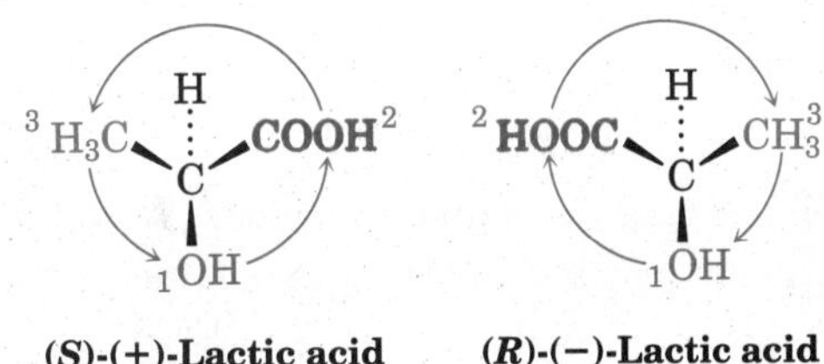

(S)-(+)-Lactic acid **(R)-(−)-Lactic acid**

Some molecules have more than one stereogenic center. Enantiomers have opposite configurations at all stereogenic centers, whereas **diastereomers** have the same configuration in at least one center and opposite configurations at the others. **Meso compounds** contain stereogenic centers but are achiral overall because they contain a plane of symmetry. **Racemates** are 50:50 mixtures of (+) and (−) enantiomers. Racemic mixtures and individual diastereomers differ in both their physical properties and their biological properties.

ADDITIONAL PROBLEMS

6.18 Cholic acid, the major steroid found in bile, was found to have a rotation of +2.22° when a 3.00 g sample was dissolved in 5.00 mL alcohol in a sample tube with a 1.00 cm path length. Calculate $[\alpha]_D$ for cholic acid.

6.19 Polarimeters are so sensitive that they can measure rotations to the thousandth of a degree, an important advantage when only small amounts of a sample are available. For example, when 7.00 mg of ecdysone, an insect hormone that controls molting in the silkworm moth, was dissolved in 1.00 mL chloroform in a cell with a 2.00 cm path length, an observed rotation of +0.087° was found. Calculate $[\alpha]_D$ for ecdysone.

6.20 Define these terms:
(a) Chirality **(b)** Stereogenic center **(c)** Diastereomer **(d)** Racemate
(e) Meso compound **(f)** Enantiomer

6.21 Which of these objects are chiral?
(a) A basketball **(b)** A wine glass **(c)** An ear **(d)** A snowflake **(e)** A coin
(f) Scissors

6.22 Which of these compounds are chiral?
(a) 2,4-Dimethylheptane **(b)** 5-Ethyl-3,3-dimethylheptane
(c) *cis*-1,3-Dimethylcyclohexane

6.23 Penicillin V is a broad-spectrum antibiotic that contains three stereogenic centers. Identify them with asterisks.

Penicillin V (antibiotic)

6.24 Draw chiral molecules that meet these descriptions:
(a) A chloroalkane, $C_5H_{11}Cl$ **(b)** An alcohol, $C_6H_{14}O$ **(c)** An alkene, C_6H_{12}
(d) An alkane, C_8H_{18}

6.25 Which of the following compounds are chiral? Label all stereogenic centers.

(a) $CH_3CH_2CH(CH_3)-C(CH_3)_2-CH_2CH_3$ **(b)** O, CH_3 **(c)** O, H_3C

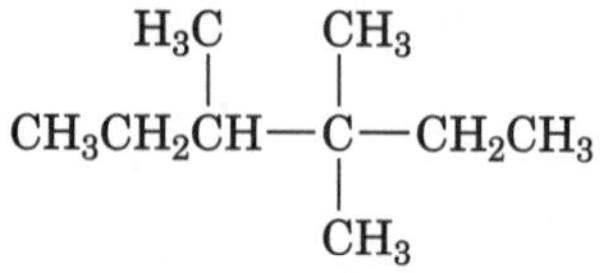

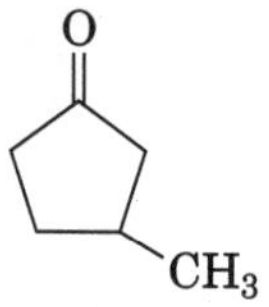

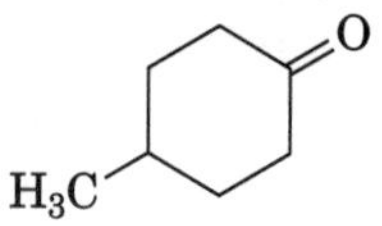

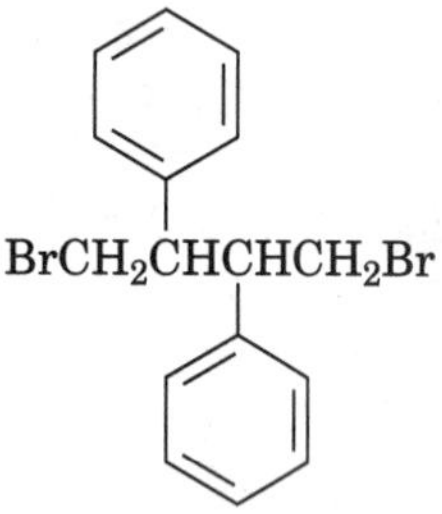

(d) $BrCH_2CHCHCH_2Br$ **(e)** CH_3, H, H_3C, H

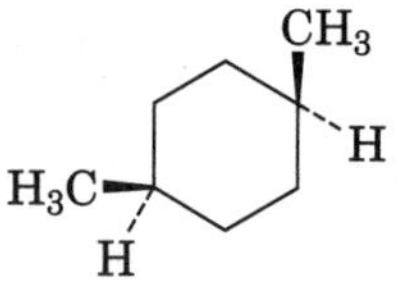

6.26 There are eight alcohols with the formula $C_5H_{12}O$. Draw them and tell which are chiral.

6.27 Propose structures for compounds that meet these descriptions:
(a) A chiral alcohol with four carbons **(b)** A chiral carboxylic acid
(c) A compound with two stereogenic centers

6.28 Assign priorities to the substituents in each of these sets:
(a) $-H$, $-OH$, $-OCH_3$, $-CH_3$ **(b)** $-Br$, $-CH_3$, $-CH_2Br$, $-Cl$
(c) $-CH{=}CH_2$, $-CH(CH_3)_2$, $-C(CH_3)_3$, $-CH_2CH_3$ **(d)** $-COOCH_3$, $-COCH_3$, $-CH_2OCH_3$, $-OCH_3$

6.29 One enantiomer of lactic acid is shown below. Is it *R* or *S*? Draw its mirror image in the standard tetrahedral representation.

COOH, C, H, OH, CH_3

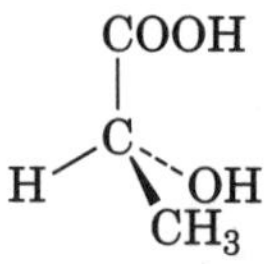

6.30 Draw tetrahedral representations of both enantiomers of the amino acid serine. Tell which of your structures is *S* and which is *R*.

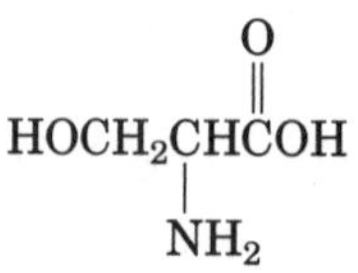

$HOCH_2CH(NH_2)COOH$ **Serine**

6.31 If naturally occurring (*S*)-serine has $[\alpha]_D = -6.83°$, what specific rotation do you expect for (*R*)-serine?

6.32 Assign *R* or *S* configuration to the stereogenic centers in these molecules:

(a) H OH **(b)** Cl H **(c)** H OCH_3, $HOCH_2$, COOH

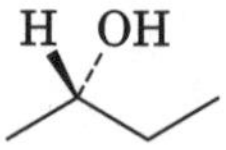

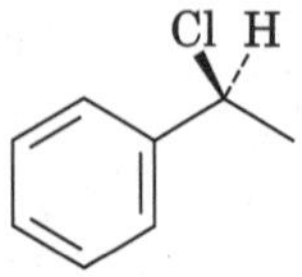

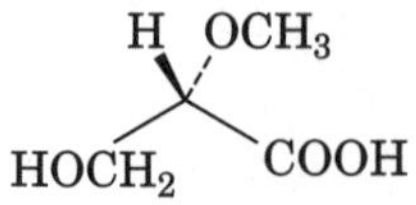

6.33 What is the relationship between the specific rotations of (2*R*,3*R*)-dihydroxypentane and (2*S*,3*S*)-dihydroxypentane? Between (2*R*,3*S*)-dihydroxypentane and (2*R*,3*R*)-dihydroxypentane?

6.34 What is the stereochemical configuration of the enantiomer of (2*S*,4*R*)-dibromooctane?

6.35 What are the stereochemical configurations of the two diastereomers of (2*S*,4*R*)-dibromooctane?

6.36 Draw examples of the following:
(a) A meso compound with the formula C_8H_{18}
(b) A compound with two stereogenic centers, one *R* and the other *S*

6.37 Draw a tetrahedral representation of (*S*)-2-butanol, $CH_3CH_2CH(OH)CH_3$.

6.38 Tell whether this Newman projection of 2-chlorobutane is *R* or *S*:

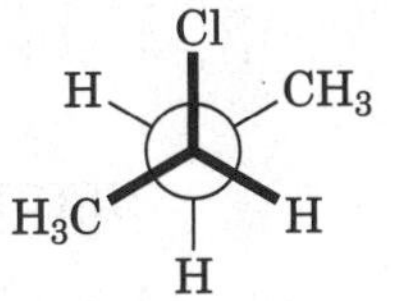

Cl
H CH3
H3C H
H

6.39 Draw a Newman projection that is enantiomeric with the one shown in Problem 6.38.

6.40 Draw a Newman projection of *meso*-tartaric acid.

6.41 Draw Newman projections of (2*R*,3*R*)- and (2*S*,3*S*)-tartaric acid and compare them to the projection you drew in Problem 6.40 for the meso form.

6.42 The sugar glucose has four stereogenic centers. How many stereoisomers of glucose are possible?

6.43 Draw a tetrahedral representation of (*R*)-3-chloro-1-pentene.

6.44 Draw all the stereoisomers of 1,2-dimethylcyclopentane. Assign *R*,*S* configurations to the stereogenic centers in all isomers, and indicate which stereoisomers are chiral and which, if any, are meso.

6.45 Assign *R* or *S* configuration to each stereogenic center in these molecules:

(a)
COOH
C
H NH2
CH3

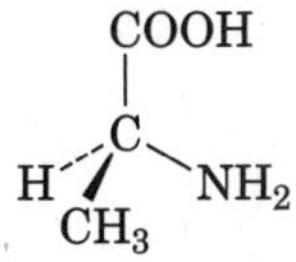

(b)
HO H
C
H2N CH3
C
H

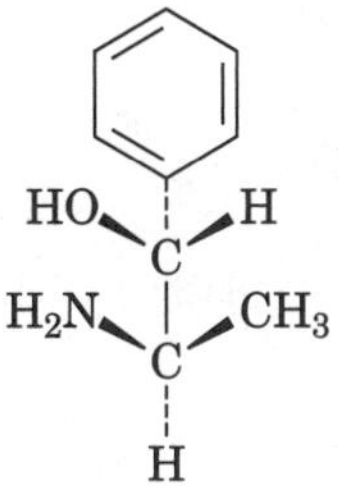

6.46 Hydroxylation of *cis*-2-butene with $KMnO_4$ yields 2,3-butanediol. What is the stereochemistry of the product? (Review Section 4.7.)

6.47 Answer Problem 6.46 for *trans*-2-butene.

6.48 How many stereoisomers of 2,4-dibromo-3-chloropentane are there? Draw them and indicate which are optically active.

6.49 Alkenes undergo reaction with peroxycarboxylic acids (RCO_3H) to give compounds called *epoxides*. For example, *cis*-2-butene gives 2,3-epoxybutane:

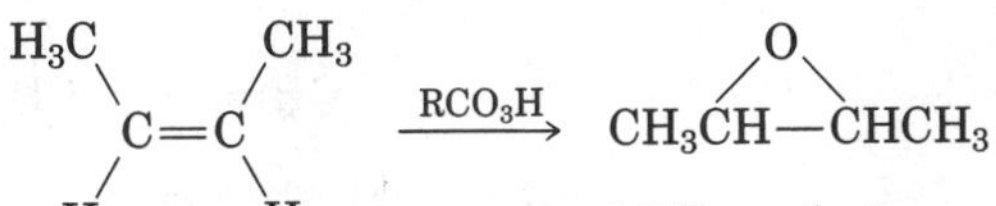

H3C CH3
C=C
H H

$\xrightarrow{RCO_3H}$

O
$CH_3CH—CHCH_3$

2,3-Epoxybutane

Assuming that both C–O bonds form from the same side of the molecule (syn stereochemistry), show the stereochemistry of the product. Is the epoxide chiral? How many stereogenic centers does it have? How would you describe the product stereochemically?

6.50 Answer Problem 6.49 assuming that the epoxidation was carried out on *trans*-2-butene.

6.51 Ribose, an essential part of ribonucleic acid (RNA), has the following structure:

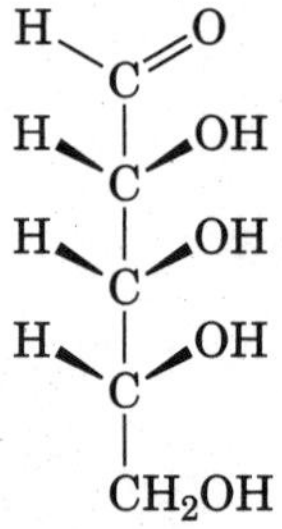

H, C=O
H–C–OH
H–C–OH
H–C–OH
CH_2OH

Ribose

How many stereogenic centers does ribose have? Identify them with asterisks. How many stereoisomers of ribose are there?

6.52 Draw the structure of the enantiomer (mirror image) of ribose (see Problem 6.51).

6.53 Draw the structure of a diastereomer of ribose (see Problem 6.51).

6.54 On catalytic hydrogenation over a platinum catalyst, ribose (see Problem 6.51) is converted into ribitol. Is ribitol optically active or inactive? Explain.

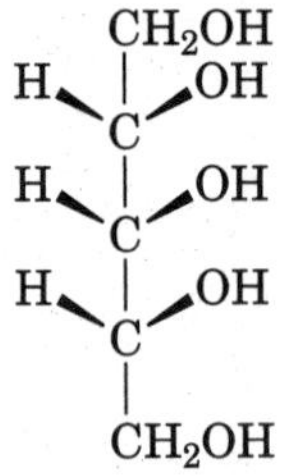

CH_2OH
H–C–OH
H–C–OH
H–C–OH
CH_2OH

Ribitol

6.55 Draw the two enantiomers of the amino acid cysteine, $HSCH_2CH(NH_2)COOH$, and identify each as *R* or *S*.

6.56 Draw the structure of (*R*)-2-methylcyclohexanone.

6.57 Compound A, C_7H_{12}, was found to be optically active. On catalytic reduction over a palladium catalyst, 2 equiv of hydrogen were absorbed, yielding compound B, C_7H_{16}. On cleavage with ozone, two fragments were obtained. One fragment was identified as acetic acid, CH_3COOH, and the other fragment, C, was found to be an optically active carboxylic acid. Formulate the reactions, and propose structures for A, B, and C.

6.58 *Allenes* are compounds with adjacent carbon–carbon double bonds. Even though they do not contain chiral carbon atoms, many allenes are chiral. For example, mycomycin, an antibiotic isolated from the bacterium *Nocardia acidophilus*, is chiral and has $[\alpha]_D = -130°$. Can you explain why mycomycin is chiral? Making a molecular model should be helpful.

$$HC{\equiv}C{-}C{\equiv}C{-}CH{=}C{=}CH{-}CH{=}CH{-}CH{=}CH{-}CH_2COOH$$

Mycomycin (an allene)

7

CHAPTER

ALKYL HALIDES

It would be difficult to study organic chemistry for long without becoming aware of halo-substituted alkanes. Among their many uses, alkyl halides are employed as industrial solvents, as inhaled anesthetics in medicine, as insecticides, and as refrigerants.

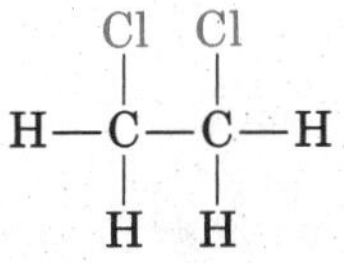

```
    Cl  Cl              F   Br              F
    |   |               |   |               |
H—C—C—H           F—C—C—H          Cl—C—F
    |   |               |   |               |
    H   H               F   Cl              Cl
```

1,2-Dichloroethane (a solvent) | **Halothane** (an inhaled anesthetic) | **Freon 12** (a refrigerant)

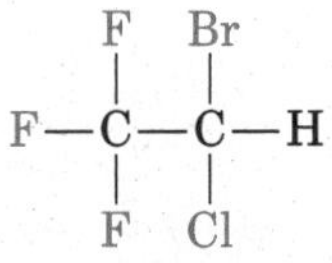

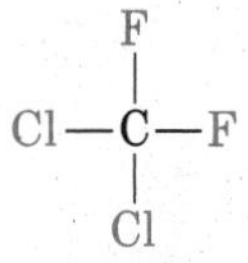

7.1 NAMING ALKYL HALIDES

Alkyl halides are named in the same way as alkanes (Section 2.3), by considering the halogen as a substituent on the parent alkane chain. There are three steps:

STEP 1 Find the longest chain and name it as the parent. If a multiple bond is present, the parent chain must contain it.

STEP 2 Number the carbons of the chain beginning at the end nearer the first substituent, regardless of whether it's alkyl or halo. Assign each substituent a number according to its position on the chain. If there are substituents the same distance from both ends, begin numbering at the end nearer the substituent with alphabetical priority.

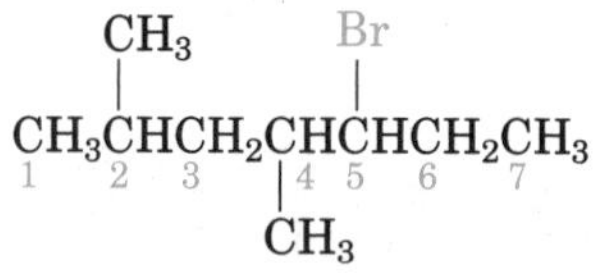

$$\overset{CH_3}{|}\quad\overset{Br}{|}$$
$$\underset{1\ \ 2\ \ 3\ \ 4\ 5\ \ 6\ \ 7}{CH_3CHCH_2CHCHCH_2CH_3}$$
$$\underset{CH_3}{|}$$

5-Bromo-2,4-dimethylheptane

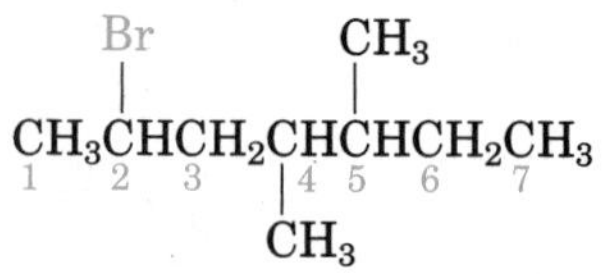

$$\overset{Br}{|}\quad\overset{CH_3}{|}$$
$$\underset{1\ \ 2\ \ 3\ \ 4\ 5\ \ 6\ \ 7}{CH_3CHCH_2CHCHCH_2CH_3}$$
$$\underset{CH_3}{|}$$

2-Bromo-4,5-dimethylheptane

STEP 3 Write the name, listing all substituents in alphabetical order and using one of the prefixes *di-*, *tri-*, and so forth, if more than one of the same substituent is present.

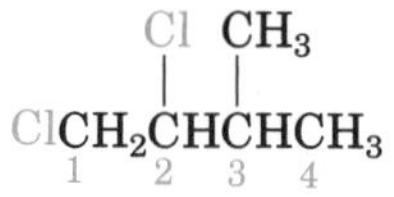

$$\overset{Cl}{|}\ \overset{CH_3}{|}$$
$$\underset{1\ \ 2\ \ 3\ \ 4}{ClCH_2CHCHCH_3}$$

1,2-Dichloro-3-methylbutane

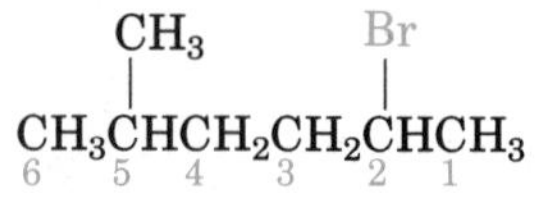

$$\overset{CH_3}{|}\qquad\overset{Br}{|}$$
$$\underset{6\ \ 5\ \ 4\ \ 3\ \ 2\ \ 1}{CH_3CHCH_2CH_2CHCH_3}$$

2-Bromo-5-methylhexane
(*NOT* 5-bromo-2-methylhexane)

In addition to their systematic names, many simple alkyl halides are also named by identifying first the alkyl group and then the halogen with an *-ide* ending. For example, CH_3I can be called either iodomethane or methyl iodide.

CH_3I

Iodomethane
(or methyl iodide)

$$\overset{Cl}{|}$$
$$CH_3CHCH_3$$

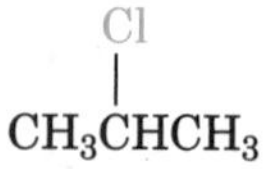

2-Chloropropane
(or isopropyl chloride)

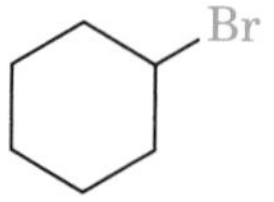

Br

Bromocyclohexane
(or cyclohexyl bromide)

PROBLEM 7.1

Give the IUPAC names of these alkyl halides:
(a) $CH_3CH_2CHBrCH_3$ (b) $CH_3CH_2CHClCH(CH_3)_2$
(c) $(CH_3)_2CHCH_2CH_2Cl$ (d) $(CH_3)_2CClCH_2CH_2Cl$
(e) $BrCH_2CH_2CH_2CH_2Cl$ (f) $CH_3CHBrCH_2CH_2CH_2Cl$

PROBLEM 7.2

Draw structures corresponding to these names:
(a) 2-Chloro-3,3-dimethylhexane (b) 3,3-Dichloro-2-methylhexane
(c) 3-Bromo-3-ethylpentane (d) 2-Bromo-5-chloro-3-methylhexane

7.2 PREPARATION OF ALKYL HALIDES: RADICAL CHLORINATION OF ALKANES

We've already seen several methods of alkyl halide preparation, including the addition reaction of HX and X_2 with alkenes (Sections 4.1 and 4.5):

$$\mathrm{>C{=}C<} + \mathrm{HCl} \longrightarrow -\overset{\mathrm{H}}{\underset{|}{\overset{|}{\mathrm{C}}}}-\overset{\mathrm{Cl}}{\underset{|}{\overset{|}{\mathrm{C}}}}-$$

$$\mathrm{>C{=}C<} + \mathrm{Br_2} \longrightarrow \mathrm{Br{-}C{-}C{-}Br}$$

Another method of alkyl halide synthesis is the reaction of alkanes with chlorine or bromine. Although inert to most reagents, alkanes react readily with chlorine in the presence of ultraviolet light ($h\nu$) to give chlorinated alkane products. For example, methane reacts with chlorine gas to give chloromethane:

$$CH_4 + Cl_2 \xrightarrow{h\nu} CH_3Cl + HCl$$

Methane **Chloromethane**

RADICAL SUBSTITUTION REACTION
Substitution that takes place by a radical mechanism

The chlorination of methane is a typical **radical substitution reaction** rather than a polar reaction of the sort we've been studying until now. Recall from Section 3.6 that radical reactions involve reagents that have an odd number of electrons. Bonds are formed in radical reactions when each partner donates one electron to the new bond, and bonds are broken when each fragment leaves with one electron.

$$\mathrm{A\cdot + \cdot B \longrightarrow A{:}B}$$ **Radical (heterogenic) bond making**

$$\mathrm{A{:}B \longrightarrow A\cdot + \cdot B}$$ **Radical (heterolytic) bond breaking**

Radical substitution reactions normally require three kinds of steps: an *initiation* step, *propagation* steps, and *termination* steps. As its name implies, the initiation step starts the reaction by producing reactive radicals. In the chlorination reaction, the relatively weak Cl–Cl bond is broken by irradiation with ultraviolet light to give two chlorine radicals.

Once chlorine radicals have been produced in small amounts, reaction of Cl_2 with methane occurs by a sequence of two propagation steps. In the first propagation step, a chlorine radical (Cl·) abstracts a hydrogen atom from methane to produce HCl and a methyl radical ($\cdot CH_3$). In the second propagation step, the methyl radical abstracts a chlorine atom from Cl_2 to yield chloromethane and a new chlorine radical, which then cycles back to the first propagation step, making the overall process a *chain reaction*. Once the reaction has been initiated, it becomes a self-sustaining cycle of endlessly repeating propagation steps 1 and 2.

Occasionally, two radicals collide and combine to form a stable product. When this kind of termination step happens, the reaction cycle is interrupted and the chain is ended. The overall mechanism of methane chlorination is shown in Figure 7.1.

Initiation step $Cl{-}Cl \xrightarrow{h\nu} 2\,Cl\cdot$

Propagation steps (a repeating cycle)

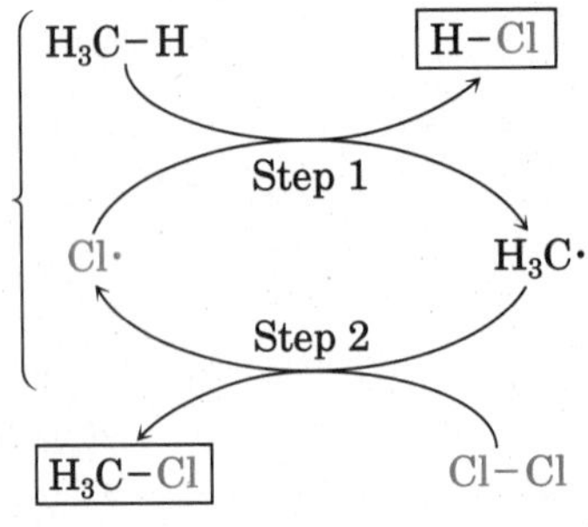

Termination steps

$$H_3C\cdot + \cdot CH_3 \longrightarrow H_3C{-}CH_3$$
$$Cl\cdot + \cdot CH_3 \longrightarrow Cl{-}CH_3$$
$$Cl\cdot + \cdot Cl \longrightarrow Cl{-}Cl$$

Overall reaction $CH_4 + Cl_2 \longrightarrow CH_3Cl + HCl$

FIGURE 7.1 Mechanism of the radical chlorination of methane.

Though interesting from a mechanistic point of view, alkane chlorination is not very useful for preparing most alkyl chlorides because mixtures of products usually result. Chlorination of methane doesn't stop cleanly at the monochlorinated stage, but continues on, giving a mixture of dichloro, trichloro, and even tetrachloro products that must be separated:

$$CH_4 + Cl_2 \longrightarrow CH_3Cl + HCl$$
$$\xrightarrow{Cl_2} CH_2Cl_2 + HCl$$
$$\xrightarrow{Cl_2} CHCl_3 + HCl$$
$$\xrightarrow{Cl_2} CCl_4 + HCl$$

The situation is even worse for chlorination of alkanes that have more than one kind of hydrogen. Chlorination of butane gives two monochlorinated products as well as several dichlorobutanes, trichlorobutanes, and so on. Of the monochloro product, 30% is 1-chlorobutane and 70% is 2-chlorobutane:

$$\underset{\textbf{Butane}}{CH_3CH_2CH_2CH_3} + Cl_2 \xrightarrow{h\nu} \underbrace{\underset{\textbf{1-Chlorobutane}}{CH_3CH_2CH_2CH_2Cl} + \underset{\textbf{2-Chlorobutane}}{CH_3CH_2\overset{\displaystyle Cl}{\overset{|}{C}}HCH_3}}_{30:70} + \text{Dichloro-, trichloro-, tetrachloro-, and so on}$$

PRACTICE PROBLEM 7.1 Draw all the monochloro products you might get from radical chlorination of 2-methylbutane.

SOLUTION Draw the structure of the starting material and begin systematically replacing each kind of hydrogen by chlorine. In this example, there are four possibilities.

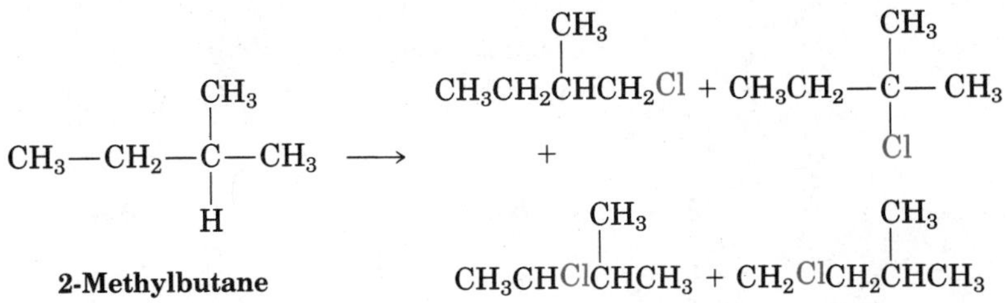

$$CH_3{-}CH_2{-}\overset{\displaystyle CH_3}{\underset{\displaystyle H}{C}}{-}CH_3 \longrightarrow CH_3CH_2\overset{CH_3}{CH}CH_2Cl + CH_3CH_2{-}\overset{\displaystyle CH_3}{\underset{\displaystyle Cl}{C}}{-}CH_3 + CH_3CHCl\overset{CH_3}{CH}CH_3 + CH_2ClCH_2\overset{CH_3}{CH}CH_3$$

2-Methylbutane

PROBLEM 7.3 Draw and name all monochloro products you might obtain from radical chlorination of 3-methylpentane. Which, if any, are chiral?

PROBLEM 7.4 Radical chlorination of pentane is a poor way to prepare 1-chloropentane, but radical chlorination of 2,2-dimethylpropane is a good way to prepare 1-chloro-2,2-dimethylpropane. Explain.

7.3 ALKYL HALIDES FROM ALCOHOLS

The best method for preparing alkyl halides is from alcohols, a reaction carried out most simply by treating the alcohol with HX:

$$R{-}OH + HX \longrightarrow R{-}X + H_2O$$

For reasons that will be discussed in Section 7.8, the reaction works best when applied to tertiary alcohols. Primary and secondary alcohols react much more slowly.

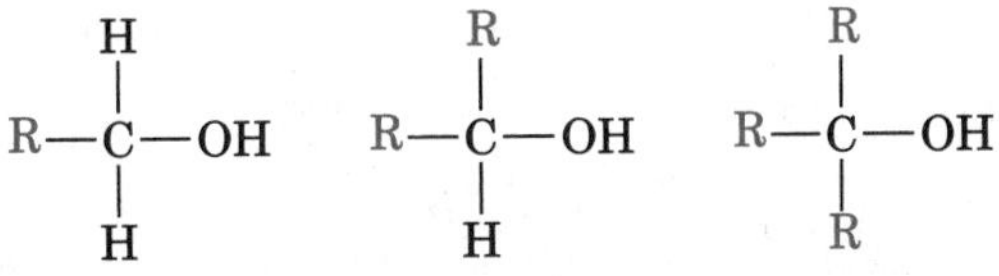

$$R{-}\overset{\displaystyle H}{\underset{\displaystyle H}{C}}{-}OH \qquad R{-}\overset{\displaystyle R}{\underset{\displaystyle H}{C}}{-}OH \qquad R{-}\overset{\displaystyle R}{\underset{\displaystyle R}{C}}{-}OH$$

Primary (1°) Secondary (2°) Tertiary (3°)

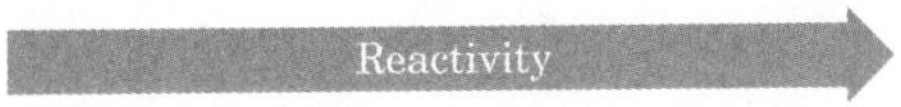

Reactivity →

Primary and secondary alcohols are converted into alkyl halides by treatment with either thionyl chloride ($SOCl_2$) or phosphorus tribromide (PBr_3). These reactions normally take place under fairly mild conditions, and yields are usually high.

$$\text{Benzoin} \xrightarrow{SOCl_2} \text{Desyl chloride (86\%)} + SO_2 + HCl$$

Benzoin **Desyl chloride (86%)**

$$3\ CH_3CH_2CH(OH)CH_3 \xrightarrow[\text{Ether, 35°C}]{PBr_3} 3\ CH_3CH_2CH(Br)CH_3 + P(OH)_3$$

2-Butanol **2-Bromobutane (86%)**

PRACTICE PROBLEM 7.2 Predict the product of this reaction:

$$C_6H_5CH(OH)CH_3 \xrightarrow{SOCl_2}$$

SOLUTION Alcohols yield alkyl chlorides on treatment with $SOCl_2$:

$$C_6H_5CH(OH)CH_3 \xrightarrow{SOCl_2} C_6H_5CH(Cl)CH_3$$

PROBLEM 7.5 How would you prepare these alkyl halides from the appropriate alcohols?

(a) 2-Chloro-2-methylpropane **(b)** 2-Bromo-4-methylpentane

(c) $BrCH_2CH_2CH_2CH_2CH(CH_3)CH_3$ **(d)** $CH_3CH_2CH(CH_3)CH_2C(Cl)(CH_3)CH_3$

PROBLEM 7.6 Predict the products of these reactions:

(a) $CH_3CH_2CH(OH)CH_2CH(CH_3)CH_3 + PBr_3 \longrightarrow ?$

(b) 1-Methylcyclohexanol (cyclohexane ring bearing CH_3 and OH) $+ HCl \longrightarrow ?$

(c) 3,3-Dimethylcyclopentanol (cyclopentane ring bearing OH and two H_3C groups) $+ SOCl_2 \longrightarrow ?$

7.4 REACTIONS OF ALKYL HALIDES: GRIGNARD REAGENTS

GRIGNARD REAGENT
Organomagnesium halide, RMgX

Alkyl halides react with metallic magnesium in ether solvent to yield organomagnesium halides, called **Grignard reagents** after their discoverer, Victor Grignard. Grignard reagents contain a carbon–metal bond and thus are *organometallic compounds.*

$$R{-}X + Mg \xrightarrow{\text{Ether}} R{-}Mg{-}X$$

where R = 1°, 2°, or 3° alkyl or aryl
X = Cl, Br, or I

For example,

$$CH_3CH_2\overset{\displaystyle Cl}{\overset{|}{C}}HCH_3 \xrightarrow[\text{Ether}]{\text{Mg}} CH_3CH_2\overset{\displaystyle MgCl}{\overset{|}{C}}HCH_3$$

2-Chlorobutane **sec-Butylmagnesium chloride**

Grignard reagents are extraordinarily useful and versatile compounds. As you might expect from the discussion of electronegativity and bond polarity in Section 1.12, a carbon–magnesium bond is strongly polarized, making the organic part strongly nucleophilic.

δ^+ MgX
δ^- C ← Nucleophilic carbon

Because of their nucleophilic character, Grignard reagents react with a wide variety of electrophiles. For example, they react with acids such as HCl or H_2O to yield hydrocarbons. The overall sequence, R–X → R–MgX → R–H, is a useful method for converting organic halides into alkanes:

$$CH_3(CH_2)_8CH_2Br \xrightarrow[\text{2. } H_2O]{\text{1. Mg}} CH_3(CH_2)_8CH_3$$

1-Bromodecane **Decane (85%)**

PRACTICE PROBLEM 7.3 By using several reactions in sequence, you can accomplish transformations that can't be done in a single step. How would you prepare the alkane methylcyclohexane from the alcohol 1-methylcyclohexanol?

$$\text{1-Methylcyclohexanol} \xrightarrow{?} \text{Methylcyclohexane}$$

1-Methylcyclohexanol Methylcyclohexane

SOLUTION We know that alcohols can be converted into alkyl halides and that alkyl halides can be converted into alkanes. Carrying out the two reactions sequentially thus converts 1-methylcyclohexanol into methylcyclohexane.

$$\text{1-Methylcyclohexanol} \xrightarrow{HBr} \text{1-Bromo-1-methylcyclohexane} \xrightarrow[2.\ H_2O]{1.\ Mg,\ ether} \text{Methylcyclohexane}$$

1-Methylcyclohexanol **1-Bromo-1-methylcyclohexane** **Methylcyclohexane**

PROBLEM 7.7 An advantage to preparing alkanes from Grignard reagents is that deuterium (D; the isotope of hydrogen with atomic weight 2) can be placed at a specific site in a molecule. How might you convert 2-bromobutane into 2-deuteriobutane?

$$CH_3\underset{|}{\overset{Br}{C}}HCH_2CH_3 \xrightarrow{?} CH_3\overset{D}{C}HCH_2CH_3$$

PROBLEM 7.8 How could you convert 4-methyl-1-pentanol into 2-methylpentane?

$$CH_3\overset{CH_3}{C}HCH_2CH_2CH_2OH$$ **4-Methyl-1-pentanol**

7.5 NUCLEOPHILIC SUBSTITUTION REACTIONS: THE DISCOVERY

In 1896, the German chemist Paul Walden reported a remarkable discovery. He found that (+)- and (−)-malic acids can be interconverted by a series of simple reactions. When Walden treated (−)-malic acid with PCl_5, he isolated (+)-chlorosuccinic acid. This, on reaction with wet Ag_2O, gave (+)-malic acid. Similarly, reaction of (+)-malic acid with PCl_5 gave (−)-chlorosuccinic acid, which was converted into (−)-malic acid when treated with wet Ag_2O. The full cycle of reactions reported by Walden is shown in Figure 7.2.

At the time, the results were astonishing. Since (−)-malic acid was converted into (+)-malic acid, *some reactions in the cycle must have occurred with a change in the configuration of the stereogenic center.* But which ones, and how?

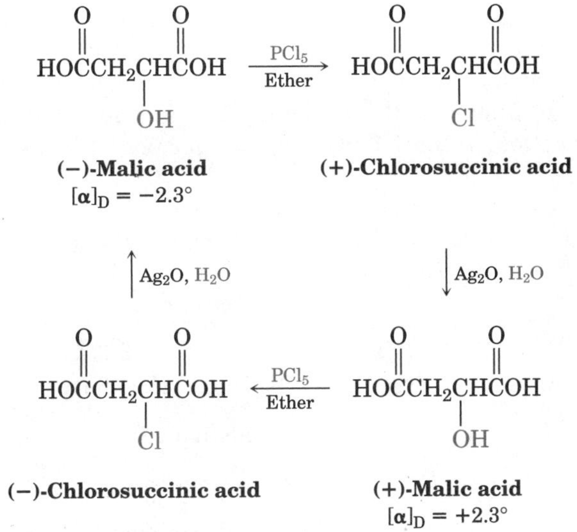

$$\mathrm{HO\overset{O}{\overset{\|}{C}}CH_2\underset{OH}{\underset{|}{C}}H\overset{O}{\overset{\|}{C}}OH \xrightarrow[\text{Ether}]{PCl_5} HO\overset{O}{\overset{\|}{C}}CH_2\underset{Cl}{\underset{|}{C}}H\overset{O}{\overset{\|}{C}}OH}$$

(−)-Malic acid $[\alpha]_D = -2.3°$ **(+)-Chlorosuccinic acid**

↑ Ag_2O, H_2O ↓ Ag_2O, H_2O

$$\mathrm{HO\overset{O}{\overset{\|}{C}}CH_2\underset{Cl}{\underset{|}{C}}H\overset{O}{\overset{\|}{C}}OH \xleftarrow[\text{Ether}]{PCl_5} HO\overset{O}{\overset{\|}{C}}CH_2\underset{OH}{\underset{|}{C}}H\overset{O}{\overset{\|}{C}}OH}$$

(−)-Chlorosuccinic acid **(+)-Malic acid** $[\alpha]_D = +2.3°$

FIGURE 7.2 Walden's cycle of reactions interconverting (+)- and (−)-malic acids.

NUCLEOPHILIC SUBSTITUTION REACTION
Substitution reaction in which one nucleophile replaces another

Today, we refer to the transformations taking place in Walden's cycle as **nucleophilic substitution reactions**, because each step involves the substitution of one nucleophile (chloride ion, Cl^-, or hydroxide ion, OH^-) for another. Nucleophilic substitution reactions are one of the most important general reaction types in organic chemistry.

7.6 KINDS OF NUCLEOPHILIC SUBSTITUTION REACTIONS

Following the work of Walden, a series of investigations was undertaken during the 1920s and 1930s to clarify the mechanism of nucleophilic substitution reactions and to find out how inversions of configuration occur. We now know that nucleophilic substitutions occur by two major paths, named the *S_N1 reaction* and the *S_N2 reaction*. In both cases, the "S_N" part of the name stands for "substitution, nucleophilic." What the 1 and the 2 stand for will become clear soon.

LEAVING GROUP
The group that is replaced in a nucleophilic substitution reaction

Regardless of mechanism, the overall change during all nucleophilic substitution reactions is the same: A **nucleophile** (Nu: or Nu:^-) reacts with a substrate R–X and substitutes for X:^- (the **leaving group**) to yield the product R–Nu. If the nucleophile is neutral (Nu:), then the product is positively charged to maintain charge conservation; if it's negatively charged (Nu:^-), the product is neutral.

$$\text{Nu:} + \text{R—X} \longrightarrow \text{R—}\overset{+}{\text{Nu}} + \text{X:}^-$$

$$\text{Nu:}^- + \text{R—X} \longrightarrow \text{R—Nu} + \text{X:}^-$$

Because of the wide scope of nucleophilic substitution reactions, many products can be prepared. In fact, we've already seen a number of nucleophilic substitution reactions in previous chapters. The reaction of acetylide anions with alkyl halides (Section 4.16), for example, is an S_N2 reaction. Table 7.1 lists some other possibilities.

TABLE 7.1 Some Nucleophilic Substitution Reactions on Bromomethane: $Nu:^- + CH_3Br \longrightarrow Nu{-}CH_3 + :Br^-$

ATTACKING NUCLEOPHILE		PRODUCT	
FORMULA	NAME	FORMULA	NAME
$CH_3\ddot{S}:^-$	Methanethiolate	CH_3SCH_3	Dimethyl sulfide
$H\ddot{S}:^-$	Hydrosulfide	$HSCH_3$	Methane thiol
$:N{\equiv}C:^-$	Cyanide	$N{\equiv}CCH_3$	Acetonitrile
$:\ddot{I}:^-$	Iodide	ICH_3	Iodomethane
$H\ddot{O}:^-$	Hydroxide	$HOCH_3$	Methanol
$CH_3\ddot{O}:^-$	Methoxide	CH_3OCH_3	Dimethyl ether
$:\ddot{N}{=}N{=}\ddot{N}:^-$	Azide	N_3CH_3	Azidomethane
$:\ddot{Cl}:^-$	Chloride	$ClCH_3$	Chloromethane
$CH_3CO_2:^-$	Acetate	$CH_3CO_2CH_3$	Methyl acetate
$H_3N:$	Ammonia	$H_3\overset{+}{N}CH_3\ Br^-$	Methylammonium bromide

PRACTICE PROBLEM 7.4 What is the substitution product from reaction of 1-chloropropane with sodium hydroxide?

SOLUTION Write the two starting materials, and identify the nucleophile (in this instance, OH^-) and the leaving group (in this instance, Cl^-). Then replace the –Cl group by –OH and write the complete equation.

$$\underset{\textbf{1-Chloropropane}}{CH_3CH_2CH_2Cl} + Na^+\ {}^-OH \longrightarrow \underset{\textbf{1-Propanol}}{CH_3CH_2CH_2OH} + Na^+\ {}^-Cl$$

PRACTICE PROBLEM 7.5 How would you prepare 1-propanethiol, $CH_3CH_2CH_2SH$, using a nucleophilic substitution reaction?

SOLUTION Since the product contains an –SH group, it might be prepared by reaction of SH^- (hydrosulfide ion) on 1-bromopropane:

$$\underset{\textbf{1-Bromopropane}}{CH_3CH_2CH_2Br} + Na^+\ {}^-SH \longrightarrow \underset{\textbf{1-Propanethiol}}{CH_3CH_2CH_2SH} + Na^+\ {}^-Br$$

PROBLEM 7.9 What substitution products would you expect to obtain from these reactions?

(a) $CH_3CH_2CHBrCH_3 + LiI \longrightarrow ?$ **(b)** $(CH_3)_2CHCH_2Cl + HS^- \longrightarrow ?$

(c) 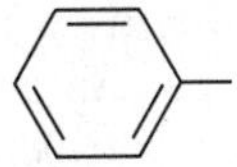 C_6H_5–$CH_2Br + NaCN \longrightarrow ?$

PROBLEM 7.10 How might you prepare these substances by using nucleophilic substitution reactions?

(a) $CH_3CH_2CH_2CH_2OH$ **(b)** $(CH_3)_2CHCH_2CH_2N_3$

7.7 THE S_N2 REACTION

S_N2 REACTION Nucleophilic substitution reaction that takes place in a single step by back-side displacement of the leaving group

An **S_N2 reaction** takes place in a single step without intermediates when the entering nucleophile attacks the substrate from a direction 180° away from the leaving group. As the nucleophile comes in on one side of the molecule, an electron pair on the nucleophile Nu:$^-$ forces out the leaving group X:$^-$, which departs from the other side of the molecule and takes with it the electron pair from the C–X bond (Figure 7.3). In the transition state for the reaction, the new Nu–C bond is partially forming at the same time the old C–X bond is partially breaking, and the negative charge is shared by both the incoming nucleophile and the outgoing leaving group.

The nucleophile Nu:$^-$ uses its lone-pair electrons to attack the alkyl halide 180° away from the halogen. This leads to a transition state with a partially formed C–Nu bond and a partially broken C–X bond.

Nu:$^-$ + C–X

↓

$[\overset{\delta-}{Nu}\cdots C \cdots \overset{\delta-}{X}]^{\ddagger}$

Transition state

The stereochemistry at carbon is inverted as the C–Nu bond forms fully and the halide departs with the electron pair from the original C–X bond.

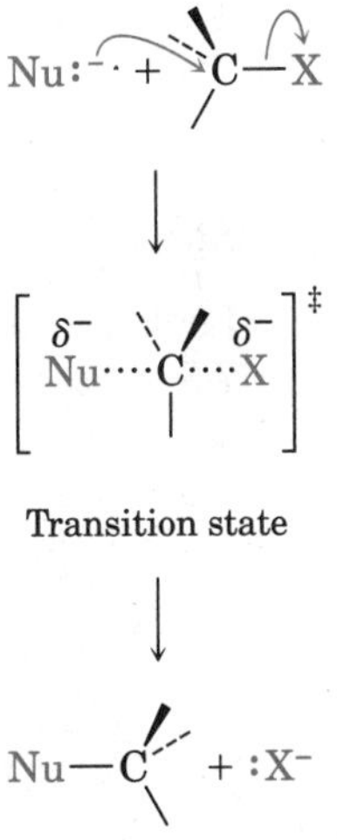

↓

Nu–C + :X$^-$

FIGURE 7.3 The mechanism of the S_N2 reaction.

Let's see what evidence there is for this mechanism and what the chemical consequences are.

Rates of S_N2 Reactions

REACTION RATE
Exact speed of a reaction under defined conditions

The exact speed with which a reaction occurs is called the **reaction rate** and is a quantity that can often be measured. The determination of reaction rates and of how those rates depend on reactant concentrations is a powerful tool for probing mechanisms. As an example, let's look at the effect of reactant concentrations on the rate of the S_N2 reaction of hydroxide ion with bromomethane to yield methanol:

$$HO:^- + CH_3{-}Br: \longrightarrow HO{-}CH_3 + :Br:^-$$

The S_N2 reaction of bromomethane with hydroxide ion takes place when substrate and nucleophile collide and react in a single step. At a given concentration of reactants, the reaction takes place at a certain rate. If we double the concentration of hydroxide ion, the frequency of encounter between the two reagents doubles, and we therefore find that the reaction rate also doubles. Similarly, if we double the concentration of bromomethane, the reaction rate doubles. Thus, the derivation of the "2" in S_N2: S_N2 reactions are said to be **bimolecular** because *two* molecules, alkyl halide and nucleophile, are involved in the step whose rate is measured.

BIMOLECULAR
Describing a reaction step that involves two molecules

PROBLEM 7.11

What effects would these changes have on the rate of the S_N2 reaction between iodomethane and sodium acetate?
(a) The CH_3I concentration is tripled.
(b) Both CH_3I and CH_3CO_2Na concentrations are doubled.

Stereochemistry of S_N2 Reactions

Look carefully at the mechanism of the S_N2 reaction shown in Figure 7.3. As the incoming nucleophile attacks the substrate and begins pushing out the leaving group on the opposite side, the stereochemistry of the molecule *inverts*. For example, treatment of (*S*)-2-bromobutane with hydroxide ion yields (*R*)-2-butanol:

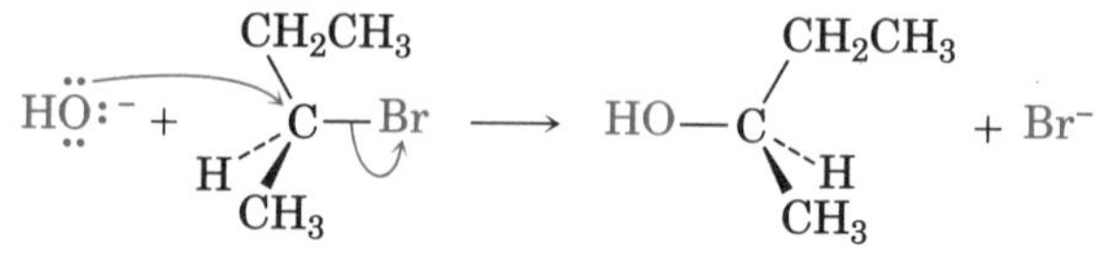

HO:⁻ + (S)-2-bromobutane [C bearing CH_2CH_3, H, CH_3, Br] ⟶ HO—C [CH_2CH_3, H, CH_3] + Br^-

(*S*)-2-Bromobutane **(*R*)-2-Butanol**

The inversion that takes place during an S_N2 reaction is similar to what happens when an umbrella turns inside out in the wind (Figure 7.4).

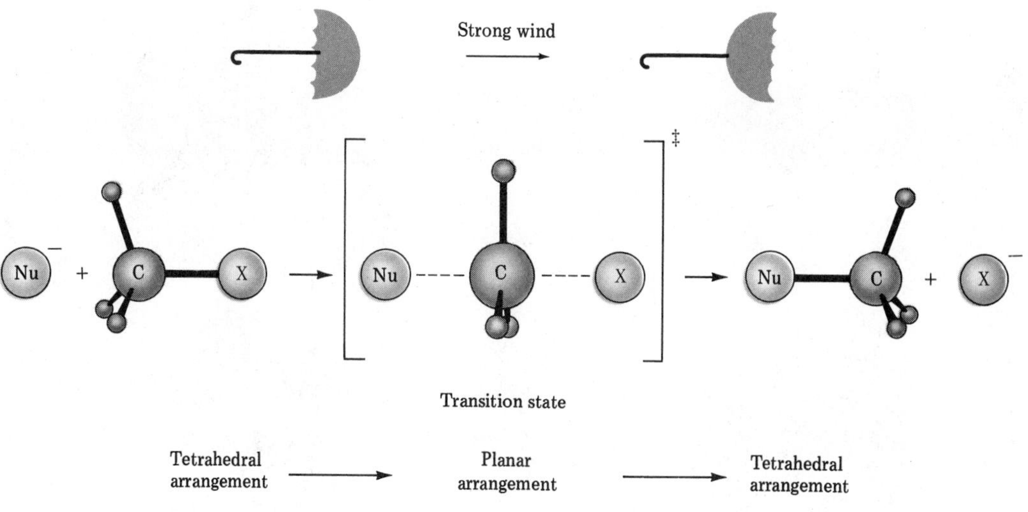

FIGURE 7.4 The stereochemistry of an S_N2 reaction. The inversion of configuration that occurs is like the inversion of an umbrella in a strong wind.

PRACTICE PROBLEM 7.6 What product would you expect to obtain from the S_N2 reaction of (*R*)-2-iodooctane with sodium cyanide, NaCN?

SOLUTION Table 7.1 shows that cyanide ion is a good nucleophile in the S_N2 reaction. We therefore expect it to displace iodide ion from (*R*)-2-iodooctane with inversion of configuration to yield (*S*)-2-methyloctanenitrile.

H I —$Na^+ \, ^-:CN$→ CN H + NaI

(*R*)-2-Iodooctane **(*S*)-2-Methyloctanenitrile**

PROBLEM 7.12 What product would you expect to obtain from the S_N2 reaction of (*S*)-2-bromohexane with sodium acetate, CH_3CO_2Na? Show the stereochemistry of both starting material and product.

PROBLEM 7.13 How can you explain the fact that treatment of (*R*)-2-bromohexane with NaBr yields *racemic* 2-bromohexane?

Steric Effects in S_N2 Reactions

Since an attacking nucleophile must approach the substrate closely to expel the leaving group in an S_N2 reaction, the ease of approach depends on the steric accessibility of the substrate. Bulky substrates in which the halide-bearing carbon atom is shielded from attack react more slowly than substrates in which the carbon is more accessible (Figure 7.5).

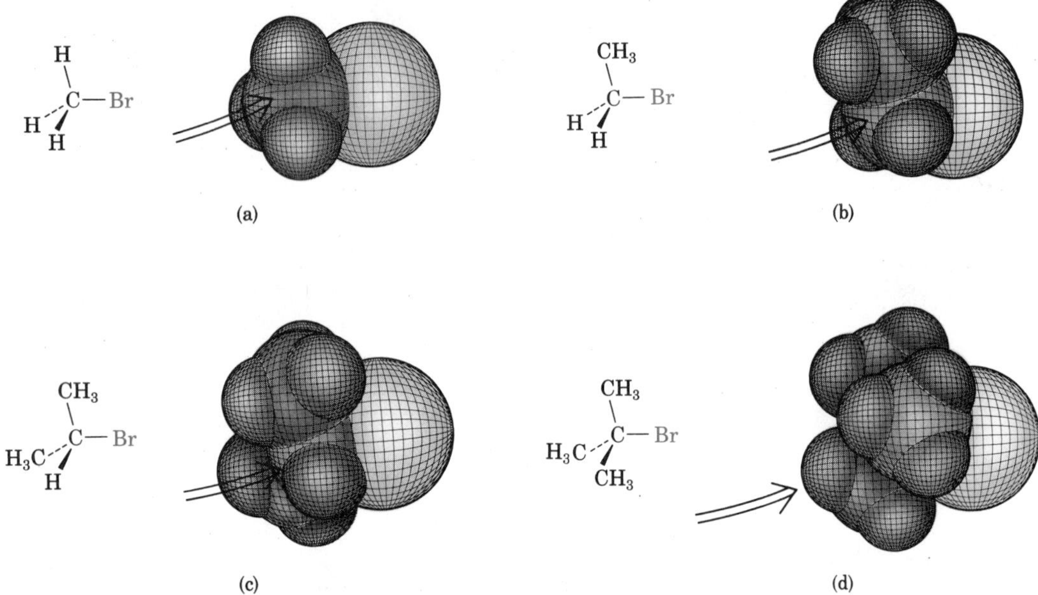

FIGURE 7.5 Steric hindrance to the S_N2 reaction. As these computer-generated models indicate, the carbon atom in (a) bromomethane is readily accessible, resulting in a fast S_N2 reaction, but the carbon atoms in (b) bromoethane, (c) 2-bromopropane, and (d) 2-bromo-2-methylpropane are successively less accessible, resulting in successively slower S_N2 reactions.

The relative reactivities of S_N2 substrates are as follows:

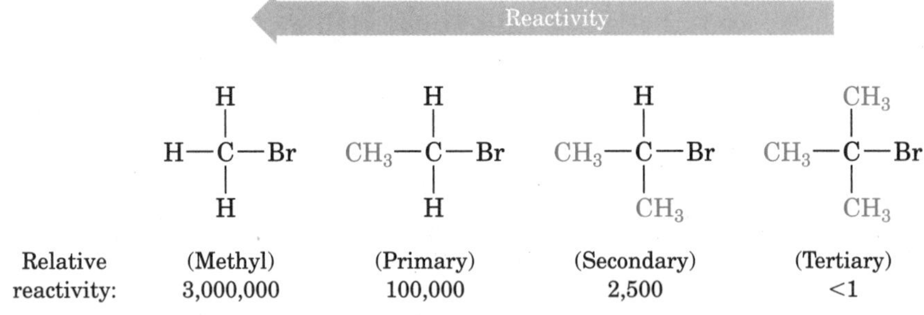

Methyl substrates (CH_3–X) are the most reactive, followed by primary alkyl (RCH_2–X) such as ethyl and propyl. Alkyl branching next to the leaving group, as in secondary substrates (R_2CH–X), slows the reaction greatly. Further branching, as in tertiary substrates (R_3C–X), effectively halts the reaction.

Although not shown in the reactivity list, vinylic (R_2C=CRX) and aryl (Ar–X) substrates are completely unreactive toward S_N2 displacements. This lack of reactivity is probably due to steric hindrance, because the incoming nucleophile would have to approach in the plane of the molecules to carry out a back-side displacement.

No S_N2 reactions with these halides:

A vinylic halide An aryl halide

PRACTICE PROBLEM 7.7 Which would you expect to be faster, the S_N2 reaction of OH^- ion with 1-bromopentane or with 2-bromopentane?

SOLUTION Since 1-bromopentane is a 1° halide and 2-bromopentane is a 2° halide, reaction with the less hindered 1-bromopentane should be faster.

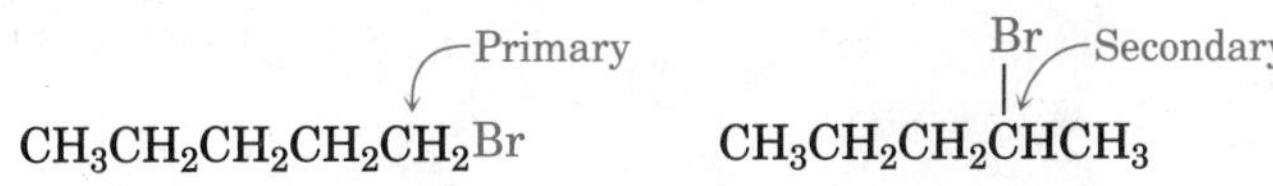

1-Bromopentane **2-Bromopentane**

PROBLEM 7.14 Which of these S_N2 reactions would you expect to be faster?

(a) Reaction of CN^- (cyanide ion) with $CH_3CHBrCH_3$ or with $CH_3CH_2CH_2Br$?

(b) Reaction of I^- with $(CH_3)_2CHCH_2Cl$ or with $H_2C{=}CHCl$?

The Leaving Group in S_N2 Reactions

Another variable that can affect the S_N2 reaction is the nature of the leaving group displaced by the attacking nucleophile. Because the leaving group is expelled with a negative charge in most S_N2 reactions, the best leaving groups are those that give the more stable anions: I^-, Br^-, and Cl^-. Of course, it's just as important to know which are poor leaving groups as to know which are good, and the reactivity list below indicates that fluoride ion (F^-), acetate ion ($CH_3CO_2^-$, or AcO^-), hydroxide ion (HO^-), alkoxide ion (RO^-), and amide ion (H_2N^-) are not displaced by nucleophiles. In other words, alkyl fluorides, esters, alcohols, ethers, and amines normally don't undergo S_N2 reactions.

← Reactivity as leaving group

$$I^- > Br^- > Cl^- \gg F^- > CH_3\overset{\overset{\displaystyle O}{\|}}{C}O^- > HO^- > CH_3O^- > H_2N^-$$

← Stability of anion

PROBLEM 7.15 Rank the following compounds in order of their expected reactivity toward S_N2 reaction: CH_3I, CH_3OH, CH_3Br, CH_3COOCH_3.

7.8 THE S_N1 REACTION

S_N1 REACTION
Nucleophilic substitution reaction that takes place in two steps through a carbocation intermediate

Most nucleophilic substitution reactions take place by the S_N2 pathway, but an alternative called the **S_N1 reaction** can also occur. In general, S_N1 reactions take place only on *tertiary* substrates and only under neutral or acidic conditions in a hydroxylic solvent such as water or alcohol. We saw in Section 7.3, for example, that alkyl halides can be prepared from alcohols by treatment with HCl or HBr. Tertiary alcohols react rapidly, but primary and secondary alcohols are much slower.

$$\text{R–OH} + \text{HX} \longrightarrow \text{R–X} + H_2O$$

$$\underset{3°}{R_3COH} > \underset{2°}{R_2CHOH} > \underset{1°}{RCH_2OH} > \underset{\text{Methanol}}{CH_3OH}$$

More reactive ← Reactivity — Less reactive

What's going on here? Clearly, a nucleophilic substitution reaction is taking place—a halide group is replacing an –OH group—yet the reactivity order 3° > 2° > 1° is backward from the normal S_N2 order. Furthermore, an –OH group is being replaced, although we said in the previous section that alcohols don't enter into S_N2 reactions. What's going on is that this is *not* an S_N2 reaction; it's an S_N1 reaction. The mechanism is shown in Figure 7.6.

Unlike an S_N2 reaction, in which the leaving group is displaced *at the same time* that the incoming nucleophile approaches, an S_N1 reaction takes place by the spontaneous loss of the leaving group *before* the incoming nucleophile approaches. Loss of the leaving group gives a carbocation intermediate that then reacts with nucleophile in a second step to yield the substitution product.

This two-step mechanism explains why tertiary alcohols react with HBr so much more rapidly than primary or secondary ones do: S_N1 reactions occur only when stable carbocation intermediates are formed. The more stable the carbocation intermediate, the faster the S_N1 reaction. Thus, the reactivity order of alcohols with HBr is the same as the stability order of carbocations (Section 4.3).

Rates of S_N1 Reactions

UNIMOLECULAR
Describing a reaction step that involves only one molecule

Unlike an S_N2 reaction, whose rate depends on the concentrations of both substrate and nucleophile, the rate of an S_N1 reaction depends only on the concentration of the substrate but is *independent of the concentration of the nucleophile*. Thus, the derivation of the "1" in S_N1: S_N1 reactions are **unimolecular** because only one molecule is involved in the step whose rate is measured. The observation that S_N1 reactions are unimolecular means that the substrate must undergo a spontaneous reaction without assistance from the nucleophile, exactly what the mechanism shown in Figure 7.6 accounts for.

$$(CH_3)_3C{-}\ddot{O}H + H{-}Br$$

The –OH group is first protonated by HBr.

$$\rightleftharpoons \left[(CH_3)_3C{-}\overset{+}{O}H_2\right] + Br^-$$

Spontaneous dissociation of the protonated alcohol occurs in a slow, rate-limiting step to yield a carbocation intermediate plus water.

Carbocation

$$\rightleftharpoons \left[(CH_3)_3C^+\right] + H_2O$$

The carbocation intermediate reacts with bromide ion in a fast step to yield the neutral substitution product.

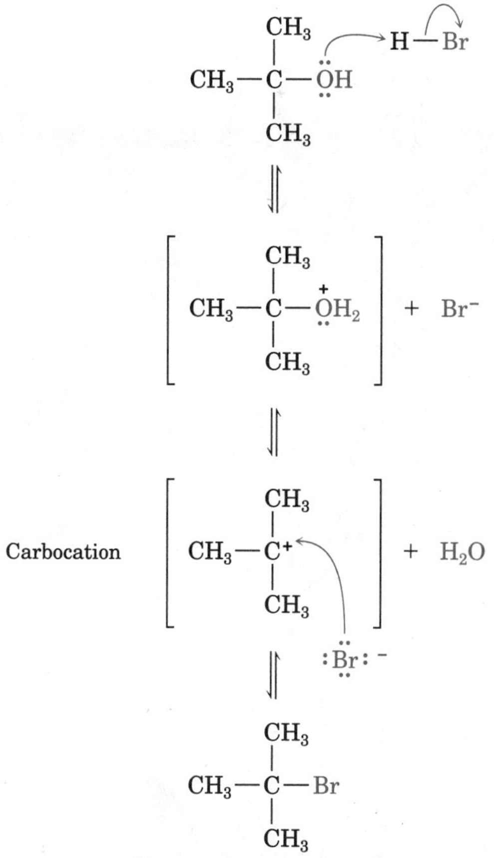

$$\xrightarrow{:\ddot{Br}:^-} (CH_3)_3C{-}Br$$

FIGURE 7.6 The mechanism of the S_N1 reaction of *tert*-butyl alcohol with HBr. Neutral water is the leaving group.

PROBLEM 7.16 What effect would the following changes have on the rate of the S_N1 reaction of *tert*-butyl alcohol with HBr?

(a) The HBr concentration is tripled.

(b) The HBr concentration is halved, and the *tert*-butyl alcohol concentration is doubled.

Stereochemistry of S_N1 Reactions

If S_N1 reactions occur through carbocation intermediates as shown in Figure 7.6, their stereochemistry should be different from S_N2 reactions. Since carbocations are planar and sp^2-hybridized, they are achiral. The positively charged carbon can therefore be attacked by a nucleophile equally well from either side, leading to a racemic mixture of enantiomers (Figure 7.7). In other words, if we carry out an S_N1 reaction on a single enantiomer of a chiral starting material, then the product must be optically inactive.

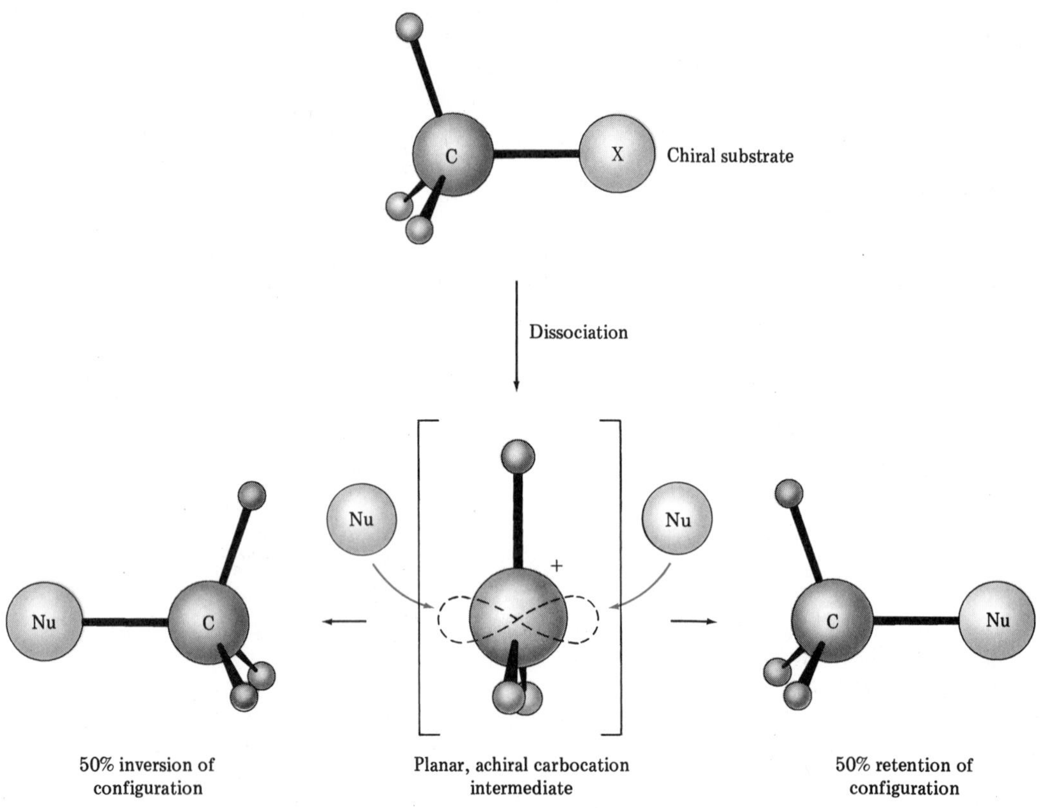

FIGURE 7.7 An S_N1 reaction on a chiral substrate. An optically active starting material gives a racemic product.

The prediction that S_N1 reactions on chiral substrates should lead to racemic products is exactly what is found. For example, reaction of optically active (*R*)-1-phenyl-1-butanol with HCl gives a racemic alkyl chloride product:

$CH_3CH_2CH_2$–C(H)(Ph)–OH + HCl ⟶ $CH_3CH_2CH_2$–C(H)(Ph)–Cl + Cl–C(H)(Ph)–$CH_2CH_2CH_3$

(*R*)-1-Phenyl-1-butanol | **(*R*)-1-Phenyl-1-chlorobutane** (50%, retention) | **(*S*)-1-Phenyl-1-chlorobutane** (50%, inversion)

PRACTICE PROBLEM 7.8 What stereochemistry would you expect in the S_N1 reaction of (*R*)-3-bromo-3-methylhexane with methanol to yield 3-methoxy-3-methylhexane?

SOLUTION First draw the starting alkyl halide, showing its correct stereochemistry. Then replace the bromine with a methoxy group ($-OCH_3$) to give the racemic product.

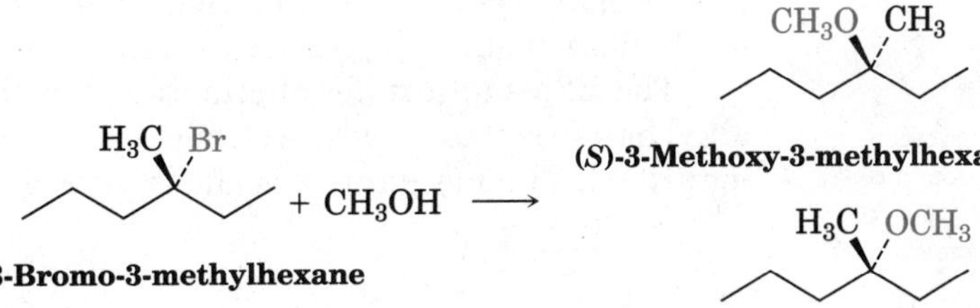

CH_3O CH_3

H_3C Br

+ CH_3OH ⟶

(*S*)-3-Methoxy-3-methylhexane (50%)

(*R*)-3-Bromo-3-methylhexane

H_3C OCH_3

(*R*)-3-Methoxy-3-methylhexane (50%)

PROBLEM 7.17 What product would you expect to obtain from the S_N1 reaction of (*S*)-3-methyl-3-octanol [(*S*)-3-hydroxy-3-methyloctane] with HBr? Show the stereochemistry of both starting material and product.

The Leaving Group in S_N1 Reactions

As with S_N2 reactions, the best leaving groups in S_N1 reactions are those that are most stable. Note that when an S_N1 reaction is carried out under acidic conditions, as when a tertiary alcohol reacts with HX to yield an alkyl halide (Figure 7.6), neutral water can act as a leaving group. The S_N1 reactivity order of leaving groups is

$$I^- > Br^- > H_2O \approx Cl^- \gg F^- > CH_3\overset{O}{\overset{\|}{C}}O^-$$

Reactivity as leaving group

7.9 ELIMINATIONS: THE E2 REACTION

Two kinds of reactions are possible when a nucleophile reacts with an alkyl halide. The nucleophile can substitute for the leaving group in an S_N1 or S_N2 reaction, or an elimination reaction can occur, leading to formation of an alkene:

Substitution: H–C–C–Br + OH^- ⟶ H–C–C–OH + Br^-

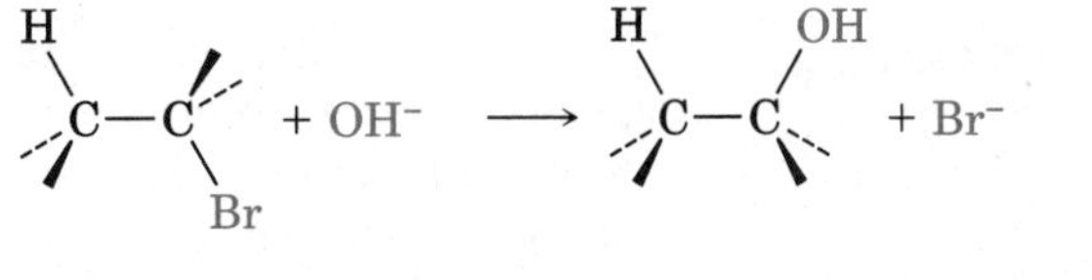

Elimination: H–C–C–Br + OH^- ⟶ C=C + H_2O + Br^-

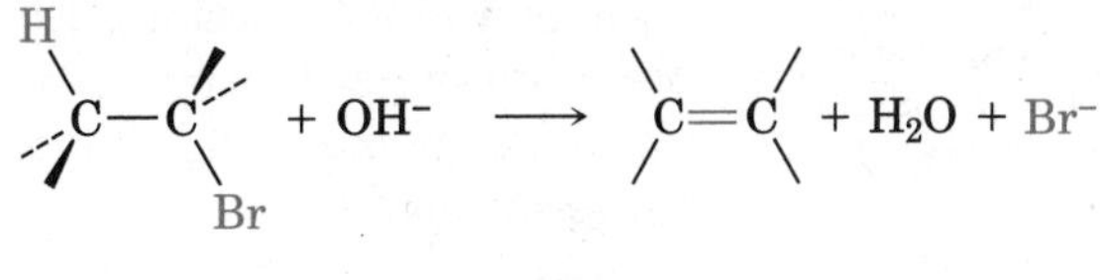

We saw in Section 4.9 that the elimination of HX from an alkyl halide is an extremely useful method for preparing alkenes. The subject is complex, though, because eliminations can take place by several different mechanistic pathways, just as substitutions can.

E2 REACTION
Elimination reaction that takes place in a single step through a bimolecular mechanism

The **E2 reaction** (for elimination, bimolecular) takes place when an alkyl halide is treated with a strong base such as hydroxide ion or alkoxide ion (RO^-). The mechanism is shown in Figure 7.8.

Base (B:) attacks a neighboring hydrogen and begins to remove the H at the same time as the alkene double bond starts to form and the X group starts to leave.

Transition state

Neutral alkene is produced when the C–H bond is fully broken and the X group has departed with the C–X bond electron pair.

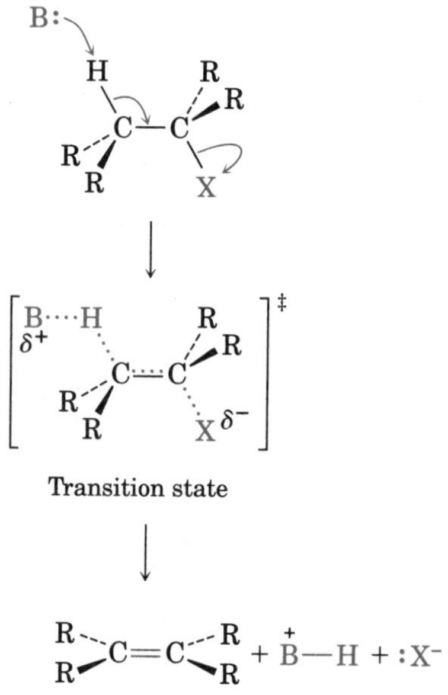

$$R_2C{=}CR_2 + \overset{+}{B}{-}H + :X^-$$

FIGURE 7.8 The mechanism of the E2 reaction. The reaction takes place in a single step, without intermediates. (Dotted lines indicate partial bonding in the transition state.)

Like the S_N2 reaction, the E2 reaction takes place in one step without intermediates. As the attacking base begins to abstract a proton from a carbon atom next to the leaving group, the C–H and C–X bonds begin to break, and the C=C double bond begins to form. When the leaving group departs, it takes with it the two electrons from the former C–X bond.

ANTI PERIPLANAR GEOMETRY
Reaction geometry in which all reacting atoms lie in a plane, with one group on top and another group on the bottom of the molecule

One good piece of evidence supporting this mechanism comes from measurements of reaction rates. Since both base and alkyl halide take part in the one step, E2 reactions show the same bimolecular behavior that S_N2 reactions do. A second piece of evidence involves the stereochemistry of E2 reactions. Eliminations almost always occur from an **anti periplanar geometry**, meaning that all reacting atoms lie in the same plane (*periplanar*) and that the H and X depart from opposite sides (*anti*) of the molecule:

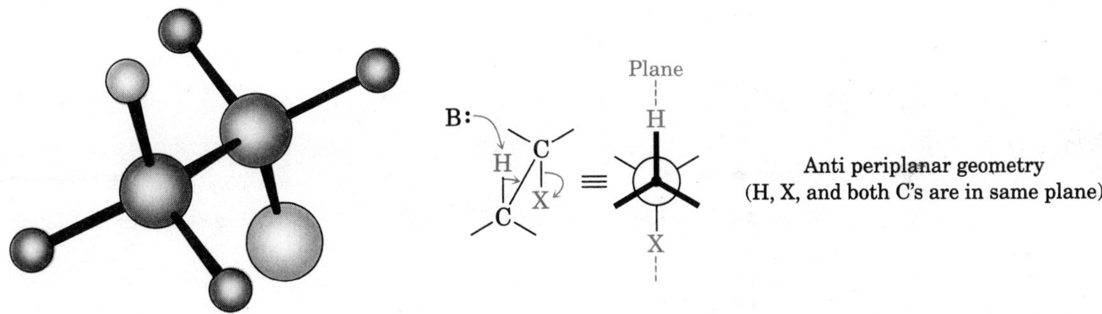

What's so special about anti periplanar geometry? Because the original C–H and C–X sp^3 sigma orbitals in the reactant must overlap and become p orbitals in the alkene product, *they must also overlap in the transition state*. This can occur only if the orbitals are in the same plane (are periplanar) to begin with (Figure 7.9).

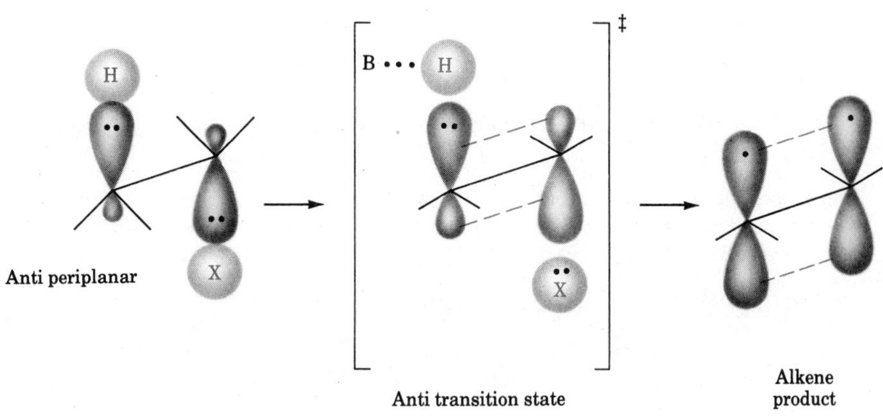

FIGURE 7.9 The transition state for an E2 reaction. Overlap of developing p orbitals in the transition state requires periplanar geometry.

Anti periplanar geometry for E2 reactions has stereochemical consequences that provide strong evidence for the proposed mechanism. To cite just one example, *meso*-1,2-dibromo-1,2-diphenylethane undergoes E2 elimination on treatment with base to give the pure E alkene, rather than a mixture of E and Z alkenes:

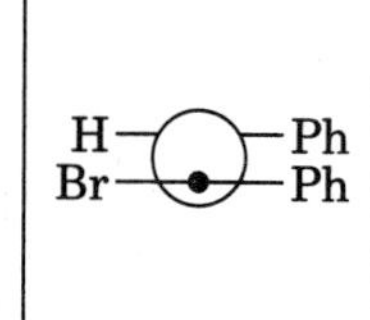

***meso*-1,2-Dibromo-1,2-diphenylethane** **(*E*)-1-Bromo-1,2-diphenylethylene**

where Ph =

PRACTICE PROBLEM 7.9 What stereochemistry would you expect for the alkene obtained by E2 elimination of (1*S*,2*S*)-1,2-dibromo-1,2-diphenylethane?

SOLUTION First, draw (1*S*,2*S*)-1,2-dibromo-1,2-diphenylethane so that you can see its stereochemistry and so that the –H and –Br groups to be eliminated are anti periplanar (molecular models are extremely helpful here). Keeping all substituents in approximately their same positions, eliminate HBr and see what alkene results. The product is (*Z*)-1-bromo-1,2-diphenylethylene.

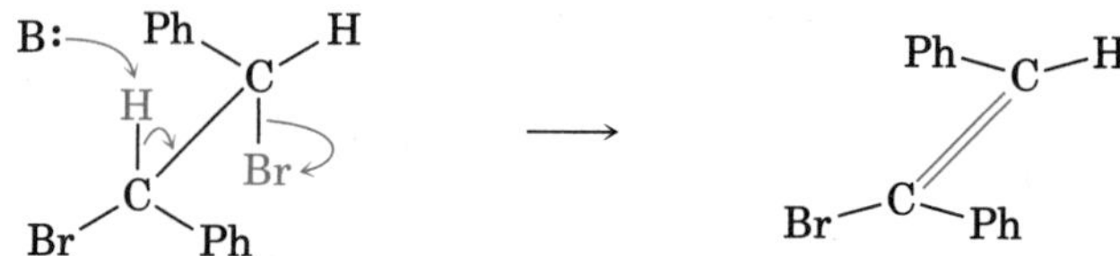

(1*S*,2*S*)-1,2-Dibromo-1,2-diphenylethane **(*Z*)-1-Bromo-1,2-diphenylethylene**

PROBLEM 7.18 Ignoring double-bond stereochemistry, what elimination products would you expect from these reactions? (Remember Zaitsev's rule; Section 4.9.)

(a) $CH_3CH_2CH(Br)CH(CH_3)CH_3$

(b) $CH_3CH(CH_3)CH_2{-}C(Cl)(CH_3){-}CH(CH_3)CH_3$

(c) Cyclohexyl–$CH(Br)CH_3$

PROBLEM 7.19 What stereochemistry would you expect for the alkene obtained by E2 elimination of (1*R*,2*R*)-1,2-dibromo-1,2-diphenylethane? Draw a Newman projection of the reacting conformation.

7.10 ▌ ELIMINATIONS: THE E1 REACTION

Just as the S_N2 reaction has an analog in the E2 reaction, the S_N1 reaction has an analog in the **E1 reaction** (elimination, unimolecular). There are numerous similarities between S_N1 and E1 processes. For example, both

E1 REACTION
Elimination reaction that takes place in two steps through a unimolecular mechanism

S_N1 and E1 reactions frequently occur when alcohols are treated with acid so that neutral H_2O is the leaving group. Furthermore, both S_N1 and E1 reactions are unimolecular processes that occur by spontaneous dissociation of the substrate to generate an intermediate carbocation, which then undergoes further reaction.

We saw in Section 4.9 that alcohols can be dehydrated by treatment with H_2SO_4 to give alkenes. Tertiary alcohols react rapidly under mild conditions, but primary and secondary alcohols react much more slowly. The acid-catalyzed dehydration of an alcohol occurs by an E1 mechanism, as shown in Figure 7.10. Acid first protonates the –OH group of the

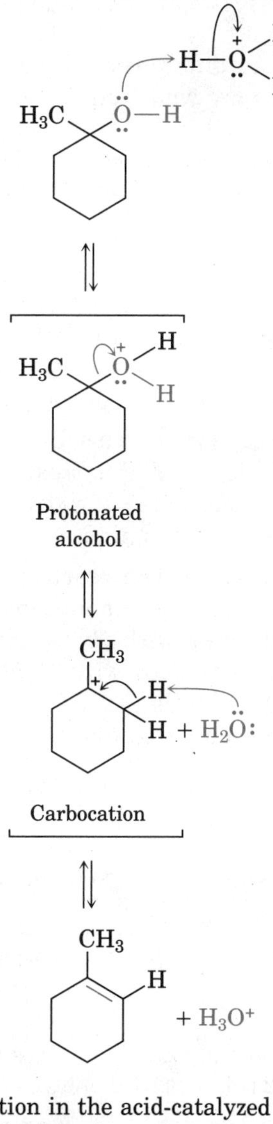

Two electrons from the oxygen atom bond to H^+, yielding a protonated alcohol intermediate.

Protonated alcohol

The carbon–oxygen bond breaks, and the two electrons from the bond stay with oxygen, leaving a carbocation intermediate.

$H + H_2\ddot{O}:$

Carbocation

Two electrons from a neighboring carbon–hydrogen bond form the alkene pi bond, and H^+ (a proton) is eliminated.

$+ H_3O^+$

FIGURE 7.10 Mechanism of the E1 reaction in the acid-catalyzed dehydration of a tertiary alcohol. Two steps are involved, and a carbocation intermediate is formed.

alcohol, the protonated alcohol spontaneously loses water to yield a carbocation intermediate, and the carbocation then loses H^+ from a neighboring carbon to give alkene product. Tertiary alcohols react faster than primary or secondary ones because they lead to more stable carbocation intermediates.

PROBLEM 7.20 What effect on the rate of an E1 reaction of 2-methyl-2-propanol would you expect if the concentration of the alcohol were tripled?

7.11 A SUMMARY OF REACTIVITY: S_N1, S_N2, E1, E2

Having seen four different kinds of nucleophilic substitution/elimination reactions, you may well wonder how to predict what will take place in any given case. Will substitution or elimination occur? Will the reaction be unimolecular or bimolecular? There are no rigid answers to these questions, but we can make some broad generalizations:

1. *Primary substrates* normally react by an S_N2 pathway because they are unhindered and because their dissociation would give unstable primary carbocations. If a primary alkyl halide is treated with a good nucleophile such as I^-, Br^-, RS^-, NH_3, or CN^-, only S_N2 substitution occurs. If a strong base such as hydroxide ion or an alkoxide ion (RO^-) is used, a small amount of competitive E2 elimination can also occur.
2. *Secondary substrates* react by either an S_N2 or an E2 pathway. If a secondary alkyl halide is treated with a good nucleophile, S_N2 substitution predominates. If a strong base such as hydroxide ion is used, E2 elimination predominates.
3. *Tertiary substrates* can react by any of three pathways, S_N1, E1, or E2, depending on the conditions and on the reactants. If a tertiary alcohol is treated with HCl, HBr, or HI, S_N1 substitution occurs; if the alcohol is treated with H_2SO_4, E1 elimination occurs; and if a tertiary alkyl halide is treated with a strong base such as hydroxide ion, E2 elimination occurs.

These generalizations are summarized in Table 7.2.

TABLE 7.2 Correlation of Structure and Reactivity for Substitution and Elimination Reactions

SUBSTRATE TYPE	S_N1	S_N2	E1	E2
RCH_2X (primary)	Does not occur	Favored	Does not occur	Can occur if strong base is used
R_2CHX (secondary)	Does not occur	Favored if good nucleophile is used	Does not occur	Favored if strong base is used
R_3CX (tertiary)	Favored in acidic solution	Does not occur	Favored in acidic solution	Favored if strong base is used

PRACTICE PROBLEM 7.10 Tell what kind of reaction this is:

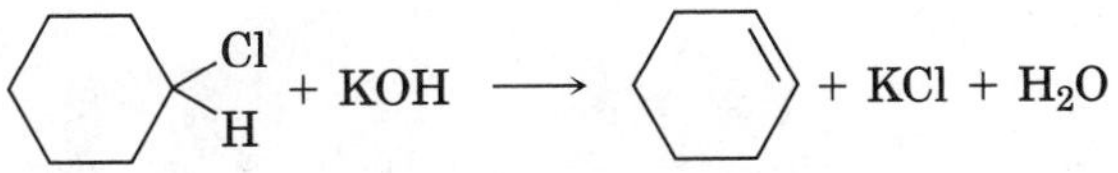

$$\text{Chlorocyclohexane (C(Cl)(H))} + KOH \longrightarrow \text{Cyclohexene} + KCl + H_2O$$

SOLUTION Because KOH is a strong base and a secondary alkyl halide is undergoing elimination of HCl, this is an E2 reaction.

PROBLEM 7.21 Tell whether these reactions are S_N1, S_N2, E1, or E2:

(a) 1-Bromobutane + NaN_3 $\longrightarrow$ 1-Azidobutane

(b) $CH_3CH_2\underset{}{\overset{Cl}{C}}HCH_2CH_3 + KOH \longrightarrow CH_3CH_2CH{=}CHCH_3$

(c) 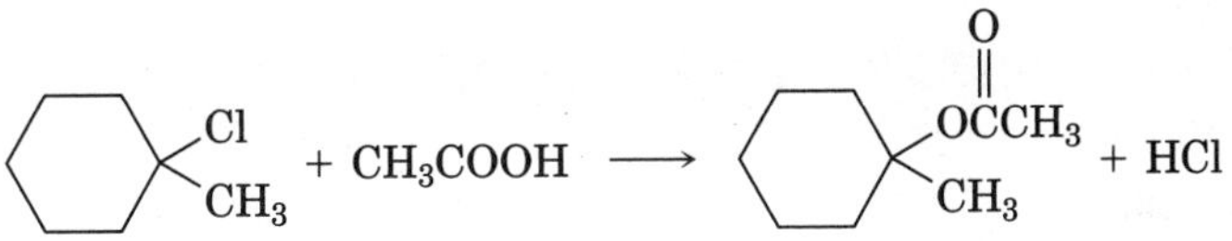 1-Chloro-1-methylcyclohexane + CH_3COOH $\longrightarrow$ 1-methylcyclohexyl acetate ($OCCH_3$ with C=O, CH_3) + HCl

7.12 BIOLOGICAL SUBSTITUTION REACTIONS

Many biological processes take place by reaction pathways analogous to those carried out in the laboratory. Thus, a number of processes that occur in living organisms involve nucleophilic substitution reactions. Perhaps the most common of all biological substitution reactions is *methylation*—the transfer of a methyl group from an electrophilic donor to a nucleophile:

$$\sim Y{-}CH_3 + :Nu^- \longrightarrow \sim Y:^- + CH_3{-}Nu$$

Methyl donor — Methylated nucleophile

Although a laboratory chemist would probably use iodomethane for a methylation reaction, living organisms operate more subtly. The large and complex molecule *S*-adenosylmethionine is the biological methyl-group donor. Since the sulfur atom of *S*-adenosylmethionine has a positive charge (a *sulfonium ion*), it is an excellent leaving group for S_N2 displacements on the methyl carbon. An example of the action of *S*-adenosylmethionine in biological methylations takes place in the adrenal medulla during the formation of adrenaline from norepinephrine (Figure 7.11).

After becoming used to simple alkyl halides like iodomethane, it's something of a shock to encounter a molecule as complex as *S*-adenosylmethionine. From the chemical standpoint, though, iodomethane and *S*-adenosylmethionine do exactly the same thing: Both transfer a methyl group in an S_N2 reaction. The same chemical principles apply to both.

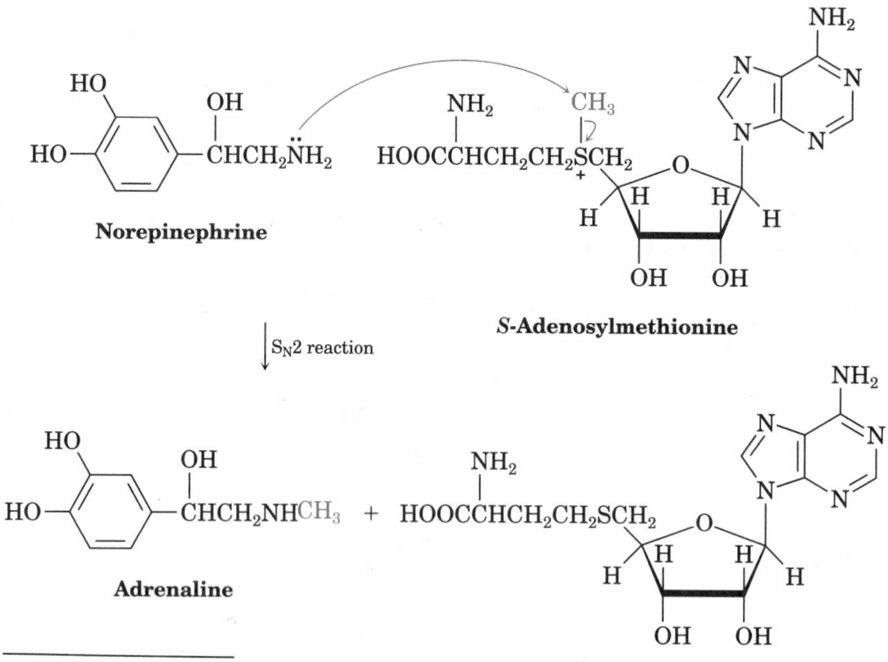
HO
HO
OH
CHCH2NH2
Norepinephrine
NH2
HOOCCHCH2CH2SCH2
CH3
NH2
N
N
N
N
O
H
H
H
H
OH OH
S-Adenosylmethionine
SN2 reaction
HO
HO
OH
CHCH2NHCH3
Adrenaline
+
NH2
HOOCCHCH2CH2SCH2

FIGURE 7.11 The biological formation of adrenaline by reaction of norepinephrine with *S*-adenosylmethionine.

INTERLUDE

ALKYL HALIDES AND THE OZONE HOLE

The aerosol can is a fixture of modern life—something we take for granted to spray our deodorants, paints, and insect repellents. In the early 1970s, though, it became apparent that the proliferation of aerosol sprays was leading to a serious environmental problem.

The volatile propellants used in aerosols at the time were various alkyl halides called *chlorofluorocarbons*, or *CFC's*, simple alkanes in which all the hydrogens have been replaced by either chlorine or fluorine. Fluorotrichloromethane (CCl_3F) and dichlorodifluoromethane (CCl_2F_2) are two of the most common CFC's. The advantage of using CFC's as aerosol propellants is that they are chemically inert and nonflammable. They don't react with the contents of the can, they leave no residue, they have no odor, and they're nontoxic. They do, however, escape into the atmosphere, where they ultimately find their way into the stratosphere.

The *ozone layer* is an atmospheric band extending from about 20 to 40 km above the earth's surface. Although ozone (O_3) is toxic in high concentrations, it is critically important in the upper atmosphere, because it acts as a shield to protect the surface of the earth from intense solar ultraviolet radiation. If the ozone layer were depleted or destroyed, more ultraviolet radiation would reach the earth, causing an increased incidence of skin cancers and eye cataracts. Unfortunately, destruction of ozone is exactly what CFC's do. Beginning around 1976, a disturbing amount of ozone depletion, the so-called ozone hole, began showing up over the south pole (Figure 7.12). Estimates of the extent of ozone destruction differ, but a recent report predicted a 5–9% depletion over the next 50 years.

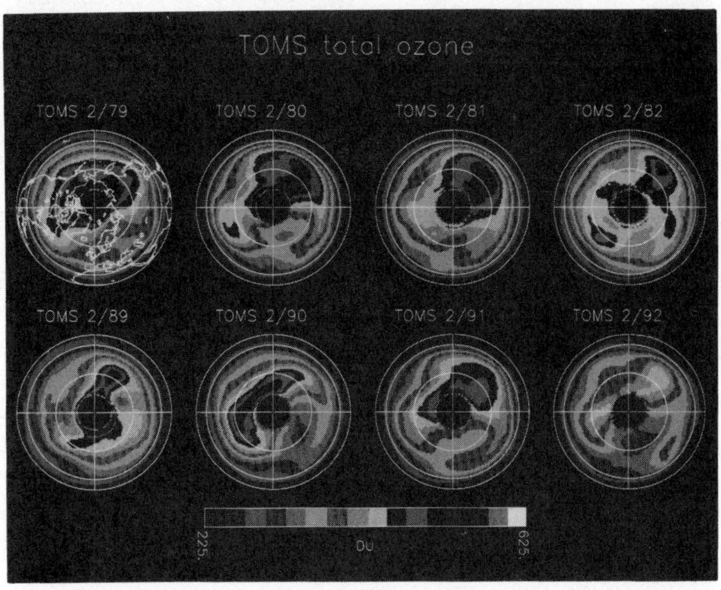

FIGURE 7.12 The development of the Antarctic ozone hole since 1979 as measured by the Total Ozone Mapping Spectrometer (TOMS). Ozone values in the hole indicated by the shaded area are up to 50% lower than normal values.

The mechanism of ozone destruction by chlorofluorocarbons involves radical reactions of the same sort we saw in the radical chlorination of methane (Section 7.2). Ultraviolet light ($h\nu$) striking a CFC molecule breaks a C–Cl bond, producing a chlorine radical. This radical then reacts with ozone to yield O_2 and ClO:

$$CCl_2F_2 \xrightarrow{h\nu} \cdot CClF_2 + \cdot Cl$$

$$\cdot Cl + O_3 \longrightarrow O_2 + ClO$$

Recognition of this problem led the U.S. government in 1980 to ban the use of CFC's for aerosol propellants. Worldwide action to reduce chlorofluorocarbon use lagged behind U.S. action, but began in September 1987, when an international agreement was reached by the European Community and 24 other nations. This agreement was extended in 1992 to call for a total ban on the release of CFC's by 1996.

SUMMARY AND KEY WORDS

Alkyl halides can be prepared by radical chlorination or bromination of alkanes, but product mixtures always result. Alkyl halides are best prepared from alcohols by treatment either with HX (for tertiary alcohols) or with $SOCl_2$ or PBr_3 (for primary and secondary alcohols). Alkyl halides react with magnesium metal to form organomagnesium halides, or **Grignard reagents**. These organometallic compounds react with acids to yield the corresponding alkanes.

Treatment of an alkyl halide with a nucleophile/base results either in substitution or in elimination. **Nucleophilic substitution reactions** occur by two mechanisms: S_N2 and S_N1. In the **S_N2 reaction**, the entering nucleophile attacks the substrate from a direction 180° away from the **leaving group**, resulting in an umbrellalike inversion of configuration at the carbon atom. S_N2 reactions are strongly inhibited by increasing steric bulk of the reagents and are favored only for primary and secondary substrates. In the **S_N1 reaction**, the substrate spontaneously dissociates to a carbocation followed by rapid attack of nucleophile. In consequence, S_N1 reactions take place with racemization of configuration at the carbon atom and are favored only for tertiary substrates.

Elimination reactions also occur by two mechanisms: E2 and E1. In the **E2 reaction**, a base abstracts a proton at the same time that the leaving group departs. The E2 reaction takes place with **anti periplanar geometry** and occurs when a substrate is treated with a strong base. In the **E1 reaction**, the substrate spontaneously dissociates to form a carbocation that can subsequently lose a neighboring proton. The reaction occurs on tertiary substrates in neutral or acidic hydroxylic solvents.

SUMMARY OF REACTIONS

1. Synthesis of alkyl halides
   (a) Radical chlorination of alkanes (Section 7.2)

$$-\overset{|}{\underset{|}{C}}-H + Cl_2 \xrightarrow{h\nu} -\overset{|}{\underset{|}{C}}-Cl + HCl$$

Reaction is very unselective

   (b) Alkyl halides from alcohols (Section 7.3)
      (1) Reaction of tertiary alcohols with HX, where X = Cl, Br

$$R_3C-OH \xrightarrow[\text{Ether}]{HX} R_3C-X + H_2O$$

(2) Reaction of primary and secondary alcohols with PBr_3 and $SOCl_2$

$$ROH + PBr_3 \longrightarrow RBr$$
$$ROH + SOCl_2 \longrightarrow RCl$$

2. Reactions of alkyl halides

(a) Formation and protonation of Grignard reagents (Section 7.4)

$$RX + Mg \longrightarrow RMgX$$
$$RMgX \longrightarrow RH$$

(b) S_N2 reaction: back-side attack of nucleophile on alkyl halide (Sections 7.6 and 7.7)

$$Nu:^- + C{-}X \longrightarrow Nu{-}C + X:^-$$

Substrate must be primary or secondary

(c) S_N1 reaction: carbocation intermediate is involved (Section 7.8)

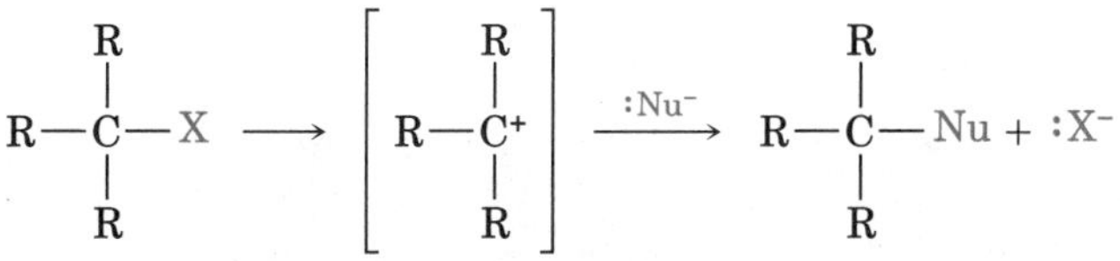

$$R_3C{-}X \longrightarrow [R_3C^+] \xrightarrow{:Nu^-} R_3C{-}Nu + :X^-$$

Substrate must be tertiary or (occasionally) secondary

(d) E2 reaction (Section 7.9)

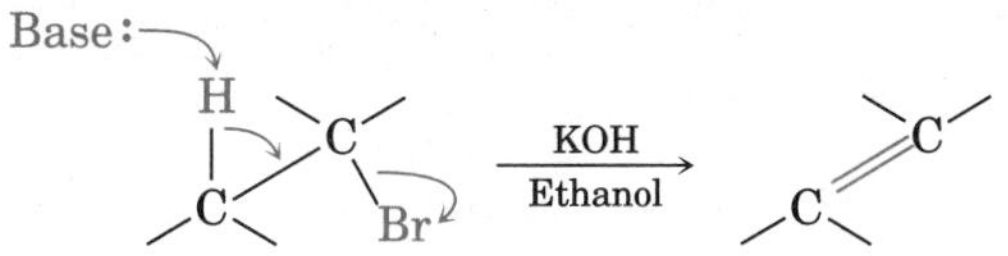

$$Base: + H{-}C{-}C{-}Br \xrightarrow[\text{Ethanol}]{\text{KOH}} C{=}C$$

Anti periplanar geometry is required

(e) E1 reaction (Section 7.10)

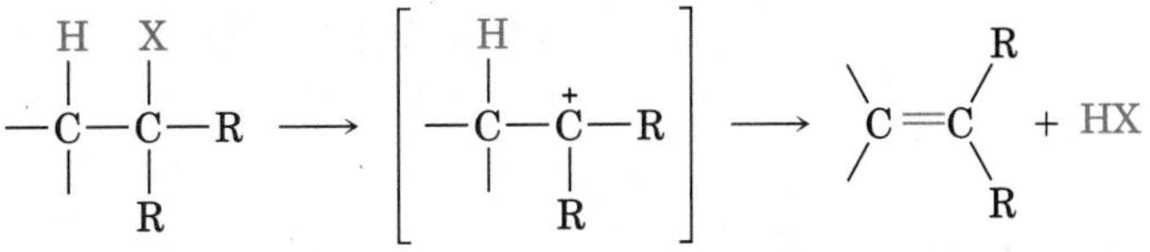

$$-\overset{H}{C}-\overset{X}{\underset{R}{C}}-R \longrightarrow \left[-\overset{H}{C}-\overset{+}{\underset{R}{C}}-R\right] \longrightarrow C{=}C(R)(R) + HX$$

Best for tertiary substrates in neutral or acidic solvents.
Carbocation intermediate is involved.

ADDITIONAL PROBLEMS

7.22 Name these alkyl halides according to IUPAC rules:
(a) $(CH_3)_2CHCHBrCHBrCH_2CH(CH_3)_2$ **(b)** $CH_3CH{=}CHCH_2CHICH_3$

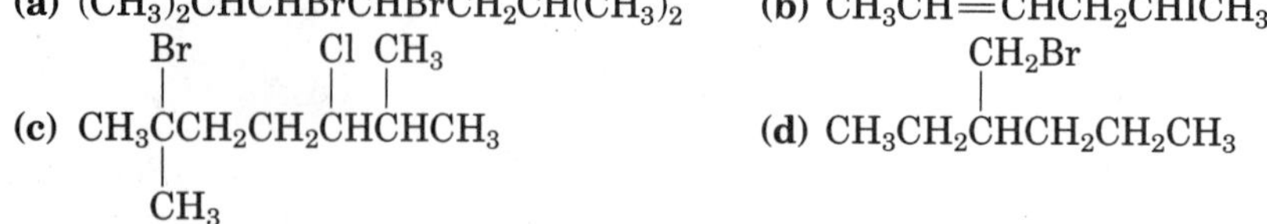

(c) $CH_3C(Br)(CH_3)CH_2CH_2CH(Cl)CH(CH_3)CH_3$ **(d)** $CH_3CH_2CH(CH_2Br)CH_2CH_2CH_3$

7.23 Draw structures corresponding to these IUPAC names:
(a) 2,3-Dichloro-4-methylhexane **(b)** 4-Bromo-4-ethyl-2-methylhexane
(c) 3-Iodo-2,2,4,4-tetramethylpentane

7.24 Draw and name the monochlorination products you might obtain by radical chlorination of 2-methylpentane. Which of the products are chiral? Are any of the products optically active?

7.25 Describe the effects of the these variables on both S_N2 and S_N1 reactions:
(a) Substrate structure **(b)** Leaving group

7.26 How would you prepare these compounds, starting with cyclopentene and any other reagents needed?
(a) Chlorocyclopentane **(b)** Cyclopentanol **(c)** Cyclopentylmagnesium chloride
(d) Cyclopentane

7.27 Which reagent in each of the following pairs is a better leaving group?
(a) F^- or Br^- **(b)** Cl^- or NH_2^- **(c)** OH^- or I^-

7.28 Predict the product(s) of these reactions:

(a) 1-methylcyclohexanol (H_3C, OH on the same ring carbon) $\xrightarrow[\text{Ether}]{\text{HBr}}$? **(b)** $CH_3CH_2CH_2CH_2OH \xrightarrow{SOCl_2}$?

(c) cyclohexanol (OH) $\xrightarrow[\text{Ether}]{PBr_3}$? **(d)** $CH_3CH_2CHBrCH_3 \xrightarrow[\text{Ether}]{\text{Mg}}$ A $\xrightarrow{H_2O}$ B

7.29 Which alkyl halide in each pair will react faster in an S_N2 reaction with hydroxide ion?
(a) Bromobenzene or benzyl bromide, $C_6H_5CH_2Br$ **(b)** CH_3Cl or $(CH_3)_3CCl$
(c) $CH_3CH{=}CHBr$ or $H_2C{=}CHCH_2Br$

7.30 How might you prepare these molecules using a nucleophilic substitution reaction at some step?

(a) CH_3CH_2Br **(b)** $CH_3CH_2CH_2CH_2CN$ **(c)** $CH_3OC(CH_3)_2CH_3$

(d) $CH_2CH_2CH_2N{=}\overset{+}{N}{=}N^-$ **(e)** CH_3CH_2SH **(f)** $CH_3\overset{O}{\overset{\|}{C}}OCH_3$

7.31 What products do you expect from reaction of 1-bromopropane with these reagents?
(a) NaI **(b)** NaCN **(c)** NaOH **(d)** Mg, then H_2O **(e)** $NaOCH_3$

7.32 Order these compounds with respect to both S_N1 and S_N2 reactivity:

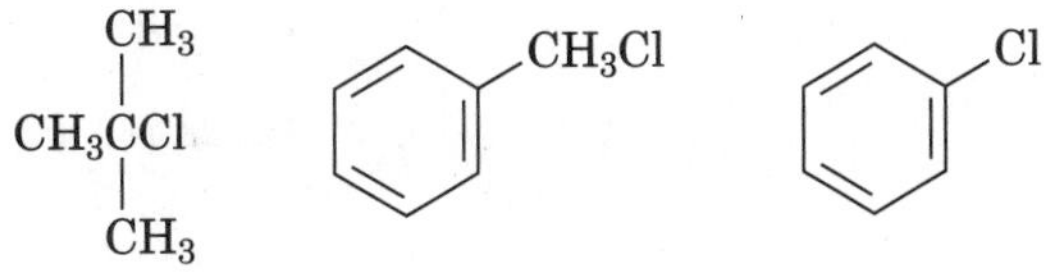

$$CH_3-\underset{CH_3}{\overset{CH_3}{C}}-Cl \qquad C_6H_5CH_2Cl \qquad C_6H_5Cl$$

7.33 Order each set of compounds with respect to S_N2 reactivity:
(a) $(CH_3)_3CCl$, $CH_3CH_2CH_2Cl$, $CH_3CH_2CHClCH_3$
(b) $(CH_3)_2CHCHBrCH_3$, $(CH_3)_2CHCH_2Br$, CH_3Br

7.34 What is wrong with each of the following reactions?

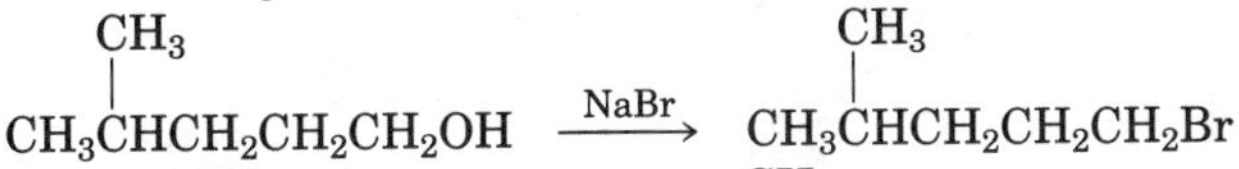

(a) 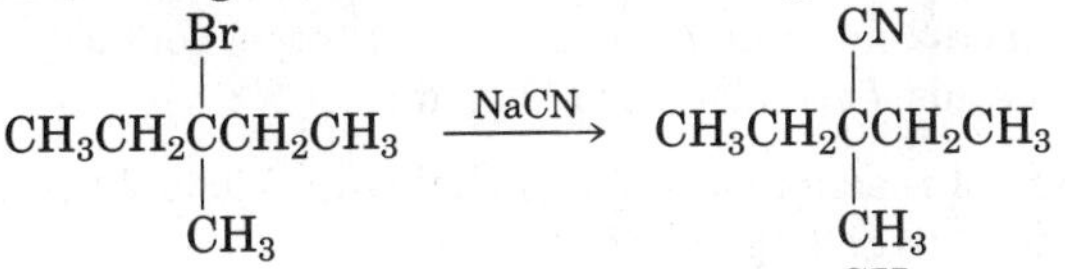 $$CH_3CH_2\underset{CH_3}{\overset{Br}{C}}CH_2CH_3 \xrightarrow{NaCN} CH_3CH_2\underset{CH_3}{\overset{CN}{C}}CH_2CH_3$$

(b) $$CH_3\overset{CH_3}{C}HCH_2CH_2CH_2OH \xrightarrow{NaBr} CH_3\overset{CH_3}{C}HCH_2CH_2CH_2Br$$

(c) $$CH_3CH_2\underset{CH_3}{\overset{OH}{C}}CH_3 \xrightarrow{HBr} CH_3CH{=}\overset{CH_3}{C}CH_3$$

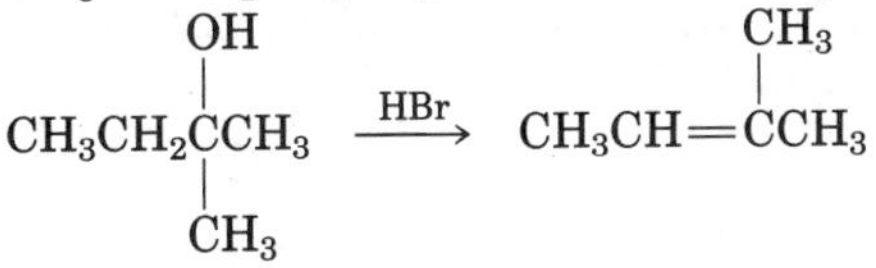

7.35 Predict the product and give the stereochemistry of reactions of these nucleophiles with (*R*)-2-bromooctane:
(a) CN^- **(b)** $CH_3CO_2^-$ **(c)** Br^-

7.36 Draw all isomers of C_4H_9Br, name them, and arrange them in order of decreasing reactivity in the S_N2 reaction.

7.37 Although radical chlorination of alkanes is usually unselective, chlorination of propene, $CH_3CH{=}CH_2$, occurs almost exclusively on the methyl group rather than on the double bond. Draw resonance structures of the allyl radical $CH_2{=}CHCH_2\cdot$ to account for this result.

7.38 Draw resonance structures of the benzyl radical $C_6H_5CH_2\cdot$ to account for the fact that radical chlorination of toluene occurs exclusively on the methyl group rather than on the ring.

7.39 Ethers can be prepared by S_N2 reaction of alkoxide ions with alkyl halides: $R{-}O^- + R'{-}Br \rightarrow R{-}O{-}R' + Br^-$. Suppose you want to prepare cyclohexyl methyl ether. Which route would be better, reaction of methoxide ion, CH_3O^-, with bromocyclohexane or reaction of cyclohexoxide with bromomethane? Explain.

7.40 How could you prepare diethyl ether, $CH_3CH_2OCH_2CH_3$, starting from ethyl alcohol and any inorganic reagents needed? (See Problem 7.39.)

7.41 How could you prepare cyclohexane starting from 3-bromocyclohexene?

7.42 The S_N2 reaction can occur *intramolecularly* (within the same molecule). What product would you expect from treatment of 4-bromo-1-butanol with base?

$$BrCH_2CH_2CH_2CH_2OH \xrightarrow{Base} [BrCH_2CH_2CH_2CH_2O^-\ Na^+] \longrightarrow\ ?$$

7.43 In light of your answer to Problem 7.42, propose a synthesis of 1,4-dioxane starting from 1,2-dibromoethane:

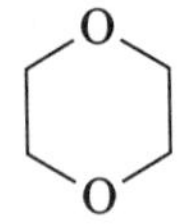

1,4-Dioxane

7.44 Propose a structure for an alkyl halide that can give a mixture of three alkenes on E2 reaction.

7.45 Heating either *tert*-butyl chloride or *tert*-butyl bromide with ethanol yields the same reaction mixture of about 80% *tert*-butyl ethyl ether [$(CH_3)_3COCH_2CH_3$] and 20% 2-methylpropene. Explain.

7.46 What effect would you expect the following changes to have on the rate of the reaction of 1-iodo-2-methylbutane with cyanide ion?

(a) CN^- concentration is halved and 1-iodo-2-methylbutane concentration is doubled.

(b) Both CN^- and 1-iodo-2-methylbutane concentrations are tripled.

7.47 What effect would you expect on the rate of reaction of ethyl alcohol with 2-iodo-2-methylbutane if the concentration of 2-iodo-2-methylbutane were tripled?

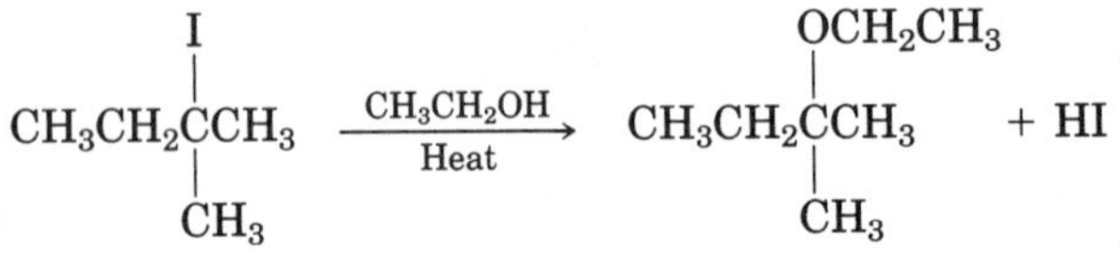

$$CH_3CH_2C(I)(CH_3)CH_3 \xrightarrow[\text{Heat}]{CH_3CH_2OH} CH_3CH_2C(OCH_2CH_3)(CH_3)CH_3 + HI$$

2-Iodo-2-methylbutane

7.48 Identify these reactions as S_N1, S_N2, E1 or E2:

(a) $C_6H_5CH(Br)CH_3 \xrightarrow{KOH} C_6H_5CH{=}CH_2$

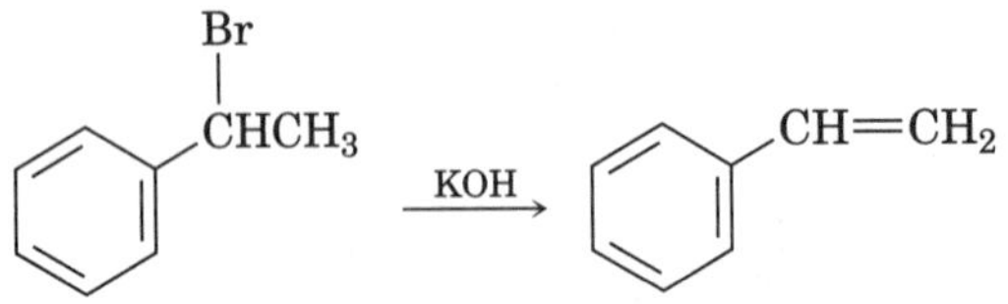

(b) $C_6H_5CH(Br)CH_3 \xrightarrow[\text{Heat}]{CH_3OH} C_6H_5CH(OCH_3)CH_3$

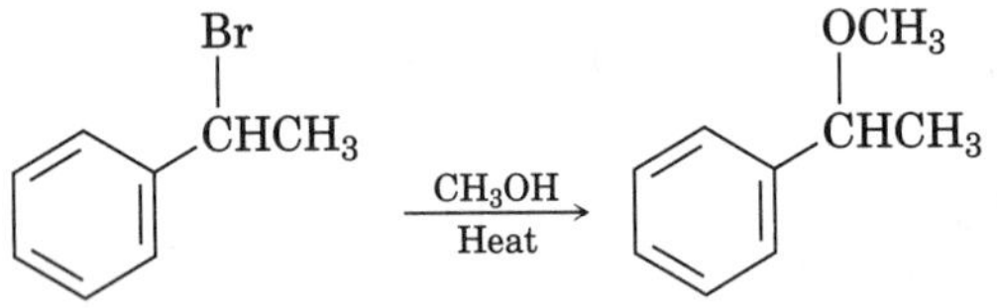

7.49 How can you explain the fact that *trans*-1-bromo-2-methylcyclohexane yields the non-Zaitsev elimination product 3-methylcyclohexene on treatment with base?

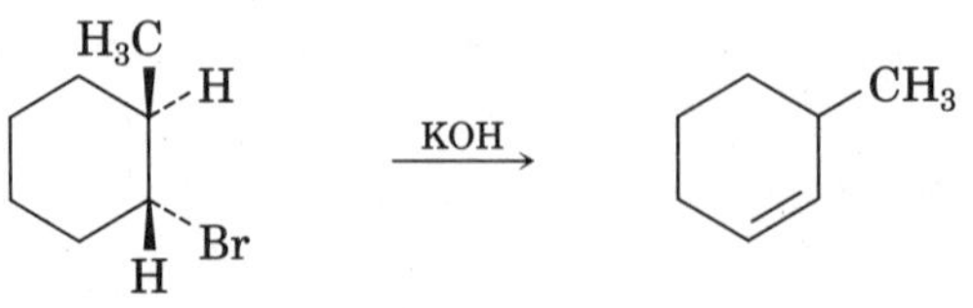

$$\text{trans-1-Bromo-2-methylcyclohexane} \xrightarrow{KOH} \text{3-Methylcyclohexene}$$

***trans*-1-Bromo-2-methylcyclohexane** **3-Methylcyclohexene**

7.50 Propose a structure for an alkyl halide that gives (*Z*)-2,3-diphenyl-2-butene on E2 reaction.

7.51 Predict the major alkene product from these eliminations:

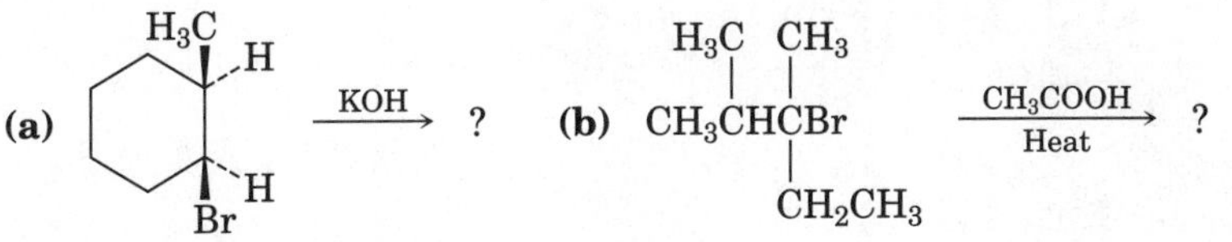

(a) [cyclohexane ring bearing H_3C and H on one carbon, Br and H on the adjacent carbon] $\xrightarrow{KOH}$?

(b) $CH_3CHC(CH_3)(CH_2CH_3)Br$ with H_3C on C2 — i.e. $CH_3CH(CH_3)C(CH_3)(CH_2CH_3)Br$ $\xrightarrow[\text{Heat}]{CH_3COOH}$?

7.52 (2*R*,3*S*)-2-bromo-3-phenylbutane undergoes E2 reaction on treatment with sodium ethoxide to yield (*Z*)-2-phenyl-2-butene.

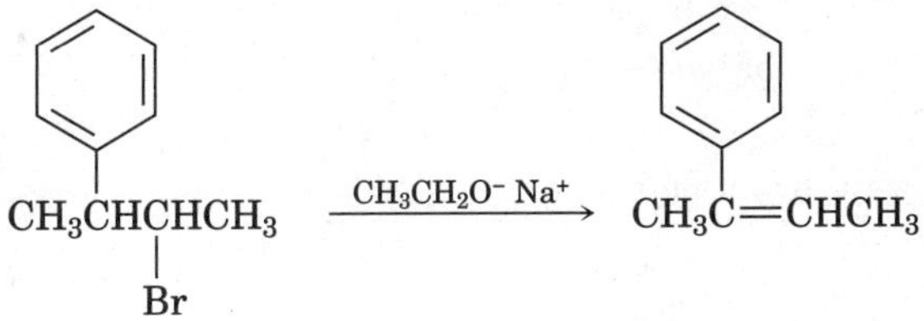

$$CH_3CH(C_6H_5)CH(Br)CH_3 \xrightarrow{CH_3CH_2O^-\ Na^+} CH_3C(C_6H_5){=}CHCH_3$$

Formulate the reaction, showing the proper stereochemistry. Explain the observed result using Newman projections.

7.53 In light of your answer to Problem 7.52, which alkene, *E* or Z, would you expect from the E2 reaction of (2*R*,3*R*)-2-bromo-3-phenylbutane?

7.54 Reaction of HBr with (*R*)-3-methyl-3-hexanol yields (±)-3-bromo-3-methylhexane. Explain.

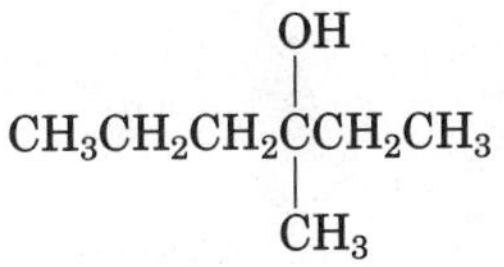

$$CH_3CH_2CH_2C(OH)(CH_3)CH_2CH_3 \quad \textbf{3-Methyl-3-hexanol}$$

7.55 (*S*)-3-Methylhexane undergoes radical chlorination to yield 3-chloro-3-methylhexane as the major product. Is the product chiral? Is it optically active? What stereoisomers are produced and in what ratio?

7.56 Draw the eight diastereomers of 1,2,3,4,5,6-hexachlorocyclohexane. One isomer loses HCl in an E2 reaction nearly 1000 times more slowly than the others. Which isomer reacts so slowly, and why?

7.57 Compound A is optically inactive and has the formula $C_{16}H_{16}Br_2$. On treatment with strong base, A gives hydrocarbon B, $C_{16}H_{14}$, which absorbs 2 equiv of hydrogen when reduced over a palladium catalyst and which reacts with ozone to give two carbonyl-containing products. One product, C, is an aldehyde with the formula C_7H_6O. The other product is glyoxal, OHCCHO. Formulate the reactions involved and suggest structures for A, B, and C. What is the stereochemistry of A?

8 CHAPTER

ALCOHOLS, ETHERS, AND PHENOLS

ALCOHOL
Compound with an –OH group bonded to a saturated, sp^3-hybridized carbon atom

Alcohols are compounds that have hydroxyl groups bonded to saturated, sp^3-hybridized carbon atoms; **phenols** have hydroxyl groups bonded to an aromatic ring; and **ethers** have an oxygen atom bonded to two organic groups. All three classes of compounds can be thought of as organic derivatives of water in which one or both of the water hydrogens are replaced by an organic substituent (H–O–H becomes R–O–H or R–O–R′).

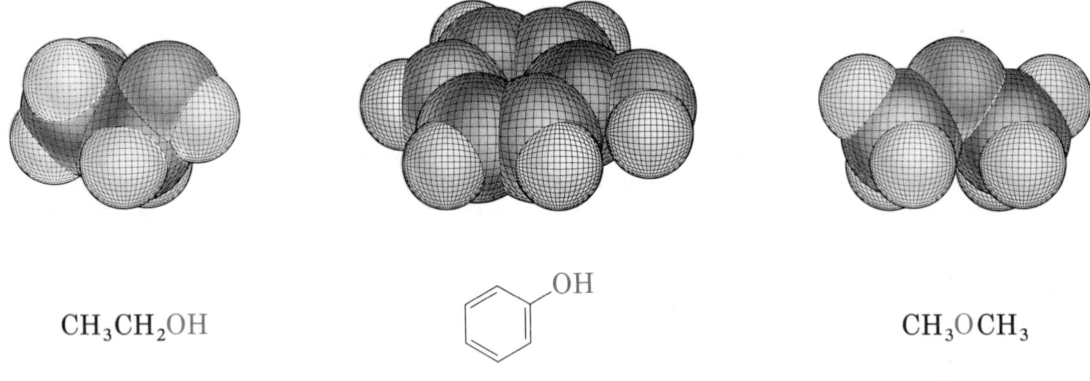

CH_3CH_2OH — **Ethanol**

Phenol (OH on benzene ring) — **Phenol**

CH_3OCH_3 — **Dimethyl ether**

PHENOL
Compound with an –OH group bonded to an aromatic ring

ETHER
Compound with two organic groups bonded to the same oxygen atom

Alcohols, phenols, and ethers occur widely in nature and have many industrial and pharmaceutical applications. Ethanol, for instance, is a fuel additive, an industrial solvent, and a beverage; menthol, an alcohol isolated from peppermint oil, is a flavoring agent; BHT (butylated hydroxytoluene) is a food additive that prolongs shelf life and protects against oxidation; and diethyl ether, the familiar "ether" of medical use, was once popular as an anesthetic agent but is now used mainly as an industrial solvent.

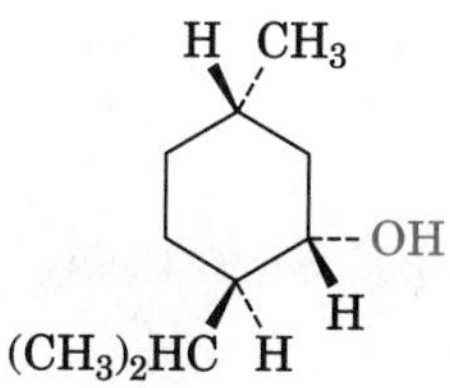

H CH_3 OH $(CH_3)_2HC$ H H

Menthol

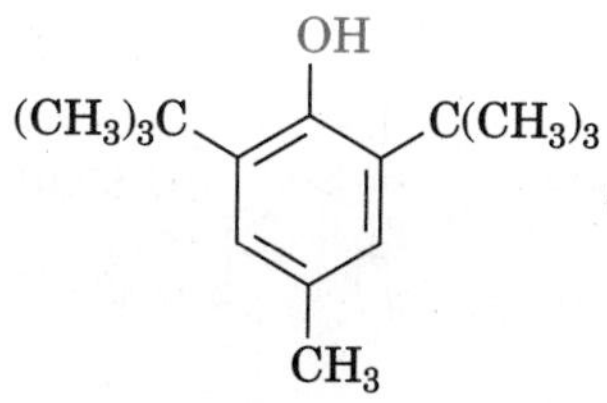

OH $(CH_3)_3C$ $C(CH_3)_3$ CH_3

BHT

$CH_3CH_2OCH_2CH_3$

Diethyl ether

8.1 NAMING ALCOHOLS, PHENOLS, AND ETHERS

Alcohols

Alcohols are classified as either primary (1°), secondary (2°), or tertiary (3°), depending on the number of carbon substituents bonded to the hydroxyl-bearing carbon:

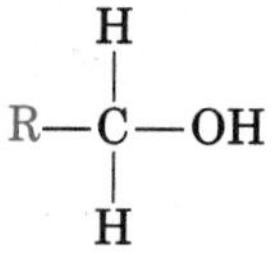

R—CH(H)—OH (H above, H below)	R—C(H)(R)—OH (H above, R below)	R—C(R)(R)—OH (R above, R below)
A primary alcohol (1°)	A secondary alcohol (2°)	A tertiary alcohol (3°)

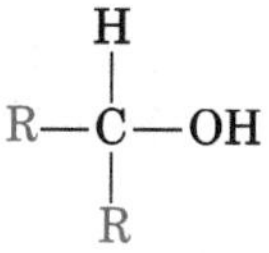

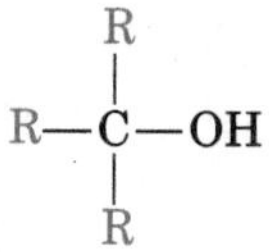

Simple alcohols are named in the IUPAC system as derivatives of the parent alkane:

STEP 1 Select as parent the longest carbon chain that contains the hydroxyl group, and replace the *-e* ending of the corresponding alkane with *-ol.*

STEP 2 Number the carbons of the parent chain beginning at the end nearer the hydroxyl group.

STEP 3 Number all substituents according to their position on the chain and write the name, listing the substituents in alphabetical order.

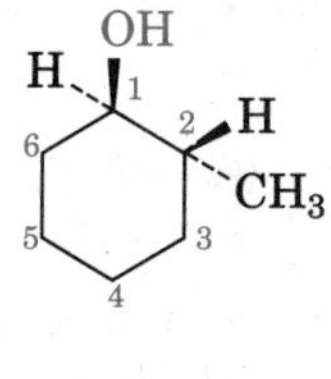

OH H 1 2 H CH_3 3 4 5 6

***trans*-2-Methylcyclohexanol**

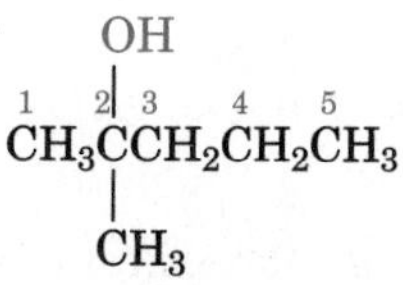

OH
1 2 3 4 5
$CH_3CCH_2CH_2CH_3$
CH_3

2-Methyl-2-pentanol

HO H 1 2 3 4 HO H

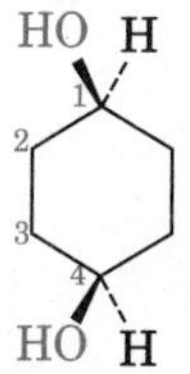

***cis*-1,4-Cyclohexanediol**

Some well-known alcohols also have common names. For example,

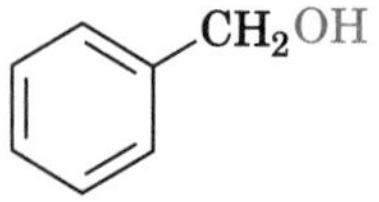

$C_6H_5CH_2OH$ — **Benzyl alcohol (phenylmethanol)**

$H_2C{=}CHCH_2OH$ — **Allyl alcohol (2-propen-1-ol)**

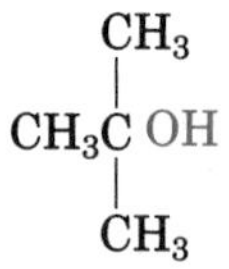

$(CH_3)_3COH$ — ***tert*-Butyl alcohol (2-methyl-2-propanol)**

$HOCH_2CH_2OH$ — **Ethylene glycol (1,2-ethanediol)**

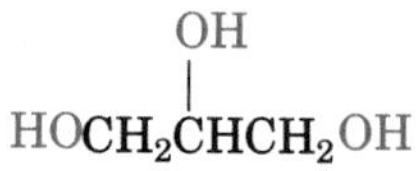

$HOCH_2CH(OH)CH_2OH$ — **Glycerol (1,2,3-propanetriol)**

Phenols

The word *phenol* is used both as the name of a specific substance (hydroxybenzene) and as the family name for all hydroxy-substituted aromatic compounds. Phenols are named as substituted aromatic compounds according to the rules discussed in Section 5.4. Note that *-phenol* is used as the parent name rather than *-benzene*. For example,

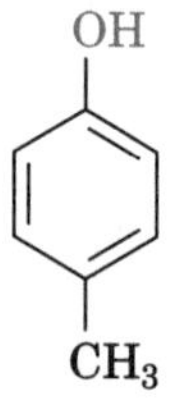

p-Methylphenol (OH and CH_3 para on benzene ring)

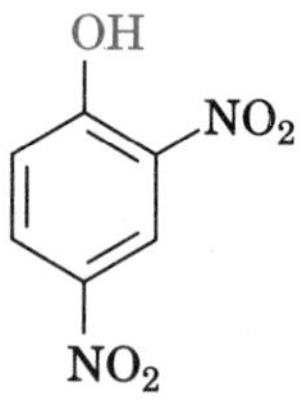

2,4-Dinitrophenol (OH with NO_2 at positions 2 and 4)

Ethers

Two systems of ether nomenclature are allowed by IUPAC rules. Simple ethers that contain no other functional groups are named by identifying the two organic groups and adding the word *ether*. For example,

$CH_3OC(CH_3)_3$ — ***tert*-Butyl methyl ether**

$CH_3CH_2OCH{=}CH_2$ — **Ethyl vinyl ether**

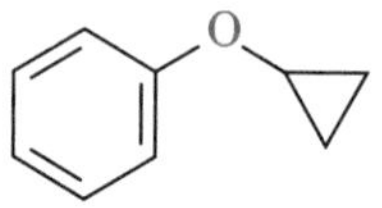

$C_6H_5O\text{-}C_3H_5$ — **Cyclopropyl phenyl ether**

If more than one ether linkage is present in the molecule or if other functional groups are present, the ether group is named as an *alkoxy* substituent on the parent compound. For example,

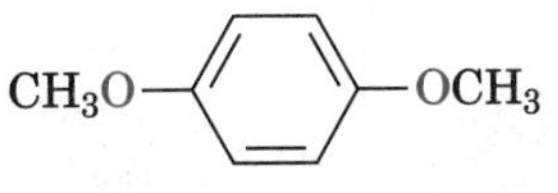

CH_3O–C6H4–OCH_3 4-cyclohexenyl–O–C(CH_3)3

***p*-Dimethoxybenzene** 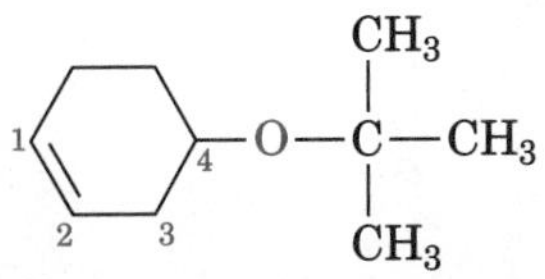 **4-*tert*-Butoxy-1-cyclohexene**

PROBLEM 8.1 Give IUPAC names for these alcohols:

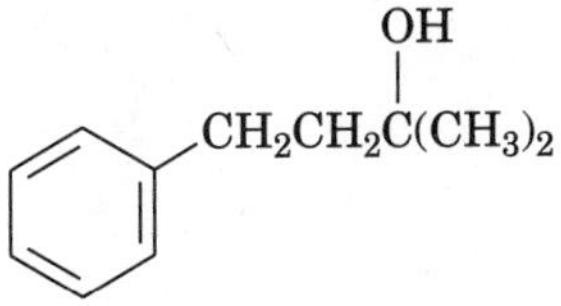

(a) $CH_3CH(OH)CH_2CH(OH)CH(CH_3)_2$

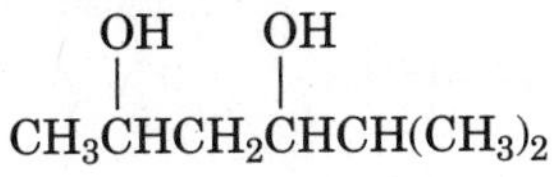

(b) $C_6H_5CH_2CH_2C(OH)(CH_3)_2$

(c) 4,4-dimethylcyclohexanol (OH; H_3C, CH_3)

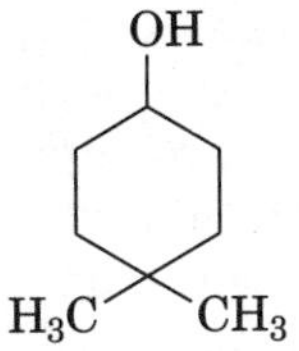

(d) Br, H, OH, H on cyclopentane

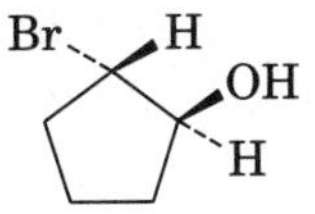

PROBLEM 8.2 Identify the alcohols in Problem 8.1 as primary, secondary, or tertiary.

PROBLEM 8.3 Draw structures corresponding to these IUPAC names:

(a) 2-Methyl-2-hexanol **(b)** 1,5-Hexanediol **(c)** 2-Ethyl-2-buten-1-ol
(d) 3-Cyclohexen-1-ol **(e)** *o*-Bromophenol **(f)** 2,4,6-Trinitrophenol

PROBLEM 8.4 Name these ethers according to IUPAC rules:

(a) $CH_3CH(CH_3)OCH(CH_3)CH_3$ **(b)** cyclopentyl–$OCH_2CH_2CH_3$

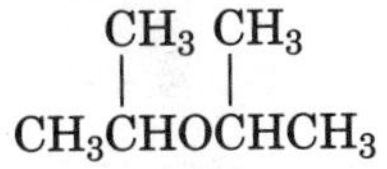

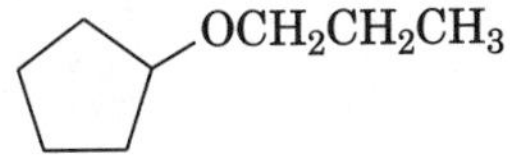

(c) Br–C6H4–OCH_3 **(d)** $(CH_3)_2CHCH_2OCH_2CH_3$

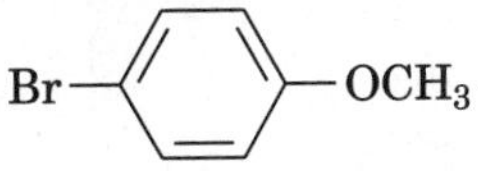

8.2 PROPERTIES OF ALCOHOLS, PHENOLS, AND ETHERS: HYDROGEN BONDING

As mentioned previously, alcohols, phenols, and ethers can be thought of as organic derivatives of water in which one or both of the hydrogens have been replaced by organic residues. Thus, all three classes of compounds have nearly the same geometry as water. The R–O–H or R–O–R′ bonds have an approximately tetrahedral bond angle (112° in dimethyl ether, for example), and the oxygen atom is sp^3-hybridized.

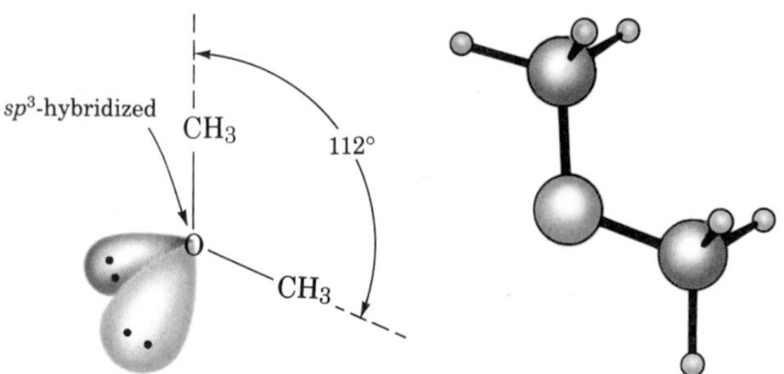

Alcohols and phenols differ significantly from the hydrocarbons and alkyl halides we've studied thus far. As shown in Figure 8.1, alcohols have higher boiling points than alkanes or haloalkanes of similar molecular weight. For example, the molecular weights of 1-propanol (mol wt = 60), butane (mol wt = 58), and chloroethane (mol wt = 65) are similar, but 1-propanol boils at 97.4°C, compared with −0.5°C for the alkane and 12.3°C for the chloroalkane. Similarly, phenols have higher boiling points than aromatic hydrocarbons. Phenol itself, for example, boils at 182°C, whereas toluene boils at 110.6°C.

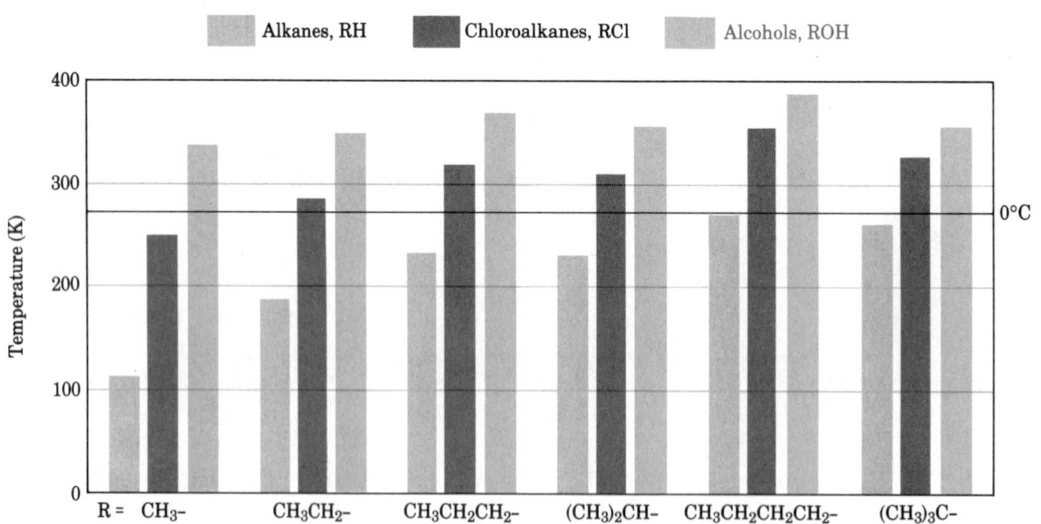

FIGURE 8.1 A comparison of boiling points for some alkanes, chloroalkanes, and alcohols.

The reason for their unusually high boiling points is that alcohols and phenols, like water, are highly associated in solution because of the formation of hydrogen bonds. The positively polarized hydroxyl hydrogen atom of one molecule forms a weak hydrogen bond to the negatively polarized oxygen atom of another molecule (Figure 8.2). Although hydrogen bonds have a strength of only about 5 kcal/mol (20 kJ/mol) versus 100 kcal/mol for a typical O–H bond, the presence of a great many hydrogen bonds in solution means that extra energy is required to break them during the boiling process. Ethers, because they lack hydroxyl groups, can't form hydrogen bonds and therefore have relatively low boiling points.

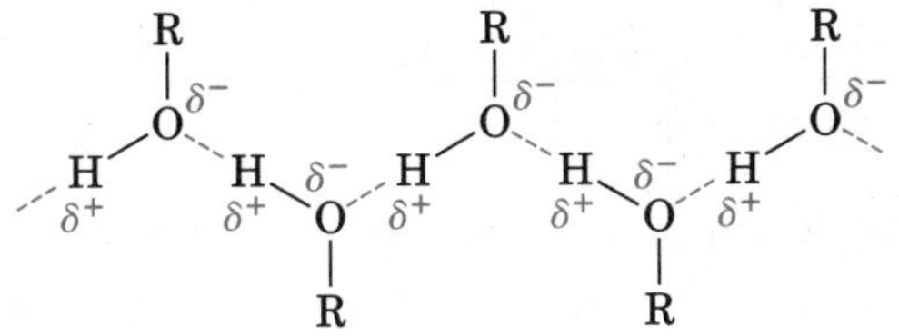

FIGURE 8.2 Hydrogen bonding in alcohols and phenols. Note that the O–H····O bond angle is approximately 180°.

8.3 PROPERTIES OF ALCOHOLS AND PHENOLS: ACIDITY

Alcohols and phenols, like water, are both weakly acidic and weakly basic. As weak Lewis bases, alcohols and phenols are reversibly protonated by acids to yield oxonium ions, ROH_2^+:

$$R{-}\ddot{O}{-}H + HX \rightleftharpoons R{-}\overset{+}{O}H_2 \; X^-$$

As weak acids, alcohols and phenols act as proton donors. In dilute aqueous solution, they dissociate to a slight extent by donating a proton to water:

$$R\ddot{O}{-}H + H_2\ddot{O}: \rightleftharpoons R\ddot{O}:^- + H_3O:^+$$

Table 8.1 gives the pK_a values of some common alcohols and phenols in comparison with water and HCl.

TABLE 8.1 Acidity Constants of Some Alcohols and Phenols

ALCOHOL OR PHENOL	pK_a	
$(CH_3)_3COH$	18.00	Weaker acid
CH_3CH_2OH	16.00	
[HOH, water][a]	[15.74]	
CH_3OH	15.54	
p-Methylphenol	10.26	
Phenol	10.00	
p-Bromophenol	9.35	
p-Nitrophenol	7.15	
[HCl, hydrochloric acid][a]	[−7.00]	Stronger acid

[a]Values for water and hydrochloric acid are shown for reference.

The data in Table 8.1 show that alcohols are about as acidic as water. They are generally much weaker, however, than carboxylic acids or mineral acids, and they don't react with bicarbonate ion. Alcohols do, however, react with alkali metals such as sodium and potassium to yield alkoxide salts that are themselves strong bases.

$$2\ CH_3OH + 2\ Na \longrightarrow 2\ CH_3O^-\ Na^+ + H_2$$

Methanol **Sodium methoxide**

$$2\ (CH_3)_3COH + 2\ K \longrightarrow 2\ (CH_3)_3CO^-\ K^+ + H_2$$

***tert*-Butyl alcohol** **Potassium *tert*-butoxide**

Phenols are much more acidic than alcohols. In fact, some nitro-substituted phenols approach or surpass the acidity of carboxylic acids. One practical consequence of this acidity is that phenols are soluble in dilute aqueous sodium hydroxide.

$$C_6H_5{-}O{-}H + NaOH \longrightarrow C_6H_5{-}O^-\ Na^+ + H_2O$$

Phenol **Sodium phenoxide**

Phenols are more acidic than alcohols because the phenoxide anion is resonance-stabilized by the aromatic ring. Sharing the negative charge increases the stability of the phenoxide anion and the acidity of the corresponding phenol.

:Ö:⁻ ⟷ :Ö:⁻ ⟷ Ö ⟷ Ö ⟷ Ö

Substituted phenols can be either more or less acidic than phenol itself, depending on their structure. Phenols with an electron-withdrawing substituent are more acidic because the substituent stabilizes the corresponding phenoxide anion. Phenols with an electron-donating substituent are less acidic because the substituent destabilizes the phenoxide anion.

EWG

Electron-withdrawing groups (EWG) stabilize phenoxide anion, resulting in increased phenol acidity

EDG

Electron-donating groups (EDG) destabilize phenoxide anion, resulting in decreased phenol acidity

PRACTICE PROBLEM 8.1 Which would you expect to be more acidic, *p*-methylphenol or *p*-cyanophenol?

SOLUTION We know from their effects on aromatic substitution (Section 5.9) that methyl is an activating group (electron donor) whereas cyano is a deactivating group (electron acceptor). Thus, *p*-cyanophenol is more acidic.

PROBLEM 8.5 Rank the compounds in each group in order of increasing acidity.

(a) Methanol, phenol, *p*-nitrophenol, *p*-methylphenol

(b) Benzyl alcohol, *p*-bromophenol, 2,4-dibromophenol, *p*-methoxyphenol

PROBLEM 8.6 Draw as many resonance structures as you can for the anion of *p*-cyanophenol.

8.4 SYNTHESIS OF ALCOHOLS

Alcohols occupy a central position in organic chemistry. They can be prepared from a variety of functional-group families (alkenes, alkyl halides, ketones, aldehydes, and esters, among others), and they can be transformed into an equally wide assortment of families. Let's review briefly some of the methods of alcohol preparation we've already seen.

Alcohols can be prepared by hydration of alkenes. Treatment of the alkene with sulfuric acid and water leads to the Markovnikov product (Section 4.4):

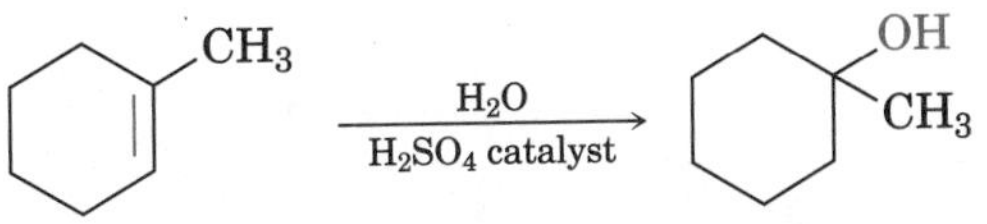

$$\xrightarrow[\text{H}_2\text{SO}_4\text{ catalyst}]{\text{H}_2\text{O}}$$

1-Methylcyclohexene **1-Methylcyclohexanol**

1,2-Diols can be prepared by direct hydroxylation of an alkene with basic potassium permanganate (Section 4.7). The reaction takes place with syn stereochemistry:

$$\text{1-Methylcyclohexene} \xrightarrow[\text{H}_2\text{O, NaOH}]{\text{KMnO}_4} \text{1-Methyl-}cis\text{-1,2-cyclohexanediol}$$

1-Methylcyclohexene **1-Methyl-*cis*-1,2-cyclohexanediol**

8.5 ALCOHOLS FROM CARBONYL COMPOUNDS

The most valuable method for preparing alcohols is by reduction of carbonyl compounds—the formal addition of H_2 to the C=O double bond:

$$\text{C=O} \xrightarrow{[\text{H}]} \text{H–C–OH}$$

where [H] is a generalized reducing agent

A carbonyl compound An alcohol

Reduction of Aldehydes and Ketones

Aldehydes and ketones are easily converted into alcohols by reduction:

$$\text{RCHO} \xrightarrow{[\text{H}]} \text{RCH}_2\text{OH} \qquad \text{RCOR}' \xrightarrow{[\text{H}]} \text{RCH(OH)R}'$$

An aldehyde A primary alcohol A ketone A secondary alcohol

Although many reducing reagents are available, sodium borohydride, $NaBH_4$, is usually chosen for ketone and aldehyde reductions because of its safety and ease of handling. Aldehydes are reduced by $NaBH_4$ to give primary alcohols, and ketones are reduced to give secondary alcohols:

$$\text{CH}_3\text{CH}_2\text{CH}_2\text{CHO} \xrightarrow[\text{2. H}_3\text{O}^+]{\text{1. NaBH}_4\text{, ethanol}} \text{CH}_3\text{CH}_2\text{CH}_2\text{CH}_2\text{OH}$$

Butanal **1-Butanol (85%)**

$$(\text{C}_6\text{H}_{11})_2\text{C=O} \xrightarrow[\text{2. H}_3\text{O}^+]{\text{1. NaBH}_4\text{, ethanol}} (\text{C}_6\text{H}_{11})_2\text{CHOH}$$

Dicyclohexyl ketone **Dicyclohexylmethanol (88%)**

Lithium aluminum hydride, $LiAlH_4$, is another reducing agent that is sometimes used to reduce ketones and aldehydes. Far more powerful and reactive than $NaBH_4$, $LiAlH_4$ is also far more dangerous since it reacts violently with water and ethanol, and decomposes explosively when heated above 120°C.

Reduction of Esters and Carboxylic Acids

Esters and carboxylic acids are reduced to primary alcohols:

$$\underset{\text{A carboxylic acid}}{R{-}\overset{O}{\overset{\|}{C}}{-}OH} \quad \text{or} \quad \underset{\text{An ester}}{R{-}\overset{O}{\overset{\|}{C}}{-}OR'} \xrightarrow{[H]} \underset{\text{A primary alcohol}}{R{-}\overset{OH}{\overset{|}{\underset{H}{\underset{|}{C}}}}{-}H}$$

Since these reactions are more difficult than the corresponding reductions of ketones and aldehydes, $LiAlH_4$ is used rather than $NaBH_4$. Note that only one hydrogen is added to the carbonyl carbon atom during reductions of ketones and aldehydes, but two hydrogens are added to the carbonyl carbon during ester and carboxylic acid reductions.

$$\underset{\textbf{Methyl 2-pentenoate}}{CH_3CH_2CH{=}CH\overset{O}{\overset{\|}{C}}OCH_3} \xrightarrow[\text{2. } H_3O^+]{\text{1. } LiAlH_4\text{, ether}} \underset{\textbf{2-Penten-1-ol (91\%)}}{CH_3CH_2CH{=}CHCH_2OH} + CH_3OH$$

$$\underset{\textbf{Oleic acid}}{CH_3(CH_2)_7CH{=}CH(CH_2)_7\overset{O}{\overset{\|}{C}}OH} \xrightarrow[\text{2. } H_3O^+]{\text{1. } LiAlH_4\text{, ether}} \underset{\textbf{9-Octadecen-1-ol (87\%)}}{CH_3(CH_2)_7CH{=}CH(CH_2)_7CH_2OH}$$

PRACTICE PROBLEM 8.2 Predict the product of this reaction:

$$CH_3CH_2CH_2\overset{O}{\overset{\|}{C}}CH_2CH_3 \xrightarrow{NaBH_4} ?$$

SOLUTION Ketones are reduced by treatment with $NaBH_4$ to yield secondary alcohols. Thus, reduction of 3-hexanone yields 3-hexanol.

$$\underset{\textbf{3-Hexanone}}{CH_3CH_2CH_2\overset{O}{\overset{\|}{C}}CH_2CH_3} \xrightarrow{NaBH_4} \underset{\textbf{3-Hexanol}}{CH_3CH_2CH_2\overset{OH}{\overset{|}{C}}HCH_2CH_3}$$

PROBLEM 8.7

How would you carry out these reactions?

(a) $CH_3C(=O)CH_2CH_2C(=O)OCH_3 \xrightarrow{?} CH_3CH(OH)CH_2CH_2C(=O)OCH_3$

(b) $CH_3C(=O)CH_2CH_2C(=O)OCH_3 \xrightarrow{?} CH_3CH(OH)CH_2CH_2CH_2OH$

PROBLEM 8.8

What carbonyl compounds give the following alcohols on reduction with $LiAlH_4$? Show all possibilities.

(a) 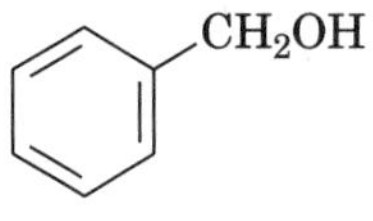 $C_6H_5CH_2OH$ (b) $C_6H_5CH(OH)CH_3$ (c) cyclohexane with OH and H

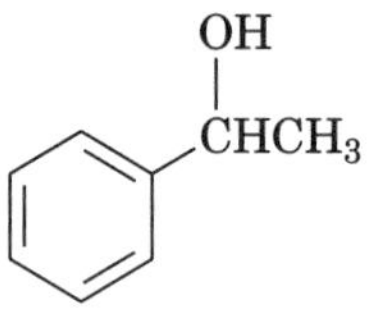

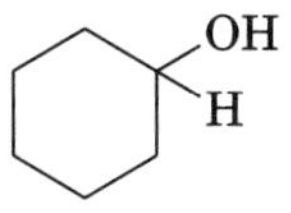

8.6 ETHERS FROM ALCOHOLS: THE WILLIAMSON ETHER SYNTHESIS

WILLIAMSON ETHER SYNTHESIS
The reaction of an alkoxide ion with an alkyl halide to yield an ether

Metal alkoxides react with alkyl halides to yield ethers, a reaction known as the **Williamson ether synthesis**. Although discovered more than 100 years ago, the Williamson synthesis is still the best method for the preparation of both symmetrical and unsymmetrical ethers.

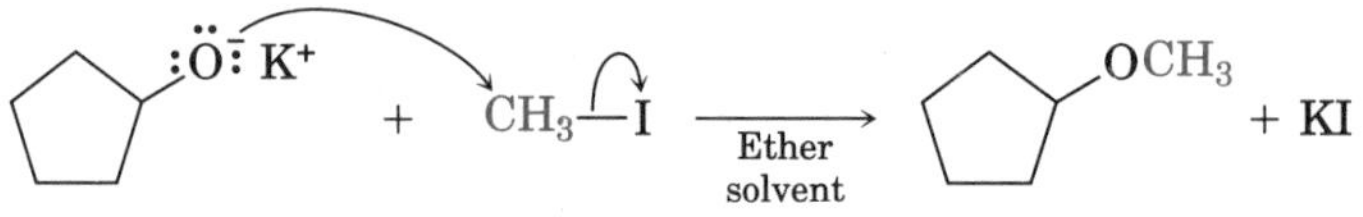

$$C_5H_9\ddot{O}:^- K^+ + CH_3{-}I \xrightarrow[\text{solvent}]{\text{Ether}} C_5H_9OCH_3 + KI$$

Potassium cyclopentoxide **Iodomethane** **Cyclopentyl methyl ether (74%)**

The alkoxide ion needed in the reaction is usually prepared by reaction of an alcohol with sodium or potassium (Section 8.3):

$$2\ ROH + 2\ Na \longrightarrow 2\ RO^-\ Na^+ + H_2$$

The Williamson synthesis is an S_N2 reaction (Section 7.7) that occurs by nucleophilic displacement of halide ion by the alkoxide ion nucleophile. Thus, the reaction is subject to all the normal S_N2 limitations. Primary alkyl halides work best because competitive E2 elimination of HX can occur with more hindered substrates. Unsymmetrical ethers are therefore best prepared by reaction of the more hindered alkoxide partner with the less hindered halide partner, rather than vice versa. For example, *tert*-butyl methyl ether is best synthesized by reaction of *tert*-butoxide ion with iodomethane, rather than by reaction of methoxide ion with 2-chloro-2-methylpropane.

$$\text{S}_\text{N}2\text{ reaction}\quad CH_3-\underset{CH_3}{\overset{CH_3}{C}}-\ddot{O}:^- + CH_3-I \longrightarrow CH_3-\underset{CH_3}{\overset{CH_3}{C}}-\ddot{O}-CH_3 + I^-$$

***tert*-Butoxide ion** **Iodomethane** ***tert*-Butyl methyl ether**

$$\text{E2 reaction}\quad CH_3\ddot{O}:^- + \overset{H}{CH_2}-\underset{CH_3}{\overset{CH_3}{C}}-Cl \longrightarrow H_2C{=}C(CH_3)_2 + CH_3\ddot{O}H + Cl^-$$

Methoxide ion **2-Chloro-2-methylpropane** **2-Methylpropene**

PROBLEM 8.9 Treatment of cyclohexanol with Na gives an alkoxide ion that undergoes reaction with iodoethane to yield cyclohexyl ethyl ether. Write the reaction showing all the steps.

PROBLEM 8.10 How would you prepare these ethers?

(a) Methyl propyl ether **(b)** Anisole (methyl phenyl ether)

(c) $C_6H_5CH_2OCH(CH_3)_2$

PROBLEM 8.11 Rank these compounds in order of their expected reactivity toward alkoxide ion nucleophiles in the Williamson ether synthesis: bromoethane, 2-bromopropane, chloroethane, 2-chloro-2-methylpropane.

8.7 REACTIONS OF ALCOHOLS

Dehydration

Alcohols can be dehydrated to give alkenes (Section 4.9). Tertiary alcohols lose water when treated with H_2SO_4 under fairly mild conditions, but primary and secondary alcohols require higher temperatures. As we saw in Section 7.10, the mechanism of this dehydration is simply an E1 reaction. Strong acid protonates the alcohol oxygen, the protonated intermediate spontaneously loses water by an E1 mechanism to generate a carbocation, and loss of a proton from a neighboring carbon atom then yields the alkene product:

$$H_3C-\underset{CH_3}{\overset{OH}{C}}-CH_3 \xrightarrow{H_2SO_4} \left[H_3C-\underset{CH_3}{\overset{+}{C}}-CH_3\right] + H_2O \xrightarrow{-H^+} H_2C{=}C(CH_3)_2$$

***tert*-Butyl alcohol** **+ H_2O** **2-Methylpropene (97%)**

Conversion into Alkyl Halides and Ethers

Alcohols can be converted into both alkyl halides (Section 7.3) and ethers. Tertiary alcohols are readily transformed into alkyl halides by an S_N1 mechanism on treatment with either HCl or HBr. Primary and secondary alcohols are much more resistant to reaction with halogen acids, however, and are best converted into halides by treatment with either $SOCl_2$ or PBr_3.

Oxidation of Alcohols

The most important reaction of alcohols is their oxidation to carbonyl compounds by a formal loss of H_2: CH–OH → C=O. Primary alcohols yield aldehydes or carboxylic acids, and secondary alcohols yield ketones, but tertiary alcohols don't normally react with oxidizing agents.

$$R-CH(OH)-H \xrightarrow{[O]} R-C(=O)-H \xrightarrow{[O]} R-C(=O)-OH$$

A primary alcohol — An aldehyde — A carboxylic acid

$$R-C(OH)(R')-H \xrightarrow{[O]} R-C(=O)-R'$$

A secondary alcohol — A ketone

where [O] is an oxidizing reagent

Primary alcohols are oxidized either to aldehydes or to carboxylic acids, depending on the reagents chosen and on the conditions used. The best method for preparing aldehydes from primary alcohols on a laboratory scale (as opposed to an industrial scale) is by use of pyridinium chlorochromate (PCC), $C_5H_6NCrO_3Cl$, in dichloromethane solvent. This reagent is too expensive for large-scale use in industry, though.

$$CH_3(CH_2)_5CH_2OH \xrightarrow[CH_2Cl_2]{PCC} CH_3(CH_2)_5CH{=}O$$

1-Heptanol — **Heptanal (78%)**

Many oxidizing agents, such as chromium trioxide (CrO_3) or sodium dichromate in aqueous acid solution, oxidize primary alcohols to carboxylic acids. Although aldehydes are intermediates in these oxidations, they usually can't be isolated because they are further oxidized too rapidly.

$$CH_3(CH_2)_8CH_2OH \xrightarrow[H_3O^+]{CrO_3} CH_3(CH_2)_8C(=O)OH$$

1-Decanol — **Decanoic acid (93%)**

Secondary alcohols are oxidized easily to produce ketones. Sodium dichromate ($Na_2Cr_2O_7$) in aqueous acetic acid is often used as the oxidant.

$$(CH_3)_3C\text{–}C_6H_{10}\text{–}OH \xrightarrow[H_2O,\ CH_3CO_2H,\ \Delta]{Na_2Cr_2O_7} (CH_3)_3C\text{–}C_6H_9{=}O$$

4-*tert*-Butylcyclohexanol **4-*tert*-Butylcyclohexanone (91%)**

PRACTICE PROBLEM 8.3 What product would you expect from reaction of benzyl alcohol with CrO_3?

$C_6H_5\text{–}CH_2OH$ **Benzyl alcohol**

SOLUTION We know that treatment of primary alcohols with CrO_3 yields carboxylic acids. Thus, oxidation of benzyl alcohol should yield benzoic acid.

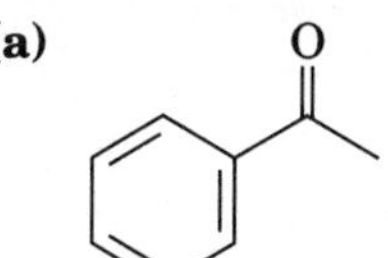

$\xrightarrow[H_3O^+]{CrO_3}$

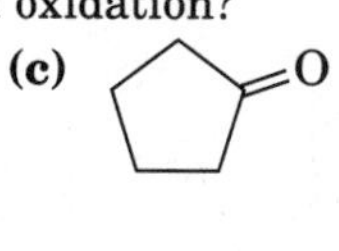

Benzyl alcohol **Benzoic acid**

PROBLEM 8.12 What alcohols would give these products on oxidation?

(a) $C_6H_5COCH_3$ **(b)** $CH_3CH(CH_3)CHO$ **(c)** Cyclopentanone ($C_5H_8{=}O$)

PROBLEM 8.13 What products would you expect to obtain from oxidation of these alcohols with CrO_3?

(a) Cyclohexanol **(b)** 1-Hexanol **(c)** 2-Hexanol

PROBLEM 8.14 What products would you expect to obtain from oxidation of the alcohols in Problem 8.13 with pyridinium chlorochromate (PCC)?

8.8 SYNTHESIS AND REACTIONS OF PHENOLS

Phenols are synthesized from aromatic starting materials by a two-step sequence. The starting compound is first sulfonated by treatment with SO_3/H_2SO_4, and the arenesulfonic acid product is then converted into a phenol by high-temperature reaction with NaOH.

$$\text{Benzene} \xrightarrow[H_2SO_4]{SO_3} \text{Benzenesulfonic acid} \xrightarrow[200°C]{NaOH} \text{Phenol (72\%)}$$

Benzene **Benzenesulfonic acid** **Phenol (72%)**

PROBLEM 8.15 *p*-Cresol (*p*-methylphenol) is used industrially both as an antiseptic and as a starting material to prepare the food additive BHT. How could you synthesize *p*-cresol from benzene?

Alcohol-Like Reactions

Phenols and alcohols are very different in spite of the fact that both have –OH groups. Phenols can't be dehydrated by treatment with acid and can't be converted into alkyl halides by treatment with HX. Phenols can, however, be converted into ethers by reaction with alkyl halides in the presence of base. Williamson ether synthesis with phenols occurs easily because phenols are more acidic than alcohols and are therefore more easily converted into their anions.

$$\text{o-Nitrophenol} + CH_3CH_2CH_2CH_2Br \xrightarrow[\text{Acetone}]{K_2CO_3} \text{Butyl o-nitrophenyl ether (80\%)}\ (OCH_2CH_2CH_2CH_3,\ NO_2)$$

***o*-Nitrophenol** **1-Bromobutane** **Butyl *o*-nitrophenyl ether (80%)**

Electrophilic Aromatic Substitution Reactions

The hydroxyl group is an activating, ortho-, para-directing substituent in electrophilic aromatic substitution reactions (Sections 5.9 and 5.10). As a result, phenols are reactive substrates for electrophilic halogenation, nitration, and sulfonation.

Phenol (OH) → with Br_2, $FeBr_3$: o-bromophenol (OH, Br); with HNO_3, H_2SO_4: o-nitrophenol (OH, NO_2); with SO_3, H_2SO_4: o-hydroxybenzenesulfonic acid (OH, SO_3H)

Oxidation of Phenols: Quinones

QUINONE Compound that contains a cyclohexadienedione functional group

Treatment of a phenol with a strong oxidizing agent such as sodium dichromate yields a cyclohexadienedione, or **quinone**:

$$\text{Phenol} \xrightarrow[H_3O^+]{Na_2Cr_2O_7} \text{Benzoquinone}$$

Phenol **Benzoquinone**

HYDROQUINONE Compound that contains a *p*-dihydroxybenzene unit

Quinones are an interesting and valuable class of compounds because of their oxidation-reduction properties. They can be easily reduced to **hydroquinones** (*p*-dihydroxybenzenes) by $NaBH_4$ or $SnCl_2$, and hydroquinones can be easily oxidized back to quinones by $Na_2Cr_2O_7$. Hydroquinone is used, among other things, as a photographic developer, because it reduces Ag^+ on film to metallic silver.

$$\text{Benzoquinone} \underset{Na_2Cr_2O_7}{\overset{NaBH_4}{\rightleftharpoons}} \text{Hydroquinone}$$

Benzoquinone **Hydroquinone**

8.9 REACTIONS OF ETHERS: ACIDIC CLEAVAGE

Ethers are relatively unreactive, a property that accounts for their common use as reaction solvents. Halogens, mild acids, bases, and nucleophiles have no effect on most ethers. In fact, ethers undergo only one reaction of general use—cleavage by strong acids. Aqueous HI is the usual reagent for cleaving ethers, though aqueous HBr also works.

Acidic ether cleavages are typical nucleophilic substitution reactions. They take place through either S_N1 or S_N2 pathways, depending on the structure of the ether. Primary and secondary alkyl ethers react by an S_N2 pathway, in which nucleophilic iodide ion attacks the protonated ether at the less highly substituted site. The ether oxygen atom stays with the more hindered alkyl group, and the iodide bonds to the less hindered group. For example, ethyl isopropyl ether yields isopropyl alcohol and iodoethane on cleavage by HI:

$$CH_3CH(CH_3)-\ddot{O}-CH_2CH_3 \xrightarrow{H-I} \left[CH_3CH(CH_3)-\overset{H}{\underset{..}{O^+}}-CH_2CH_3\right] \xrightarrow{:\ddot{I}:^-} CH_3CH(CH_3)OH + CH_3CH_2I$$

Ethyl isopropyl ether **Isopropyl alcohol** **Iodoethane**

Tertiary ethers cleave by either an S_N1 or E1 mechanism since they can produce stable intermediate carbocations. These reactions are often fast and take place at room temperature or below.

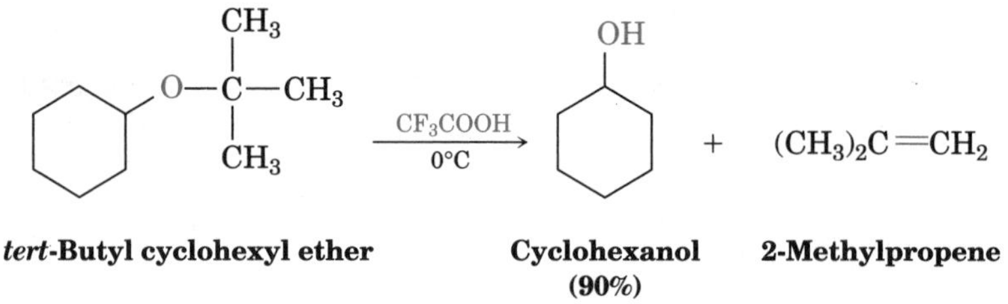

$$\text{tert-Butyl cyclohexyl ether (cyclohexyl}-O-C(CH_3)_3) \xrightarrow[0°C]{CF_3COOH} \text{Cyclohexanol (90\%)} + (CH_3)_2C{=}CH_2 \text{ (2-Methylpropene)}$$

***tert*-Butyl cyclohexyl ether** **Cyclohexanol (90%)** **2-Methylpropene**

PRACTICE PROBLEM 8.4 What products would you expect from the reaction of methyl cyclopentyl ether with HI?

SOLUTION Iodide ion attacks the methyl group rather than the secondary cyclopentyl group in an S_N2 reaction, giving iodomethane and cyclopentanol:

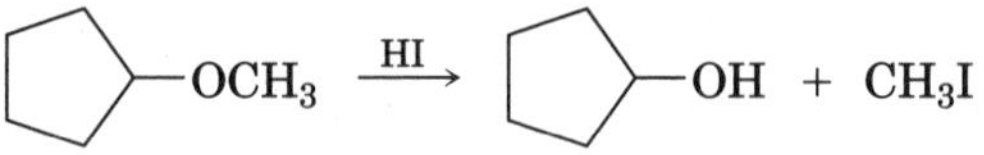

$$\text{cyclopentyl}-OCH_3 \xrightarrow{HI} \text{cyclopentyl}-OH + CH_3I$$

PROBLEM 8.16 What products do you expect from the reaction of these ethers with HI?
(a) $CH_3CH_2OCH_2CH_3$ **(b)** Cyclohexyl ethyl ether **(c)** $(CH_3)_3COCH_2CH_3$

PROBLEM 8.17 Write the mechanism of the cleavage of *tert*-butyl cyclohexyl ether with CF_3COOH to yield cyclohexanol and 2-methylpropene. What kind of reaction is occurring?

8.10 CYCLIC ETHERS: EPOXIDES

For the most part, cyclic ethers behave like acyclic ethers. The chemistry of the ether functional group is the same whether it's in an open chain or in a ring. Thus, common cyclic ethers such as tetrahydrofuran (THF) and dioxane are often used as solvents because of their inertness.

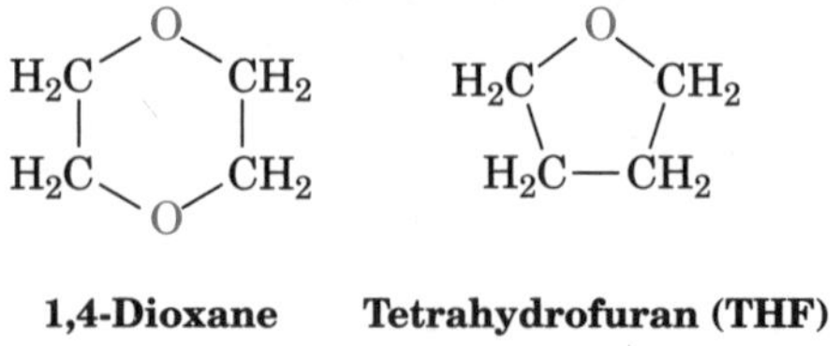

H_2C—O—CH_2, H_2C—O—CH_2 (six-membered ring); H_2C—O—CH_2, H_2C—CH_2 (five-membered ring)

1,4-Dioxane **Tetrahydrofuran (THF)**

EPOXIDE
A three-membered, oxygen-containing ring

OXIRANE
Alternative name for an epoxide

One group of cyclic ethers that behaves differently from open-chain compounds is the group called **epoxides**, or **oxiranes**. These are cyclic ethers that have a three-membered ring. Epoxides are usually prepared by treatment of an alkene with a peroxyacid, RCO_3H. *m*-Chloroperoxybenzoic acid is often used because it is more stable and more easily handled than most other peroxyacids.

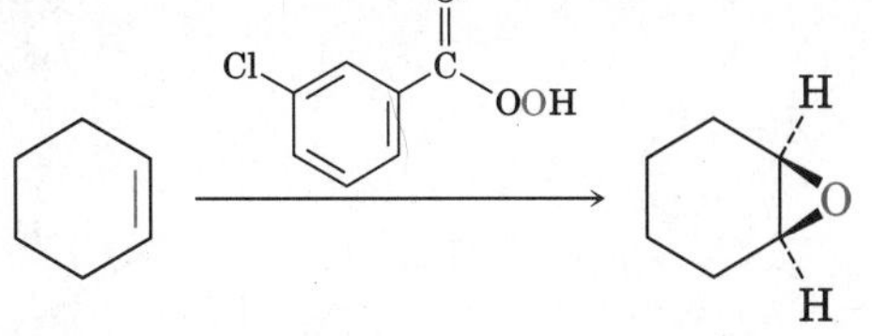

Cyclohexene 1,2-Epoxycyclohexane (85%)

PROBLEM 8.18 What product do you expect from reaction of *cis*-2-butene with *m*-chloroperoxybenzoic acid, assuming syn stereochemistry?

PROBLEM 8.19 Reaction of *trans*-2-butene with *m*-chloroperoxybenzoic acid yields a different epoxide from that obtained by reaction of the cis isomer (see Problem 8.18). How can you account for this?

8.11 RING-OPENING REACTIONS OF EPOXIDES

Epoxide rings are opened by treatment with acid in much the same way that other ethers are cleaved. The major difference is that epoxides react under much milder conditions because of the strain of the small ring. Dilute aqueous mineral acid at room temperature converts epoxides to 1,2-diols (also called *vicinal glycols*). Two million tons of ethylene glycol, most of it used as automobile antifreeze, are produced every year by acid-catalyzed hydration of ethylene oxide.

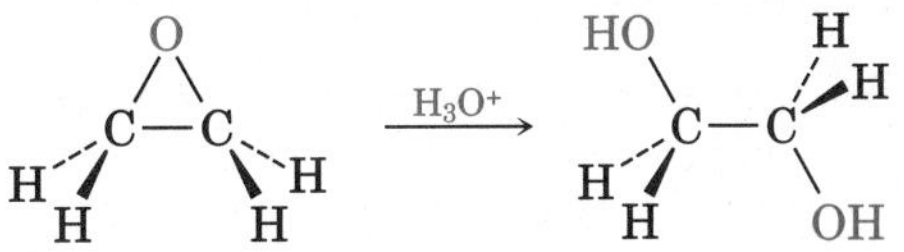

$\xrightarrow{H_3O^+}$

Ethylene oxide Ethylene glycol (1,2-ethanediol)

Acid-induced epoxide ring-opening takes place by S_N2 attack of a nucleophile on the protonated epoxide, in a manner analogous to the final step of alkene bromination in which a three-membered-ring bromonium ion is opened by nucleophilic attack (Section 4.5).

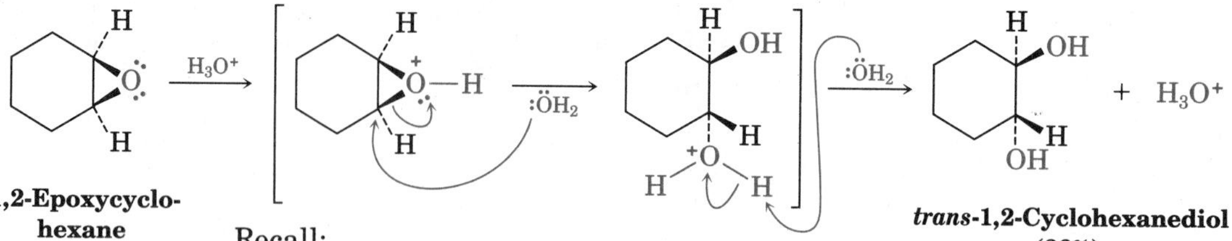

1,2-Epoxycyclohexane

H_3O^+

$:\ddot{O}H_2$

trans-1,2-Cyclohexanediol (86%)

$+ H_3O^+$

Recall:

Br_2

$:\ddot{Br}:^-$

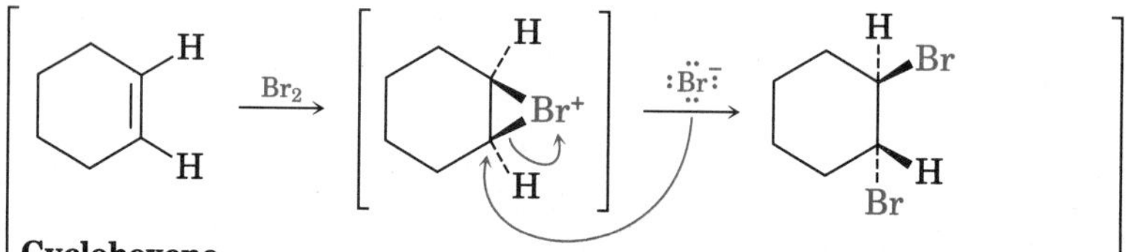

Cyclohexene

trans-1,2-Dibromocyclohexane

Epoxide opening is also involved in the mechanism by which the polycyclic aromatic hydrocarbons (PAH's) in chimney soot and cigarette smoke cause cancer. Benzo[a]pyrene, one of the best-studied PAH's, is converted by metabolic oxidation into a diol epoxide. In the body, the epoxide ring reacts with an amino group in cellular DNA to give an altered DNA that is covalently bound to the PAH. With its DNA thus altered, the cell is unable to function normally.

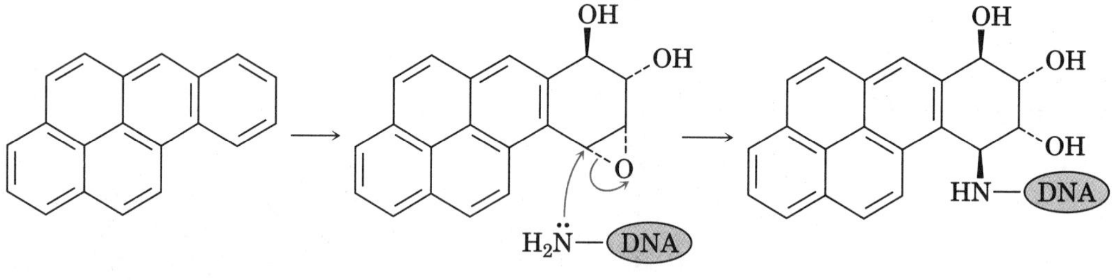

$H_2\ddot{N}$—DNA

HN—DNA

Benzo[a]pyrene

A diol epoxide

PROBLEM 8.20 Show the steps involved in the reaction of *cis*-2,3-epoxybutane with aqueous acid to yield 2,3-butanediol. What is the stereochemistry of the product if the ring opening takes place by normal back-side S_N2 attack?

PROBLEM 8.21 Answer Problem 8.20 for the reaction of *trans*-2,3-epoxybutane. Is the same product formed?

8.12 THIOLS AND SULFIDES

THIOL
Compound with the –SH functional group

SULFIDE
Compound that has two organic groups bonded to a sulfur atom

Sulfur is the element just below oxygen in the periodic table, and many oxygen-containing organic compounds have sulfur analogs. For instance, **thiols, R–SH**, are sulfur analogs of alcohols, and **sulfides, R–S–R′**, are sulfur analogs of ethers. Thiols are named in the same way as alcohols, with the suffix *-thiol* used in place of *-ol*. The –SH group itself is referred to as a **mercapto group**.

MERCAPTO GROUP
Alternative name for the thiol group, –SH

CH_3CH_2SH

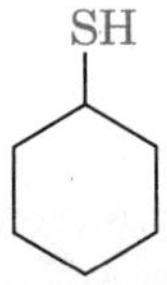

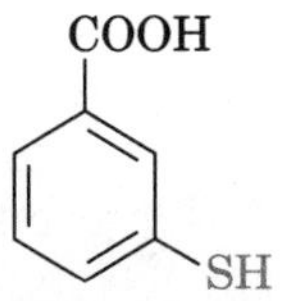

Ethanethiol **Cyclohexanethiol** ***m*-Mercaptobenzoic acid**

Sulfides are named in the same way as ethers, with *sulfide* used in place of *ether* for simple compounds and with *alkylthio* used in place of *alkoxy* for more complex substances.

$CH_3—S—CH_3$

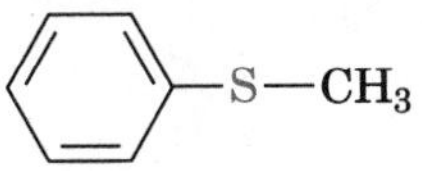

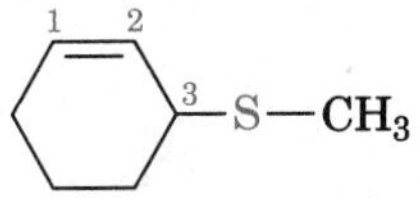

Dimethyl sulfide **Methyl phenyl sulfide** **3-(Methylthio)cyclohexene**

Thiols are usually prepared from the corresponding alkyl halide by S_N2 displacement with a sulfur nucleophile such as hydrosulfide anion, SH^-:

$$CH_3(CH_2)_6CH_2Br + Na^+ :\ddot{S}H^- \longrightarrow CH_3(CH_2)_6CH_2SH + NaBr$$

1-Bromooctane **Sodium hydrosulfide** **1-Octanethiol**

Sulfides are prepared by treating a primary or secondary alkyl halide with a thiolate ion, RS^-. Reaction occurs by an S_N2 mechanism that is analogous to the Williamson ether synthesis (Section 8.6). Thiolate anions are among the best nucleophiles known, and product yields are usually high in these sulfide-forming reactions.

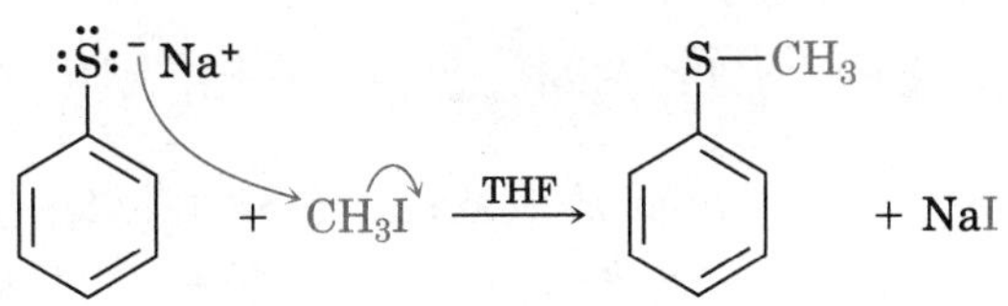

$$+ CH_3I \xrightarrow{THF} + NaI$$

Sodium benzenethiolate **Methyl phenyl sulfide (96%)**

The most unforgettable characteristic of thiols is their appalling odor. Skunk scent is due primarily to the simple thiols 3-methyl-1-butanethiol and 2-butene-1-thiol. Thiols can be oxidized by mild reagents such as bromine to yield disulfides, R–S–S–R, and disulfides can be reduced back to thiols by treatment with zinc metal and acetic acid:

$$2\ R—SH \underset{Zn,\ H^+}{\overset{Br_2}{\rightleftharpoons}} R—S—S—R + 2\ HBr$$

A thiol A disulfide

We'll see in Section 15.5 that the thiol–disulfide interconversion is extremely important in biochemistry, because disulfide "bridges" form the cross-links between protein chains that help stabilize the three-dimensional conformations of proteins.

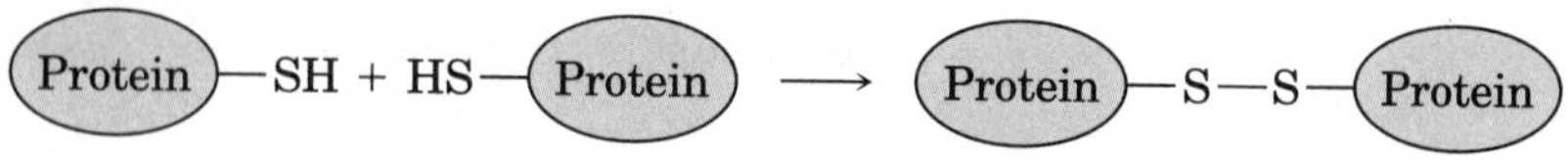

$$\text{Protein}-SH + HS-\text{Protein} \longrightarrow \text{Protein}-S-S-\text{Protein}$$

A cross-linked protein

PROBLEM 8.22 Name these thiols by IUPAC rules:
(a) $CH_3CH_2CH(SH)CH_3$ (b) $(CH_3)_3CCH_2CH(SH)CH_2CH(CH_3)_2$
(c) SH

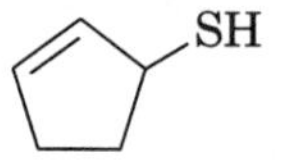

PROBLEM 8.23 Name these compounds by IUPAC rules:
(a) $CH_3CH_2SCH_3$ (b) $(CH_3)_3CSCH_2CH_3$ (c) SCH_3 SCH_3

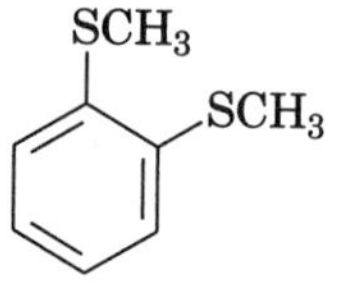

PROBLEM 8.24 2-Butene-1-thiol is one component of skunk spray. How would you synthesize this substance from 2-buten-1-ol? From methyl 2-butenoate, $CH_3CH{=}CHCOOCH_3$? More than one step is required in both instances.

INTERLUDE

ETHANOL AS CHEMICAL, DRUG, AND POISON

The production of ethanol by fermentation of grains and sugars is one of the oldest known organic reactions, going back at least to the ancient Greeks. Fermentation is carried out by adding yeast to an aqueous sugar solution, where enzymes break down carbohydrates into ethanol and CO_2:

$$\underset{\text{A carbohydrate}}{C_6H_{12}O_6} \xrightarrow{\text{Yeast}} 2\ CH_3CH_2OH + 2\ CO_2$$

Nearly 300 million gallons of ethanol are produced each year in the United States, primarily for use as a solvent. Only about 5% of this industrial ethanol comes from fermentation, though; most is obtained by acid-catalyzed hydration of ethylene:

$$H_2C{=}CH_2 + H_2O \xrightarrow[\text{catalyst}]{\text{Acid}} CH_3CH_2OH$$

Ethanol is classified for medical purposes as a central nervous system (CNS) depressant. Its effects (that is, being drunk) resemble the human response to anesthetics. There is an initial excitability and increase in sociable behavior, but this results from depression of inhibition rather than from stimulation. At a blood alcohol concentration of 0.1–0.3%, motor coordination is affected, accompanied by loss of balance, slurred speech, and amnesia. When blood alcohol concentration rises to 0.3–0.4%, nausea and loss of consciousness occur. Above 0.6%, spontaneous respiration and cardiovascular regulation are affected, ultimately leading to death. The LD_{50} (Chapter 1 Interlude) of ethanol is 10.6 g/kg.

The passage of ethanol through the body begins with its absorption in the stomach and small intestine, followed by rapid distribution to all body fluids and organs. In the pituitary gland, ethanol inhibits the production of a hormone that regulates urine flow, causing increased urine production and dehydration. In the stomach, ethanol stimulates production of acid. Throughout the body, ethanol causes blood vessels to dilate, resulting in flushing of the skin and a sensation of warmth as blood moves into capillaries beneath the surface. The result is not a warming of the body, though, but an increased loss of heat at the surface.

The metabolism of ethanol occurs mainly in the liver and proceeds by oxidation in two steps, first to acetaldehyde (CH_3CHO) and then to acetic acid (CH_3COOH). In chronic alcoholics, ethanol and acetaldehyde are toxic, leading to devastating physical and metabolic deterioration. The liver usually suffers the worst damage since it is the major site of alcohol metabolism.

The quick and uniform distribution of ethanol in body fluids, the ease with which it crosses lung membranes, and its ready oxidizability provide the basis for simple tests for blood alcohol concentration. The *Breathalyzer test* measures alcohol concentration in expired air by the color change that occurs when the chromium in the bright orange oxidizing agent potassium dichromate ($K_2Cr_2O_7$) is reduced to blue-green chromium(III). In most states, a blood alcohol level above 0.10% is sufficient for a charge of driving while intoxicated.

SUMMARY AND KEY WORDS

Ethers, alcohols, and **phenols** are organic derivatives of water in which one or both of the water hydrogens have been replaced by organic groups. Ethers have two organic groups bonded to an oxygen. The groups may be alkyl, alkenyl, or aryl, and the oxygen atom may be in a ring or in an open chain. Ethers are prepared by S_N2 reaction of an alkoxide ion with a primary alkyl halide—the **Williamson synthesis**. Ethers are inert to most reagents, but are attacked by strong acids to give cleavage products. **Epoxides** differ from other ethers in their ease of cleavage. The high reactivity of the strained three-membered ether ring allows epoxide rings to be opened easily.

Alcohols are compounds that have a hydroxyl group bonded to an alkyl residue. They can be prepared in many ways, including hydroxylation of alkenes and hydration of alkenes. The most important method of alcohol synthesis involves *reduction* of carbonyl compounds. Aldehydes, esters, and carboxylic acids yield primary alcohols on reduction; ketones yield secondary alcohols.

Alcohols are weak acids and can be converted into their *alkoxide anions* on treatment with strong base or with alkali metals. Alcohols can also be dehydrated to yield alkenes, can be transformed into alkyl halides by treatment with PBr_3 or $SOCl_2$, and can be converted into ethers by reaction of their anions with alkyl halides. The most important reaction of alcohols is their *oxidation* to yield carbonyl compounds. Primary alcohols give either aldehydes or carboxylic acids, secondary alcohols yield ketones, and tertiary alcohols are not oxidized.

Phenols are aromatic counterparts of alcohols and are prepared by reaction of arenesulfonic acids with NaOH at high temperature. Although similar to alcohols in some respects, phenols are more acidic because phenoxide anions are stabilized by resonance. Phenols undergo electrophilic aromatic substitution and can be oxidized to **quinones**.

Sulfides (R–S–R) and **thiols** (R–SH) are sulfur analogs of ethers and alcohols. Thiols are prepared by S_N2 reaction of an alkyl halide with HS^-, and sulfides are prepared by further alkylation of the thiol with a second molecule of alkyl halide.

SUMMARY OF REACTIONS

1. Synthesis of alcohols (Section 8.5)

   (a) Reduction of aldehydes to yield primary alcohols

$$\mathrm{R\overset{\overset{O}{\|}}{C}H} \xrightarrow{\mathrm{NaBH_4}} \mathrm{RCH_2OH}$$

   (b) Reduction of ketones to yield secondary alcohols

$$\mathrm{R\overset{\overset{O}{\|}}{C}R'} \xrightarrow{\mathrm{NaBH_4}} \mathrm{R\overset{\overset{OH}{|}}{C}HR'}$$

   (c) Reduction of esters to yield primary alcohols

$$\mathrm{R\overset{\overset{O}{\|}}{C}OR'} \xrightarrow{\mathrm{LiAlH_4}} \mathrm{RCH_2OH}$$

(d) Reduction of carboxylic acids to yield primary alcohols

$$RCOOH \xrightarrow{LiAlH_4} RCH_2OH$$

2. Synthesis of ethers (Section 8.6)

$$RO^- \ Na^+ + R'Br \xrightarrow[\text{reaction}]{S_N2} ROR'$$

3. Synthesis of phenols (Section 8.8)

$$C_6H_5SO_3H \xrightarrow{KOH} C_6H_5OH$$

4. Synthesis of epoxides (Section 8.10)

$$>C=C< \xrightarrow{RCO_3H} \text{epoxide}$$

5. Synthesis of thiols (Section 8.12)

$$Na^+ \ {}^-SH + RBr \xrightarrow[\text{reaction}]{S_N2} RSH$$

6. Synthesis of sulfides (Section 8.12)

$$RS^- \ Na^+ + R'Br \xrightarrow[\text{reaction}]{S_N2} RSR'$$

7. Reactions of alcohols

(a) Conversion into ethers (Section 8.6)

$$2\ ROH + 2\ Na \longrightarrow 2\ RO^- \ Na^+ + H_2$$

$$RO^- \ Na^+ + R'Br \xrightarrow[\text{reaction}]{S_N2} ROR'$$

(b) Dehydration to yield alkenes (Section 8.7)

$$-\underset{|}{\overset{H}{\overset{|}{C}}}-\underset{|}{\overset{OH}{\overset{|}{C}}}- \xrightarrow{H_2SO_4} >C=C< + H_2O$$

(c) Oxidation to yield carbonyl compounds (Section 8.7)

$$RCH_2OH \xrightarrow[\text{chlorochromate}]{\text{Pyridinium}} RCH(=O) \quad \text{An aldehyde}$$

$$RCH_2OH \xrightarrow[H_3O^+]{CrO_3} RCOH(=O) \quad \text{A carboxylic acid}$$

$$RCH(OH)R' \xrightarrow[\text{chlorochromate}]{\text{Pyridinium}} RCR'(=O) \quad \text{A ketone}$$

8. Reactions of ethers; acidic cleavage (Section 8.9)

$$ROR' + HI \longrightarrow ROH + R'I$$

ADDITIONAL PROBLEMS

8.25 Draw structures corresponding to these IUPAC names:
(a) Ethyl isopropyl ether (b) 3,4-Dimethoxybenzoic acid (c) 2-Methyl-2,5-heptanediol
(d) *trans*-3-Ethylcyclohexanol (e) 4-Allyl-2-methoxyphenol (eugenol, from oil of cloves)

8.26 Name these compounds according to IUPAC rules:

(a) 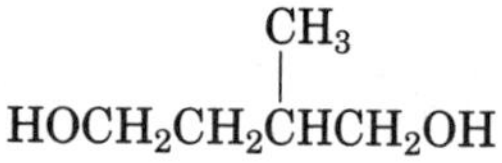 $HOCH_2CH_2CH(CH_3)CH_2OH$ (b) 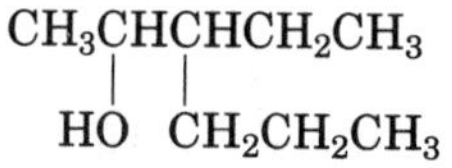 $CH_3CH(HO)CH(CH_2CH_2CH_3)CH_2CH_3$ (c) 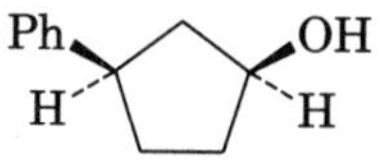 Ph, H / OH, H cyclopentane

(d) 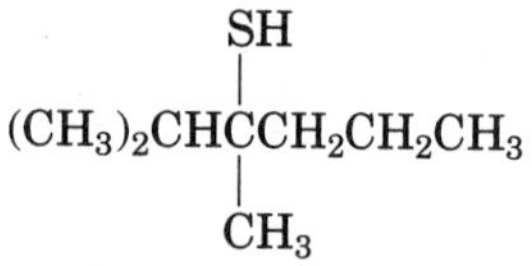 $(CH_3)_2CHC(SH)(CH_3)CH_2CH_2CH_3$

8.27 Draw and name the eight isomeric alcohols with the formula $C_5H_{12}O$.

8.28 Which of the eight alcohols you identified in Problem 8.27 would react with aqueous acidic $Na_2Cr_2O_7$? Show the products you would expect from each reaction.

8.29 Which of the eight alcohols you identified in Problem 8.27 are chiral?

8.30 Draw and name the six ethers that are isomeric with the alcohols you drew in Problem 8.27. Which are chiral?

8.31 Show the HI cleavage products of the ethers you drew in Problem 8.30.

8.32 Predict the likely products of these cleavage reactions:

(a) 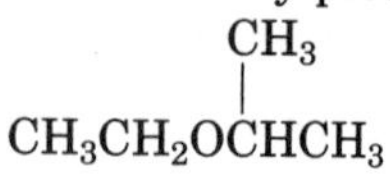 $CH_3CH_2OCH(CH_3)CH_3 \xrightarrow{HI,\ H_2O}$ (b) $(CH_3)_3CCH_2OCH_3 \xrightarrow{HI,\ H_2O}$

8.33 What reagents would you use to carry out these transformations?

(a) 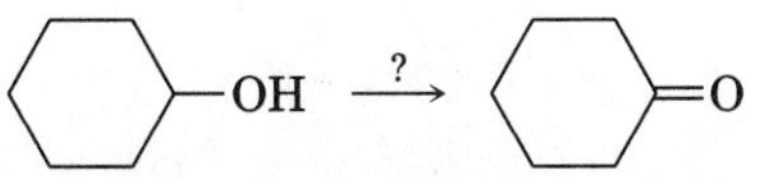 Cyclohexanol $\xrightarrow{?}$ Cyclohexanone (b) 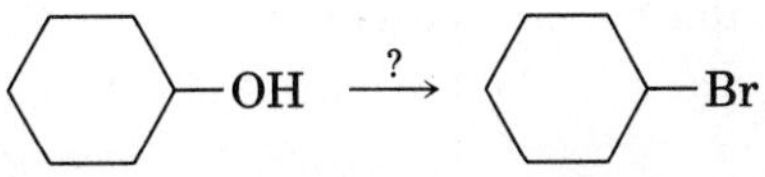 Cyclohexanol $\xrightarrow{?}$ Bromocyclohexane

(c) $CH_3CH_2CH_2OH \xrightarrow{?} CH_3CH_2CHO$ (d) $CH_3CH_2CH_2OH \xrightarrow{?} CH_3CH_2COOH$

(e) $CH_3CH_2CH_2OH \xrightarrow{?} CH_3CH_2CH_2O^- \ Na^+$ (f) $CH_3CH_2CH_2OH \xrightarrow{?} CH_3CH_2CH_2Cl$

8.34 How would you prepare these compounds from 2-phenylethanol?
(a) Benzoic acid (b) Ethylbenzene (c) 2-Bromo-1-phenylethane
(d) Phenylacetic acid ($C_6H_5CH_2COOH$) (e) Phenylacetaldehyde ($C_6H_5CH_2CHO$)

8.35 Give the structures of the major products you would obtain from reaction of phenol with these reagents:
(a) Br_2 (1 mol) (b) Br_2 (3 mol) (c) NaOH, then CH_3I (d) $Na_2Cr_2O_7$, H_3O^+

8.36 What products would you obtain from reaction of 1-butanol with these reagents?
(a) PBr_3 (b) CrO_3, H_3O^+ (c) Na (d) Pyridinium chlorochromate

8.37 What products would you obtain from reaction of 1-methylcyclohexanol with these reagents?
(a) HBr (b) H_2SO_4 (c) CrO_3 (d) Na (e) Product from part (d), then CH_3I

8.38 What alcohols would you oxidize to obtain these products?

(a) 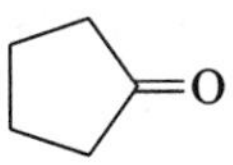 Cyclopentanone (=O) (b) 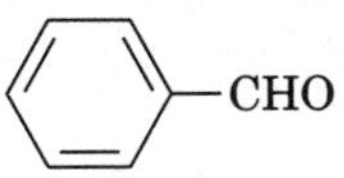 Benzene ring–CHO (c) $CH_3CH(CH_3)COOH$

8.39 Show the alcohols you would obtain by reduction of these carbonyl compounds:

(a) $CH_3CH(CH_3)CH_2CHO$ (b) 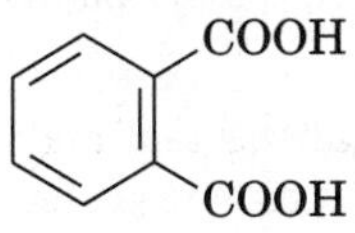 Benzene ring with two adjacent COOH groups (c) $CH_3CH_2C(=O)CH_2CH(CH_3)CH_3$

8.40 Reduction of 2-butanone with $NaBH_4$ yields 2-butanol. Is the product chiral? Is it optically active? Explain.

$CH_3C(=O)CH_2CH_3$ **2-Butanone**

8.41 When 4-chloro-1-butanol is treated with a strong base such as sodium hydride, NaH, tetrahydrofuran is produced. Suggest a mechanism for this reaction.

$$ClCH_2CH_2CH_2CH_2OH \xrightarrow[\text{Ether}]{\text{NaH}} \text{(tetrahydrofuran)} + H_2 + NaCl$$

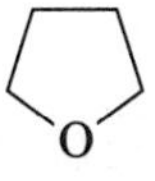

8.42 Which is more acidic, *p*-methylphenol or *p*-bromophenol? Explain.

8.43 Why can't the Williamson ether synthesis be used to prepare diphenyl ether?

8.44 Rank these substances in order of increasing acidity:

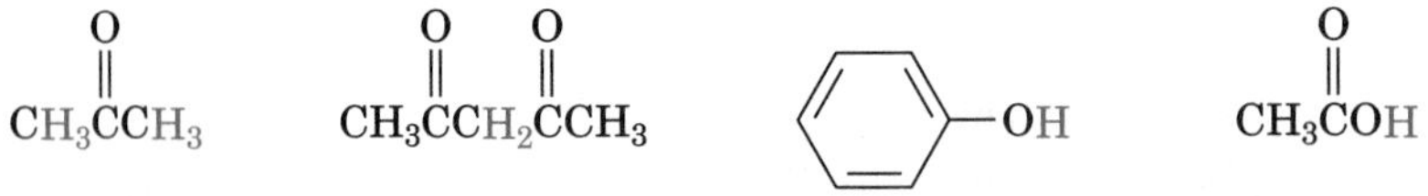

CH_3COCH_3	$CH_3COCH_2COCH_3$	C_6H_5OH	CH_3COOH
Acetone	**2,4-Pentanedione**	**Phenol**	**Acetic acid**
$pK_a = 19$	$pK_a = 9$	$pK_a = 10$	$pK_a = 4.7$

8.45 Which, if any, of the substances in Problem 8.44 are strong enough acids to react substantially with NaOH? (The pK_a of H_2O is 15.7.)

8.46 Is *tert*-butoxide anion a strong enough base to react with water? In other words, does the following reaction take place as written? (The pK_a of *tert*-butyl alcohol is 18.)

$$(CH_3)_3CO^- \, Na^+ + H_2O \xrightarrow{?} (CH_3)_3COH + NaOH$$

8.47 Sodium bicarbonate, $NaHCO_3$, is the sodium salt of carbonic acid (H_2CO_3), $pK_a \approx 6.4$. Which of the substances shown in Problem 8.44 will react with sodium bicarbonate?

8.48 Assume that you have two unlabeled bottles, one that contains phenol ($pK_a = 10$) and one that contains acetic acid ($pK_a = 4.7$). In light of your answer to Problem 8.47, propose a simple way to tell what is in each bottle.

8.49 Starting from benzene, how would you prepare benzyl phenyl ether, $C_6H_5OCH_2C_6H_5$? (More than one step is required.)

8.50 Since all hamsters look pretty much alike, pairing and mating is governed by chemical means of communication. Investigations have shown that dimethyl disulfide, CH_3SSCH_3, is secreted by female hamsters as a sex attractant for males. How would you synthesize dimethyl disulfide in the laboratory if you wanted to trick your hamster?

8.51 *p*-Nitrophenol ($pK_a = 7.15$) is much more acidic than phenol ($pK_a = 10.0$). Draw as many resonance structures as you can for the *p*-nitrophenoxide anion.

8.52 The herbicide 2,4,5-T (2,4,5-trichlorophenoxyacetic acid) can be prepared by heating a mixture of 2,4,5-trichlorophenol and $ClCH_2COOH$ with NaOH. Show the mechanism of the reaction.

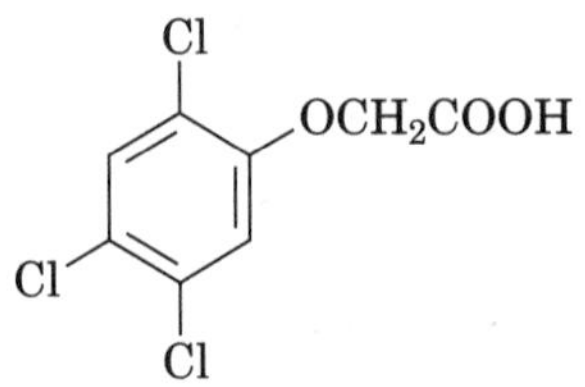

Cl, OCH_2COOH, Cl, Cl (benzene ring)

2,4,5-T

8.53 *tert*-Butyl ethers can be prepared by the reaction of an alcohol with 2-methylpropene in the presence of an acid catalyst. Propose a mechanism for this reaction.

8.54 How would you prepare these ethers?

(a) 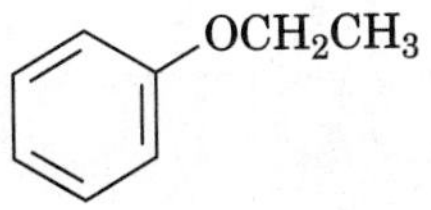 (b) 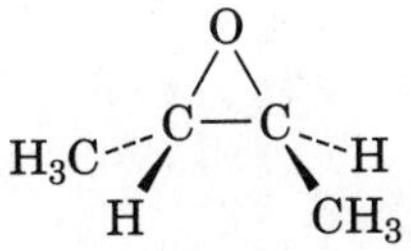

8.55 What cleavage product would you expect from reaction of tetrahydrofuran with hot aqueous HI?

Tetrahydrofuran

8.56 Methyl phenyl ether can be cleaved to yield iodomethane and lithium phenoxide upon heating with LiI. Propose a mechanism for this reaction.

8.57 The Zeisel method, a procedure for determining the number of methoxyl groups (CH_3O-) in a compound, involves heating a weighed amount of compound with HI. Ether cleavage occurs, and the iodomethane formed is distilled off and passed into a solution of $AgNO_3$. The silver iodide that precipitates is then weighed, and the percentage of methoxy groups in the sample is thereby determined. For example, 1.06 g of vanillin, the material responsible for the characteristic odor of vanilla, yields 1.60 g AgI. If vanillin has a molecular weight of 152, how many methoxyls does it contain?

9 CHAPTER

ALDEHYDES AND KETONES: NUCLEOPHILIC ADDITION REACTIONS

CARBONYL GROUP
The carbon–oxygen double bond, C=O

In this and the next two chapters, we'll discuss the most important functional group in organic chemistry—the **carbonyl group, C=O**. Carbonyl compounds are everywhere in nature. Most biologically important molecules contain carbonyl groups, as do many pharmaceutical agents and many of the synthetic chemicals that touch our everyday lives. Acetic acid (the chief component of vinegar), acetaminophen (an over-the-counter headache remedy), and Dacron (the polyester material used in clothing) all contain different kinds of carbonyl groups.

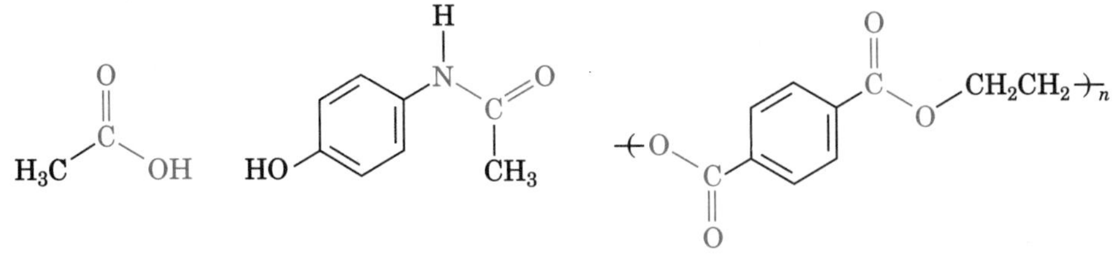

$H_3C-C(=O)-OH$ — **Acetic acid** (a carboxylic acid)

HO–C₆H₄–NH–C(=O)–CH_3 — **Acetaminophen** (an amide)

$-\!\!(O-C(=O)-C_6H_4-C(=O)-O-CH_2CH_2)\!\!-_n$ — **Dacron** (a polyester)

9.1 KINDS OF CARBONYL COMPOUNDS

Table 9.1 shows some of the many kinds of carbonyl compounds. All contain an acyl part, $R-\overset{O}{\overset{\|}{C}}-$, bonded to another atom, which may be a carbon, hydrogen, oxygen, halogen, sulfur, or other substituent.

TABLE 9.1 Some Types of Carbonyl Compounds

NAME	GENERAL FORMULA	NAME ENDING
Aldehyde	$R{-}C({=}O){-}H$	*-al*
Ketone	$R{-}C({=}O){-}R'$	*-one*
Carboxylic acid	$R{-}C({=}O){-}O{-}H$	*-oic acid*
Acid chloride	$R{-}C({=}O){-}Cl$	*-yl* or *-oyl chloride*
Acid anhydride	$R{-}C({=}O){-}O{-}C({=}O){-}R'$	*-oic anhydride*
Ester	$R{-}C({=}O){-}O{-}R'$	*-oate*
Lactone (cyclic ester)	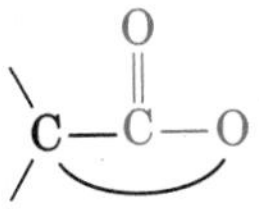 $>C{-}C({=}O){-}O$ (ring)	None
Amide	$R{-}C({=}O){-}N<$	*-amide*

It turns out to be very useful to classify carbonyl compounds into two general categories, based on the kinds of chemistry they undergo:

Aldehydes (RCHO)
Ketones (R_2CO)

The acyl groups in these two families are bonded to substituents (–H and –R, respectively) that *can't stabilize a negative charge and therefore can't act as leaving groups*. Aldehydes and ketones behave similarly and undergo many of the same reactions.

Carboxylic acids (RCOOH)
Esters (RCOOR′)
Acid chlorides (RCOCl)
Acid anhydrides (RCOOCOR′)
Amides ($RCONH_2$)

The acyl groups in carboxylic acids and their derivatives are bonded to substitutents (oxygen, halogen, nitrogen) that *can stabilize a negative charge and can serve as leaving groups in substitution reactions*. The chemistry of these compounds is therefore similar.

PROBLEM 9.1 Propose structures for molecules that meet these descriptions:
(a) A ketone, $C_5H_{10}O$ (b) An aldehyde, $C_6H_{10}O$
(c) A keto aldehyde, $C_6H_{10}O_2$ (d) A cyclic ketone, C_5H_8O

9.2 STRUCTURE AND PROPERTIES OF CARBONYL GROUPS

The carbon–oxygen double bond of carbonyl groups is similar in some respects to the carbon–carbon double bond of alkenes (Figure 9.1). The carbonyl carbon atom is sp^2-hybridized and forms three sigma bonds. The fourth valence electron remains in a carbon *p* orbital and forms a pi bond to oxygen by overlap with an oxygen *p* orbital. The oxygen also has two nonbonding pairs of electrons, which occupy its remaining two orbitals. Like alkenes, carbonyl compounds are planar about the double bond and have bond angles of approximately 120°.

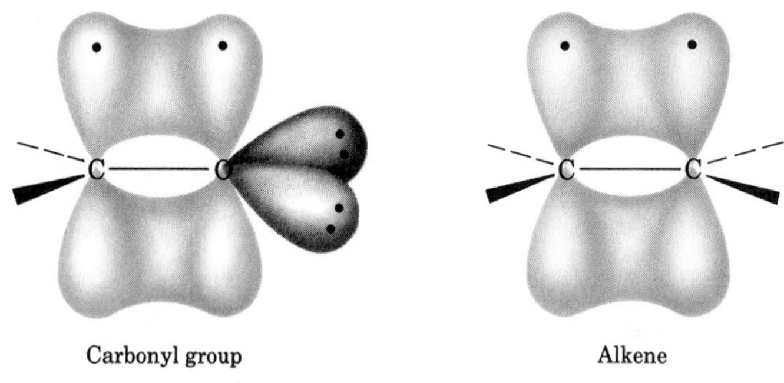

FIGURE 9.1 Electronic structure of the carbonyl group.

Carbon–oxygen double bonds are polarized because of the high electronegativity of oxygen relative to carbon. Since the carbonyl carbon is positively polarized, it is an electrophilic (Lewis acidic) site and is attacked by nucleophiles. Conversely, the carbonyl oxygen is negatively polarized and is a nucleophilic (Lewis basic) site. We'll see in this and the next two chapters that most carbonyl-group reactions can be understood in terms of bond polarities.

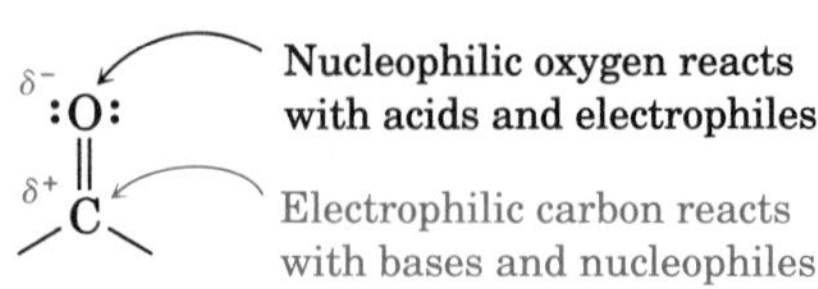

9.3 NAMING ALDEHYDES AND KETONES

Systematic names for aldehydes are obtained by replacing the terminal *-e* of the alkane name with *-al*. The longest chain selected for the base name must contain the –CHO group, and the CHO carbon is always numbered as carbon 1. For example,

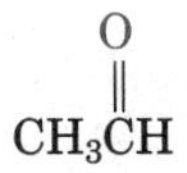

$$CH_3\overset{O}{\overset{\|}{C}}H \qquad CH_3CH_2\overset{O}{\overset{\|}{C}}H \qquad \underset{5}{C}H_3\underset{4}{\overset{CH_3}{\overset{|}{C}}}H\underset{3}{C}H_2\underset{\underset{CH_2CH_3}{|}}{\overset{2}{C}}H\underset{1}{\overset{O}{\overset{\|}{C}}}H$$

Ethanal (acetaldehyde) **Propanal (propionaldehyde)** **2-Ethyl-4-methylpentanal**

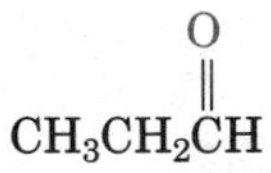

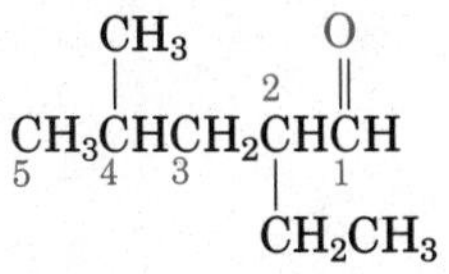

Note that the longest chain in 2-ethyl-4-methylpentanal is a hexane, but this chain doesn't include the –CHO group.

For more complex aldehydes in which the –CHO group is attached to a ring, the suffix *-carbaldehyde* is used:

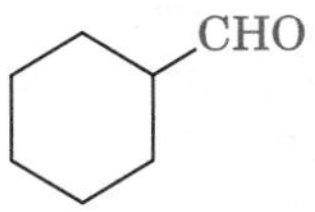

Cyclohexanecarbaldehyde 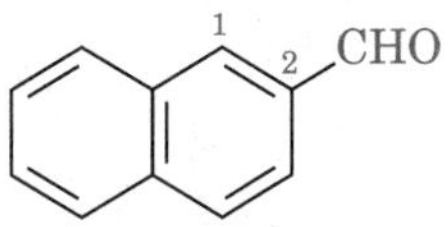 **2-Naphthalenecarbaldehyde**

Some simple and well-known aldehydes also have common names, as indicated in Table 9.2.

TABLE 9.2 Common Names of Some Simple Aldehydes

FORMULA	COMMON NAME	SYSTEMATIC NAME
$HCHO$	Formaldehyde	Methanal
CH_3CHO	Acetaldehyde	Ethanal
CH_3CH_2CHO	Propionaldehyde	Propanal
$CH_3CH_2CH_2CHO$	Butyraldehyde	Butanal
$CH_3CH_2CH_2CH_2CHO$	Valeraldehyde	Pentanal
$H_2C{=}CHCHO$	Acrolein	2-Propenal
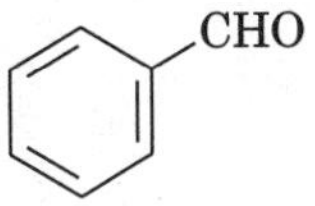 C_6H_5–CHO	Benzaldehyde	Benzenecarbaldehyde

Ketones are named by replacing the terminal *-e* of the corresponding alkane name with *-one* (pronounced **oan**). The chain selected for the base name is the longest one that contains the ketone group, and the numbering begins at the end nearer the carbonyl carbon. For example,

$CH_3C(=O)CH_3$ — **Propanone (acetone)**

$CH_3CH_2C(=O)CH_2CH_2CH_3$ (numbered 1 2 3 4 5 6) — **3-Hexanone**

$CH_3CH{=}CHCH_2C(=O)CH_3$ (numbered 6 5 4 3 2 1) — **4-Hexen-2-one**

A few ketones also have common names:

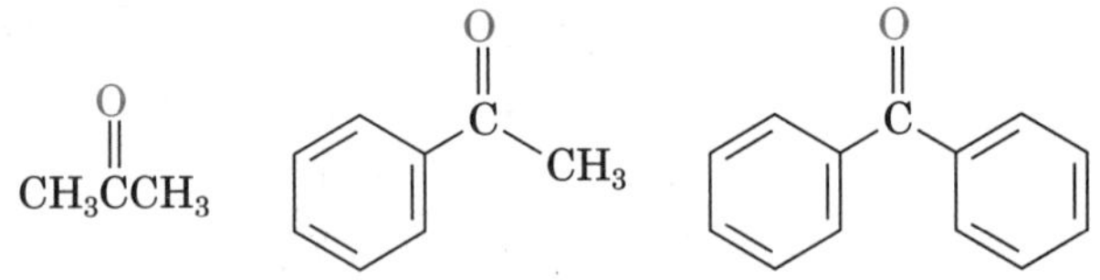

$CH_3C(=O)CH_3$ — **Acetone**

$C_6H_5C(=O)CH_3$ — **Acetophenone**

$C_6H_5C(=O)C_6H_5$ — **Benzophenone**

When it is necessary to refer to the $-COR$ group as a substituent, the term *acyl* (**a**-sil) is used. Similarly, $-COCH_3$ is an *acetyl* group, $-CHO$ is called a *formyl* group, $-COAr$ is an *aroyl* group, and $-COC_6H_5$ is a *benzoyl* group.

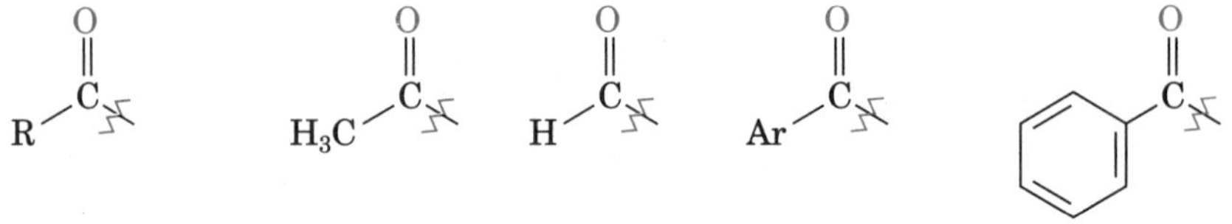

$R-C(=O)-$ — An acyl group (R = alkyl, alkenyl)

$H_3C-C(=O)-$ — Acetyl

$H-C(=O)-$ — Formyl

$Ar-C(=O)-$ — Aroyl (Ar = aromatic)

$C_6H_5-C(=O)-$ — Benzoyl

Occasionally, the doubly bonded oxygen must be considered a substituent, and the prefix *oxo-* is used. For example,

$CH_3CH_2CH_2C(=O)CH_2C(=O)OCH_3$ (numbered 6 5 4 3 2 1) — **Methyl 3-oxohexanoate**

PROBLEM 9.2 Name these aldehydes and ketones:

(a) $CH_3CH_2C(=O)CH(CH_3)_2$

(b) 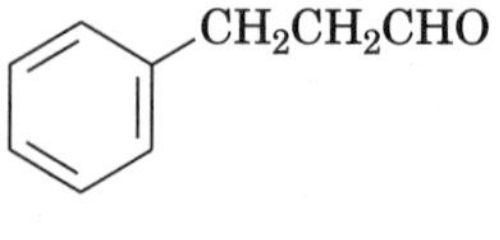 $C_6H_5CH_2CH_2CHO$

(c) $CH_3\overset{O}{\overset{\|}{C}}CH_2CH_2CH_2\overset{O}{\overset{\|}{C}}CH_2CH_3$ (d) 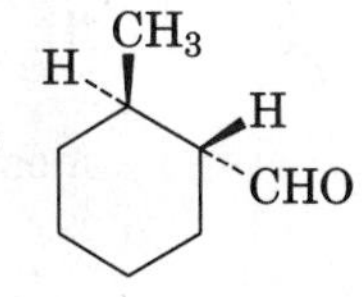 [structure: cyclohexane with H, CH_3 and H, CHO substituents]

(e) $OHCCH_2CH_2CH_2CHO$ (f) 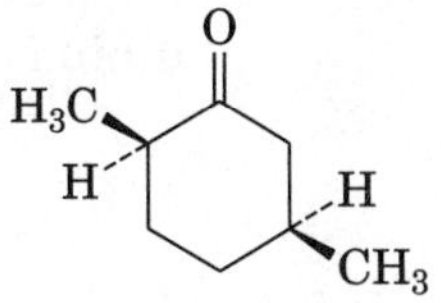 [structure: cyclohexanone with H_3C, H and H, CH_3 substituents]

PROBLEM 9.3

Draw structures corresponding to these IUPAC names:

(a) 3-Methylbutanal **(b)** 3-Methyl-3-butenal
(c) 4-Chloro-2-pentanone **(d)** Phenylacetaldehyde
(e) 2,2-Dimethylcyclohexanecarbaldehyde **(f)** 1,3-Cyclohexanedione

9.4 SYNTHESIS OF ALDEHYDES

We've already discussed two good methods of aldehyde synthesis: oxidation of primary alcohols and cleavage of alkenes. Let's review briefly:

1. Primary alcohols can be oxidized to give aldehydes (Section 8.7). The reaction is often carried out using pyridinium chlorochromate (PCC) in dichloromethane solution.

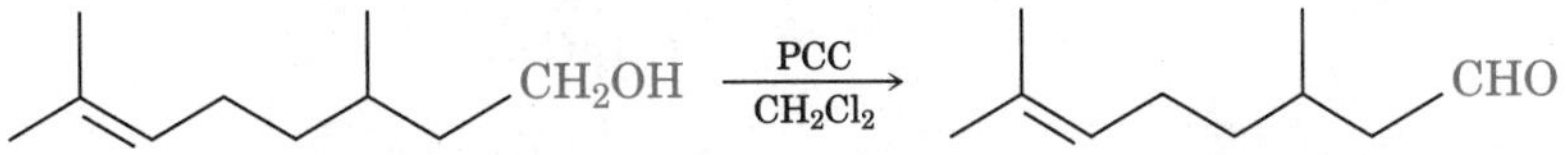

$$\text{Citronellol } (\ldots CH_2OH) \xrightarrow[CH_2Cl_2]{PCC} \text{Citronellal } (\ldots CHO)$$

Citronellol **Citronellal (82%)**

2. Alkenes with at least one vinylic proton (*remember*: *vinylic* means "on the double-bond carbon") undergo oxidative cleavage when treated with ozone to yield aldehydes (Section 4.7).

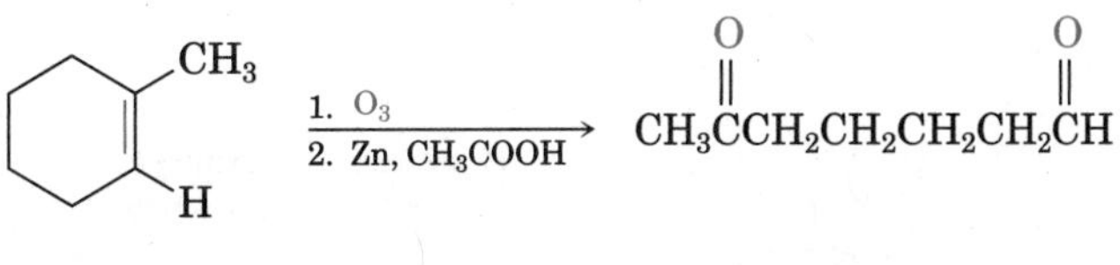

$$\text{1-Methylcyclohexene} \xrightarrow[2.\ Zn,\ CH_3COOH]{1.\ O_3} CH_3\overset{O}{\overset{\|}{C}}CH_2CH_2CH_2CH_2\overset{O}{\overset{\|}{C}}H$$

1-Methylcyclohexene **6-Oxoheptanal (86%)**

PROBLEM 9.4

Show how you might prepare pentanal from these starting materials:

(a) 1-Pentanol **(b)** 1-Hexene **(c)** 5-Decene
(d) $CH_3CH_2CH_2CH_2COOH$

9.5 SYNTHESIS OF KETONES

For the most part, methods of ketone synthesis are similar to those for aldehydes:

1. Secondary alcohols are oxidized to give ketones (Section 8.7). Pyridinium chlorochromate (PCC), CrO_3, and $Na_2Cr_2O_7$ are all effective.

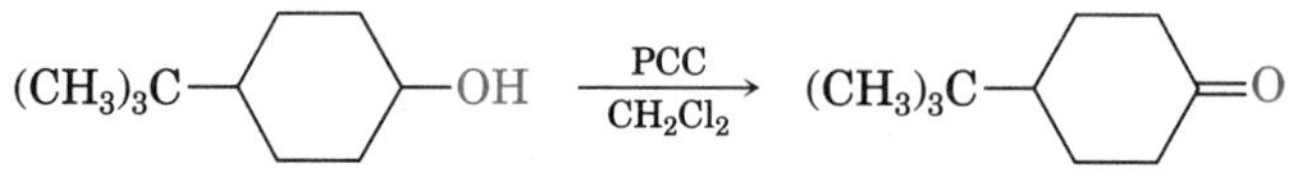

4-*tert*-Butylcyclohexanol **4-*tert*-Butylcyclohexanone (90%)**

2. Ozonolysis of alkenes yields ketones if one of the double-bond carbon atoms is disubstituted (Section 4.7).

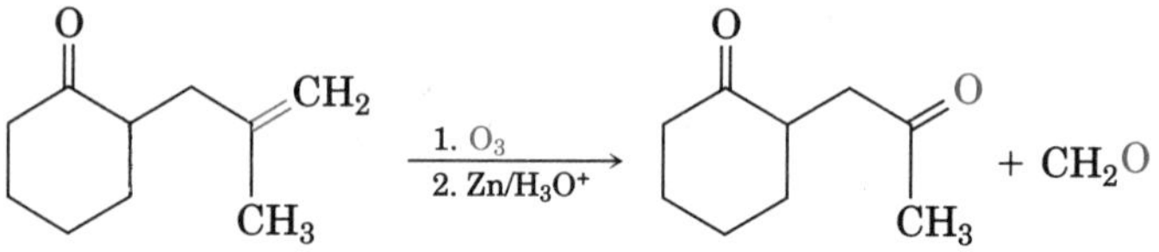

70%

3. Terminal alkynes undergo hydration to yield methyl ketones. The reaction is catalyzed by mercuric sulfate (Section 4.15).

$$CH_3(CH_2)_3C{\equiv}CH \xrightarrow[Hg(OAc)_2]{H_3O^+} CH_3(CH_2)_3\overset{\overset{O}{\|}}{C}{-}CH_3$$

1-Hexyne **2-Hexanone (78%)**

4. Aromatic rings undergo Friedel–Crafts acylation with an acid chloride to yield alkyl aryl ketones (Section 5.8).

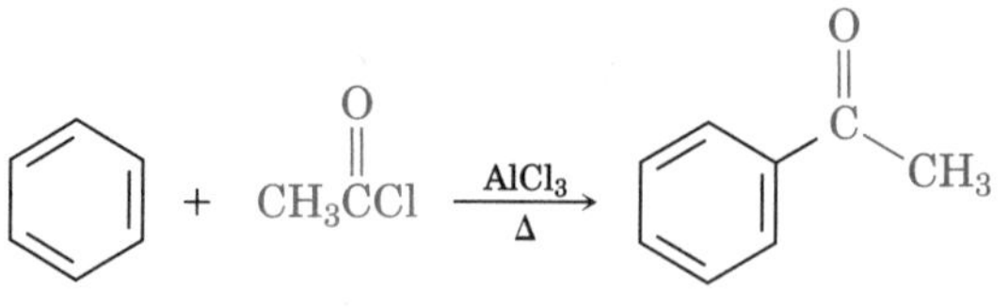

Benzene **Acetyl chloride** **Acetophenone (95%)**

PROBLEM 9.5 Show how you could prepare 2-hexanone from these starting materials:
(a) 2-Hexanol (b) 1-Hexyne (c) 2-Methyl-1-hexene

PROBLEM 9.6 How would you carry out the following reactions? (More than one step may be required.)
(a) 3-Hexene ⟶ 3-Hexanone (b) Benzene ⟶ 1-Phenylethanol

9.6 OXIDATION OF ALDEHYDES

Aldehydes are easily oxidized to yield carboxylic acids, RCHO → RCOOH, but ketones are unreactive toward oxidation. This reactivity difference is a consequence of the structural difference between the two functional groups: Aldehydes have a –CHO proton that can be removed during oxidation, but ketones do not.

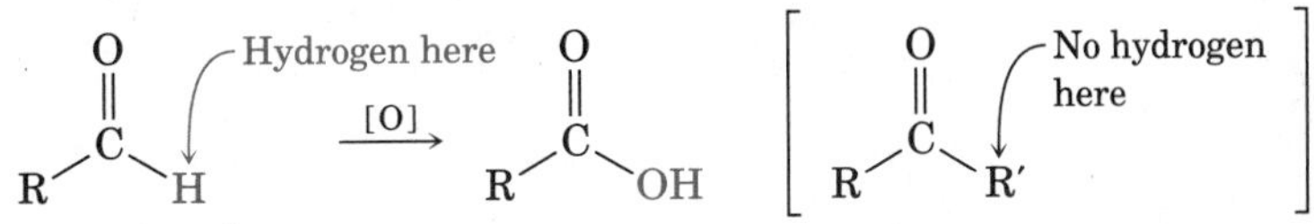

$$\underset{\text{An aldehyde}}{R-\overset{O}{\overset{\|}{C}}-H} \xrightarrow{[O]} R-\overset{O}{\overset{\|}{C}}-OH \qquad \left[\underset{\text{A ketone}}{R-\overset{O}{\overset{\|}{C}}-R'}\right]$$

Hydrogen here

No hydrogen here

TOLLENS' REAGENT Solution of Ag^+ in aqueous NH_3; useful for oxidizing aldehydes to carboxylic acids

One of the simplest methods for oxidizing an aldehyde is to use silver ion, Ag^+, in dilute aqueous ammonia (**Tollens' reagent**). As the oxidation proceeds, silver metal is deposited on the walls of the reaction flask as a shiny mirror. In fact, Tollens' reagent can be used to detect the presence of an aldehyde functional group in a sample of unknown structure. A small amount of the unknown is dissolved in ethanol in a test tube, and a few drops of Tollens' reagent are added. If the test tube becomes silvery, the unknown is an aldehyde.

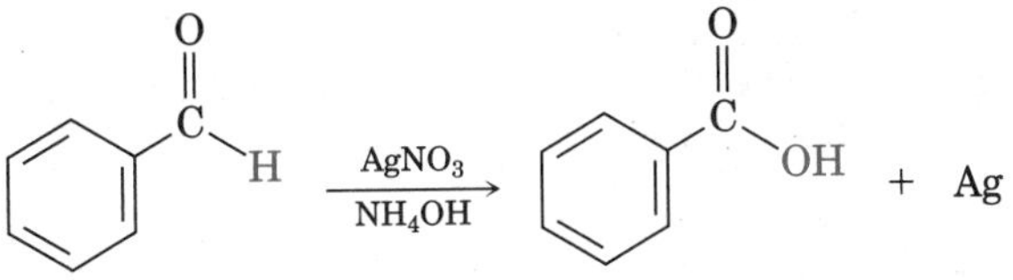

$$\underset{\textbf{Benzaldehyde}}{C_6H_5-\overset{O}{\overset{\|}{C}}-H} \xrightarrow[NH_4OH]{AgNO_3} \underset{\textbf{Benzoic acid}}{C_6H_5-\overset{O}{\overset{\|}{C}}-OH} + Ag$$

PRACTICE PROBLEM 9.1 What product would you obtain from the reaction of 3-methylbutanal with Tollens' reagent?

SOLUTION Write the structure of the aldehyde starting material and then replace the hydrogen bonded to the carbonyl group by –OH:

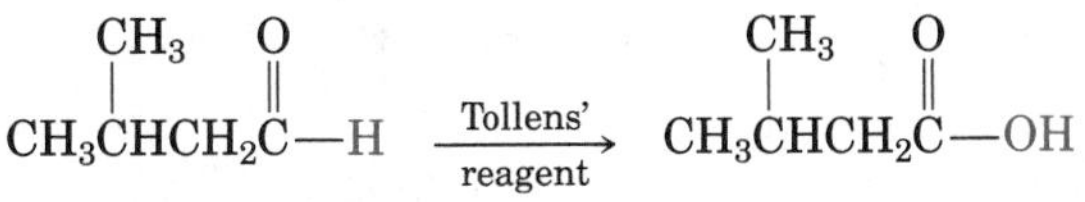

$$\underset{\textbf{3-Methylbutanal}}{CH_3\overset{CH_3}{\overset{|}{C}}HCH_2\overset{O}{\overset{\|}{C}}-H} \xrightarrow[\text{reagent}]{\text{Tollens'}} \underset{\textbf{3-Methylbutanoic acid}}{CH_3\overset{CH_3}{\overset{|}{C}}HCH_2\overset{O}{\overset{\|}{C}}-OH}$$

PROBLEM 9.7 Predict the products of the reaction of these substances with Tollens' reagent:
(a) Pentanal **(b)** 2,2-Dimethylhexanal **(c)** Cyclohexanone

9.7 REACTIONS OF ALDEHYDES AND KETONES: NUCLEOPHILIC ADDITIONS

NUCLEOPHILIC ADDITION REACTION
Reaction that involves the addition of a nucleophile to a carbonyl group

CARBANION
Substance that has a trivalent, negatively charged carbon atom

The most important reaction of ketones and aldehydes is the **nucleophilic addition reaction**. As the name implies, a nucleophilic addition reaction involves the addition of a nucleophile (:Nu or $:Nu^-$) to the carbonyl group. Hydroxide ion (OH^-), hydride ion (H^-), carbon anions (**carbanions,** R_3C^-), water (H_2O), ammonia (H_3N), and alcohols (ROH) are several of many possibilities.

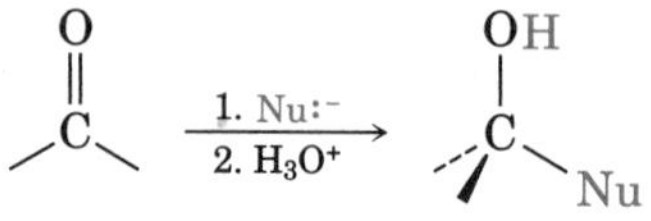

1. Nu:⁻ 2. H_3O^+

The general mechanism of nucleophilic addition is shown in Figure 9.2. As indicated, the nucleophile uses its electrons to form a new bond to the carbonyl-group carbon, the carbon–oxygen double bond breaks, and a proton bonds to the oxygen.

An electron pair from the nucleophile attacks the electrophilic carbonyl carbon, pushing an electron pair from the C=O bond out onto oxygen. The carbonyl carbon rehybridizes from sp^2 to sp^3.

Aldehyde or ketone

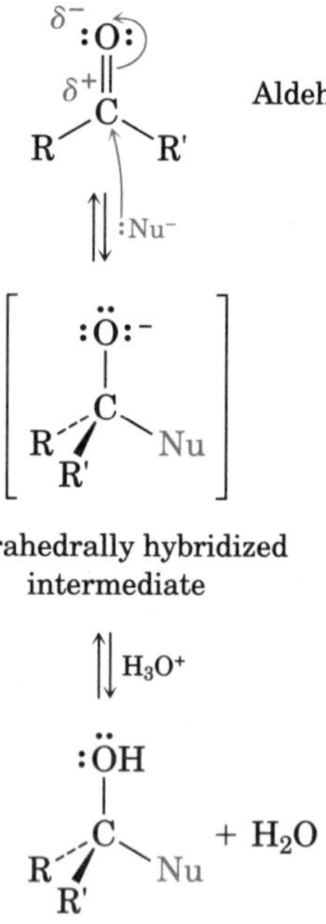

:Nu⁻

Tetrahedrally hybridized intermediate

Protonation of the anion resulting from nucleophilic attack yields the neutral alcohol addition product.

H_3O^+

$+ H_2O$

FIGURE 9.2 General mechanism of a nucleophilic addition reaction.

Aldehydes are generally more reactive than ketones in nucleophilic additions. The presence of two relatively large substituents in ketones versus only one large substituent in aldehydes means that attacking nucleophiles are able to approach aldehydes more readily (Figure 9.3).

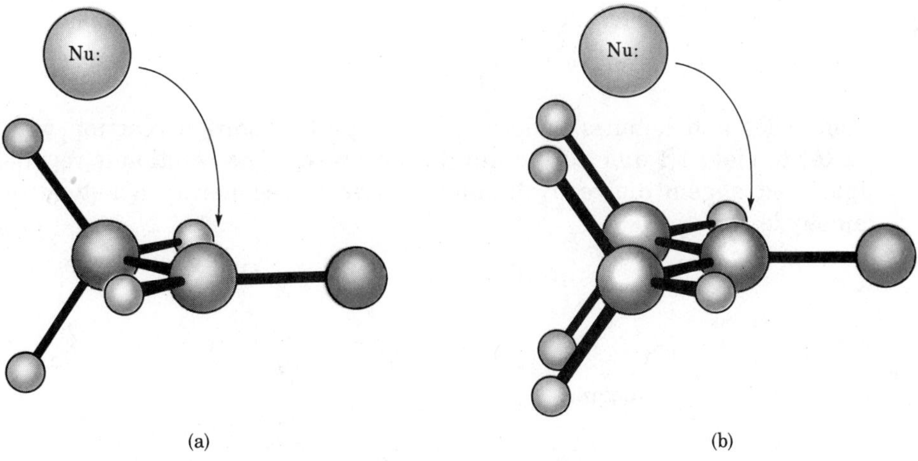

FIGURE 9.3 Nucleophilic attack on an aldehyde (a) is relatively unhindered, but attack on a ketone (b) is sterically hindered because of the two relatively large substituents attached to the carbonyl-group carbon.

PRACTICE PROBLEM 9.2 What product would you expect from nucleophilic addition of aqueous hydroxide ion to acetaldehyde?

SOLUTION Hydroxide ion adds to the carbonyl-group carbon atom, giving an alkoxide ion intermediate that is protonated to yield a 1,1-dialcohol:

$$CH_3CHO \xrightleftharpoons{:\ddot{O}H^-} \left[CH_3CH(O^-)OH \right] \xrightarrow{H_3O^+} CH_3CH(OH)OH$$

Acetaldehyde

PROBLEM 9.8 What product would you expect if the nucleophile cyanide ion, CN^-, were to add to acetone, and the intermediate were to be protonated?

PROBLEM 9.9 The reduction of a ketone to a secondary alcohol on treatment with $NaBH_4$ (Section 8.5) is a nucleophilic addition reaction in which the nucleophile hydride ion (H^-) adds to the carbonyl group, and the alkoxide ion intermediate is then protonated. Show the mechanism of this reduction.

PROBLEM 9.10 Which would you expect to be more reactive toward nucleophilic additions, propanal or 2,2-dimethylpropanal? Explain.

9.8 NUCLEOPHILIC ADDITION OF WATER: HYDRATION

GEMINAL
Referring to two groups attached to the same carbon atom

Aldehydes and ketones undergo a nucleophilic addition reaction with water to yield 1,1-diols, or **geminal (gem) diols**. The reaction is reversible, and a gem diol can eliminate water to regenerate a ketone or aldehyde:

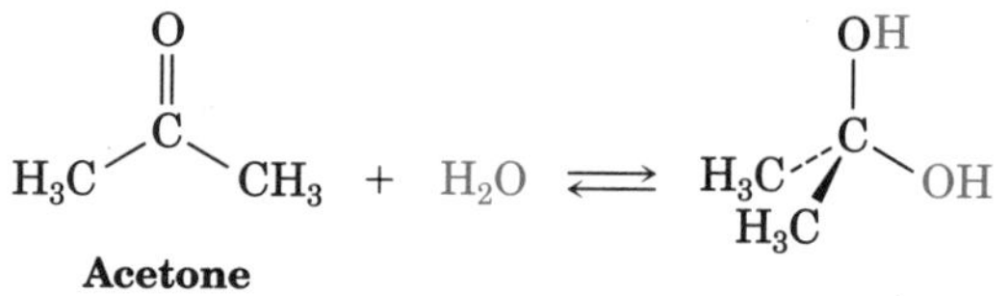

$$H_3C{-}C({=}O){-}CH_3 + H_2O \rightleftharpoons (H_3C)_2C(OH)_2$$

Acetone **Acetone hydrate** (a gem diol)

The position of the equilibrium between gem diols and ketones/aldehydes depends on the structure of the carbonyl compound. Although the equilibrium strongly favors the carbonyl compound in most cases, the gem diol is favored for a few simple aldehydes. For example, an aqueous solution of acetone consists of about 0.1% gem diol and 99.9% ketone, whereas an aqueous solution of formaldehyde consists of 99.9% gem diol and 0.1% aldehyde.

The nucleophilic addition of water to ketones and aldehydes is slow in pure water but is catalyzed by both acid and base. Although these catalysts don't change the *position* of the equilibrium, they strongly affect the rate at which the hydration reaction occurs.

The base-catalyzed reaction takes place in several steps, as shown in Figure 9.4. The attacking nucleophile is the negatively charged hydroxide ion.

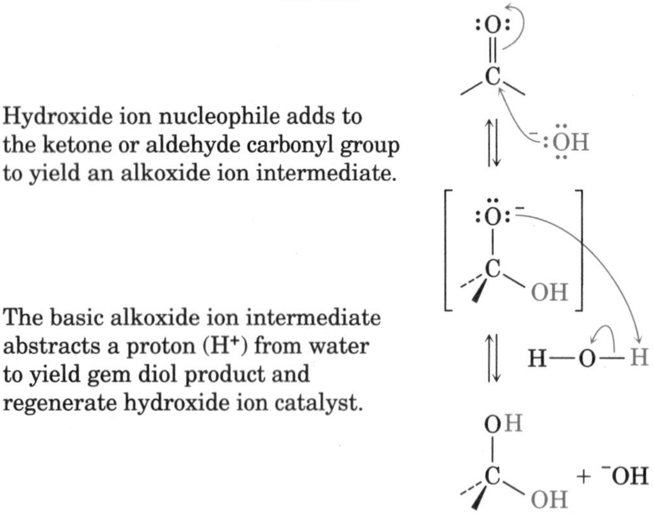

Hydroxide ion nucleophile adds to the ketone or aldehyde carbonyl group to yield an alkoxide ion intermediate.

$$C{=}O + {}^-{:}\ddot{O}H \rightleftharpoons \left[{}^-{:}\ddot{O}{-}C{-}OH\right]$$

The basic alkoxide ion intermediate abstracts a proton (H^+) from water to yield gem diol product and regenerate hydroxide ion catalyst.

$$\left[{}^-{:}\ddot{O}{-}C{-}OH\right] + H{-}O{-}H \rightleftharpoons HO{-}C{-}OH + {}^-OH$$

FIGURE 9.4 Mechanism of the base-catalyzed hydration reaction of a ketone or aldehyde. The base is a more reactive nucleophile than neutral water.

Acid catalyst protonates the basic carbonyl oxygen atom, making the ketone or aldehyde a much better acceptor of nucleophiles.

Nucleophilic addition of neutral water yields a protonated gem diol.

Loss of a proton regenerates the acid catalyst and gives neutral gem diol product.

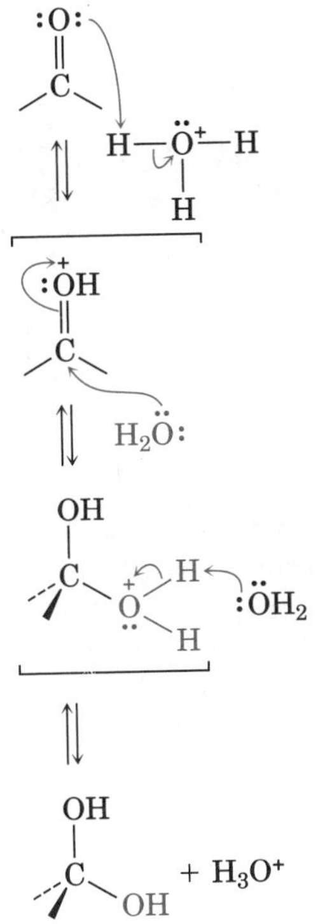

FIGURE 9.5 Mechanism of the acid-catalyzed hydration reaction of a ketone or aldehyde. The acid catalyst protonates the carbonyl starting material, thus making it more electrophilic and more reactive.

The acid-catalyzed hydration reaction also takes place in several steps (Figure 9.5). The acid catalyst first protonates the Lewis-basic oxygen atom of the carbonyl group, and subsequent nucleophilic addition of water yields a protonated gem diol. Loss of a proton then gives the neutral gem diol product.

Note the differences between the acid-catalyzed and base-catalyzed processes. The *base*-catalyzed reaction takes place rapidly because hydroxide ion is a much better nucleophilic *donor* than neutral water. The *acid*-catalyzed reaction takes place rapidly because the carbonyl compound is converted by protonation into a much better electrophilic *acceptor*. Most nucleophilic addition reactions can be similarly catalyzed by either acid or base.

PROBLEM 9.11 When dissolved in water, trichloroacetaldehyde (chloral, CCl_3CHO) exists primarily as the gem diol, chloral hydrate, $CCl_3CH(OH)_2$ (better known by the non-IUPAC name of "knockout drops"). Show the structure of chloral hydrate.

PROBLEM 9.12 The oxygen in water is primarily (99.8%) ^{16}O, but water enriched with the heavy isotope ^{18}O is also available. When a ketone or aldehyde is dissolved in $H_2{}^{18}O$, the isotopic label becomes incorporated into the carbonyl group: $R_2C{=}O + H_2O^* \rightarrow R_2C{=}O^* + H_2O$ (where $O^* = {}^{18}O$). Explain.

9.9 NUCLEOPHILIC ADDITION OF ALCOHOLS: ACETAL FORMATION

ACETAL
Compound that has two –OR groups bonded to the same carbon atom

Ketones and aldehydes react with alcohols in the presence of an acid catalyst to yield **acetals, $R_2C(OR')_2$.**

Ketone/aldehyde + 2 R'OH ⇌ (Acid cataylst) An acetal + H_2O

HEMIACETAL
Compound that has one –OR group and one –OH group attached to the same carbon atom

Acetal formation involves the acid-catalyzed nucleophilic addition of an alcohol to the carbonyl group in a manner similar to that of the acid-catalyzed hydration we saw in the previous section. The initial nucleophilic addition step yields a hydroxy ether called a **hemiacetal**, which reacts further with a second equivalent of alcohol to yield the acetal plus water. For example, reaction of cyclohexanone with methanol yields the dimethyl acetal.

A ketone ⇌ (CH_3OH, H^+ catalyst) [A hemiacetal (OH, OCH_3)] ⇌ (CH_3OH, H^+ catalyst) An acetal (OCH_3, OCH_3) + H_2O

As with hydration (Section 9.8), all the steps during acetal formation are reversible, and the reaction can be made to go either forward (from carbonyl compound to acetal) or backward (from acetal to carbonyl compound) by changing the reaction conditions. The forward reaction is favored by conditions that remove water from the medium and thus drive the equilibrium to the right. The backward reaction is favored by treating the acetal with mineral acid in the presence of a large excess of water.

PROTECTING GROUP
Group that is temporarily introduced into a molecule to protect a functional group from reaction elsewhere in the molecule

Acetals are valuable to organic chemists because they can serve as **protecting groups** for ketones and aldehydes. To see what this means, imagine that you are faced with having to reduce the keto ester methyl 4-oxopentanoate to obtain the keto alcohol 5-hydroxy-2-pentanone. This reaction can't be done in a single step because of the presence in the molecule of the ketone carbonyl group. If you were to treat methyl 4-oxopentanoate with $LiAlH_4$, both ester and ketone groups would be reduced.

$$CH_3\overset{O}{\overset{\|}{C}}CH_2CH_2\overset{O}{\overset{\|}{C}}OCH_3 \xrightarrow{?} CH_3\overset{O}{\overset{\|}{C}}CH_2CH_2CH_2OH$$

Methyl 4-oxopentanoate **5-Hydroxy-2-pentanone**

This situation isn't unusual. It often happens that one functional group in a complex molecule interferes with intended chemistry on another functional group elsewhere in the molecule. In such situations, you can often circumvent the problem by *protecting* the interfering functional group to render it unreactive, carrying out the desired reaction, and then removing the protecting group.

Ketones and aldehydes can be protected by converting them into acetals. Acetals, like other ethers, are stable to bases, reducing agents, and other nucleophiles, but they are acid-sensitive. Thus, you can selectively reduce the ester group in methyl 4-oxopentanoate by converting the keto group into an acetal, treating the compound with $LiAlH_4$ in ether, and then removing the acetal protecting group by treatment with aqueous acid.

$$CH_3\overset{O}{\overset{\|}{C}}CH_2CH_2\overset{O}{\overset{\|}{C}}OCH_3 \xrightarrow[H^+ \text{ catalyst}]{2\ CH_3OH} CH_3C(OCH_3)_2CH_2CH_2\overset{O}{\overset{\|}{C}}OCH_3$$

Methyl 4-oxopentanoate

$$\downarrow \text{1. } LiAlH_4 \quad \text{2. } H_2O$$

$$2\ CH_3OH + CH_3\overset{O}{\overset{\|}{C}}CH_2CH_2CH_2OH \xleftarrow{H_3O^+} CH_3C(OCH_3)_2CH_2CH_2CH_2OH$$

5-Hydroxy-2-pentanone

PRACTICE PROBLEM 9.3 What product would you obtain from the acid-catalyzed reaction of 2-methylcyclopentanone with methanol?

SOLUTION Replace the oxygen of the ketone with two $-OCH_3$ groups from the alcohol:

$$\text{2-methylcyclopentanone (ring C=O, }CH_3\text{)} \xrightarrow[H^+ \text{ catalyst}]{CH_3OH} \text{1,1-dimethoxy-2-methylcyclopentane (}OCH_3, OCH_3, CH_3\text{)} + H_2O$$

PROBLEM 9.13 What product would you expect from the acid-catalyzed reaction of cyclohexanone and ethanol?

PROBLEM 9.14 Show the mechanism of the acid-catalyzed formation of a *cyclic* acetal from ethylene glycol and acetone.

$$H_3C\overset{O}{\overset{\|}{C}}CH_3 \xrightarrow[H^+ \text{ catalyst}]{HOCH_2CH_2OH} \text{(cyclic acetal: } H_3C-C(-O-CH_2CH_2-O-)-CH_3\text{)} + H_2O$$

PROBLEM 9.15

Show how you might carry out the following transformation. (A protection step is needed.)

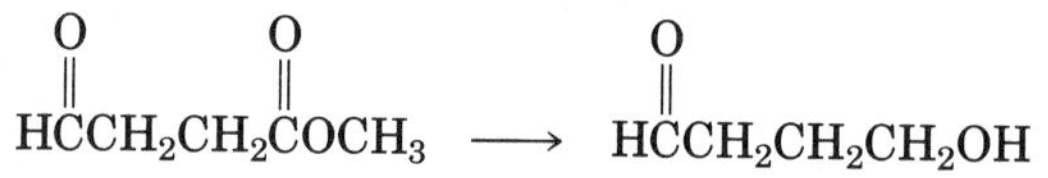

$$\overset{O}{\overset{\|}{H C}}CH_2CH_2\overset{O}{\overset{\|}{C}}OCH_3 \longrightarrow \overset{O}{\overset{\|}{H C}}CH_2CH_2CH_2OH$$

9.10 NUCLEOPHILIC ADDITION OF AMINES: IMINE FORMATION

IMINE
Compound with a C=N functional group

Ammonia and primary amines, $R'NH_2$, add to aldehydes and ketones to yield **imines, $R_2C{=}NR'$**, in a reaction analogous to hydration (Section 9.8). Imines are formed by nucleophilic addition to the carbonyl group by the nucleophilic amine, followed by loss of water from the amino alcohol addition product.

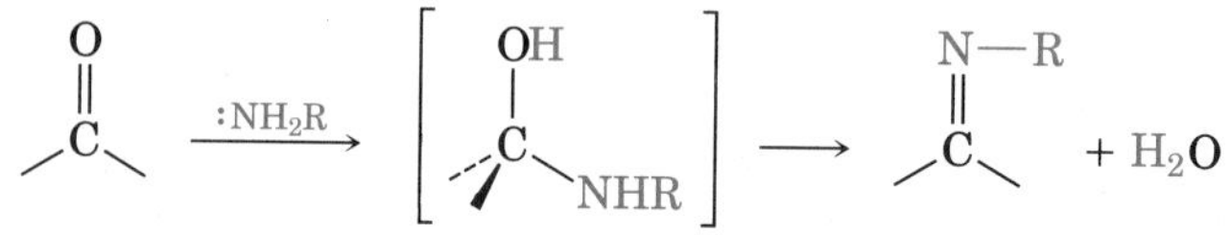

$$R_2C{=}O \xrightarrow{:NH_2R} \left[R_2C(OH)NHR\right] \longrightarrow R_2C{=}N{-}R + H_2O$$

A ketone or aldehyde — Amino alcohol intermediate — An imine

Imine derivatives, such as *oximes*, and *2,4-dinitrophenylhydrazones* (abbreviated as 2,4-DNP's) are also easily prepared by reaction of a ketone or aldehyde with the appropriate $H_2N{-}Y$ compound. Such imines, which are usually crystalline, easy-to-handle materials, are often prepared as a means of converting liquid ketones or aldehydes into solid derivatives.

Oxime

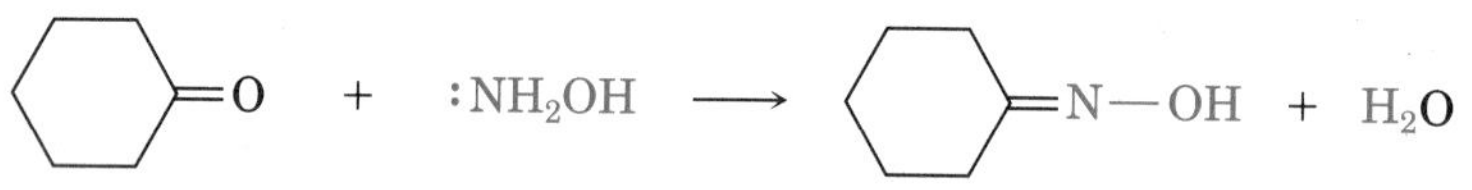

$$C_6H_{10}{=}O + :NH_2OH \longrightarrow C_6H_{10}{=}N{-}OH + H_2O$$

Cyclohexanone — **Hydroxylamine** — **Cyclohexanone oxime** (mp 90°C)

2,4-Dinitrophenylhydrazone

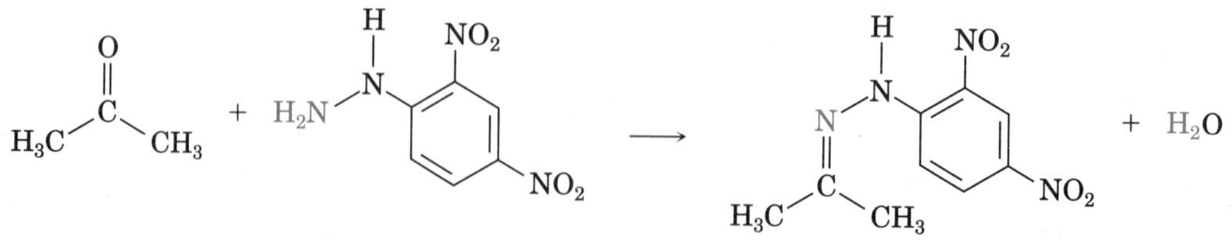

$$(H_3C)_2C{=}O + H_2N{-}NH{-}C_6H_3(NO_2)_2 \longrightarrow (H_3C)_2C{=}N{-}NH{-}C_6H_3(NO_2)_2 + H_2O$$

Acetone — **2,4-Dinitrophenylhydrazine** — **Acetone 2,4-dinitrophenylhydrazone** (mp 126°C)

WOLFF–KISHNER REACTION
Process for reducing a ketone or aldehyde to an alkane by reaction with KOH and hydrazine

An important variant of imine formation involves the treatment of a ketone or aldehyde with hydrazine, $H_2N{-}NH_2$, in the presence of strong base. Called the **Wolff–Kishner reaction** after its discoverers, the process is a useful method for converting ketones or aldehydes into alkanes, $R_2C{=}O \rightarrow R_2CH_2$. The reaction involves initial formation of an imine intermediate called a *hydrazone*, followed by loss of nitrogen and formation of the alkane product.

$$C_6H_5C(=O)CH_2CH_3 \xrightarrow[KOH]{H_2NNH_2} \left[C_6H_5C(=N{-}NH_2)CH_2CH_3 \right] \longrightarrow C_6H_5CH_2CH_2CH_3 + N_2$$

Propiophenone A hydrazone **Propylbenzene (82%)**

PRACTICE PROBLEM 9.4 What product do you expect from the reaction of 2-butanone with hydroxylamine, NH_2OH?

SOLUTION Take oxygen from the ketone and two hydrogens from the amine to form water, and then join the fragments that remain:

$$CH_3CH_2\overset{O}{\overset{\|}{C}}CH_3 + H_2NOH \longrightarrow CH_3CH_2\overset{NOH}{\overset{\|}{C}}CH_3 + H_2O$$

PROBLEM 9.16 Write the products you would obtain from treatment of cyclohexanone with these reagents:
(a) H_2NOH **(b)** 2,4-Dinitrophenylhydrazine **(c)** N_2H_4, KOH
(d) $NaBH_4$

PROBLEM 9.17 Show how you could prepare butylbenzene from benzene by carrying out a Friedel–Crafts acylation reaction (Section 5.8) followed by a Wolff–Kishner reduction.

9.11 NUCLEOPHILIC ADDITION OF GRIGNARD REAGENTS: ALCOHOL FORMATION

Grignard reagents, RMgX, react with ketones and aldehydes to yield alcohols in the same way that $NaBH_4$ and $LiAlH_4$ do (Section 8.5).

Ketone/aldehyde — RMgX → [alkoxide, R] — H_3O^+ → alcohol (OH, R)

Ketone/aldehyde — $NaBH_4$ → [alkoxide, H] — H_3O^+ → alcohol (OH, H)

As we saw in Section 7.4, Grignard reagents are prepared by reaction of alkyl, aryl, or vinylic halides with magnesium in ether.

$$\underset{\text{An organohalide}}{R{-}X} \xrightarrow[\text{Ether}]{\text{Mg}} \underset{\text{A Grignard reagent}}{\overset{\delta^-}{R}{-}\overset{\delta^+}{MgX}}$$

The carbon–magnesium bond of Grignard reagents is polarized so that the carbon atom is both nucleophilic and basic. Grignard reagents therefore react as though they were carbanions, R^-, and they undergo nucleophilic addition to ketones and aldehydes just as water and alcohols do. Unlike the addition of water and alcohols, though, the nucleophilic addition of a Grignard reagent is irreversible. The reaction first produces a tetrahedrally hybridized magnesium alkoxide intermediate, which is then protonated to yield the neutral alcohol on treatment with aqueous acid.

Carbonyl — $:R^-\ ^+MgX$, Nucleophilic addition → Tetrahedral intermediate ($:\ddot{O}:^-\ ^+MgX$, R) — H_3O^+ → Alcohol (OH, R) + HOMgX

A great many alcohols can be obtained from Grignard reactions, depending on the reagents used. For example, formaldehyde, CH_2O, reacts with Grignard reagents to give primary alcohols, RCH_2OH:

$$\underset{\text{Cyclohexylmagnesium bromide}}{C_6H_{11}MgBr} \xrightarrow[\text{2. } H_3O^+]{\text{1. } CH_2O \text{ (formaldehyde)}} \underset{\text{Cyclohexylmethanol (65\%) (a primary alcohol)}}{C_6H_{11}CH_2OH}$$

Cyclohexylmagnesium bromide

Cyclohexylmethanol (65%)
(a primary alcohol)

Aldehydes react with Grignard reagents to give secondary alcohols, and ketones react similarly to yield tertiary alcohols:

$$\mathrm{CH_3CH(CH_3)CH_2CHO} + \mathrm{C_6H_5MgBr} \longrightarrow \mathrm{CH_3CH(CH_3)CH_2CH(OH)C_6H_5}$$

3-Methylbutanal **Phenylmagnesium bromide** **3-Methyl-1-phenyl-1-butanol** (a secondary alcohol)

$$\text{Cyclohexanone} \xrightarrow[\text{2. } H_3O^+]{\text{1. } CH_3CH_2MgBr} \text{1-ethylcyclohexanol (OH, } CH_2CH_3\text{)}$$

Cyclohexanone **1-Ethylcyclohexanol (89%)** (a tertiary alcohol)

Although useful, the Grignard reaction also has limitations. For example, Grignard reagents can't be prepared from organohalides if there are other reactive functional groups in the same molecule. A compound that is both an alkyl halide and a ketone won't form a Grignard reagent—it reacts with itself instead. Similarly, a compound that is both an alkyl halide and a carboxylic acid, alcohol, or amine can't form a Grignard reagent because the acidic **RCOOH**, **ROH**, or **RNH$_2$** protons in the molecule simply react with the basic Grignard reagent as it's formed.

In general, Grignard reagents can't be prepared from compounds that have these functional groups in the molecule:

Br—(Molecule)—FG

where FG = –OH, –NH$_2$, –SH, –COOH, –NO$_2$, –CHO, –COR, –CN, or –CONH$_2$.

PRACTICE PROBLEM 9.5 How can you use the addition of a Grignard reagent to a ketone to synthesize 2-phenyl-2-propanol?

SOLUTION First, draw the structure of the product and identify the groups bonded to the alcohol carbon atom. In this instance, there are two methyl groups (–CH$_3$) and one phenyl (–C$_6$H$_5$). One of the three must come from a Grignard reagent, and the remaining two must come from a ketone. Thus, the possibilities are addition of methylmagnesium bromide to acetophenone and addition of phenylmagnesium bromide to acetone:

$$\mathrm{C_6H_5COCH_3} \xrightarrow[\text{2. } H_3O^+]{\text{1. } CH_3MgBr} \mathrm{C_6H_5C(OH)(CH_3)CH_3} \xleftarrow[\text{2. } H_3O^+]{\text{1. } C_6H_5MgBr} \mathrm{H_3CCOCH_3}$$

Acetophenone **2-Phenyl-2-propanol** **Acetone**

PROBLEM 9.18 Show the products obtained from addition of methylmagnesium bromide to these compounds:
(a) Cyclopentanone **(b)** Benzophenone (diphenyl ketone)
(c) 3-Hexanone

PROBLEM 9.19 How might you use a Grignard addition reaction to prepare these alcohols?
(a) 2-Methyl-2-propanol **(b)** 1-Methylcyclohexanol
(c) 3-Methyl-3-pentanol

9.12 SOME BIOLOGICAL NUCLEOPHILIC ADDITION REACTIONS

Nature synthesizes the molecules of life using many of the reactions that chemists use in the laboratory. This is particularly true of carbonyl-group reactions, where nucleophilic addition steps are an important part of the biosynthesis of many vital molecules.

One of the pathways by which amino acids are made involves a nucleophilic addition reaction of α-keto acids. To choose a specific case, the bacterium *Bacillus subtilis* synthesizes alanine from pyruvic acid and ammonia. The key step in this biological transformation is the nucleophilic addition of ammonia to the ketone carbonyl group of pyruvic acid to give an imine that is further reduced by enzymes.

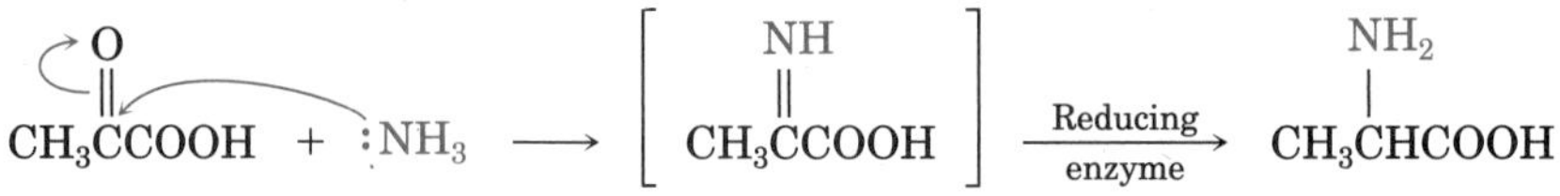

$$CH_3\overset{O}{\overset{\|}{C}}COOH + :NH_3 \longrightarrow \left[CH_3\overset{NH}{\overset{\|}{C}}COOH \right] \xrightarrow[\text{enzyme}]{\text{Reducing}} CH_3\overset{NH_2}{\overset{|}{C}}HCOOH$$

Pyruvic acid An imine **Alanine**

Other examples of nucleophilic carbonyl addition occur frequently in carbohydrate chemistry. For example, the six-carbon sugar glucose acts in some respects as if it were an aldehyde. It can, for example, be oxidized to yield a carboxylic acid. Spectroscopic examination of glucose shows, however, that no aldehyde group is present; instead, glucose exists as a *cyclic hemiacetal*. The hydroxyl group at carbon 5 adds to the aldehyde at carbon 1 in an internal nucleophilic addition step.

H–C1(=O); H–C2–OH; HO–C3–H; H–C4–OH; H–C5–OH; C6H2OH = [ring with HO, HO, CH2OH (6), O–H, C1=O, OH] ⇌ [ring with HO, HO, CH2OH, O, OH, OH]

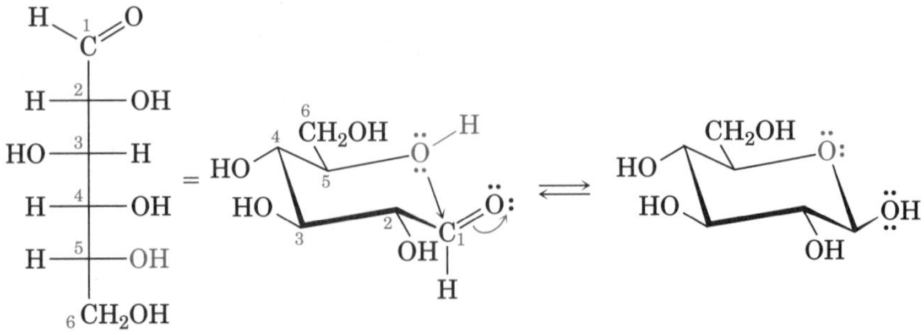

Glucose (open form) **Glucose** (hemiacetal form)

Further reaction between molecules of glucose leads to the carbohydrate polymer *cellulose*, which constitutes the major building block of plant cell walls. Cellulose consists simply of glucose units joined by acetal linkages between carbon 1 of one glucose and the hydroxyl group at carbon 4 of another glucose:

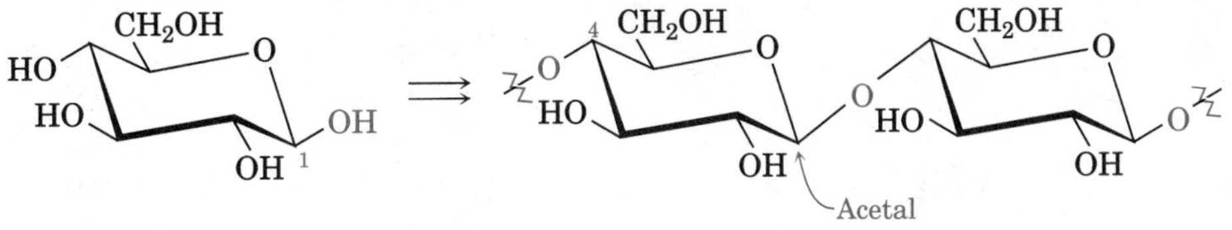

CH_2OH HO HO OH OH 1 → 4 CH_2OH O HO OH Acetal O HO CH_2OH OH O

Glucose **Cellulose**

We'll study this and other reactions of carbohydrates in Chapter 14.

INTERLUDE

CHEMICAL WARFARE IN NATURE

Among many known nucleophilic additions of ketones and aldehydes is their reaction with HCN (hydrogen cyanide) to yield cyano alcohols, or *cyanohydrins* [RCH(OH)CN]:

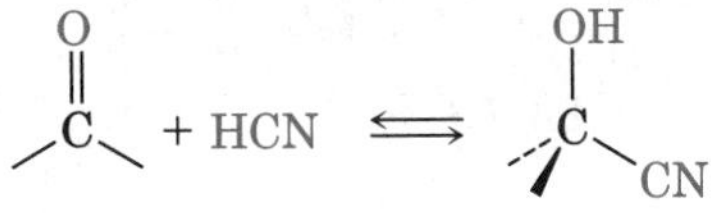

C=O + HCN ⇌ C(OH)CN

A ketone or aldehyde   A cyanohydrin

The formation of cyanohydrins from ketones and aldehydes is of more than just chemical interest, since cyanohydrins also play an interesting role in the chemical defense mechanisms of certain plants and insects against predators. For example, when the millipede *Apheloria corrugata* is attacked by ants, it secretes mandelonitrile and an enzyme that catalyzes the decomposition of mandelonitrile into benzaldehyde and HCN. The millipede actually protects itself by discharging poisonous HCN at would-be attackers!

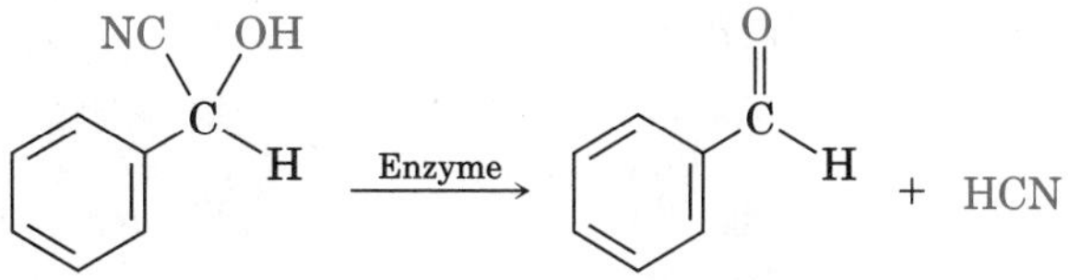

NC OH C H —Enzyme→ C(=O)H + HCN

Mandelonitrile
(from *Apheloria corrugata*)

In a similar vein, the pits of apricots and peaches contain a group of substances called *cyanogenic glycosides*. These compounds, one of which (amygdalin, or Laetrile) is notorious because of its claimed anticancer activity, consist of benzaldehyde cyanohydrin bonded to simple sugars such as glucose. When eaten, the sugar unit is cleaved off, and HCN is released. Predators soon learn to avoid these seeds.

NC O—(Glucose)$_2$ C H

Amygdalin (Laetrile)

SUMMARY AND KEY WORDS

Carbonyl compounds can be classified into two general categories:

R_2CO **RCHO**	**Ketones** and **aldehydes** are similar in their reactivity and are distinguished by the fact that the substituents on the acyl carbon can't act as leaving groups.
RCOOH **RCOOR′** **$RCONH_2$** **RCOCl** **RCOOCOR′**	**Carboxylic acids** and their derivatives—**esters, amides, acid anhydrides,** and **acid chlorides**—are distinguished by the fact that the substituents on the acyl carbon *can* act as leaving groups.

Structurally, a carbon–oxygen double bond is similar to a carbon–carbon double bond. The carbonyl carbon atom is sp^2-hybridized and forms both an sp^2 sigma bond and a p pi bond to oxygen. Carbonyl groups are strongly polarized because of the electronegativity of oxygen. Thus, carbonyl carbons are strongly electrophilic.

Aldehydes are usually prepared by oxidative cleavage of alkenes or by oxidation of primary alcohols. Ketones are similarly prepared by oxidative cleavage of alkenes or by oxidation of secondary alcohols.

Ketones and aldehydes behave similarly in much of their chemistry, though aldehydes are generally more reactive than ketones. Both undergo **nucleophilic addition reactions**, and a variety of product types can be prepared. For example, ketones and aldehydes are reduced by $NaBH_4$ or $LiAlH_4$ to yield secondary and primary alcohols, respectively. Addition of Grignard reagents also leads to alcohols. Primary amines add to ketones and aldehydes to give **imines, $R_2C{=}NR'$**. If hydrazine, H_2NNH_2, is used as the amine nucleophile, the initially formed imine intermediate undergoes further reaction with base to yield an alkane (the **Wolff–Kishner reaction**). Alcohols add to ketones and aldehydes to yield **acetals, $R_2C(OR')_2$**, which are valuable as carbonyl **protecting groups**.

SUMMARY OF REACTIONS

1. Reaction of ketones and aldehydes with alcohols to yield acetals (Section 9.9)

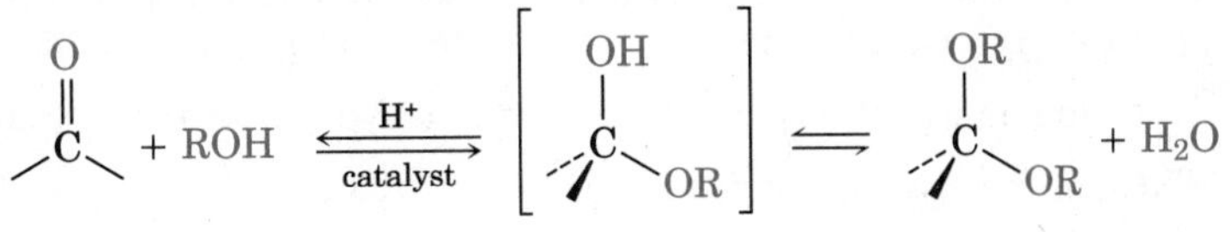

2. Reaction of ketones and aldehydes with amines to yield imines (Section 9.10)

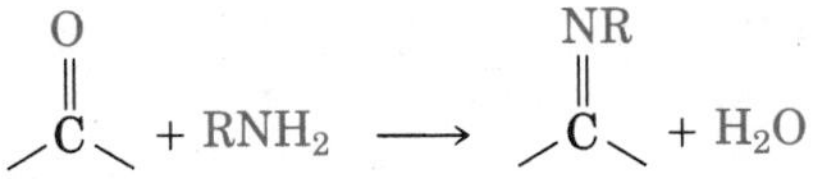

3. Wolff–Kishner reaction to yield alkanes (Section 9.10)

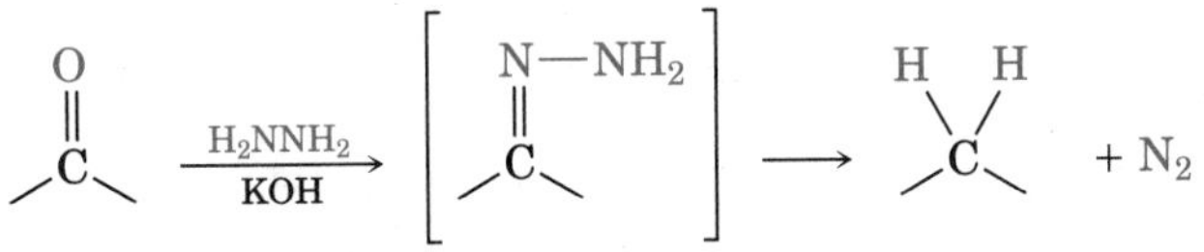

4. Reaction of ketones and aldehydes with Grignard reagents to yield alcohols (Section 9.11)

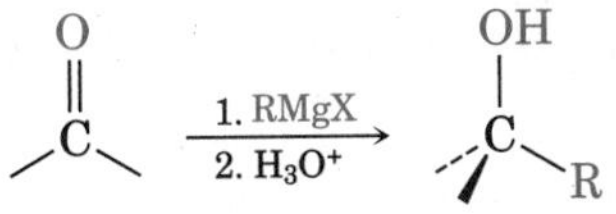

ADDITIONAL PROBLEMS

9.20 Identify the different kinds of carbonyl groups in these molecules:

(a)

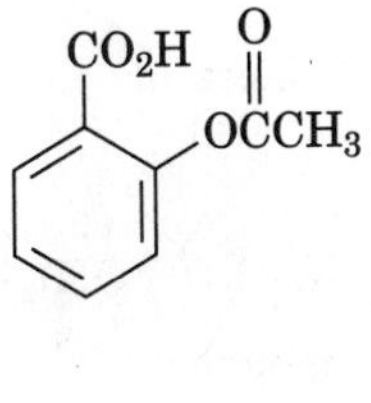

Aspirin

(b)

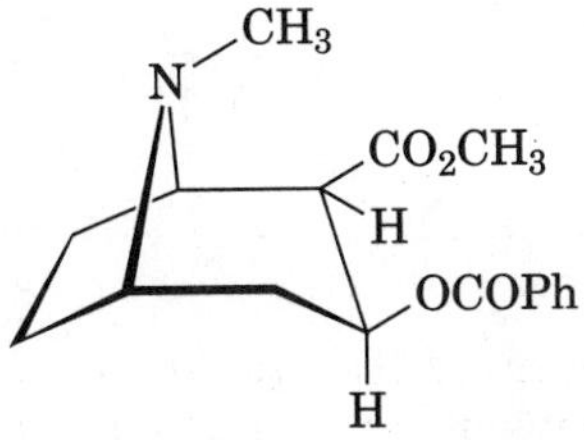

Cocaine

(c)

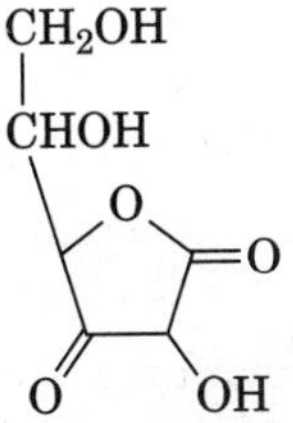

Ascorbic acid (vitamin C)

9.21 What is the structural difference between aldehydes and ketones?

9.22 Draw structures corresponding to these names:
(a) Bromoacetone **(b)** 3-Methyl-2-butanone
(c) 3,5-Dinitrobenzaldehyde **(d)** 3,5-Dimethylcyclohexanone
(e) 2,2,4,4-Tetramethyl-3-pentanone **(f)** Butanedial
(g) (*S*)-2-Hydroxypropanal **(h)** 3-Phenyl-2-propenal

9.23 Draw and name the seven ketones and aldehydes with the formula $C_5H_{10}O$.

9.24 Which of the compounds you identified in Problem 9.23 are chiral?

9.25 Draw structures of molecules that meet these descriptions:
(a) A cyclic ketone, C_6H_8O **(b)** A diketone, $C_6H_{10}O_2$ **(c)** An aryl ketone, C_9H_{10}
(d) A 2-bromoaldehyde, C_5H_9BrO

9.26 Give IUPAC names for these structures:

(a) 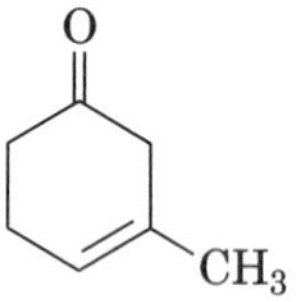 O, CH_3

(b) 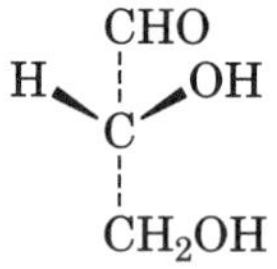 CHO, H, OH, C, CH_2OH

(c) O

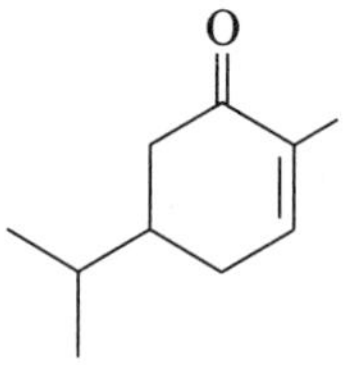

(d) 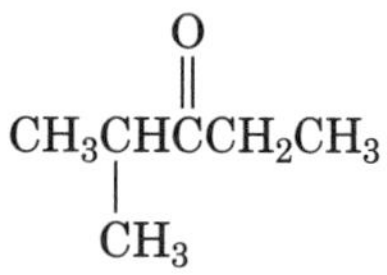 $CH_3CH(CH_3)C(=O)CH_2CH_3$

(e) 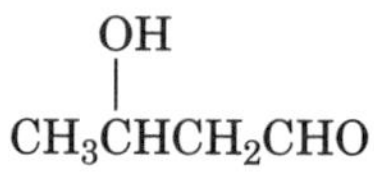 $CH_3CH(OH)CH_2CHO$

(f) 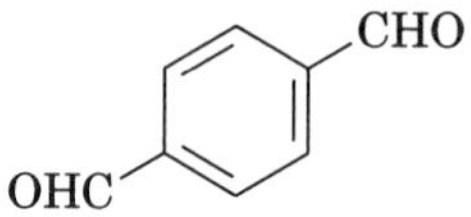 OHC, CHO

9.27 Give an example of each of the following:
(a) An acetal **(b)** A cyanohydrin **(c)** A gem diol **(d)** An oxime **(e)** An imine
(f) A hemiacetal

9.28 Predict the products of the reaction of phenylacetaldehyde, $C_6H_5CH_2CHO$, with these reagents:
(a) $NaBH_4$, then H_3O^+ **(b)** Tollens' reagent **(c)** NH_2OH
(d) CH_3MgBr, then H_3O^+ **(e)** CH_3OH, H^+ catalyst **(f)** H_2NNH_2, KOH

9.29 Answer Problem 9.28 for the reaction of acetophenone, $C_6H_5COCH_3$, with the same reagents.

9.30 Reaction of 2-butanone with HCN yields a cyanohydrin product having a new stereogenic center. What stereochemistry would you expect the product to have? (Review Section 6.13.)

9.31 In light of your answer to Problem 9.30, what stereochemistry would you expect the product from the reaction of phenylmagnesium bromide with 2-butanone to have?

9.32 Starting from 2-cyclohexenone and any other reagents needed, how would you prepare these substances? (More than one step may be required.)
(a) Cyclohexene **(b)** 1-Methylcyclohexanol **(c)** Cyclohexanol
(d) 1-Phenyl-2-cyclohexen-1-ol

9.33 How can you explain the observation that the S_N2 reaction of (dibromomethyl)benzene, $C_6H_5CHBr_2$, with NaOH yields benzaldehyde rather than (dihydroxymethyl)benzene, $C_6H_5CH(OH)_2$?

9.34 Use a Grignard reaction on a ketone or aldehyde to synthesize these compounds:
(a) 2-Pentanol **(b)** 2-Phenyl-2-butanol **(c)** 1-Ethylcyclohexanol **(d)** Diphenylmethanol

9.35 Show the products that result from the reaction of phenylmagnesium bromide with these reagents:
(a) CH_2O **(b)** Benzophenone ($C_6H_5COC_6H_5$) **(c)** 3-Pentanone

9.36 Show how you could make these alcohols using a Grignard reaction:

(a) $CH_3CH(CH_3)CH_2CH_2CH_2OH$ (b) 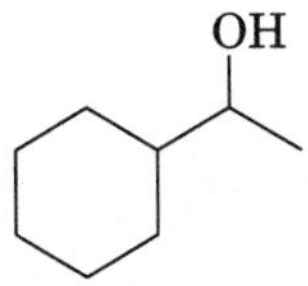 [cyclohexyl–CH(OH)–CH₃ structure] (c) $CH_3CH_2CH(OH)CH{=}CHCH_3$

9.37 How could you convert bromobenzene into benzoic acid, C_6H_5COOH? (More than one step is required.)

9.38 Show the structures of the intermediate hemiacetals and the final acetals that result from these reactions:

(a) 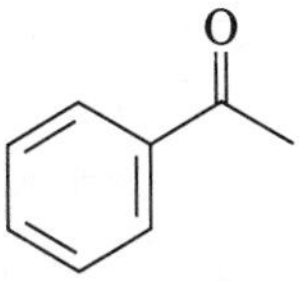 [acetophenone structure] + $CH_3CH(OH)CH_3 \xrightarrow[\text{catalyst}]{H^+}$ (b) $CH_3CH_2C(=O)CH_2CH_3$ + [cyclopentanol structure] $\xrightarrow[\text{catalyst}]{H^+}$

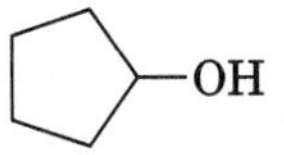

9.39 Show the structures of the starting alcohols and ketones or aldehydes you would use to make these acetals:

(a) $CH_3CH_2CH(CH_3)CH_2CH(OCH_3)OCH_3$ (b) 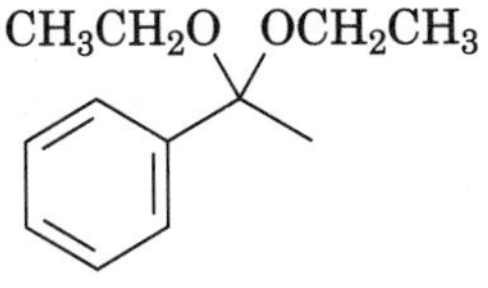 CH_3CH_2O OCH_2CH_3 (c)

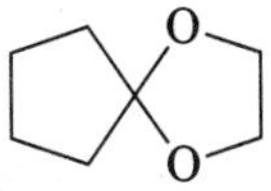

9.40 Show the products from the reaction of 2-pentanone with these reagents:
(a) NH_2OH (b) 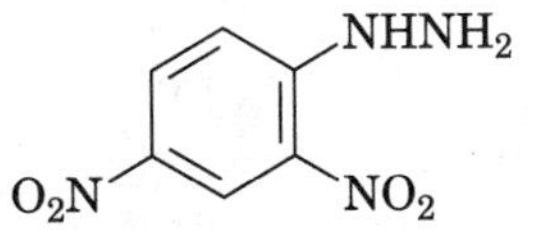 [2,4-dinitrophenylhydrazine structure: $NHNH_2$, O_2N, NO_2] (c) H_2NNH_2, KOH

9.41 How would you synthesize the following compounds from cyclohexanone?
(a) 1-Methylcyclohexene (b) *cis*-1,2-Cyclohexanediol (c) 1-Bromo-1-methylcyclohexane
(d) 1-Cyclohexylcyclohexanol

9.42 How can you explain the observation that treatment of 4-hydroxycyclohexanone with 1 equiv of methylmagnesium bromide yields none of the expected addition product, whereas treatment with an excess of Grignard reagent leads to a good yield of 1-methyl-1,4-cyclohexanediol?

9.43 Carvone is the major constituent of spearmint oil. What products would you expect from the reaction of carvone with the following reagents?

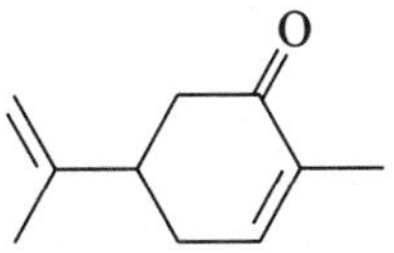

Carvone

(a) $LiAlH_4$, then H_3O^+ (b) C_6H_5MgBr, then H_3O^+ (c) H_2, Pd catalyst (d) CH_3OH, H^+

9.44 When 4-hydroxybutanal is treated with methanol in the presence of an acid catalyst, 2-methoxytetrahydrofuran is obtained. Propose a mechanism to account for this result.

$$HOCH_2CH_2CH_2CHO \xrightarrow[H^+]{CH_3OH}$$ [2-methoxytetrahydrofuran structure: O, OCH_3]

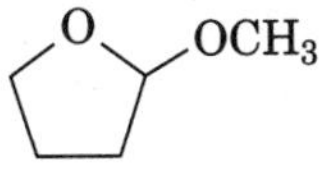

9.45 Using your knowledge of the reactivity differences between aldehydes and ketones, show how the following two selective reductions might be carried out. One of the schemes requires a protection step.

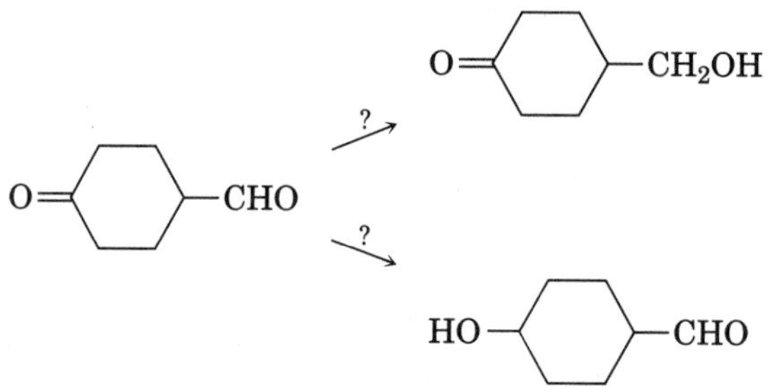

O= cyclohexane —CHO →? O= cyclohexane —CH_2OH

O= cyclohexane —CHO →? HO— cyclohexane —CHO

9.46 Treatment of a ketone or aldehyde with a thiol in the presence of an acid catalyst yields a thioacetal, $R_2C(SR')_2$. To what other reaction is this thioacetal formation analogous?

9.47 When crystals of pure α-glucose are dissolved in water, isomerization slowly occurs to produce β-glucose. How does this isomerization occur?

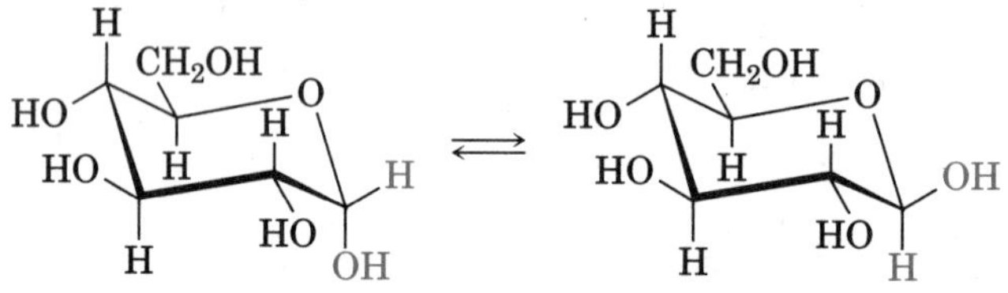

α-Glucose ⇌ **β-Glucose**

9.48 Ketones react with dimethylsulfonium methylide to yield epoxides by a mechanism that involves an initial nucleophilic addition followed by an intramolecular S_N2 substitution. Formulate the mechanism.

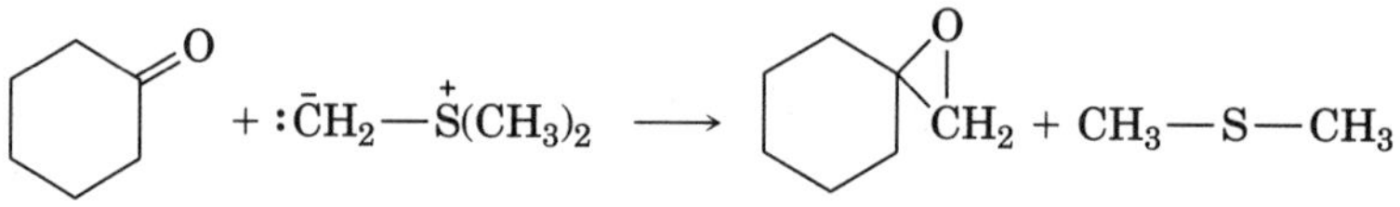

$$\text{cyclohexanone} + :\bar{C}H_2—\overset{+}{S}(CH_3)_2 \longrightarrow \text{epoxide } (CH_2) + CH_3—S—CH_3$$

Dimethylsulfonium methylide

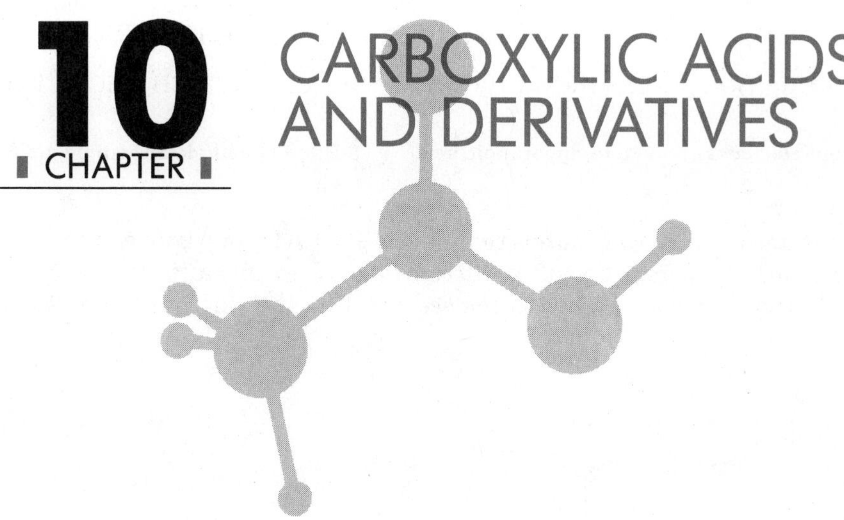

10 CHAPTER CARBOXYLIC ACIDS AND DERIVATIVES

Carboxylic acids and their derivatives are carbonyl compounds in which the acyl group is bonded to an electronegative atom such as oxygen, halogen, nitrogen, or sulfur. Although there are many different kinds of carboxylic acid derivatives, we'll be concerned only with four of the most common types in addition to the acids themselves: acid halides, acid anhydrides, esters, and amides. In contrast to aldehydes and ketones, these compounds contain an acyl group bonded to a substituent that can act as a leaving group in substitution reactions. Also in this chapter, we'll discuss *nitriles*, a class of compounds closely related to carboxylic acids.

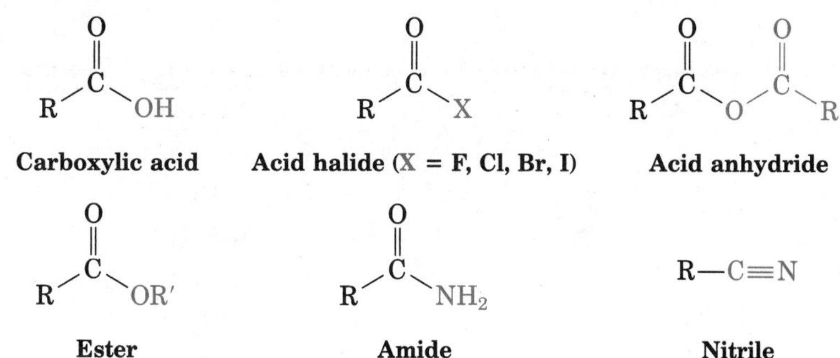

10.1 NAMING CARBOXYLIC ACIDS AND DERIVATIVES

Carboxylic Acids: RCOOH

IUPAC rules allow for two systems of nomenclature. Simple open-chain carboxylic acids are named by replacing the terminal *-e* of the alkane name with *-oic acid*. The carboxyl carbon atom is always numbered C1.

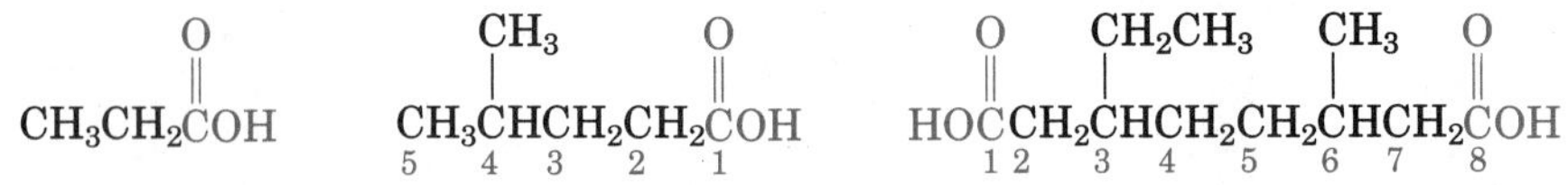

CH_3CH_2COOH (with C=O) — **Propanoic acid**

$CH_3CH(CH_3)CH_2CH_2COOH$ (carbons numbered 5 4 3 2 1) — **4-Methylpentanoic acid**

$HOOCCH_2CH(CH_2CH_3)CH_2CH_2CH(CH_3)CH_2COOH$ (carbons numbered 1 2 3 4 5 6 7 8) — **3-Ethyl-6-methyloctanedioic acid**

Alternatively, compounds that have a –COOH group bonded to a ring are named by using the suffix *-carboxylic acid.* In this alternate system, the carboxylic acid carbon is *attached to* C1 on the ring but is not itself numbered.

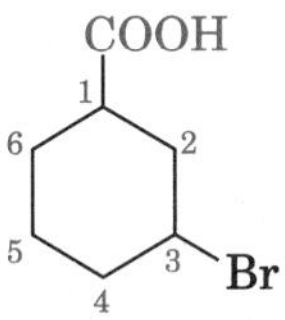

COOH on C1 of cyclohexane ring, Br on C3 (ring numbered 1–6) — **3-Bromomcyclohexanecarboxylic acid**

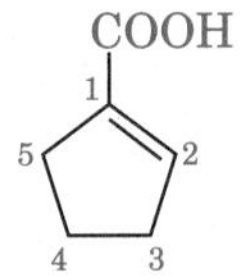

COOH on C1 of cyclopentene ring, double bond C1=C2 (ring numbered 1–5) — **1-Cyclopentenecarboxylic acid**

Because many carboxylic acids were among the first organic compounds to be isolated and purified, there are a large number of common names for acids (Table 10.1). We'll use systematic names in this book, with the exception of formic (methanoic) acid, HCOOH, and acetic (ethanoic) acid, CH_3COOH, whose names are so well known that it makes little sense to refer to them in any other way.

TABLE 10.1 Some Common Names of Carboxylic Acids and Acyl Groups

CARBOXYLIC ACID		ACYL GROUP	
Structure	Name	Name	Structure
HCOOH	Formic	Formyl	HCO—
CH_3COOH	Acetic	Acetyl	CH_3CO—
CH_3CH_2COOH	Propionic	Propionyl	CH_3CH_2CO—
$CH_3CH_2CH_2COOH$	Butyric	Butyryl	$CH_3(CH_2)_2CO$—
HOOCCOOH	Oxalic	Oxalyl	—OCCO—
$HOOCCH_2COOH$	Malonic	Malonyl	—$OCCH_2CO$—
$HOOCCH_2CH_2COOH$	Succinic	Succinyl	—$OC(CH_2)_2CO$—
$H_2C{=}CHCOOH$	Acrylic	Acryloyl	$H_2C{=}CHCO$—
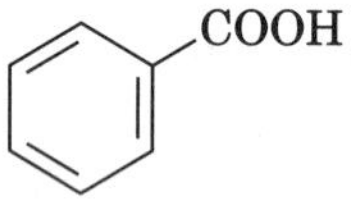 C_6H_5COOH (benzene ring–COOH)	Benzoic	Benzoyl	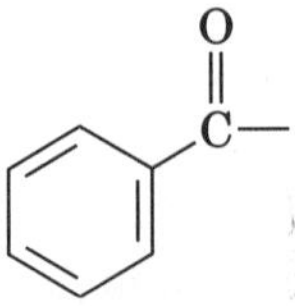 C_6H_5C(=O)— (benzene ring–C(=O)—)

PROBLEM 10.1 Give IUPAC names for these compounds:

(a) $(CH_3)_2CHCH_2COOH$ **(b)** $CH_3CHBrCH_2CH_2COOH$

(c) $CH_3CH{=}CHCH_2CH_2COOH$

(d) $CH_3CH_2\underset{}{\overset{COOH}{\overset{|}{C}}}HCH_2CH_2CH_3$

(e) (cyclohexane ring bearing COOH and H on one carbon, and CH_3 and H on the adjacent carbon)

COOH
H
CH_3
H

PROBLEM 10.2 Draw structures corresponding to these names:

(a) 2,3-Dimethylhexanoic acid **(b)** 4-Methylpentanoic acid

(c) *o*-Hydroxybenzoic acid **(d)** *trans*-1,2-Cyclobutanedicarboxylic acid

Acid Halides: RCOX

Carboxylic acid halides are named by identifying the acyl group and then the halide. The acyl group name is derived from the acid name by replacing the *-ic acid* ending with *-yl*, or the *-carboxylic acid* ending with *-carbonyl*. For example,

$CH_3\overset{O}{\overset{\|}{C}}Cl$

Acetyl chloride
(from acetic acid)

$C_6H_5\overset{O}{\overset{\|}{C}}Br$

Benzoyl bromide
(from benzoic acid)

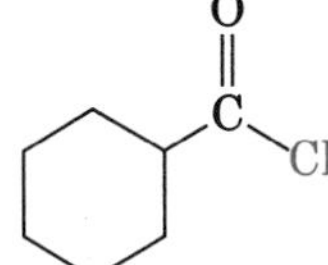

Cyclohexanecarbonyl chloride
(from cyclohexanecarboxylic acid)

Acid Anhydrides: RCO_2COR

Symmetrical anhydrides of simple carboxylic acids and cyclic anhydrides of dicarboxylic acids are named by replacing the word *acid* with *anhydride*:

$$CH_3\overset{O}{\overset{\|}{C}}{-}O{-}\overset{O}{\overset{\|}{C}}CH_3 \qquad CH_3(CH_2)_5\overset{O}{\overset{\|}{C}}{-}O{-}\overset{O}{\overset{\|}{C}}(CH_2)_5CH_3$$

Acetic anhydride **Heptanoic anhydride**

Amides: $RCONH_2$

Amides with an unsubstituted $-NH_2$ group are named by replacing the *-oic acid* or *-ic acid* ending with *-amide*, or by replacing the *-carboxylic acid* ending with *-carboxamide*:

$CH_3\overset{O}{\overset{\|}{C}}NH_2$

Acetamide
(from acetic acid)

$CH_3(CH_2)_4\overset{O}{\overset{\|}{C}}NH_2$

Hexanamide
(from hexanoic acid)

$C_5H_9\overset{O}{\overset{\|}{C}}{-}NH_2$

Cyclopentanecarboxamide
(from cyclopentanecarboxylic acid)

If the nitrogen atom is substituted, the amide is named by first identifying the substituent group and then citing the parent name. The substituents are preceded by the letter *N* to identify them as being directly attached to nitrogen.

$CH_3CH_2\overset{O}{\overset{\|}{C}}NHCH_3$ $\qquad$ $\square\!-\!\overset{O}{\overset{\|}{C}}-N(CH_2CH_3)_2$

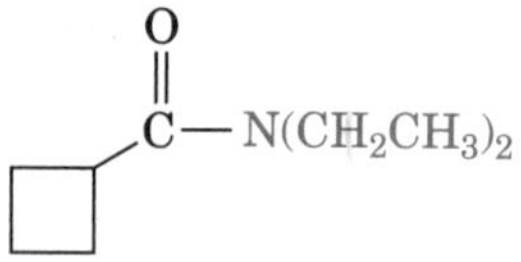

***N*-Methylpropanamide** $\qquad$ ***N,N*-Diethylcyclobutanecarboxamide**

Esters: RCO_2R'

Systematic names for esters are derived by first giving the name of the alkyl group attached to oxygen and then identifying the carboxylic acid. In so doing, the *-ic acid* ending is replaced by *-ate*:

$CH_3\overset{O}{\overset{\|}{C}}OCH_2CH_3$ $\qquad$ $H_2C(\overset{O}{\overset{\|}{C}}-OCH_3)_2$ $\qquad$ $C_6H_{11}-\overset{O}{\overset{\|}{C}}-OC(CH_3)_3$

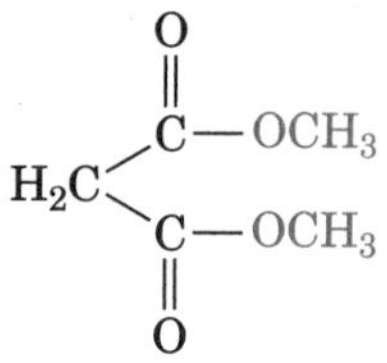

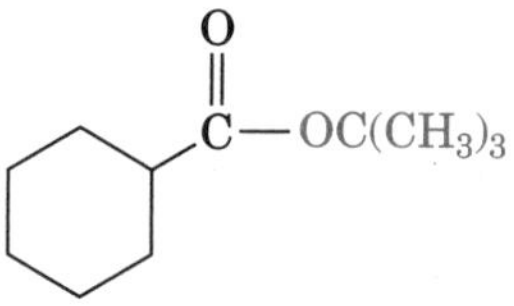

Ethyl acetate (the ethyl ester of acetic acid)

Dimethyl malonate (the dimethyl ester of malonic acid)

***tert*-Butyl cyclohexanecarboxylate** (the *tert*-butyl ester of cyclohexanecarboxylic acid)

Nitriles: R–C≡N

NITRILE Compound with a C≡N functional group

Compounds containing the –C≡N functional group are known as **nitriles**. Simple acyclic nitriles are named by adding *-nitrile* as a suffix to the alkane name, with the nitrile carbon numbered C1.

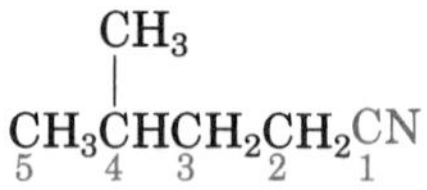

$\underset{5}{CH_3}\underset{4}{\overset{CH_3}{\overset{|}{C}}H}\underset{3}{C}H_2\underset{2}{C}H_2\underset{1}{C}N$ $\qquad$ **4-Methylpentanenitrile**

More complex nitriles are named as derivatives of carboxylic acids by replacing the *-ic acid* or *-oic acid* ending with *-onitrile*, or by replacing the *-carboxylic acid* ending with *-carbonitrile*. In this system, the nitrile carbon atom is attached to C1 but is not itself numbered:

$CH_3C{\equiv}N$ $\qquad$ $C_6H_5-C{\equiv}N$ $\qquad$ (cyclohexane ring numbered 1–6; CN on C1; two CH_3 on C2)

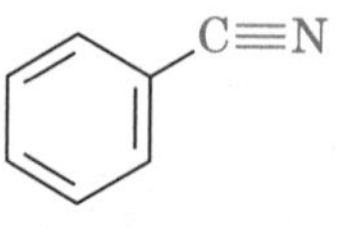

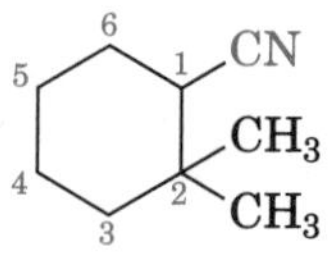

Acetonitrile (from acetic acid)

Benzonitrile (from benzoic acid)

2,2-Dimethylcyclohexanecarbonitrile (from 2,2-dimethylcyclohexanecarboxylic acid)

PROBLEM 10.3 Give IUPAC names for these structures:

(a) $(CH_3)_2CHCH_2CH_2COCl$ (b) $CH_3CH_2CH(CH_3)CN$

(c) $H_2C{=}CHCH_2CH_2CONH_2$ (d) $(CH_3CH_2)_2CHCN$

(e) $O_2CCH(CH_3)_2$ (f) CH_3 COCl C=C CH_3 CH_3

(g) O C ⟶ ½ O (h) $CO_2CH(CH_3)_2$

PROBLEM 10.4 Draw structures corresponding to these names:

(a) 2,2-Dimethylpropanoyl chloride (b) *N*-Methylbenzamide

(c) 5,5-Dimethylhexanenitrile (d) *tert*-Butyl butanoate

(e) *trans*-2-Methylcyclohexanecarboxamide (f) *p*-Methylbenzoic anhydride

(g) *cis*-3-Methylcyclohexanecarbonyl bromide (h) *p*-Bromobenzonitrile

10.2 OCCURRENCE, STRUCTURE, AND PROPERTIES OF CARBOXYLIC ACIDS

Carboxylic acids occupy a central place among carbonyl compounds, both in nature and in the laboratory. Vinegar, for example, is simply a dilute solution of acetic acid, CH_3COOH; butanoic acid, $CH_3CH_2CH_2COOH$, is responsible for the rancid odor of sour butter; and hexanoic acid (caproic acid), $CH_3(CH_2)_4COOH$, is partially responsible for the unmistakable aroma of goats (Latin *caper*, "goat").

Since the carboxylic acid functional group, –COOH, is structurally similar to both ketones and alcohols, we might expect to see similar properties. As in ketones, the carboxyl carbon is sp^2-hybridized. Carboxylic acid groups are therefore planar, with C–C–O and O–C–O bond angles of approximately 120°. The structure of acetic acid is shown in Figure 10.1.

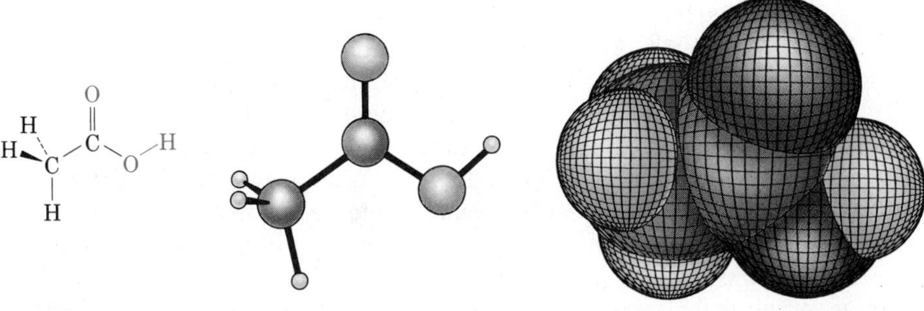

FIGURE 10.1 The structure of acetic acid.

Like alcohols, carboxylic acids are strongly associated because of intermolecular hydrogen bonding. Most carboxylic acids exist as dimers held together by two hydrogen bonds:

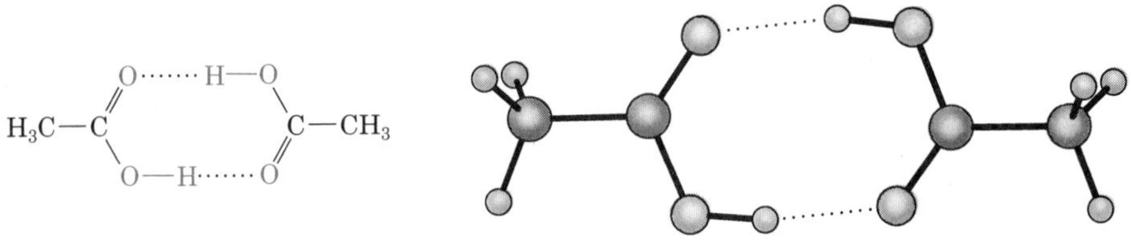

$$H_3C-C(=O\cdots H-O)(-O-H\cdots O=)C-CH_3$$

Acetic acid dimer

This strong hydrogen bonding has a noticeable effect on boiling points. Carboxylic acids normally boil at much higher temperatures than alkanes or alkyl halides of similar molecular weight. Acetic acid, for example, boils at 118°C, whereas chloropropane boils at 46.6°C.

10.3 ACIDITY OF CARBOXYLIC ACIDS

As their name implies, carboxylic acids are *acidic*. They therefore react with bases such as sodium hydroxide to give metal carboxylate salts. Although carboxylic acids with more than six carbon atoms are only slightly soluble in water, alkali metal salts of carboxylic acids are generally quite water-soluble because of their ionic nature. It's often possible to take advantage of this solubility to purify acids by extracting their salts into aqueous base, then reacidifying and extracting the pure acid back into an organic solvent.

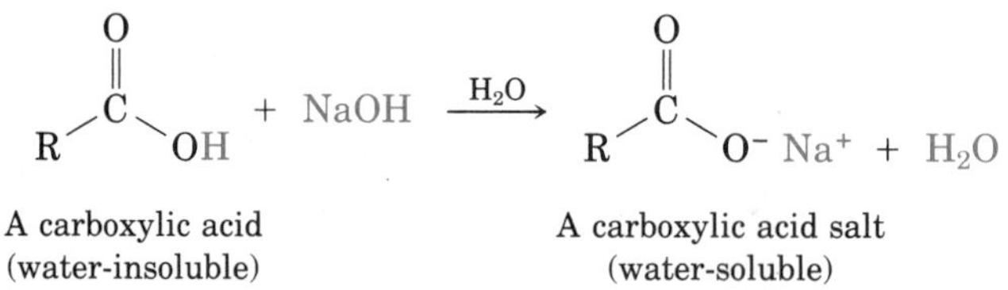

$$RCOOH + NaOH \xrightarrow{H_2O} RCOO^- \ Na^+ + H_2O$$

A carboxylic acid (water-insoluble) — A carboxylic acid salt (water-soluble)

For most carboxylic acids, the acidity constant K_a (Section 1.13) is on the order of 10^{-5}. Acetic acid, for example, has $K_a = 1.76 \times 10^{-5}$, which corresponds to a pK_a of 4.75. In practical terms, a K_a value near 10^{-5} means that only about 1% of the molecules in a 0.1 M aqueous solution are dissociated, as opposed to the 100% dissociation observed for strong mineral acids like HCl and H_2SO_4. As indicated by the list of K_a values in Table 10.2, there is a considerable range in the strengths of various carboxylic acids. Trichloroacetic acid ($K_a = 0.20$), for example, is more than 12,000 times as strong as acetic acid ($K_a = 1.76 \times 10^{-5}$). How can we account for such differences?

TABLE 10.2 Acid Strengths of Some Carboxylic Acids

NAME	K_a	pK_a	
HCl (hydrochloric acid)[a]	(10^7)	(−7)	Stronger acid
CCl_3COOH	0.20	0.70	↑
$CHCl_2COOH$	3.3×10^{-2}	1.48	
$CH_2ClCOOH$	1.4×10^{-3}	2.85	
$HCOOH$	1.77×10^{-4}	3.75	
C_6H_5COOH	6.46×10^{-5}	4.19	
$H_2C{=}CHCOOH$	5.6×10^{-5}	4.25	
CH_3COOH	1.76×10^{-5}	4.75	
CH_3CH_2OH (ethanol)[a]	(10^{-16})	(16)	Weaker acid

[a]Values for HCl and ethanol are shown for reference.

Because the dissociation of a carboxylic acid is an equilibrium process, any substituent that stabilizes the carboxylate anion by withdrawing electrons favors increased dissociation and increases acidity. Thus, introducing an electron-withdrawing chlorine atom makes chloroacetic acid stronger than acetic acid by a factor of 75. Introducing two electronegative substituents makes dichloroacetic acid some 3000 times as strong as acetic acid, and introducing three substituents makes trichloroacetic acid more than 12,000 times as strong.

Although weaker than mineral acids, carboxylic acids are nevertheless much stronger acids than alcohols. Ethanol, for example, has a K_a of approximately 10^{-16}, making ethanol a weaker acid than acetic acid by a factor of 10^{11}.

CH_3CH_2OH	CH_3COOH	$CH_2ClCOOH$	$CHCl_2COOH$	CCl_3COOH	HCl
$pK_a = 16$	$pK_a = 4.75$	$pK_a = 2.85$	$pK_a = 1.48$	$pK_a = 0.70$	$pK_a = -7$

Weaker acid → Stronger acid

Why are carboxylic acids so much more acidic than alcohols even though both contain OH groups? The easiest way to answer this question is to look at the relative stability of carboxylate anions versus alkoxide anions. In alkoxides, the negative charge is *localized* on the one oxygen atom. In carboxylate anions, however, the negative charge is *delocalized*, or spread out over both oxygen atoms, resulting in resonance stabilization (Section 4.12) of the ion. In other words, a carboxylate anion is a stabilized resonance hybrid of two equivalent line-bond structures.

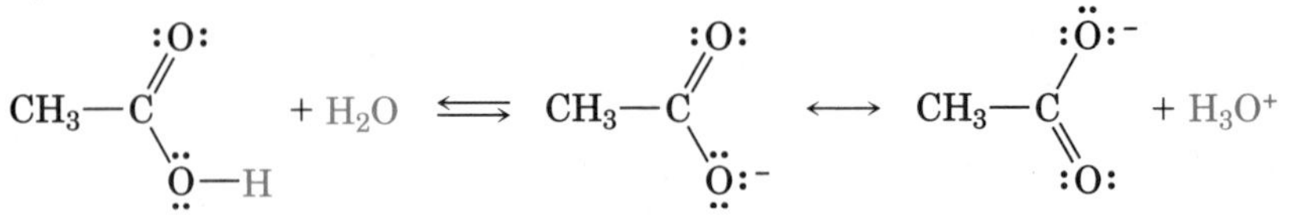

$$CH_3-C(=O)-OH + H_2O \rightleftharpoons CH_3-C(=O)-O^- \longleftrightarrow CH_3-C(-O^-)=O + H_3O^+$$

Carboxylic acid

Resonance-stabilized carboxylate ion (two equivalent resonance forms)

PRACTICE PROBLEM 10.1 Which would you expect to be the stronger acid, benzoic acid or *p*-nitrobenzoic acid?

SOLUTION We know from its effect on aromatic substitution (Section 5.9) that the nitro group is electron-withdrawing and can stabilize a negative charge. Thus, *p*-nitrobenzoic acid is stronger than benzoic acid.

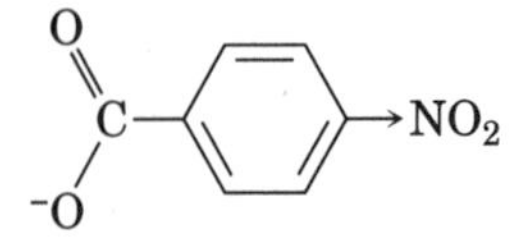

$$^-O-C(=O)-C_6H_4 \rightarrow NO_2$$

Nitro group withdraws electrons from ring and stabilizes negative charge.

PROBLEM 10.5 Draw structures for the products of these reactions:
(a) Benzoic acid + $NaOCH_3 \longrightarrow$? **(b)** $(CH_3)_3CCOOH + KOH \longrightarrow$?

PROBLEM 10.6 Rank these compounds in order of increasing acidity: sulfuric acid, methanol, phenol, *p*-nitrophenol, acetic acid.

PROBLEM 10.7 Rank these compounds in order of increasing acidity:
(a) CH_3CH_2COOH, $BrCH_2COOH$, $BrCH_2CH_2COOH$
(b) Benzoic acid, ethanol, *p*-cyanobenzoic acid

10.4 SYNTHESIS OF CARBOXYLIC ACIDS

We've already seen most of the common methods for preparing carboxylic acids, but let's review briefly:

1. Oxidation of substituted alkylbenzenes with potassium permanganate or sodium dichromate gives substituted benzoic acids (Section 5.11):

$$O_2N\text{–}C_6H_4\text{–}CH_3 \xrightarrow[H_2O,\ 95°C]{KMnO_4} O_2N\text{–}C_6H_4\text{–}\overset{O}{\overset{\|}{C}}OH$$

p-Nitrotoluene **p-Nitrobenzoic acid (88%)**

2. Oxidation of primary alcohols and aldehydes yields carboxylic acids (Sections 8.7 and 9.6). Primary alcohols are often oxidized with chromium trioxide or sodium dichromate; aldehydes are oxidized with Tollens' reagent ($AgNO_3$ in NH_4OH).

$$CH_3(CH_2)_8CH_2OH \xrightarrow[H_2O,\ H_2SO_4]{CrO_3} CH_3(CH_2)_8\overset{O}{\overset{\|}{C}}OH$$

1-Decanol **Decanoic acid (93%)**

$$CH_3CH_2CH_2CH_2CH_2\overset{O}{\overset{\|}{C}}H \xrightarrow[NH_4OH]{AgNO_3} CH_3CH_2CH_2CH_2CH_2\overset{O}{\overset{\|}{C}}OH$$

Hexanal **Hexanoic acid (85%)**

Hydrolysis of Nitriles

Carboxylic acids can be prepared from nitriles, R–C≡N, by reaction with aqueous acid or base (*hydrolysis*). Since nitriles themselves are usually prepared by S_N2 reaction between an alkyl halide and cyanide ion, CN^-, the two-step sequence of cyanide ion displacement followed by nitrile hydrolysis is an excellent method for converting an alkyl halide into a carboxylic acid (RBr → RC≡N → RCOOH). A good example of the reaction occurs in the commercial synthesis of the antiarthritis drug, Fenoprofen:

$$C_6H_5\text{–O–}C_6H_4\text{–}\underset{Br}{\underset{|}{C}}HCH_3 \xrightarrow{\text{1. NaCN; 2. }^-OH/H_2O\text{; 3. }H_3O^+} C_6H_5\text{–O–}C_6H_4\text{–}\underset{CH_3}{\underset{|}{C}}HCOOH$$

Fenoprofen
(an antiarthritic agent)

As with all S_N2 reactions, the method works best with primary alkyl halides, but secondary alkyl halides often can be used as well (Section 7.7).

Carboxylation of Grignard Reagents

CARBOXYLATION
The addition of CO_2 to a molecule

Yet another method of preparing carboxylic acids is by reaction of Grignard reagents, RMgX, with carbon dioxide. This **carboxylation** reaction is carried out either by pouring a solution of the Grignard reagent over dry ice (solid CO_2) or by bubbling a stream of dry CO_2 gas through the Grignard reagent solution.

Br, H_3C, CH_3, CH_3 (ring) $\xrightarrow[\text{Ether}]{\text{Mg}}$ MgBr, H_3C, CH_3, CH_3 (ring) $\xrightarrow[\text{2. } H_3O^+]{\text{1. } CO_2\text{, ether}}$ COOH, H_3C, CH_3, CH_3 (ring)

1-Bromo-2,4,6-trimethylbenzene

2,4,6-Trimethylbenzoic acid (87%)

PRACTICE PROBLEM 10.2 How would you convert 2-chloro-2-methylpropane into 2,2-dimethylpropanoic acid?

SOLUTION Since 2-chloro-2-methylpropane is a tertiary alkyl halide, it won't undergo S_N2 substitution with cyanide ion. Thus, the only way you could carry out the desired reaction is to convert the alkyl halide into a Grignard reagent and then add CO_2.

$$(CH_3)_3CCl + Mg \longrightarrow (CH_3)_3CMgCl \xrightarrow[\text{2. } H_3O^+]{\text{1. } CO_2} (CH_3)_3C\overset{\overset{\displaystyle O}{\|}}{C}OH$$

PROBLEM 10.8 Predict the products of these reactions:

(a) Bromobenzene (Br on benzene ring) $\xrightarrow[\text{2. } CO_2 \text{ 3. } H_3O^+]{\text{1. Mg, ether}}$?

(b) $CH_3\underset{}{\overset{CH_3}{|}}{CH}CH_2CH_2Br$ $\xrightarrow[\text{2. NaOH, } H_2O \text{ 3. } H_3O^+]{\text{1. NaCN}}$?

PROBLEM 10.9 Show the steps in the conversion of iodomethane to acetic acid by the nitrile hydrolysis route. Would a similar route work for the conversion of iodobenzene to benzoic acid? Explain.

PROBLEM 10.10 Show all the steps in the conversion of iodobenzene to benzoic acid by the Grignard carboxylation route. Would a similar route work for the conversion of iodomethane to acetic acid?

10.5 NUCLEOPHILIC ACYL SUBSTITUTION REACTIONS

NUCLEOPHILIC ACYL SUBSTITUTION REACTION
Substitution reaction that replaces one nucleophile bonded to a carbonyl group by another

We saw in Chapter 9 that the addition of nucleophiles to the polar C=O bond is a general feature of ketone and aldehyde chemistry. Carboxylic acids and their derivatives also react with nucleophiles, but the initially formed intermediate expels the substituent originally bonded to the carbonyl carbon, leading to the formation of a new carbonyl compound by a **nucleophilic acyl substitution reaction** (Figure 10.2).

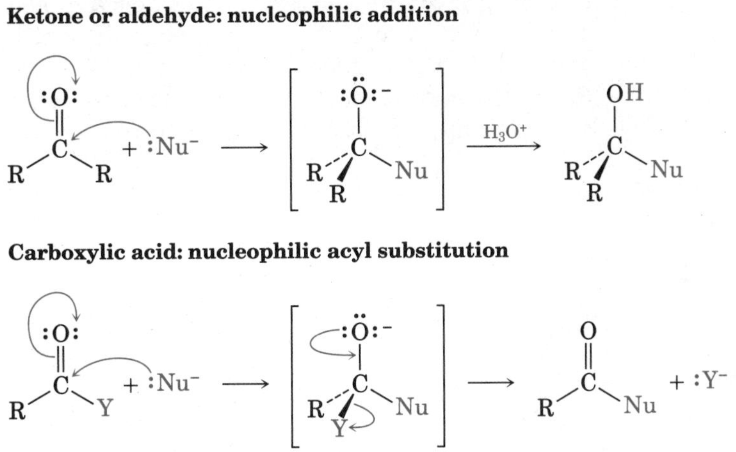

Ketone or aldehyde: nucleophilic addition

:O:, C, R, R + :Nu⁻ ⟶ [:Ö:⁻, C, R, R, Nu] —H_3O^+⟶ OH, C, R, R, Nu

Carboxylic acid: nucleophilic acyl substitution

:O:, C, R, Y + :Nu⁻ ⟶ [:Ö:⁻, C, R, Y, Nu] ⟶ O, C, R, Nu + :Y⁻

FIGURE 10.2 The general mechanisms of nucleophilic addition and nucleophilic acyl substitution reactions.

The different behaviors of ketones/aldehydes and carboxylic acid derivatives are a consequence of structure. Carboxylic acid derivatives have an acyl function bonded to a group –Y that can leave as a stable anion. As soon as addition of a nucleophile occurs, the group –Y leaves and a new carbonyl compound forms. Ketones and aldehydes have no such leaving group, however, and therefore don't undergo substitution.

Note that the overall nucleophilic substitution reaction of acyl derivatives is superficially similar to what occurs during S_N2 reactions of alkyl halides (Section 7.7) in that a leaving group is replaced by an incoming nucleophile. The *mechanisms* of the two reactions are very different, however: S_N2 reactions occur in a single step by back-side displacement of a leaving group, whereas nucleophilic acyl substitutions take place in two steps through a tetrahedrally hybridized intermediate.

In comparing the reactivity of different acyl derivatives, the more highly polar a compound is, the more reactive it is. Thus, acid chlorides are the most reactive compounds because the electronegative chlorine atom strongly polarizes the carbonyl group, whereas amides are the least reactive compounds:

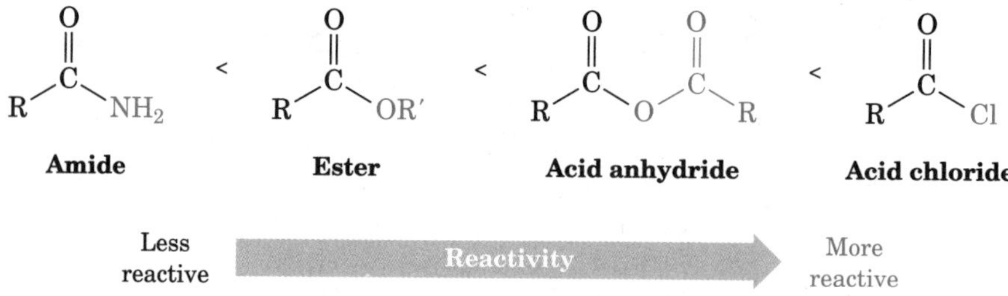

An important consequence of these reactivity differences is that it's usually possible to convert a more reactive acid derivative into a less reactive one. Acid chlorides, for example, can be converted into esters and amides, but amides and esters can't be converted into acid chlorides. Remembering the reactivity order is therefore a useful way to keep track of a large number of reactions (Figure 10.3).

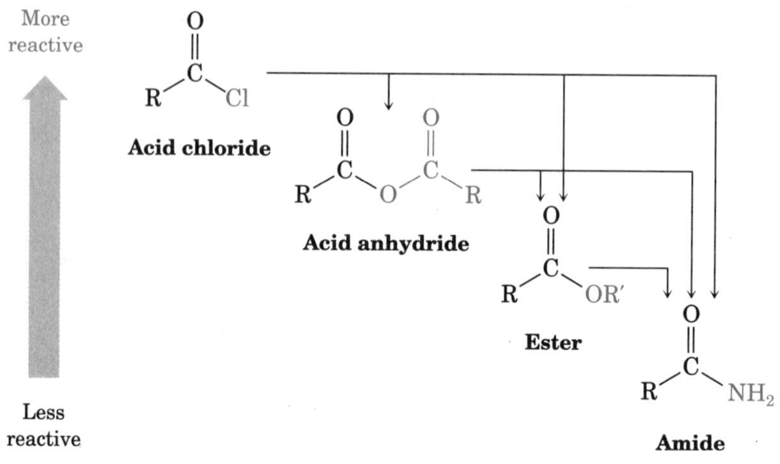

FIGURE 10.3 Interconversions of carboxylic acid derivatives.

PRACTICE PROBLEM 10.3 Which is more reactive in a nucleophilic acyl substitution reaction with hydroxide ion, CH_3CONH_2 or CH_3COCl?

SOLUTION Since Cl is more electronegative than N, the carbonyl group of an acid chloride is more polar than the carbonyl group of an amide, and acid chlorides are more reactive than amides.

PROBLEM 10.11 Which of the following compounds are more reactive in nucleophilic acyl substitution reactions?

(a) CH_3COCl or CH_3COOCH_3
(b) $(CH_3)_2CHCONH_2$ or $CH_3CH_2COOCH_3$
(c) CH_3COOCH_3 or $CH_3COOCOCH_3$
(d) CH_3COOCH_3 or CH_3CHO

PROBLEM 10.12 How can you account for the fact that methyl trifluoroacetate, CF_3COOCH_3, is more reactive than methyl acetate, CH_3COOCH_3, in nucleophilic acyl substitution reactions?

10.6 REACTIONS OF CARBOXYLIC ACIDS

Reduction: Conversion of Acids into Alcohols

We saw in Section 8.5 that carboxylic acids are reduced by lithium aluminum hydride ($LiAlH_4$) to yield primary alcohols:

$$CH_3(CH_2)_7CH{=}CH(CH_2)_7\overset{\displaystyle O}{\overset{\|}{C}}OH \xrightarrow[2.\ H_3O^+]{1.\ LiAlH_4} CH_3(CH_2)_7CH{=}CH(CH_2)_7CH_2OH$$

Oleic acid **_cis_-9-Octadecen-1-ol (87%)**

Conversion of Acids into Acid Chlorides

The most important reactions of carboxylic acids are those that convert the carboxyl group into other acid derivatives by nucleophilic acyl substitution. Acid chlorides, anhydrides, esters, and amides can all be prepared from carboxylic acids.

Acid chlorides are usually prepared by treatment of carboxylic acids with thionyl chloride, $SOCl_2$. The net effect is substitution of the acid –OH group by –Cl. For example,

$$H_3C\text{–}C_6H_2(CH_3)_2\text{–}\overset{\displaystyle O}{\overset{\|}{C}}\text{–}OH \xrightarrow[CHCl_3]{SOCl_2} H_3C\text{–}C_6H_2(CH_3)_2\text{–}\overset{\displaystyle O}{\overset{\|}{C}}\text{–}Cl + HCl + SO_2$$

2,4,6-Trimethylbenzoic acid **2,4,6-Trimethylbenzoyl chloride (90%)**

Conversion of Acids into Acid Anhydrides

Acid anhydrides are formally derived from two molecules of carboxylic acid by removing one molecule of water. Anhydrides are difficult to prepare directly from the corresponding acids, however, and only acetic anhydride is commercially available.

$$H_3C\text{–}\overset{\displaystyle O}{\overset{\|}{C}}\text{–}O\text{–}\overset{\displaystyle O}{\overset{\|}{C}}\text{–}CH_3$$

Acetic anhydride

Conversion of Acids into Esters

One of the most important reactions of carboxylic acids is their conversion into esters. Among the many methods for accomplishing this transformation is the S_N2 reaction between a carboxylate anion nucleophile and a primary alkyl halide (Section 7.7):

$$CH_3CH_2CH_2\overset{O}{\overset{\|}{C}}\ddot{\underset{..}{O}}{:}^- \; Na^+ + CH_3{-}I \xrightarrow[\text{reaction}]{S_N2} CH_3CH_2CH_2\overset{O}{\overset{\|}{C}}OCH_3 + NaI$$

Sodium butanoate **Methyl butanoate, an ester (97%)**

FISCHER ESTERIFICATION REACTION
Conversion of a carboxylic acid into an ester by reaction with alcohol and an acid catalyst

Alternatively, esters can be synthesized by a nucleophilic acyl substitution reaction of a carboxylic acid with an alcohol. Called the **Fischer esterification reaction**, this method involves heating the carboxylic acid with an acid catalyst in an alcohol solvent.

$$C_6H_5{-}\overset{O}{\overset{\|}{C}}{-}OH + CH_3CH_2OH \xrightarrow[\text{catalyst}]{H_2SO_4} C_6H_5{-}\overset{O}{\overset{\|}{C}}{-}OCH_2CH_3 + H_2O$$

Benzoic acid **Ethyl benzoate (91%)**

The Fischer esterification reaction, whose mechanism is shown in Figure 10.4, is a nucleophilic acyl substitution process. The catalyst first protonates an oxygen atom of the –COOH group, which makes the carboxylic acid much more reactive toward nucleophiles. An alcohol molecule then adds to the protonated carboxylic acid, and subsequent loss of water yields the ester product.

All steps in the Fischer esterification reaction are reversible, and the position of the equilibrium can be driven to either side depending on the reaction conditions. Ester formation is favored when alcohol is used as solvent, but carboxylic acid is favored when water is used as solvent.

PRACTICE PROBLEM 10.4 How might you prepare the following ester using a Fischer esterification reaction?

$$C_6H_5{-}\overset{O}{\overset{\|}{C}}{-}OCH_2CH_2CH_3$$

SOLUTION The trick is to identify the two parts of the ester. The target molecule is propyl benzoate, so it can be prepared by treating benzoic acid with 1-propanol.

$$C_6H_5{-}\overset{O}{\overset{\|}{C}}{-}OH + HOCH_2CH_2CH_3 \xrightarrow[\text{catalyst}]{H^+} C_6H_5{-}\overset{O}{\overset{\|}{C}}{-}OCH_2CH_2CH_3$$

Benzoic acid **1-Propanol** **Propyl benzoate**

Protonation of the carbonyl oxygen activates the carboxylic acid...

...toward nucleophilic attack by alcohol, yielding a tetrahedral intermediate.

Transfer of a proton from one oxygen atom to another yields a second tetrahedral intermediate and converts the OH group into a good leaving group.

Loss of a proton regenerates the acid catalyst and gives the ester product

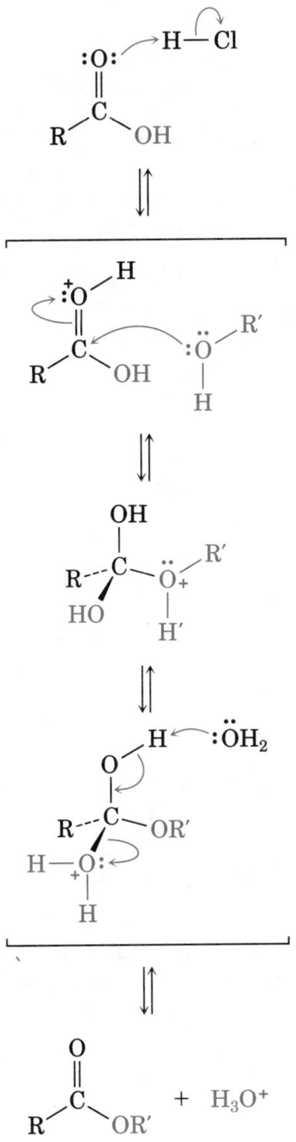

$$RCOOH + HCl \rightleftharpoons RC(=\overset{+}{O}H)OH \xrightarrow{R'OH} R\text{–}C(OH)_2\text{–}\overset{+}{O}HR' \rightleftharpoons R\text{–}C(OH)(OR')\text{–}\overset{+}{O}H_2 \rightleftharpoons RCOOR' + H_3O^+$$

FIGURE 10.4 Mechanism of the Fischer esterification reaction of a carboxylic acid to yield an ester.

PROBLEM 10.13 What products would you obtain by treating benzoic acid with the following reagents? Formulate the reactions.
(a) $SOCl_2$ (b) CH_3OH, HCl (c) $LiAlH_4$ (d) NaOH

PROBLEM 10.14 Show how you might prepare these esters using Fischer esterification reactions:
(a) Butyl acetate (b) Methyl butanoate

PROBLEM 10.15 If 5-hydroxypentanoic acid is treated with an acid catalyst, an intramolecular esterification reaction occurs. What is the structure of the product? (*Intramolecular* means within the same molecule.)

Conversion of Acids into Amides

Amides are carboxylic acid derivatives in which the acid hydroxyl group has been replaced by a nitrogen substituent, $-NH_2$, $-NHR$, or $-NR_2$. Amides are difficult to prepare directly from acids because amines are bases (see Section 12.3), which convert acidic carboxyl groups into their carboxylate anions:

$$RC(=O)OH + :NH_3 \longrightarrow RC(=O)O^- \; NH_4^+$$

Since the carboxylate anion has a negative charge, it no longer undergoes attack by nucleophiles except at high temperatures. We'll see a method for carrying out the transformation in Section 15.8 in connection with the synthesis of proteins from amino acids.

10.7 CHEMISTRY OF ACID HALIDES

Synthesis of Acid Halides

Acid chlorides are prepared from carboxylic acids by reaction with thionyl chloride, $SOCl_2$, as we saw in the previous section:

$$RC(=O)OH \xrightarrow{SOCl_2} RC(=O)Cl$$

Reactions of Acid Halides

Acid halides are among the most reactive of the various carboxylic acid derivatives and can be converted into many other kinds of substances. (We'll discuss only the reactions of acid chlorides in this chapter, but similar chemistry also applies to other acid halides.) Most reactions of acid halides occur by nucleophilic acyl substitution mechanisms. As illustrated in Figure 10.5, the halide ion can be replaced by $-OH$ to yield an acid, by $-OR$ to yield an ester, or by $-NH_2$ to yield an amide. Although Figure 10.5 illustrates these reactions only for acid chlorides, they also take place with other acid halides.

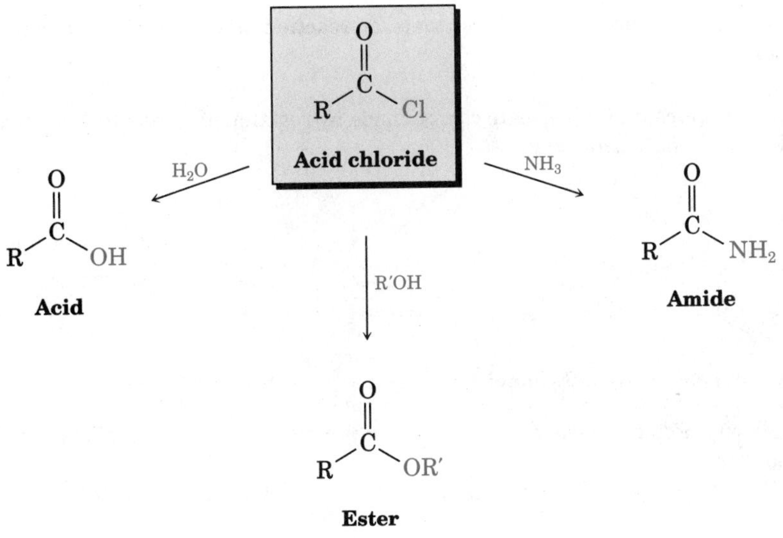

FIGURE 10.5 Some nucleophilic acyl substitution reactions of acid chlorides.

HYDROLYSIS: CONVERSION OF ACID CHLORIDES INTO ACIDS Acid chlorides react with water to yield carboxylic acids. This hydrolysis reaction is a typical nucleophilic acyl substitution process that is initiated by attack of the nucleophile water on the acid chloride carbonyl group. The initially formed tetrahedral intermediate undergoes loss of HCl to yield the product:

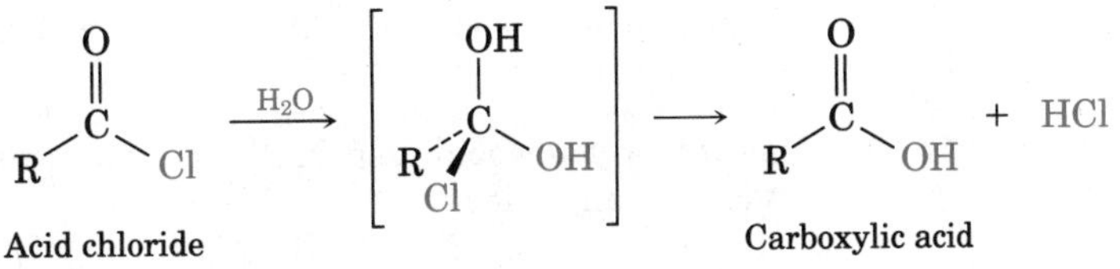

$$\underset{\text{Acid chloride}}{RCOCl} \xrightarrow{H_2O} \left[RC(OH)_2Cl \right] \longrightarrow \underset{\text{Carboxylic acid}}{RCOOH} + HCl$$

ALCOHOLYSIS: CONVERSION OF ACID HALIDES INTO ESTERS Acid chlorides react with alcohols to yield esters in a reaction analogous to their reaction with water to yield acids:

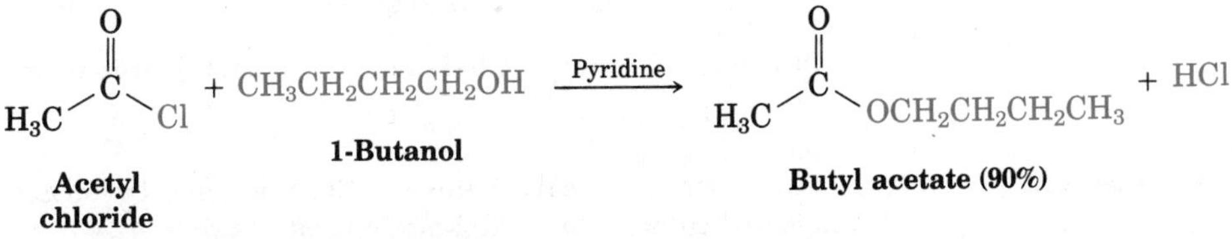

$$\underset{\text{Acetyl chloride}}{H_3CCOCl} + \underset{\text{1-Butanol}}{CH_3CH_2CH_2CH_2OH} \xrightarrow{\text{Pyridine}} \underset{\text{Butyl acetate (90\%)}}{H_3CCOOCH_2CH_2CH_2CH_3} + HCl$$

Since HCl is generated as a by-product of alcoholysis, the reaction is usually carried out in the presence of an amine base such as pyridine (see Section 12.7), which reacts with the HCl as it's formed and prevents it from causing side reactions.

PRACTICE PROBLEM 10.5 Show how you could prepare ethyl benzoate by reaction of an acid chloride with an alcohol.

SOLUTION As its name implies, ethyl benzoate can be made by reaction of ethyl alcohol with the acid chloride of benzoic acid:

$$C_6H_5COCl + CH_3CH_2OH \xrightarrow[\text{solvent}]{\text{Pyridine}} C_6H_5COOCH_2CH_3$$

Benzoyl chloride **Ethanol** **Ethyl benzoate**

PROBLEM 10.16 How could you prepare these esters using the reaction of an acid chloride with an alcohol?
(a) $CH_3CH_2COOCH_3$ (b) $CH_3COOCH_2CH_3$ (c) Cyclohexyl acetate

AMINOLYSIS: CONVERSION OF ACID CHLORIDES INTO AMIDES Acid chlorides react rapidly with ammonia and with amines to give amides. Both mono- and disubstituted amines can be used. For example, 2-methylpropanamide is prepared by reaction of 2-methylpropanoyl chloride with ammonia. Note that one extra equivalent of ammonia is added to react with the HCl generated.

$$(CH_3)_2CHCCl(=O) + 2\ NH_3 \xrightarrow{H_2O} (CH_3)_2CHCNH_2(=O) + \overset{+}{N}H_4\ \overset{-}{Cl}$$

2-Methylpropanoyl chloride **2-Methylpropanamide (83%)**

PRACTICE PROBLEM 10.6 Show how you would prepare *N*-methylpropanamide by reaction of an acid chloride with an amine.

SOLUTION Reaction of methylamine with propanoyl chloride gives *N*-methylpropanamide:

$$CH_3CH_2CCl(=O) + 2\ CH_3NH_2 \longrightarrow CH_3CH_2CNHCH_3(=O) + CH_3NH_3^+\ Cl^-$$

Propanoyl chloride **Methylamine** ***N*-Methylpropanamide**

PROBLEM 10.17 Write the steps in the mechanism of the reaction between ammonia and 2-methylpropanoyl chloride to yield 2-methylpropanamide.

PROBLEM 10.18 What amines would react with what acid chlorides to give the following amide products?
(a) $CH_3CH_2CONH_2$ (b) $(CH_3)_2CHCH_2CONHCH_3$
(c) *N,N*-Dimethylpropanamide (d) *N,N*-Diethylbenzamide

10.8 CHEMISTRY OF ACID ANHYDRIDES

Synthesis of Acid Anhydrides

The best method of preparing acid anhydrides is by a nucleophilic acyl substitution reaction of an acid chloride with a carboxylic acid anion. Both symmetrical and unsymmetrical acid anhydrides can be prepared in this way.

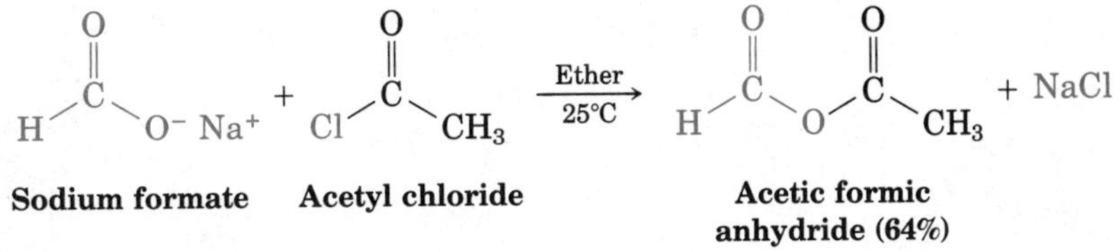

$$HCO_2^- Na^+ + ClCOCH_3 \xrightarrow[25°C]{\text{Ether}} HCOOCOCH_3 + NaCl$$

Sodium formate **Acetyl chloride** **Acetic formic anhydride (64%)**

Reactions of Acid Anhydrides

The chemistry of acid anhydrides is similar to that of acid chlorides. Although anhydrides react more slowly than acid chlorides, the kinds of reactions the two functional groups undergo are the same. Thus, acid anhydrides react with water to form acids, with alcohols to form esters, and with amines to form amides (Figure 10.6).

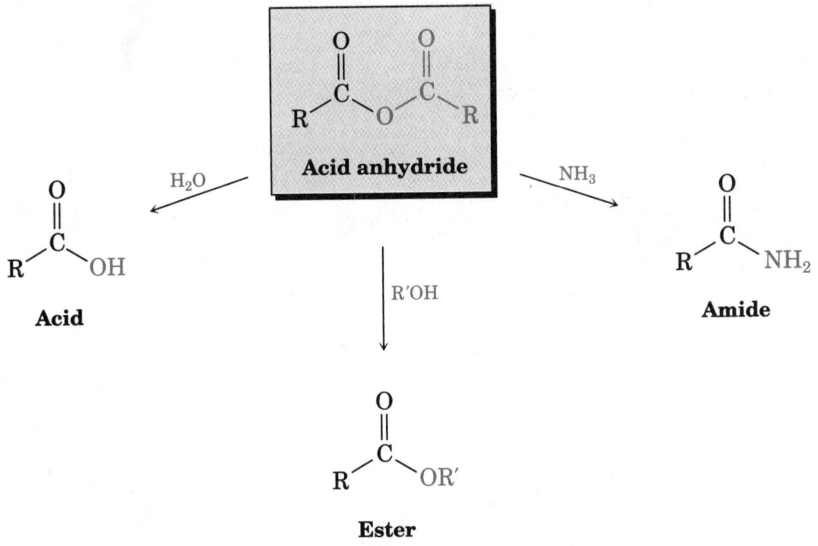

Acid anhydride ($RCOOCOR$)

H_2O → **Acid** ($RCOOH$)

$R'OH$ → **Ester** ($RCOOR'$)

NH_3 → **Amide** ($RCONH_2$)

FIGURE 10.6 Some reactions of acid anhydrides.

Acetic anhydride is often used to prepare acetate esters of complex alcohols and to prepare substituted acetamides from amines. For example, aspirin (an ester) is prepared by reaction of acetic anhydride with *o*-hydroxybenzoic acid. Similarly, acetaminophen (an amide) is prepared by reaction of acetic anhydride with *p*-hydroxyaniline.

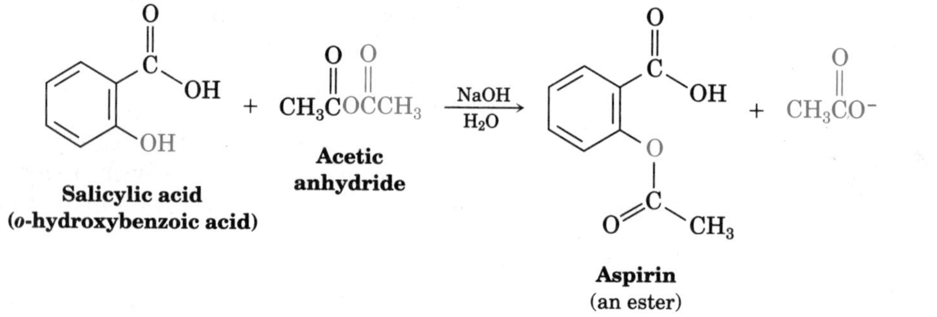

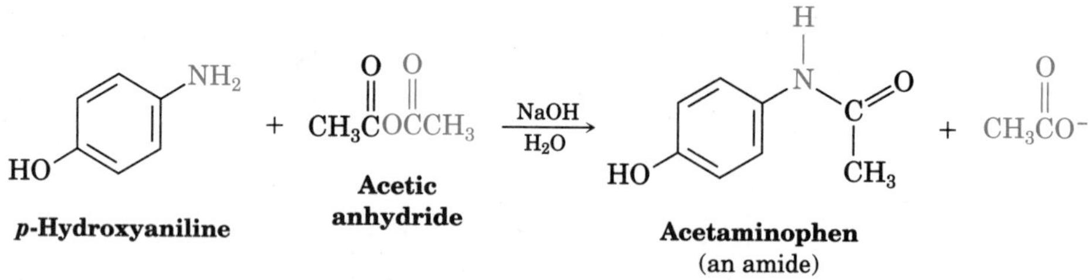

Salicylic acid (o-hydroxybenzoic acid) + CH_3COCCH_3 (Acetic anhydride) $\xrightarrow[H_2O]{NaOH}$ Aspirin (an ester) + CH_3CO^-

p-Hydroxyaniline + CH_3COCCH_3 (Acetic anhydride) $\xrightarrow[H_2O]{NaOH}$ Acetaminophen (an amide) + CH_3CO^-

Notice in both of these examples that only "half" of the anhydride molecule is used; the other half acts as the leaving group during the nucleophilic acyl substitution step and produces carboxylate anion as a by-product. Thus, anhydrides are inefficient, and acid chlorides are normally preferred for introducing acyl substituents other than acetyl groups.

PRACTICE PROBLEM 10.7 What is the product of the following reaction?

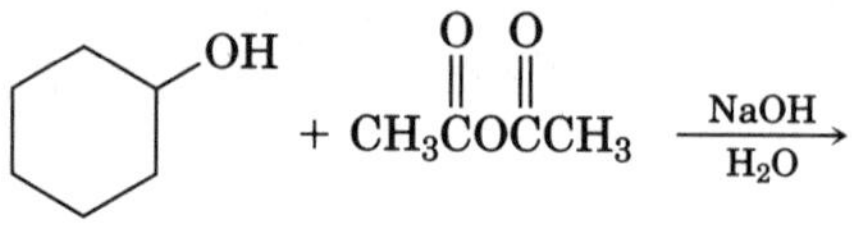

Cyclohexanol (OH) + CH_3COCCH_3 $\xrightarrow[H_2O]{NaOH}$

SOLUTION Reaction of cyclohexanol with acetic anhydride yields cyclohexyl acetate by nucleophilic acyl substitution of the alcohol group by the acetate group of the anhydride.

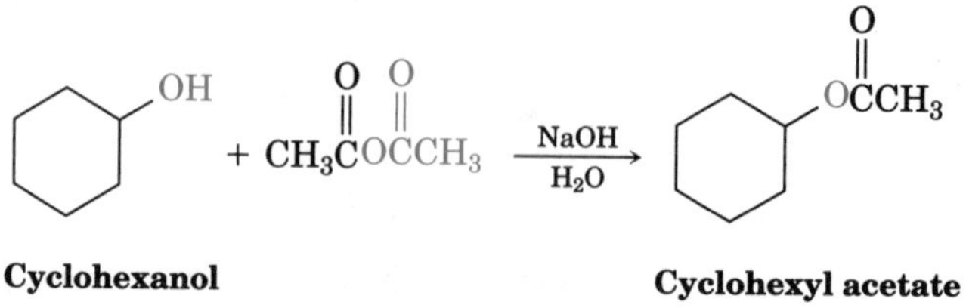

Cyclohexanol + CH_3COCCH_3 $\xrightarrow[H_2O]{NaOH}$ Cyclohexyl acetate ($OCCH_3$)

PROBLEM 10.19 Write the steps in the mechanism of the reaction between p-hydroxyaniline and acetic anhydride to prepare acetaminophen.

PROBLEM 10.20 What product would you expect to obtain from the reaction of 1 equiv of methanol with a cyclic anhydride such as phthalic anhydride?

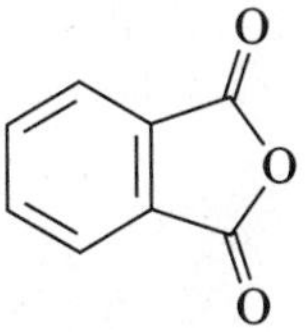

Phthalic anhydride

10.9 CHEMISTRY OF ESTERS

Esters are among the most widespread of naturally occurring compounds. Many simple esters are pleasant-smelling liquids that are responsible for the fragrant odors of fruits and flowers. For example, methyl butanoate has been isolated from pineapple oil, and isopentyl acetate has been found in banana oil. The ester linkage is also present in animal fats and other biologically important molecules.

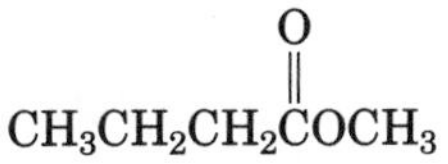

$CH_3CH_2CH_2COCH_3$ (C=O)

Methyl butanoate
(from pineapples)

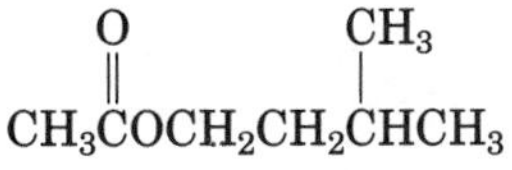

$CH_3COCH_2CH_2CHCH_3$ (C=O; CH_3 branch)

Isopentyl acetate
(from bananas)

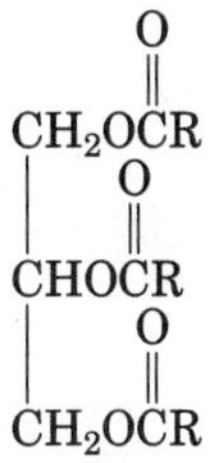

CH_2OCR / $CHOCR$ / CH_2OCR (each C=O)

A fat
(R = C_{12-18} chains)

Synthesis of Esters

Esters are usually prepared either from acids or from acid chlorides by the methods already discussed. Thus, carboxylic acids are converted directly into esters either by S_N2 reaction of a carboxylate salt with a primary alkyl halide or by reaction of the acid with an alcohol (Section 10.6). Acid chlorides are converted into esters by reaction with an alcohol in the presence of base (Section 10.7).

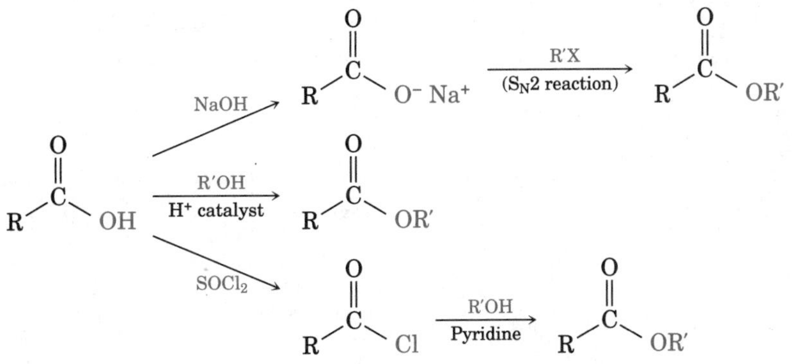

$$RCOOH \xrightarrow{NaOH} RCOO^- Na^+ \xrightarrow[(S_N2\ reaction)]{R'X} RCOOR'$$

$$RCOOH \xrightarrow[H^+\ catalyst]{R'OH} RCOOR'$$

$$RCOOH \xrightarrow{SOCl_2} RCOCl \xrightarrow[Pyridine]{R'OH} RCOOR'$$

Reactions of Esters

Esters show the same kinds of chemistry we've seen for other acyl derivatives, but they're less reactive toward nucleophiles than acid chlorides or anhydrides. Figure 10.7 shows some general reactions of esters.

Ester

H_2O → Acid

NH_3 → Amide

[H] → 1° Alcohol

$R''MgX$ → 3° Alcohol

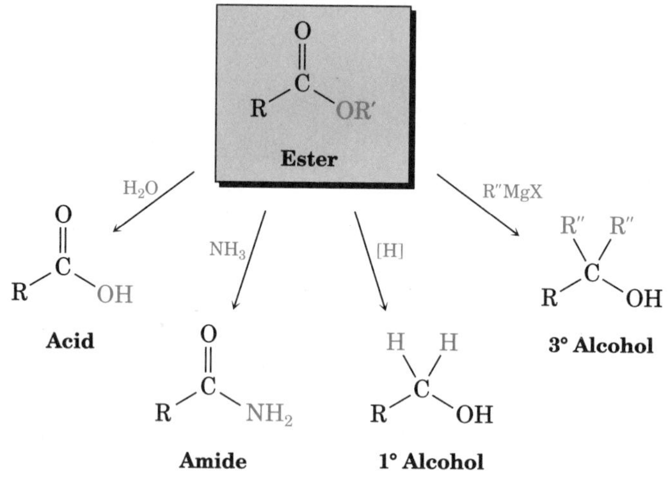

FIGURE 10.7 Some general reactions of esters.

HYDROLYSIS: CONVERSION OF ESTERS INTO ACIDS Esters are hydrolyzed either by aqueous base or by aqueous acid to yield a carboxylic acid plus an alcohol:

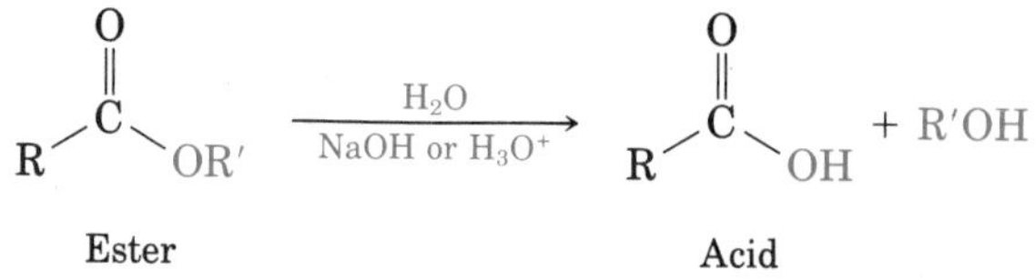

$$\underset{\text{Ester}}{RCOOR'} \xrightarrow[\text{NaOH or } H_3O^+]{H_2O} \underset{\text{Acid}}{RCOOH} + R'OH$$

SAPONIFICATION Base-induced hydrolysis of an ester to yield a carboxylate anion

Hydrolysis in basic solution is called **saponification** (Latin *sapo*, "soap"). (As we'll see in Section 16.3, the boiling of animal fat with alkali to make soap is a saponification since fats have ester linkages.) Ester hydrolysis occurs by a typical nucleophilic acyl substitution pathway in which hydroxide ion nucleophile adds to the ester carbonyl group, yielding a tetrahedral intermediate. Loss of alkoxide ion then gives a carboxylic acid, which is deprotonated to give the acid salt:

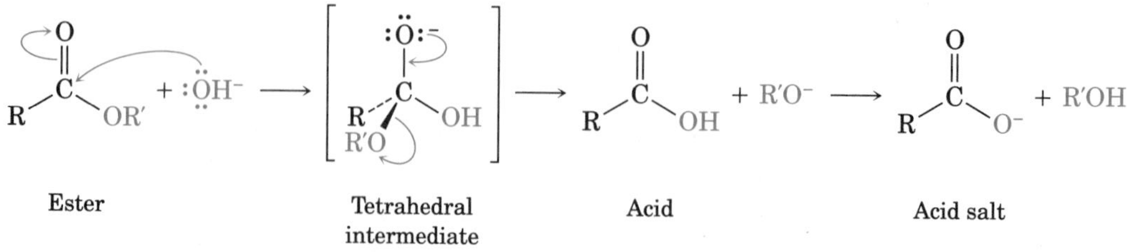

$$\underset{\text{Ester}}{RCOOR'} + :\ddot{O}H^- \longrightarrow \underset{\text{Tetrahedral intermediate}}{\left[R(R'O)C(\ddot{O}:^-)OH \right]} \longrightarrow \underset{\text{Acid}}{RCOOH} + R'O^- \longrightarrow \underset{\text{Acid salt}}{RCOO^-} + R'OH$$

PRACTICE PROBLEM 10.8 Write the products of the following saponification reaction:

$$\underset{\text{Ethyl 3-methylbutanoate}}{CH_3\overset{CH_3}{\overset{|}{C}}HCH_2\overset{O}{\overset{\|}{C}}OCH_2CH_3} \xrightarrow[2.\ H_3O^+]{1.\ NaOH,\ H_2O} ?$$

SOLUTION Esters are cleaved by aqueous base into their acid and alcohol components by breaking the bond between the carbonyl carbon and the alcohol oxygen:

$$\underset{\text{Ethyl 3-methylbutanoate}}{CH_3\overset{CH_3}{\overset{|}{C}}HCH_2\overset{O}{\overset{\|}{C}}OCH_2CH_3} \xrightarrow[2.\ H_3O^+]{1.\ NaOH,\ H_2O} \underset{\text{3-Methylbutanoic acid}}{CH_3\overset{CH_3}{\overset{|}{C}}HCH_2\overset{O}{\overset{\|}{C}}OH} + \underset{\text{Ethanol}}{CH_3CH_2OH}$$

PROBLEM 10.21 Show the products of hydrolysis of these esters:
(a) Isopropyl acetate (b) Methyl cyclohexanecarboxylate

PROBLEM 10.22 Why do you suppose saponification of esters is not reversible? In other words, why doesn't treatment of a carboxylic acid with alkoxide ion give an ester?

AMINOLYSIS: CONVERSION OF ESTERS INTO AMIDES Esters react with ammonia and amines to yield amides. The reaction is not often used, however, because higher yields are usually obtained starting from the acid chloride rather than from the ester.

$$\underset{\text{Methyl benzoate}}{C_6H_5\overset{O}{\overset{\|}{C}}OCH_3} \xrightarrow[\text{Ether}]{NH_3} \underset{\text{Benzamide}}{C_6H_5\overset{O}{\overset{\|}{C}}NH_2} + CH_3OH$$

REDUCTION: CONVERSION OF ESTERS INTO ALCOHOLS Esters are easily reduced by treatment with $LiAlH_4$ to yield primary alcohols (Section 8.5):

$$\underset{\text{Ethyl-2-pentenoate}}{CH_3CH_2CH{=}CH\overset{O}{\overset{\|}{C}}OCH_2CH_3} \xrightarrow[2.\ H_3O^+]{1.\ LiAlH_4,\ \text{ether}} \underset{\text{2-Penten-1-ol (91\%)}}{CH_3CH_2CH{=}CHCH_2OH} + CH_3CH_2OH$$

Hydride ion first adds to the carbonyl group, followed by elimination of alkoxide ion to yield an aldehyde intermediate. Further reduction of the aldehyde gives the primary alcohol.

Ester + :H⁻ → [Tetrahedral intermediate → R'O⁻ + Aldehyde intermediate] —1. $LiAlH_4$, 2. H_3O^+→ 1° Alcohol + R'OH

PRACTICE PROBLEM 10.9 What products would you obtain by reduction of propyl benzoate with $LiAlH_4$?

SOLUTION Reduction of esters with $LiAlH_4$ yields two molecules of alcohol product, one from the acyl portion of the ester and one from the alkoxy portion. Thus, reduction of propyl benzoate yields benzyl alcohol (from the acyl group) and 1-propanol (from the alkoxyl group).

$C_6H_5COCH_2CH_2CH_3$ —1. $LiAlH_4$, 2. H_3O^+→ $C_6H_5CH_2OH$ + $HOCH_2CH_2CH_3$

Propyl benzoate **Benzyl alcohol** **1-Propanol**

PROBLEM 10.23 Show the products you would obtain by reduction of these esters with $LiAlH_4$:
(a) $CH_3CH_2CH_2CH(CH_3)COOCH_3$ **(b)** Phenyl benzoate

PROBLEM 10.24 What product would you expect from the reaction of a cyclic ester such as butyrolactone with $LiAlH_4$?

Butyrolactone

REACTION OF ESTERS WITH GRIGNARD REAGENTS Grignard reagents react with esters to yield tertiary alcohols in which two of the substituents are identical. For example, methyl benzoate reacts with 2 equiv of methylmagnesium bromide to yield 2-phenyl-2-propanol. The reaction occurs by addition of a Grignard reagent to the ester, elimination of methoxide ion to give an intermediate ketone, and further addition to the ketone to yield the tertiary alcohol (Figure 10.8).

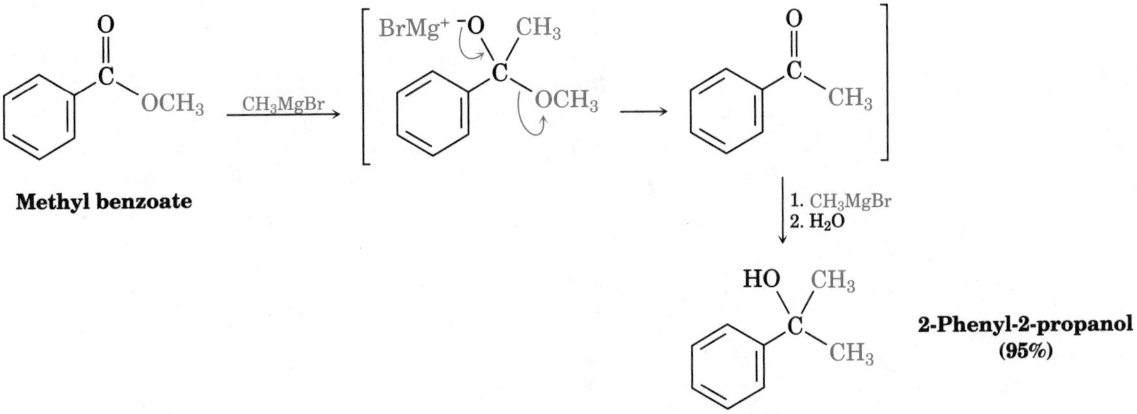

$$\text{C}_6\text{H}_5\text{C(=O)OCH}_3 \xrightarrow{\text{CH}_3\text{MgBr}} \left[\text{C}_6\text{H}_5\text{C}(\text{O}^-\ \text{BrMg}^+)(\text{CH}_3)(\text{OCH}_3) \longrightarrow \text{C}_6\text{H}_5\text{C(=O)CH}_3 \right] \xrightarrow[\text{2. } \text{H}_2\text{O}]{\text{1. } \text{CH}_3\text{MgBr}} \text{C}_6\text{H}_5\text{C(OH)(CH}_3)_2$$

Methyl benzoate **2-Phenyl-2-propanol (95%)**

FIGURE 10.8 Mechanism of the reaction of Grignard reagents with esters to yield tertiary alcohols. A ketone intermediate is involved.

PRACTICE PROBLEM 10.10 How could you use the reaction of a Grignard reagent with an ester to prepare 1,1-diphenyl-1-propanol?

SOLUTION The product of the reaction between a Grignard reagent and an ester is a tertiary alcohol in which the alcohol carbon and one of the attached groups have come from the ester, and the remaining two groups bonded to the alcohol carbon have come from the Grignard reagent. Since 1,1-diphenyl-1-propanol has two phenyl groups and one ethyl group bonded to the alcohol carbon, it must have been prepared from the reaction of a phenylmagnesium halide with an ester of propanoic acid:

$$2\ C_6H_5MgBr + CH_3CH_2\overset{O}{\overset{\|}{C}}OCH_3 \longrightarrow CH_3CH_2-\underset{C_6H_5}{\underset{|}{\overset{OH}{\overset{|}{C}}}}-C_6H_5$$

1,1-Diphenyl-1-propanol

PROBLEM 10.25 What ester and what Grignard reagent might you use to prepare these alcohols?
(a) 2-Phenyl-2-propanol **(b)** 1,1-Diphenylethanol **(c)** 3-Ethyl-3-heptanol

10.10 CHEMISTRY OF AMIDES

Synthesis of Amides

Amides are usually prepared by reaction of an acid chloride with an amine, as we saw in Section 10.7. Ammonia, monosubstituted amines, and disubstituted amines all undergo this reaction.

$$\mathrm{RCOCl} \xrightarrow{NH_3} \mathrm{RCONH_2} \qquad \mathrm{RCOCl} \xrightarrow{R'NH_2} \mathrm{RCONHR'} \qquad \mathrm{RCOCl} \xrightarrow{R'_2NH} \mathrm{RCONR'_2}$$

Reactions of Amides

Amides are much less reactive than acid chlorides, acid anhydrides, and esters. We'll see in Chapter 15, for example, that the amide linkage is stable enough to serve as the basic unit from which proteins are made.

$$H_2N{-}CH(R){-}C(=O){-}OH \longrightarrow {-}NH{-}CH(R){-}C(=O){-}NH{-}CH(R'){-}C(=O){-}NH{-}CH(R''){-}C(=O){-}$$

Amino acids — A protein (a polyamide)

HYDROLYSIS: CONVERSION OF AMIDES INTO ACIDS Amides undergo hydrolysis to yield carboxylic acids plus amine upon heating in either aqueous acid or base. Although the reaction is difficult and requires prolonged heating, the overall transformation is a typical nucleophilic acyl substitution of $-OH$ for $-NH_2$.

$$\underset{\text{Amide}}{\mathrm{RC(=O)NH_2}} \xrightarrow[\text{Heat}]{H_3O^+ \text{ or } HO^-,\ H_2O} \underset{\text{Acid}}{\mathrm{RC(=O)OH}} + NH_3$$

REDUCTION: CONVERSION OF AMIDES INTO AMINES Like other carboxylic acid derivatives, amides are reduced by lithium aluminum hydride. The product of this reduction, however, is an *amine* rather than an alcohol:

$$\underset{\textbf{Benzamide}}{\mathrm{C_6H_5C(=O)NH_2}} \xrightarrow[\text{2. } H_2O]{\text{1. } LiAlH_4\text{, ether}} \underset{\textbf{Benzylamine (93\%)}}{\mathrm{C_6H_5CH_2NH_2}}$$

The effect of amide reduction is to convert the amide carbonyl group into a methylene group (C=O → CH_2). This kind of reaction is specific for amides and doesn't occur with other carboxylic acid derivatives.

PRACTICE PROBLEM 10.11 How could you prepare *N*-ethylaniline by reduction of an amide with $LiAlH_4$?

SOLUTION Since reduction of an amide with $LiAlH_4$ yields an amine, the starting material for synthesis of *N*-ethylaniline must have been *N*-phenylacetamide.

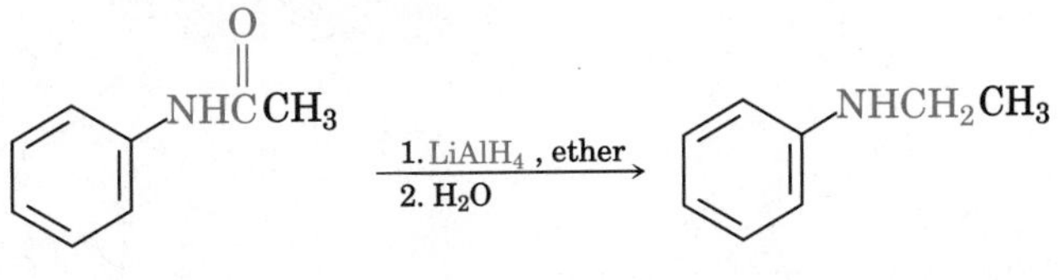

$$C_6H_5NHCOCH_3 \xrightarrow[2.\ H_2O]{1.\ LiAlH_4,\ ether} C_6H_5NHCH_2CH_3$$

***N*-Phenylacetamide** ***N*-Ethylaniline**

PROBLEM 10.26 How would you convert *N*-ethylbenzamide into these products?
(a) Benzoic acid (b) Benzyl alcohol
(c) *N*-Ethylbenzylamine, $C_6H_5CH_2NHCH_2CH_3$

PROBLEM 10.27 The lithium aluminum hydride reduction of amides to yield amines is equally effective with both acyclic and cyclic amides (*lactams*). What product would you obtain from reduction of 5,5-dimethyl-2-pyrrolidone with $LiAlH_4$?

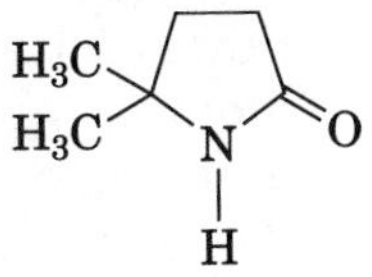

5,5-Dimethyl-2-pyrrolidone
(a lactam)

10.11 CHEMISTRY OF NITRILES

Nitriles, R–C≡N, are not related to carboxylic acids in the same sense that acyl derivatives are, but the chemistries of nitriles and carboxylic acids are so similar that the two classes of compounds can be considered together.

Synthesis of Nitriles

The simplest method of preparing nitriles is by the S_N2 displacement reaction of cyanide ion on a primary alkyl halide (Section 7.7):

$$RCH_2Br + Na^+\ CN^- \xrightarrow[\text{reaction}]{S_N2} RCH_2CN + NaBr$$

Reactions of Nitriles

The chemistry of nitriles is similar in many respects to the chemistry of carbonyl compounds, and nitriles undergo many of the same kinds of reactions as carboxylic acid derivatives (Figure 10.9).

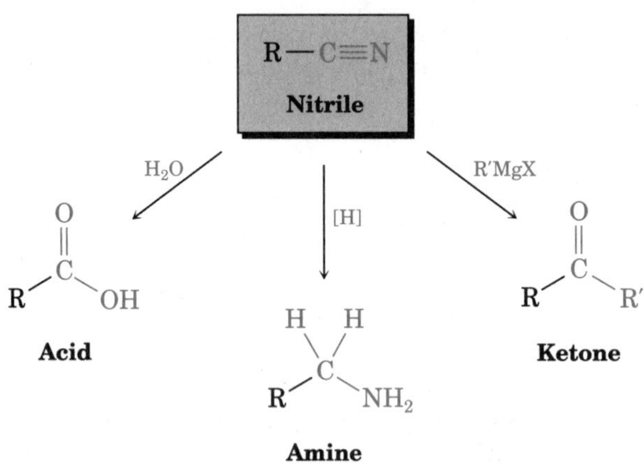

FIGURE 10.9 Some reactions of nitriles.

Like carbonyl groups, nitriles are strongly polarized. Thus, the nitrile-group carbon atom is electrophilic and undergoes attack by nucleophiles to yield an sp^2-hybridized intermediate imine anion that is analogous to the sp^3-hybridized intermediate alkoxide anion formed by addition of a nucleophile to a carbonyl group. Once formed, the intermediate imine anion can then go on to yield further products.

Carbonyl compound: $R_2C{=}O$ (δ^+ C, δ^- O) $\xrightarrow{:Nu^-}$ $[R_2C(Nu)O^-]$ $\longrightarrow$ Products

Nitrile: $R{-}\overset{\delta^+}{C}{\equiv}\overset{\delta^-}{N}$ $\xrightarrow{:Nu^-}$ $[RC(Nu){=}N^-]$ $\longrightarrow$ Products

HYDROLYSIS: CONVERSION OF NITRILES INTO CARBOXYLIC ACIDS Nitriles are hydrolyzed in either acidic or basic solution to yield carboxylic acids and ammonia (or an amine).

$$R{-}C{\equiv}N \xrightarrow[H_2O,\ NaOH]{H_3O^+\ \text{or}} RCOOH + NH_3$$

REDUCTION: CONVERSION OF NITRILES INTO AMINES Reduction of nitriles with $LiAlH_4$ gives primary amines, just as reduction of esters gives primary alcohols.

$$o\text{-}CH_3C_6H_4C{\equiv}N \xrightarrow[2.\ H_2O]{1.\ LiAlH_4,\ ether} o\text{-}CH_3C_6H_4CH_2NH_2$$

o-Methylbenzonitrile **o-Methylbenzylamine (88%)**

REACTION OF NITRILES WITH GRIGNARD REAGENTS Grignard reagents, RMgX, add to nitriles to give intermediate imine anions that can be hydrolyzed to yield ketones:

$$R{-}C{\equiv}N: + \ddot{R}'^{-}\ \overset{+}{M}gX \longrightarrow \left[R{-}\underset{}{\overset{:\ddot{N}^{-}\ \overset{+}{M}gX}{\overset{\|}{C}}}{-}R' \right] \xrightarrow{H_3O^+} R{-}\overset{O}{\overset{\|}{C}}{-}R' + NH_3$$

Nitrile Imine anion Ketone

For example, benzonitrile reacts with ethylmagnesium bromide to give propiophenone in high yield:

$$C_6H_5C{\equiv}N \xrightarrow[2.\ H_3O^+]{1.\ CH_3CH_2MgBr,\ ether} C_6H_5{-}\overset{O}{\overset{\|}{C}}{-}CH_2CH_3$$

Benzonitrile **Propiophenone (89%)**

PRACTICE PROBLEM 10.12 Show how you could prepare 2-hexanone by reaction of a Grignard reagent on a nitrile.

SOLUTION There are two ways to prepare a ketone from a nitrile by Grignard addition:

$$\left.\begin{array}{l} CH_3C{\equiv}N\ +\ CH_3CH_2CH_2CH_2MgBr \\ CH_3MgBr\ +\ N{\equiv}CCH_2CH_2CH_2CH_3 \end{array}\right\rangle \longrightarrow CH_3\overset{O}{\overset{\|}{C}}CH_2CH_2CH_2CH_3$$

PROBLEM 10.28 What nitrile would you react with what Grignard reagent to prepare these ketones?

(a) $CH_3CH_2COCH_2CH_3$ **(b)** $CH_3CH_2COCH(CH_3)_2$
(c) Acetophenone (methyl phenyl ketone) **(d)** (dicyclohexyl ketone structure: two cyclohexane rings joined by C=O)

PROBLEM 10.29

How would you prepare the following molecules from the indicated starting materials? More than one step is required in each case.

(a) $(CH_3)_2CHCH_2CH_2NH_2$ from $(CH_3)_2CHCH_2I$

(b) 1-Phenyl-2-butanone from benzyl bromide, $C_6H_5CH_2Br$

10.12 NYLONS AND POLYESTERS: STEP-GROWTH POLYMERS

CHAIN-GROWTH POLYMER Polymer produced by chain reaction of a monofunctional monomer

STEP-GROWTH POLYMER Polymer produced by stepwise reactions between two difunctional monomers

There are two main classes of synthetic polymers: chain-growth polymers and step-growth polymers. **Chain-growth polymers**, such as polyethylene and the other alkene polymers we saw in Section 4.8, are prepared by chain-reaction processes in which an initiator first adds to the double bond of an alkene monomer to produce a reactive intermediate. This intermediate adds to a second alkene monomer unit, and the polymer chain lengthens as more monomer units add successively to the end of the growing chain.

Step-growth polymers are produced by polymerization reactions between two difunctional molecules. Each new bond is formed in a discrete step, independent of all other bonds in the polymer, and chain reactions aren't involved. The key bond-forming step is often a carbonyl nucleophilic acyl substitution.

A large number of step-growth polymers have been made; some of the commercially more important ones are shown in Table 10.3.

TABLE 10.3 Some Important Step-Growth Polymers and Their Uses

MONOMER NAME	FORMULA	TRADE OR COMMON NAME OF POLYMER	USES
Adipic acid Hexamethylenediamine	$HOOC(CH_2)_4COOH$ $H_2N(CH_2)_6NH_2$	→ Nylon 66	Fibers, clothing, tire cord, bearings
Ethylene glycol Dimethyl terephthalate	$HOCH_2CH_2OH$ $CH_3OOC{-}C_6H_4{-}COOCH_3$ (para; benzene ring with COOCH₃ at top and bottom)	→ Dacron, Terylene, Mylar	Fibers, clothing, tire cord, film
Caprolactam	seven-membered ring lactam (C=O, N—H)	→ Nylon 6, Perlon	Fibers, large cast articles

Nylons

NYLONS
Polyamides prepared by reaction between a diacid and a diamine.

The nylons, first synthesized by Wallace Carothers at the Du Pont Company, are the best-known step-growth polymers. **Nylons** are polyamides, usually prepared by reaction between a diamine and a diacid. For example, nylon 66 is prepared by heating the six-carbon adipic acid (hexanedioic acid) with the six-carbon hexamethylenediamine (1,6-hexanediamine) at 280°C:

$$H_2N(CH_2)_6NH_2 \text{ (Hexamethylenediamine)} + HO\overset{O}{\overset{\|}{C}}(CH_2)_4\overset{O}{\overset{\|}{C}}OH \text{ (Adipic acid)} \longrightarrow -\!\!\left[\overset{O}{\overset{\|}{C}}(CH_2)_4\overset{O}{\overset{\|}{C}}-NH(CH_2)_6NH\right]\!\!-_n + n\,H_2O$$

Hexamethylenediamine

Adipic acid

Nylon 66

Nylons are used in engineering applications and in making fibers. A combination of high-impact strength and abrasion resistance makes nylon an excellent metal substitute for bearings and gears. As fibers, nylon is used in a variety of applications, from clothing to Aramid tire cord to carpets to Perlon mountaineering ropes.

Polyesters

POLYESTERS
Polymers prepared by reaction between a diacid and a dialcohol

Just as polyamides (nylons) are made by reaction between diacids and diamines, **polyesters** are made by reaction between diacids and dialcohols. The most generally useful polyester is made by a nucleophilic acyl substitution reaction between dimethyl terephthalate (dimethyl 1,4-benzenedicarboxylate) and ethylene glycol. The product is used under the trade name Dacron to make clothing fiber and tire cord, and under the name Mylar to make plastic film and recording tape. The tensile strength of polyester film is nearly equal to that of steel.

$$CH_3O\overset{O}{\overset{\|}{C}}-C_6H_4-\overset{O}{\overset{\|}{C}}OCH_3 + HOCH_2CH_2OH \longrightarrow -\!\!\left[\overset{O}{\overset{\|}{C}}-C_6H_4-\overset{O}{\overset{\|}{C}}-OCH_2CH_2O\right]\!\!-_n + 2n\,CH_3OH$$

Dimethyl terephthalate

Ethylene glycol

A polyester (Dacron, Mylar)

PRACTICE PROBLEM 10.13 Draw the structure of Qiana, a polyamide fiber made by reaction of hexanedioic acid with 1,4-cyclohexanediamine.

SOLUTION

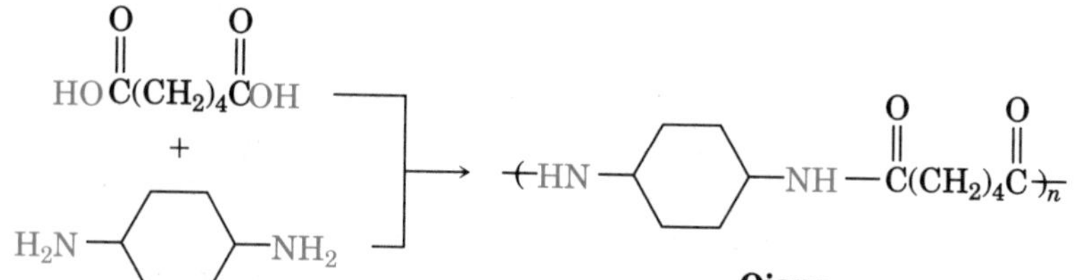

$$HO\overset{O}{\overset{\|}{C}}(CH_2)_4\overset{O}{\overset{\|}{C}}OH + H_2N\text{—}C_6H_{10}\text{—}NH_2 \longrightarrow \left(HN\text{—}C_6H_{10}\text{—}NH\text{—}\overset{O}{\overset{\|}{C}}(CH_2)_4\overset{O}{\overset{\|}{C}}\right)_n$$

Qiana

PROBLEM 10.30 Kevlar, a nylon polymer used to make bulletproof vests, is made by reaction of 1,4-benzenedicarboxylic acid with 1,4-benzenediamine. Show the structure of Kevlar.

INTERLUDE

THIOL ESTERS: BIOLOGICAL CARBOXYLIC ACID DERIVATIVES

Nucleophilic acyl substitution reactions take place in living organisms just as they take place in the chemical laboratory; the same principles apply in both cases. Nature, however, uses *thiol esters*, RCOSR′, rather than acid chlorides or acid anhydrides as its reagents. The pK_a of a typical alkanethiol (RSH) is about 10, placing thiols midway in acid strength between carboxylic acids (pK_a ≈ 5) and alcohols (pK_a ≈16). As a result, thiolate anions (RS^-) can act as leaving groups in nucleophilic acyl substitution reactions. Thiol esters aren't so reactive that they hydrolyze rapidly like anhydrides, yet they are more reactive than normal esters.

Acetyl coenzyme A, usually abbreviated as acetyl CoA, is the most common thiol ester in nature. Acetyl CoA is a much more complex molecule than acetyl chloride or acetic anhydride, yet it serves exactly the same purpose as either of these simpler reagents. Nature uses acetyl CoA as a reactive acetylating agent in nucleophilic acyl substitution reactions (an *acetylating* agent introduces an acetyl group, CH_3CO):

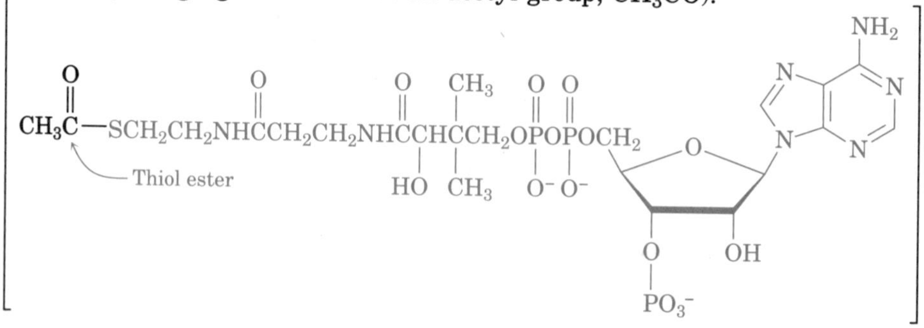

$$CH_3\overset{O}{\overset{\|}{C}}\text{—}SCH_2CH_2NH\overset{O}{\overset{\|}{C}}CH_2CH_2NH\overset{O}{\overset{\|}{C}}\underset{HO}{\underset{|}{C}H}\overset{CH_3}{\underset{CH_3}{C}}CH_2O\underset{O^-}{\overset{O}{P}}O\underset{O^-}{\overset{O}{P}}OCH_2\text{—(ribose, 3′-}OPO_3^-\text{, OH; adenine with } NH_2\text{)}$$

Thiol ester

Acetyl CoA

$$H_3C\overset{O}{\overset{\|}{C}}SCoA + :Nu^- \longrightarrow H_3C\overset{O}{\overset{\|}{C}}Nu + {}^-SCoA$$

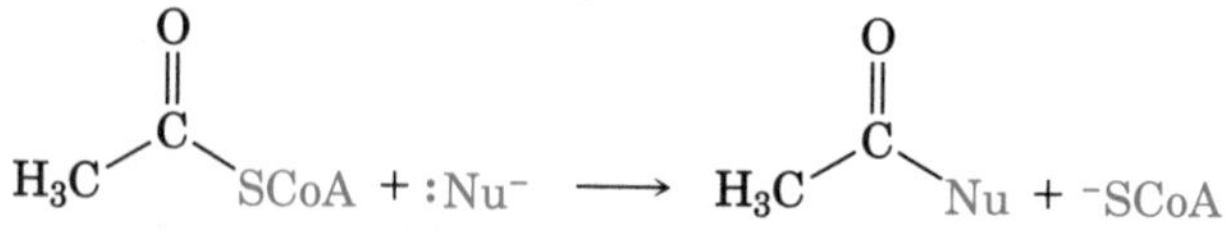

Acetyl CoA

As an example, *N*-acetylglucosamine, an important constituent of cell-surface membranes in mammals, is synthesized in nature by a reaction between glucosamine and acetyl CoA:

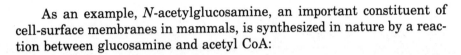

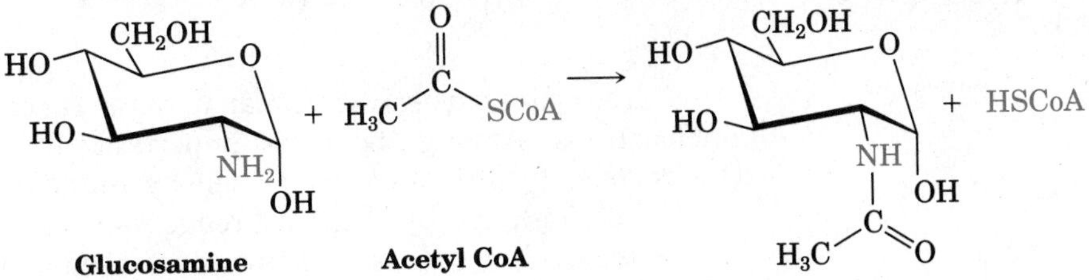

$$\text{Glucosamine} + H_3C{-}C({=}O){-}SCoA \longrightarrow N\text{-Acetylglucosamine} + HSCoA$$

Glucosamine
(an amine)

Acetyl CoA

***N*-Acetylglucosamine**
(an amide)

SUMMARY AND KEY WORDS

Carboxylic acids are among the most important building blocks for synthesizing other molecules, both in nature and in the chemical laboratory. The distinguishing characteristic of carboxylic acids is their acidity. Although weaker than mineral acids like HCl, carboxylic acids are much more acidic than alcohols. The reason for this difference is that carboxylate ions are stabilized by resonance between two equivalent forms:

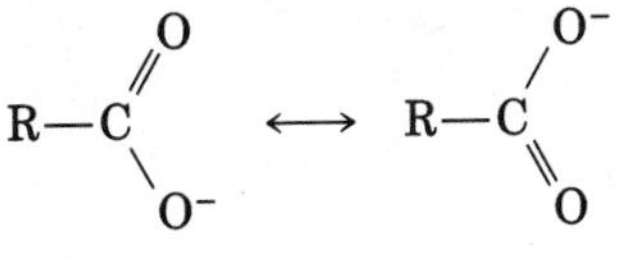

$$R{-}C({=}O){-}O^- \longleftrightarrow R{-}C(O^-){=}O$$

Most carboxylic acids have pK_a values near 5, but the exact acidity of an acid depends on its structure. Carboxylic acids substituted by an electron-withdrawing group are more acidic (have a lower pK_a) because their carboxylate ions are stabilized.

Carboxylic acids can be transformed into a variety of **acyl derivatives** in which the acid –OH group has been replaced by other substituents. **Acid chlorides, acid anhydrides, esters**, and **amides** are the most important acyl derivatives. The chemistries of these different acyl derivatives are similar and are dominated by a single general reaction type: the **nucleophilic acyl substitution reaction**. These substitutions take place by addition of a nucleophile to the polar carbonyl group of the acid derivative, followed by expulsion of the leaving group.

$$\mathrm{R{-}C({=}O){-}Y} + :\mathrm{Nu}^- \longrightarrow \left[\mathrm{R{-}C(O^-)(Y){-}Nu}\right] \longrightarrow \mathrm{R{-}C({=}O){-}Nu} + :\mathrm{Y}^-$$

Carboxylic acid derivatives can undergo reaction with many different nucleophiles. Among the most important are substitution by water (hydrolysis), by alcohols (alcoholysis), by amines (aminolysis), by hydride ion (reduction), and by Grignard reagents.

Nitriles, R–C≡N, can also be considered as carboxylic acid derivatives because they undergo nucleophilic additions to the polar C≡N bond in the same way as carbonyl compounds. The most important reactions of nitriles are their hydrolysis to yield carboxylic acids, their reduction to yield primary amines, and their reaction with Grignard reagents to yield ketones.

SUMMARY OF REACTIONS

1. Reactions of carboxylic acids (Section 10.6)
   (a) Conversion into acid chlorides

$$\mathrm{R{-}C({=}O){-}OH} \xrightarrow{\mathrm{SOCl_2}} \mathrm{R{-}C({=}O){-}Cl}$$

   (b) Conversion into esters

$$\mathrm{R{-}C({=}O){-}OH} + \mathrm{R'OH} \xrightarrow[\text{catalyst}]{\mathrm{H^+}} \mathrm{R{-}C({=}O){-}OR'}$$

2. Reactions of acid halides (Section 10.7)
   (a) Conversion into carboxylic acids

$$\mathrm{R{-}C({=}O){-}Cl} + \mathrm{H_2O} \longrightarrow \mathrm{R{-}C({=}O){-}OH}$$

   (b) Conversion into esters

$$\mathrm{R{-}C({=}O){-}Cl} + \mathrm{R'OH} \xrightarrow{\text{Pyridine}} \mathrm{R{-}C({=}O){-}OR'}$$

(c) Conversion into amides

$$\mathrm{R{-}\overset{\overset{\large O}{\|}}{C}{-}Cl} + NH_3 \xrightarrow{\text{NaOH}} \mathrm{R{-}\overset{\overset{\large O}{\|}}{C}{-}NH_2}$$

3. Reactions of acid anhydrides (Section 10.8)

(a) Conversion into esters

$$\mathrm{R{-}\overset{\overset{\large O}{\|}}{C}{-}O{-}\overset{\overset{\large O}{\|}}{C}{-}R} + R'OH \xrightarrow{\text{NaOH}} \mathrm{R{-}\overset{\overset{\large O}{\|}}{C}{-}OR'}$$

(b) Conversion into amides

$$\mathrm{R{-}\overset{\overset{\large O}{\|}}{C}{-}O{-}\overset{\overset{\large O}{\|}}{C}{-}R} + NH_3 \xrightarrow{\text{NaOH}} \mathrm{R{-}\overset{\overset{\large O}{\|}}{C}{-}NH_2}$$

4. Reactions of esters (Section 10.9)

(a) Conversion into acids

$$\mathrm{R{-}\overset{\overset{\large O}{\|}}{C}{-}OR'} + H_2O \xrightarrow[\text{NaOH}]{\text{H}^+\text{ or}} \mathrm{R{-}\overset{\overset{\large O}{\|}}{C}{-}OH}$$

(b) Conversion into amides

$$\mathrm{R{-}\overset{\overset{\large O}{\|}}{C}{-}OR'} + NH_3 \longrightarrow \mathrm{R{-}\overset{\overset{\large O}{\|}}{C}{-}NH_2}$$

(c) Conversion into primary alcohols by reduction

$$\mathrm{R{-}\overset{\overset{\large O}{\|}}{C}{-}OR'} \xrightarrow{\text{LiAlH}_4} RCH_2OH$$

(d) Conversion into tertiary alcohols by Grignard reaction

$$\mathrm{R{-}\overset{\overset{\large O}{\|}}{C}{-}OR'} \xrightarrow[\text{2. H}_3\text{O}^+]{\text{1. R'MgX}} \mathrm{R{-}\underset{\underset{\large R'}{|}}{\overset{\overset{\large OH}{|}}{C}}{-}R'}$$

5. Reactions of amides (Section 10.10)
(a) Conversion into carboxylic acids

$$RC(=O)NH_2 + H_2O \xrightarrow[\text{NaOH}]{\text{H}^+ \text{ or}} RC(=O)OH$$

(b) Conversion into amines by reduction

$$RC(=O)NH_2 \xrightarrow[\text{2. } H_2O]{\text{1. } LiAlH_4} RCH_2NH_2$$

6. Reactions of nitriles (Section 10.11)
(a) Conversion into carboxylic acids

$$R{-}C{\equiv}N + H_2O \xrightarrow[\text{NaOH}]{\text{H}^+ \text{ or}} RC(=O)OH$$

(b) Conversion into amines by reduction

$$R{-}C{\equiv}N \xrightarrow[\text{2. } H_2O]{\text{1. } LiAlH_4} RCH_2NH_2$$

(c) Conversion into ketones by Grignard reaction

$$R{-}C{\equiv}N \xrightarrow[\text{2. } H_2O]{\text{1. } R'MgX} RC(=O)R'$$

ADDITIONAL PROBLEMS

10.31 Give IUPAC names for these carboxylic acids:

(a) $CH_3CH(COOH)CH_2CH_2CH(COOH)CH_3$ (COOH groups on the second and fifth carbons)

(b) $(CH_3)_3CCOOH$

(c) $CH_3CH_2CH_2CH(CH_2CH_2CH_3)CH_2COOH$

(d) Benzene ring bearing COOH, with NO_2 para (p-nitro substituted benzoic acid structure)

(e) Cyclododecene ring with COOH on a double-bond carbon

(f) $BrCH_2CHBrCH_2CH_2COOH$

10.32 Give IUPAC names for these carboxylic acid derivatives:

(a) 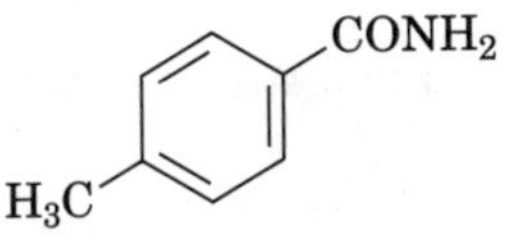 H_3C–C_6H_4–$CONH_2$ (para) **(b)** $(CH_3CH_2)_2CHCH{=}CHCN$ **(c)** $CH_3O_2CCH_2CH_2CO_2CH_3$

(d) 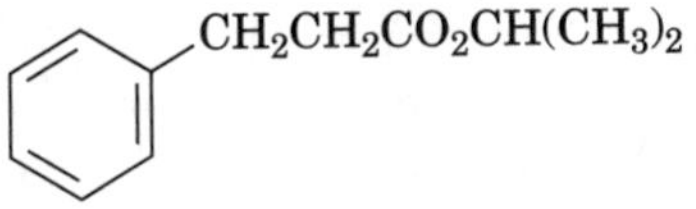 C_6H_5–$CH_2CH_2CO_2CH(CH_3)_2$ **(e)** 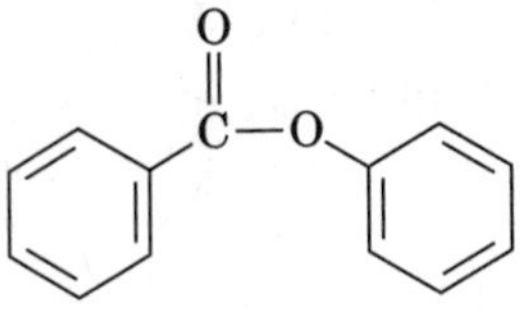 C_6H_5–$\overset{O}{\overset{\|}{C}}$–O–$C_6H_5$

(f) $CH_3CHBrCH_2CONHCH_3$ **(g)** 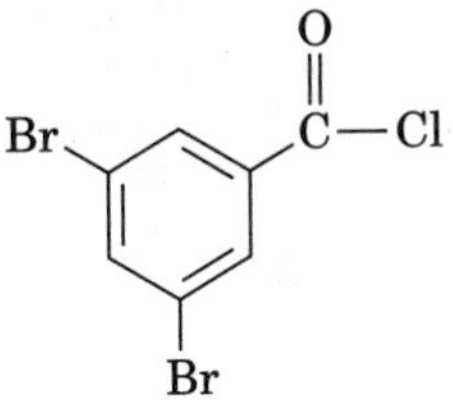 3,5-Br$_2C_6H_3$–$\overset{O}{\overset{\|}{C}}$–Cl **(h)** 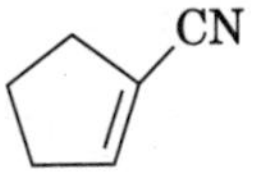 cyclopentene with CN

10.33 Draw structures corresponding to these IUPAC names:
(a) 4,5-Dimethylheptanoic acid **(b)** *cis*-1,2-Cyclohexanedicarboxylic acid
(c) Heptanedioic acid **(d)** Triphenylacetic acid
(e) 2,2-Dimethylhexanamide **(f)** Phenylacetamide
(g) 2-Cyclobutenecarbonitrile **(h)** Ethyl cyclohexanecarboxylate

10.34 Acetic acid boils at 118°C, but its ethyl ester boils at 77°C. Why is the boiling point of the acid so much higher, even though it has a lower molecular weight?

10.35 Draw and name the eight carboxylic acids with formula $C_6H_{12}O_2$.

10.36 Draw and name compounds that meet these descriptions:
(a) Three acid chlorides, C_6H_9ClO **(b)** Three amides, $C_7H_{11}NO$ **(c)** Three nitriles, C_5H_7N
(d) Three esters, $C_5H_8O_2$

10.37 The following reactivity order has been found for the saponification of alkyl acetates by hydroxide ion:

$$CH_3COOCH_3 > CH_3COOCH_2CH_3 > CH_3COOCH(CH_3)_2 > CH_3COOC(CH_3)_3$$

How can you explain this reactivity order?

10.38 Citric acid has pK_a = 3.14, and tartaric acid has pK_a = 2.98. Which acid is stronger?

10.39 Order the compounds in each set with respect to increasing acidity:
(a) Acetic acid, chloroacetic acid, trifluoroacetic acid
(b) Benzoic acid, *p*-bromobenzoic acid, *p*-nitrobenzoic acid
(c) Acetic acid, phenol, cyclohexanol

10.40 How can you explain the fact that 2-chlorobutanoic acid has pK_a = 2.86, 3-chlorobutanoic acid has pK_a = 4.05, 4-chlorobutanoic acid has pK_a = 4.52, and butanoic acid itself has pK_a = 4.82?

10.41 Rank these compounds in order of their reactivity toward nucleophilic acyl substitution:
(a) CH_3COOCH_3 **(b)** CH_3COCl **(c)** CH_3CONH_2 **(d)** $CH_3CO_2COCH_3$

10.42 How can you prepare acetophenone (methyl phenyl ketone) from the following starting materials? (More than one step may be required.)
(a) Benzonitrile **(b)** Bromobenzene **(c)** Methyl benzoate **(d)** Benzene

10.43 How might you prepare the following products starting with butanoic acid? (More than one step may be required.)
(a) 1-Butanol **(b)** Butanal **(c)** 1-Bromobutane **(d)** Pentanenitrile **(e)** 1-Butene
(f) Butylamine, $CH_3CH_2CH_2CH_2NH_2$

10.44 Predict the product of the reaction of *p*-methylbenzoic acid with each of these reagents:
(a) $LiAlH_4$ **(b)** CH_3OH, HCl **(c)** $SOCl_2$ **(d)** NaOH, then CH_3I

10.45 A chemist in need of 2,2-dimethylpentanoic acid decided to synthesize some by reaction of 2-chloro-2-methylpentane with NaCN, followed by hydrolysis of the product. After carrying out the reaction sequence, however, none of the desired product could be found. What do you suppose went wrong?

10.46 Which method of carboxylic acid synthesis, Grignard carboxylation or nitrile hydrolysis, would you use for each of the following reactions? Explain the reasons for each choice.

(a) o-(CH_2Br)phenol (benzene ring with CH_2Br and OH) $\longrightarrow$ benzene ring with CH_2COOH and OH

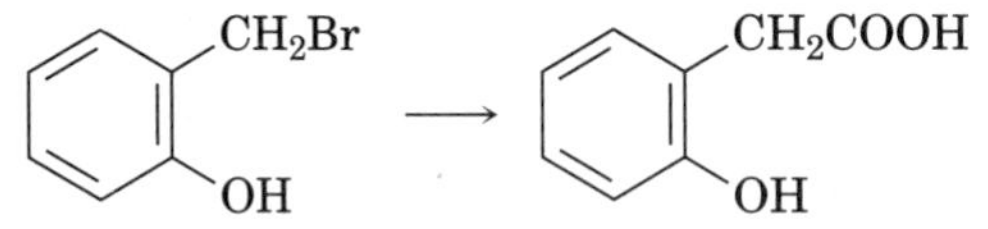

(b) $CH_3CH_2CHBrCH_3 \longrightarrow CH_3CH_2CH(CH_3)COOH$

(c) $CH_3C(=O)CH_2CH_2CH_2I \longrightarrow CH_3C(=O)CH_2CH_2CH_2COOH$
(d) $HOCH_2CH_2CH_2Br \longrightarrow HOCH_2CH_2CH_2COOH$

10.47 How can you explain the observation that an attempted Fischer esterification of 2,4,6-trimethylbenzoic acid with methanol/HCl is unsuccessful? No ester is obtained, and the starting acid is recovered unchanged.

10.48 Acid chlorides undergo reduction with $LiAlH_4$ in the same way that esters do to yield primary alcohols. What are the products of these reactions?

(a) $CH_3CH(CH_3)CH_2CH_2C(=O)Cl \xrightarrow[2.\ H_2O]{1.\ LiAlH_4}$? **(b)** cyclopentane bearing COCl and CH_3 on the same carbon $\xrightarrow[2.\ H_2O]{1.\ LiAlH_4}$?

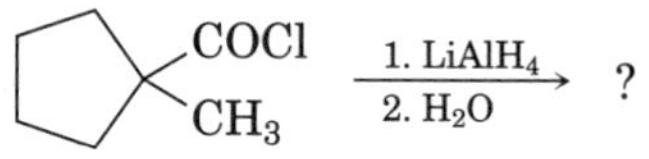

10.49 The reaction of an acid chloride with $LiAlH_4$ to yield a primary alcohol (Problem 10.48) takes place in two steps. The first step is a nucleophilic acyl substitution of H^- for Cl^- to yield an aldehyde, and the second step is nucleophilic addition of H^- to the aldehyde to yield an alcohol. Write the mechanism of the reduction of CH_3COCl.

10.50 Acid chlorides undergo reaction with Grignard reagents at −78°C to yield ketones. Propose a mechanism for the reaction.

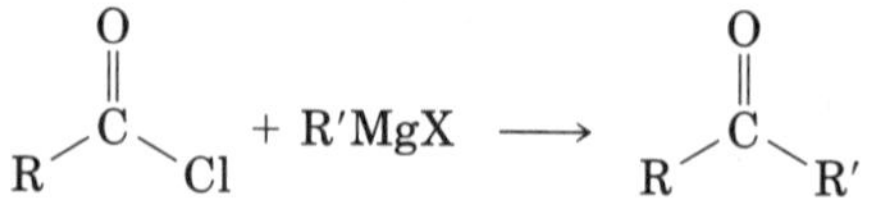

$$RC(=O)Cl + R'MgX \longrightarrow RC(=O)R'$$

10.51 If the reaction of an acid chloride with a Grignard reagent (Problem 10.50) is carried out at room temperature, a tertiary alcohol is formed.
(a) Propose a mechanism for this reaction.
(b) What are the products of the reaction of methylmagnesium bromide, CH_3MgBr, with the acid chlorides given in Problem 10.48?

10.52 When dimethyl carbonate, $CH_3OCOOCH_3$, is treated with phenylmagnesium bromide, triphenylmethanol is formed. Explain how this occurs.

10.53 Predict the product, if any, of reaction between propanoyl chloride and the following reagents. (See Problems 10.49 and 10.50.)

(a) Excess CH_3MgBr in ether **(b)** NaOH in H_2O **(c)** Methylamine
(d) $LiAlH_4$ **(e)** Cyclohexanol **(f)** Sodium acetate

10.54 Answer Problem 10.53 for reaction between methyl propanoate and the listed reagents.

10.55 What esters and what Grignard reagents would you use to make these alcohols?

(a) $CH_3CH_2CH_2C(OH)(CH_3)CH_3$ **(b)** $(C_6H_5)_2C(OH)CH_3$

10.56 Show two ways to make these esters:

(a) $CH_3CH(CH_3)CH_2CH_2C(=O)OCH_2CH_3$ **(b)** cyclopentyl–$CH_2OC(=O)CH_3$

10.57 What products would you obtain on saponification of these esters?

(a) p-Br–C_6H_4–$C(=O)OCH(CH_3)CH_3$ **(b)** Cyclohexyl propanoate

10.58 When *methyl* acetate is heated in pure ethanol containing a small amount of HCl catalyst, *ethyl* acetate results. Propose a mechanism for this reaction.

10.59 *tert*-Butoxycarbonyl azide, an important reagent used in protein synthesis, is prepared by treating *tert*-butoxycarbonyl chloride with sodium azide. Propose a mechanism for this reaction.

$$(CH_3)_3COCOCl + NaN_3 \longrightarrow (CH_3)_3COCON_3 + NaCl$$

10.60 What product would you expect to obtain upon treatment of the cyclic ester, butyrolactone, with excess phenylmagnesium bromide?

Butyrolactone

10.61 *N,N*-Diethyl-*m*-toluamide (DEET) is the active ingredient in many insect repellents. How might you synthesize this substance from *m*-bromotoluene?

$$H_3C-C_6H_4-\overset{O}{\overset{\|}{C}}-N(CH_2CH_3)_2$$

***N,N*-Diethyl-*m*-toluamide**

10.62 In the iodoform reaction, a triiodomethyl ketone reacts with aqueous base to yield a carboxylate ion and iodoform (triiodomethane). Propose a mechanism for this reaction.

$$R-\overset{O}{\overset{\|}{C}}-CI_3 \xrightarrow{NaOH,\ H_2O} R-\overset{O}{\overset{\|}{C}}-O^- + CHI_3$$

10.63 The K_a for bromoacetic acid is approximately 1×10^{-3}. What percentage of the acid is dissociated in a 0.10 M aqueous solution?

11 CHAPTER

CARBONYL ALPHA-SUBSTITUTION REACTIONS AND CONDENSATION REACTIONS

Much of the chemistry of carbonyl compounds can be explained by just four fundamental reactions. We've already looked in detail at two of the four: the nucleophilic addition reaction (Chapter 9) and the nucleophilic acyl substitution reaction (Chapter 10). In this chapter, we'll look at the other two: the alpha-substitution reaction and the carbonyl condensation reaction.

ALPHA-SUBSTITUTION REACTION
Reaction that results in substitution of a hydrogen on the α-carbon of a carbonyl compound

ALPHA POSITION
Position next to a carbonyl carbon

Alpha-substitution reactions occur at the position *next to* the carbonyl group—the **alpha (α) position**—and involve the substitution of an α-hydrogen atom by some other group:

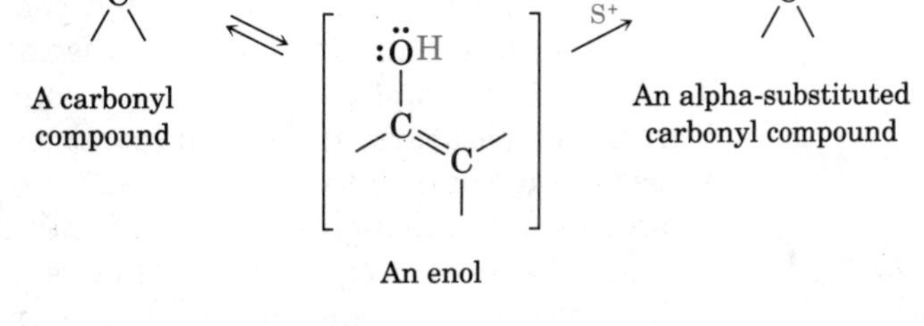

A carbonyl compound

An enolate ion

An enol

S^+

An alpha-substituted carbonyl compound

CARBONYL CONDENSATION REACTION
Reaction between two carbonyl compounds in which the α-carbon of one partner bonds to the carbonyl carbon of the other

Carbonyl condensation reactions take place when *two* carbonyl compounds react with each other in such a way that the α-carbon of one partner becomes bonded to the carbonyl carbon of the second partner:

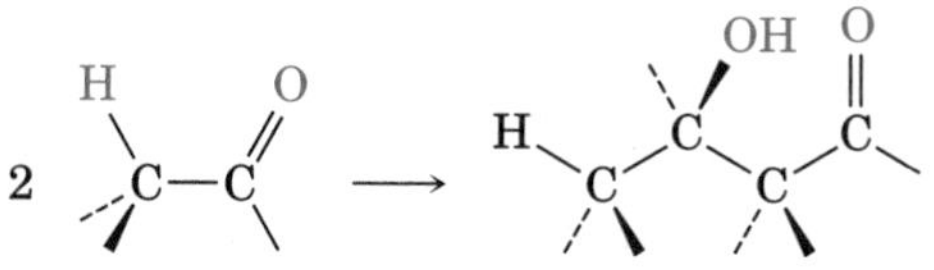

The key feature of both α-substitution reactions and carbonyl condensation reactions is that they take place by the formation of either *enol* or *enolate ion* intermediates. Let's begin our study by learning more about these two species.

11.1 KETO–ENOL TAUTOMERISM

ENOL
Vinylic alcohol

TAUTOMERS
Easily interconvertible constitutional isomers

Carbonyl compounds that have hydrogen atoms on their α-carbons are easily interconverted with their corresponding **enol** (*ene* + *ol*; unsaturated alcohol) isomers. As we saw in Section 4.15, this interconversion between keto and enol forms is a special kind of isomerism called *tautomerism* (Greek *tauto*, "the same," and *meros*, "part"). The individual isomers are called **tautomers**.

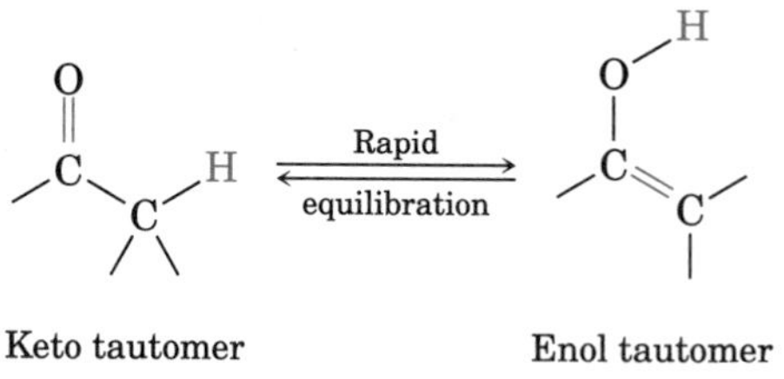

Rapid equilibration

Keto tautomer   Enol tautomer

Note that two isomers must be *easily* interconvertible to be considered tautomers. Thus, keto and enol isomers of carbonyl compounds are tautomers, but two alkene isomers such as 1-butene and 2-butene are not, because they don't interconvert rapidly.

At equilibrium, most carbonyl compounds exist almost entirely in the keto form, and it's difficult to isolate the pure enol form. Cyclohexanone, for example, contains only about 0.000 1% of its enol tautomer at room temperature, and acetone contains only about 0.000 001% enol. The percentage of enol tautomer is even less for carboxylic acids, esters, and amides. Even though enols are difficult to isolate and are present to only a small extent at equilibrium, they're nevertheless extremely important intermediates in much of the chemistry of carbonyl compounds.

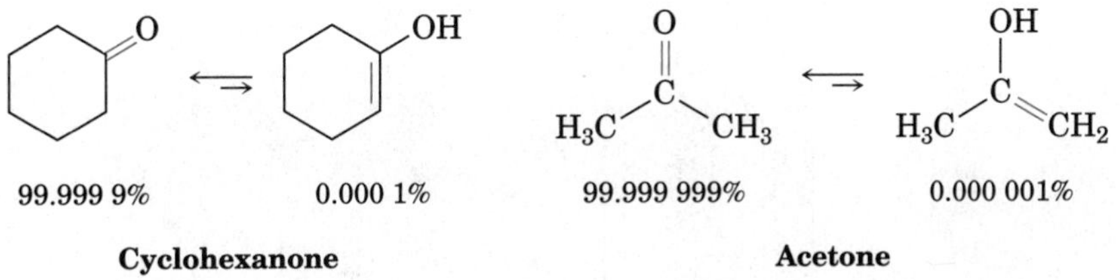

Cyclohexanone		Acetone	
99.999 9%	0.000 1%	99.999 999%	0.000 001%

Keto–enol tautomerism of carbonyl compounds is catalyzed by both acids and bases. Acid catalysis involves protonation of the carbonyl oxygen atom (a Lewis base) to give an intermediate cation that can then lose a proton from the α-carbon to yield the enol (Figure 11.1). This proton loss from the positively charged intermediate is analogous to what occurs during an E1 reaction when a carbocation loses a proton from the neighboring carbon to form an alkene (Section 7.10).

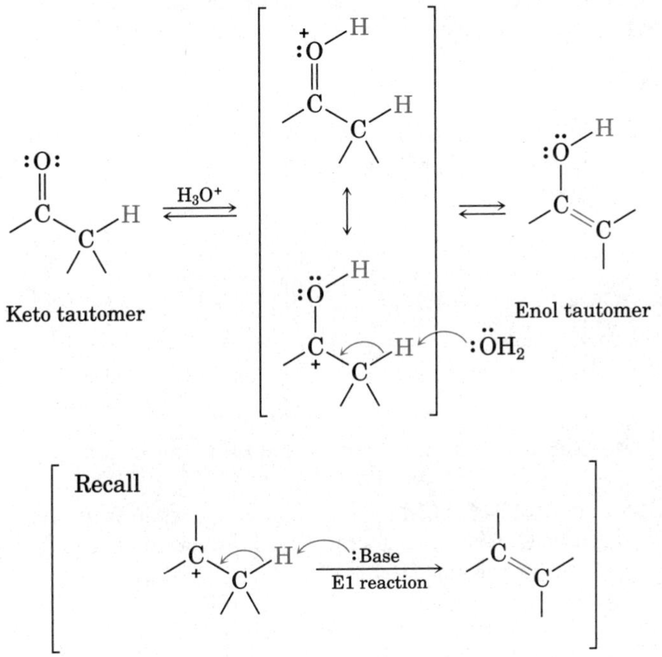

Keto tautomer $\xrightarrow{H_3O^+}$ Enol tautomer

Recall

:Base / E1 reaction

FIGURE 11.1 Mechanism of acid-catalyzed enol formation.

ENOLATE ION
Anion of an enol; a resonance-stabilized α-keto carbanion

Base-catalyzed enol formation occurs because the presence of a carbonyl group makes the hydrogens on the α-carbon slightly acidic. Thus, a carbonyl compound can act as a weak acid and lose one of its α-hydrogens to the base. The resultant resonance-stabilized anion, an **enolate ion**, is then reprotonated to yield a neutral compound. If protonation of the enolate ion takes place on the α-carbon, the keto tautomer is regenerated, and no net change occurs. If, however, protonation takes place on the oxygen atom, then an enol tautomer is formed (Figure 11.2).

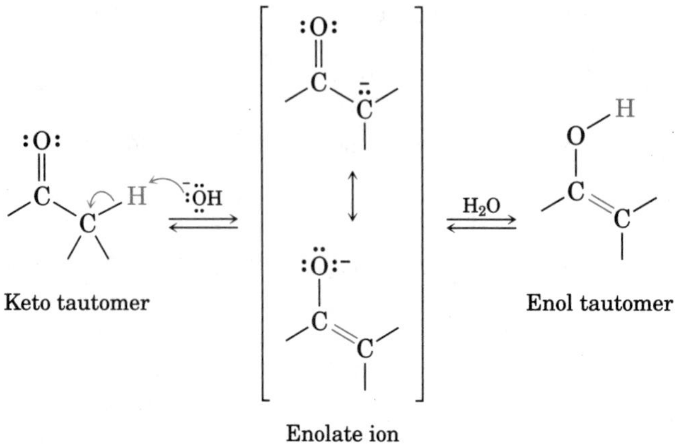

Keto tautomer

H_2O

Enolate ion

Enol tautomer

FIGURE 11.2 Mechanism of base-catalyzed enol formation. The intermediate enolate anion, a resonance hybrid of two forms, can be protonated either on carbon to regenerate the starting ketone or on oxygen to give an enol.

Note that only the protons on the α position of carbonyl compounds are acidic. The protons at beta (β), gamma (γ), delta (δ), and other positions aren't acidic, because the resulting anions can't be resonance-stabilized by the carbonyl group.

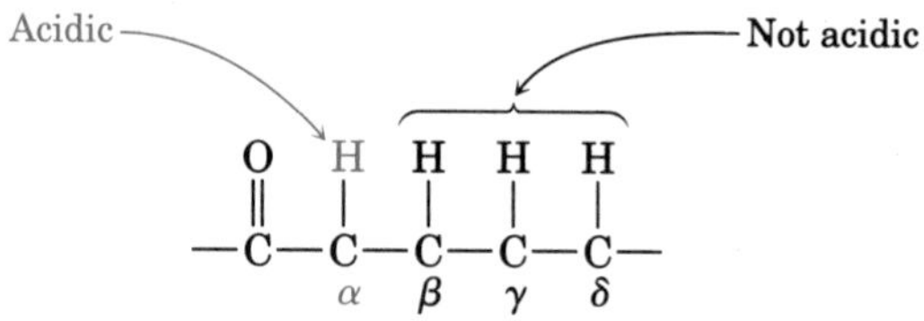

Acidic

Not acidic

α β γ δ

PRACTICE PROBLEM 11.1 Show the structure of the enol tautomer of propanal.

SOLUTION Enols are formed by removing a hydrogen from the carbon next to the carbonyl carbon, forming a double bond between the two carbons, and replacing the hydrogen on the carbonyl oxygen:

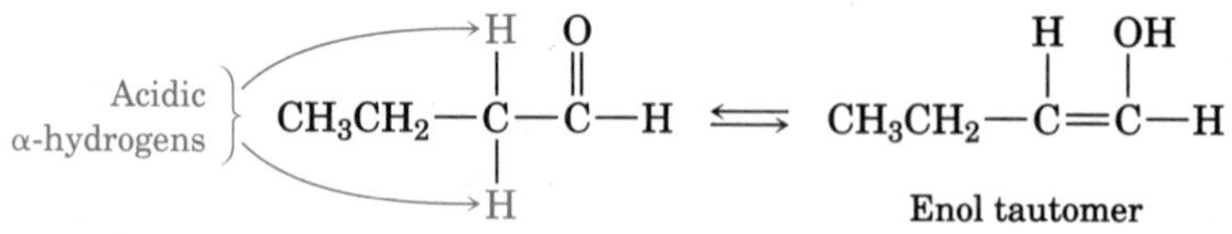

Acidic α-hydrogens

$$CH_3CH_2—CH_2—C(=O)—H \rightleftharpoons CH_3CH_2—CH=C(OH)—H$$

Enol tautomer

PROBLEM 11.1 Draw structures for the enol tautomers of these compounds:
(a) Cyclopentanone **(b)** Acetyl chloride **(c)** Ethyl acetate
(d) Acetic acid **(e)** Acetophenone (methyl phenyl ketone)

PROBLEM 11.2 How many acidic hydrogens does each of the molecules listed in Problem 11.1 have? Identify them.

PROBLEM 11.3 Account for the fact that 2-methylcyclohexanone can form two enol tautomers. Show the structures of both.

11.2 REACTIVITY OF ENOLS: THE MECHANISM OF ALPHA-SUBSTITUTION REACTIONS

What kind of chemistry might we expect of enols? Since their double bonds are electron-rich, enols behave as nucleophiles and react with electrophiles in much the same way as alkenes (Section 4.1). Because of electron donation from the oxygen lone-pair electrons, however, enols are even more reactive than alkenes.

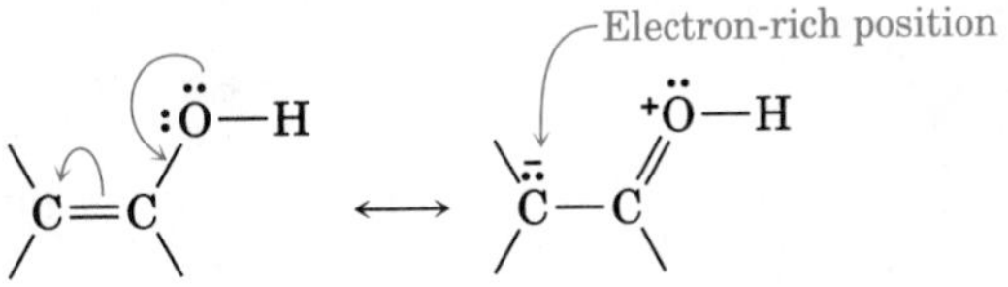

Electron-rich position

When an alkene reacts with an electrophile, the intermediate carbocation reacts with a nucleophile to give the addition product. When an enol reacts with an electrophile, however, the intermediate cation can lose the hydroxyl proton to regenerate a carbonyl compound. The net result of the reaction is α substitution (Figure 11.3).

Acid-catalyzed enol formation occurs by the usual mechanism.

Acid catalyst

An electron pair from the enol oxygen attacks an electrophile, forming a new bond and leaving a cation intermediate that is stabilized by resonance between two forms.

Base

Loss of a proton from oxygen yields the neutral alpha-substitution product as a new C=O bond is formed.

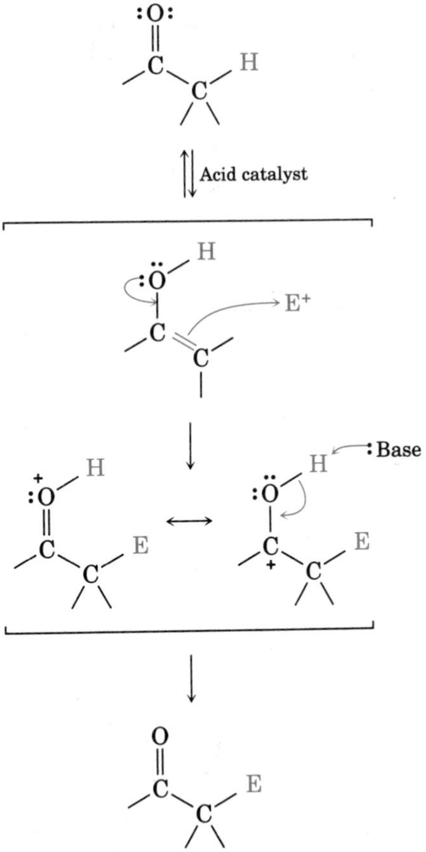

FIGURE 11.3 The general mechanism of a carbonyl α-substitution reaction.

11.3 ALPHA HALOGENATION OF KETONES AND ALDEHYDES

Ketones and aldehydes are halogenated at their alpha positions by reaction with chlorine, bromine, or iodine in acidic solution. Bromine is most often used, and acetic acid is often employed as solvent. The reaction is a typical α-substitution process that proceeds through an enol intermediate.

$$C_6H_5COCH_3 + Br_2 \xrightarrow[\text{solvent}]{CH_3COOH} C_6H_5COCH_2Br + HBr$$

Acetophenone **α-Bromoacetophenone (72%)**

α-Bromo ketones are useful substances because they undergo elimination of HBr on treatment with base to yield α,β-unsaturated ketones. For example, 2-bromo-2-methylcyclohexanone gives 2-methyl-2-cyclohexenone in 62% yield when heated in pyridine. The reaction takes place by the normal E2 elimination pathway (Section 7.9) and is an excellent way of introducing carbon–carbon double bonds into molecules.

$$\text{2-Bromo-2-methylcyclohexanone} \xrightarrow[\Delta]{\text{pyridine}} \text{2-Methyl-2-cyclohexenone} + HBr$$

2-Bromo-2-methylcyclohexanone **2-Methyl-2-cyclohexenone (62%)** (an α,β-unsaturated ketone)

PRACTICE PROBLEM 11.2 What product would you obtain from the reaction of cyclopentanone with Br_2 in acetic acid?

SOLUTION Locate the α-hydrogens in the starting ketone and replace one of them by bromine to carry out an α-substitution reaction:

$$\text{Cyclopentanone} + Br_2 \xrightarrow[\text{solvent}]{CH_3COOH} \text{2-Bromocyclopentanone} + HBr$$

Cyclopentanone **2-Bromocyclopentanone**

PROBLEM 11.4 Show the products of these reactions:

(a) 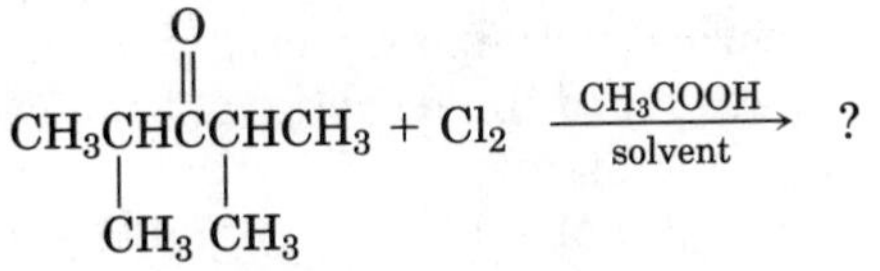 $CH_3CH(CH_3)C(=O)CH(CH_3)CH_3 + Cl_2 \xrightarrow[\text{solvent}]{CH_3COOH} ?$

(b) 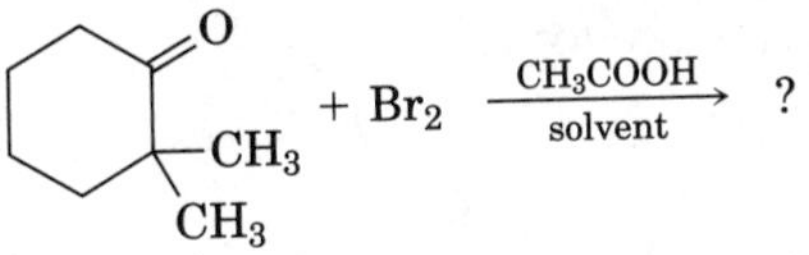 2,2-Dimethylcyclohexanone $+ Br_2 \xrightarrow[\text{solvent}]{CH_3COOH} ?$

PROBLEM 11.5 Show how you might prepare 1-penten-3-one from 3-pentanone:

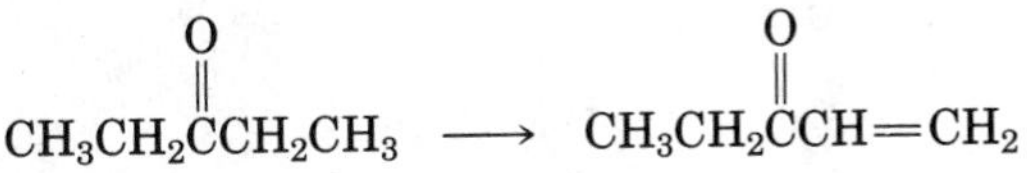

$$CH_3CH_2\overset{O}{\overset{\|}{C}}CH_2CH_3 \longrightarrow CH_3CH_2\overset{O}{\overset{\|}{C}}CH{=}CH_2$$

PROBLEM 11.6 When optically active (*R*)-3-phenyl-2-butanone is exposed to aqueous acid, a loss of optical activity occurs, and racemic 3-phenyl-2-butanone is produced. Explain how this loss of optical activity takes place. (Review Section 6.13.)

11.4 ACIDITY OF α-HYDROGEN ATOMS: ENOLATE ION FORMATION

During the discussion of base-catalyzed enol formation in Section 11.1, we said that carbonyl compounds act as weak acids. Strong bases can abstract acidic α protons from carbonyl compounds to form resonance-stabilized enolate ions:

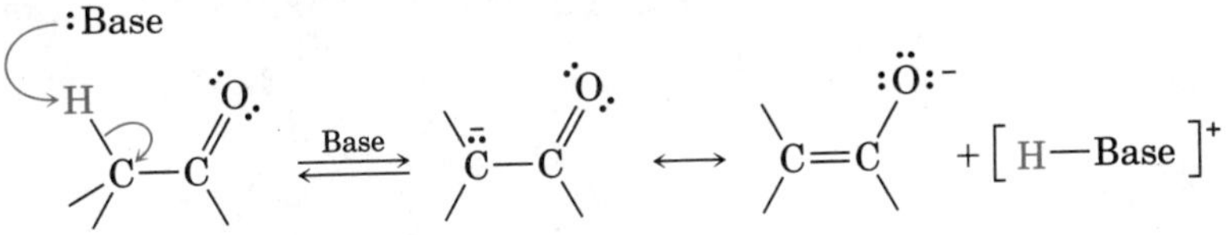

:Base + H–C–C=O $\underset{}{\overset{\text{Base}}{\rightleftharpoons}}$ $^{-}$:C–C=O $\longleftrightarrow$ C=C–O:$^{-}$ + [H—Base]$^{+}$

An enolate ion

Why are carbonyl compounds acidic? If we compare acetone, $pK_a = 19.3$, with ethane, $pK_a \approx 60$, we find that the presence of the carbonyl group increases the acidity of the neighboring C–H by a factor of 10^{40}.

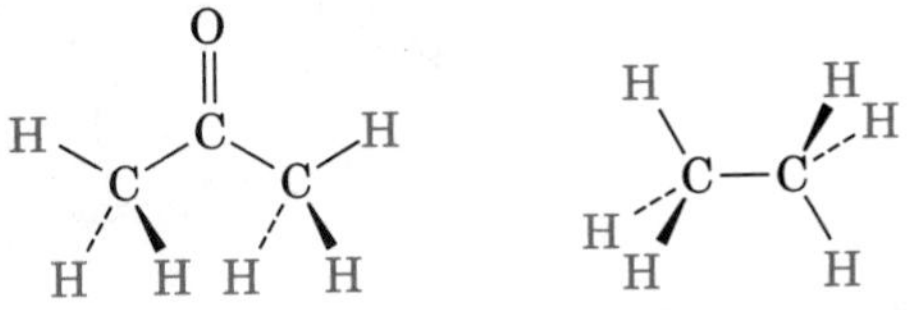

Acetone (pK_a = 19.3) **Ethane ($pK_a \approx$ 60)**

The best way to understand the acidity of carbonyl compounds is to look at an orbital picture of an enolate ion (Figure 11.4). Proton abstraction from a carbonyl compound occurs when the alpha C–H sigma bond is oriented parallel to the *p* orbitals of the carbonyl group. The α-carbon of the product enolate ion is sp^2-hybridized and has a *p* orbital that overlaps the carbonyl *p* orbitals. Thus, the negative charge is shared by the electronegative oxygen atom, and the enolate ion is stabilized by resonance between two forms.

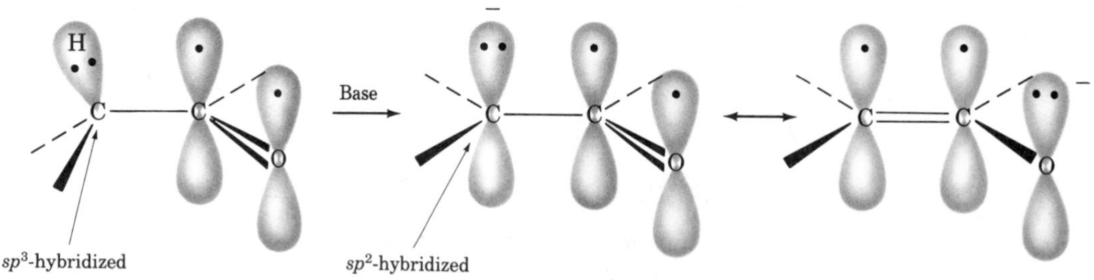

FIGURE 11.4 Mechanism of enolate ion formation by abstraction of an acidic alpha proton from a carbonyl compound.

Carbonyl compounds are more acidic than alkanes for the same reason that carboxylic acids are more acidic than alcohols (Section 10.3). In both cases, the anions are stabilized by resonance. Enolate ions differ from carboxylate ions, though, because their two resonance forms aren't equivalent. The resonance form with the negative charge on the enolate oxygen atom is lower in energy than the form with the charge on carbon. The principle behind resonance stabilization is the same in both cases, however.

CH_3CH_3 versus

Ethane
($pK_a \approx 60$)

Acetone
($pK_a = 19.3$)

Nonequivalent resonance forms

CH_3OH versus

Methanol
($pK_a = 15.5$)

Acetic acid
($pK_a = 4.7$)

Equivalent resonance forms

Since α-hydrogen atoms of carbonyl compounds are only weakly acidic compared with mineral acids or carboxylic acids, strong bases must be used to form enolate ions. If sodium ethoxide is used, ionization of acetone takes place only to the extent of about 0.1% since ethanol (pK_a = 16) is a stronger acid than acetone (pK_a = 19.3). If, however, a very powerful base such as sodium amide ($NaNH_2$, the sodium salt of ammonia) or sodium hydride (NaH, the sodium salt of H_2) is used, then a carbonyl compound is completely converted into its enolate ion.

NaH / THF ; $+ H_2$; $O^- Na^+$

Cyclohexanone **Cyclohexanone enolate (100%)**

All types of carbonyl compounds, including aldehydes, ketones, esters, acid chlorides, and amides, are much more acidic than alkanes. Table 11.1 lists the approximate pK_a values of various kinds of carbonyl compounds and shows how these values compare with other common acids.

TABLE 11.1 Acidity Constants for Some Organic Compounds

COMPOUND TYPE	COMPOUND	pK_a	
Carboxylic acid	CH_3COOH	5	Stronger acid
1,3-Diketone	$CH_2(COCH_3)_2$	9	↑
1,3-Keto ester	$CH_3COCH_2CO_2C_2H_5$	11	
1,3-Diester	$CH_2(CO_2C_2H_5)_2$	13	
Water	HOH	15.74	
Primary alcohol	CH_3CH_2OH	16	
Acid chloride	CH_3COCl	16	
Aldehyde	CH_3CHO	17	
Ketone	CH_3COCH_3	19	
Ester	$CH_3CO_2C_2H_5$	25	
Nitrile	CH_3CN	25	
Dialkylamide	$CH_3CON(CH_3)_2$	30	
Ammonia	NH_3	35	Weaker acid

When a C–H bond is flanked by two carbonyl groups, acidity is enhanced even more. Thus, Table 11.1 shows that 1,3-diketones (called β-diketones), 1,3-keto esters (β-keto esters), and 1,3-diesters are more acidic than water. The enolate ions derived from these β-dicarbonyl compounds are stabilized by delocalization of the negative charge onto both of the neighboring carbonyl oxygens. For example, there are three resonance forms for the enolate ion from 2,4-pentanedione:

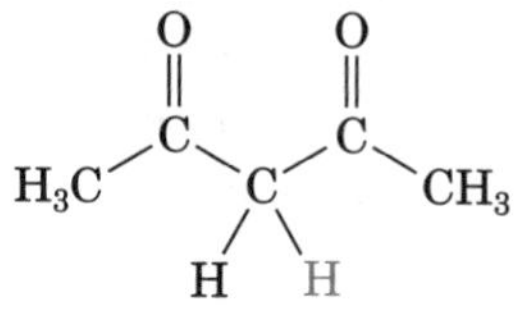

2,4-Pentanedione
(a β-diketone; $pK_a = 9$)

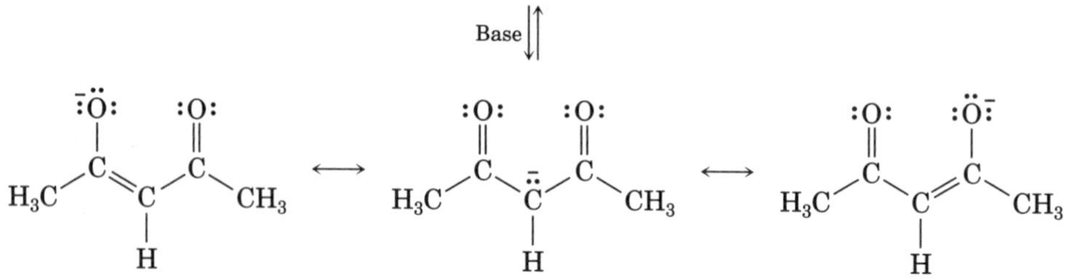

Base

PRACTICE PROBLEM 11.3 Draw structures of the two enolate ions you could obtain by deprotonation of 3-methylcyclohexanone.

SOLUTION Locate the acidic hydrogens and then remove them one at a time to generate the possible enolate ions. In this case, 3-methylcyclohexanone can be deprotonated either at C_2 or at C_6.

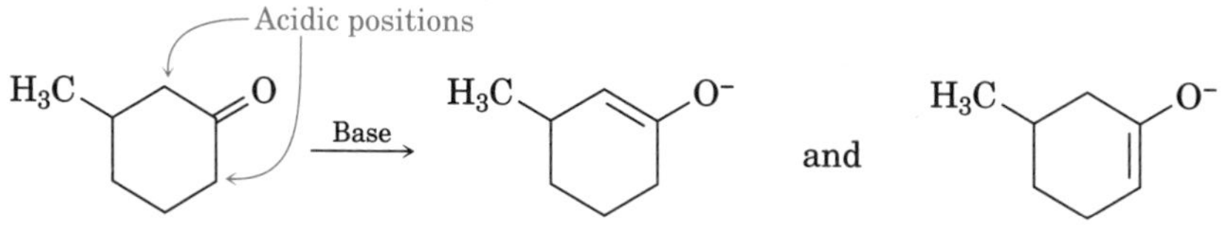

Acidic positions

Base

and

PROBLEM 11.7 Identify the acidic hydrogens in these molecules:
(a) CH_3CH_2CHO **(b)** $(CH_3)_3CCOCH_3$ **(c)** CH_3COOH
(d) $CH_3CH_2CH_2C{\equiv}N$ **(e)** 1,3-Cyclohexanedione

PROBLEM 11.8 Show the enolate ions you would obtain by deprotonation of these carbonyl compounds:
(a) Butanal **(b)** 2-Butanone **(c)** 2-Methylcyclohexanone

PROBLEM 11.9 Draw three resonance forms for the most stable enolate ion you would obtain by deprotonation of methyl 3-oxobutanoate.

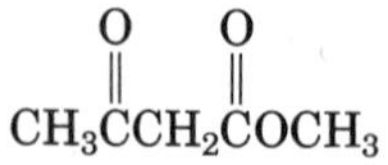

$CH_3\overset{O}{\overset{\|}{C}}CH_2\overset{O}{\overset{\|}{C}}OCH_3$ **Methyl 3-oxobutanoate**

11.5 REACTIVITY OF ENOLATE IONS

Enolate ions are more useful than enols for two reasons. First, pure enols can't normally be isolated. Instead, enols are usually generated only as transient intermediates in low concentration. By contrast, stable solutions of pure enolate ions are easily prepared from many carbonyl compounds by treatment with a strong base. Second, enolate ions are more reactive than enols. Whereas enols are neutral, enolate ions have a negative charge that makes them much better nucleophiles. Thus, the alpha position of enolate ions is highly reactive toward electrophiles.

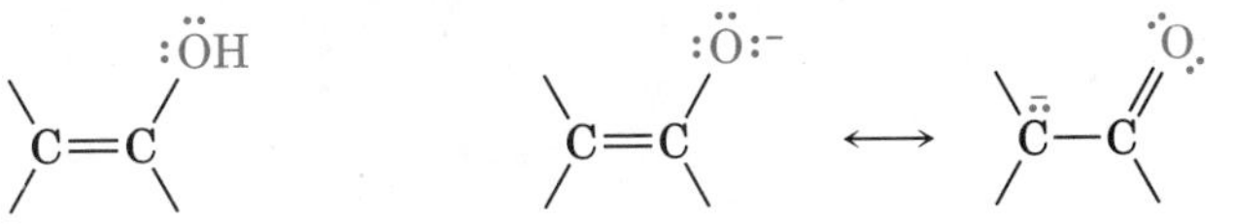

Enol: neutral, moderately reactive, very difficult to isolate

Enolate: negatively charged, very reactive, easily prepared

Since enolate ions are resonance hybrids of two nonequivalent forms, they can be thought of either as α-keto carbanions ($^{-}$C–C=O) or as vinylic alkoxides (C=C–O^{-}). Thus, enolate ions can react with electrophiles either on oxygen or on carbon. Reaction on oxygen yields an enol derivative, whereas reaction on carbon yields an α-substituted carbonyl compound (Figure 11.5). Both kinds of reactivity are known, but reaction on carbon is the more commonly observed pathway.

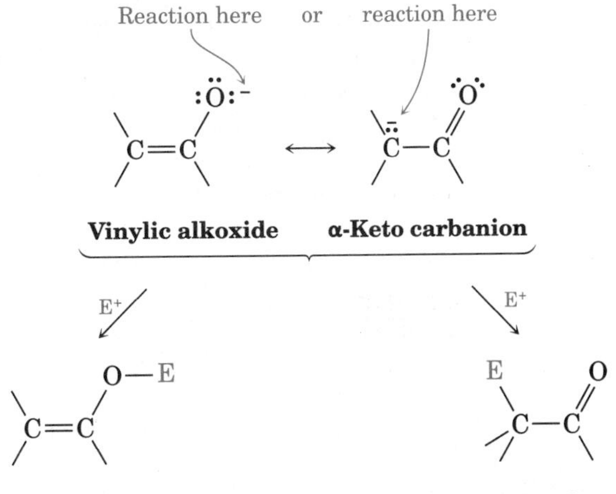

Reaction here or reaction here

Vinylic alkoxide **α-Keto carbanion**

E^+ E^+

An enol derivative (E^+ = an electrophile) An α-substituted carbonyl compound

FIGURE 11.5 Two modes of enolate ion reactivity. Reaction on carbon to yield an α-substituted carbonyl product is the more commonly followed pathway.

11.6 ALKYLATION OF ENOLATE IONS

ALKYLATION
Alpha substitution of a carbonyl compound by reaction of an enolate ion with an alkyl halide

One of the most important reactions of enolate ions is their **alkylation** by treatment with an alkyl halide. The alkylation reaction is extremely useful because it forms a new carbon–carbon bond, thereby joining two smaller pieces into one larger molecule. Alkylation occurs when the nucleophilic enolate ion reacts with the electrophilic alkyl halide in an S_N2 reaction, displacing the halide ion by back-side attack:

Enolate ion + Alkyl halide $\xrightarrow[\text{reaction}]{S_N2}$ product $+ :X^-$

Like all S_N2 reactions (Section 7.7), alkylations with R–X are successful only when the alkyl group *R* is primary or methyl. Otherwise, a competing E2 elimination occurs if a secondary or tertiary halide is used. The leaving group X can be chloride, bromide, or iodide.

MALONIC ESTER SYNTHESIS
Method for forming alpha-substituted acetic acids by reaction of diethyl malonate with an alkyl halide, followed by decarboxylation

The **malonic ester synthesis**, one of the best-known carbonyl alkylation reactions, is an excellent method for preparing substituted acetic acids from alkyl halides:

$$\underset{\text{Alkyl halide}}{R{-}X} \xrightarrow[\text{ester synthesis}]{\text{Via malonic}} \underset{\alpha\text{-Substituted acetic acid}}{R{-}CH_2\overset{\overset{O}{\|}}{C}OH}$$

Diethyl propanedioate, commonly called diethyl malonate or malonic ester, is relatively acidic ($pK_a = 13$) because its α-hydrogen atoms are flanked by two carbonyl groups. Thus, malonic ester is easily converted into its enolate ion by reaction with sodium ethoxide in ethanol. The enolate ion, in turn, is readily alkylated by treatment with an alkyl halide, yielding an α-substituted malonic ester.

$$\underset{\text{Malonic ester}}{H{-}C(H)(CO_2CH_2CH_3){-}CO_2CH_2CH_3} \xrightarrow[CH_3CH_2OH]{Na^+\ ^-OCH_2CH_3} \underset{\text{Sodio malonic ester}}{Na^+\ ^-{:}C(H)(CO_2CH_2CH_3){-}CO_2CH_2CH_3} \xrightarrow{RX} \underset{\text{An alkylated malonic ester}}{R{-}C(H)(CO_2CH_2CH_3){-}CO_2CH_2CH_3} + NaX$$

The product of a malonic ester alkylation has one acidic α-hydrogen left, and the alkylation process can therefore be repeated a second time to yield a dialkylated malonic ester:

$$\text{R—}\underset{\text{H}}{\overset{\text{CO}_2\text{CH}_2\text{CH}_3}{\text{C}}}\text{—CO}_2\text{CH}_2\text{CH}_3 \xrightarrow[\text{2. R}'\text{X}]{\text{1. Na}^+\ {}^-\text{OCH}_2\text{CH}_3} \text{R—}\underset{\text{R}'}{\overset{\text{CO}_2\text{CH}_2\text{CH}_3}{\text{C}}}\text{—CO}_2\text{CH}_2\text{CH}_3 + \text{NaX}$$

A dialkylated malonic ester

DECARBOXYLATION Loss of carbon dioxide from a molecule

Once formed, alkylated malonic esters can be hydrolyzed and decarboxylated (**decarboxylation** means the loss of carbon dioxide, CO_2) when heated with aqueous HCl. The product is a substituted carboxylic acid. Note that decarboxylation is not a general reaction of carboxylic acids but is a unique feature of compounds like malonic acids that have a second carbonyl group two atoms away.

$$\text{R—}\underset{\text{R}'}{\overset{\text{CO}_2\text{CH}_2\text{CH}_3}{\text{C}}}\text{—CO}_2\text{CH}_2\text{CH}_3 \xrightarrow[\text{Heat}]{\text{H}_3\text{O}^+} \text{R—}\underset{\text{R}'}{\overset{\text{H}}{\text{C}}}\text{—}\overset{\text{O}}{\overset{\|}{\text{C}}}\text{OH} + \text{CO}_2 + 2\ \text{CH}_3\text{CH}_2\text{OH}$$

The overall effect of the malonic ester synthesis is to convert an alkyl halide into a carboxylic acid and to lengthen the carbon chain by two atoms ($RX \rightarrow RCH_2COOH$). For example,

$$\underset{\textbf{1-Bromobutane}}{\text{CH}_3\text{CH}_2\text{CH}_2\text{CH}_2\text{Br}} + \text{Na}^+\ {}^-\text{:CH(COOCH}_3)_2 \longrightarrow \text{CH}_3(\text{CH}_2)_2\text{CH}_2\text{CH(COOCH}_3)_2$$

$$\xrightarrow{\text{H}_3\text{O}^+,\ \text{heat}} \underset{\textbf{Hexanoic acid (75\%)}}{\text{CH}_3(\text{CH}_2)_2\text{CH}_2\text{CH}_2\text{COOH}} + \text{CO}_2 + 2\ \text{CH}_3\text{OH}$$

PRACTICE PROBLEM 11.4 How would you prepare heptanoic acid by a malonic ester synthesis?

SOLUTION The malonic ester synthesis converts an alkyl halide into a carboxylic acid with two more carbons in its chain. Thus, a seven-carbon acid chain must be derived from a five-carbon alkyl halide such as 1-bromopentane.

$$\text{CH}_3\text{CH}_2\text{CH}_2\text{CH}_2\text{CH}_2\text{Br} + \text{CH}_2(\text{COOCH}_2\text{CH}_3)_2 \xrightarrow[\text{2. H}_3\text{O}^+,\ \text{heat}]{\text{1. Na}^+\ {}^-\text{OCH}_2\text{CH}_3} \text{CH}_3\text{CH}_2\text{CH}_2\text{CH}_2\text{CH}_2\text{CH}_2\text{COOH}$$

PROBLEM 11.10 What alkyl halide would you use to prepare these compounds by a malonic ester synthesis?
(a) Butanoic acid **(b)** 3-Phenylpropanoic acid **(c)** 5-Methylhexanoic acid

PROBLEM 11.11 Show how you could use a malonic ester synthesis to prepare these compounds:
(a) 4-Methylpentanoic acid **(b)** 2-Methylpentanoic acid

11.7 CARBONYL CONDENSATION REACTIONS

We've seen now that carbonyl compounds can behave either as electrophiles or as nucleophiles. In nucleophilic addition reactions and nucleophilic acyl substitution reactions, the carbonyl group behaves as an electrophile by accepting electrons from an attacking nucleophile. In α-substitution reactions, however, the carbonyl compound behaves as a nucleophile when it is converted into an enol or enolate ion.

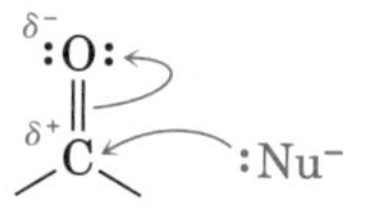

δ^- :O: δ^+ C :Nu$^-$ — :Ö:$^-$ C=C E$^+$

Electrophilic carbonyl group is attacked by nucleophiles

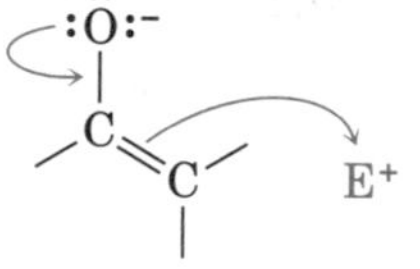

Nucleophilic enolate ion attacks electrophiles

Carbonyl condensation reactions, the fourth and last general category of carbonyl-group reactions we'll study, involve both kinds of reactivity. These reactions take place between two carbonyl components and involve a combination of nucleophilic addition and α-substitution steps. One component acts as an electron donor and undergoes an α-substitution process, while the other component acts as an electron acceptor and undergoes a nucleophilic addition process. There are numerous variations of carbonyl condensation reactions, depending on the two carbonyl components, but the general mechanism shown in Figure 11.6 remains the same.

11.8 CONDENSATION OF ALDEHYDES AND KETONES: THE ALDOL REACTION

When acetaldehyde is treated in an alcoholic solvent with a basic catalyst such as sodium hydroxide or sodium ethoxide, a rapid and reversible condensation reaction occurs. The product is a β-hydroxy aldehyde product known commonly as *aldol* (*ald*ehyde + alcoh*ol*).

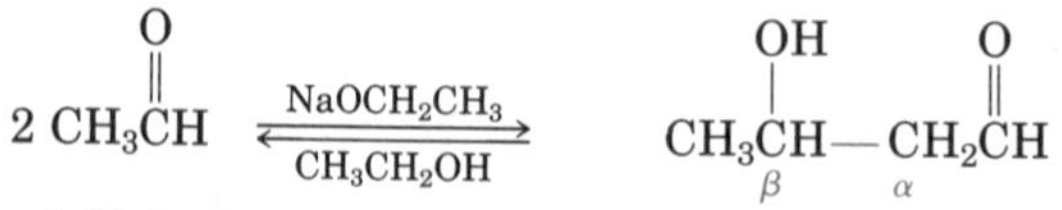

$$2\ CH_3\overset{O}{\overset{\|}{C}}H \underset{CH_3CH_2OH}{\overset{NaOCH_2CH_3}{\rightleftharpoons}} CH_3\overset{OH}{\overset{|}{C}}H_{\beta}-CH_2{}_{\alpha}\overset{O}{\overset{\|}{C}}H$$

Acetaldehyde **Aldol (a β-hydroxy aldehyde)**

ALDOL REACTION Carbonyl condensation between two ketones or aldehydes, leading to a β-hydroxy ketone or aldehyde product

Called the **aldol reaction**, base-catalyzed dimerization is a general reaction of all ketones and aldehydes with α-hydrogen atoms. If the ketone or aldehyde does not have an α-hydrogen atom, aldol condensation can't occur. The exact position of the aldol equilibrium depends both on reaction conditions and on substrate structure. As the following examples indicate, the aldol equilibrium generally favors the condensation product for mono-

One carbonyl component with an α-hydrogen atom is converted by base into its enolate ion.

This enolate ion acts as a nucleophilic donor and adds to the electrophilic carbonyl group of the acceptor component.

Nucleophilic donor

Electrophilic acceptor

Protonation of the tetrahedral alkoxide ion intermediate gives the neutral condensation product.

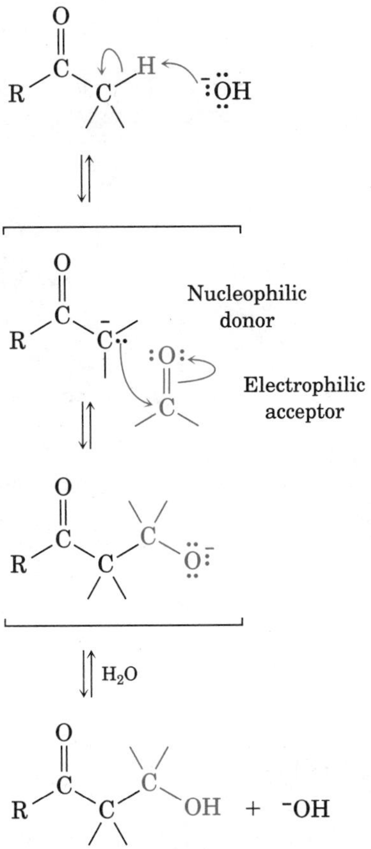

H_2O

FIGURE 11.6 The general mechanism of a carbonyl condensation reaction. One component (the donor) acts as a nucleophile, while the other component (the acceptor) acts as an electrophile.

substituted acetaldehydes (RCH_2CHO) but favors the starting material for disubstituted acetaldehydes (R_2CHCHO) and for ketones.

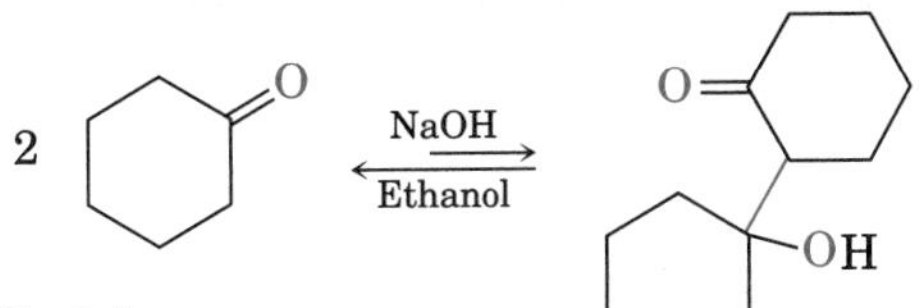

2 Cyclohexanone $\xrightarrow[\text{Ethanol}]{\text{NaOH}}$ 22%

2 Phenylacetaldehyde $\xrightarrow[\text{Ethanol}]{\text{NaOH}}$ 90%

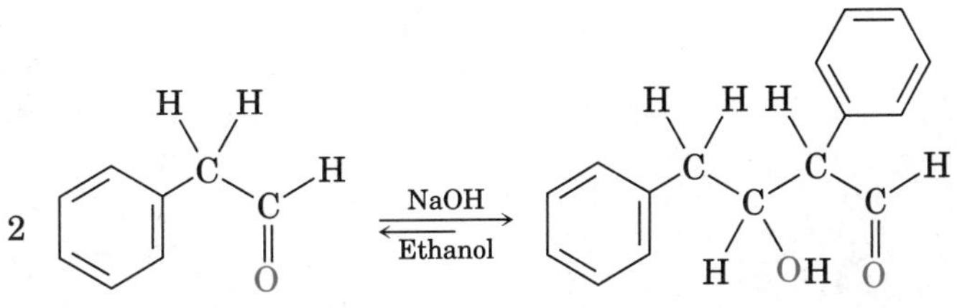

PRACTICE PROBLEM 11.5 What is the structure of the aldol product derived from propanal?

SOLUTION An aldol reaction combines two molecules of starting material, forming a bond between the α-carbon of one partner and the carbonyl carbon of the second partner:

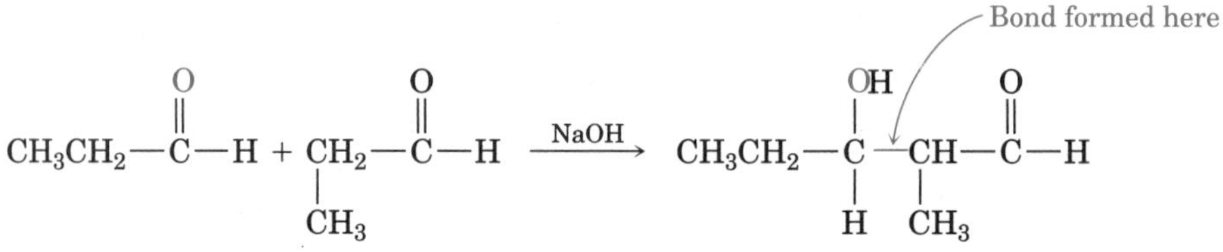

$$CH_3CH_2-\overset{O}{\overset{\|}{C}}-H + \underset{CH_3}{\underset{|}{CH_2}}-\overset{O}{\overset{\|}{C}}-H \xrightarrow{NaOH} CH_3CH_2-\underset{H}{\underset{|}{\overset{OH}{\overset{|}{C}}}}-\underset{CH_3}{\underset{|}{CH}}-\overset{O}{\overset{\|}{C}}-H$$

Bond formed here

PROBLEM 11.12 Which of the following compounds can undergo the aldol reaction and which cannot? Explain.
(a) Cyclohexanone (b) Benzaldehyde
(c) 2,2,6,6-Tetramethylcyclohexanone (d) Formaldehyde

PROBLEM 11.13 Show the product of the aldol reaction of these compounds:
(a) Butanal (b) Cyclopentanone (c) Acetophenone

11.9 DEHYDRATION OF ALDOL PRODUCTS: SYNTHESIS OF ENONES

ENONE
Unsaturated ketone

The β-hydroxy ketones and β-hydroxy aldehydes formed in aldol reactions are easily dehydrated to yield conjugated **enones** (*ene* + *one*). In fact, it's this loss of water that gives the aldol *condensation* its name, since water condenses out of the reaction.

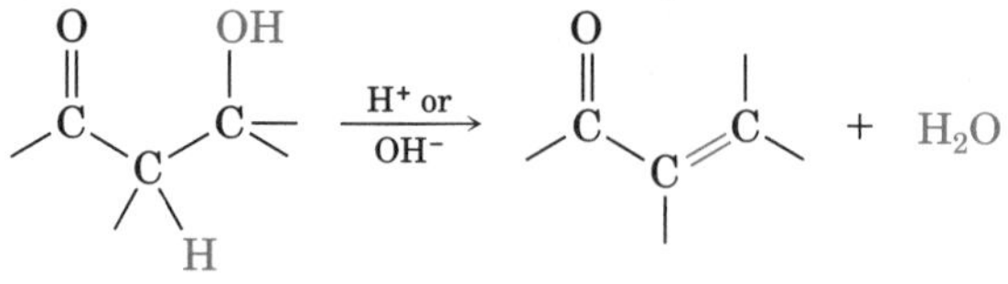

$$\overset{O}{\overset{\|}{C}}-\underset{H}{\underset{|}{C}}-\overset{OH}{\overset{|}{C}} \xrightarrow[OH^-]{H^+ \text{ or}} \overset{O}{\overset{\|}{C}}-\underset{|}{C}=\overset{|}{C} + H_2O$$

A β-hydroxy ketone or aldehyde — A conjugated enone

Although most alcohols are resistant to dehydration by dilute acid or base (Section 8.8), hydroxyl groups that are beta to a carbonyl group are special. Under basic conditions, an acidic α-hydrogen is abstracted, and the resultant enolate ion expels hydroxide ion. Under acidic conditions, the hydroxyl group is protonated and then expelled by the neighboring enol.

Base-catalyzed

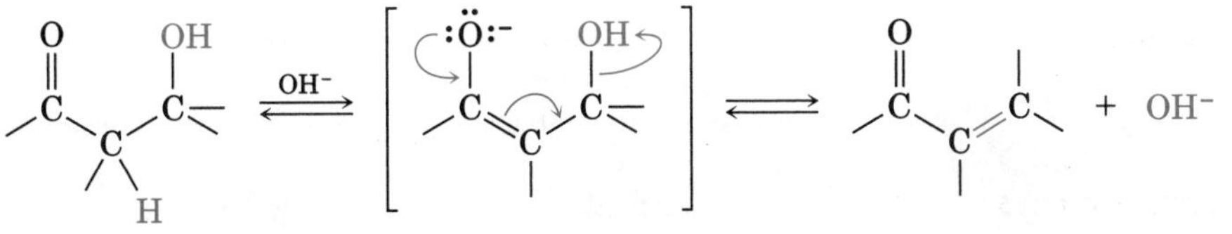

$$\text{O=C(}\!-\!)\text{–C(H)–C(OH)} \xrightleftharpoons{OH^-} \left[\text{:\ddot{O}:}^- \text{–C=C–C–OH} \right] \rightleftharpoons \text{O=C–C=C} + OH^-$$

Enolate ion

Acid-catalyzed

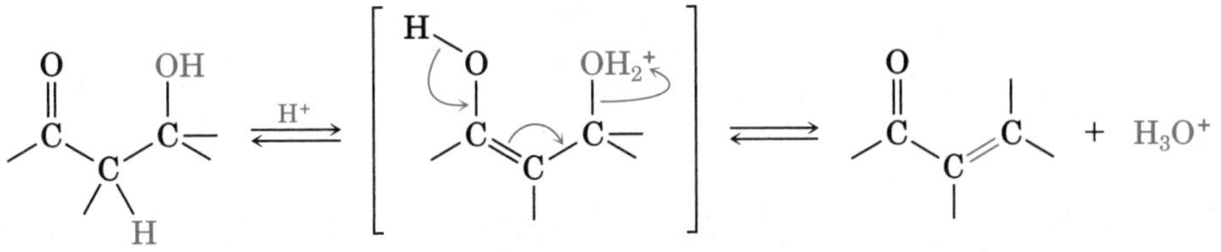

$$\text{O=C–C(H)–C(OH)} \xrightleftharpoons{H^+} \left[\text{H–O–C=C–C–}OH_2^+ \right] \rightleftharpoons \text{O=C–C=C} + H_3O^+$$

Enol

The conditions needed to cause aldol dehydration are often only a bit more vigorous (slightly higher temperature, for example) than the conditions needed for the aldol dimerization itself. As a result, conjugated enones are often obtained directly from aldol reactions; the intermediate β-hydroxy carbonyl compounds usually aren't even isolated.

Conjugated enones are more stable than nonconjugated enones for the same reasons that conjugated dienes are more stable than nonconjugated dienes (Section 4.10). Interaction between the π electrons of the carbon–carbon double bond and the π electrons of the carbonyl group allows delocalization of the π electrons over all four atomic centers.

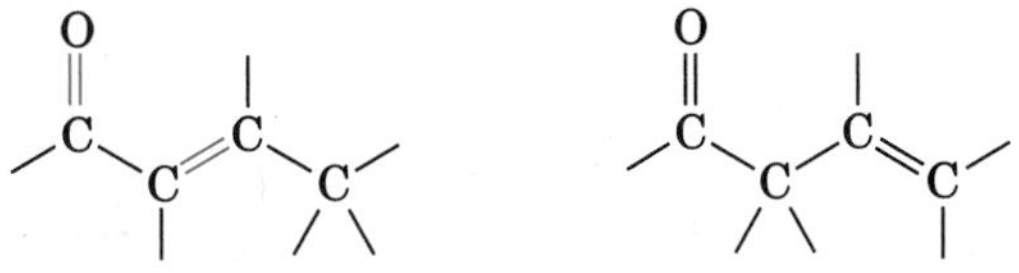

Conjugated enone (more stable)

Nonconjugated enone (less stable)

PRACTICE PROBLEM 11.6 What is the structure of the enone obtained from aldol condensation of acetaldehyde?

SOLUTION In the aldol reaction, H_2O is eliminated by removing two hydrogens from the acidic α position of one partner and the oxygen from the second partner:

$$H_3C\text{—}\overset{H}{\overset{|}{C}}\text{=O} + H_2\overset{H}{\overset{|}{C}}\text{—CHO} \xrightarrow{NaOH} H_3C\text{—}\overset{H}{\overset{|}{C}}\text{=}\overset{H}{\overset{|}{C}}\text{—CHO} + H_2O$$

2-Butenal

PROBLEM 11.14 Write the structures of the enone products you would obtain from aldol condensation of these compounds:
(a) Acetone **(b)** Cyclopentanone **(c)** Acetophenone **(d)** Propanal

PROBLEM 11.15 Aldol condensation of 2-butanone leads to a mixture of two enones (ignoring double-bond stereochemistry). Draw the two enones.

11.10 CONDENSATION OF ESTERS: THE CLAISEN CONDENSATION REACTION

CLAISEN CONDENSATION REACTION
A carbonyl condensation reaction between two esters, leading to formation of a β-keto ester product

Esters, like aldehydes and ketones, are weakly acidic. When an ester with an α-hydrogen is treated with 1 equiv of a base such as sodium ethoxide, a reversible condensation reaction yields a β-keto ester product. For example, ethyl acetate yields ethyl acetoacetate on base treatment. This reaction between two ester components is known as the **Claisen condensation reaction**.

$$2\ CH_3\overset{O}{\overset{\|}{C}}OCH_2CH_3 \xrightarrow[2.\ H_3O^+]{1.\ Na^+\ ^-OCH_2CH_3,\ ethanol} CH_3\overset{O}{\overset{\|}{\underset{\beta}{C}}}-\underset{\alpha}{CH_2}\overset{O}{\overset{\|}{C}}OCH_2CH_3 + CH_3CH_2OH$$

Ethyl acetate

Ethyl acetoacetate
(a β-keto ester; 75%)

The mechanism of the Claisen reaction is similar to that of the aldol reaction, involving the nucleophilic addition of an ester enolate ion donor to the carbonyl group of a second ester molecule (Figure 11.7). From the point of view of the donor component, the Claisen condensation is simply an α-substitution reaction. From the point of view of the acceptor component, the Claisen condensation is a nucleophilic acyl substitution reaction.

The only difference between an aldol condensation and a Claisen condensation involves the fate of the initially formed tetrahedral intermediate. The tetrahedral intermediate in the aldol reaction is protonated to give a stable alcohol product, exactly the behavior previously seen for ketones (Section 9.7). The tetrahedral intermediate in the Claisen reaction, however, expels an alkoxide leaving group to yield an acyl substitution product, exactly the behavior previously seen for esters (Section 10.5).

Methoxide ion base abstracts an acidic α-hydrogen atom from an ester molecule, yielding an ester enolate ion.

$$CH_3\overset{O}{\overset{\|}{C}}OCH_3 \underset{}{\overset{^-OCH_3}{\rightleftharpoons}} \left[^-:CH_2\overset{O}{\overset{\|}{C}}OCH_3 \right] + HOCH_3$$

The enolate ion adds to a second ester molecule by nucleophilic addition, yielding a tetrahedral intermediate.

$$\rightleftharpoons \left[CH_3\overset{:\ddot{O}:^-}{\overset{|}{C}}(OCH_3)-CH_2\overset{O}{\overset{\|}{C}}OCH_3 \right]$$

Loss of methoxide ion from the tetrahedral intermediate yields methyl acetoacetate and regenerates the basic catalyst.

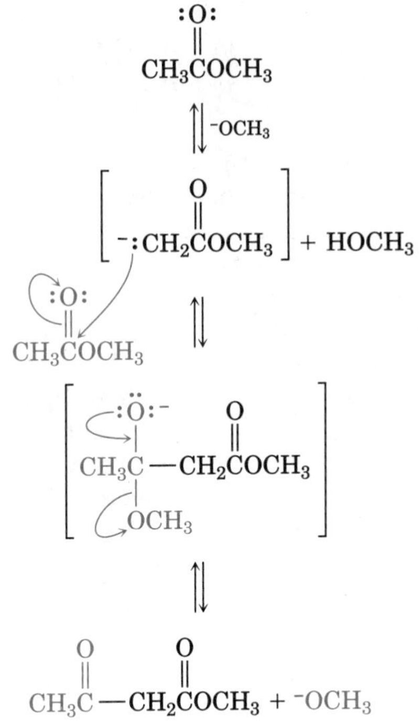

$$\rightleftharpoons CH_3\overset{O}{\overset{\|}{C}}-CH_2\overset{O}{\overset{\|}{C}}OCH_3 + {}^-OCH_3$$

FIGURE 11.7 The mechanism of the Claisen condensation reaction.

PRACTICE PROBLEM 11.7 What product would you obtain from Claisen condensation of methyl propanoate?

SOLUTION The Claisen condensation of an ester results in the loss of one molecule of alcohol and the formation of a product in which an acyl group of one reactant bonds to the α-carbon of the second reactant:

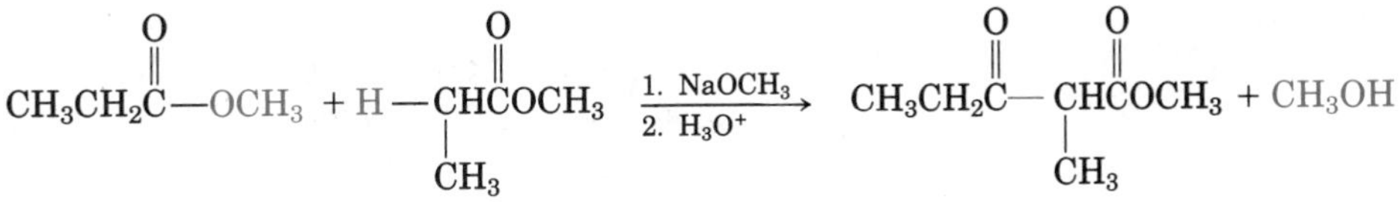

$$CH_3CH_2\overset{O}{\overset{\|}{C}}-OCH_3 + H-\underset{CH_3}{\underset{|}{C}}H\overset{O}{\overset{\|}{C}}OCH_3 \xrightarrow[2.\ H_3O^+]{1.\ NaOCH_3} CH_3CH_2\overset{O}{\overset{\|}{C}}-\underset{CH_3}{\underset{|}{C}}H\overset{O}{\overset{\|}{C}}OCH_3 + CH_3OH$$

Methyl propanoate (2 molecules) → **Methyl 2-methyl-3-oxopentanoate**

PROBLEM 11.16 Which of the following esters can't undergo Claisen condensation? Explain.
(a) Methyl formate **(b)** Methyl propenoate **(c)** Methyl propanoate

PROBLEM 11.17 Show the products you would obtain by Claisen condensation of these esters:
(a) $(CH_3)_2CHCH_2COOCH_3$ **(b)** Methyl phenylacetate
(c) Methyl cyclohexylacetate

INTERLUDE

BIOLOGICAL CARBONYL CONDENSATION REACTIONS

Carbonyl condensation reactions are used in nature for the biological synthesis of innumerable different molecules. Fats, amino acids, steroid hormones, and many other kinds of compounds are biosynthesized by plants and animals using carbonyl condensation reactions as the key step.

Nature uses the two-carbon acetate fragment of acetyl coenzyme A as the major building block for synthesis. We saw in the Chapter 10 Interlude that acetyl CoA can serve as an electrophilic acceptor for attack of nucleophiles at the acyl carbon. In addition, it can serve as a nucleophilic donor by loss of its acidic α proton to generate an enolate ion. The enolate ion of acetyl CoA can then add to another carbonyl group in a condensation reaction. For example, citric acid is biosynthesized by addition of acetyl CoA to the ketone carbonyl group of oxaloacetic acid (2-oxobutanedioic acid) in a type of aldol reaction.

$$CH_3\overset{O}{\overset{\|}{C}}SCoA \longrightarrow \text{“}:\bar{C}H_2\overset{O}{\overset{\|}{C}}SCoA\text{”}$$

Acetyl CoA, a thiol ester

$$O{=}C(COOH)(CH_2COOH) + :\bar{C}H_2\overset{O}{\overset{\|}{C}}SCoA \rightleftharpoons HO{-}C(COOH)(CH_2COOH){-}CH_2COOH$$

Oxaloacetic acid **Citric acid**

Acetyl CoA is also involved in the biosynthesis of steroids, fats, and other lipids. The key step in these biosyntheses is a Claisen-like condensation of acetyl CoA to yield acetoacetyl CoA.

$$H_3C\overset{:O:}{\overset{\|}{C}}SCoA + H_2\ddot{C}^{-}\overset{O}{\overset{\|}{C}}SCoA \longrightarrow \left[H_3C{-}C(:\ddot{O}:^{-})(SCoA){-}CH_2\overset{O}{\overset{\|}{C}}SCoA \right] \longrightarrow CH_3\overset{O}{\overset{\|}{C}}CH_2\overset{O}{\overset{\|}{C}}SCoA + {}^{-}SCoA$$

Acetyl CoA **Acetyl CoA** **Acetoacetyl CoA**

SUMMARY AND KEY WORDS

Alpha substitutions and **carbonyl condensations** are two of the four fundamental reaction types in carbonyl-group chemistry. Alpha-substitution reactions, which take place via **enol** or **enolate ion** intermediates, result in the replacement of an α-hydrogen atom by some other substituent.

Carbonyl compounds are in equilibrium with their enols, a process known as **tautomerism**. Enol tautomers are normally present to only a small extent and usually can't be isolated in pure form. Nevertheless, enols react rapidly with a variety of electrophiles. For examples, ketones and aldehydes are rapidly halogenated at the **alpha position** by reaction with chlorine, bromine, or iodine in acetic acid solution.

Alpha-hydrogen atoms of carbonyl compounds are acidic and can be abstracted by bases to yield enolate ions. Ketones, aldehydes, esters, amides, and nitriles can all be deprotonated. The most important reaction of enolate ions is their S_N2 **alkylation** by alkyl halides. The nucleophilic enolate ion attacks an alkyl halide from the back side, displacing the leaving halide group and yielding an α-alkylated carbonyl product. The **malonic ester synthesis**, which involves alkylation of diethyl malonate with alkyl halides, provides a method for preparing monoalkylated or dialkylated acetic acids.

A carbonyl condensation reaction takes place between two carbonyl components and involves a combination of nucleophilic addition and α-substitution steps. One carbonyl component (the donor) is converted into a nucleophilic enolate ion, which then adds to the electrophilic carbonyl group of the second component (the acceptor).

The **aldol reaction** is a carbonyl condensation that occurs between two ketone or aldehyde components. Aldol reactions are reversible, leading first to β-hydroxy ketone products and then to α,β-unsaturated ketones. The **Claisen condensation reaction** is a carbonyl condensation that occurs between two ester components and that leads to a β-keto ester product.

SUMMARY OF REACTIONS

1. Halogenation of ketones and aldehydes (Section 11.3)

$$\text{C(=O)–C–H} + X_2 \xrightarrow{CH_3COOH} \text{C(=O)–C–X} \quad \text{where X = Cl, Br, or I}$$

2. Malonic ester synthesis (Section 11.6)

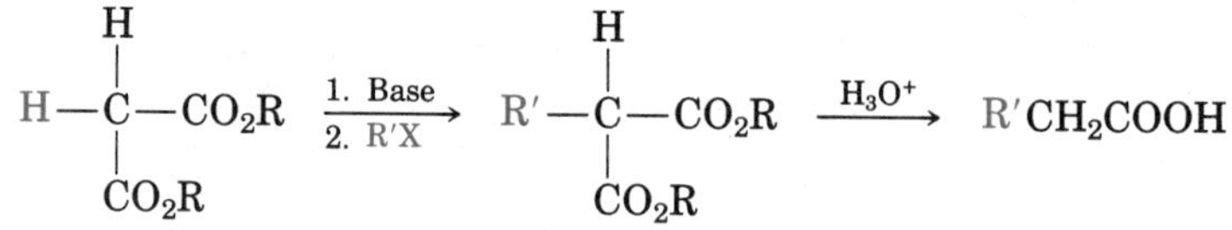

$$\mathrm{H{-}\overset{H}{\underset{CO_2R}{C}}{-}CO_2R \xrightarrow[\text{2. R'X}]{\text{1. Base}} R'{-}\overset{H}{\underset{CO_2R}{C}}{-}CO_2R \xrightarrow{H_3O^+} R'CH_2COOH}$$

3. Aldol reaction of ketones and aldehydes (Section 11.8)

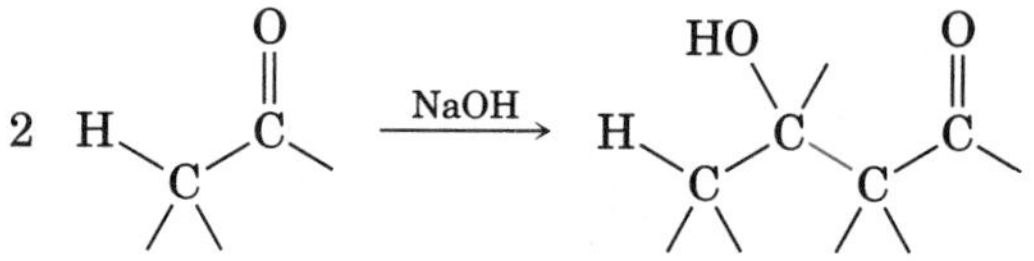

$$2\ \mathrm{H{-}\overset{|}{\underset{/\ \backslash}{C}}{-}\overset{O}{\overset{\|}{C}}{-}} \xrightarrow{\text{NaOH}} \mathrm{H{-}\underset{/\ \backslash}{C}{-}\overset{HO}{C}{-}\underset{/\ \backslash}{C}{-}\overset{O}{\overset{\|}{C}}{-}}$$

4. Claisen condensation reaction of esters (Section 11.10)

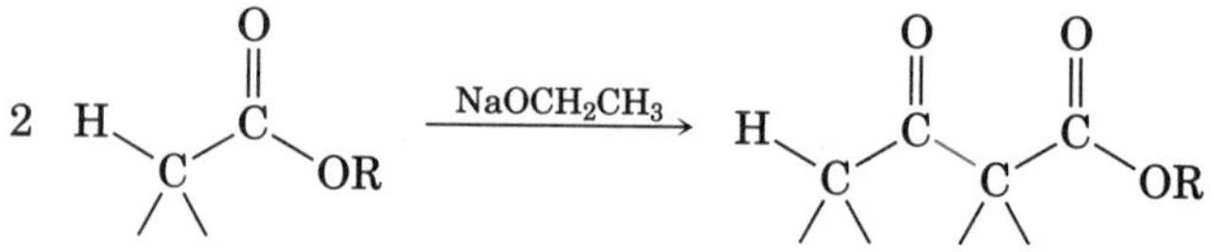

$$2\ \mathrm{H{-}\underset{/\ \backslash}{C}{-}\overset{O}{\overset{\|}{C}}{-}OR} \xrightarrow{\text{NaOCH}_2\text{CH}_3} \mathrm{H{-}\underset{/\ \backslash}{C}{-}\overset{O}{\overset{\|}{C}}{-}\underset{/\ \backslash}{C}{-}\overset{O}{\overset{\|}{C}}{-}OR}$$

ADDITIONAL PROBLEMS

11.18 Indicate the acidic hydrogen atoms in these molecules:

(a) $HOCH_2\overset{O}{\overset{\|}{C}}CH_3$ (b) $HOCH_2CH_2\overset{O}{\overset{\|}{C}}C(CH_3)_3$ (c) 1,3-Cyclopentanedione
(d) $CH_3CH{=}CHCHO$

11.19 Draw structures for the possible monoenol tautomers of 1,3-cyclohexanedione. How many enol forms are possible, and which would you expect to be most stable? Explain.

11.20 Rank these compounds in order of increasing acidity: CH_3CH_2COOH, CH_3COCH_3, CH_3CH_2OH, $CH_3COCH_2COCH_3$.

11.21 Why do you suppose acetone is enolized only to the extent of about 0.000 1% at equilibrium, whereas 2,4-pentanedione is 76% enolized?

11.22 Write resonance structures for these anions:

(a) $CH_3\overset{O}{\overset{\|}{C}}\overset{\ddot{-}}{C}H\overset{O}{\overset{\|}{C}}CH_3$ (b) $:\overset{-}{C}H_2C{\equiv}N$ (c) $CH_3CH{=}CH\overset{\ddot{-}}{C}H\overset{O}{\overset{\|}{C}}CH_3$ (d) $N{\equiv}C\overset{\ddot{-}}{C}HCO_2C_2H_5$

11.23 When acetone is treated with acid in deuterated water, D_2O, deuterium becomes incorporated into the molecule. Propose a mechanism to account for this reaction.

$$CH_3\overset{O}{\overset{\|}{C}}CH_3 + D_2O \overset{DCl}{\rightleftharpoons} CH_3\overset{O}{\overset{\|}{C}}CH_2D$$

11.24 Why is an enolate ion generally more reactive than an enol?

11.25 How do the mechanisms of base-catalyzed enolization and acid-catalyzed enolization differ?

11.26 When optically active (*R*)-2-methylcyclohexanone is treated with aqueous HCl or NaOH, racemic 2-methylcyclohexanone is produced. Explain.

11.27 When optically active (*R*)-3-methylcyclohexanone is treated with aqueous HCl or NaOH, no racemization occurs. Instead, the optically active ketone is recovered unchanged. How can you reconcile this observation with your answer to Problem 11.26?

11.28 Monoalkylated acetic acids (RCH_2COOH) and dialkylated acetic acids ($R_2CHCOOH$) can be prepared by malonic ester synthesis, but trialkylated acetic acids (R_3CCOOH) can't be prepared in this way. Explain.

11.29 Which of the following compounds would you expect to undergo aldol condensation? Draw the product in each case.
(a) 2,2-Dimethylpropanal **(b)** Cyclobutanone **(c)** Benzophenone (diphenyl ketone)
(d) Decanal

11.30 Which of the following esters can be prepared by a malonic ester synthesis? Show what reagents you would use.
(a) Ethyl pentanoate **(b)** Ethyl 3-methylbutanoate **(c)** Ethyl 2-methylbutanoate
(d) Ethyl 2,2-dimethylpropanoate

11.31 The aldol condensation reaction can be carried out intramolecularly by treatment of a diketone with base. What diketone would you start with to prepare 3-methyl-2-cyclohexenone? Show the reaction.

11.32 Nonconjugated β,γ-unsaturated ketones such as 3-cyclohexenone are in an acid-catalyzed equilibrium with their conjugated α,β-unsaturated isomers. Propose a mechanism for the acid-catalyzed interconversion of the two isomers.

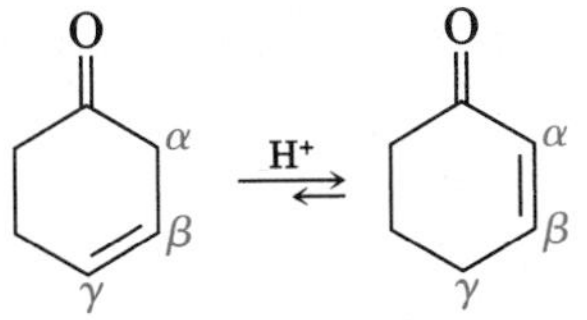

O α β γ $\underset{}{\overset{H^+}{\rightleftharpoons}}$ O α β γ

11.33 The α,β to β,γ interconversion of unsaturated ketones (see Problem 11.32) is also catalyzed by base. Propose a mechanism for the reaction.

11.34 One consequence of the base-catalyzed α,β to β,γ isomerization of unsaturated ketones (see Problem 11.33) is that C5-substituted 2-cyclopentenones can be interconverted with C2-substituted 2-cyclopentenones. Propose a mechanism to account for this isomerization.

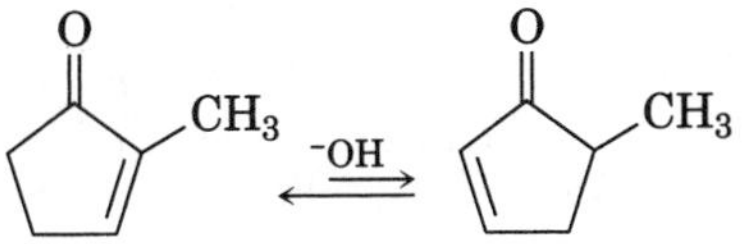

O CH_3 $\overset{^-OH}{\rightleftharpoons}$ O CH_3

11.35 If a 1:1 mixture of ethyl acetate and ethyl propanoate is treated with base under Claisen condensation conditions, a mixture of four β-keto ester products is obtained. Show their structures.

11.36 If a mixture of ethyl acetate and ethyl benzoate is treated with base, a mixture of two Claisen condensation products is obtained. Explain.

11.37 Cinnamaldehyde, the aromatic constituent of cinnamon oil, can be synthesized by a mixed aldol-like reaction between benzaldehyde and acetaldehyde. Formulate the reaction. What other product would you expect to obtain?

CHO

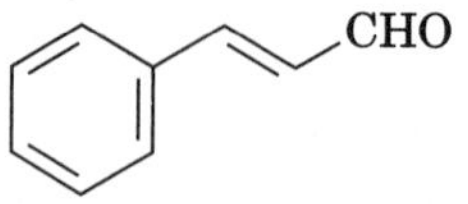

Cinnamaldehyde

11.38 How might you prepare these compounds using aldol condensation reactions?
(a) $C_6H_5C(CH_3){=}CHCOC_6H_5$ **(b)** 4-Methyl-3-penten-2-one

11.39 1-Butanol is synthesized commercially, starting from acetaldehyde, by a three-step route that involves an aldol reaction followed by two reductions. How might you carry out this transformation?

$$CH_3CHO \xrightarrow{\text{3 steps}} CH_3CH_2CH_2CH_2OH$$

11.40 By starting with a dihalide, cyclic compounds can be prepared using the malonic ester synthesis. What product would you expect to obtain from the reaction between diethyl malonate, 1,4-dibromobutane, and 2 equiv of base?

11.41 The aldol reaction can sometimes take place internally if a dicarbonyl compound is treated with base. What product would you expect to obtain from aldol cyclization of hexanedial, $OHCCH_2CH_2CH_2CH_2CHO$?

11.42 How can you account for the fact that *cis*- and *trans*-4-*tert*-butyl-2-methylcyclohexanone are interconverted by base treatment? Which of the two isomers do you think is more stable, and why? (See Section 2.11.)

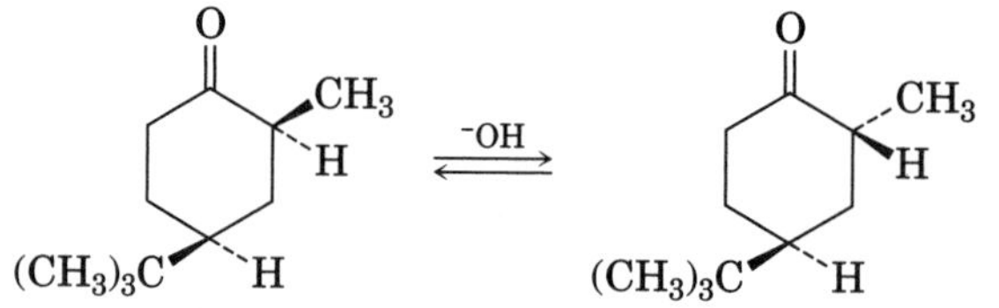

O, CH_3, H, $(CH_3)_3C$, H $\xrightleftharpoons{^-OH}$ O, CH_3, H, $(CH_3)_3C$, H

11.43 Show how you might convert geraniol, the chief constituent of rose oil, into ethyl geranylacetate.

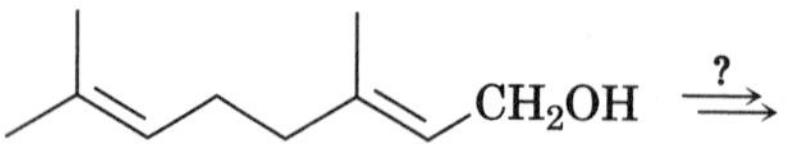

CH_2OH $\xrightarrow{?}$ O, C, OCH_2CH_3

Geraniol 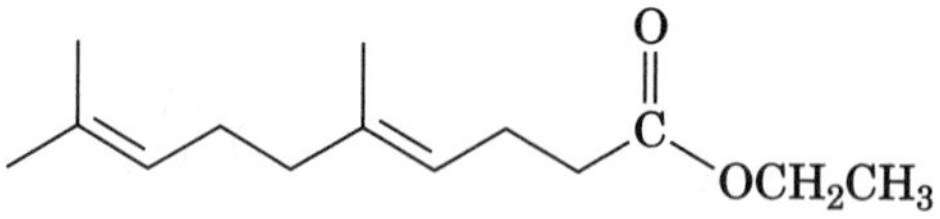 **Ethyl geranylacetate**

11.44 The *acetoacetic ester synthesis* is closely related to the malonic ester synthesis, but involves alkylation with the anion of ethyl acetoacetate rather than diethyl malonate. Treatment of the ethyl acetoacetate anion with an alkyl halide, followed by decarboxylation, yields a ketone product:

$$CH_3\overset{O}{\overset{\|}{C}}CH_2\overset{O}{\overset{\|}{C}}OCH_2CH_3 \xrightarrow[\text{2. RX} \quad \text{3. } H_3O^+]{\text{1. } Na^+\ ^-OCH_2CH_3} CH_3\overset{O}{\overset{\|}{C}}CH_2{-}R + CO_2 + HOCH_2CH_3$$

How would you prepare these compounds using an acetoacetic ester synthesis?
(a) 4-Phenyl-2-butanone **(b)** 5-Methyl-2-hexanone **(c)** 3-Methyl-2-hexanone

11.45 Which of the following compounds can't be prepared by an acetoacetic ester synthesis (see Problem 11.44)? Explain.
(a) 2-Butanone **(b)** Phenylacetone **(c)** Acetophenone **(d)** 3,3-Dimethyl-2-butanone

11.46 The Claisen condensation is reversible. That is, a β-keto ester can be cleaved by base into two fragments. Show the mechanism by which the following cleavage occurs:

$$C_6H_5\overset{O}{\overset{\|}{C}}CH_2\overset{O}{\overset{\|}{C}}OCH_2CH_3 \xrightarrow[H_2O]{NaOH} C_6H_5\overset{O}{\overset{\|}{C}}O^-\ Na^+ + CH_3\overset{O}{\overset{\|}{C}}OCH_2CH_3$$

12 CHAPTER AMINES

AMINE
Organic derivative of ammonia

PRIMARY AMINE
Amine with one organic substituent on nitrogen, RNH_2

SECONDARY AMINE
Amine with two organic substituents on nitrogen, R_2NH

TERTIARY AMINE
Amine with three organic substituents on nitrogen, R_3N

Amines are organic derivatives of ammonia in the same way that alcohols and ethers are organic derivatives of water. Amines are classified either as **primary (RNH_2), secondary (R_2NH),** or **tertiary (R_3N)**, depending on the number of organic substituents attached to nitrogen. For example, methylamine (CH_3NH_2) is a primary amine and trimethylamine [$(CH_3)_3N$] is a tertiary amine. Note that this usage of the terms *primary*, *secondary*, and *tertiary* is different from our previous usage. When we speak of a tertiary alcohol or alkyl halide, we refer to the degree of substitution at the alkyl *carbon* atom, but when we speak of a tertiary amine, we refer to the degree of substitution at the *nitrogen* atom.

$$CH_3-\underset{CH_3}{\overset{CH_3}{\underset{|}{\overset{|}{C}}}}-OH \qquad CH_3-\underset{CH_3}{\overset{CH_3}{\underset{|}{\overset{|}{N}}}} \qquad CH_3-\underset{CH_3}{\overset{CH_3}{\underset{|}{\overset{|}{C}}}}-NH_2$$

***tert*-Butyl alcohol** (a tertiary alcohol) — **Trimethylamine** (a tertiary amine) — ***tert*-Butylamine** (a primary amine)

Compounds with four groups attached to nitrogen are also known, but the nitrogen atom must carry a positive charge. Such compounds are called **quaternary ammonium salts**.

QUATERNARY AMMONIUM SALT
Compound with four organic substituents attached to a positively charged nitrogen, $R_4N^+ X^-$

$$R-\underset{R}{\overset{R}{\underset{|}{\overset{|}{N^+}}}}-R \quad X^- \qquad \text{A quaternary ammonium salt}$$

ALKYLAMINE
Amine that has its nitrogen atom bonded to a saturated, alkyl-group carbon

ARYLAMINE
Amine that has its nitrogen atom bonded to an aromatic ring

Amines can be either alkyl-substituted (**alkylamines**) or aryl-substituted (**arylamines**). Although much of the chemistry of the two classes is similar, we'll soon see that there are also important differences.

$CH_3CH_2\ddot{N}H_2$ — **Ethylamine** (an alkylamine)

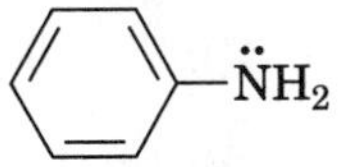

$C_6H_5\text{—}\ddot{N}H_2$ — **Aniline** (an arylamine)

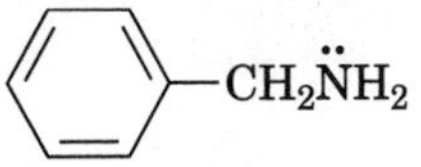

$C_6H_5\text{—}CH_2\ddot{N}H_2$ — **Benzylamine** (an alkylamine)

PRACTICE PROBLEM 12.1 Classify these amines as either primary, secondary, or tertiary:

(a) $CH_3CH_2CH(CH_3)NH_2$ **(b)** pyrrolidine ring with N—H **(c)** $C_6H_5N(CH_3)_2$

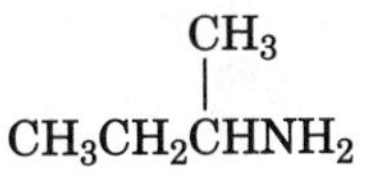

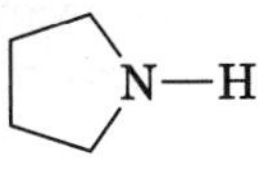

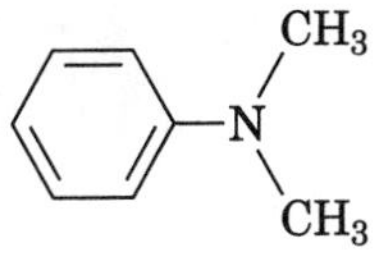

SOLUTION Amine (a) is primary, (b) is secondary, and (c) is tertiary.

PROBLEM 12.1 Classify the following compounds as either primary, secondary, or tertiary amines or as quaternary ammonium salts:

(a) $(CH_3)_2CHNH_2$ **(b)** $(CH_3CH_2)_2NH$

(c) cyclohexyl—$N(CH_3)_2$ **(d)** $C_6H_5\text{—}CH_2\overset{+}{N}(CH_3)_3\ I^-$

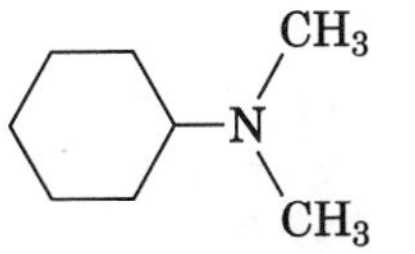

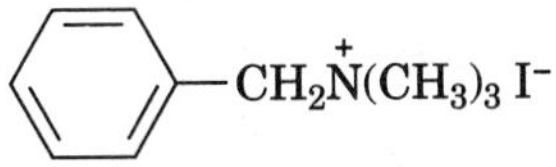

PROBLEM 12.2 Draw structures of compounds that meet these descriptions:

(a) A secondary amine with one isopropyl group

(b) A tertiary amine with one phenyl group and one ethyl group

(c) A quaternary ammonium salt with four different groups bonded to nitrogen

12.1 NAMING AMINES

Primary amines, RNH_2, are named in the IUPAC system in either of two ways. For simple amines, the suffix *-amine* is added to the name of the organic substituent:

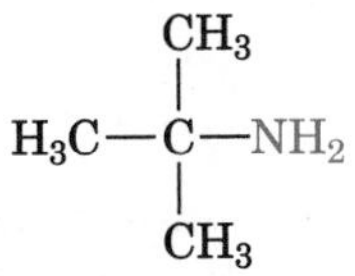

$H_3C\text{—}C(CH_3)_2\text{—}NH_2$ — ***tert*-Butylamine**

cyclohexyl—NH_2 — **Cyclohexylamine**

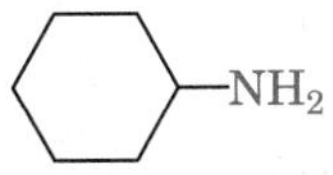

$H_2NCH_2CH_2CH_2CH_2NH_2$ — **1,4-Butanediamine**

Amines that also have other functional groups are named by considering the $-NH_2$ as an *amino* substituent on the parent molecule:

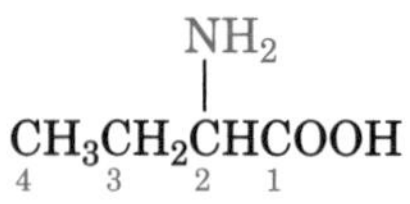

$$\underset{4\ \ 3\ \ \ 2\ \ 1}{CH_3CH_2\overset{\overset{NH_2}{|}}{C}HCOOH}$$

2-Aminobutanoic acid

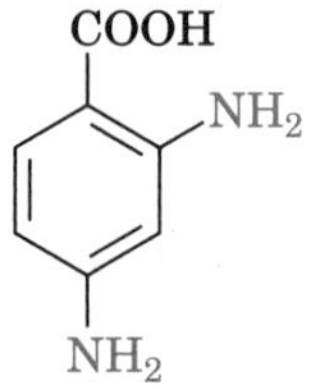

2,4-Diaminobenzoic acid

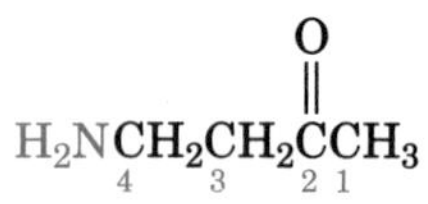

$$\underset{4\ \ \ 3\ \ \ 2\ 1}{H_2NCH_2CH_2\overset{\overset{O}{\|}}{C}CH_3}$$

4-Amino-2-butanone

Symmetrical secondary and tertiary amines are named by adding the prefix *di-* or *tri-* to the alkyl group:

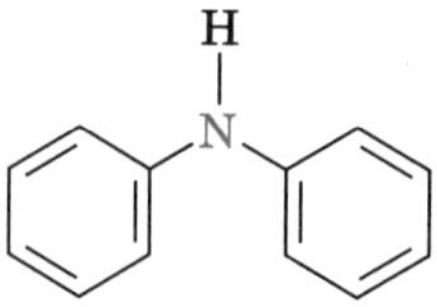

Diphenylamine

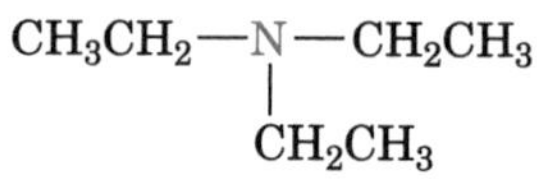

$$CH_3CH_2-\underset{\underset{CH_2CH_3}{|}}{N}-CH_2CH_3$$

Triethylamine

Unsymmetrically substituted secondary and tertiary amines are named as *N*-substituted primary amines. The largest organic group is chosen as the parent, and the other groups are considered as *N*-substituents on the parent (*N* since they're attached to nitrogen).

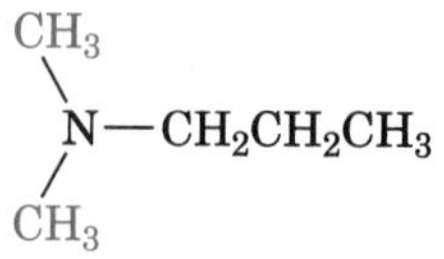

$$(CH_3)_2N-CH_2CH_2CH_3$$

***N,N*-Dimethylpropylamine**
(propylamine is the parent name; the two methyl groups are substituents on nitrogen)

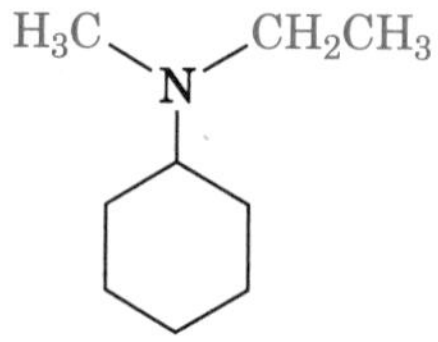

***N*-Ethyl-*N*-methylcyclohexylamine**
(cyclohexylamine is the parent name; methyl and ethyl are *N*-substituents)

There are few common names for simple amines, but phenylamine is usually called *aniline.*

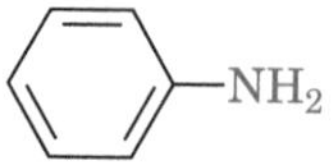

Aniline

HETEROCYCLIC AMINE Amine in which the nitrogen atom occurs in a ring

Heterocyclic amines, compounds in which the nitrogen atom occurs as part of a ring, are also common, and each different heterocyclic ring system is given its own parent name. In all cases, the nitrogen atom is numbered as position 1.

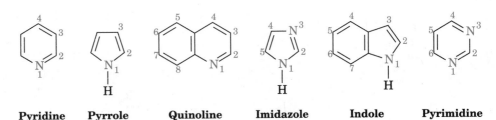

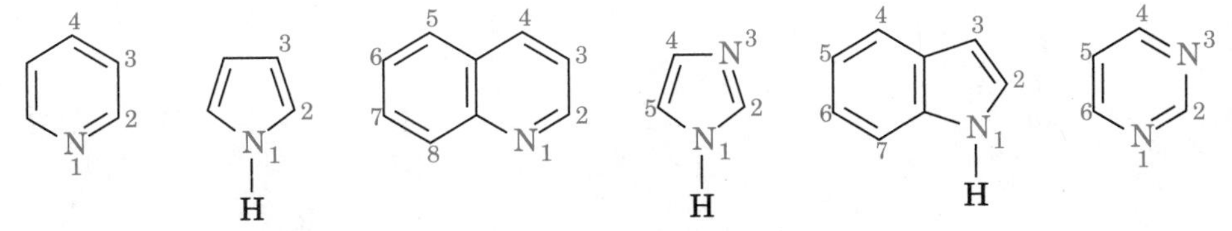

Pyridine **Pyrrole** **Quinoline** **Imidazole** **Indole** **Pyrimidine**

PROBLEM 12.3

Name these compounds by IUPAC rules:

(a) $CH_3NHCH_2CH_3$ **(b)**

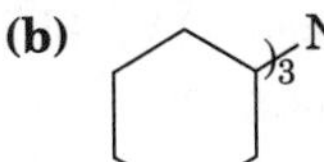

(c) N—CH_3

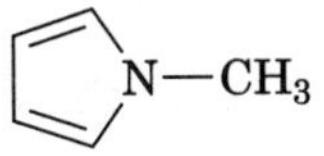

(d) CH_3 / N \ $CH_2CH_2CH_3$ **(e)** $H_2NCH_2CH_2\overset{CH_3}{\overset{|}{C}}HNH_2$

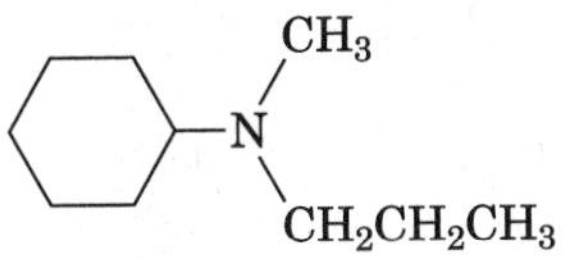

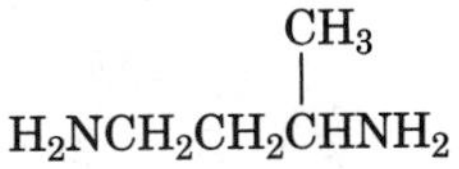

PROBLEM 12.4

Draw structures corresponding to these IUPAC names:

(a) Triethylamine **(b)** *N*-Methylaniline

(c) Tetraethylammonium bromide **(d)** *p*-Bromoaniline

(e) *N*-Ethyl-*N*-methylcyclopentylamine

12.2 STRUCTURE AND PROPERTIES OF AMINES

The bonding in amines is similar to the bonding in ammonia. The nitrogen atom is sp^3-hybridized, with the three substituents occupying three corners of a tetrahedron and the nitrogen's nonbonding lone pair of electrons occupying the fourth corner. As expected, the C–N–C bond angles are very close to the 109° tetrahedral value. For trimethylamine, the C–N–C angle is 108°, and the C–N bond length is 1.47 Å.

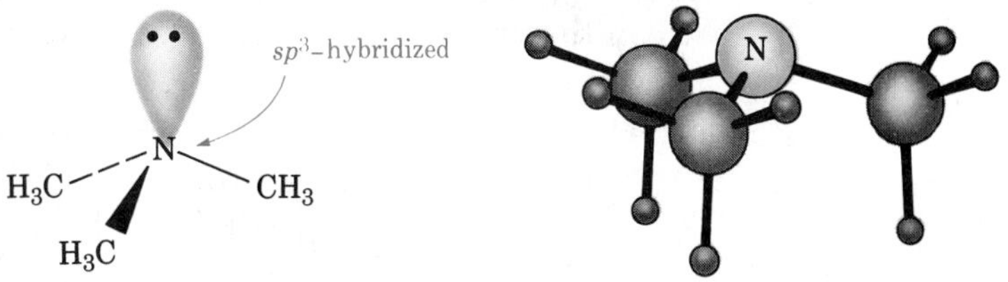

Trimethylamine

Like alcohols, amines are highly polar, and those with fewer than five carbon atoms are generally water-soluble. Also like alcohols, primary and secondary amines form hydrogen bonds and therefore have higher boiling points than alkanes of similar molecular weight.

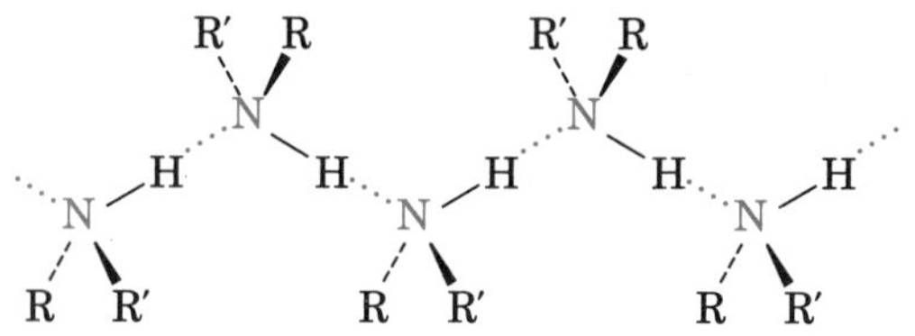

One other characteristic property of amines is their odor. Low-molecular-weight amines have a characteristic fishlike aroma, while diamines such as putrescine (1,4-butanediamine) have odors that are as putrid as their names suggest.

12.3 AMINE BASICITY

The chemistry of amines is dominated by the nitrogen lone pair of electrons. Because of the lone pair, amines are both basic and nucleophilic. They react with Lewis acids to form acid–base salts, and they react with electrophiles in many of the polar reactions seen in previous chapters.

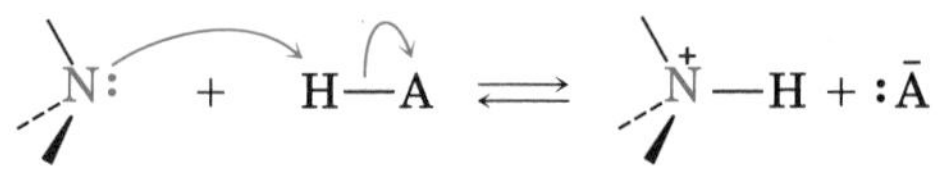

An amine (a Lewis base) An acid A salt

BASICITY CONSTANT, K_b Value that measures the strength of a base in water solution

Amines are much more basic than alcohols, ethers, or water. When an amine is dissolved in water, an equilibrium is established in which water acts as a protic acid and donates a proton to the amine. By finding the equilibrium constant for the reaction, we can define a **basicity constant, K_b,** that measures the ability of an amine to accept a proton, and we can thereby establish a relative order of base strength.

$$RNH_2 + H_2O \rightleftharpoons RNH_3^+ + OH^-$$

$$K_b = \frac{[RNH_3^+][OH^-]}{[RNH_2]} \quad \text{and} \quad pK_b = -\log K_b$$

The larger the K_b (and the smaller the pK_b), the stronger the base; the smaller the K_b (and the larger the pK_b), the weaker the base. Table 12.1 gives the pK_b values of some common amines and indicates that substitution has relatively little effect on alkylamine basicity. Most simple alkylamines have pK_b's in the narrow range 3–4, regardless of their exact structure.

TABLE 12.1 Base Strengths of Some Common Amines

NAME	STRUCTURE	pK_b	
Diethylamine	$(CH_3CH_2)_2\ddot{N}H$	3.06	More basic
Triethylamine	$(CH_3CH_2)_3N:$	3.21	
Ethylamine	$CH_3CH_2\ddot{N}H_2$	3.25	
Dimethylamine	$(CH_3)_2\ddot{N}H$	3.27	
Methylamine	$CH_3\ddot{N}H_2$	3.36	
Trimethylamine	$(CH_3)_3N:$	4.21	
Ammonia	$:NH_3$	4.74	
Aniline	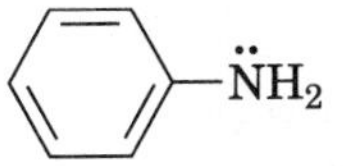 $C_6H_5\text{—}\ddot{N}H_2$	9.37	Less basic

The most important conclusion from Table 12.1 is that arylamines such as aniline are weaker bases than alkylamines by a factor of about 10^6. The nitrogen lone-pair electrons of arylamines are delocalized by orbital overlap with the pi orbitals of the aromatic ring, and are therefore less available for bonding to an acid. In resonance terms, arylamines are more stable than alkylamines because of their five resonance structures:

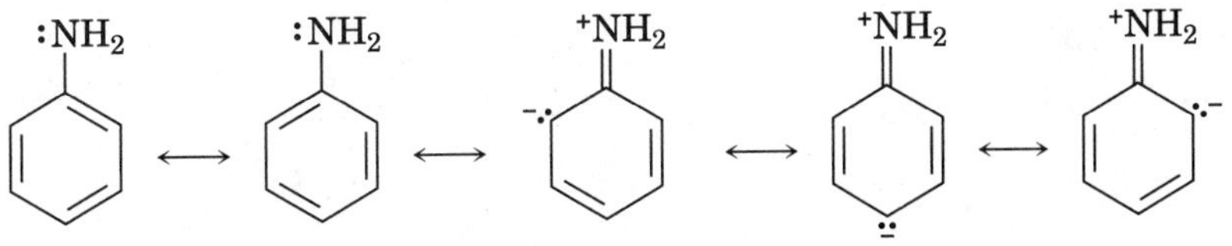

$:NH_2 \longleftrightarrow :NH_2 \longleftrightarrow {}^{+}NH_2 \longleftrightarrow {}^{+}NH_2 \longleftrightarrow {}^{+}NH_2$

In contrast to amines, *amides* ($RCONH_2$) are nonbasic. Amides don't form salts when treated with acids, and their aqueous solutions are neutral. The main reason for the decreased basicity of amides relative to amines is that the nitrogen lone-pair electrons are delocalized by orbital overlap with the neighboring carbonyl-group pi orbital. The electrons are therefore much less available for bonding to an acid. In resonance terms, amides are more stable and less reactive than amines because they are hybrids of two resonance forms:

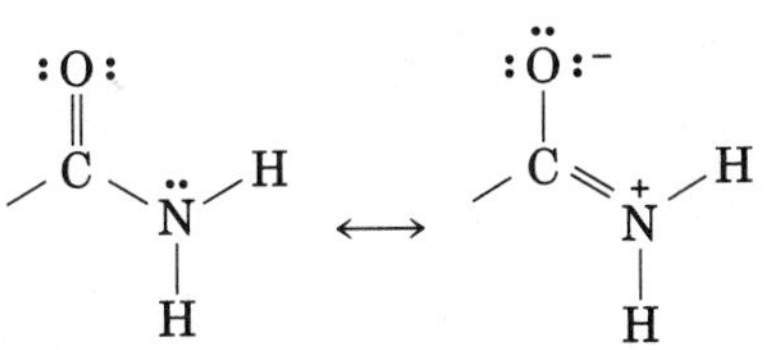

$:O: \quad C \quad \ddot{N} \quad H \quad H \longleftrightarrow :\ddot{O}:^- \quad C \quad \overset{+}{N} \quad H \quad H$

We can take advantage of the basicity of amines to purify them. For example, if we have a mixture of an amine (basic), a ketone (neutral), and a carboxylic acid (acidic), we can dissolve the mixture in an organic solvent and add aqueous HCl. The basic amine dissolves in the acidic water as its ammonium ion, while the neutral ketone and the carboxylic acid remain in the organic solvent. Separation of the water layer and neutralization of the ammonium ion by addition of NaOH then provides the pure amine. Meanwhile, addition of an aqueous solution of NaOH to the organic solvent converts the carboxylic acid into its water-soluble carboxylate ion, leaving the neutral ketone in the ether. Neutralization of the carboxylate ion by addition of HCl gives the pure carboxylic acid (Figure 12.1).

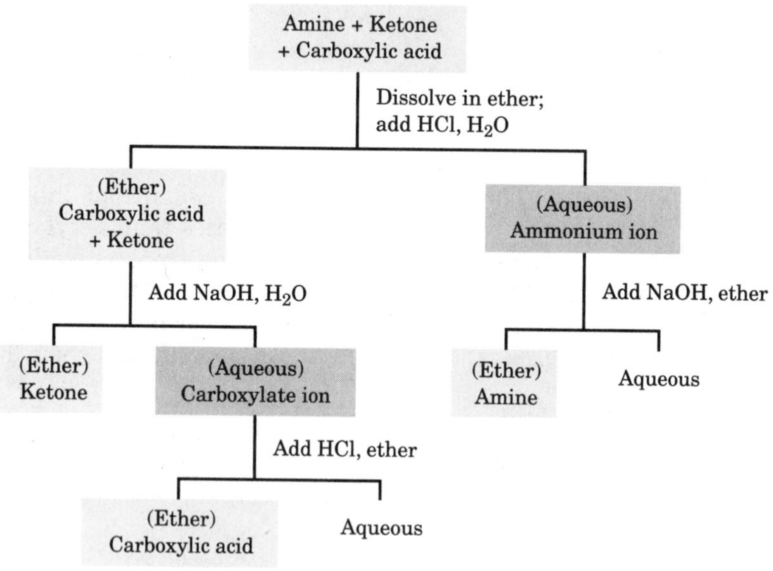

FIGURE 12.1 Separation and purification of a basic amine from a mixture of acidic and neutral components.

PRACTICE PROBLEM 12.2 Which would you expect to be the stronger base, aniline or *p*-nitroaniline?

SOLUTION Since a nitro group on a benzene ring is strongly electron-withdrawing (Section 5.9), it pulls electrons from the $-NH_2$ group, making them less available for donation to acids and making *p*-nitroaniline a weaker base than aniline.

PRACTICE PROBLEM 12.3 Predict the product of this reaction:

$$CH_3CH_2NHCH_3 + HCl \longrightarrow ?$$

SOLUTION Amines are protonated by acids to yield ammonium salts.

$$CH_3CH_2NHCH_3 + HCl \longrightarrow CH_3CH_2\overset{+}{N}H_2CH_3\ Cl^-$$

PROBLEM 12.5 Which would you expect to be the stronger base, aniline or *p*-methylaniline? Explain.

PROBLEM 12.6 Predict the product of this reaction:

$$\text{C}_5\text{H}_9\text{–N(CH}_3\text{)(H)} + \text{HBr} \longrightarrow ?$$

PROBLEM 12.7 Which compound in each of the following pairs is more basic?

(a) $CH_3CH_2NH_2$ or $CH_3CH_2CONH_2$ **(b)** NaOH or $C_6H_5NH_2$
(c) CH_3NHCH_3 or $CH_3NHC_6H_5$ **(d)** CH_3OCH_3 or $(CH_3)_3N$

12.4 RESOLUTION OF ENANTIOMERS BY USE OF AMINE SALTS

RESOLUTION Separation of a racemic mixture into two pure enantiomers

Amine basicity is often used to carry out the separation (**resolution**) of a racemic carboxylic acid mixture into its two pure enantiomers. [Recall from Section 6.10 that a racemic mixture is a 50:50 mixture of (+) and (−) enantiomers.]

$$50\%\ (+) : 50\%\ (-) \xrightarrow{\text{Resolve}} \text{Pure } (+) \text{ and Pure } (-)$$

Racemic mixture of enantiomers — Pure separate enantiomers

Historically, Louis Pasteur was the first person to resolve a racemic mixture when he was able to crystallize a salt of (±)-tartaric acid and separate two kinds of crystals by hand (Section 6.5). Pasteur's method isn't generally applicable, though, since few racemic mixtures crystallize into separate mirror-image forms. The most commonly used method of resolution makes use of an acid–base reaction between a racemic mixture of carboxylic acids and a chiral amine.

To understand how this method of resolution works, let's see what happens when a racemic mixture of chiral acids such as (+)- and (−)-lactic acids reacts with an achiral amine base such as methylamine (Figure 12.2). Stereochemically, the situation is analogous to what happens when left and right hands (chiral) pick up a ball (achiral). Both left and right hands pick up the ball equally well, and the products—ball in right hand versus ball in left hand—are mirror images. In the same way, both (+)- and (−)-lactic acid react with methylamine equally well, and the product is a mixture of two salts—methylammonium (+)-lactate and methylammonium (−)-lactate. Just as with the chiral hands and achiral ball, the two salts are mirror images and we still have a racemic mixture.

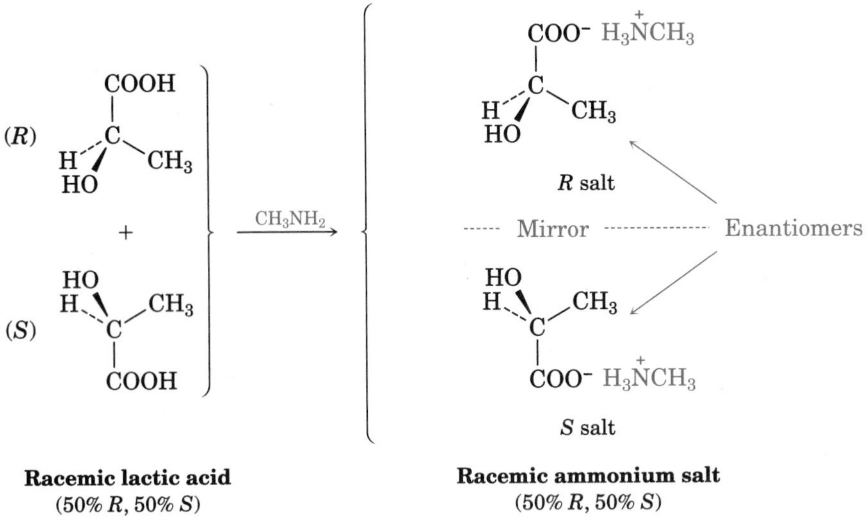

FIGURE 12.2 Reaction of racemic lactic acid with achiral methylamine leads to a racemic mixture of salts.

Now let's see what happens when the racemic mixture of (+)- and (−)-lactic acids reacts with a single enantiomer of a chiral amine base such as (*R*)-1-phenylethylamine (Figure 12.3). Stereochemically, this situation is analogous to what happens when a hand (a chiral reagent) puts on a glove (*also a chiral reagent*). *Left and right hands don't put on the same glove in the same way.* The products—right hand in right glove

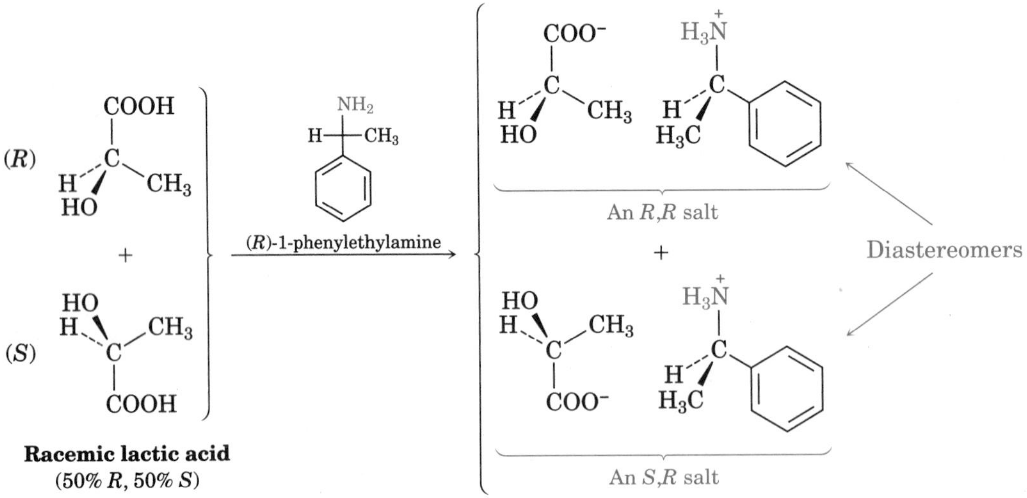

FIGURE 12.3 Reaction of racemic lactic acid with optically pure (*R*)-1-phenylethylamine leads to a mixture of diastereomeric salts.

versus left hand in right glove—are not mirror images; they're altogether different.

In the same way, (+)- and (−)-lactic acid react with (*R*)-1-phenylethylamine to give different products. (*R*)-Lactic acid reacts with (*R*)-1-phenylethylamine to give the *R*,*R* salt, whereas (*S*)-lactic acid reacts with the same *R* amine to give the *S*,*R* salt. *These two salts are diastereomers* (Section 6.7). They are different compounds and have different chemical and physical properties. It may therefore be possible to separate them by fractional crystallization or by some other laboratory technique. Once separated, acidification of the two diastereomeric salts with mineral acid then allows us to recover and isolate the two pure enantiomers of lactic acid.

PRACTICE PROBLEM 12.4 Suppose that racemic lactic acid reacts with methanol to form the ester, methyl lactate. What stereochemistry would you expect the products to have? What is the relationship of the products?

SOLUTION Reaction of a racemic acid with an achiral alcohol such as methanol yields a racemic mixture of mirror-image ester products:

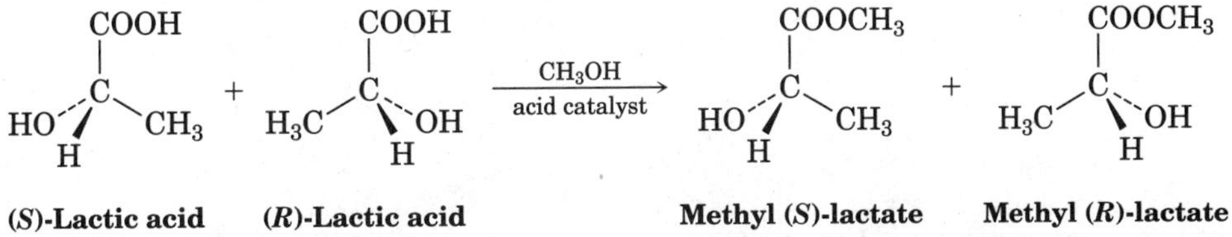

COOH, HO, C, CH_3, H + COOH, H_3C, C, OH, H $\xrightarrow[\text{acid catalyst}]{CH_3OH}$ $COOCH_3$, HO, C, CH_3, H + $COOCH_3$, H_3C, C, OH, H

(*S*)-Lactic acid **(*R*)-Lactic acid** **Methyl (*S*)-lactate** **Methyl (*R*)-lactate**

PROBLEM 12.8 Suppose that racemic lactic acid reacts with (*S*)-2-butanol to form an ester. What stereochemistry would you expect the products to have? What is the relationship of the products?

PROBLEM 12.9 How might you use the reaction described in Problem 12.8 to resolve (±)-lactic acid?

12.5 SYNTHESIS OF AMINES

S_N2 Reactions of Alkyl Halides

Alkylamines are excellent nucleophiles in S_N2 reactions (Section 7.7). As a result, the simplest method of amine synthesis is by S_N2 reaction of ammonia or an alkylamine with an alkyl halide. If ammonia is used, a primary amine results; if a primary amine is used, a secondary amine results; and so on. Even tertiary amines react with alkyl halides to yield quaternary ammonium salts, $R_4N^+ X^-$.

Ammonia	$\ddot{N}H_3 + R{-}X$	$\longrightarrow$	$RNH_3^+ X^-$	$\xrightarrow{NaOH}$ RNH_2	Primary
Primary	$R\ddot{N}H_2 + R{-}X$	$\longrightarrow$	$R_2NH_2^+ X^-$	$\xrightarrow{NaOH}$ R_2NH	Secondary
Secondary	$R_2\ddot{N}H + R{-}X$	$\longrightarrow$	$R_3NH^+ X^-$	$\xrightarrow{NaOH}$ R_3N	Tertiary
Tertiary	$R_3\ddot{N} + R{-}X$	$\longrightarrow$	$R_4N^+ X^-$	Quaternary ammonium salt	

S_N2 reaction

Unfortunately, these reactions don't stop cleanly after a single alkylation has occurred. Since primary, secondary, and tertiary amines all have similar reactivity, the initially formed monoalkylated product often undergoes further reaction to yield a mixture of products. For example, treatment of 1-bromooctane with a twofold excess of ammonia leads to a mixture containing only a 45% yield of octylamine. A nearly equal amount of dioctylamine is produced by double alkylation, along with smaller amounts of trioctylamine and tetraoctylammonium bromide.

$$CH_3(CH_2)_6CH_2Br + :NH_3 \longrightarrow CH_3(CH_2)_6CH_2\ddot{N}H_2 + [CH_3(CH_2)_6CH_2]_2\ddot{N}H$$

1-Bromooctane **Octylamine (45%)** **Dioctylamine (43%)**

$$+ [CH_3(CH_2)_6CH_2]_3N: + [CH_3(CH_2)_6CH_2]_4\overset{+}{N}\overset{-}{Br}$$

Trace Trace

PRACTICE PROBLEM 12.5 How could you prepare diethylamine starting from ammonia?

SOLUTION Look at the starting material (NH_3) and the product $(CH_3CH_2)_2NH$, and note the difference. Since two ethyl groups have become bonded to the nitrogen atom, the reaction must involve ammonia and two molecules of an ethyl halide:

$$2\ CH_3CH_2Br + NH_3 \longrightarrow (CH_3CH_2)_2NH$$

PROBLEM 12.10 Show how you could prepare these amines from ammonia:
(a) Triethylamine **(b)** Tetramethylammonium bromide

Reduction of Nitriles and Amides

We've already seen how amines can be prepared by reduction of nitriles (Section 10.11) and amides (Section 10.10) with $LiAlH_4$. The two-step sequence of S_N2 reaction of an alkyl halide with cyanide ion, followed by reduction, is a good method for converting an alkyl halide into a primary amine having one more carbon atom. Amide reduction provides a method for converting carboxylic acids into amines with the same number of carbon atoms.

$$\underset{\textbf{Alkyl halide}}{RX} \xrightarrow{NaCN} RCN \xrightarrow[\text{2. } H_2O]{\text{1. } LiAlH_4\text{, ether}} \underset{1^\circ \text{ amine}}{RCH_2NH_2}$$

$$\underset{\textbf{Carboxylic acid}}{R{-}\overset{O}{\overset{\|}{C}}{-}OH} \xrightarrow[\text{2. } NH_3]{\text{1. } SOCl_2} R{-}\overset{O}{\overset{\|}{C}}{-}NH_2 \xrightarrow[\text{2. } H_2O]{\text{1. } LiAlH_4\text{, ether}} \underset{1^\circ \text{ amine}}{RCH_2NH_2}$$

PRACTICE PROBLEM 12.6 What amide would you start with to prepare *N*-ethylcyclohexylamine?

SOLUTION Reduction of an amide with $LiAlH_4$ yields an amine in which the amide carbonyl group has been replaced by a methylene ($-CH_2-$) unit, $RCONR_2 \rightarrow RCH_2NR_2$. Since *N*-ethylcyclohexylamine has only one $-CH_2-$ carbon attached to its nitrogen (the ethyl group), the product must come from reduction of *N*-cyclohexylacetamide:

$$\underset{\textbf{\textit{N}-Cyclohexylacetamide}}{C_6H_{11}{-}NH{-}\overset{O}{\overset{\|}{C}}{-}CH_3} \xrightarrow[\text{2. } H_2O]{\text{1. } LiAlH_4} \underset{\textbf{\textit{N}-Ethylcyclohexylamine}}{C_6H_{11}{-}NHCH_2CH_3}$$

PRACTICE PROBLEM 12.7 What nitrile would yield butylamine on reaction with $LiAlH_4$?

SOLUTION Reduction of a nitrile with $LiAlH_4$ yields a primary amine whose $-CH_2NH_2$ part comes from the $-C{\equiv}N$ group. Thus, butylamine must have come from butanenitrile:

$$\underset{\textbf{Butanenitrile}}{CH_3CH_2CH_2C{\equiv}N} \xrightarrow[\text{2. } H_2O]{\text{1. } LiAlH_4} \underset{\textbf{Butylamine}}{CH_3CH_2CH_2CH_2NH_2}$$

PROBLEM 12.11 Propose structures for amides that might be precursors of these amines:
(a) Propylamine **(b)** Dipropylamine **(c)** Benzylamine, $C_6H_5CH_2NH_2$

PROBLEM 12.12 Propose structures for nitriles that might be precursors of these amines:
(a) $CH_3\overset{CH_3}{\overset{|}{C}}HCH_2CH_2NH_2$ **(b)** Benzylamine, $C_6H_5CH_2NH_2$

Reduction of Nitroarenes

Arylamines are prepared by nitration of an aromatic starting material, followed by reduction of the nitro group. The reduction step can be carried out in different ways, depending on the circumstances. Catalytic hydrogenation over platinum works well, but is sometimes incompatible with the presence elsewhere in the molecule of other reducible groups such as carbon–carbon double bonds. Iron, zinc, tin, and stannous chloride ($SnCl_2$) are also effective when used in aqueous acid.

$$p\text{-}(CH_3)_3C{-}C_6H_4{-}NO_2 \xrightarrow[\text{or 1. Fe, }H_3O^+\text{; 2. NaOH}]{H_2\text{, Pt catalyst}} p\text{-}(CH_3)_3C{-}C_6H_4{-}NH_2$$

***p-tert*-Butylnitrobenzene** → ***p-tert*-Butylaniline**

PRACTICE PROBLEM 12.8 How could you synthesize *p*-methylaniline starting from benzene? More than one step is required.

SOLUTION A methyl group is introduced onto a benzene ring by a Friedel–Crafts reaction with $CH_3Cl/AlCl_3$ (Section 5.8), and an amino group is introduced onto a ring by nitration and reduction. The overall sequence is

$$C_6H_6 \xrightarrow[AlCl_3]{CH_3Cl} C_6H_5CH_3 \xrightarrow[H_2SO_4]{HNO_3} p\text{-}CH_3C_6H_4NO_2 \xrightarrow[\text{2. NaOH}]{\text{1. Fe, }H_3O^+} p\text{-}CH_3C_6H_4NH_2$$

Benzene → **Toluene** → ***p*-Nitrotoluene** → ***p*-Methylaniline**

PROBLEM 12.13 How could you synthesize the following amines starting from benzene? More than one step is required in each case.
(a) *m*-Aminobenzoic acid **(b)** 2,4,6-Tribromoaniline

12.6 REACTIONS OF AMINES

We've already seen the two most important reactions of alkylamines: alkylation and acylation. As we saw in the previous section, primary, secondary, and tertiary amines can be alkylated by reaction with alkyl halides. Primary and secondary (but not tertiary) amines can also be acylated by nucleophilic acyl substitution reactions with acid chlorides or acid anhydrides (Sections 10.7 and 10.8). The products are amides.

$$NH_3 \xrightarrow[\text{Pyridine}]{RCOCl} R{-}\overset{O}{\overset{\|}{C}}{-}NH_2 + HCl$$

$$R'NH_2 \xrightarrow[\text{Pyridine}]{RCOCl} R{-}\overset{O}{\overset{\|}{C}}{-}NHR' + HCl$$

$$R'_2NH \xrightarrow[\text{Pyridine}]{RCOCl} R{-}\overset{O}{\overset{\|}{C}}{-}NR'_2 + HCl$$

PROBLEM 12.14 Write an equation for the reaction of diethylamine with acetyl chloride to yield *N,N*-diethylacetamide.

Diazonium Salts: The Sandmeyer Reaction

DIAZOTIZATION Reaction of a primary amine with nitrous acid to yield a diazonium salt, $RN_2^+ X^-$

Primary amines react with nitrous acid, HNO_2, in a **diazotization** reaction to yield diazonium salts, $R{-}\overset{+}{N}{\equiv}N\ X^-$. Although alkyl diazonium salts are too reactive to be isolated, aryl diazonium salts, $Ar{-}\overset{+}{N}{\equiv}N\ X^-$, are more stable.

$$C_6H_5NH_2 + HNO_2 + H_2SO_4 \longrightarrow C_6H_5\overset{+}{N}{\equiv}N\ HSO_4^- + 2\ H_2O$$

Aniline **Benzenediazonium bisulfate**

Aryl diazonium salts are extremely useful compounds, because the diazonio group (N_2^+) can be replaced by nucleophiles in a substitution reaction:

$$C_6H_5\overset{+}{N}{\equiv}N\ HSO_4^- + :Nu^- \longrightarrow C_6H_5Nu + N_2$$

Many different nucleophiles react with arenediazonium salts, and many different substituted benzenes can be prepared by using this reaction. The overall sequence of (1) nitration, (2) reduction, (3) diazotization, and (4) nucleophilic replacement, is probably the single most versatile method for preparing substituted aromatic rings (Figure 12.4, page 364).

SANDMEYER REACTION Conversion of an arenediazonium salt into an aryl halide by reaction with a cuprous halide

Aryl chlorides and bromides are prepared by reaction of an aryl diazonium salt with HX in the presence of a small amount of cuprous halide (CuX) catalyst, a process called the **Sandmeyer reaction**. Aryl iodides are prepared by reaction with sodium iodide.

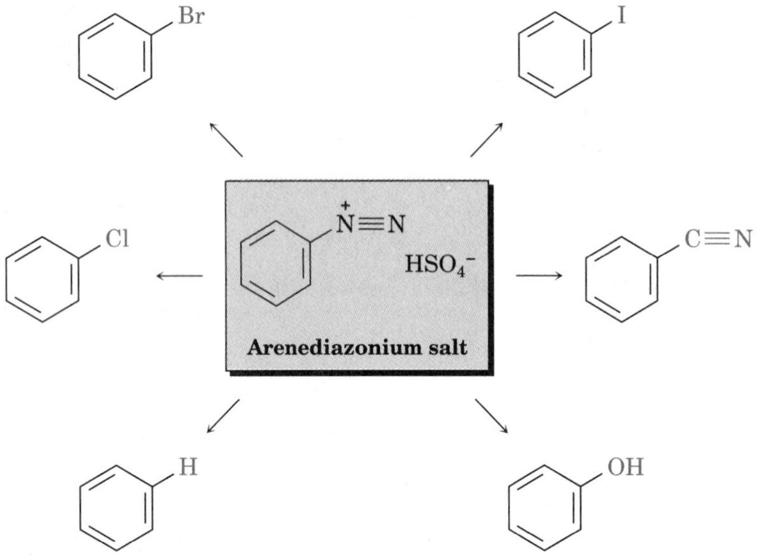

FIGURE 12.4 Preparation of substituted aromatic compounds by diazonio replacement reactions.

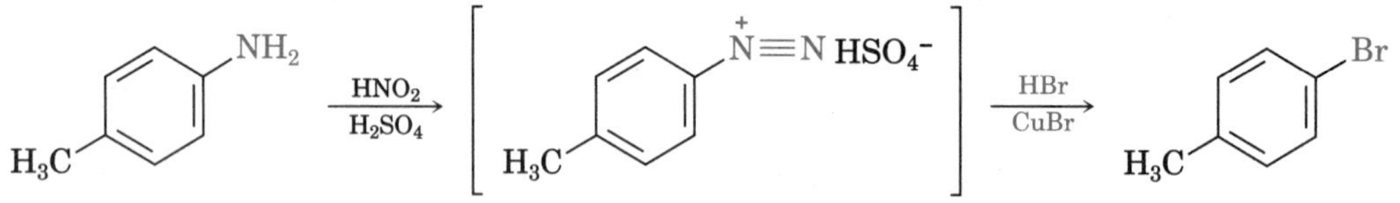

***p*-Methylaniline** **_p_-Bromotoluene (73%)**

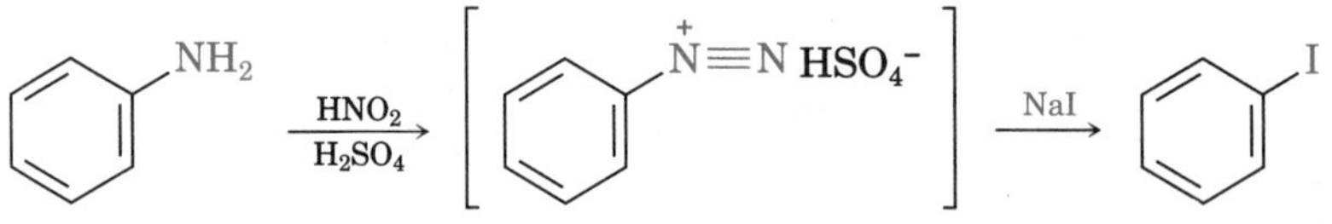

Aniline **Iodobenzene (67%)**

Treatment of an arenediazonium salt with KCN and a small amount of cuprous cyanide, CuCN, yields a nitrile, ArCN. This reaction is particularly useful because it allows the replacement of a nitrogen substituent by a carbon substituent. The nitrile can then be elaborated into other functional groups such as $-COOH$ or $-CH_2NH_2$. For example, hydrolysis of *o*-methylbenzonitrile, produced by a Sandmeyer-type reaction of *o*-methylbenzenediazonium bisulfate with cuprous cyanide, yields *o*-methylbenzoic acid:

$$\text{o-Methylaniline} \xrightarrow[H_2SO_4]{HNO_2} \text{o-Methylbenzenediazonium bisulfate} \xrightarrow[CuCN]{KCN} \text{o-Methylbenzonitrile} \xrightarrow{H_3O^+} \text{o-Methylbenzoic acid}$$

o-Methylaniline **o-Methylbenzene-diazonium bisulfate** **o-Methylbenzonitrile** **o-Methylbenzoic acid**

The diazonio group can also be replaced by –OH or by –H. Phenols are prepared by addition of the aryl diazonium salt to hot aqueous acid. For example,

$$\text{m-Nitroaniline} \xrightarrow[H_2SO_4]{HNO_2} \overset{+}{N}{\equiv}N\ HSO_4^- \xrightarrow{H_3O^+} \text{m-Nitrophenol (86\%)}$$

m-Nitroaniline **m-Nitrophenol (86%)**

Replacement of the diazonio group by –H is accomplished by reaction of the diazonium salt with hypophosphorous acid, H_3PO_2. For example, *p*-methylaniline can be converted into 3,5-dibromotoluene by a sequence involving bromination, diazotization, and hypophosphorous acid treatment:

$$\text{p-Methylaniline} \xrightarrow{2\ Br_2} \text{2,6-dibromo-4-methylaniline} \xrightarrow[H_2SO_4]{HNO_2} \overset{+}{N}{\equiv}N\ HSO_4^- \xrightarrow{H_3PO_2} \text{3,5-Dibromotoluene}$$

p-Methylaniline **3,5-Dibromotoluene**

PRACTICE PROBLEM 12.9 How would you prepare *p*-methylphenol from benzene using a diazonio replacement reaction?

SOLUTION Working backward, the immediate precursor of the target molecule might be *p*-methylbenzenediazonium ion, which could be prepared from *p*-nitrotoluene. *p*-Nitrotoluene, in turn, could be prepared by nitration of toluene, which could be prepared by Friedel–Crafts methylation of benzene.

$$C_6H_6 \xrightarrow[AlCl_3]{CH_3Cl} C_6H_5{-}CH_3 \xrightarrow[H_2SO_4]{HNO_3} O_2N{-}C_6H_4{-}CH_3$$

$$O_2N{-}C_6H_4{-}CH_3 \xrightarrow{Fe,\ HCl} H_2N{-}C_6H_4{-}CH_3 \xrightarrow[2.\ H_3O^+]{1.\ HNO_2} HO{-}C_6H_4{-}CH_3$$

PROBLEM 12.15 How would you prepare *p*-bromobenzonitrile from bromobenzene using a diazonio replacement reaction?

PROBLEM 12.16 How would you prepare these compounds from benzene?
(a) *m*-Bromobenzoic acid **(b)** *m*-Bromochlorobenzene

Diazonium Coupling Reactions

AZO COMPOUND
Compound with the –N=N– functional group

In addition to their reactivity in Sandmeyer-type substitution reactions, arenediazonium salts undergo a coupling reaction with activated aromatic rings, such as phenols and substituted anilines, to yield brightly colored **azo compounds, Ar–N=N–Ar′,** which are widely used as dyes. *p*-(Dimethylamino)azobenzene, for example, is a bright yellow substance that was once used as a coloring agent in margarine.

$$C_6H_5\overset{+}{N}{\equiv}N\ HSO_4^- + C_6H_5\ddot{N}(CH_3)_2 \longrightarrow C_6H_5N{=}N{-}C_6H_4{-}N(CH_3)_2$$

Benzenediazonium bisulfate — ***N,N*-Dimethylaniline** — ***p*-(Dimethylamino)azobenzene** (yellow crystals, mp 127°C)

Diazonium coupling reactions are typical electrophilic aromatic substitutions (Sections 5.6–5.8) in which the positively charged diazonium ion is the electrophile that reacts with the electron-rich ring of a phenol or an arylamine. Reaction almost always occurs at the para position, although ortho attack can take place if the para position is blocked.

PROBLEM 12.17 Propose a synthesis of *p*-(dimethylamino)azobenzene from benzene.

PROBLEM 12.18 Show the mechanism of the azo coupling reaction between phenol and benzenediazonium sulfate.

12.7 HETEROCYCLIC AMINES

CARBOCYCLE
Cyclic molecule that contains only carbon in its ring

HETEROCYCLE
Cyclic molecule that contains more than one kind of atom in its ring

Cyclic organic compounds can be classed either as *carbocycles* or as *heterocycles.* **Carbocycles** contain only carbon atoms in their rings, but **heterocycles** contain one or more different atoms in addition to carbon. Heterocyclic amines are particularly common in organic chemistry, and many have important biological properties. For example, the antiulcer agent cimetidine and the sedative phenobarbital are heterocyclic amines:

Cimetidine (an antiulcer agent) **Phenobarbital (a sedative)**

For the most part, heterocyclic amines have the same chemistry as their open-chain counterparts. In certain cases, though, particularly when the ring is unsaturated, heterocycles have unique and interesting properties. Let's look at several examples.

Pyrrole, a Five-Membered Aromatic Heterocycle

Pyrrole, a five-membered heterocyclic amine, has two double bonds and one nitrogen. Though pyrrole is both an amine and a conjugated diene, its chemistry is not consistent with either of these structural features. Unlike most amines, pyrrole isn't basic; unlike most conjugated dienes, pyrrole doesn't undergo electrophilic addition reactions. How can we explain these observations?

In fact, pyrrole is aromatic. Even though it has a five-membered ring, pyrrole has six pi electrons in a cyclic conjugated pi orbital system, just as benzene does (Section 5.3). Each of the four carbon atoms contributes one pi electron, and the sp^2-hybridized nitrogen atom contributes two more (its lone pair). The six pi electrons occupy *p* orbitals with lobes above and below the plane of the flat ring, as shown in Figure 12.5.

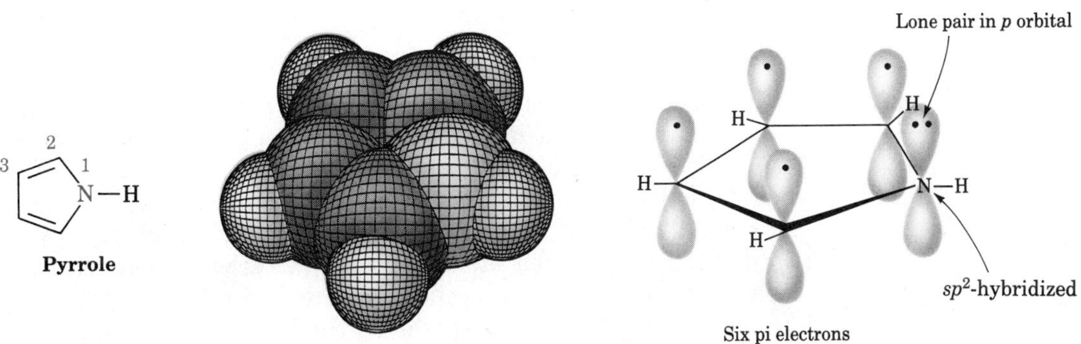

FIGURE 12.5 Pyrrole, an aromatic heterocycle.

Like benzene, pyrrole undergoes substitution of a ring hydrogen atom on reaction with electrophiles. Substitution normally occurs at the position next to nitrogen, as the following nitration shows:

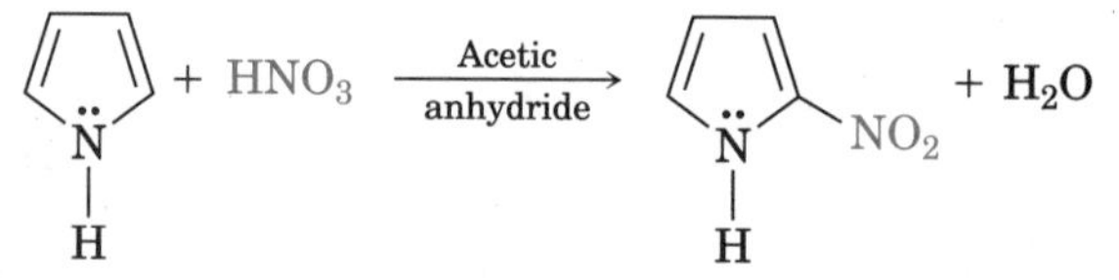

$$\text{Pyrrole} + HNO_3 \xrightarrow[\text{anhydride}]{\text{Acetic}} \text{2-Nitropyrrole (83\%)} + H_2O$$

Pyrrole **2-Nitropyrrole (83%)**

Substituted pyrrole rings form the basic building blocks from which a number of important plant and animal pigments are constructed. Among these is *heme*, an iron-containing tetrapyrrole in blood.

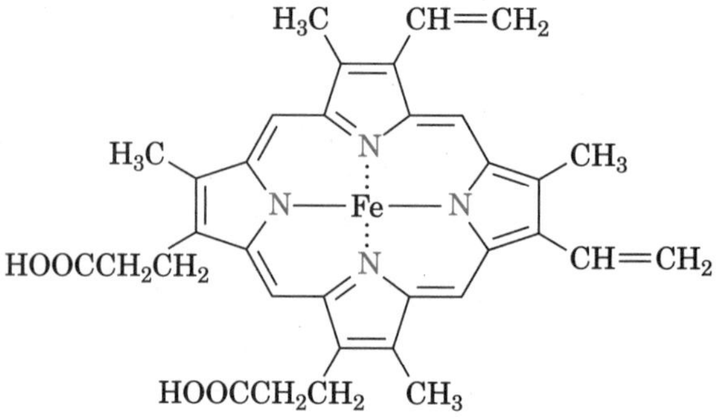

Heme

PROBLEM 12.19 Pyrrole undergoes other typical electrophilic substitution reactions in addition to nitration. What products would you expect to obtain from reaction of *N*-methylpyrrole with these reagents?
(a) Br_2 (b) CH_3Cl, $AlCl_3$ (c) CH_3COCl, $AlCl_3$

PROBLEM 12.20 Review the mechanism of the bromination of benzene (Section 5.6) and then propose a mechanism for the nitration of pyrrole.

Pyridine, a Six-Membered Aromatic Heterocycle

Pyridine is the nitrogen-containing heterocyclic analog of benzene. Like benzene, pyridine is a flat molecule with bond angles of approximately 120° and with carbon–carbon bond lengths of 1.39 Å, intermediate between normal single and double bonds. Also like benzene, pyridine is aromatic, with six pi electrons in a cyclic, conjugated pi orbital system. The sp^2-hybridized nitrogen atom and the five carbon atoms each contribute one pi electron to the cyclic conjugated pi orbitals of the ring. Unlike the situation in pyrrole, however, the lone-pair electrons on the pyridine nitrogen atom are not part of the pi orbital system but instead occupy an sp^2 orbital in the plane of the ring (Figure 12.6).

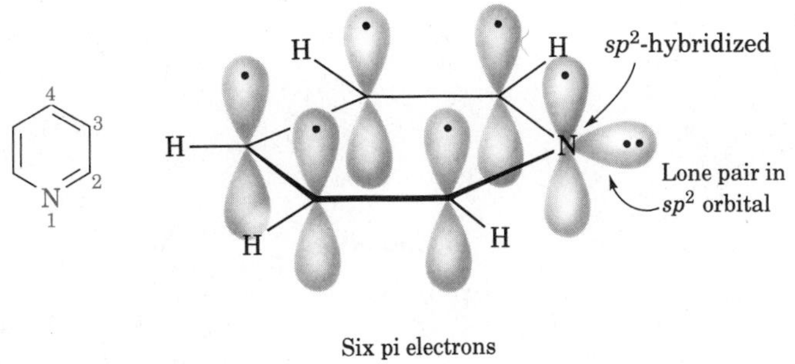

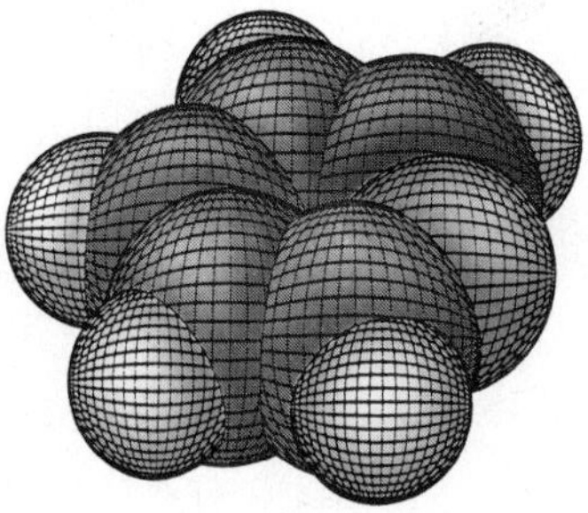

FIGURE 12.6 Electronic structure of pyridine.

Substituted pyridines such as the B_6 complex vitamins pyridoxal and pyridoxine are important biologically. Present in yeast, cereal, and other foodstuffs, the B_6 vitamins play an important role in the synthesis of some amino acids.

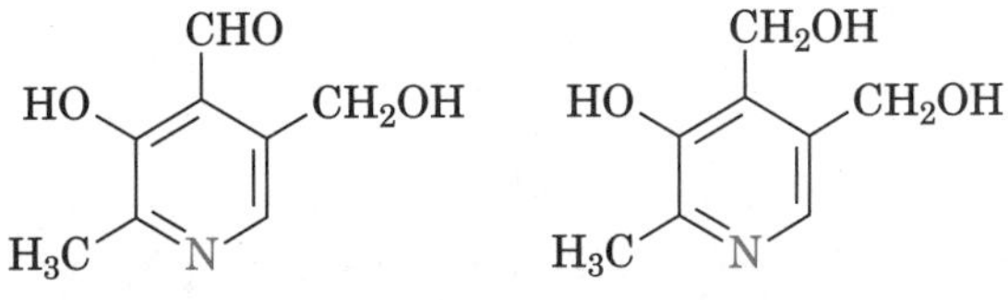

Pyridoxal **Pyridoxine**

PROBLEM 12.21 The five-membered heterocycle imidazole contains two nitrogen atoms, one "pyrrole-like" and one "pyridine-like." Draw an orbital picture of imidazole and indicate in which orbital each nitrogen has its lone-pair electrons.

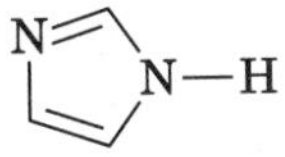

Imidazole

Fused-Ring Aromatic Heterocycles

Quinoline, isoquinoline, and indole are *fused-ring* heterocycles that contain both a benzene ring and a heterocyclic aromatic ring sharing a common bond. All three ring systems occur widely in nature, and many members of the class have useful biological activity. Thus, quinine, a quinoline derivative found in the bark of the South American cinchona tree, is an important antimalarial drug. Lysergic acid, an indole derivative found in the ergot fungus that grows on rotting grain, is the parent acid from which the psychoactive drug LSD is derived.

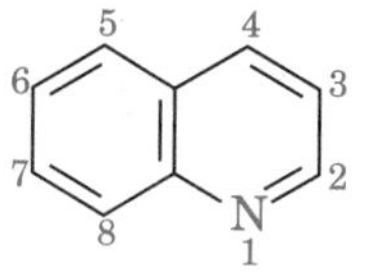

Quinoline 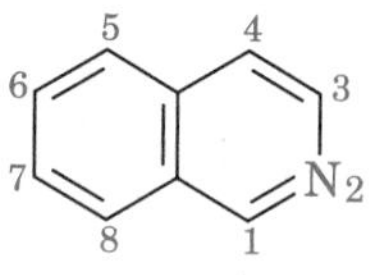 Isoquinoline 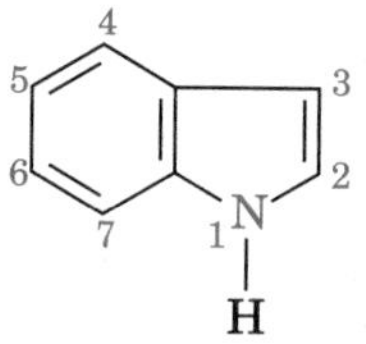 Indole

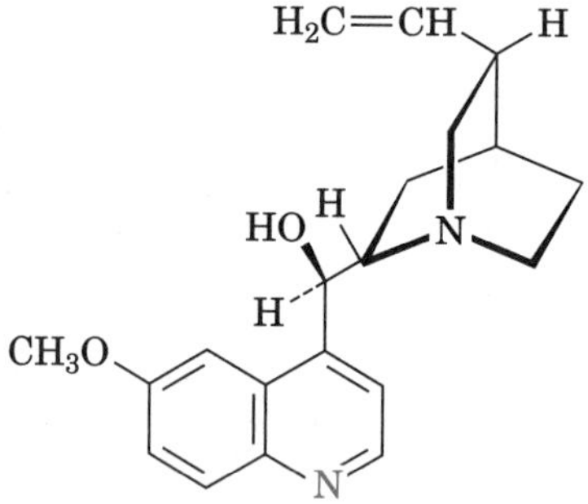

Quinine 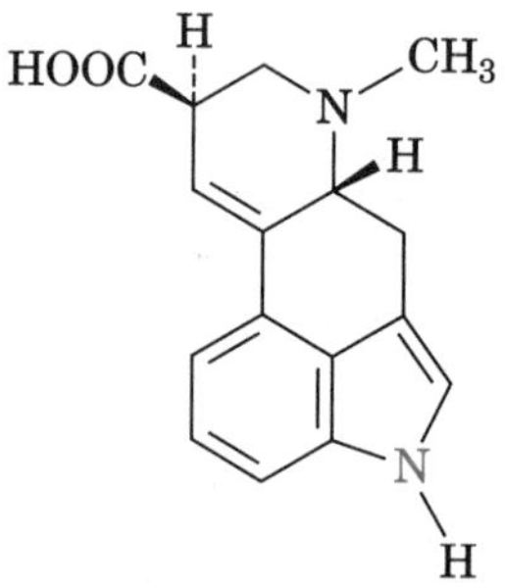 Lysergic acid

12.8 NATURALLY OCCURRING AMINES: MORPHINE ALKALOIDS

ALKALOID
Naturally occurring amine

Amines were among the first organic compounds to be isolated, and an enormous variety of amines is found in both plants and animals. Morphine, for example, is a powerful analgesic (painkiller) isolated from the opium poppy, *Papaver somniferum.* Once known as "vegetable alkali" because their water solutions are basic, naturally occurring amines such as morphine are now referred to as **alkaloids**.

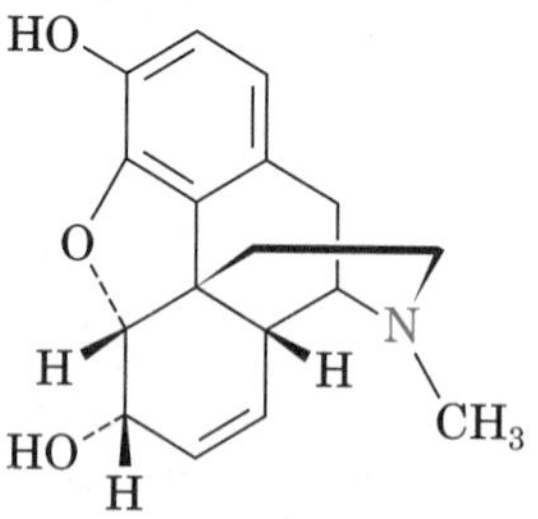

Morphine (an analgesic)

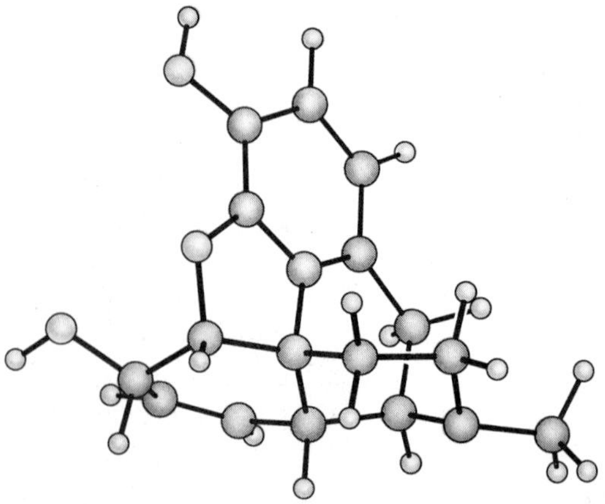

The medical uses of the morphine family of alkaloids have been known at least since the seventeenth century, when crude extracts of the opium poppy were used for the relief of pain. Morphine was the first pure alkaloid to be isolated from the poppy, but its close relative, codeine, also occurs naturally. Codeine, which is simply the methyl ether of morphine, is used in prescription cough medicines. Heroin, another close relative of morphine, does not occur naturally but is synthesized by diacetylation of morphine.

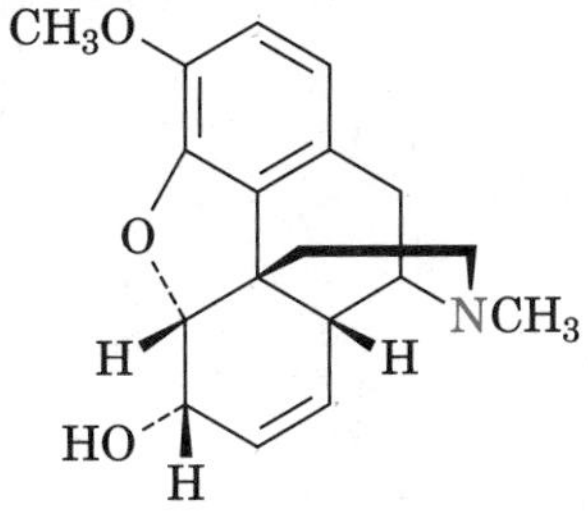

Codeine 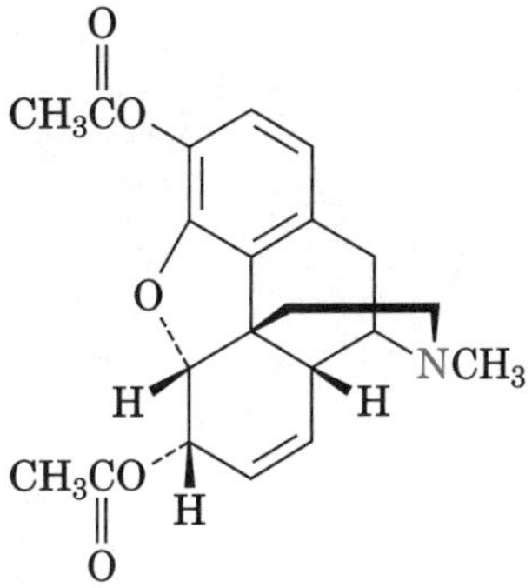 **Heroin**

Morphine and its relatives are extremely useful pharmaceutical agents, yet they also pose an enormous societal problem because of their addictive properties. Much effort has gone into studying how morphine works and into developing modified morphine analogs that retain the desired painkilling activity without causing addiction. Our present understanding is that morphine binds to opiate receptor sites in the brain, changing the brain's perception of pain. Hundreds of morphine-like molecules havé been synthesized and tested for their analgesic properties.

Studies have shown that not all of the complex framework of morphine is necessary for biological activity. According to the "morphine rule," biological activity requires: (1) an aromatic ring attached to (2) a quaternary carbon atom, and (3) a tertiary amine situated (4) two carbon atoms farther away. For example, meperidine (Demerol) is widely used as a painkiller, and methadone has been used as an antagonist in the treatment of heroin addiction to reverse the undesirable side effects of morphine.

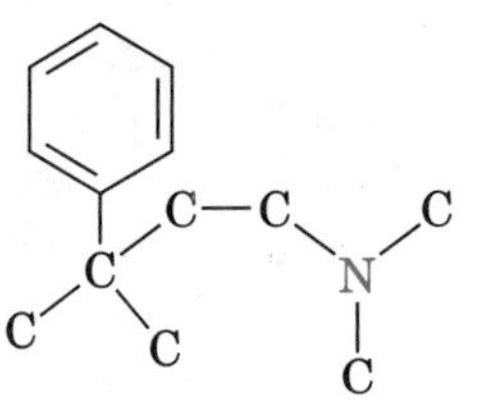

The morphine rule:
an aromatic ring, attached to a quarternary carbon, attached to two or more carbons, attached to a tertiary amine

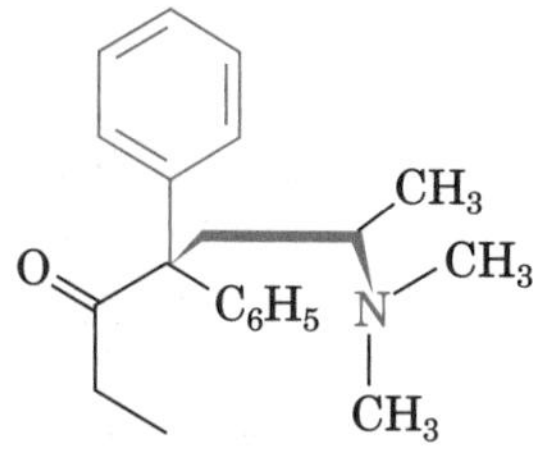

Methadone

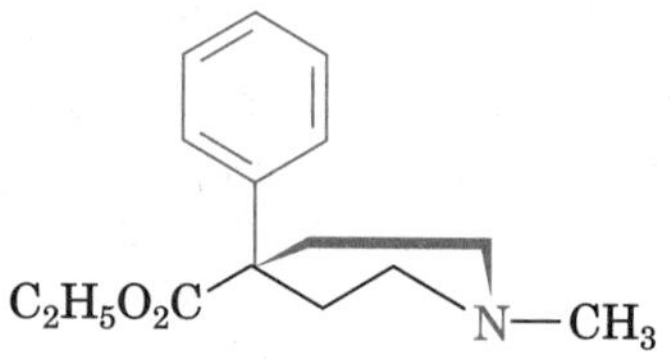

Meperidine

PROBLEM 12.22 Show how the morphine rule fits the structure of dextromethorphan, a common cough remedy.

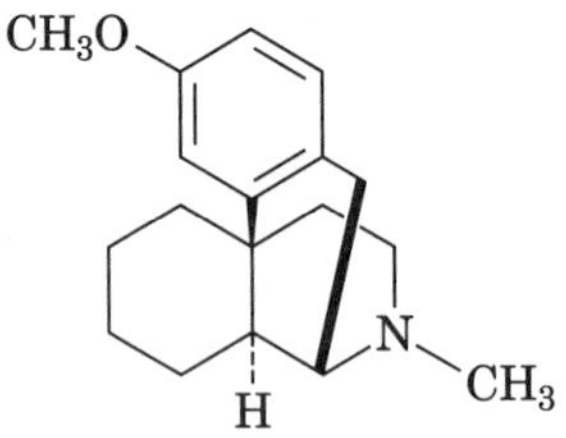

Dextromethorphan

INTERLUDE

ORGANIC DYES AND THE CHEMICAL INDUSTRY

The founding of the modern organic chemical industry can be traced to the need for a single organic compound—aniline—and to the activities of one person, Sir William Henry Perkin. Perkin, a student at the Royal College of Chemistry in London, spent his free time working in an improvised home laboratory, and decided during Easter vacation in 1856 to examine the oxidation of aniline with potassium dichromate. Although the reaction appeared unpromising at first, yielding a tarry black product, Perkin was able by careful extraction with methanol to isolate a few percent yield of a beautiful purple pigment with the properties of a dye.

Since the only dyes known at the time were naturally occurring vegetable dyes like indigo, Perkin's synthetic purple dye, which he named *mauve*, created a sensation. Realizing the commercial possibilities, Perkin did what any young entrepreneur would do today. He resigned his post at the Royal College and, at the age of 18, formed a company to exploit his remarkable discovery. Since there had never before been a need for synthetic chemicals, no chemical industry existed at the time. Large-scale chemical manufacture was unknown, and Perkin therefore devised a procedure for preparing the needed quantities of aniline by nitration of benzene.

Subsequent work showed that Perkin's original mauve was in fact not derived from aniline at all but from a small amount of methylaniline impurity in his starting material. Pure aniline yields a similar dye, however, which came to be marketed under the name *pseudomauveine.*

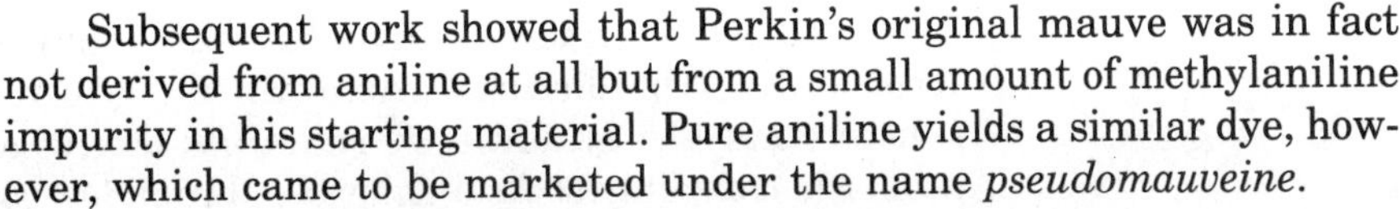

Perkin's mauve
(pseudomauveine has no methyl groups)

Today, dyestuff manufacture is a thriving and important part of the chemical industry, and many pigments such as *p*-(dimethylamino)azobenzene, used at one time as a food-coloring agent under the name "butter yellow," are derived from aniline (Section 12.6).

SUMMARY AND KEY WORDS

Amines are organic derivatives of ammonia. They are named in the IUPAC system either by adding the suffix *-amine* to the names of the alkyl substituents or by considering the amino group as a substituent on a more complex parent molecule.

Bonding in amines is similar to that in ammonia. The nitrogen atom is sp^3-hybridized, the three substituents are directed to three corners of a tetrahedron, and the lone pair of nonbonding electrons occupies the fourth corner of the tetrahedron.

The chemistry of amines is dominated by the presence of the lone-pair electrons on nitrogen. Thus, amines are both basic and nucleophilic. **Arylamines** are generally weaker bases than **alkylamines,** because their lone-pair electrons are delocalized by orbital overlap with the aromatic pi electron system.

The simplest method of amine synthesis involves S_N2 reaction of ammonia or an amine with an alkyl halide. Alkylation of ammonia yields a **primary amine**; alkylation of a primary amine yields a **secondary amine**; and so on. Amines can also be prepared from amides and nitriles by reduction with $LiAlH_4$. Arylamines are prepared by nitration of an aromatic ring, followed by reduction of the nitro group.

Many of the reactions that amines undergo are familiar from previous chapters. Thus, amines react with alkyl halides in S_N2 reactions and with acid chlorides in nucleophilic acyl substitution reactions. Arylamines are converted by treatment with nitrous acid into **aryl diazonium salts, $Ar{-}N_2^+\ X^-$**. The diazonio group can then be replaced by many other substituents (the **Sandmeyer reaction**) to give a variety of substituted aromatic compounds. Aryl chlorides, bromides, iodides, and nitriles can be prepared, as can phenols and arenes.

Heterocyclic amines, compounds in which the nitrogen atom is in a ring, are greatly diverse in their structures and properties. For example, pyrrole, pyridine, indole, and quinoline all show aromatic properties.

SUMMARY OF REACTIONS

1. Synthesis of amines (Section 12.5)

(a) Alkylamines by S_N2 reaction

$$NH_3 + RX \longrightarrow RNH_2$$
$$RNH_2 + RX \longrightarrow R_2NH$$
$$R_2NH + RX \longrightarrow R_3N$$
$$R_3N + RX \longrightarrow R_4N^+\ X^-$$

(b) Arylamines by reduction of nitroarenes

$$C_6H_5NO_2 \xrightarrow{Fe,\ H_3O^+} C_6H_5NH_2$$

2. Reactions of amines (Section 12.6)

(a) Formation and reactions of arenediazonium salts

$$C_6H_5NH_2 \xrightarrow[H_2SO_4]{HNO_2} C_6H_5\overset{+}{N}{\equiv}N\ HSO_4^-$$

- $\xrightarrow{HCl,\ CuCl}$ C_6H_5Cl
- $\xrightarrow{HBr,\ CuBr}$ C_6H_5Br
- $\xrightarrow{NaI}$ C_6H_5I
- $\xrightarrow{KCN,\ CuCN}$ C_6H_5CN
- $\xrightarrow{H_3O^+}$ C_6H_5OH
- $\xrightarrow{H_3PO_2}$ C_6H_5H

(b) Diazo coupling

$$C_6H_5\overset{+}{N}{\equiv}N\ HSO_4^- + C_6H_5OH\ (\text{or } {-}NR_2) \longrightarrow C_6H_5{-}N{=}N{-}C_6H_4{-}OH\ (\text{or } {-}NR_2)$$

ADDITIONAL PROBLEMS

12.23 Classify each of the amine (not amide) nitrogen atoms in the following substances as either primary, secondary, or tertiary:

(a) $(C_2H_5)_2N-C(=O)$ ... $N-CH_3$ ... N—H

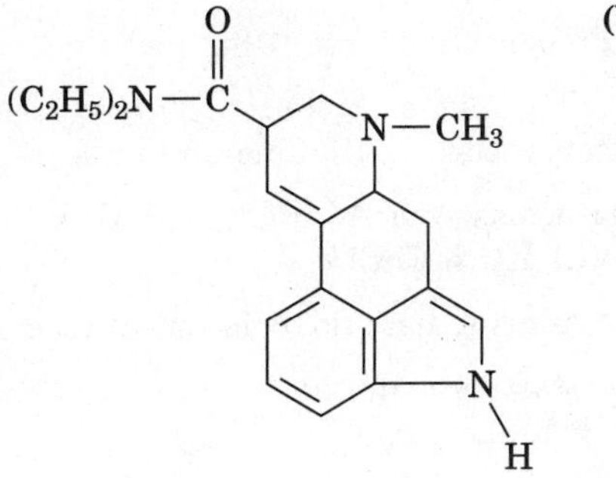

Lysergic acid diethylamide

(b) H_3C, N, O, CH_3, N, N, O, N, CH_3

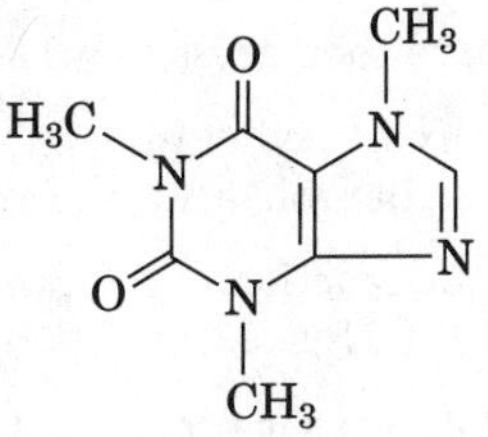

Caffeine

12.24 Draw structures corresponding to these IUPAC names:
(a) *N,N*-Dimethylaniline (b) *N*-Methylcyclohexylamine
(c) (Cyclohexylmethyl)amine (d) (2-Methylcyclohexyl)amine
(e) 3-(*N,N*-Dimethylamino)propanoic acid

12.25 Name these compounds according to IUPAC rules:
(a) NH_2, Br, Br
(b) cyclopentyl$-CH_2CH_2NH_2$
(c) cyclopentyl$-NHCH_2CH_3$
(d) cyclopentyl$-N(CH_3)_2$
(e) pyrrolidine $N-CH_2CH_2CH_3$
(f) $H_2NCH_2CH_2CH_2CN$

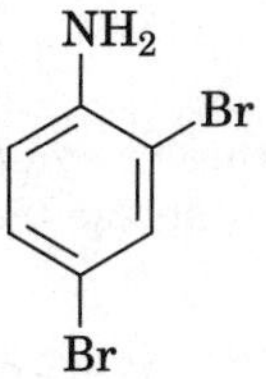

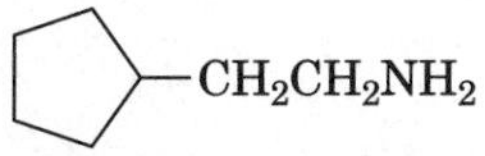

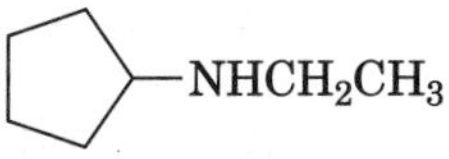

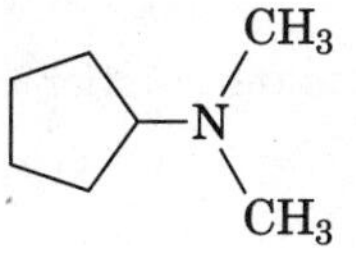

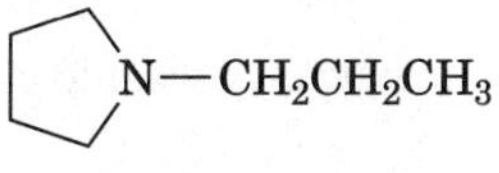

12.26 How can you explain the fact that trimethylamine (bp 3°C) has a lower boiling point than dimethylamine (bp 7°C)?

12.27 Mescaline, a powerful hallucinogen derived from the peyote cactus, has the systematic name 2-(3,4,5-trimethoxyphenyl)ethylamine. Draw its structure.

12.28 There are eight isomeric amines with the formula $C_4H_{11}N$. Draw them, name them, and classify each as primary, secondary, or tertiary.

12.29 Propose structures for amines that fit these descriptions:
(a) A secondary arylamine (b) A 1,3,5-trisubstituted arylamine
(c) An achiral quaternary ammonium salt (d) A five-membered heterocyclic amine

12.30 Show the products of these reactions:
(a) $CH_3CH_2CH_2NH_2 + CH_3Br \longrightarrow ?$ (b) Cyclohexylamine + HBr $\longrightarrow$?
(c) $CH_3CH_2CONH_2 + LiAlH_4 \longrightarrow ?$ (d) Benzonitrile + $LiAlH_4 \longrightarrow$?

12.31 How might you prepare these amines, starting from ammonia and any alkyl halides needed?
(a) Hexylamine **(b)** Benzylamine **(c)** Tetramethylammonium iodide
(d) *N*-Methylcyclohexylamine

12.32 How might you prepare each of these amines from 1-bromobutane?
(a) Butylamine **(b)** Dibutylamine **(c)** Pentylamine

12.33 How might you prepare each of the amines in Problem 12.32 from 1-butanol?

12.34 How would you prepare benzylamine, $C_6H_5CH_2NH_2$, from each of these starting materials?
(a) Benzamide **(b)** Benzoic acid **(c)** Nitrobenzene **(d)** Chlorobenzene

12.35 Write equations for the reaction of *p*-bromobenzenediazonium bisulfate with these reagents:
(a) H_3O^+ **(b)** HBr, CuBr **(c)** H_3PO_2 **(d)** KCN, CuCN

12.36 Show how you might prepare benzoic acid from aniline. A diazonio replacement reaction is needed.

12.37 How might you prepare pentylamine from these starting materials?
(a) Pentanamide **(b)** Pentanenitrile **(c)** Pentanoic acid

12.38 Which compound is more basic, $CH_3CH_2NH_2$ or $CF_3CH_2NH_2$? Explain.

12.39 Which compound is more basic, *p*-aminobenzaldehyde or aniline?

12.40 Which compound is more basic, triethylamine or aniline? Does the following reaction proceed as written?

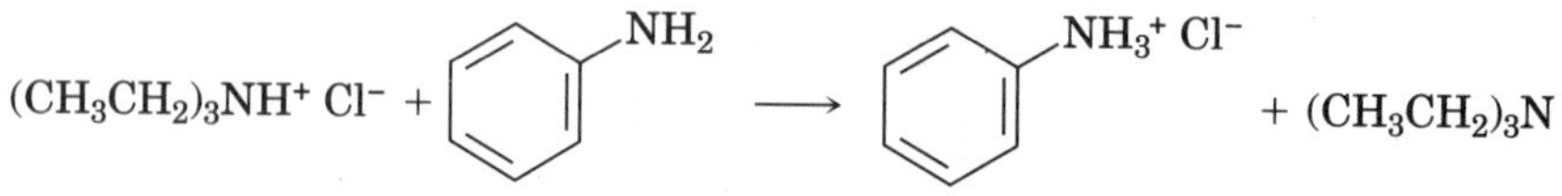

$$(CH_3CH_2)_3NH^+\ Cl^- + C_6H_5NH_2 \longrightarrow C_6H_5NH_3^+\ Cl^- + (CH_3CH_2)_3N$$

12.41 1,6-Hexanediamine, one of the starting materials used for the manufacture of nylon 66, can be synthesized by a route that begins with the addition of chlorine to 1,3-butadiene (Section 4.10). How would you carry out the complete synthesis?

12.42 Another method for making 1,6-hexanediamine (see Problem 12.41) starts from adipic acid (hexanedioic acid). How would you carry out the synthesis?

12.43 Give the structures of the major organic products you would obtain from the reaction of *m*-methylaniline with these reagents:
(a) Br_2 (1 mol) **(b)** CH_3I (excess) **(c)** CH_3COCl, pyridine

12.44 Suppose that you were given a mixture of toluene, aniline, and phenol. Describe how you would separate the mixture into its three pure components.

12.45 Would you expect diphenylamine to be more basic or less basic than aniline? Explain.

12.46 Draw structures for these amines:
(a) 2-Ethylpyrrole **(b)** 2,3-Dimethylaniline **(c)** 3-Methylindole

12.47 Furan, the ether analog of pyrrole, is aromatic in the same way that pyrrole is. Draw an orbital picture of furan and show how it has six electrons in its cyclic conjugated pi orbitals.

12.48 By analogy with the chemistry of pyrrole, what product would you expect from the reaction of furan with Br_2 (see Problem 12.47)?

12.49 How would you synthesize 1,3,5-tribromobenzene from benzene? (A diazonio replacement reaction is needed.)

12.50 We've seen that amines are basic and amides are neutral. *Imides*, compounds with two carbonyl groups flanking an N–H, are actually acidic. Show by drawing resonance structures of the anion why imides are acidic.

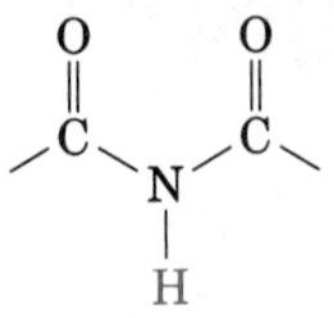

An imide

12.51 Tyramine is an alkaloid found, among other places, in mistletoe and in ripe cheese. How would you prepare tyramine from toluene?

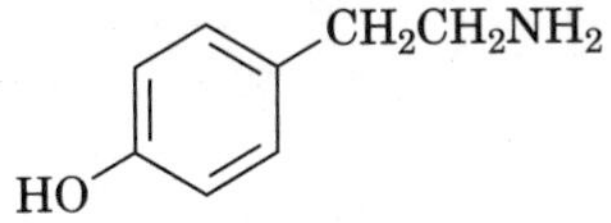

Tyramine

12.52 Atropine, $C_{17}H_{23}NO_3$, is a poisonous alkaloid isolated from the leaves and roots of *Atropa belladonna*, the deadly nightshade. In low doses, atropine acts as a muscle relaxant: 0.5 ng (1 nanogram = 10^{-9} g) is sufficient to cause pupil dilation. On reaction with aqueous NaOH, atropine yields tropic acid, $C_6H_5CH(CH_2OH)COOH$, and tropine, $C_8H_{15}NO$. Tropine, an optically inactive alcohol, yields tropidene on dehydration. Propose a structure for atropine.

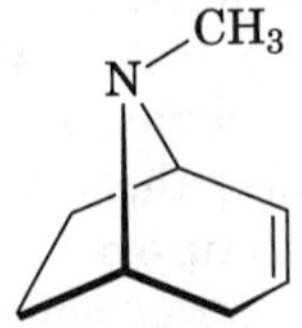

Tropidene

12.53 Choline, a component of the phospholipids in cell membranes, can be prepared by S_N2 reaction of trimethylamine with ethylene oxide. Propose a mechanism for the reaction.

12.54 Methyl orange is an azo dye that is widely used as a pH indicator. How would you synthesize methyl orange from benzene?

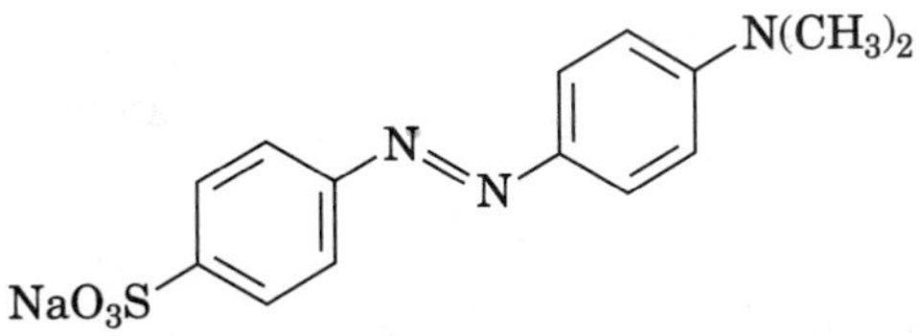

Methyl orange

13 CHAPTER STRUCTURE DETERMINATION

Structure determination is at the center of organic chemistry. Every time a reaction is run, the products have to be isolated, purified, and identified. In the nineteenth and early twentieth centuries, determining the structure of an organic molecule was a time-consuming process requiring skill and patience. In the past few decades, though, extraordinary advances have been made in chemical instrumentation. Sophisticated instruments are now available that greatly simplify structure determination.

What are the instruments for determining structures, and how are they used? We'll answer these questions by looking at three of the most useful methods of structure determination—infrared spectroscopy, ultraviolet spectroscopy, and nuclear magnetic resonance spectroscopy—each of which yields a different kind of structural information.

Infrared spectroscopy	What functional groups are present?
Ultraviolet spectroscopy	Is a conjugated pi electron system present?
Nuclear magnetic resonance spectroscopy	What carbon–hydrogen framework is present?

13.1 INFRARED SPECTROSCOPY AND THE ELECTROMAGNETIC SPECTRUM

Infrared (IR) spectroscopy is a method of structure determination that depends on the interaction of molecules with infrared radiant energy. Before beginning a study of infrared spectroscopy, however, we need to look into the nature of radiant energy and the electromagnetic spectrum.

ELECTROMAGNETIC RADIATION
Different kinds of radiant energy that make up the electromagnetic spectrum

ELECTROMAGNETIC SPECTRUM
The total range of electromagnetic radiation

Visible light, X rays, microwaves, radio waves, and so forth, are all different kinds of **electromagnetic radiation**. Collectively, they make up the **electromagnetic spectrum**, shown in Figure 13.1. The electromagnetic spectrum is loosely divided into regions, with the familiar visible region accounting for only a small portion of the overall spectrum (from 3.8×10^{-5} to 7.8×10^{-5} cm in wavelength). The visible region is flanked by the infrared and ultraviolet regions.

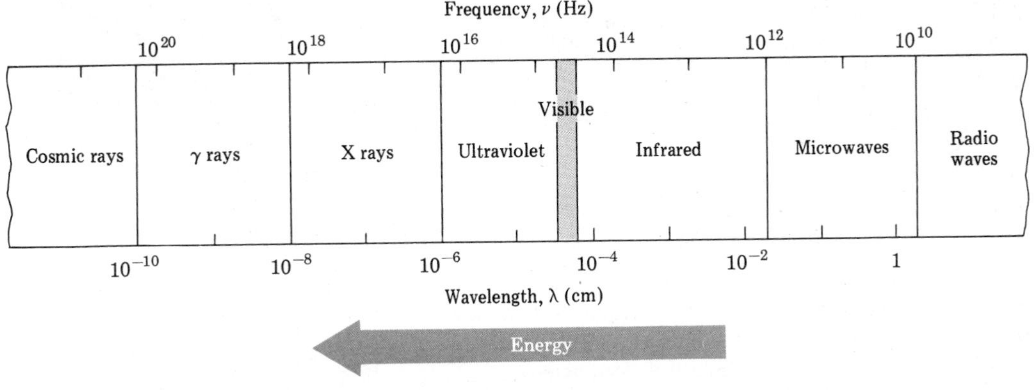

FIGURE 13.1 The electromagnetic spectrum.

FREQUENCY (ν)
Number of waves passing a fixed point per unit of time

HERTZ (Hz)
Unit of frequency measurement; reciprocal seconds, s^{-1}

WAVELENGTH (λ)
Length of a complete wave cycle

AMPLITUDE
Height of a wave from midpoint to peak

Electromagnetic radiation can be thought of as having dual behavior. In some respects it has the properties of a particle (called a *photon*), yet in other respects it behaves as a wave traveling at the speed of light. Like all waves, electromagnetic radiation is characterized by a *frequency*, a *wavelength*, and an *amplitude* (see Figure 13.2, page 380). The **frequency, ν** (Greek nu), of a wave is the number of peaks that pass by a fixed point per unit of time, usually expressed in reciprocal seconds (s^{-1}), or **hertz, Hz** ($1 \text{ Hz} = 1 \text{ s}^{-1}$). The **wavelength, λ** (Greek lambda), of the wave is the length from one wave maximum to the next. The **amplitude** of the wave is its height measured from the midpoint between peak and trough. (The intensity of radiant energy, whether a feeble beam or a blinding glare, is proportional to the square of the wave's amplitude.)

Multiplying the wavelength of a wave in centimeters (cm) by its frequency in reciprocal seconds (s^{-1}) gives the speed of the wave in centimeters per second (cm/s). The rate of travel of all electromagnetic radiation in a vacuum is a constant value, commonly called the "speed of light" and abbreviated *c*. It is one of the most accurately known of all physical constants, with a numerical value of $2.997\,924\,58 \times 10^{10}$ cm/s, usually rounded off to 3.00×10^{10} cm/s.

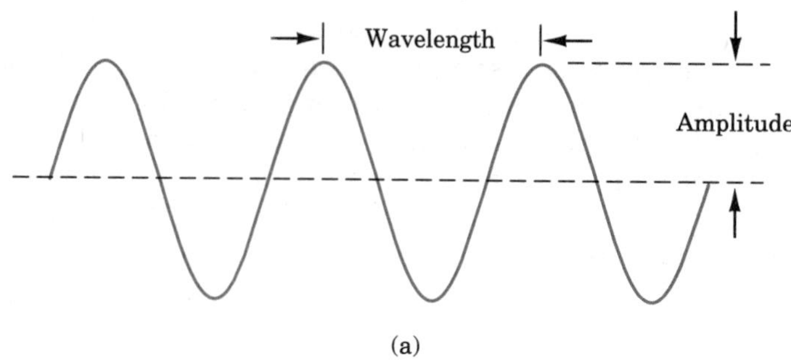

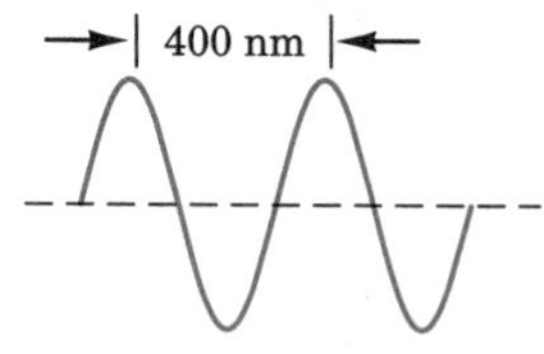

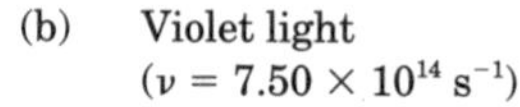

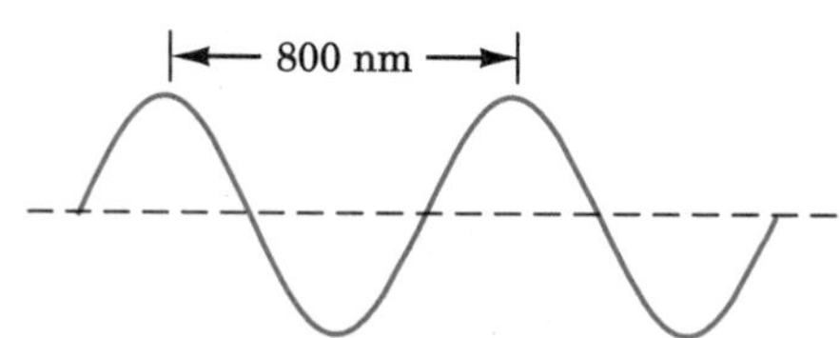

(c) Infrared radiation
($\nu = 3.75 \times 10^{14}\ s^{-1}$)

FIGURE 13.2 (a) Wavelength (λ) is the distance between two successive wave maxima. Amplitude is the height of the wave measured from the center. (b)–(c) What we perceive as different kinds of electromagnetic radiation are simply waves with different wavelengths and frequencies.

$$\text{Wavelength} \times \text{frequency} = \text{speed}$$
$$\lambda\ (\text{cm}) \times \nu\ (\text{s}^{-1}) = c\ (\text{cm/s})$$

This can be rewritten as:

$$\lambda = \frac{c}{\nu} \quad \text{or} \quad \nu = \frac{c}{\lambda}$$

Electromagnetic energy is transmitted only in discrete energy bundles, called *quanta*. The amount of energy ε corresponding to 1 quantum of energy (or 1 photon) of a given frequency ν is expressed by the equation

$$\varepsilon = h\nu = \frac{hc}{\lambda}$$

where

ε = Energy of 1 photon (1 quantum)

h = Planck's constant (6.62×10^{-34} J · s = 1.58×10^{-34} cal · s)

ν = Frequency (s^{-1})

λ = Wavelength (cm)

c = Speed of light (3×10^{10} cm/s)

This equation says that the energy of a given photon varies *directly* with its frequency ν but *inversely* with its wavelength λ. High frequencies and short wavelengths correspond to high-energy radiation such as gamma rays; low frequencies and long wavelengths correspond to low-energy radiation such as radio waves.

When an organic compound is exposed to electromagnetic radiation, it absorbs energy of certain wavelengths and transmits energy of other wavelengths. Thus, if we irradiate an organic compound with energy of many wavelengths and determine which are absorbed and which are transmitted, we can determine the **absorption spectrum** of the compound. The results are displayed on a graph that plots the wavelength versus the amount of radiation transmitted through the sample.

ABSORPTION SPECTRUM
Plot of wavelength versus absorption that results when a compound is irradiated with electromagnetic radiation

The IR spectrum of ethanol is shown in Figure 13.3. The horizontal axis shows the wavelength in micrometers (μm), and the vertical axis shows the intensity of the corresponding energy absorptions in percent transmittance. The baseline corresponding to 0% absorption (or 100% transmittance) runs along the top of the chart, and a downward spike means that energy absorption has occurred at that wavelength.

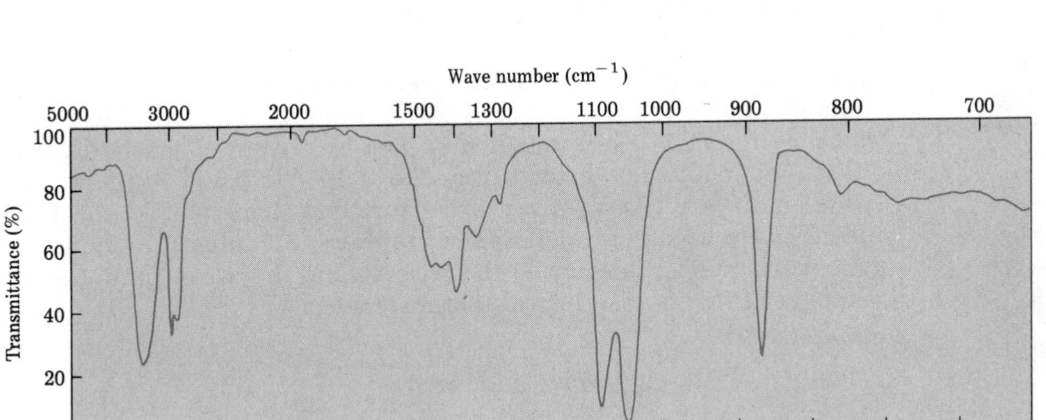

FIGURE 13.3 The infrared spectrum of ethanol. A transmittance of 100% means that all the energy is passing through the sample. A lower transmittance means that some energy is being absorbed. Thus, each downward spike corresponds to an energy absorption.

PRACTICE PROBLEM 13.1 Which is higher in energy, FM radio waves with a frequency of 1.015×10^8 Hz (101.5 MHz) or visible light with a frequency of 5×10^{14} Hz?

SOLUTION The equation $\varepsilon = h\nu$ says that energy increases as frequency increases. Thus, visible light is higher in energy than radio waves.

PRACTICE PROBLEM 13.2 What is the wavelength of visible light with a frequency of 4.5×10^{14} Hz?

SOLUTION Frequency and wavelength are related by the equation $\lambda = c/\nu$, where c is the speed of light (3.0×10^{10} cm/s):

$$\lambda = \frac{3.0 \times 10^{10} \text{ cm/s}}{4.5 \times 10^{14} \text{ s}^{-1}} = 6.7 \times 10^{-5} \text{ cm}$$

PROBLEM 13.1 How does the energy of infrared radiation with $\lambda = 10^{-4}$ cm compare with that of an X ray with $\lambda = 3 \times 10^{-7}$ cm?

PROBLEM 13.2 Which is higher in energy, radiation with $\nu = 4 \times 10^9$ Hz or radiation with $\lambda = 9 \times 10^{-4}$ cm?

13.2 INFRARED SPECTROSCOPY OF ORGANIC MOLECULES

WAVE NUMBER Unit of frequency measurement equal to the reciprocal of the wavelength ($\tilde{\nu} = 1/\lambda$ cm)

The infrared region of the electromagnetic spectrum covers the range from just above the visible (7.8×10^{-5} cm) to approximately 10^{-2} cm, but only the middle of the region is used by organic chemists (Figure 13.4). This midportion extends from 2.5×10^{-3} to 2.5×10^{-4} cm, and wavelengths are usually given in micrometers (1 μm $= 10^{-6}$ m). Frequencies are usually given in **wave numbers ($\tilde{\nu}$)**, rather than in hertz. The wave number is equal to the reciprocal of the wavelength, and is expressed in units of reciprocal centimeters (cm^{-1}):

$$\text{Wave number } (\tilde{\nu}) = \frac{1}{\lambda \text{ (cm)}}$$

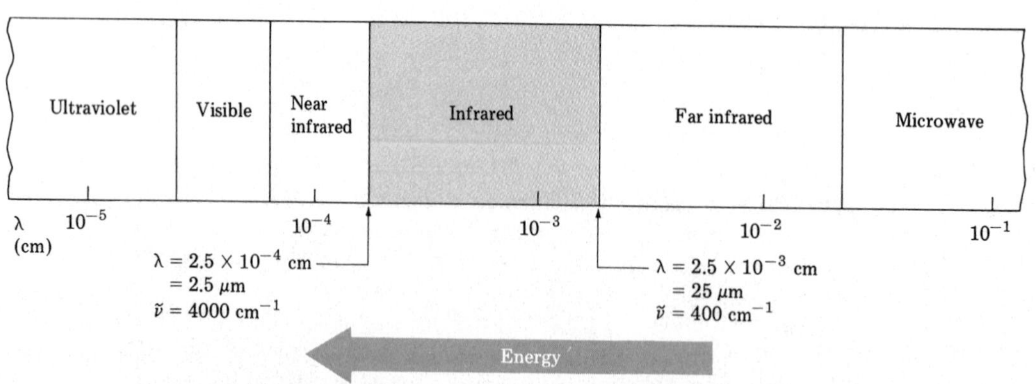

FIGURE 13.4 The infrared (IR) region of the electromagnetic spectrum.

Why does a molecule absorb some wavelengths of infrared light but not others? All molecules have a certain amount of energy that causes bonds to stretch and bend, atoms to wag and rock, and other molecular vibrations to occur. The amount of energy a molecule contains is not continuously variable, though, but is *quantized.* That is, a molecule can bend or vibrate only at specific frequencies that correspond to specific energy levels. When the molecule is irradiated with infrared light, *the vibrating bond will absorb energy only if the frequencies of the light and of the vibration are the same.*

Since each infrared absorption corresponds to a specific molecular motion, we can see what kinds of motions a molecule has by observing its infrared spectrum. By then working backward and interpreting the molecule's motions, we can find out what kinds of bonds (functional groups) are present in the molecule.

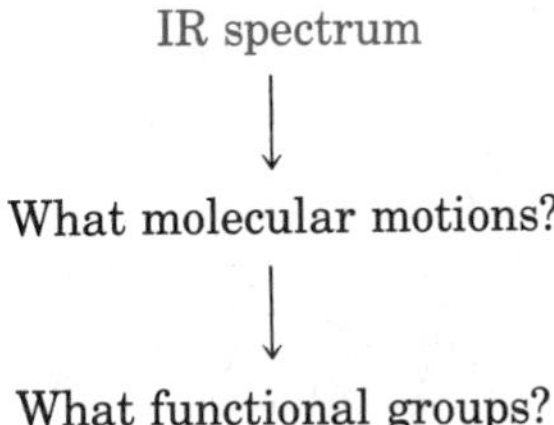

The full interpretation of an IR spectrum is difficult because most organic molecules have dozens of different bond stretching and bending motions. Thus, an IR spectrum usually contains dozens of absorptions. Fortunately, we don't have to interpret an IR spectrum fully to get useful information, because *functional groups have characteristic infrared absorptions that don't change from one compound to another*. The C=O absorption of ketones is almost always in the range 1690–1750 cm^{-1}; the O–H absorption of alcohols is almost always in the range 3200-3600 cm^{-1}; and the C=C absorptions of alkenes is almost always in the range 1640–1680 cm^{-1}. By learning to recognize where characteristic functional-group absorptions occur, it's possible to interpret infrared spectra.

Look at Figure 13.5 (p. 384) to see how IR spectra can be used. Although the IR spectra of 1-hexanol and 2-hexanone both contain many peaks, the characteristic absorptions of the different functional groups allow the compounds to be distinguished. 1-Hexanol shows a characteristic alcohol O–H absorption at 3300 cm^{-1} and a C–O absorption at 1060 cm^{-1}; 2-hexanone shows a characteristic ketone C=O peak at 1715 cm^{-1}.

One other point about infrared spectroscopy: It's also possible to obtain structural information from an IR spectrum by noticing which absorptions are *not* present. If the spectrum of an unknown has no absorption near 3400 cm^{-1}, the unknown isn't an alcohol; if the spectrum has no absorption near 1715 cm^{-1}, the unknown isn't a ketone; and so on. Table 13.1 (p. 385) lists the characteristic IR absorption frequencies of some of the most common functional groups.

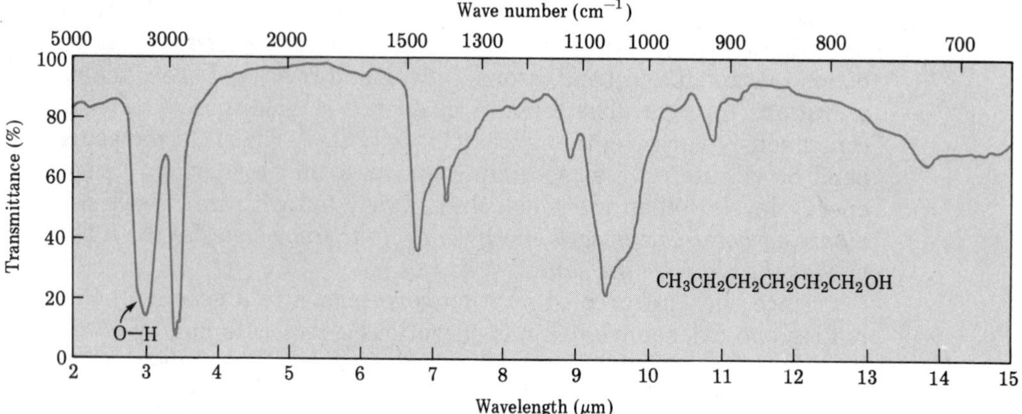

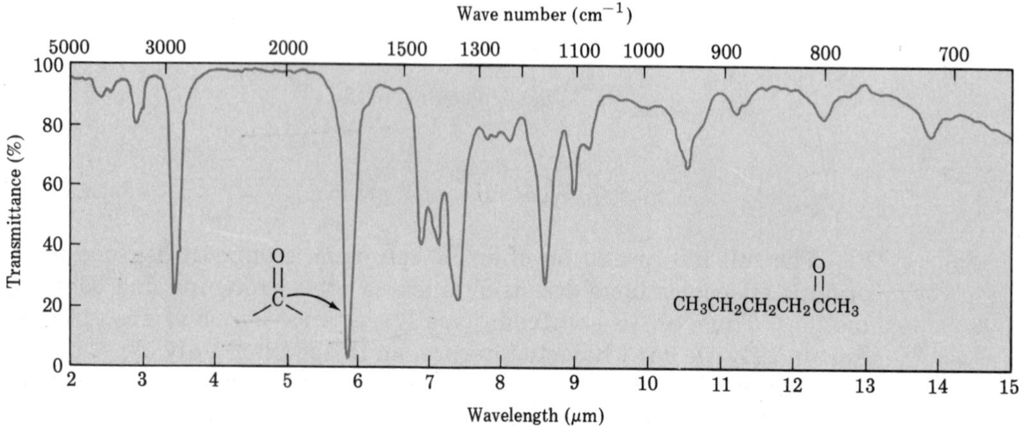

FIGURE 13.5 Infrared spectra of 1-hexanol (top) and 2-hexanone (bottom). Such spectra are easily obtained with 1–2 mg samples in a few minutes with commercial instruments.

It helps in remembering the positions of various IR absorptions to divide the infrared range from 4000 to 200 cm^{-1} into four regions, as shown in Figure 13.6.

Wave number (cm^{-1})

4000	2500	2000	1500	400
C—H O—H N—H	C≡C C≡N	C=C C=O C=N	Fingerprint region	

FIGURE 13.6 Regions in the infrared spectrum.

TABLE 13.1 Characteristic Infrared Absorptions of Some Functional Groups

FUNCTIONAL-GROUP CLASS	BAND POSITION (cm^{-1})	INTENSITY OF ABSORPTION
Alkanes, alkyl groups		
C—H	2850–2960	Medium to strong
Alkenes		
=C—H	3020–3100	Medium
C=C	1650–1670	Medium
Alkynes		
≡C—H	3300	Strong
—C≡C—	2100–2260	Medium
Alkyl halides		
C—Cl	600–800	Strong
C—Br	500–600	Strong
C—I	500	Strong
Alcohols		
O—H	3200–3600	Strong, broad
C—O	1050–1150	Strong
Aromatics		
C—H (aromatic ring)	3030	Medium
C=C (aromatic ring)	1600, 1500	Strong
Amines		
N—H	3310–3500	Medium
C—N	1030, 1230	Medium
Carbonyl compounds[a]		
C=O	1670–1780	Strong
Carboxylic acids		
O—H	2500–3100	Strong, very broad
Nitriles		
C≡N	2210–2260	Medium
Nitro compounds		
NO_2	1540	Strong

[a]Acids, esters, aldehydes, and ketones.

1. The region from 4000 to 2500 cm^{-1} corresponds to N–H, C–H, and O–H bond stretching motions. Both N–H and O–H bonds absorb in the 3300–3600 cm^{-1} range, whereas C–H bond stretching occurs near 3000 cm^{-1}. Since almost all organic compounds have C–H bonds, almost all IR spectra have an intense absorption in this region.

2. The region from 2500 to 2000 cm^{-1} is where triple-bond stretching occurs. Both nitriles (RC≡N) and alkynes (RC≡CR) show peaks here.
3. The region from 2000 to 1500 cm^{-1} contains C=O, C=N, and C=C double-bond absorptions. Carbonyl groups generally absorb from 1670 to 1780 cm^{-1}, and alkene stretching normally occurs in the narrow range from 1640 to 1680 cm^{-1}. The exact position of a C=O absorption is often diagnostic of the exact kind of carbonyl group in the molecule. Esters usually absorb at 1735 cm^{-1}, aldehydes at 1725 cm^{-1}, and open-chain ketones at 1715 cm^{-1}.
4. The region below 1500 cm^{-1} is the so-called *fingerprint region*. A large number of absorptions due to various C–O, C–C, and C–N single-bond vibrations occur here, forming a unique pattern that acts as an identifying "fingerprint" of each organic molecule.

PRACTICE PROBLEM 13.3 Refer to Table 13.1 and make an educated guess about the functional groups that cause these IR absorptions:
(a) 1735 cm^{-1} **(b)** 3500 cm^{-1}

SOLUTION **(a)** An absorption at 1735 cm^{-1} is in the carbonyl-group region of the IR spectrum.
(b) An absorption at 3500 cm^{-1} is in the –OH (alcohol) region.

PRACTICE PROBLEM 13.4 Acetone and 2-propen-1-ol ($H_2C{=}CHCH_2OH$) are isomers. How could you distinguish them by IR spectroscopy?

SOLUTION Acetone has a strong ketone carbonyl absorption at 1710 cm^{-1}. 2-Propen-1-ol has a hydroxyl absorption at 3500 cm^{-1} and an alkene absorption at 1660 cm^{-1}.

PROBLEM 13.3 What functional groups might molecules contain if they show IR absorptions at these frequencies?
(a) 1715 cm^{-1} **(b)** 1540 cm^{-1} **(c)** 2210 cm^{-1}
(d) 1720 and 2500–3100 cm^{-1} **(e)** 3500 and 1735 cm^{-1}

PROBLEM 13.4 How might you use IR spectroscopy to help you distinguish between these pairs of isomers?
(a) Ethanol and dimethyl ether **(b)** Cyclohexane and 1-hexene
(c) Propanoic acid and 3-hydroxypropanal

13.3 ULTRAVIOLET SPECTROSCOPY

The ultraviolet (UV) region of the electromagnetic spectrum extends from the low-wavelength end of the visible region (4×10^{-5} cm) to 10^{-6} cm. The portion of greatest interest to organic chemists, though, is the narrow range from 2×10^{-5} cm to 4×10^{-5} cm. Absorptions in this region are measured in nanometers (nm; 1 nm = 10^{-9} m = 10^{-7} cm. Thus, the ultraviolet range of interest is from 200 to 400 nm (Figure 13.7).

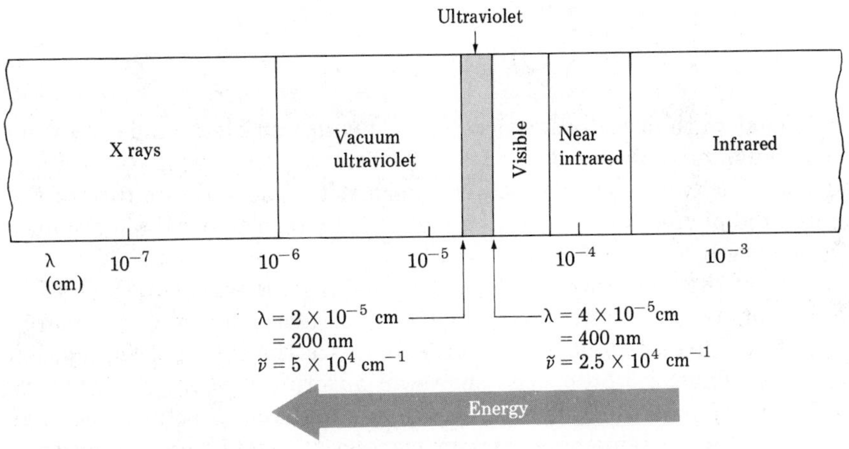

FIGURE 13.7 The ultraviolet (UV) region of the electromagnetic spectrum.

We've seen that an organic molecule either absorbs or transmits electromagnetic radiation, depending on the radiation's energy level. With IR radiation, the energies absorbed correspond to the amounts necessary to increase molecular motions of functional groups. With ultraviolet irradiation, the energies absorbed correspond to the amounts necessary to raise the energy levels of pi electrons.

Ultraviolet spectra are recorded by irradiating a sample with UV light of continuously changing wavelength. When the wavelength of light corresponds to the energy level required to excite a pi electron to a higher level, energy is absorbed. The absorption is detected and displayed on a chart that plots wavelength versus percent radiation absorbed.

A typical UV spectrum—that of 1,3-butadiene—is shown in Figure 13.8. Unlike IR spectra, which generally show many sharp lines, UV spectra are usually quite simple. Often, there's only a single broad peak, which is identified by noting the wavelength at the very top of the peak (λ_{max}). For 1,3-butadiene, λ_{max} = 217 nm.

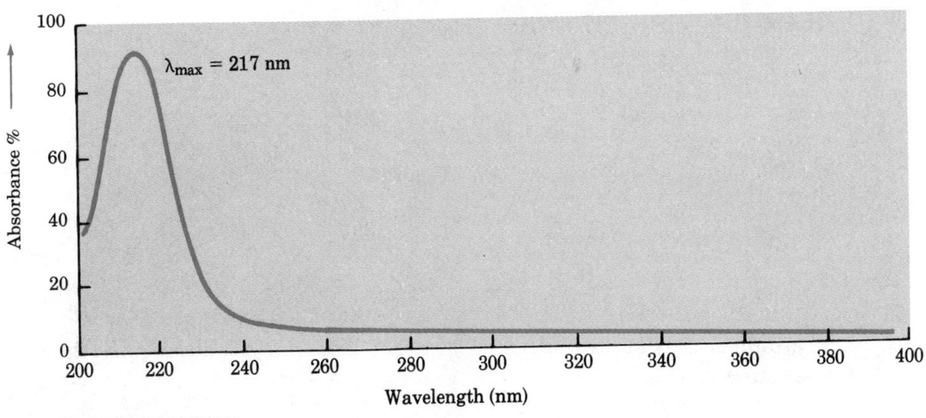

FIGURE 13.8 Ultraviolet spectrum of 1,3-butadiene.

13.4 INTERPRETING ULTRAVIOLET SPECTRA: THE EFFECT OF CONJUGATION

The wavelength of radiation necessary to cause an electronic excitation in a conjugated molecule depends on the nature of the conjugated system. Working backward, we can obtain information about the nature of the conjugated pi electron system in a molecule by measuring the molecule's UV spectrum.

One of the most important factors affecting the wavelength of a UV absorption is the extent of conjugation. It turns out that the energy required for an electronic transition decreases as the extent of conjugation increases. Thus, 1,3-buta*diene* shows an absorption at $\lambda_{max} = 217$ nm, 1,3,5-hexa*triene* absorbs at $\lambda_{max} = 258$ nm, and 1,3,5,7-octa*tetra*ene has $\lambda_{max} = 290$ nm. (Remember: Longer wavelength means lower energy.)

Other kinds of conjugated pi electron systems besides dienes and polyenes also show ultraviolet absorptions. Conjugated enones such as 3-butene-2-one and aromatic molecules such as benzene also have characteristic UV absorptions that aid in structure determination. The UV absorption maxima of some representative conjugated molecules are given in Table 13.2.

TABLE 13.2 Ultraviolet Absorption Maxima of Some Conjugated Molecules

NAME	STRUCTURE	λ_{max} (nm)
Ethylene	$H_2C{=}CH_2$	171
2-Methyl-1,3-butadiene	$H_2C{=}C(CH_3){-}CH{=}CH_2$	220
1,3-Cyclohexadiene	[cyclohexadiene ring]	256
1,3,5-Hexatriene	$H_2C{=}CH{-}CH{=}CH{-}CH{=}CH_2$	258
3-Buten-2-one	$H_2C{=}CH{-}C(CH_3){=}O$	219
Benzene	[benzene ring]	254

PRACTICE PROBLEM 13.5 1,5-Hexadiene and 1,3-hexadiene are isomers. How can you distinguish them by UV spectroscopy?

SOLUTION 1,3-Hexadiene is a conjugated diene, but 1,5-hexadiene is nonconjugated. Only the conjugated isomer shows a UV absorption above 200 nm.

PROBLEM 13.5 Which of the following show UV absorptions in the range 200-400 nm?

(a) 1,3-Cyclohexadiene (b) 1,4-Cyclohexadiene
(c) Methyl propenoate (d) *p*-Bromotoluene
(e) 2-Methylcyclohexanone (f) 2-Methyl-2-cyclohexenone

PROBLEM 13.6 How can you distinguish between 1,3-hexadiene and 1,3,5-hexatriene by UV spectroscopy?

13.5 NUCLEAR MAGNETIC RESONANCE SPECTROSCOPY

Of all techniques available for determining structure, nuclear magnetic resonance (NMR) spectroscopy is the most valuable. It's the method that organic chemists turn to first for information. We've seen that IR spectroscopy provides information about a molecule's functional groups and that UV spectroscopy provides information about a molecule's conjugated pi electron system. NMR spectroscopy doesn't replace or duplicate either of these techniques; rather, it complements them by providing a "map" of the carbon–hydrogen framework of an organic molecule. Taken together, IR, UV, and NMR spectroscopies often make it possible to find the structures of extremely complex molecules.

How does NMR spectroscopy work? Many kinds of nuclei, including ^{1}H and ^{13}C, behave like a child's top spinning about an axis. Since they're positively charged, these spinning nuclei act like tiny magnets and interact with an external magnetic field (denoted H_0). In the absence of an external magnetic field, the nuclear spins of magnetic nuclei are oriented randomly. When a sample containing these nuclei is placed between the poles of a strong magnet, however, the nuclei adopt specific orientations, much as a compass needle orients itself in the earth's magnetic field.

A spinning ^{1}H or ^{13}C nucleus can orient itself so that its own tiny magnetic field is aligned either with (*parallel to*) or against (*antiparallel to*) the external field. The two orientations don't have the same energy and therefore aren't present in equal amounts in the sample. The parallel orientation is slightly lower in energy, making this spin state slightly favored over the antiparallel orientation (Figure 13.9, p. 390).

If the magnetically oriented nuclei are irradiated with electromagnetic radiation of the proper frequency, energy absorption occurs, and the lower-energy state "spin-flips" to the higher-energy state. When this spin-flip occurs, the nuclei are said to be in resonance with the applied radiation—hence the name, *nuclear magnetic resonance*.

The amount of radio-frequency (rf) energy necessary for resonance depends both on the strength of the external magnetic field and on the chemical identity of the nuclei being irradiated. If a very strong magnetic field is applied, the energy difference between the two spin states is large, and higher-energy (higher-frequency) radiation is required. If a weaker magnetic field is applied, less energy is required to effect the transition between nuclear spin states.

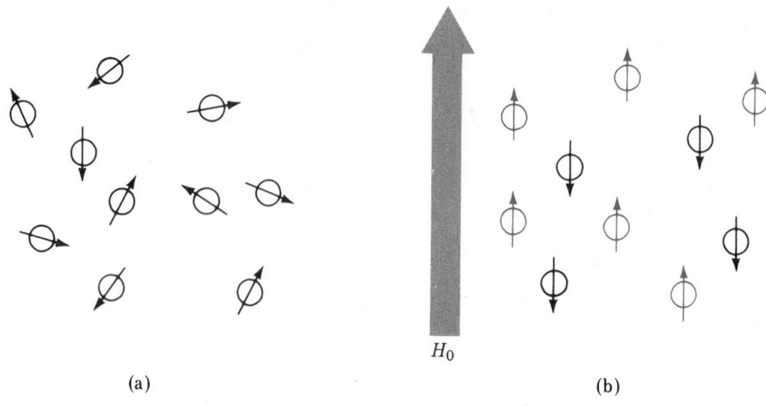

FIGURE 13.9 Nuclear spins are oriented randomly in the absence of a strong external magnetic field (a), but have a specific orientation in the presence of an external magnetic field H_0 (b). Note that some of the spins (color) are aligned parallel to the external field, and others are antiparallel. The parallel spin state is lower in energy.

In practice, superconducting magnets producing enormously powerful fields up to 140,000 gauss (G) are sometimes used, but a field strength of 14,100 G is more common. At this magnetic field strength, energy in the 60 MHz range (1 MHz = 1 megahertz = 10^6 s^{-1}) is required to bring a ^1H nucleus into resonance, and energy of 15 MHz is required to bring a ^{13}C nucleus into resonance.

PROBLEM 13.7 NMR spectroscopy uses electromagnetic radiation with a frequency of 6×10^7 Hz. Is this a greater or lesser amount of energy than that used by IR spectroscopy?

13.6 THE NATURE OF NMR ABSORPTIONS

From the description thus far, you might expect all protons (^1H nuclei) in a molecule to absorb energy at the same frequency and all ^{13}C nuclei to absorb at the same frequency. If this were true, we would observe only a single NMR absorption peak in the ^1H or ^{13}C spectrum of a molecule, a situation that would be of little use for structure determination. In fact, the absorption frequency is not the same for all ^1H or all ^{13}C nuclei.

All nuclei are surrounded by electron clouds. When an external magnetic field is applied to a sample molecule, the electron clouds around nuclei set up tiny local magnetic fields of their own. These local fields act in opposition to the applied field, so that the effective field actually felt by the nucleus is a bit weaker than the applied field:

SHIELDING
An effect in NMR spectroscopy in which electrons around nuclei diminish the effect of the applied magnetic field

$$H_{\text{effective}} = H_{\text{applied}} - H_{\text{local}}$$

In describing this effect of local fields, we say that the carbon and hydrogen nuclei are **shielded** from the applied field by the electron clouds that surround them. Since each different kind of ^1H or ^{13}C nucleus in a

13.7 CHEMICAL SHIFTS

NMR spectra are displayed on charts that show the applied field strength increasing from left to right (Figure 13.10). Thus, the left side of the chart is the low-field (or *downfield*) side, and the right side is the high-field (or *upfield*) side. To define the position of an absorption, the NMR chart is calibrated and a reference point is used. In practice, a small amount of tetramethylsilane [TMS, $(CH_3)_4Si$] is added to the sample so that a reference absorption line is produced when the spectrum is run. TMS is used as a reference for both ^1H and ^{13}C spectra, because it gives rise to a single peak that occurs at a higher field (farther right on the chart) than other absorptions that are normally found in organic molecules.

CHEMICAL SHIFT
Position of an absorption on the NMR chart

DELTA SCALE
Arbitrary scale used for defining the position of NMR absorptions; $1\ \delta = 1$ ppm of spectrometer frequency

The place on the chart at which a nucleus absorbs is called its **chemical shift**. By convention, the chemical shift of TMS is called zero, and other peaks normally occur at lower fields (to the left on the chart). NMR charts are calibrated in units of frequency using an arbitrary scale called the **delta scale**, where 1 delta unit (δ) is equal to 1 part per million (ppm) of the spectrometer operating frequency. For example, if we were using a 60 MHz instrument to measure the ^1H NMR spectrum of a substance, $1\ \delta$ would be 1 ppm of 60,000,000 Hz, or 60 Hz. Similarly, if we were measuring the spectrum with a 100 MHz instrument, then $1\ \delta = 100$ Hz.

Although this method of calibrating NMR charts may seem complex, there's a good reason for it. There are many different kinds of NMR spectrometers operating at many different frequencies and magnetic field strengths. By using a system of measurement in which NMR absorptions are expressed in relative terms (parts per million relative to spectrometer frequency) rather than in absolute terms (Hz), comparisons of spectra obtained on different instruments are possible. *The chemical shift of an NMR absorption given in parts per million or δ units is constant, regardless of the operating frequency of the instrument.* A ^1H nucleus that absorbs at $2.0\ \delta$ on a 60 MHz instrument ($2.0 \text{ ppm} \times 60 \text{ MHz} = 120$ Hz to the left of TMS) also absorbs at $2.0\ \delta$ on a 300 MHz instrument ($2.0 \text{ ppm} \times 300 \text{ MHz} = 600$ Hz to the left of TMS).

PRACTICE PROBLEM 13.7 Cyclohexane shows an absorption at $1.43\ \delta$ in its ^1H NMR spectrum. How many hertz away from TMS is this on a spectrometer operating at 60 MHz? On a spectrometer operating at 220 MHz?

SOLUTION On a 60 MHz spectrometer, $1\ \delta = 60$ Hz. Thus, $1.43\ \delta = 86$ Hz away from the TMS reference peak. On a 220 MHz spectrometer, $1\ \delta = 220$ Hz and $1.43\ \delta = 315$ Hz.

PROBLEM 13.10 When the ^1H NMR spectrum of acetone is recorded on a 60 MHz instrument, a single sharp resonance line at $2.1\ \delta$ is observed.

(a) How far away from TMS (in hertz) does the acetone absorption occur?
(b) What is the position of the acetone absorption in δ units on a 100 MHz instrument?
(c) How many hertz away from TMS does the absorption in the 100 MHz spectrum correspond to?

PROBLEM 13.11 The following ^1H NMR resonances were recorded on a spectrometer operating at 60 MHz. Convert each into δ units.
(a) $CHCl_3$, 436 Hz **(b)** CH_3Cl, 183 Hz **(c)** CH_3OH, 208 Hz
(d) CH_2Cl_2, 318 Hz

13.8 CHEMICAL SHIFTS IN ^1H NMR SPECTRA

Everything we've said thus far about NMR spectroscopy applies to both ^1H and ^{13}C spectra. Now let's focus only on ^1H NMR spectroscopy to see how it can be used in organic structure determination. Most ^1H NMR absorptions occur in the range from 0 to 8 δ, which can be divided into five regions that are characteristic of certain kinds of protons (Figure 13.11). Once the regions are memorized, it's possible to tell at a glance what kinds of protons a molecule contains.

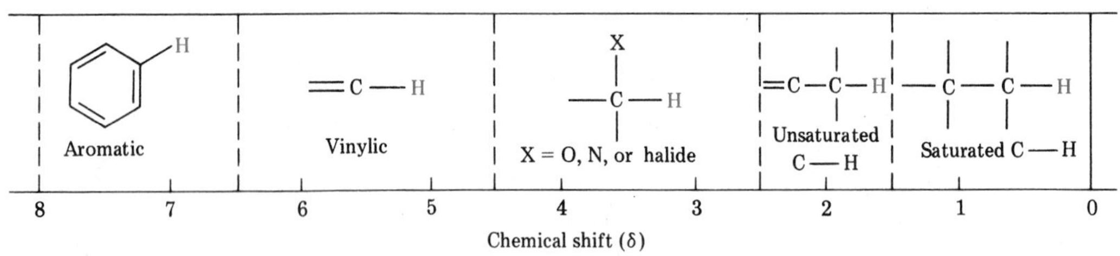

FIGURE 13.11 Chemical shifts of different kinds of protons.

Table 13.3 shows the correlation of ^1H chemical shift with environment in more detail. In general, protons bonded to saturated sp^3 carbons absorb at higher fields (right-hand side of the spectrum), whereas protons bonded to sp^2 carbons absorb at lower fields (left-hand side of the spectrum). Protons on carbons that are bonded to electronegative atoms such as N, O, or halogen also absorb at lower fields.

PRACTICE PROBLEM 13.8 Methyl 2,2-dimethylpropanoate $(CH_3)_3COOCH_3$ has two peaks in its ^1H NMR spectrum. At what approximate chemical shifts do they come?

SOLUTION The CH_3O– protons absorb around 3.5–4.0 δ since they are on carbon bonded to oxygen. The $(CH_3)_3C$– protons absorb around 1.0 δ since they are typical alkane-like protons. (See Figure 13.12.)

PROBLEM 13.12 Each of the following compounds exhibits a single ^1H NMR peak. Approximately where would you expect each compound to absorb?
(a) Ethane **(b)** Acetone **(c)** Benzene **(d)** Trimethylamine

TABLE 13.3 Correlation of ^{1}H Chemical Shift with Environment

TYPE OF PROTON	FORMULA	CHEMICAL SHIFT (δ)	TYPE OF PROTON	FORMULA	CHEMICAL SHIFT (δ)
Reference peak	$(CH_3)_4Si$	0	Alkyl iodide	$I-\overset{\vert}{\underset{\vert}{C}}-H$	2.0–4.0
Saturated primary	$-CH_3$	0.7–1.3	Alcohol, ether	$-O-\overset{\vert}{\underset{\vert}{C}}-H$	3.3–4.0
Saturated secondary	$-CH_2-$	1.2–1.4	Alkynyl	$-C\equiv C-H$	2.5–2.7
Saturated tertiary	$-\overset{\vert}{\underset{\vert}{C}}-H$	1.4–1.7	Vinylic	$>C=\underset{\vert}{C}-H$	5.0–6.5
Allylic primary	$>C=\underset{\vert}{C}-CH_3$	1.6–1.9	Aromatic	$Ar-H$	6.5–8.0
Methyl ketones	$-\overset{O}{\overset{\Vert}{C}}-CH_3$	2.1–2.4	Aldehyde	$-\overset{O}{\overset{\Vert}{C}}-H$	9.7–10.0
Aromatic methyl	$Ar-CH_3$	2.5–2.7	Carboxylic acid	$-\overset{O}{\overset{\Vert}{C}}-O-H$	11.0–12.0
Alkyl chloride	$Cl-\overset{\vert}{\underset{\vert}{C}}-H$	3.0–4.0	Alcohol	$-\overset{\vert}{\underset{\vert}{C}}-O-H$	Extremely variable (2.5–5.0)
Alkyl bromide	$Br-\overset{\vert}{\underset{\vert}{C}}-H$	2.5–4.0			

13.9 INTEGRATION OF ^{1}H NMR SPECTRA: PROTON COUNTING

Look at the ^{1}H NMR spectrum of methyl 2,2-dimethylpropanoate in Figure 13.12 (p. 396). There are two peaks, corresponding to the two kinds of protons, but the peaks aren't the same size. The peak at 1.2 δ, due to the $(CH_3)_3C-$ protons, is larger than the peak at 3.7 δ, due to the $-OCH_3$ protons.

INTEGRATION

A means of electronically measuring the ratios of the number of nuclei responsible for each peak in an NMR spectrum

The area under each peak is proportional to the number of protons causing that peak. By electronically measuring (**integrating**) the area under each peak, it's possible to measure the relative number of each kind of proton in a molecule. Integrated peak areas are superimposed over the spectrum in a "stair-step" manner, with the height of each step proportional to the area of the peak, and therefore proportional to the relative number of protons causing the peak. For example, the two peaks in methyl 2,2-dimethylpropanoate are found to have a 1:3 (or 3:9) ratio when integrated—exactly what we expect since the three $-OCH_3$ protons are equivalent and the nine $(CH_3)_3C-$ protons are equivalent.

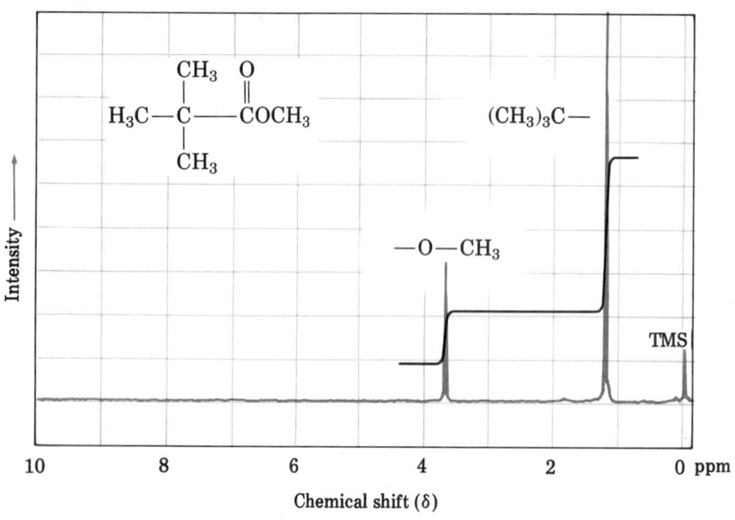

FIGURE 13.12 The ^1H NMR spectrum of methyl 2,2-dimethylpropanoate. Integrating the peaks in a "stair-step" manner shows that they have a 1:3 ratio, corresponding to the ratio of the numbers of protons responsible for each peak.

PROBLEM 13.13 How many peaks would you expect to see in the ^1H NMR spectrum of *p*-dimethylbenzene (*p*-xylene)? What ratio of peak areas would you expect to find on integration of the spectrum? Refer to Table 13.3 for approximate chemical shift values, and sketch what the spectrum might look like.

13.10 SPIN–SPIN SPLITTING IN ^1H NMR SPECTRA

In the ^1H NMR spectra we've seen thus far, each different kind of proton in a molecule has given rise to a single peak. It often happens, though, that the absorption of a proton splits into multiple peaks. For example, the ^1H NMR spectrum of chloroethane in Figure 13.13 indicates that the $-CH_2Cl$ protons appear as four peaks (a *quartet*) at 3.6 δ, and the $-CH_3$ protons appear as a *triplet* at 1.5 δ.

SPIN–SPIN SPLITTING Splitting of an NMR absorption into multiple peaks because of the interaction of neighboring nuclear spins

COUPLING Interaction of neighboring nuclear spins that results in spin–spin splitting

Called **spin–spin splitting**, the phenomenon of multiple absorptions is due to the fact that the nuclear spin of one atom interacts, or **couples**, with the nuclear spin of nearby atoms. In other words, the tiny magnetic field of one nucleus affects the magnetic field felt by a neighboring nucleus.

To understand the reasons for spin–spin splitting, look at the $-CH_3$ protons in chloroethane (Figure 13.13). The three equivalent $-CH_3$ protons are neighbored by two magnetic nuclei, the $-CH_2Cl$ protons. Each of the $-CH_2Cl$ protons has its own nuclear spin, which can align either with or against the applied magnetic field, producing a tiny effect that is felt by the neighboring $-CH_3$ protons.

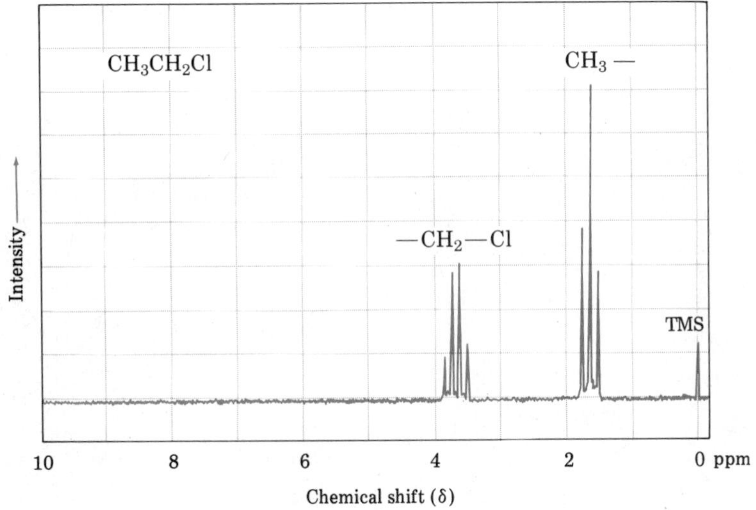

FIGURE 13.13 The ^{1}H NMR spectrum of chloroethane, CH_3CH_2Cl.

There are three ways in which the two $-CH_2Cl$ protons can align. If both protons align *with* the applied magnetic field, the total effective field felt by the neighboring $-CH_3$ protons is slightly larger than it would otherwise be. Consequently, the applied field necessary to cause resonance is slightly reduced. Alternatively, if one $-CH_2Cl$ proton aligns *with* and one aligns *against* the applied field (two possible ways), there is no effect on the neighboring $-CH_3$ protons. Finally, if both $-CH_2Cl$ protons align against the applied field, the effective field felt by the $-CH_3$ protons is slightly smaller than it would otherwise be, and the applied field needed for resonance must be slightly increased. Each of the three possible alignments of $-CH_2Cl$ spins is adopted by a certain fraction of molecules in the sample, causing the neighboring $-CH_3$ protons to appear as three peaks with a 1:2:1 ratio in the NMR spectrum. Figure 13.14 (p. 398) shows schematically how spin–spin splitting arises.

In the same way that the $-CH_3$ protons of chloroethane are split into a triplet in the NMR spectrum, the $-CH_2Cl$ protons are split into a quartet. The three spins of the neighboring $-CH_3$ protons align in four combinations: all three with the applied field, two with and one against (three possibilities), one with and two against (three possibilities), or all three against. Thus, four peaks are produced in a 1:3:3:1 ratio.

n + 1 RULE

The signal of a proton with n neighboring protons splits into $n + 1$ peaks in the NMR spectrum

As a general rule (the ***n* + 1 rule**), protons that have n neighboring protons show $n + 1$ peaks in their NMR spectrum. For example, the $-CHBr-$ proton in 2-bromopropane ($CH_3CHBrCH_3$) appears as a seven-line multiplet (a *septet*) in the NMR spectrum (Figure 13.15, p.398), because its signal is split by the six neighboring protons ($n + 1 = 7$ when $n = 6$). Similarly, the $-CH_3$ protons of 2-bromopropane appear as a doublet, because their signal is split only by the single neighboring $-CHBr-$ proton.

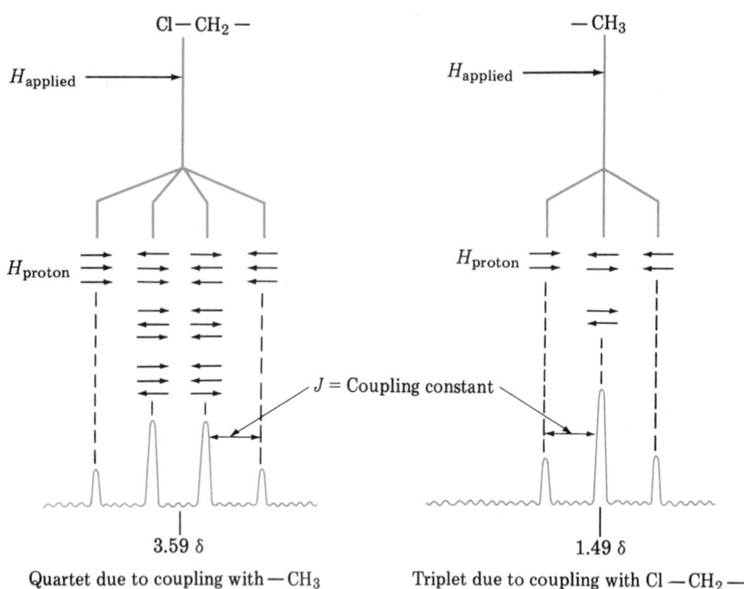

FIGURE 13.14 The origin of spin–spin splitting in chloroethane. The nuclear spins of neighboring protons (indicated by horizontal arrows) align either with or against the applied field, causing the splitting of absorptions into multiplets.

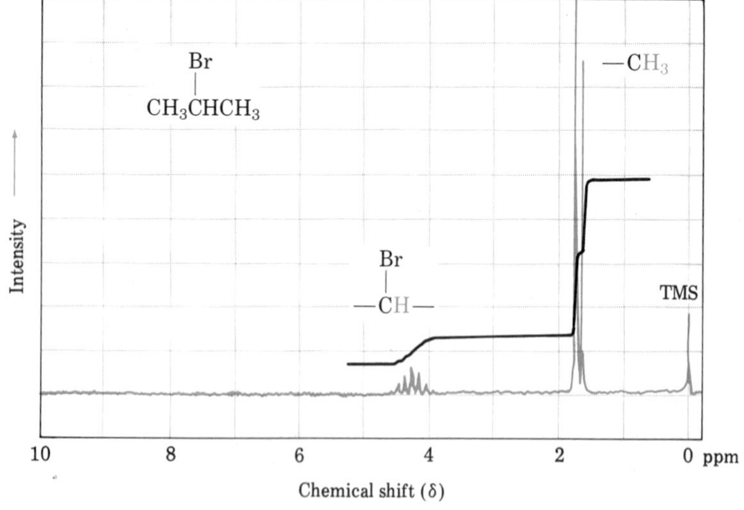

FIGURE 13.15 The ^1H NMR spectrum of 2-bromopropane. The $-CH_3$ proton signal is split into a doublet, and the $-CHBr-$ proton signal is split into a septet.

COUPLING CONSTANT (J) Measure of the amount of coupling between two neighboring nuclei

The distance between peaks in a multiplet is called the **coupling constant**, denoted J. Coupling constants are measured in hertz and generally fall in the range 0–18 Hz. Though the exact value depends on the geometry of the molecule, a typical value for an open-chain alkane is J = 6–8 Hz. Note that the same coupling constant is shared by both groups of hydrogens whose spins are coupled. In chloroethane, for instance, the $-CH_2Cl$ protons are coupled to the $-CH_3$ protons with coupling constant J = 7 Hz. The $-CH_3$ protons are similarly coupled to the $-CH_2Cl$ protons with the same J = 7 Hz coupling constant.

Three important rules about spin–spin splitting are illustrated by the spectra of chloroethane in Figure 13.13 and 2-bromopropane in Figure 13.15:

RULE 1 Chemically equivalent protons don't show spin–spin splitting. The equivalent protons can be on the same carbon or on different carbons, but their signals still appear as singlets and don't split.

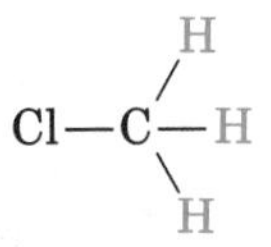

```
      H              H     H
     /                \   /
Cl—C—H           Cl—C—C—Cl
     \                /   \
      H              H     H
```

Three C–H protons are chemically equivalent; no splitting occurs

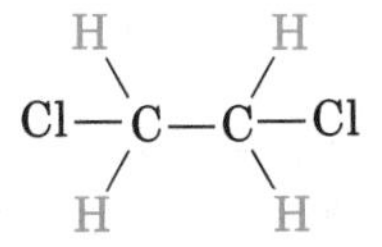

Four C–H protons are chemically equivalent; no splitting occurs

RULE 2 A proton with n equivalent neighboring protons gives a signal that is split into a multiplet of $n + 1$ peaks with coupling constant J. Protons that are more than two carbon atoms apart usually don't split each other.

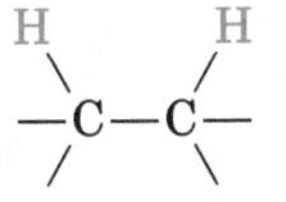

```
H     H          H         H
 \   /            \   |   /
 —C—C—            —C—C—C—
 /   \            /   |   \
```

Splitting observed

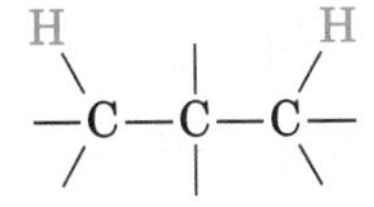

Splitting not usually observed

RULE 3 Two groups of protons coupled to each other have the same coupling constant J.

PRACTICE PROBLEM 13.9 Predict the splitting pattern for each kind of hydrogen in isopropyl propanoate, $CH_3CH_2COOCH(CH_3)_2$.

SOLUTION First, find how many different kinds of protons are present (there are four). Then find out how many neighboring protons each kind has, and apply the $n + 1$ rule to determine the splitting patterns:

(Triplet) (Quartet) (Septet) (Doublet)

$$CH_3{-}CH_2{-}\overset{\displaystyle O}{\overset{\|}{C}}{-}O{-}\underset{\displaystyle CH_3}{\overset{\displaystyle H}{C}}{-}CH_3$$

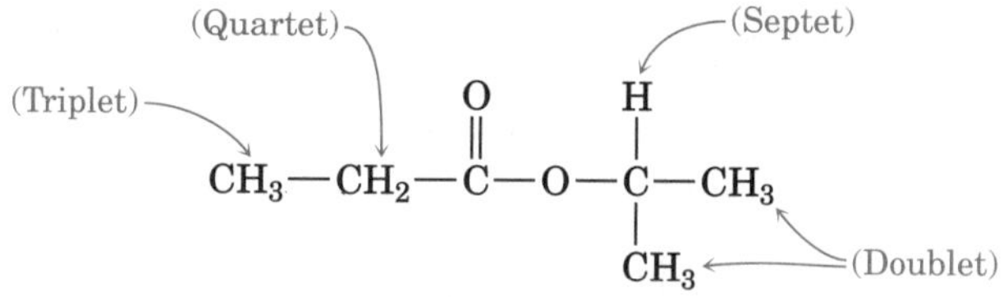

Isopropyl propanoate

PROBLEM 13.14 Predict the splitting patterns you would expect for each proton in these molecules:
(a) $(CH_3)_3CH$ **(b)** CH_3CHBr_2 **(c)** $CH_3OCH_2CH_2Br$
(d) $CH_3CH_2COOCH_3$ **(e)** $ClCH_2CH_2CH_2Cl$ **(f)** $(CH_3)_2CHCOOCH_3$

PROBLEM 13.15 Propose structures for compounds that show these ^1H NMR spectra:
(a) C_2H_6O; one singlet **(b)** $C_3H_6O_2$; two singlets
(c) C_3H_7Cl; one doublet and one septet

13.11 USES OF ^1H NMR SPECTRA

^1H NMR spectroscopy is used to help identify the product of nearly every reaction run in the laboratory. For example, we said in Section 4.2 that addition of HCl to alkenes occurs with Markovnikov orientation; that is, the more highly substituted chloroalkane is formed. With the help of ^1H NMR, we can now prove this statement.

Does addition of HCl to 1-methylcyclohexene yield 1-chloro-1-methylcyclohexane or 1-chloro-2-methylcyclohexane?

CH_3 | $\xrightarrow[\text{Ether}]{\text{HCl}}$ | H_3C Cl | or | H CH_3 H Cl ?

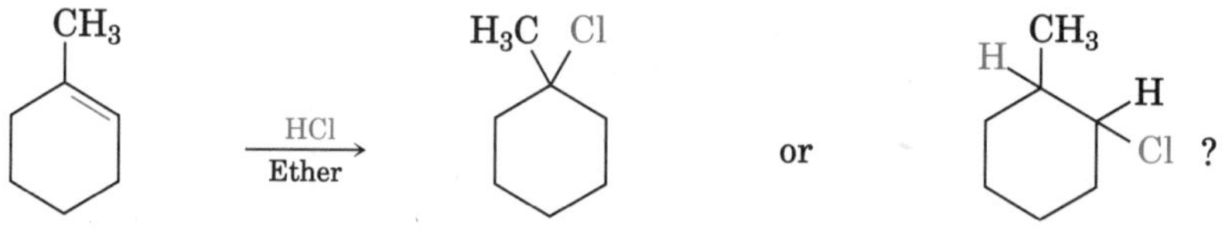

1-Methylcyclohexene **1-Chloro-methylcyclohexane** (Markovnikov) **1-Chloro-2-methylcyclohexane** (non-Markovnikov)

The ^1H NMR spectrum of the reaction product is shown in Figure 13.16. Although many of the ring protons overlap into a broad, poorly defined multiplet centered around 1.6 δ, the spectrum also shows a large singlet absorption in the saturated methyl region at 1.5 δ, indicating that the product has a methyl group bonded to a quaternary carbon ($R_3C{-}CH_3$). Furthermore, the spectrum shows no absorptions in the range 4–5 δ, where we would expect the signal of a R_2CHCl proton to occur. Thus, the reaction product must be 1-chloro-1-methylcyclohexane.

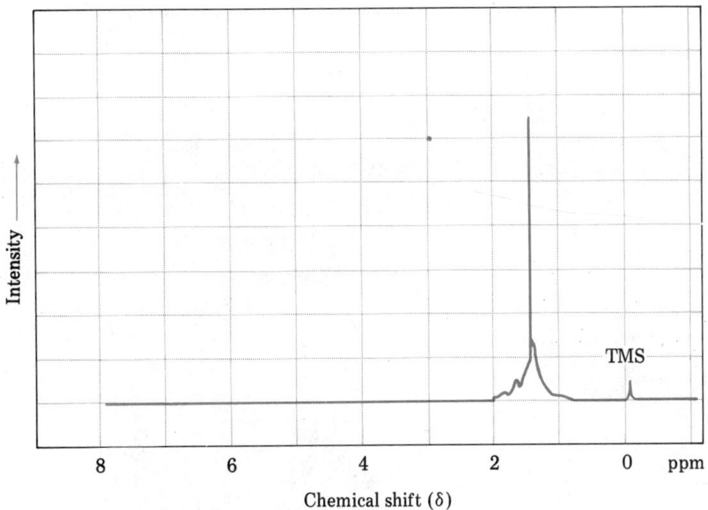

FIGURE 13.16 The ^1H NMR spectrum of the reaction product from HCl and 1-methylcyclohexene. The presence of the $-CH_3$ absorption at 1.6 δ and the absence of any absorptions near 4 δ identify the product as 1-chloro-1-methylcyclohexane.

13.12 ^{13}C NMR SPECTROSCOPY

In some ways, it's surprising that carbon NMR is even possible. After all, ^{12}C, the most abundant carbon isotope, has no nuclear spin and is not observable by NMR. The only naturally occurring carbon isotope with a magnetic moment is ^{13}C, but its natural abundance is only 1.1%. Thus, only about 1 of every 100 carbon atoms in organic molecules is observable by NMR. Fortunately, the technical problems caused by this low abundance have been overcome by improved electronics and computer techniques, and ^{13}C NMR has now become a routine structural tool.

At its simplest, ^{13}C NMR makes it possible to count the number of carbon atoms in a molecule. In addition, it's possible to get information about the environment of each carbon by observing its chemical shift. As illustrated by the ^{13}C NMR spectrum of methyl acetate shown earlier (Figure 13.10), we normally observe a single, sharp resonance line for each kind of carbon atom in a molecule. Thus, methyl acetate has three nonequivalent carbon atoms and three peaks in its ^{13}C NMR spectrum. (Coupling between adjacent carbon atoms isn't seen, because the low natural abundance of ^{13}C makes it unlikely that two such nuclei would be next to each other in a molecule.)

Most ^{13}C resonances are between 0 and 250 δ, with the exact chemical shift dependent on a carbon's environment in the molecule. Figure 13.17 shows how environment and chemical shift are correlated.

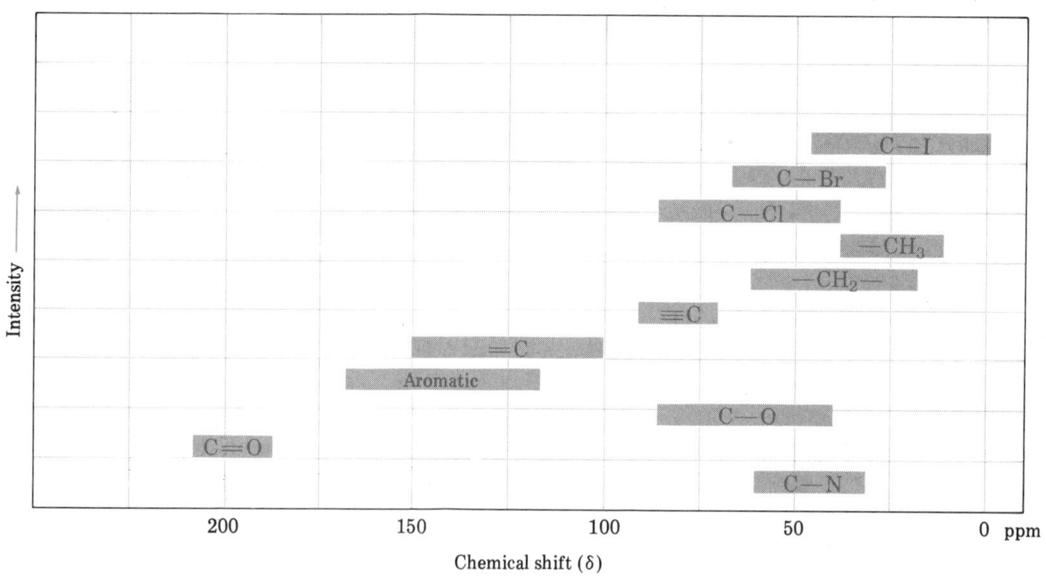

FIGURE 13.17 Chemical shift correlations for ^{13}C NMR.

Although the factors that determine chemical shifts are complex, it's possible to make some generalizations. One rule is that carbons bonded to electronegative atoms like oxygen, nitrogen, and halogen absorb downfield (to the left) of normal alkane carbons. Another general rule is that sp^3-hybridized carbons absorb in the range 0–100 δ, and sp^2 carbons absorb in the range 100–200 δ. Carbonyl-group carbons are particularly distinct in the ^{13}C NMR spectrum and are easily observed at the extreme low-field side of the chart, in the range 170–210 δ. For example, the ^{13}C NMR spectrum of *p*-bromoacetophenone in Figure 13.18 shows an absorption for the carbonyl carbon at 197 δ.

The ^{13}C NMR spectrum of *p*-bromoacetophenone is interesting for another reason as well. Note that only six absorptions are observed even though the molecule has eight carbons. *p*-Bromoacetophenone has a symmetry plane that makes carbons 4 and 4′, and carbons 5 and 5′, equivalent. Thus, the six ring carbons show only four absorptions in the range 128–137 δ. In addition, the $-CH_3$ carbon absorbs at 26 δ.

PRACTICE PROBLEM 13.10 How many resonance lines would you expect to see in the ^{13}C NMR spectrum of methylcyclopentane?

SOLUTION Methylcyclopentane has a symmetry plane. Thus, it has only four kinds of carbons and only four peaks in its ^{13}C NMR spectrum.

PROBLEM 13.16 How many resonance lines would you expect to observe in the ^{13}C NMR spectra of these compounds?

(a) Cyclopentane **(b)** 1,3-Dimethylbenzene **(c)** 1,2-Dimethylbenzene
(d) 1-Methylcyclohexene

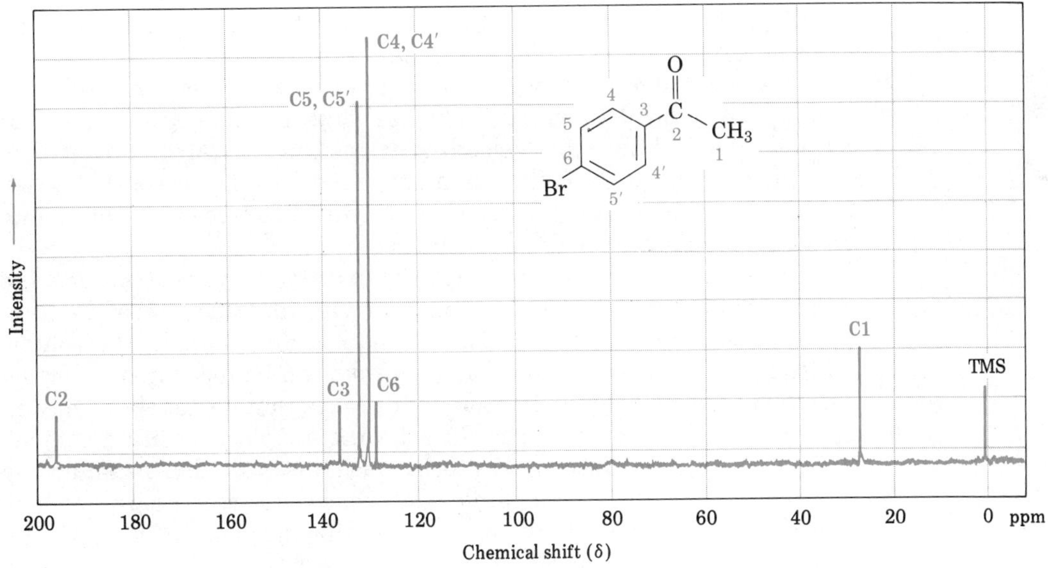

FIGURE 13.18 The ^{13}C NMR spectrum of *p*-bromoacetophenone, $BrC_6H_4COCH_3$.

PROBLEM 13.17 Propose structures for compounds whose ^{13}C NMR spectra fit these descriptions:

(a) A hydrocarbon with seven peaks in its spectrum
(b) A six-carbon compound with only five peaks in its spectrum
(c) A four-carbon compound with three peaks in its spectrum

INTERLUDE

MAGNETIC RESONANCE IMAGING (MRI)

As practiced by organic chemists, NMR spectroscopy is a valued method of structure determination. A small amount of sample, typically 10 mg or less, is dissolved in 1 mL or so of a suitable solvent, the solution is placed in a thin glass tube, and the tube is placed into the narrow (1–2 cm) gap between the poles of a strong magnet. Imagine, though, that a much larger NMR instrument were available. Instead of a few milligrams, the sample size could be tens of kilograms; instead of a narrow gap between magnet poles, the gap could be large enough for a whole person to climb into so that an NMR spectrum of body parts could be obtained. What you've just imagined is an instrument for *magnetic resonance imaging (MRI)*, a new and valuable diagnostic technique that has created enormous excitement in the medical community because of its advantages over X-ray or radioactive imaging methods. (Fortunately, you don't need to be dissolved in solvent nor sit in a thin glass tube while undergoing MRI.)

Like NMR spectroscopy, MRI takes advantage of the magnetic properties of certain nuclei and of the signals emitted when those nuclei are stimulated by radio-frequency energy. Unlike what happens in NMR spectroscopy, though, MRI instruments use powerful computers and data manipulation techniques to look at the three-dimensional *location* of magnetic nuclei in the body, rather than at the chemical nature of the nuclei. Most MRI instruments currently look at hydrogen, present in abundance wherever there is water or fat in the body.

The signals produced vary with the density of hydrogen atoms and with the nature of their surroundings, allowing identification of different types of tissue and the visualization of motion. For example, the volume of blood leaving the heart in a single stroke can be measured, allowing observation of the heart in motion. Soft tissues that do not show up well on X rays can also be seen clearly, allowing diagnosis of brain tumors, strokes, and other conditions. The technique is also valuable in diagnosing knee damage, because it is a painless alternative to arthroscopy, in which an endoscope is physically introduced into the knee joint (Figure 13.19).

Several types of atoms in addition to hydrogen can be detected by MRI, and the applications of images based on ^{31}P atoms are being explored. The technique holds great promise for studies of metabolism.

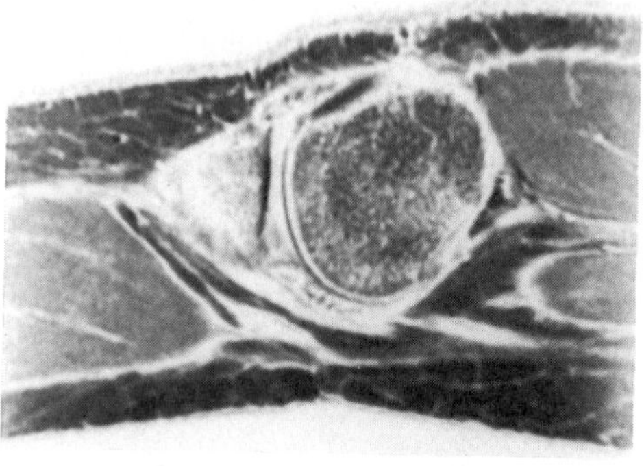

FIGURE 13.19 An MRI image of a normal knee joint.

SUMMARY AND KEY WORDS

Three main spectroscopic methods are used to determine the structures of organic molecules. Each of the three gives a different kind of information:

Infrared spectroscopy	What functional groups are present?
Ultraviolet spectroscopy	Is a conjugated pi electron system present?
Nuclear magnetic resonance spectroscopy	What carbon–hydrogen framework is present?

When an organic molecule is irradiated with **infrared (IR)** radiation, frequencies of light corresponding to the energy levels of molecular bending and stretching motions are absorbed. Each kind of functional group has a characteristic set of infrared absorptions that allows it to be identified. For example, an alkene C=C bond absorbs in the range 1650–1670 cm^{-1}, a saturated ketone absorbs near 1715 cm^{-1}, and a nitrile absorbs near 2230 cm^{-1}. By observing which frequencies of IR radiation are absorbed by a molecule *and which are not*, the functional groups in a molecule can be identified.

Ultraviolet (UV) spectroscopy is applicable to conjugated pi electron systems. When a conjugated molecule is irradiated with ultraviolet light, energy absorption occurs, leading to excitation of pi electrons to higher energy levels. The greater the extent of conjugation, the longer the wavelength needed for excitation.

Nuclear magnetic resonance (NMR) spectroscopy is the most important of the common spectroscopic techniques. When ^1H and ^{13}C nuclei are placed in a magnetic field, their spins orient themselves either with or against the field. On irradiation with radio-frequency (rf) waves, energy is absorbed and the nuclear spins flip from the lower-energy state to the higher-energy state. This absorption of energy is detected, amplified, and displayed as an NMR spectrum. NMR spectra display four general features:

1. **Number of resonance lines.** Each nonequivalent kind of ^1H or ^{13}C nucleus in a molecule gives rise to a different resonance line.
2. **Chemical shift.** The exact position of each peak is called its chemical shift and is correlated to the chemical nature of each ^1H or ^{13}C nucleus. Most ^1H absorptions fall in the range 0–10 δ downfield from the TMS reference signal.
3. **Integration.** The area under each peak can be electronically integrated to determine the relative number of protons responsible for each peak.
4. **Spin–spin splitting.** Neighboring nuclear spins can **couple**, splitting NMR absorptions into multiplets. The NMR signal of a ^1H nucleus neighbored by n adjacent protons splits into $n + 1$ peaks with **coupling constant J**.

ADDITIONAL PROBLEMS

13.18 What kinds of functional groups might compounds contain if they show the following IR absorptions?
(a) 1670 cm^{-1} **(b)** 1735 cm^{-1} **(c)** 1540 cm^{-1}
(d) 1715 cm^{-1} and 2500–3100 cm^{-1} (broad)

13.19 At what approximate positions might the following compounds show IR absorptions?
(a) Benzoic acid **(b)** Methyl benzoate **(c)** *p*-Hydroxybenzonitrile
(d) 3-Cyclohexenone **(e)** Methyl 4-oxopentanoate

13.20 The following ^{1}H NMR absorptions, determined on a spectrometer operating at 60 MHz, are given in hertz downfield from the TMS standard. Convert the absorptions to δ units.
(a) 131 Hz (b) 287 Hz (c) 451 Hz

13.21 At what positions, in hertz downfield from TMS standard, would the NMR absorptions in Problem 13.20 appear on a spectrometer operating at 100 MHz?

13.22 The following NMR absorptions, given in δ units, were obtained on a spectrometer operating at 80 MHz. Convert the chemical shifts from δ units into hertz downfield from TMS.
(a) 2.1 δ (b) 3.45 δ (c) 6.30 δ

13.23 If C–O single-bond stretching occurs at 1000 cm^{-1} and C=O double-bond stretching occurs at 1700 cm^{-1}, which of the two requires more energy? How does your answer correlate with the relative strengths of single and double bonds?

13.24 Tell what is meant by each of these terms:
(a) Chemical shift (b) Coupling constant (c) λ_{max} (d) Spin–spin splitting
(e) Wave number (f) Applied magnetic field

13.25 When measured on a spectrometer operating at 60 MHz, chloroform ($CHCl_3$) shows a single sharp absorption at 7.3 δ.
(a) How many parts per million downfield from TMS does chloroform absorb?
(b) How many hertz downfield from TMS does chloroform absorb if the measurement is carried out on a spectrometer operating at 360 MHz?
(c) What is the position of the chloroform absorption in δ units if it is measured on a 360 MHz spectrometer?

13.26 How many absorptions would you expect in the ^{13}C NMR spectra of these compounds?
(a) 1,1-Dimethylcyclohexane (b) Ethyl methyl ether (c) Cyclohexanone
(d) 2-Methyl-2-butene (e) *cis*-2-Pentene (f) *trans*-2-Pentene

13.27 How many types of nonequivalent protons are there in each of the molecules listed in Problem 13.26?

13.28 Describe the ^{1}H NMR spectra you would expect for these compounds:
(a) CH_3CHCl_2 (b) $CH_3COOCH_2CH_3$ (c) $(CH_3)_3CCH_2CH_3$

13.29 The following compounds all show a single line in their ^{1}H NMR spectra. List them in order of expected increasing chemical shift: CH_4, CH_2Cl_2, cyclohexane, CH_3COCH_3, $H_2C{=}CH_2$, benzene.

13.30 Propose structures for compounds that meet these descriptions:
(a) C_5H_8, with IR absorptions at 3300 and 2150 cm^{-1}
(b) C_4H_8O, with a strong IR absorption at 3400 cm^{-1}
(c) C_4H_8O, with a strong IR absorption at 1715 cm^{-1}
(d) C_8H_{10}, with IR absorptions at 1600 and 1500 cm^{-1}

13.31 How would you use infrared spectroscopy to distinguish between these pairs of isomers?
(a) $(CH_3)_3N$ and $CH_3CH_2NHCH_3$ (b) CH_3COCH_3 and $CH_2{=}CHCH_2OH$
(c) CH_3COCH_3 and CH_3CH_2CHO

13.32 How would you use ^{1}H NMR spectroscopy to distinguish between the isomer pairs shown in Problem 13.31?

13.33 How could you use ^{13}C NMR spectroscopy to distinguish between the isomer pairs shown in Problem 13.31?

13.34 Assume that you're carrying out the dehydration of 1-methylcyclohexanol to yield 1-methylcyclohexene. How could you use IR spectroscopy to determine when the reaction is complete? What characteristic absorptions would you expect for both starting material and product?

13.35 The infrared spectrum of a compound with the formula C_4H_8O is shown. Propose a likely structure for the substance.

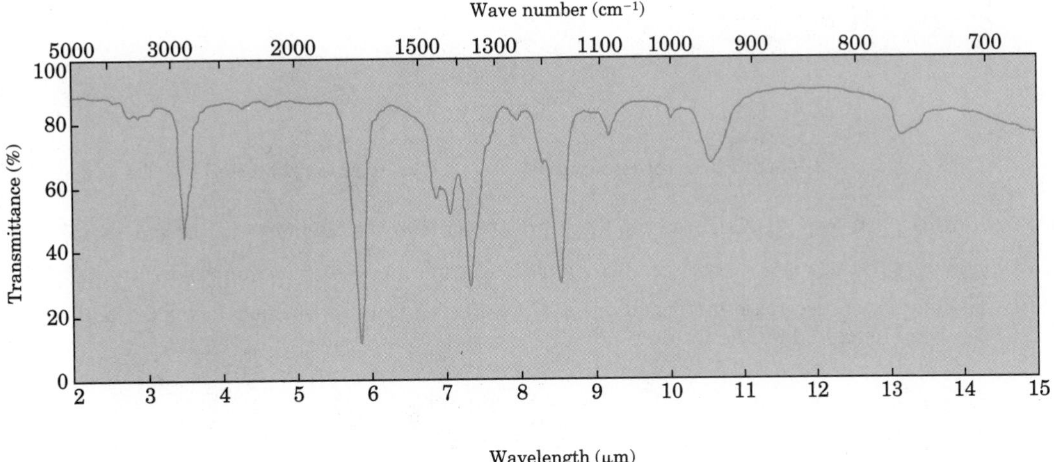

13.36 Dehydration of 1-methylcyclohexanol might lead to either of two isomeric alkenes, 1-methylcyclohexene or methylenecyclohexane. How could you use NMR spectroscopy (both ^1H and ^{13}C) to determine the structure of the product?

13.37 3,4-Dibromohexane can undergo base-induced double dehydrobromination to yield either 3-hexyne or 2,4-hexadiene. How could you use UV spectroscopy to help identify the product? How could you use ^1H NMR spectroscopy?

13.38 Describe the ^1H and ^{13}C NMR spectra you expect for these compounds:
(a) $ClCH_2CH_2CH_2Cl$ **(b)** $CH_3COCH_2CH_2Cl$

13.39 Propose structures for compounds with the following formulas that show only one peak in their ^1H NMR spectra:
(a) C_5H_{12} **(b)** C_5H_{10} **(c)** $C_4H_8O_2$

13.40 Assume that you have a compound with formula C_3H_6O.
(a) Propose as many structures as you can that fit the molecular formula (there are seven).
(b) If your compound has an IR absorption at 1715 cm^{-1}, what can you conclude?
(c) If your compound has a single ^1H NMR absorption at 2.1 δ, what is its structure?

13.41 Propose structures for compounds that fit these ^1H NMR data:

(a) $C_5H_{10}O$
6 H doublet at 0.95 δ, $J = 7$ Hz
3 H singlet at 2.10 δ
1 H multiplet at 2.43 δ

(b) C_3H_5Br
3 H singlet at 2.32 δ
1 H singlet at 5.25 δ
1 H singlet at 5.54 δ

13.42 How can you use ^1H and ^{13}C NMR to help distinguish among these four isomers?

$$\begin{array}{l} CH_2-CH_2 \\ | \quad\quad\;\; | \\ CH_2-CH_2 \end{array} \qquad H_2C{=}CHCH_2CH_3 \qquad CH_3CH{=}CHCH_3 \qquad \begin{array}{c} CH_3 \\ | \\ CH_3C{=}CH_2 \end{array}$$

13.43 How can you use ^1H NMR to help distinguish between the following isomers?

13.44 How can you use ^{13}C NMR to help distinguish between the isomers in Problem 13.43?

13.45 How can you use UV spectroscopy to help distinguish between the isomers in Problem 13.43?

13.46 The energy of electromagnetic radiation, E, expressed in units of kilocalories per mole (kcal/mol), can be determined by the formula

$$E = \frac{2.86 \times 10^{-3}}{\lambda \text{ (cm)}} \text{ kcal/mol}$$

What is the energy of infrared radiation of wavelength 10^{-4} cm?

13.47 Using the formula given in Problem 13.46, calculate the energy required to effect the electronic excitation of 1,3-butadiene (λ_{max} = 217 nm).

13.48 Using the equation given in Problem 13.46, calculate the amount of energy required to spin-flip a proton in a spectrometer operating at 100 MHz. Does increasing the spectrometer frequency from 60 MHz to 100 MHz increase or decrease the amount of energy necessary for resonance?

13.49 The ^1H NMR spectrum of compound A, $C_3H_6Br_2$, is shown. Propose a plausible structure for A and explain how the peaks in the spectrum fit your structure.

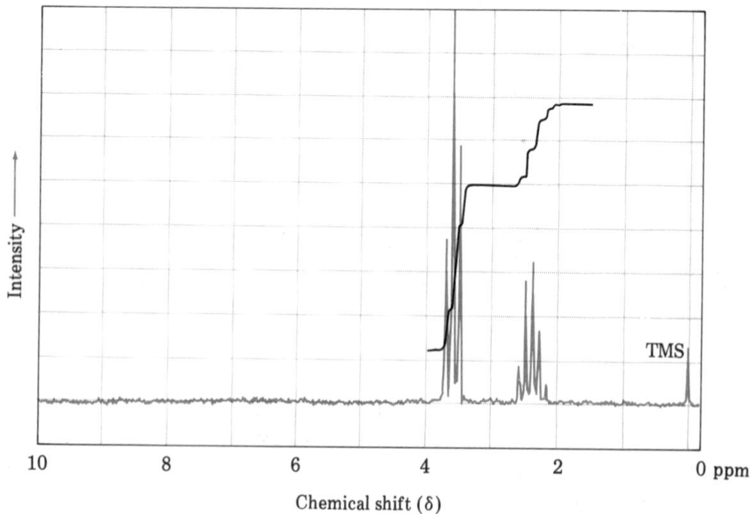

13.50 The compound whose ^1H NMR spectrum is shown has the formula $C_4H_7O_2Cl$ and has an IR absorption peak at 1740 cm^{-1}. Propose a plausible structure.

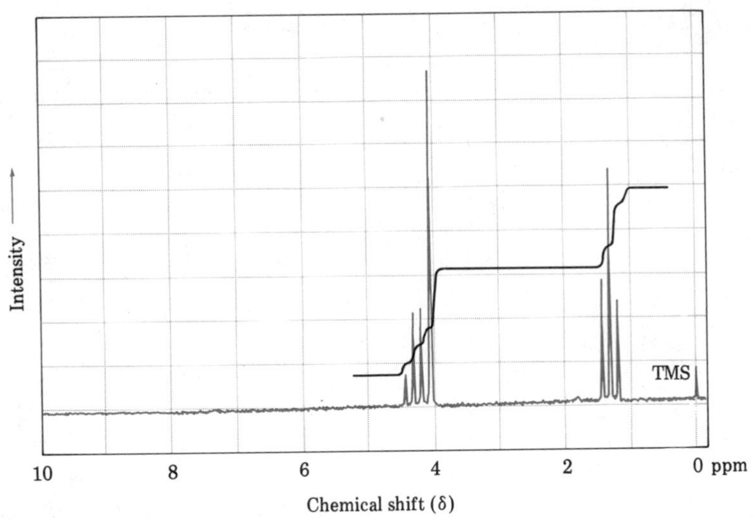

13.51 Propose a structure for a compound with formula C_4H_9Br that has the following ^1H NMR spectrum:

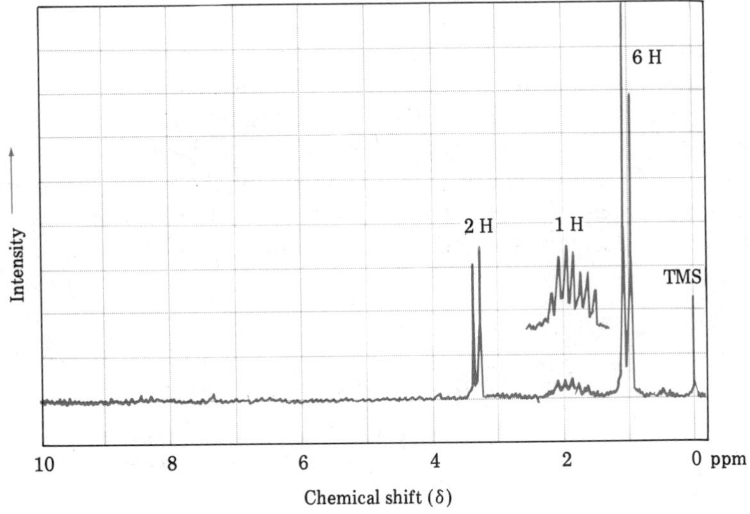

13.52 Propose structures for compounds that fit the following ^1H NMR data:

(a) $C_4H_6Cl_2$
- 3 H singlet at 2.18 δ
- 2 H doublet at 4.16 δ, $J = 7$ Hz
- 1 H triplet at 5.71 δ, $J = 7$ Hz

(b) $C_{10}H_{14}$
- 9 H singlet at 1.30 δ
- 5 H singlet at 7.30 δ

14 CHAPTER
BIOMOLECULES: CARBOHYDRATES

Carbohydrates are everywhere in nature; they occur in every living organism and are essential to life. The sugar and starch in food, and the cellulose in wood, paper, and cotton are nearly pure carbohydrate. Modified carbohydrates form part of the coating around living cells, other carbohydrates are found in the DNA that carries genetic information, and still others are invaluable as medicines.

CARBOHYDRATE
A straight-chain polyhydroxy ketone or aldehyde such as glucose

The word **carbohydrate** derives historically from the fact that glucose, the first simple carbohydrate to be obtained pure, has the molecular formula $C_6H_{12}O_6$ and was originally thought to be a "hydrate of carbon," $C_6(H_2O)_6$. This view was soon abandoned, but the name persisted. Today, the term *carbohydrate* is used to refer loosely to the broad class of polyhydroxylated aldehydes and ketones commonly called sugars.

```
H   O
 \ //
  C
  |
H—C—OH
  |
HO—C—H
  |
H—C—OH
  |
H—C—OH
  |
 CH2OH
```

Glucose (also called dextrose), a pentahydroxyhexanal

Carbohydrates are synthesized by green plants during photosynthesis, a complex process during which carbon dioxide is converted into glucose. Many molecules of glucose are then chemically linked for storage by the plant in the form of either cellulose or starch. It has been estimated that more than 50% of the dry weight of the earth's biomass—all plants and animals—consists of glucose polymers. When eaten and then metabolized, carbohydrates provide the major source of energy required by organisms. Thus, carbohydrates act as the chemical intermediaries by which solar energy is stored and used to support life.

$$6\ CO_2 + 6\ H_2O \xrightarrow{\text{Sunlight}} 6\ O_2 + \underset{\textbf{Glucose}}{C_6H_{12}O_6} \longrightarrow \text{Cellulose, starch}$$

14.1 CLASSIFICATION OF CARBOHYDRATES

SIMPLE SUGAR A carbohydrate like glucose that can't be hydrolyzed into smaller molecules

MONOSACCHARIDE Simple sugar

COMPLEX CARBOHYDRATE Carbohydrate composed of two or more simple sugars linked together by an acetal bond

DISACCHARIDE Complex carbohydrate having two simple sugars bonded together

POLYSACCHARIDE Complex carbohydrate having many simple sugars bonded together

ALDOSE Simple sugar with an aldehyde carbonyl group

KETOSE Simple sugar with a ketone carbonyl group

Carbohydrates are generally classed into two groups, *simple* and *complex*. **Simple sugars**, or **monosaccharides**, are carbohydrates like glucose and fructose that can't be hydrolyzed into smaller molecules. **Complex carbohydrates** are composed of two or more simple sugars linked together. Sucrose (table sugar), for example, is a **disaccharide** (two sugars) made up of one glucose molecule linked to one fructose molecule. Similarly, cellulose is a **polysaccharide** (many sugars) made up of several thousand glucose molecules linked together. Hydrolysis of polysaccharides breaks them down into their constituent monosaccharide units.

$$1\ \text{Sucrose} \xrightarrow{H_3O^+} 1\ \text{Glucose} + 1\ \text{Fructose}$$

$$\text{Cellulose} \xrightarrow{H_3O^+} {\sim}3000\ \text{Glucose}$$

Monosaccharides are further classified as either **aldoses** or **ketoses**. The *-ose* suffix is used as the family name ending for carbohydrates, and the *aldo-* and *keto-* prefixes identify the nature of the carbonyl group, whether aldehyde or ketone. The number of carbon atoms in the monosaccharide is indicated by using *tri-*, *tetr-*, *pent-*, *hex-*, and so forth, in the parent name. For example, glucose is an *aldohexose*, a six-carbon aldehydo sugar; fructose is a *ketohexose*, a six-carbon keto sugar; and ribose is an *aldopentose*, a five-carbon aldehydo sugar. Most of the commonly occurring simple sugars are either aldopentoses or aldohexoses.

```
   H   O           CH2OH           H   O
    \\//             |              \\//
     C             C=O              C
     |               |              |
  H—C—OH        HO—C—H          H—C—OH
     |               |              |
 HO—C—H          H—C—OH         H—C—OH
     |               |              |
  H—C—OH         H—C—OH         H—C—OH
     |               |              |
  H—C—OH           CH2OH          CH2OH
     |
   CH2OH
```

Glucose (an aldohexose) **Fructose (a ketohexose)** **Ribose (an aldopentose)**

PRACTICE PROBLEM 14.1 Classify this monosaccharide:

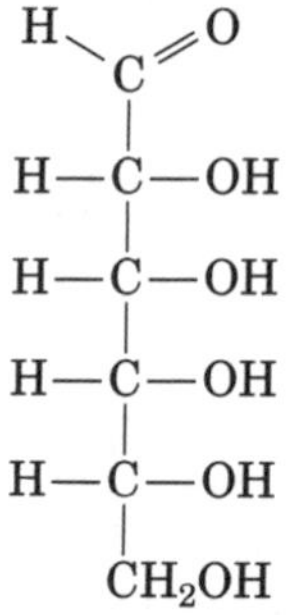

$$\begin{array}{c} \mathrm{H{\diagdown}C{=}O} \\ | \\ \mathrm{H{-}C{-}OH} \\ | \\ \mathrm{H{-}C{-}OH} \\ | \\ \mathrm{H{-}C{-}OH} \\ | \\ \mathrm{H{-}C{-}OH} \\ | \\ \mathrm{CH_2OH} \end{array}$$

Allose

SOLUTION Since allose has six carbons and an aldehyde carbonyl group, it is an aldohexose.

PROBLEM 14.1 Classify each of the following monosaccharides:

(a) 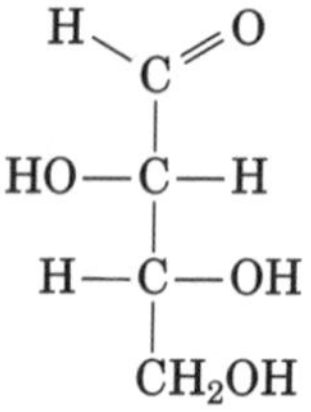
$$\begin{array}{c} \mathrm{H{\diagdown}C{=}O} \\ | \\ \mathrm{HO{-}C{-}H} \\ | \\ \mathrm{H{-}C{-}OH} \\ | \\ \mathrm{CH_2OH} \end{array}$$
Threose

(b) 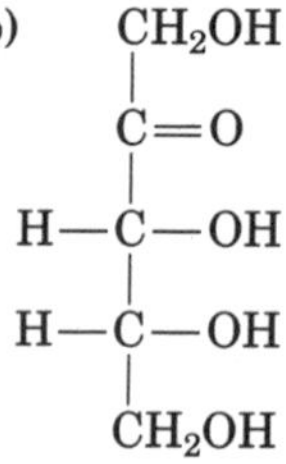
$$\begin{array}{c} \mathrm{CH_2OH} \\ | \\ \mathrm{C{=}O} \\ | \\ \mathrm{H{-}C{-}OH} \\ | \\ \mathrm{H{-}C{-}OH} \\ | \\ \mathrm{CH_2OH} \end{array}$$
Ribulose

(c) 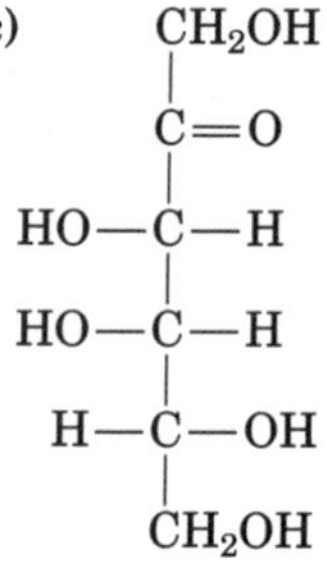
$$\begin{array}{c} \mathrm{CH_2OH} \\ | \\ \mathrm{C{=}O} \\ | \\ \mathrm{HO{-}C{-}H} \\ | \\ \mathrm{HO{-}C{-}H} \\ | \\ \mathrm{H{-}C{-}OH} \\ | \\ \mathrm{CH_2OH} \end{array}$$
Tagatose

(d) 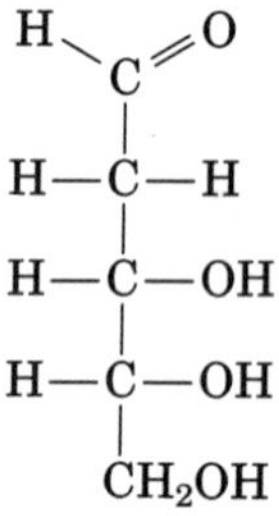
$$\begin{array}{c} \mathrm{H{\diagdown}C{=}O} \\ | \\ \mathrm{H{-}C{-}H} \\ | \\ \mathrm{H{-}C{-}OH} \\ | \\ \mathrm{H{-}C{-}OH} \\ | \\ \mathrm{CH_2OH} \end{array}$$
2-Deoxyribose

14.2 CONFIGURATIONS OF MONOSACCHARIDES: FISCHER PROJECTIONS

FISCHER PROJECTION Method for depicting stereochemistry at a chiral carbon by showing a tetrahedral carbon as two perpendicular crossed lines

Since all carbohydrates have chiral carbon atoms, it was recognized long ago that a standard method of representation is needed to describe carbohydrate stereochemistry. In 1891, Emil Fischer suggested a method based on the projection of a tetrahedral carbon atom onto a flat surface. These **Fischer projections** were soon adopted and are now a standard means of depicting stereochemistry at stereogenic centers.

A tetrahedral carbon atom in a Fischer projection is represented by two perpendicular lines. The horizontal lines represent bonds coming out of the page, and the vertical lines represent bonds going into the page:

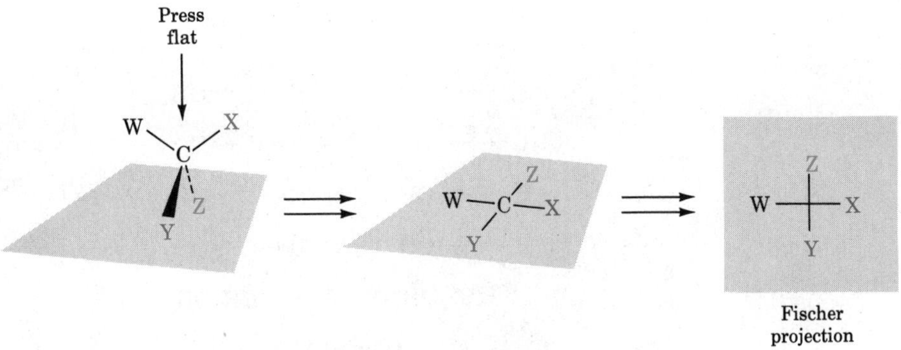

Press flat

W, X, C, Z, Y

W—C—X, Z, Y

Fischer projection

By convention, the carbonyl carbon is placed at or near the top in Fischer projections. Thus, (R)-glyceraldehyde, the simplest monosaccharide, can be represented as shown in Figure 14.1.

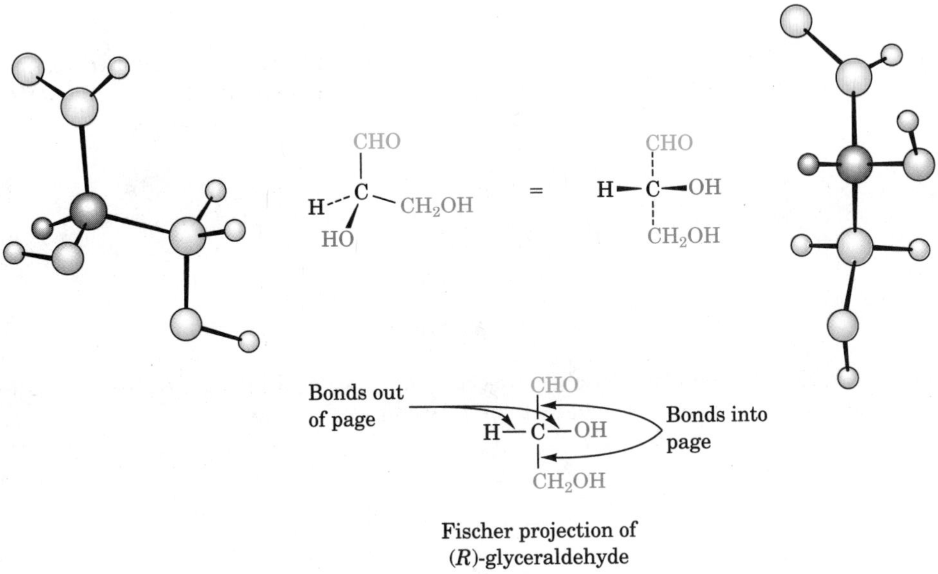

CHO, H, C, CH_2OH, HO = CHO, H—C—OH, CH_2OH

Bonds out of page

CHO, H—C—OH, CH_2OH

Bonds into page

Fischer projection of (R)-glyceraldehyde

FIGURE 14.1 Fischer projection of (R)-glyceraldehyde.

Carbohydrates with more than one stereogenic center are shown by stacking the atoms, one on top of the other. Once again, the carbonyl carbon is at or near the top of the Fischer projection. Glucose, for example, has four stereogenic centers stacked on top of each other in a Fischer projection:

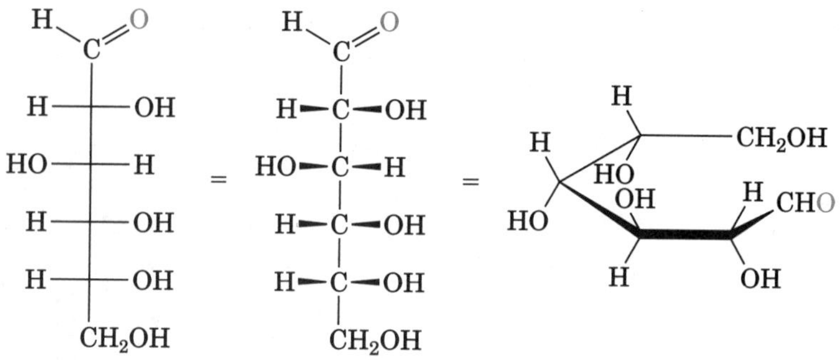

H–C=O; H—C—OH; HO—C—H; H—C—OH; H—C—OH; CH_2OH = (Fischer and wedge forms) = (tetrahedral chain form with CH_2OH, CHO, H, HO, OH labels)

Glucose
(carbonyl group at top)

PRACTICE PROBLEM 14.2 Convert the following tetrahedral representation of (*R*)-2-butanol into a Fischer projection:

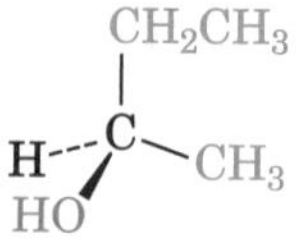

CH_2CH_3 / H---C—CH_3 / HO   **(*R*)-2-Butanol**

SOLUTION Orient the molecule so that two horizontal bonds are facing you and two vertical bonds are receding away from you. Then press the molecule flat into the paper, indicating the chiral carbon as the intersection of two crossed lines:

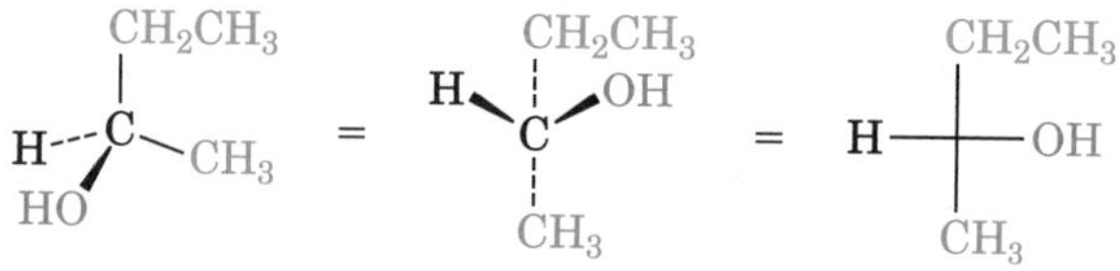

CH_2CH_3 / H---C—CH_3 / HO = H, OH on C, CH_2CH_3 and CH_3 dashed = H—+—OH with CH_2CH_3 above and CH_3 below

(*R*)-2-Butanol

PRACTICE PROBLEM 14.3 Convert the following Fischer projection of lactic acid into a tetrahedral representation, and indicate whether the molecule is (*R*) or (*S*):

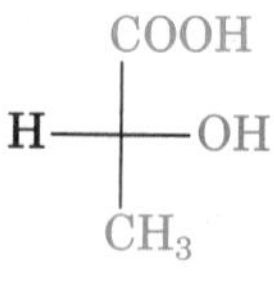

COOH / H—+—OH / CH_3 **Lactic acid**

SOLUTION After placing a carbon atom at the intersection of the two crossed lines, imagine that the two horizontal bonds are coming toward you and the two vertical bonds are receding away from you. The projection represents (*R*)-lactic acid.

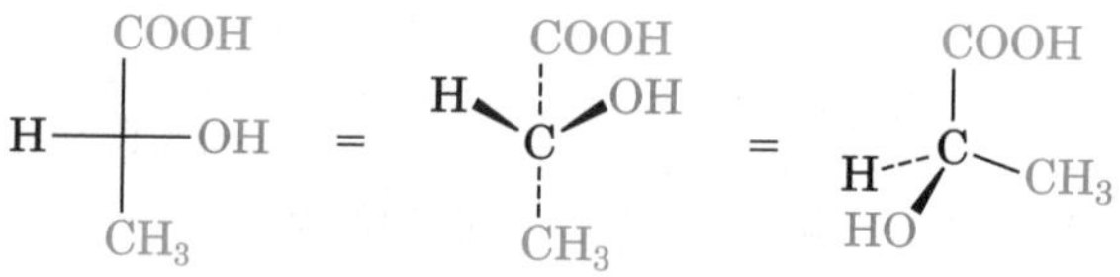

COOH / H—+—OH / CH_3 = H, OH on C, COOH and CH_3 dashed = COOH / H---C—CH_3 / HO

(*R*)-Lactic acid

PROBLEM 14.2 Convert this tetrahedral representation of (*S*)-glyceraldehyde into a Fischer projection:

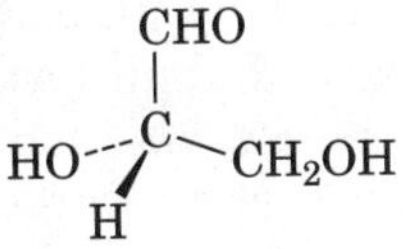

CHO
HO---C---CH₂OH (wedge H)

(S)-Glyceraldehyde

PROBLEM 14.3 Draw Fischer projections of both (*R*)-2-chlorobutane and (*S*)-2-chlorobutane.

PROBLEM 14.4 Convert these Fischer projections into tetrahedral representations, and assign *R* or *S* stereochemistry to each:

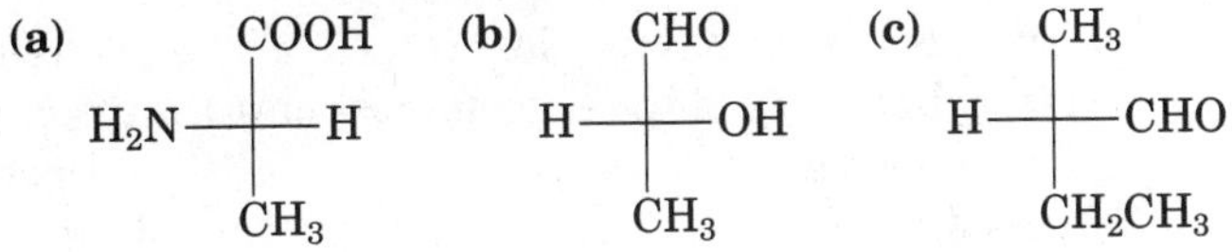

(a) COOH on top; H_2N on the left, H on the right; CH_3 on the bottom

(b) CHO on top; H on the left, OH on the right; CH_3 on the bottom

(c) CH_3 on top; H on the left, CHO on the right; CH_2CH_3 on the bottom

14.3 D,L SUGARS

Glyceraldehyde has one chiral carbon atom and therefore has two enantiomeric (mirror-image) forms, but only the dextrorotatory enantiomer occurs naturally. That is, a sample of naturally occurring glyceraldehyde placed in a polarimeter (see Section 6.4) rotates plane-polarized light in a clockwise direction, denoted (+). Since (+)-glyceraldehyde is known to have the *R* configuration at C2, it can be represented as in Figure 14.2. For historical reasons dating from long before the adoption of the *R*,*S* system, (*R*)-(+)-glyceraldehyde is also referred to as D-*glyceraldehyde*, where the D stands for dextrorotatory. The other enantiomer, (*S*)-(−)-glyceraldehyde, is known as L-*glyceraldehyde* (L for levorotatory).

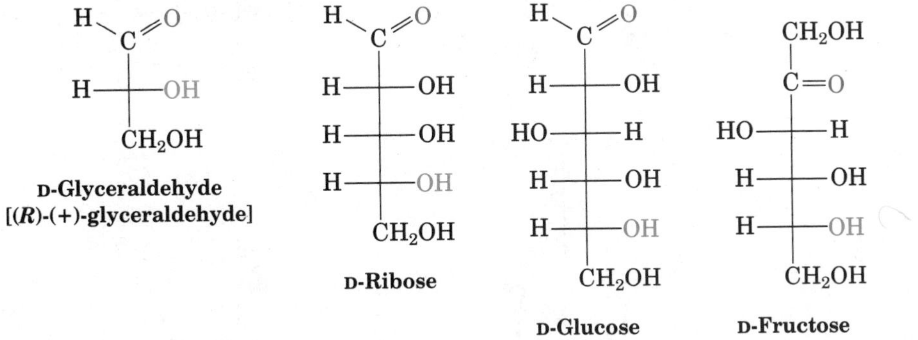

D-Glyceraldehyde [(*R*)-(+)-glyceraldehyde]: CHO (H–C=O) on top; H on the left, OH on the right; CH_2OH on the bottom

D-Ribose: CHO (H–C=O) on top; then H on the left, OH on the right; H on the left, OH on the right; H on the left, OH on the right; CH_2OH on the bottom

D-Glucose: CHO (H–C=O) on top; then H on the left, OH on the right; HO on the left, H on the right; H on the left, OH on the right; H on the left, OH on the right; CH_2OH on the bottom

D-Fructose: CH_2OH on top; then C=O; HO on the left, H on the right; H on the left, OH on the right; H on the left, OH on the right; CH_2OH on the bottom

FIGURE 14.2 Some naturally occurring D sugars. The hydroxyl at the lowest stereogenic center is on the right in Fischer projections.

Because of the way that monosaccharides are synthesized in nature, glucose, fructose, ribose, and most other naturally occurring monosaccharides have the same stereochemical configuration as D-glyceraldehyde at the chiral carbon atom farthest from the carbonyl group. In Fischer projections, therefore, most naturally occurring sugars have the hydroxyl group at the lowest chiral carbon atom pointing to the *right* (Figure 14.2). Such compounds are referred to as **D sugars**.

D SUGAR
Sugar whose hydroxyl group at the chiral carbon farthest from the carbonyl group points to the right when the molecule is drawn in Fischer projection

L SUGAR
Sugar whose hydroxyl group at the chiral carbon farthest from the carbonyl group points to the left when the molecule is drawn in Fischer projection

In contrast to the D sugars, all **L sugars** have the hydroxyl group at the stereogenic center farthest from the carbonyl group on the *left* in Fischer projections. Thus, L sugars are mirror images (enantiomers) of D sugars. Note that the D and L notations have no relation to the direction in which a given sugar rotates plane-polarized light. A D sugar may be either dextrorotatory or levorotatory. The prefix D indicates only that the stereochemistry of the bottommost chiral carbon atom is the same as that of D-glyceraldehyde and is to the right in Fischer projection when the molecule is drawn in the standard way with the carbonyl group at or near the top.

Mirror

H–C=O; HO—H; CH_2OH

L-Glyceraldehyde
[(S)-(−)-glyceraldehyde]

H–C=O; HO—H; H—OH; HO—H; HO—H; CH_2OH

L-Glucose
(not naturally occurring)

O=C–H; H—OH; HO—H; H—OH; H—OH; CH_2OH

D-Glucose

PRACTICE PROBLEM 14.4 Look at the Fischer projection of D-fructose in Figure 14.2, and then draw a Fischer projection of L-fructose.

SOLUTION Since L-fructose is the enantiomer (mirror image) of D-fructose, we simply take the structure of D-fructose and reverse the configuration at each stereogenic center:

Mirror

CH_2OH; =O; HO—H; H—OH; H—OH; CH_2OH

D-Fructose

CH_2OH; =O; H—OH; HO—H; HO—H; CH_2OH

L-Fructose

PROBLEM 14.5 Which of the following are L sugars and which are D sugars?

(a)

$$\begin{array}{ccc} & CHO & \\ HO & — & H \\ HO & — & H \\ & CH_2OH & \end{array}$$

(b)

$$\begin{array}{ccc} & CHO & \\ H & — & OH \\ HO & — & H \\ H & — & OH \\ & CH_2OH & \end{array}$$

(c)

$$\begin{array}{ccc} & CH_2OH & \\ & C{=}O & \\ HO & — & H \\ H & — & OH \\ & CH_2OH & \end{array}$$

PROBLEM 14.6 Draw the enantiomers (mirror images) of the carbohydrates shown in Problem 14.5 and identify each as a D sugar or as an L sugar.

14.4 CONFIGURATIONS OF ALDOSES

Aldotetroses are four-carbon sugars with two stereogenic centers. Thus, there are $2^2 = 4$ possible stereoisomeric aldotetroses, or two D,L pairs of enantiomers, named *erythrose* and *threose.*

Aldopentoses have three stereogenic centers and a total of $2^3 = 8$ possible stereoisomers, or four D,L pairs of enantiomers. These four pairs are named *ribose*, *arabinose*, *xylose*, and *lyxose*. All except lyxose occur widely in nature. D-ribose is an important part of RNA (ribonucleic acid), L-arabinose is found in many plants, and D-xylose is found in wood.

Aldohexoses have four stereogenic centers, for a total of $2^4 = 16$ possible stereoisomers, or eight D,L pairs of enantiomers. The names of the eight pairs are *allose*, *altrose*, *glucose*, *mannose*, *gulose*, *idose*, *galactose*, and *talose*. Of the eight, only D-glucose, from starch and cellulose, and D-galactose, from gums and fruit pectins, are widely distributed in nature. D-Mannose and D-talose also occur naturally, but in lesser abundance.

Fischer projections of the four-, five-, and six-carbon aldoses are shown in Figure 14.3 for the D series. Starting from D-glyceraldehyde, we can imagine constructing the two D-aldotetroses by inserting a new chiral carbon atom just below the aldehyde carbon. Each of the two D-aldotetroses then leads to two D-aldopentoses (four total), and each of the four D-aldopentoses leads to two D-aldohexoses (eight total).

PROBLEM 14.7 Write Fischer projections for the following L sugars. Remember that an L sugar is the mirror image of the corresponding D sugar shown in Figure 14.3.
(a) L-Arabinose **(b)** L-Threose **(c)** L-Galactose

PROBLEM 14.8 How many aldoheptoses are possible? How many of them are D sugars and how many are L sugars?

PROBLEM 14.9 Draw Fischer projections for the two D-aldoheptoses (Problem 14.8) whose stereochemistry at C3, C4, C5, and C6 is the same as that of glucose at C2, C3, C4, and C5.

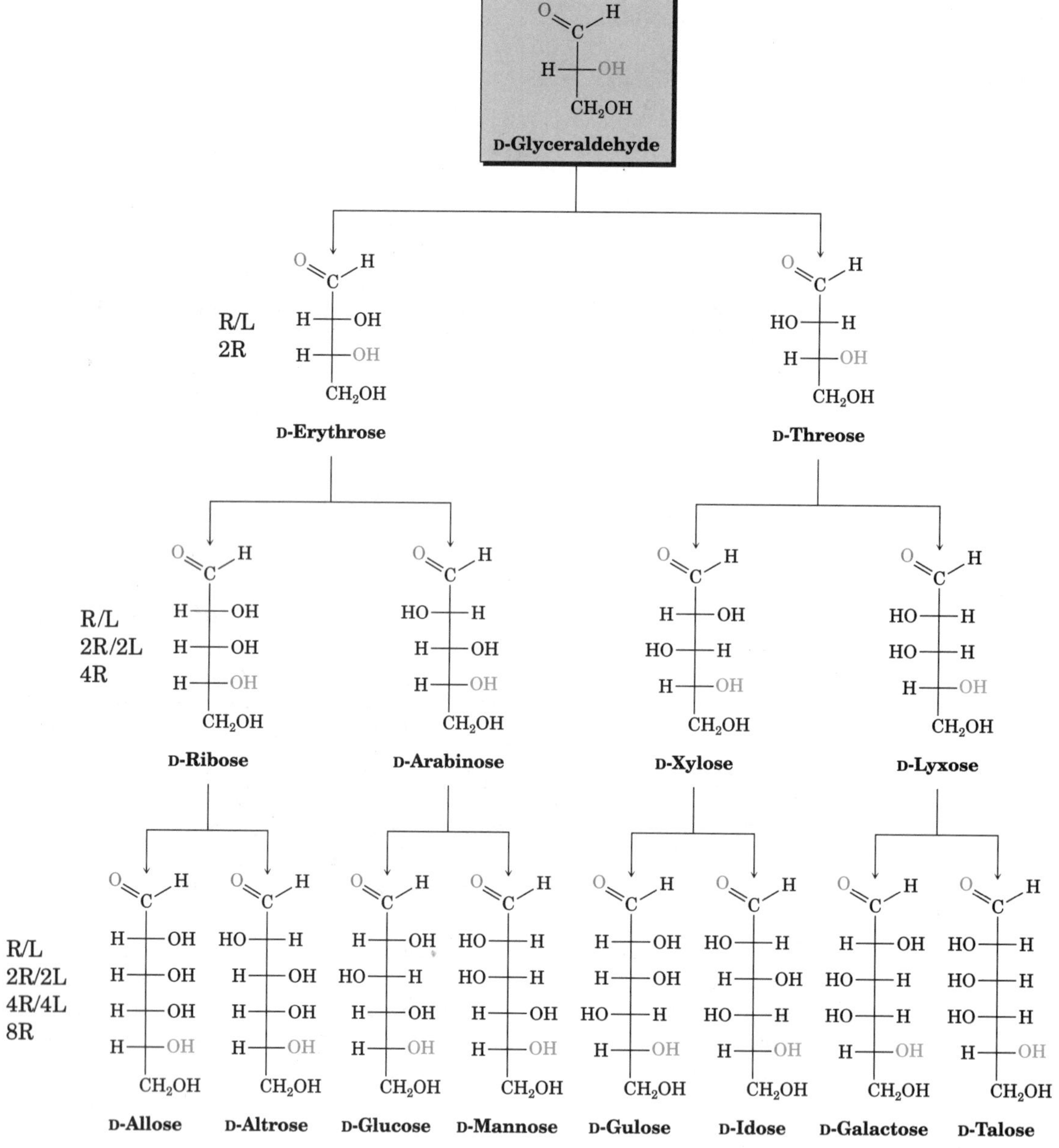

FIGURE 14.3 Configurations of D-aldoses: The structures are arranged in order from left to right so that the hydroxyl groups on C2 alternate right/left (R/L) in going across a series; the hydroxyl groups at C3 alternate two right/two left (2R/2L); the hydroxyl groups at C4 alternate four right/four left (4R/4L); and the hydroxyl groups at C5 are to the right in all eight (8R).

14.5 CYCLIC STRUCTURES OF MONOSACCHARIDES: HEMIACETAL FORMATION

We said during the discussion of carbonyl-group chemistry in Section 9.11 that alcohols undergo a rapid and reversible nucleophilic addition reaction with ketones and aldehydes to form hemiacetals:

$$\underset{\text{An aldehyde}}{\mathrm{R{-}C({=}O){-}H}} + \mathrm{R'OH} \xrightleftharpoons[\text{catalyst}]{\mathrm{H^+}} \underset{\text{A hemiacetal}}{\mathrm{H{-}C(OH)(R){-}OR'}}$$

If both the hydroxyl and the carbonyl group are in the same molecule, an *intramolecular* nucleophilic addition can take place, leading to the formation of a *cyclic* hemiacetal. Five- and six-membered cyclic hemiacetals form particularly easily, and many carbohydrates therefore exist in an equilibrium between open-chain and cyclic forms. For example, glucose exists in aqueous solution primarily as the six-membered (**pyranose**) ring formed by intramolecular nucleophilic addition of the hydroxyl group at C5 to the C1 aldehyde group. Fructose, on the other hand, exists to the extent of about 20% as the five-membered (**furanose**) ring formed by addition of the hydroxyl group at C5 to the C2 ketone. The terms *pyranose* for a six-membered ring and *furanose* for a five-membered ring are derived from the names of the simple cyclic ethers pyran and furan. The cyclic forms of glucose and fructose are shown in Figure 14.4 (p. 420).

PYRANOSE
Six-membered ring structure of a simple sugar

FURANOSE
Five-membered ring structure of a simple sugar

Pyranose and furanose rings are often represented using the **Haworth projections** shown in Figure 14.4, rather than Fischer projections. In a Haworth projection, the hemiacetal ring is drawn as if it were flat and is viewed edge-on, with the oxygen atom at the upper right. Although convenient, this view isn't really accurate because pyranose rings are actually chair-shaped like cyclohexane (Section 2.9), rather than flat. Nevertheless, Haworth projections are widely used because they make it possible to see at a glance the cis–trans relationships among hydroxyl groups on the ring.

HAWORTH PROJECTION
View of a furanose or pyranose sugar in which the ring is flat and viewed edge-on, with the oxygen atom at the upper right

When converting from one kind of projection to the other, remember that a hydroxyl on the *right* in a Fischer projection is *down* in a Haworth projection. Conversely, a hydroxyl on the *left* in a Fischer projection is *up* in a Haworth projection. For D sugars, the terminal $-CH_2OH$ group is always up in Haworth projections, whereas for L sugars, the $-CH_2OH$ group is down. Figure 14.5 illustrates the conversion for glucose.

PRACTICE PROBLEM 14.5 D-Mannose differs from D-glucose in its stereochemistry at C2. Draw a Haworth projection of D-mannose in its pyranose form.

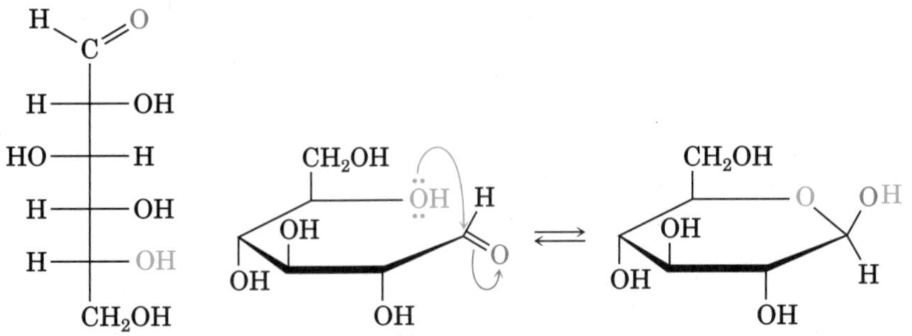

D-Glucose (Fischer)

D-Glucose, pyranose form (Haworth)

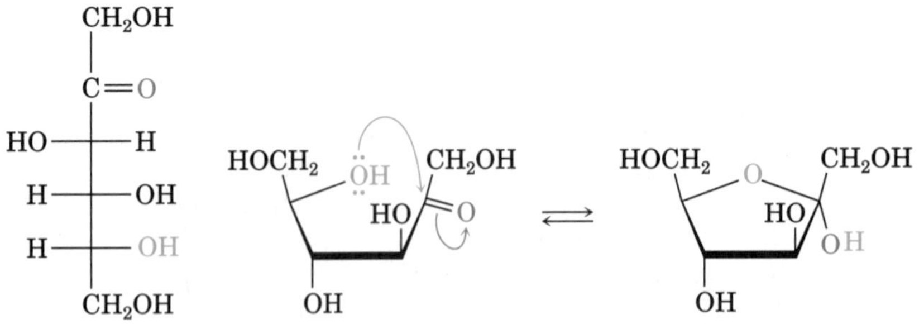

D-Fructose (Fischer)

D-Fructose, furanose form (Haworth)

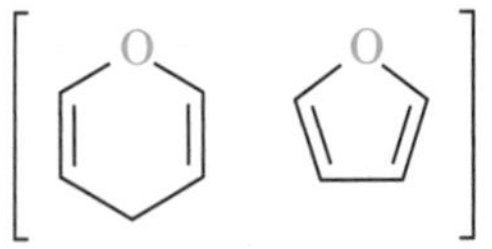

Pyran Furan

FIGURE 14.4 Glucose and fructose in their cyclic pyranose and furanose forms.

SOLUTION First, draw a Fischer projection of D-mannose. Then lay it on its side and curl it around so that the aldehyde group (C1) is toward the front and the CH_2OH (C6) is toward the rear. Now connect the hydroxyl at C5 to the C1 carbonyl group to form a pyranose ring.

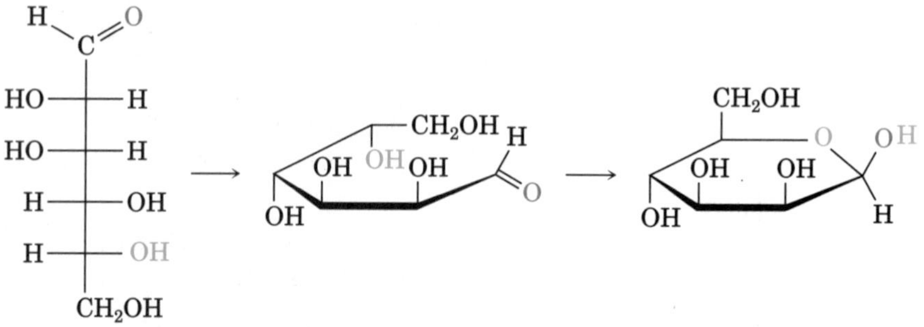

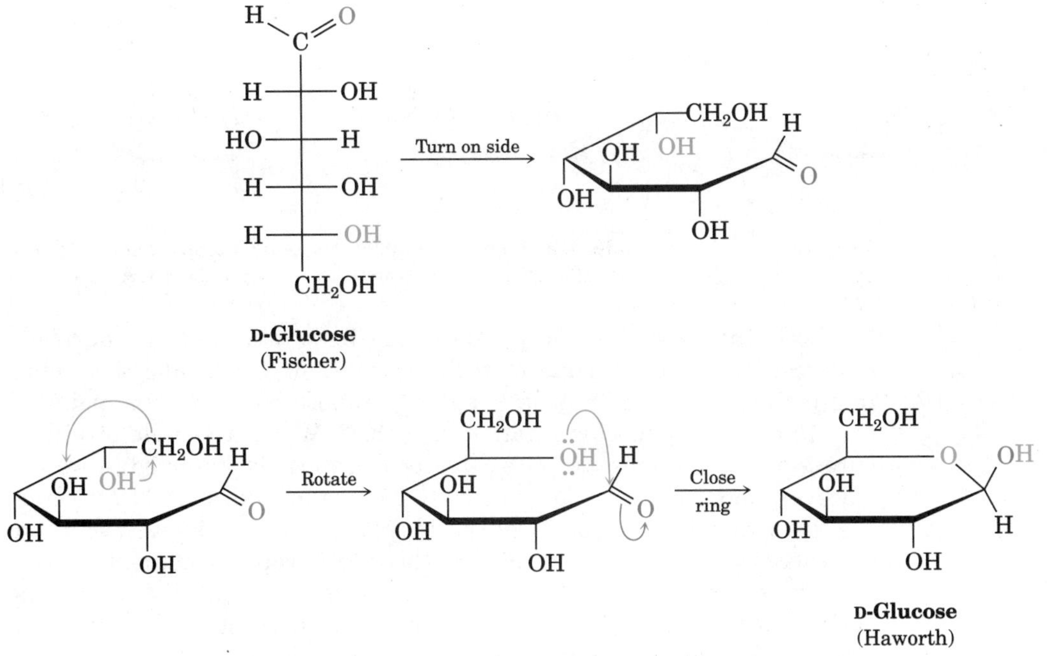

FIGURE 14.5 Interconversion of Fischer and Haworth projections of D-glucose.

PROBLEM 14.10 D-Galactose differs from D-glucose in its stereochemistry at C4. Draw a Haworth projection of D-galactose in its pyranose form.

PROBLEM 14.11 Ribose exists largely in a furanose form produced by addition of the C4 hydroxyl group to the C1 aldehyde. Draw a Haworth projection of D-ribose in its furanose form.

14.6 MONOSACCHARIDE ANOMERS: MUTAROTATION

ANOMER
Pyranose or furanose sugar whose hydroxyl group at C1 is either up (α) or down (β)

ANOMERIC CENTER
The hemiacetal carbon in a pyranose or furanose sugar

When an open-chain monosaccharide cyclizes to a furanose or pyranose form, a new stereogenic center is formed at what used to be the carbonyl carbon. Two diastereomers called **anomers** are produced, with the hemiacetal carbon referred to as the **anomeric center**. For example, glucose cyclizes reversibly in aqueous solution to yield a 36:64 mixture of two anomers. The minor anomer with the C1 –OH group trans to the $-CH_2OH$ substituent at C5 (and therefore *down* in a Haworth projection) is called the **alpha (α) anomer**; its complete name is α-D-glucopyranose. The major anomer with the C1 –OH group cis to the $-CH_2OH$ substituent at C5 (and therefore *up* in a Haworth projection) is called the **beta (β) anomer**; its complete name is β-D-glucopyranose.

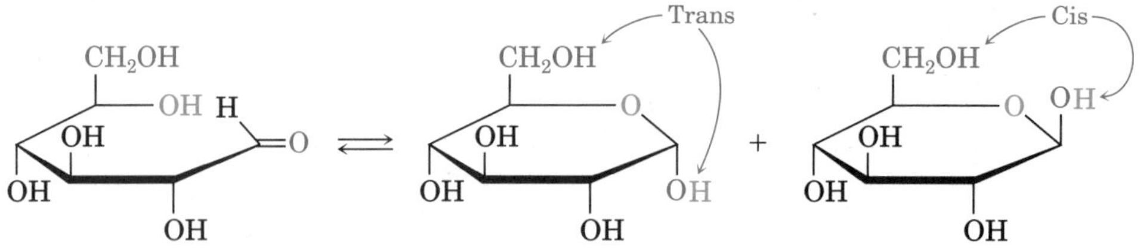

D-Glucose | **α-D-Glucopyranose (36%)** (OH and CH_2OH are trans) | **β-D-Glucopyranose (64%)** (OH and CH_2OH are cis)

Both anomers of D-glucopyranose can be crystallized and purified. Pure α-D-glucopyranose has a melting point of 146°C and a specific rotation $[\alpha]_D$ of +112.2°; pure β-D-glucopyranose has a melting point of 148–155°C and a specific rotation of +18.7°. When a sample of either pure α-D-glucopyranose or pure β-D-glucopyranose is dissolved in water, however, both optical rotations slowly change and ultimately converge to a constant value of +52.6°. The specific rotation of the α anomer solution decreases from +112.2° to +52.6°, and the specific rotation of the β anomer solution increases from +18.7° to +52.6°. Known as **mutarotation**, this phenomenon is due to the slow conversion of the pure α and β enantiomers into the 36:64 equilibrium mixture.

MUTAROTATION Change in optical rotation observed when a solution of a single sugar anomer equilibrates to a mixture of anomers

Mutarotation occurs by a reversible ring opening of each anomer to the open-chain aldehyde form, followed by reclosure. Although equilibration is slow at neutral pH, it is catalyzed by both acid and base.

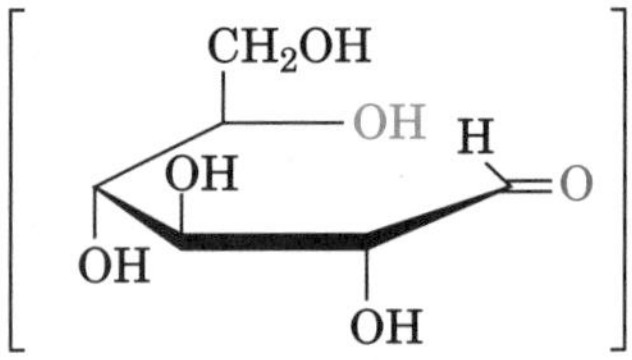

D-Glucose

α-D-Glucopyranose (36%) $[\alpha]_D = +112.2°$

β-D-Glucopyranose (64%) $[\alpha]_D = +18.7°$

PRACTICE PROBLEM 14.6 Draw Haworth projections of the two pyranose anomers of D-galactose, and identify each as α or β.

SOLUTION The α anomer has the –OH group at C1 pointing down, and the β anomer has the –OH group at C1 pointing up.

α-D-Galactopyranose **β-D-Galactopyranose**

PROBLEM 14.12 Draw the two anomers of D-fructose in their furanose forms.

PROBLEM 14.13 If the specific rotation of pure α-D-glucopyranose is $+112.2°$ and the specific rotation of pure β-D-glucopyranose is $+18.7°$, show how the equilibrium percentages of α and β anomers can be calculated from the equilibrium specific rotation of $+52.6°$.

PROBLEM 14.14 Many other sugars besides glucose exhibit mutarotation. For example, α-D-galactopyranose has $[\alpha]_D = +150.7°$, and β-D-galactopyranose has $[\alpha]_D = +52.8°$. If either anomer is dissolved in water and allowed to reach equilibrium, the specific rotation of the solution is $+80.2°$. What are the percentages of each anomer at equilibrium?

14.7 CONFORMATIONS OF MONOSACCHARIDES

Although Haworth projections are easy to draw, they don't give an accurate three-dimensional picture of a molecule. Pyranose rings, like cyclohexane rings (Section 2.10), have a chairlike geometry with axial and equatorial substituents. Any substituent that is up in a Haworth projection is also up in a chair conformation, and any substituent that is down in a Haworth projection is also down in the chair conformation. Haworth projections can be converted into chair representations by following three steps:

STEP 1 Draw the Haworth projection with the ring oxygen atom at the upper right.

STEP 2 Raise the leftmost carbon atom (C4) *above* the ring plane.

STEP 3 Lower the anomeric carbon atom (C1) *below* the ring plane.

Figure 14.6 shows how this is done for α-D-glucopyranose and β-D-glucopyranose.

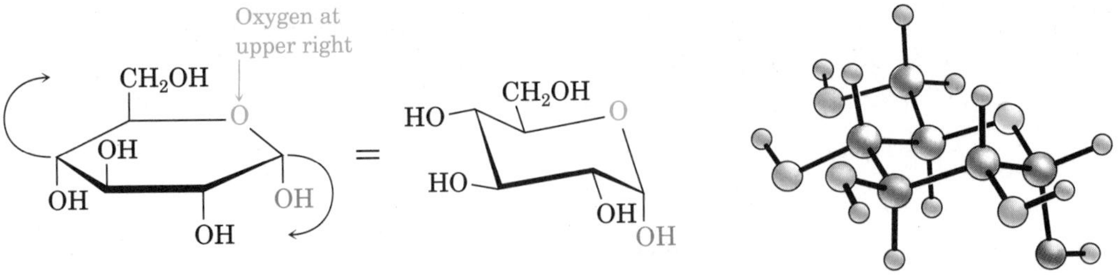

Oxygen at upper right

α-D-Glucopyranose

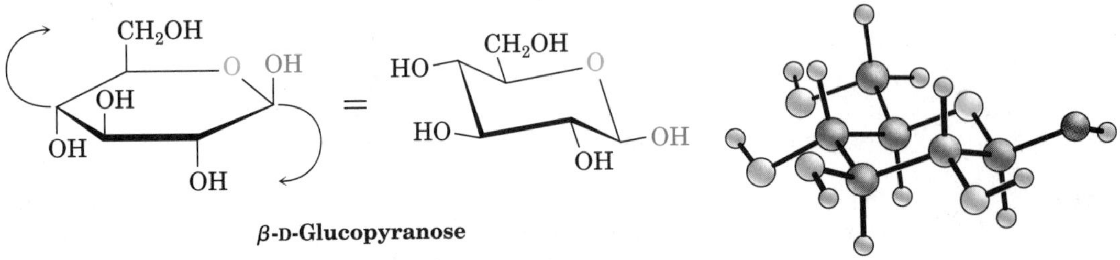

β-D-Glucopyranose

FIGURE 14.6 Chair representations of α-D-glucopyranose and β-D-glucopyranose.

Note that in β-D-glucopyranose, all the substituents on the ring are equatorial. Thus, β-D-glucopyranose is the least sterically crowded and most stable of the eight D-aldohexoses.

PROBLEM 14.15 Draw β-D-galactopyranose in its chair conformation. Label all the ring substituents as axial or equatorial.

PROBLEM 14.16 Draw β-D-mannopyranose in its chair conformation, and label all substituents as axial or equatorial. Which would you expect to be more stable, mannose or galactose (Problem 14.15)?

14.8 REACTIONS OF MONOSACCHARIDES

Since monosaccharides contain only two kinds of functional groups, carbonyls and hydroxyls, most of the chemistry of monosaccharides is the now familiar chemistry of these two groups.

Ester and Ether Formation

Monosaccharides behave as simple alcohols in much of their chemistry. For example, carbohydrate hydroxyl groups can be converted into esters and ethers. Ester and ether derivatives of carbohydrates are often much easier to work with than the free sugars. Because of their many hydroxyl groups, monosaccharides are usually soluble in water but insoluble in organic solvents such as ether. Ester and ether derivatives, however, are soluble in organic solvents and are easily crystallized.

Esterification is carried out by treating the carbohydrate with an acid chloride or acid anhydride in the presence of a base (Sections 10.7 and 10.8). All the hydroxyl groups react, including the anomeric one. For example, β-D-glucopyranose is converted into its pentaacetate by treatment with acetic anhydride in pyridine solution:

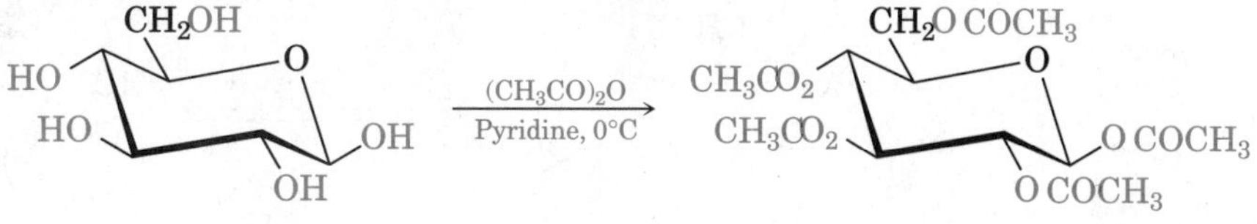

$$\xrightarrow[\text{Pyridine, 0°C}]{(CH_3CO)_2O}$$

β-D-Glucopyranose → **Penta-O-acetyl-β-D-glucopyranose (91%)**

Carbohydrates are converted into ethers by treatment with an alkyl halide in the presence of base (the Williamson ether synthesis; Section 8.6). Silver oxide is a particularly mild and useful base for this reaction, since hydroxide and alkoxide bases tend to degrade the sensitive sugar molecules. For example, α-D-glucopyranose is converted into its pentamethyl ether in 85% yield on reaction with iodomethane and silver oxide:

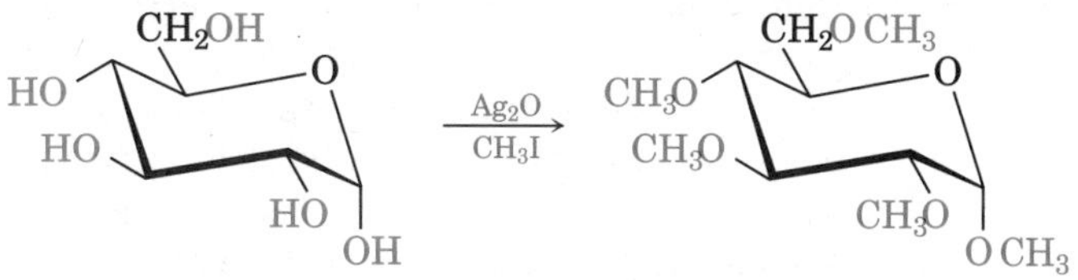

$$\xrightarrow[CH_3I]{Ag_2O}$$

α-D-Glucopyranose → **α-D-Glucopyranose pentamethyl ether (85%)**

PROBLEM 14.17 Draw the products you would obtain by reaction of β-D-ribofuranose with:
(a) CH_3I, Ag_2O **(b)** $(CH_3CO)_2O$, pyridine

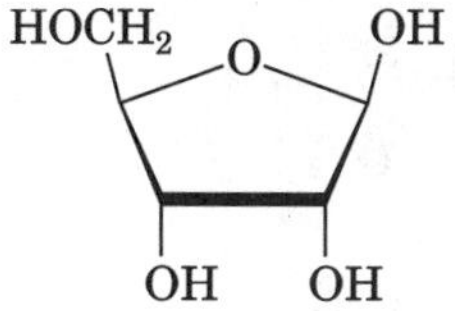

β-D-Ribofuranose

Glycoside Formation

We saw in Section 9.9 that treatment of a hemiacetal with an alcohol and an acid catalyst yields an acetal:

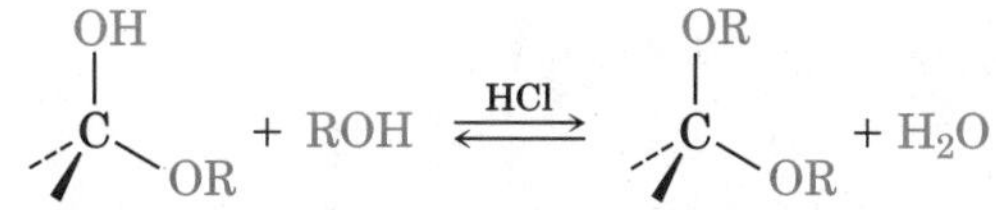

$$\text{A hemiacetal} + ROH \underset{}{\overset{HCl}{\rightleftharpoons}} \text{An acetal} + H_2O$$

In the same way, treatment of a monosaccharide hemiacetal with an alcohol and an acid catalyst yields an acetal in which the anomeric hydroxyl group has been replaced by an alkoxy group. For example, reaction of glucose with methanol gives methyl β-D-glucopyranoside:

β-D-Glucopyranose (a hemiacetal)

Methyl β-D-Glucopyranoside (an acetal)

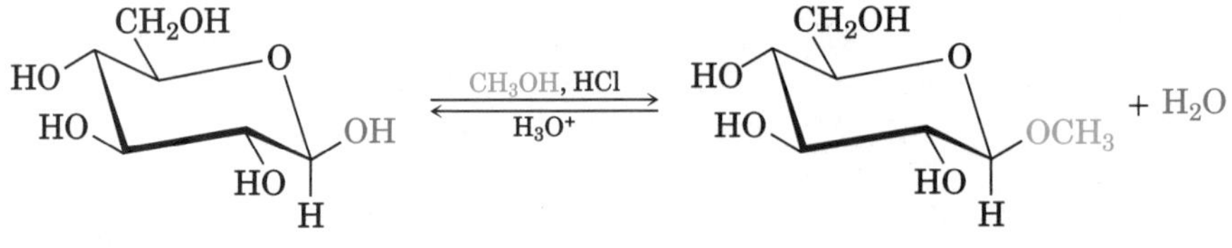

GLYCOSIDE Carbohydrate acetal formed by reaction of a carbohydrate with an alcohol

Carbohydrate acetals are called **glycosides** and are named by first citing the alkyl group and then replacing the *-ose* ending of the sugar with *-oside*. Glycosides, like all acetals, are stable to water. They aren't in equilibrium with an open-chain form and they don't show mutarotation. They can, however, be converted back to the free monosaccharide by hydrolysis with aqueous acid.

Glycosides are widespread in nature, and a great many biologically active molecules contain glycosidic linkages. For example, digitoxin, the active component of the digitalis preparations used for treatment of heart disease, is a glycoside consisting of a complex steroid alcohol linked to a trisaccharide (Figure 14.7). Note that the three sugars are also linked to each other by glycoside bonds.

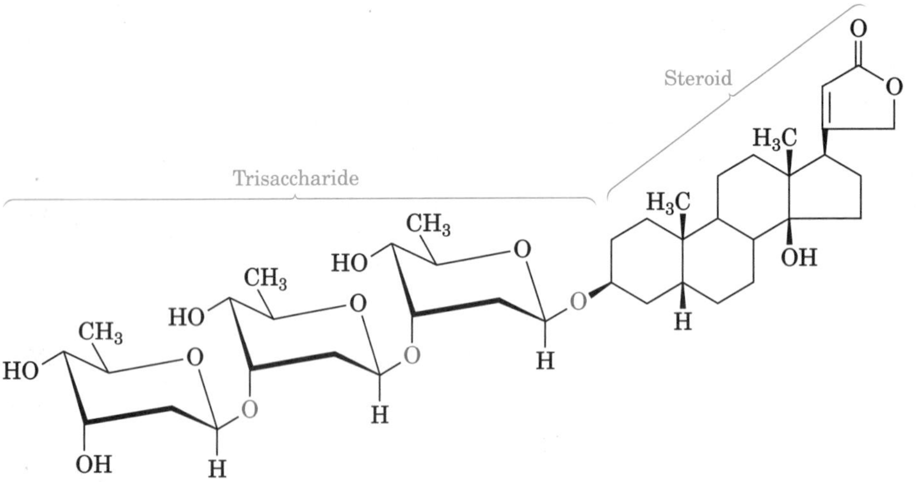

FIGURE 14.7 The structure of digitoxin, a complex glycoside.

PRACTICE PROBLEM 14.7 What product would you expect from the acid-catalyzed reaction of β-D-ribofuranose with methanol?

SOLUTION The acid-catalyzed reaction of a monosaccharide with an alcohol yields a glycoside in which the anomeric –OH group is replaced by the –OR group of the alcohol:

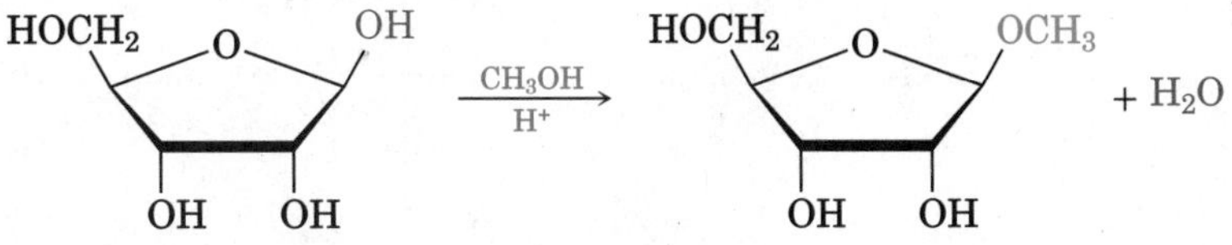

$$\beta\text{-D-Ribofuranose} \xrightarrow[H^+]{CH_3OH} \text{Methyl } \beta\text{-D-ribofuranoside} + H_2O$$

β-D-Ribofuranose

Methyl β-D-ribofuranoside

PROBLEM 14.18 Draw the product you would obtain from the acid-catalyzed reaction of β-D-galactopyranose with ethanol.

Reduction of Monosaccharides

ALDITOL
Polyalcohol formed by reduction of a ketose or aldose

Treatment of an aldose or ketose with sodium borohydride reduces it to a polyalcohol called an **alditol**. The reaction occurs by interception of the open-chain form present in the aldehyde ⇄ hemiacetal equilibrium.

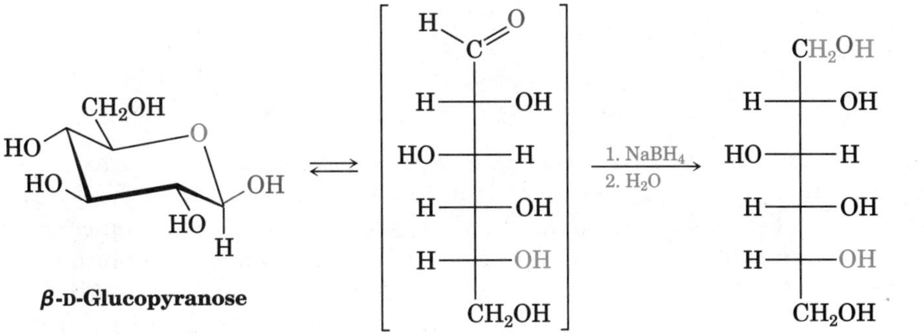

$$\beta\text{-D-Glucopyranose} \rightleftharpoons [\text{D-Glucose}] \xrightarrow[2.\ H_2O]{1.\ NaBH_4} \text{D-Glucitol}$$

β-D-Glucopyranose

D-Glucose

D-Glucitol (D-sorbitol)
an alditol

D-Glucitol, the alditol produced on reduction of D-glucose, is itself a naturally occurring substance that has been isolated from many fruits and berries. It is used under the name D-sorbitol as a sweetener and sugar substitute in many foods.

PRACTICE PROBLEM 14.8 Show the structure of the alditol you would obtain from reduction of D-galactose.

SOLUTION First, draw D-galactose in its open-chain form. Then convert the –CHO group at C1 into a $-CH_2OH$ group.

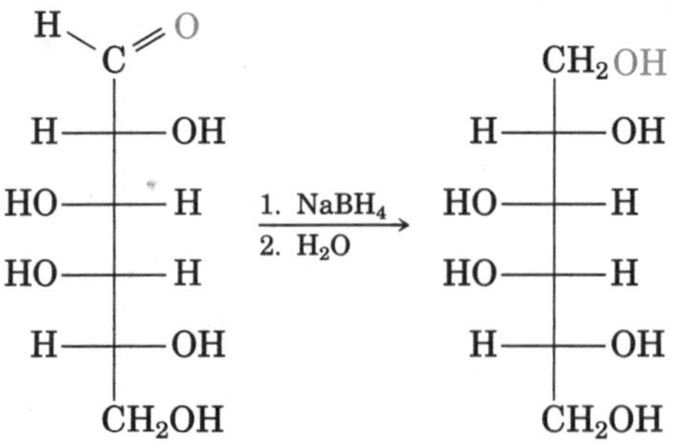

1. $NaBH_4$
2. H_2O

D-Galactose **D-Galactitol**

PROBLEM 14.19 How can you account for the fact that reduction of D-glucose leads to an optically active alditol (D-glucitol) whereas reduction of D-galactose leads to an optically inactive alditol (see Section 6.8)?

PROBLEM 14.20 Reduction of L-gulose with $NaBH_4$ leads to the same alditol (D-glucitol) as reduction of D-glucose. Explain.

Oxidation of Monosaccharides

ALDONIC ACID
Polyhydroxy monocarboxylic acid formed by oxidation of an aldose

REDUCING SUGAR
Sugar that reduces Tollens' reagent or Fehling's reagent, because it contains a hemiacetal group

Like other aldehydes, aldoses are easily oxidized, yielding carboxylic acids called **aldonic acids**. Aldoses react with Tollens' reagent (Ag^+ in aqueous ammonia), Fehling's reagent (Cu^{2+} with aqueous sodium tartrate), and Benedict's reagent (Cu^{2+} with aqueous sodium citrate) to yield the oxidized sugar and a reduced metallic species. All three reactions serve as simple chemical tests for what are called **reducing sugars** (*reducing*, because the sugar reduces the metallic oxidizing agent).

If Tollens' reagent is used, metallic silver is produced as a shiny mirror on the walls of the reaction flask or test tube (Section 9.6). If Fehling's or Benedict's reagent is used, a reddish precipitate of cuprous oxide signals a positive result. Some diabetes self-test kits sold in drugstores for home use employ Benedict's test. As little as 0.1% glucose in urine gives a positive test.

All aldoses are reducing sugars since they contain aldehyde carbonyl groups, but glycosides are nonreducing. Glycosides don't react with Tollens' or Fehling's reagents because the acetal group can't open to an aldehyde under basic conditions.

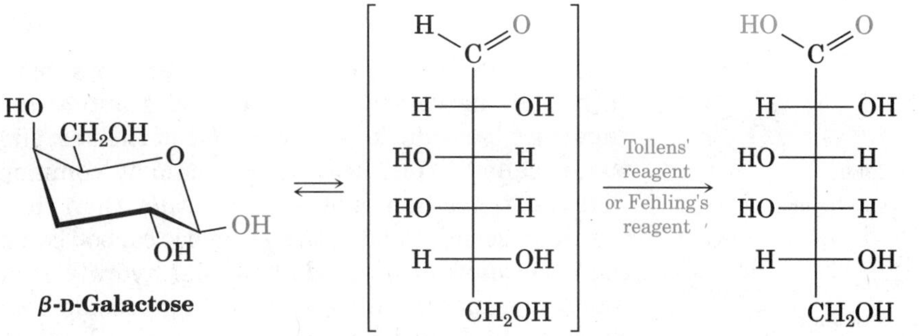

β-D-Galactose

D-Galactonic acid (an aldonic acid)

ALDARIC ACID Polyhydroxy dicarboxylic acid formed by oxidation of an aldose

If warm dilute nitric acid is used as the oxidizing agent, aldoses are oxidized to dicarboxylic acids called **aldaric acids**. Both the aldehyde carbonyl and the terminal $-CH_2OH$ group are oxidized in this reaction.

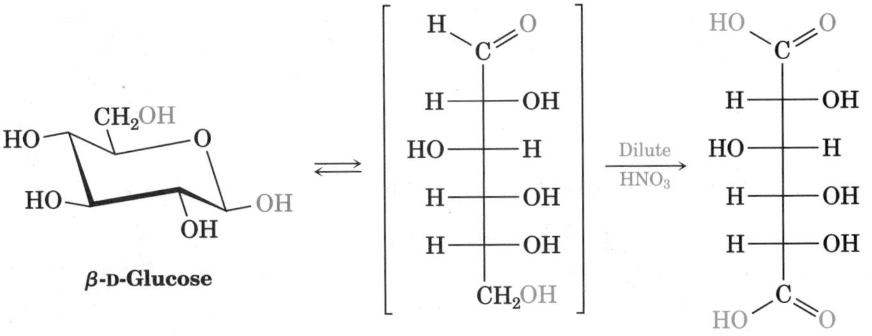

β-D-Glucose

D-Glucaric acid (an aldaric acid)

PROBLEM 14.21 D-Glucose yields an optically active aldaric acid on treatment with nitric acid, but D-allose yields an optically inactive aldaric acid. Explain.

PROBLEM 14.22 Which of the other six D-aldohexoses yield optically active aldaric acids, and which yield optically inactive aldaric acids? (See Problem 14.21.)

14.9 DISACCHARIDES

We saw in the previous section that reaction of a monosaccharide hemiacetal yields a glycoside in which the anomeric hydroxyl group is replaced by an alkoxyl substituent. If the alcohol is another sugar, the glycoside product is a disaccharide.

Cellobiose and Maltose

1,4′ LINK
Glycosidic link between the C1 carbonyl group of one sugar and the C4 hydroxyl group of another sugar

Disaccharides can contain a glycosidic acetal bond between the anomeric carbon of one sugar and a hydroxyl group at *any* position on the other sugar. A glycosidic link between C1 of the first sugar and C4 of the second sugar, called a **1,4′ link**, is particularly common. (The prime indicates that the 4′ position is on a sugar other than the nonprime 1 position.)

A glycosidic bond to the anomeric carbon can be either α or β. *Maltose*, the disaccharide obtained by partial hydrolysis of starch, consists of two D-glucopyranoses joined by a 1,4′-α-glycoside bond. *Cellobiose*, the disaccharide obtained by partial hydrolysis of cellulose, consists of two D-glucopyranoses joined by a 1,4′-β-glycoside bond.

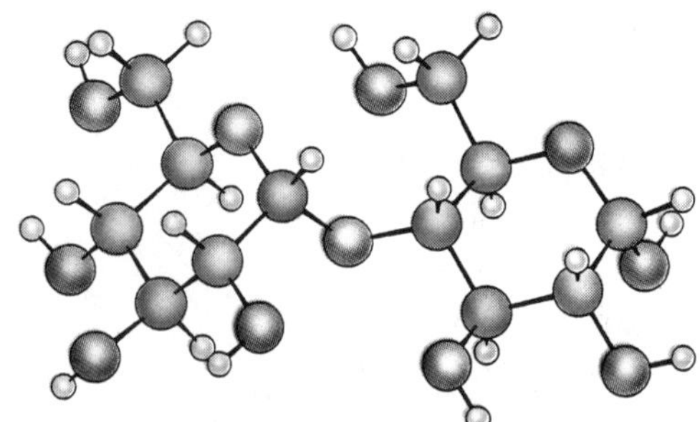

Maltose, a 1,4′-α-glycoside
[4-*O*-(α-D-glucopyranosyl)-α-D-glucopyranose]

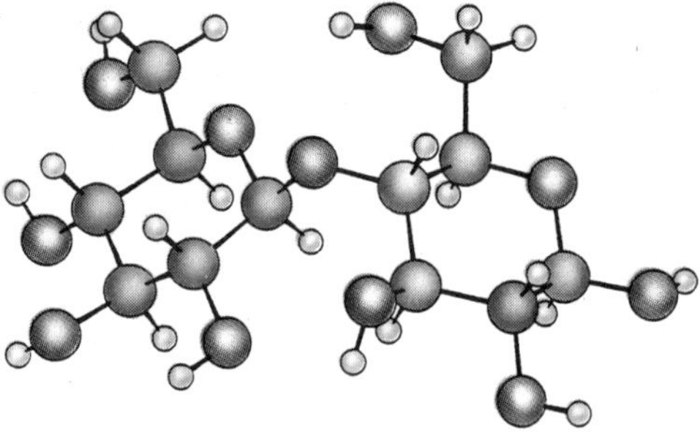

Cellobiose, a 1,4′-β-glycoside
[4-*O*-(β-D-glucopyranosyl)-β-D-glucopyranose]

Both maltose and cellobiose are reducing sugars because the right-hand saccharide units have hemiacetal groups. Both are therefore in equilibrium with aldehyde forms, which can reduce Tollens' or Fehling's reagent. For a similar reason, both maltose and cellobiose exhibit mutarotation of the α and β anomers of the glucopyranose unit on the right.

Despite the similarities of their structures, maltose and cellobiose are dramatically different biologically. Cellobiose can't be digested by humans and can't be fermented by yeast. Maltose, however, is digested without difficulty and is readily fermented.

PROBLEM 14.23 Draw the structures of the products obtained from reaction of cellobiose with these reagents:
(a) $NaBH_4$ **(b)** $AgNO_3$, H_2O, NH_3

Sucrose

Sucrose, ordinary table sugar, is probably the most abundant pure organic chemical in the world. Whether from sugar cane (20% by weight) or from sugar beets (15% by weight), and whether raw or refined, all table sugar is sucrose.

Sucrose is a disaccharide that yields 1 equiv of glucose and 1 equiv of fructose on hydrolysis of its glycoside link. This 1:1 mixture of glucose and fructose is often referred to as *invert sugar*, because the sign of optical rotation changes (inverts) during the hydrolysis from sucrose, $[\alpha]_D = +66.5°$, to a glucose/fructose mixture, $[\alpha]_D = -22°$. Certain insects such as honeybees have enzymes called *invertases* that catalyze the hydrolysis of sucrose to glucose + fructose. Honey, in fact, is primarily a mixture of glucose, fructose, and sucrose.

Unlike most other disaccharides, sucrose isn't a reducing sugar and doesn't exhibit mutarotation. These observations imply that sucrose has no hemiacetal group and also suggest that the glucose and fructose units both must be glycosides. This can happen only if the two sugars are joined by a glycoside link between anomeric carbons, C1 of glucose and C2 of fructose.

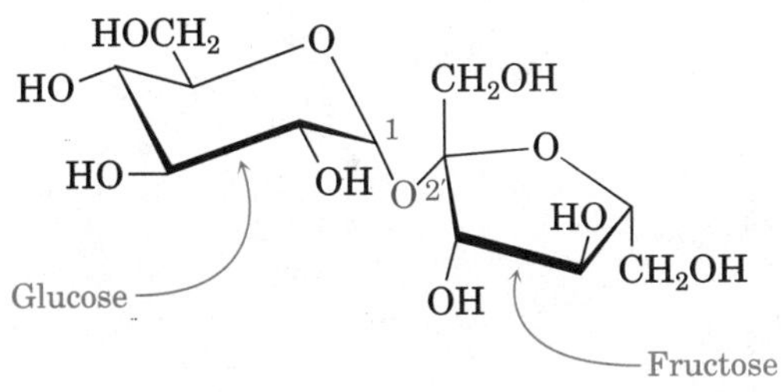

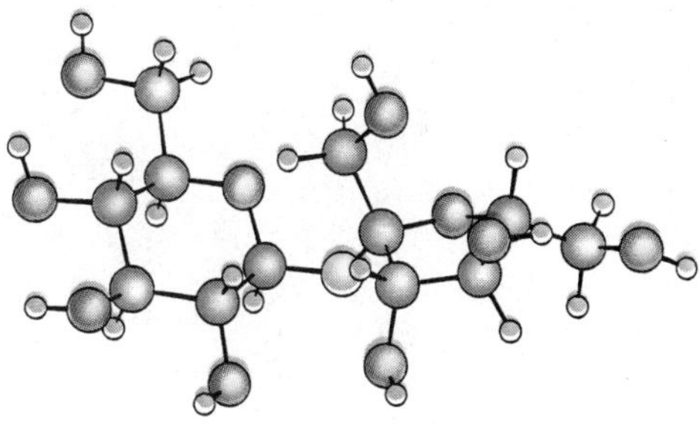

Sucrose, a 1,2′-glycoside
[2-*O*-(α-D-glucopyranosyl)-β-D-fructofuranoside]

14.10 POLYSACCHARIDES

Polysaccharides are carbohydrates in which tens, hundreds, or even thousands of simple sugars are linked by glycoside bonds. Since these compounds have no free anomeric hydroxyls (except for one at the end of the chain), they aren't reducing sugars and don't show mutarotation. Cellulose and starch are the two most widely occurring polysaccharides.

Cellulose

Cellulose consists of D-glucose units linked by the 1,4′-β-glycoside bonds we saw in cellobiose. Several thousand glucose units are linked to form one large molecule, and different molecules can then interact to form a large aggregate structure held together by hydrogen bonds.

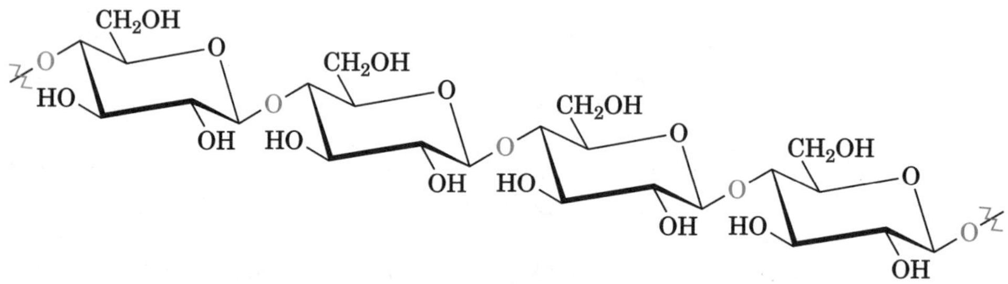

Cellulose, a 1,4′-*O*-(β-D-glucopyranoside) polymer

Nature uses cellulose primarily as a structural material to impart strength and rigidity to plants. Wood, leaves, grasses, and cotton are primarily cellulose. Cellulose also serves as a raw material for the manufacture of cellulose acetate, known commercially as acetate rayon.

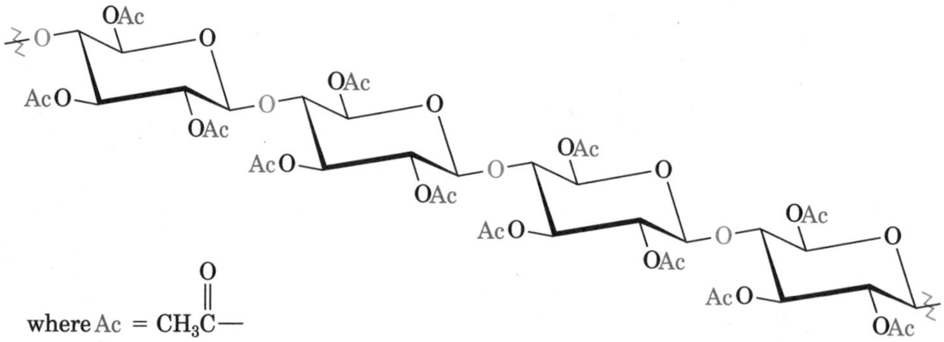

where Ac $= CH_3\overset{O}{\overset{\|}{C}}—$

A segment of cellulose acetate (acetate rayon)

Starch and Glycogen

Starch, a glucose polymer whose monosaccharide units are linked by the 1,4′-α-glycoside bonds we saw in maltose, can be separated into two fractions called *amylopectin* and *amylose*. Amylose, which accounts for about 20% by weight of starch, consists of several hundred glucose molecules linked by 1,4′-α-glycoside bonds:

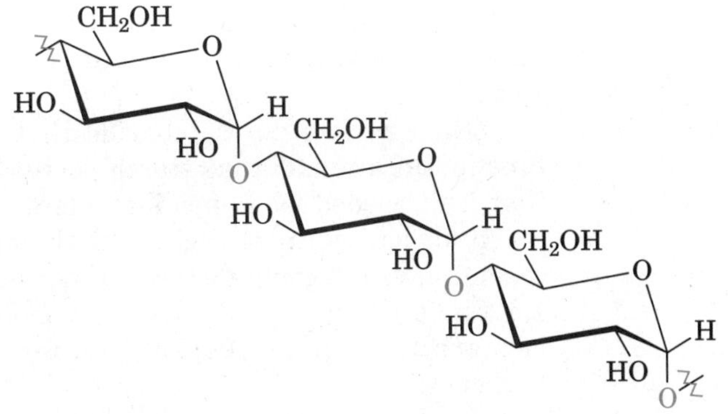

Amylose, a 1,4′-*O*-(α-D-glucopyranoside) polymer

Amylopectin, which accounts for the remaining 80% of starch, is more complex in structure than amylose. Unlike cellulose or amylose, which are linear polymers, amylopectin contains 1,6′-α-glycoside *branches* approximately every 25 glucose units. As a result, amylopectin has an exceedingly complex three-dimensional structure (Figure 14.8).

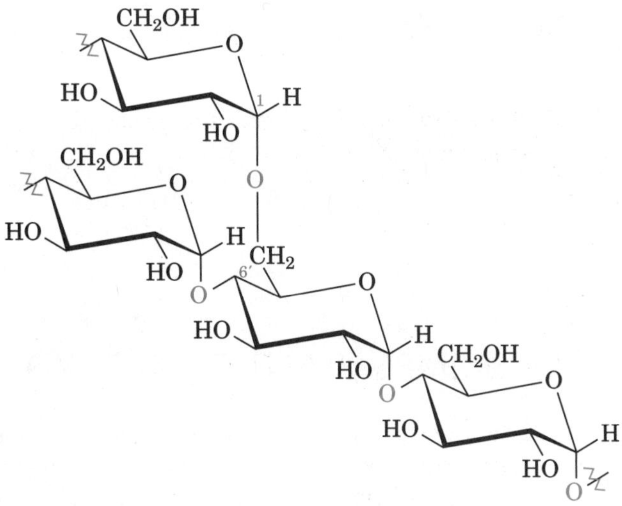

FIGURE 14.8 A 1,6′-α-glycoside branch in amylopectin.

Nature uses starch as the means by which plants store energy for later use. Potatoes, corn, and cereal grains contain large amounts of starch. When eaten, starch is digested in the mouth and stomach by enzymes called *glycosidases*, which catalyze the hydrolysis of glycoside bonds and release individual molecules of glucose. Like most enzymes, glycosidases are highly selective in their action. They hydrolyze only the α-glycoside links in starch and leave the β-glycoside links in cellulose untouched. Thus, humans can eat potatoes and grains but not grass and wood.

Glycogen is a polysaccharide that serves the same energy-storage function in animals that starch serves in plants. Dietary carbohydrate that isn't needed for immediate energy is converted by the body to glycogen for long-term storage. Like the amylopectin found in starch, glycogen contains a complex three-dimensional structure with both 1,4′ and 1,6′ links (Figure 14.9). Glycogen molecules are larger than those of amylopectin—up to 100,000 glucose units—and contain even more branches.

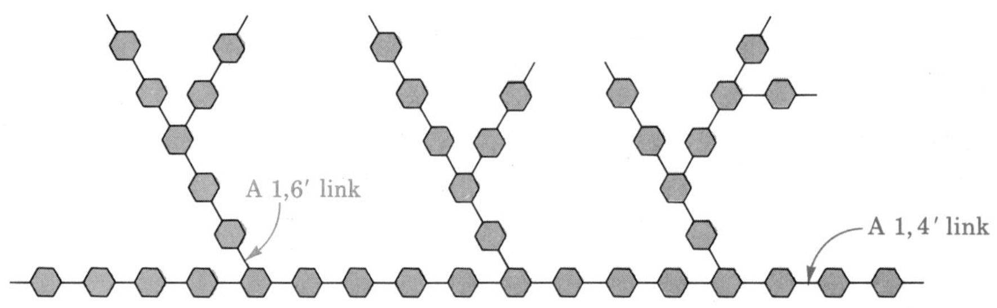

FIGURE 14.9 A representation of the structure of glycogen. The hexagons represent glucose units linked by 1,4′ and 1,6′ acetal bonds.

14.11 OTHER IMPORTANT CARBOHYDRATES

In addition to the common carbohydrates mentioned in previous sections, there are a variety of important carbohydrate-derived materials whose structures have been chemically modified. Their structural resemblance to sugars is clear, but they aren't simple aldoses or ketoses.

Deoxy Sugars

DEOXY SUGAR
Sugar with an –OH group missing from one carbon

Deoxy sugars differ from normal sugars by having one of their oxygen atoms "missing." In other words, an –OH group is replaced by an –H. 2-Deoxyribose, a sugar found in DNA (deoxyribonucleic acid), is the most important deoxy sugar. Note that 2-deoxyribose adopts a furanose (five-membered) form:

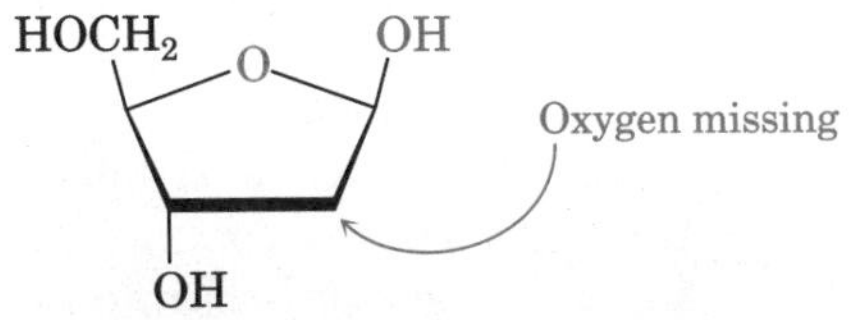

HOCH$_2$ O OH
Oxygen missing
OH

2-Deoxyribose

Amino Sugars

AMINO SUGAR
Sugar with an –OH group on one carbon replaced by an $-NH_2$ group

Amino sugars, such as D-glucosamine, have one of their –OH groups replaced by an $-NH_2$. The *N*-acetyl amide derived from D-glucosamine is the monosaccharide unit from which *chitin*, the hard crust that protects insects and shellfish, is built. Still other amino sugars are found in antibiotics such as streptomycin and gentamicin.

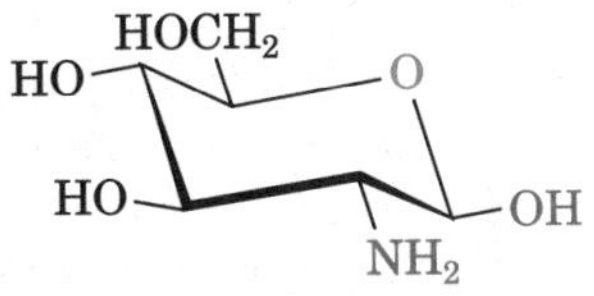

β-D-Glucosamine (an amino sugar)

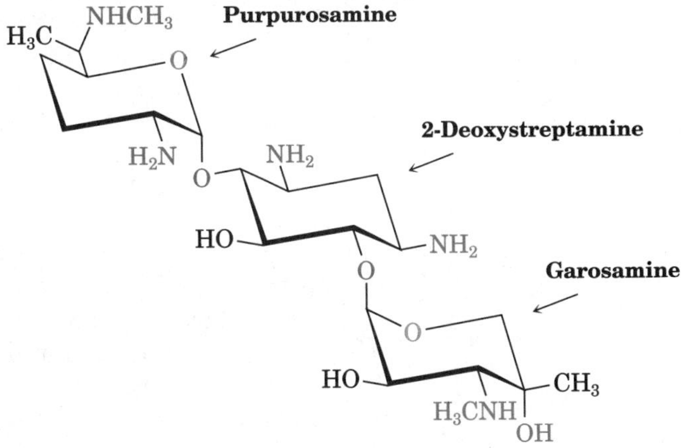

Purpurosamine
2-Deoxystreptamine
Garosamine

Gentamicin (an antibiotic)

14.12 CELL-SURFACE CARBOHYDRATES

For many years, carbohydrates were thought to be uninteresting compounds whose only biological purposes were to serve as structural materials and as energy sources. Although carbohydrates do indeed fill these two roles, recent work has shown that they perform many other important biochemical functions as well. Polysaccharides are now known to be centrally involved in the critical process by which one cell type recognizes another. Small polysaccharide chains, covalently bound by glycoside links to hydroxyl groups on proteins (*glycoproteins*), act as biochemical labels on cell surfaces, as illustrated by the human blood-group antigens.

It has been known for over 80 years that human blood can be classified into four blood-group types, A, B, AB, and O, and that blood from a donor of one type can't be transfused into a recipient with another type unless the two types are compatible (Table 14.1). If an incompatible mix is made, the red blood cells clump together, or *agglutinate*.

TABLE 14.1 Human Blood-Group Compatibilities

	ACCEPTOR BLOOD TYPE			
DONOR BLOOD TYPE	A	B	AB	O
A	○	×	○	×
B	×	○	○	×
AB	×	×	○	×
O	○	○	○	○

○ = compatible; × = incompatible

The agglutination of incompatible red blood cells, which indicates that the recipient's immune system has recognized the presence of foreign cells in the body and has formed antibodies to them, results from the presence of polysaccharide markers on the surface of the cells. Type A, B, and O red blood cells each have characteristic markers called *antigenic determinants*; type AB cells have both type A and type B markers. The structures of all three blood-group determinants have been elucidated and are shown in Figure 14.10. All three contain *N*-acetylamino sugars as well as the unusual monosaccharide L-fucose.

The antigenic determinant of blood group O is a trisaccharide, whereas the determinants of blood groups A and B are tetrasaccharides. Type A and B determinants differ only in the substitution of an acetylamino group ($-NHCOCH_3$) for a hydroxyl in the terminal galactose unit.

INTERLUDE

SWEETNESS

Say the word *sugar* and most people immediately think of sweet-tasting candies, donuts, and such. In fact, most of the simple carbohydrates we've discussed in this chapter do taste sweet, but the degree of sweetness varies greatly from one sugar to another. With sucrose (table sugar) as a reference point, fructose is nearly twice as sweet, but lactose is only about one-sixth as sweet. Comparisons are difficult, though, because sweetness is simply a matter of taste, and the ranking of sugars is a matter of personal opinion. Nevertheless, the ordering in Table 14.2 is generally accepted.

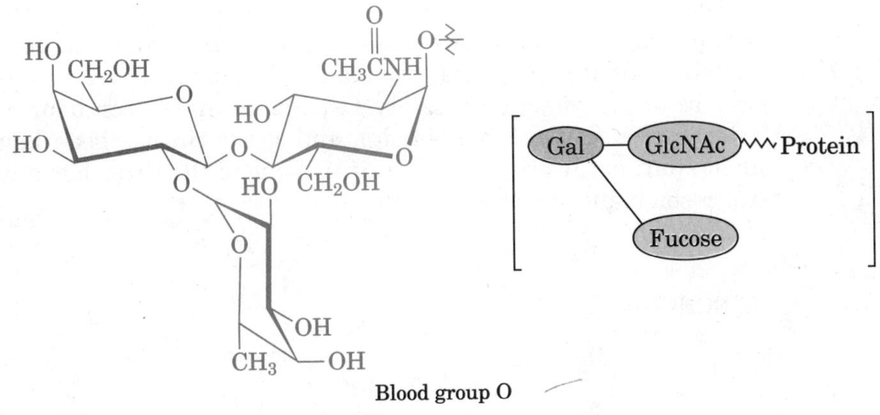

Blood group O

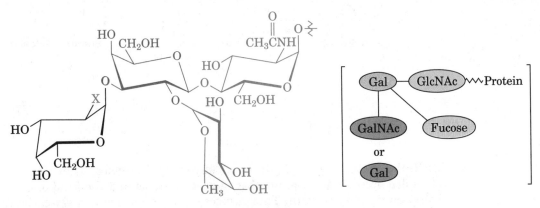

Blood group A, X = $NHCOCH_3$
Blood group B, X = OH

FIGURE 14.10 Structures of the A, B, and O blood-group antigenic determinants. (Gal is D-galactose; GlcNAc is *N*-acetylglucosamine; GalNAc is *N*-acetylgalactosamine.)

TABLE 14.2 Sweetness of Some Sugars and Sugar Substitutes

NAME	TYPE	SWEETNESS
Lactose	Disaccharide	0.16
Glucose	Monosaccharide	0.75
Sucrose	Disaccharide	1.00
Fructose	Monosaccharide	1.75
Cyclamate	Synthetic	300
Aspartame	Synthetic	1500
Saccharin	Synthetic	3500

The desire of many people to cut their caloric intake has led to the development of synthetic sweeteners such as aspartame, saccharin, and cyclamate. All are far sweeter than natural sugars, but doubts have been raised as to their long-term safety. Cyclamates have been banned in the United States (but not in Canada), and saccharin has been banned in Canada (but not in the United States). None of the three has any structural resemblance to carbohydrates.

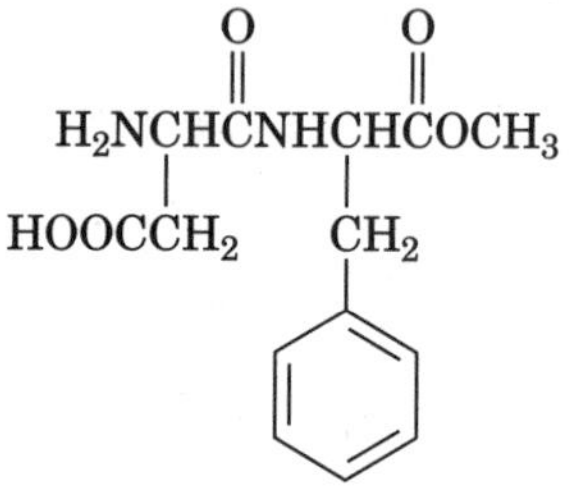

$H_2NCH(CH_2COOH)C(=O)NHCH(CH_2C_6H_5)C(=O)OCH_3$

Aspartame

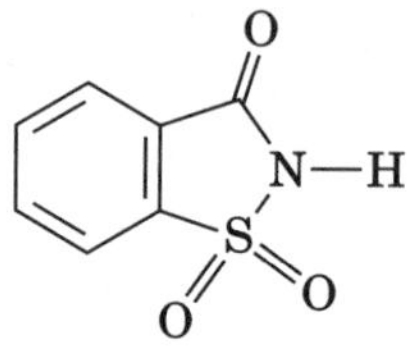

Saccharin

C_6H_{11}–$NHSO_2O^-$ Na^+

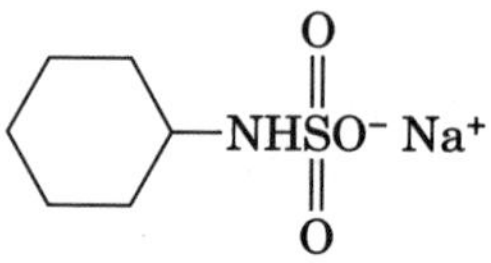

Sodium cyclamate

SUMMARY AND KEY WORDS

Carbohydrates are polyhydroxy aldehydes and ketones. They can be classified according to the number of carbon atoms and the kind of carbonyl group they contain. Thus, glucose is an *aldohexose*, a six-carbon aldehydo sugar. **Monosaccharides** are further classified as either **D** or **L sugars**, depending on the stereochemistry of the chiral carbon atom farthest from the carbonyl group. Most naturally occurring sugars are in the D series.

Monosaccharides normally exist as cyclic hemiacetals rather than as open-chain aldehydes or ketones. The hemiacetal linkage results from reaction of the carbonyl group with a hydroxyl group three or four carbon atoms away. A five-membered ring hemiacetal is a **furanose**; a six-membered ring hemiacetal is a **pyranose**. Cyclization leads to the formation of a new stereogenic center (the **anomeric center**) and to production of two diastereomeric hemiacetals called **alpha (α)** and **beta (β) anomers**.

Stereochemical relationships among monosaccharides are portrayed in several ways. **Fischer projections** display chiral carbon atoms as a pair of perpendicular crossed lines. These projections are useful in allowing us to quickly relate one sugar to another, but cyclic **Haworth projections** provide a better view. Any group to the right in a Fischer projection is down in a Haworth projection.

Much of the chemistry of monosaccharides is the familiar chemistry of alcohol and carbonyl functional groups. Thus, the hydroxyl groups of carbohydrates form esters and ethers in the normal way. The carbonyl group of a monosaccharide can be reduced with sodium borohydride to yield an **alditol**, can be oxidized with Tollens' or Fehling's reagent to yield an **aldonic acid**, can be oxidized with warm nitric acid to yield an **aldaric acid**, and can be treated with an alcohol in the presence of acid catalyst to yield a **glycoside**.

Disaccharides are complex carbohydrates in which two simple sugars are linked by a glycoside bond between the anomeric carbon of one unit and a hydroxyl of the second unit. The two sugars can be the same, as in maltose and cellobiose, or different, as in sucrose. The glycoside bond can be either α (maltose) or β (cellobiose) and can involve any hydroxyl of the second sugar. A **1,4′ link** is most common (cellobiose, maltose), but other links, such as 1,6′ (amylopectin) and 1,2′ (sucrose) also occur. **Polysaccharides**, such as cellulose, starch, and glycogen, are used in nature both as structural materials and for long-term energy storage.

ADDITIONAL PROBLEMS

14.24 Classify the following sugars by type (for example, glucose is an aldohexose):

(a) CH_2OH–C=O–CH_2OH

(b) CH_2OH; H–C–OH; C=O; H–C–OH; CH_2OH

(c) CHO; H–C–OH; HO–C–H; H–C–OH; HO–C–H; H–C–OH; CH_2OH

14.25 Write open-chain structures for a ketotetrose and a ketopentose.

14.26 Write an open-chain structure for a deoxyaldohexose.

14.27 Write an open-chain structure for a five-carbon amino sugar.

14.28 The structure of ascorbic acid (vitamin C) is shown. Does ascorbic acid have a D or an L configuration?

OH; HO; C=C; C=O; H–C–O; HO–C–H; CH_2OH

Ascorbic acid

14.29 Draw a Haworth projection of ascorbic acid (see Problem 14.28).

14.30 Define the following terms, and give an example of each:
(a) Monosaccharide (b) Anomeric center (c) Haworth projection (d) Fischer projection
(e) Glycoside (f) Reducing sugar (g) Pyranose form (h) 1,4′ Link
(i) D-Sugar

14.31 The following cyclic structure is that of gulose. Is this a furanose or pyranose form? Is it an α or β anomer? Is it a D sugar or an L sugar?

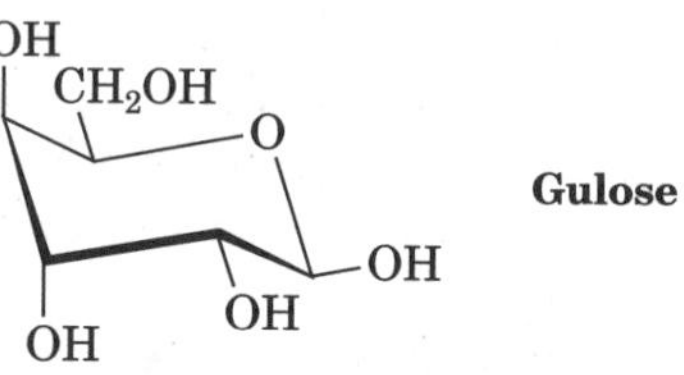

Gulose

14.32 Uncoil gulose (see Problem 14.31) and write it in its open-chain form.

14.33 Draw D-ribulose in its five-membered cyclic β hemiacetal form.

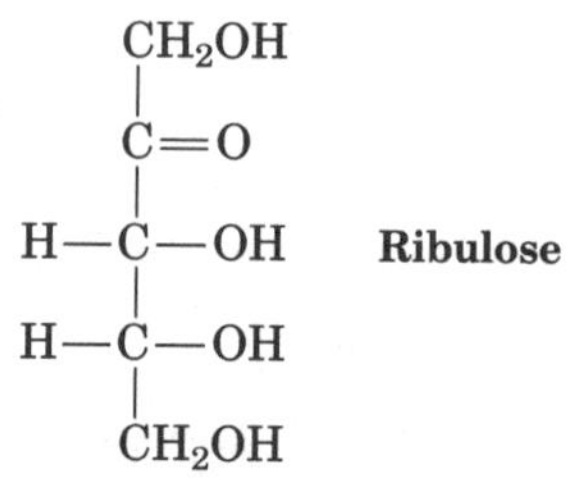

Ribulose

14.34 Look up the structure of D-talose in Figure 14.3, and draw the β anomer in its pyranose form. Identify the ring substituents as axial or equatorial.

14.35 Draw structures for the products you would expect to obtain from the reaction of β-D-talopyranose (see Problem 14.34) with each of the following reagents:
(a) $NaBH_4$ (b) Warm dilute HNO_3 (c) $AgNO_3$, NH_3, H_2O
(d) CH_3CH_2OH, H^+ (e) CH_3I, Ag_2O (f) $(CH_3CO)_2O$, pyridine

14.36 What is the stereochemical relationship of D-allose to L-allose? What generalizations can you make about the following properties of the two sugars?
(a) Melting point (b) Solubility in water (c) Specific rotation (d) Density

14.37 What is the stereochemical relationship of D-ribose to L-xylose? What generalizations can you make about the following properties of the two sugars?
(a) Melting point (b) Solubility in water (c) Specific rotation (d) Density

14.38 How many D-2-ketohexoses are there? Draw them.

14.39 One of the D-2-ketohexoses (see Problem 14.38) is called *sorbose*. On treatment with $NaBH_4$, sorbose yields a mixture of gulitol and iditol. What is the structure of D-sorbose? (Gulitol and iditol are the alditols obtained by reduction of gulose and idose.)

14.40 Another D-2-ketohexose, *psicose*, yields a mixture of allitol and altritol when reduced with $NaBH_4$. What is the structure of psicose?

14.41 Fructose exists at equilibrium as an approximately 2:1 mixture of β-D-fructopyranose and β-D-fructofuranose. Draw both forms in Haworth projection.

14.42 Draw Fischer projections of these substances:
(a) (*R*)-2-Methylbutanoic acid **(b)** (*S*)-3-Methyl-2-pentanone

14.43 Convert these Fischer projections into tetrahedral representations:

(a) Fischer projection: top Br; left H; right OCH_3; bottom CH_3

(b) Fischer projection: top CH_3; left H; right NH_2; bottom CH_2CH_3

14.44 Which of the eight D-aldohexoses yield optically inactive (meso) alditols on reduction with $NaBH_4$?

14.45 What other D-aldohexose would give the same alditol as D-talose? (See Problem 14.44.)

14.46 Which of the eight D-aldohexoses give the same aldaric acids as their L enantiomers?

14.47 Which of the other three D-aldopentoses gives the same aldaric acid as D-lyxose?

14.48 The *Ruff degradation* is a method used to shorten an aldose chain by one carbon atom. The original C1 carbon atom is cleaved off, and the original C2 carbon atom becomes the aldehyde of the chain-shortened aldose. For example, D-glucose, an aldohexose, is converted by Ruff degradation into D-arabinose, an aldopentose. What other D-aldohexose would also yield D-arabinose on Ruff degradation?

14.49 D-Galactose and D-talose yield the same aldopentose on Ruff degradation (see Problem 14.48). What does this tell you about the stereochemistry of galactose and of talose? Which D-aldopentose is obtained?

14.50 The aldaric acid obtained by nitric acid oxidation of D-erythrose, one of the D-aldotetroses, is optically inactive. The aldaric acid obtained from oxidation of the other D-aldotetrose, D-threose, however, is optically active. How does this information allow you to assign structures to the two D-aldotetroses?

14.51 Gentiobiose is a rare disaccharide found in saffron and gentian. It is a reducing sugar and forms only glucose on hydrolysis with aqueous acid. If gentiobiose contains a 1,6′-β-glycoside link, what is its structure?

14.52 The *cyclitols* are a group of carbocyclic sugar derivatives with the general formula 1,2,3,4,5,6-cyclohexanehexaol—that is, a cyclohexane ring with one hydroxyl on each carbon. Draw the structures of the nine stereoisomeric cyclitols in Haworth projection.

14.53 Raffinose, a trisaccharide found in sugar beets, is formed by a 1,6′ α linkage of D-galactose to the glucose unit of sucrose. Draw the structure of raffinose.

14.54 Is raffinose (see Problem 14.53) a reducing sugar? Explain.

15 CHAPTER

BIOMOLECULES: AMINO ACIDS, PEPTIDES, AND PROTEINS

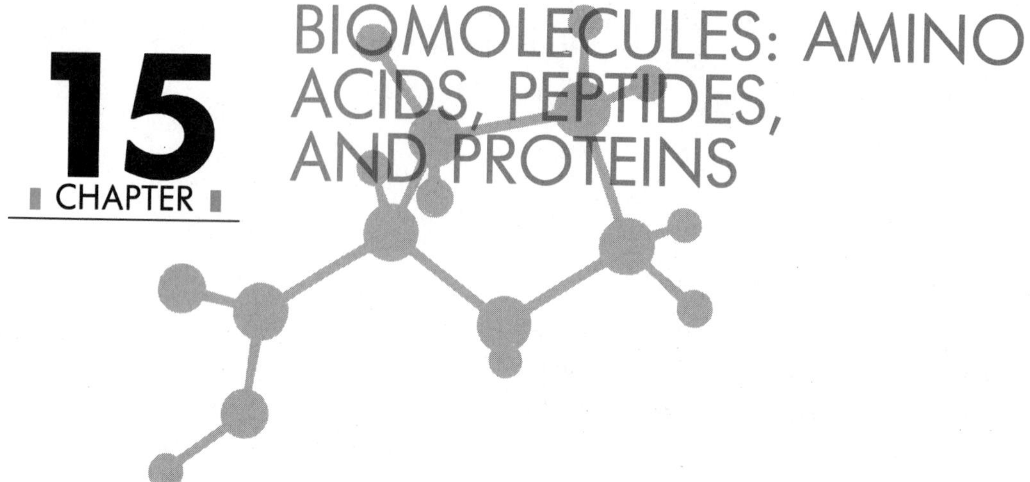

PROTEIN
Large biological polymer made up of many amino acid units linked together by amide bonds

AMINO ACID
Compound that contains both an amino group and a carboxylic acid group

Proteins are large biomolecules that occur in every living organism. They are of many types and have many biological functions. The keratin of skin and fingernails, the insulin that regulates glucose metabolism in the body, and the DNA polymerase that catalyzes the synthesis of DNA in cells are all proteins. Regardless of their appearance or function, all proteins are chemically similar. All are made up of many *amino acid* units linked together by amide bonds in a long chain.

Amino acids, as their name implies, are difunctional. They contain both a basic amino group and an acidic carboxyl group:

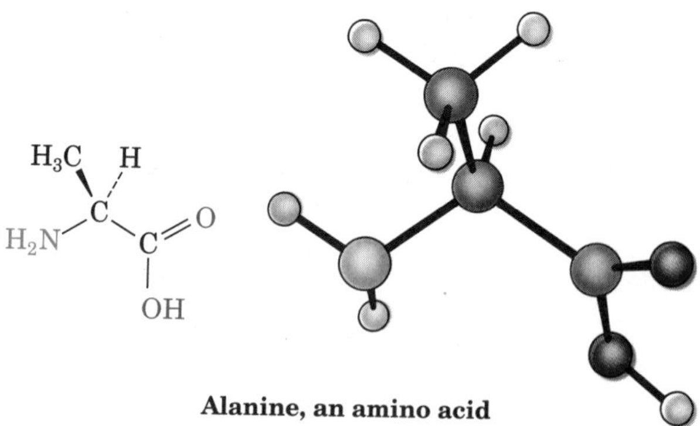

Alanine, an amino acid

PEPTIDE BOND
Amide bond linking amino acids together in peptides and proteins

DIPEPTIDE
Compound that has two amino acid units linked together

TRIPEPTIDE
Compound that has three amino acid units linked together

Their value as building blocks for proteins derives from the fact that amino acids can link together into *peptides* by forming amide, or **peptide, bonds**. A **dipeptide** results when the $-NH_2$ of one amino acid is linked to the $-COOH$ of a second amino acid by an amide bond, a **tripeptide** results from linkage of three amino acids by two amide bonds, and so on.

POLYPEPTIDE
Compound that has up to 50 amino acid units linked together

Any number of amino acids can link together to form a long chain. Chains with fewer than 50 amino acids are usually called **polypeptides**, while the term *protein* is used for longer chains.

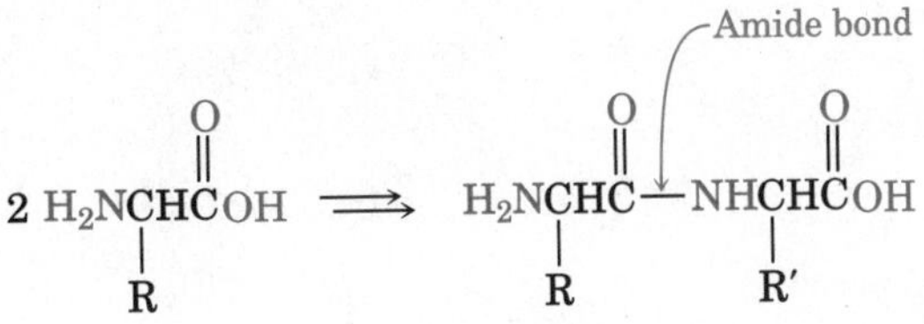

$$2\ H_2N\underset{\displaystyle R}{\underset{|}{C}}H\overset{\displaystyle O}{\overset{\|}{C}}OH \Longrightarrow H_2N\underset{\displaystyle R}{\underset{|}{C}}H\overset{\displaystyle O}{\overset{\|}{C}}\text{—}NH\underset{\displaystyle R'}{\underset{|}{C}}H\overset{\displaystyle O}{\overset{\|}{C}}OH$$

Amide bond

A dipeptide (one amide bond)

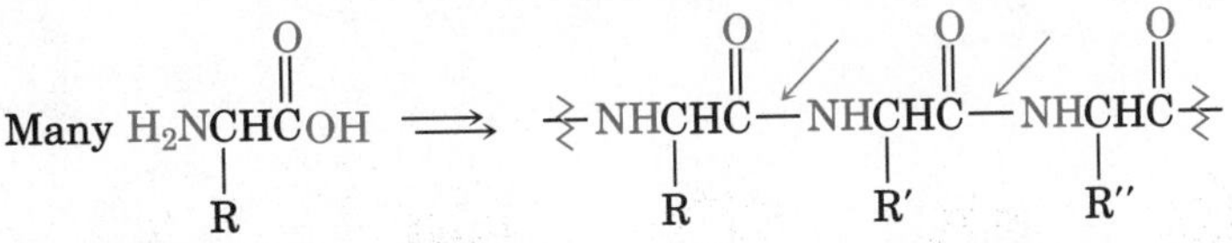

$$\text{Many } H_2N\underset{\displaystyle R}{\underset{|}{C}}H\overset{\displaystyle O}{\overset{\|}{C}}OH \Longrightarrow \text{⌇}NH\underset{\displaystyle R}{\underset{|}{C}}H\overset{\displaystyle O}{\overset{\|}{C}}\text{—}NH\underset{\displaystyle R'}{\underset{|}{C}}H\overset{\displaystyle O}{\overset{\|}{C}}\text{—}NH\underset{\displaystyle R''}{\underset{|}{C}}H\overset{\displaystyle O}{\overset{\|}{C}}\text{⌇}$$

A polypeptide (many amide bonds)

15.1 STRUCTURES OF AMINO ACIDS

α-AMINO ACID
Compound in which the amino group is attached to the carbon atom next to the carboxyl group

The structures of the 20 amino acids commonly found in proteins are shown in Table 15.1. All 20 are ***α*-amino acids**; that is, the amino group in each is attached to the carbon atom α to (next to) the carbonyl group. Note that 19 of the 20 amino acids are primary amines, RNH_2, and differ only in the nature of their side chains. Proline, however, is a secondary amine whose nitrogen and α-carbon atom are part of a five-membered ring. Proline can still form amide bonds like the other 19 α-amino acids, however.

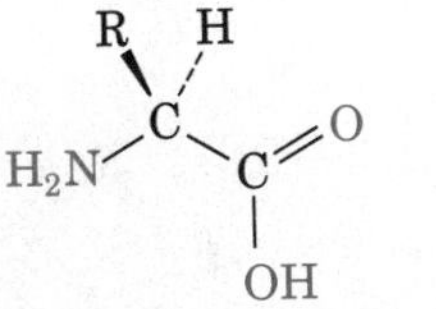

R H H₂N C C=O OH

A primary α-amino acid (R = a side chain)

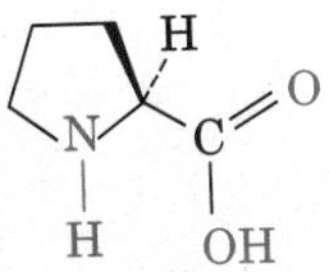

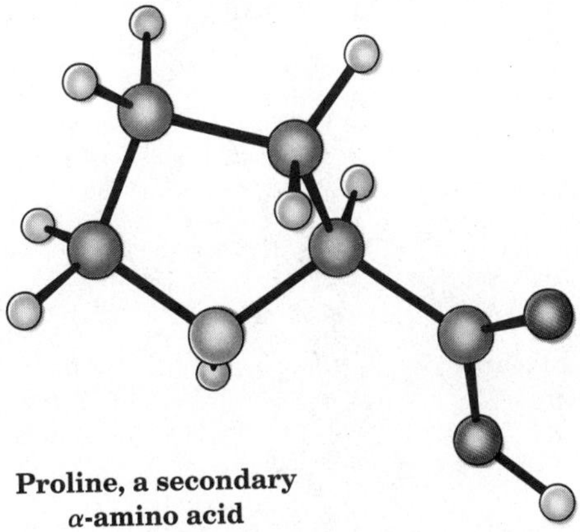

H N H C=O OH

Proline, a secondary α-amino acid

TABLE 15.1 Structures of the 20 Common Amino Acids Found in Proteins; Amino Acids Essential to Human Diets Are Shown in Red

NAME	ABBREVIATIONS	MOLECULAR WEIGHT	STRUCTURE	ISOELECTRIC POINT	pK_a α-COOH	pK_a α-NH_3^+
NEUTRAL AMINO ACIDS						
Alanine	Ala (A)	89	$CH_3CH(NH_2)C(=O)OH$	6.00	2.34	9.69
Asparagine	Asn (N)	132	$H_2NC(=O)CH_2CH(NH_2)C(=O)OH$	5.41	2.02	8.80
Cysteine	Cys (C)	121	$HSCH_2CH(NH_2)C(=O)OH$	5.07	1.96	10.28
Glutamine	Gln (Q)	146	$H_2NC(=O)CH_2CH_2CH(NH_2)C(=O)OH$	5.65	2.17	9.13
Glycine	Gly (G)	75	$CH_2(NH_2)C(=O)OH$	5.97	2.34	9.60
Isoleucine	Ile (I)	131	$CH_3CH_2CH(CH_3)CH(NH_2)C(=O)OH$	6.02	2.36	9.60
Leucine	Leu (L)	131	$CH_3CH(CH_3)CH_2CH(NH_2)C(=O)OH$	5.98	2.36	9.60
Methionine	Met (M)	149	$CH_3SCH_2CH_2CH(NH_2)C(=O)OH$	5.74	2.28	9.21
Phenylalanine	Phe (F)	165	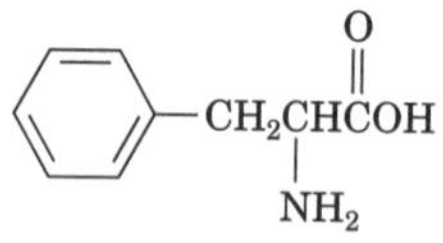 C_6H_5–$CH_2CH(NH_2)C(=O)OH$	5.48	1.83	9.13
Proline	Pro (P)	115	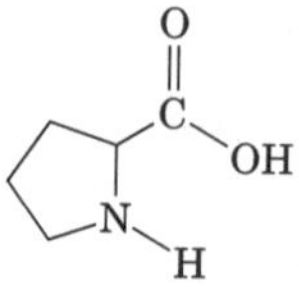 pyrrolidine ring (N–H) with C(=O)OH	6.30	1.99	10.60
Serine	Ser (S)	105	$HOCH_2CH(NH_2)C(=O)OH$	5.68	2.21	9.15

NAME	ABBREVIATIONS	MOLECULAR WEIGHT	STRUCTURE	ISOELECTRIC POINT	pK_a α-COOH	pK_a α-NH_3^+
Threonine	Thr (T)	119	$CH_3CH(OH)CH(NH_2)COOH$	5.60	2.09	9.10
Tryptophan	Trp (W)	204	indol-3-yl–$CH_2CH(NH_2)COOH$	5.89	2.83	9.39
Tyrosine	Tyr (Y)	181	HO–C_6H_4–$CH_2CH(NH_2)COOH$	5.66	2.20	9.11
Valine	Val (V)	117	$CH_3CH(CH_3)CH(NH_2)COOH$	5.96	2.32	9.62
ACIDIC AMINO ACIDS						
Aspartic acid	Asp (D)	133	$HOOCCH_2CH(NH_2)COOH$	2.77	1.88	9.60
Glutamic acid	Glu (E)	147	$HOOCCH_2CH_2CH(NH_2)COOH$	3.22	2.19	9.67
BASIC AMINO ACIDS						
Arginine	Arg (R)	174	$H_2NC(=NH)NHCH_2CH_2CH_2CH(NH_2)COOH$	10.76	2.17	9.04
Histidine	His (H)	155	imidazol-4-yl–$CH_2CH(NH_2)COOH$	7.59	1.82	9.17
Lysine	Lys (K)	146	$H_2NCH_2CH_2CH_2CH_2CH(NH_2)COOH$	9.74	2.18	8.95

Note also that each of the amino acids in Table 15.1 is referred to by a three-letter shorthand code: Ala for alanine, Gly for glycine, and so on. In addition, a one-letter code, shown in parentheses in the table, is frequently used.

With the exception of glycine, H_2NCH_2COOH, the α carbons of the amino acids are stereogenic centers. Two enantiomeric forms of each amino acid are therefore possible, but nature uses only a single enantiomer to build proteins. In Fischer projections, naturally occurring amino acids are represented by placing the carboxyl group at the top as if drawing a carbohydrate (Section 14.2) and then placing the amino group on the left. Because of their stereochemical similarity to L sugars (Section 14.3), the naturally occurring α-amino acids are often referred to as L-amino acids.

COOH / H_2N—C—H / CH_3

(*S*)-Alanine (L-alanine)

COOH / H_2N—C—H / CH_2—C_6H_5

(*S*)-Phenylalanine (L-phenylalanine)

COOH / H_2N—C—H / CH_2OH

(*S*)-Serine (L-serine)

(CHO / HO—C—H / CH_2OH

Stereochemically similiar to L-glyceraldehyde)

The 20 common amino acids can be further classified as either neutral, basic, or acidic, depending on the nature of their side chains. Fifteen of the 20 have neutral side chains, but 2 (aspartic acid and glutamic acid) have an extra carboxylic acid function in their side chains, and 3 (lysine, arginine, and histidine) have basic amino groups in their side chains.

All 20 of the amino acids are necessary for protein synthesis, but humans are thought to be able to synthesize only 10 of the 20. The other 10 are called **essential amino acids** because they must be obtained from dietary sources. Failure to include an adequate dietary supply of any of these essential amino acids leads to poor growth and general failure to thrive.

ESSENTIAL AMINO ACID
Amino acid that humans must obtain in their diet

PROBLEM 15.1 Look carefully at the 20 amino acids in Table 15.1. How many contain aromatic rings? How many contain sulfur? How many are alcohols? How many have hydrocarbon side chains?

PROBLEM 15.2 Eighteen of the 19 L-amino acids have the *S* configuration at the α-carbon. Cysteine is the only L-amino acid that has an *R* configuration. Explain.

PROBLEM 15.3 Draw L-alanine in the standard three-dimensional format using solid, wedged, and dashed lines.

15.2 DIPOLAR STRUCTURE OF AMINO ACIDS

ZWITTERION
Dipolar substance that has both plus and minus charges

Since amino acids contain both acidic and basic groups in the same molecule, they undergo an internal acid–base reaction and exist primarily in the form of a dipolar ion, or **zwitterion** (German *zwitter*, "hybrid").

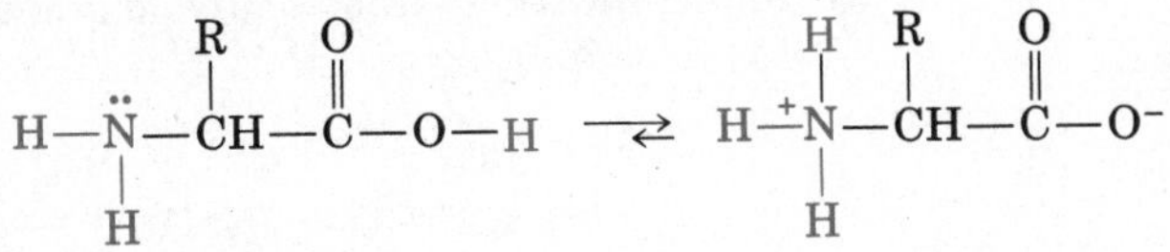

$$H_2\ddot{N}{-}CHR{-}C(=O){-}O{-}H \rightleftharpoons H_3\overset{+}{N}{-}CHR{-}C(=O){-}O^-$$

A zwitterion

Amino acid zwitterions are a kind of internal salt and therefore have many of the physical properties associated with salts. Thus, amino acids are crystalline, with high melting points, and are soluble in water but insoluble in hydrocarbons. In addition, amino acids are *amphoteric*: They can react either as acids or as bases, depending on the circumstances. In aqueous acid solution, an amino acid zwitterion *accepts* a proton to yield a cation; in aqueous basic solution, the zwitterion *loses* a proton to form an anion.

In acid solution

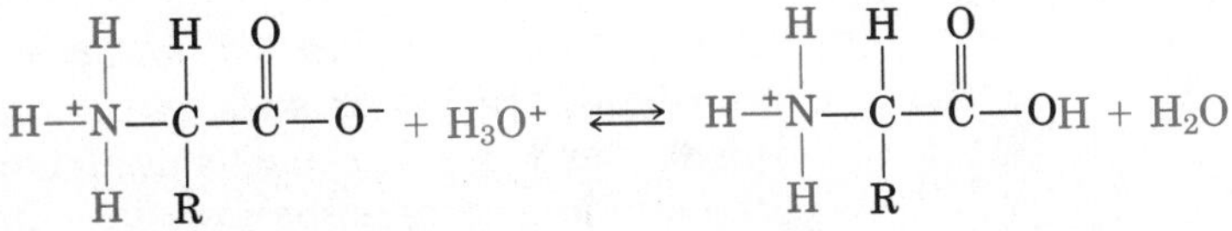

$$H_3\overset{+}{N}{-}CHR{-}C(=O){-}O^- + H_3O^+ \rightleftharpoons H_3\overset{+}{N}{-}CHR{-}C(=O){-}OH + H_2O$$

In base solution

$$H_3\overset{+}{N}{-}CHR{-}C(=O){-}O^- + {}^-OH \rightleftharpoons H_2N{-}CHR{-}C(=O){-}O^- + H_2O$$

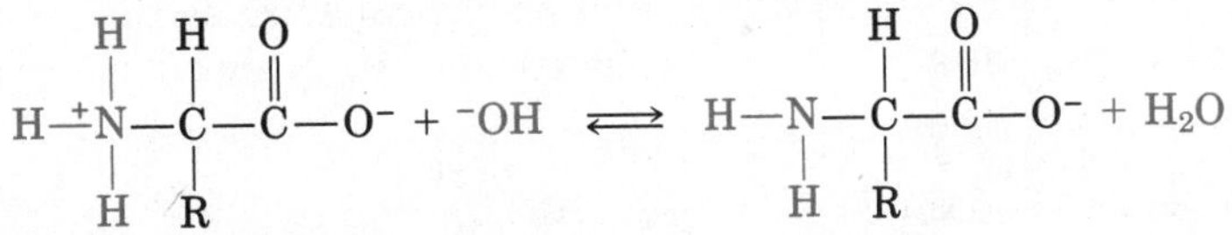

Note that it's the carboxylate anion, $-COO^-$, rather than the amino group that acts as the basic site in the zwitterion and accepts the proton in acid solution. Similarly, it's the ammonium cation, $-NH_3^+$, rather than the carboxyl group that acts as the acidic site and donates a proton in basic solution.

PRACTICE PROBLEM 15.1 Write an equation for the reaction of glycine hydrochloride with:
(a) 1 equiv NaOH **(b)** 2 equiv NaOH

SOLUTION Glycine hydrochloride has the structure

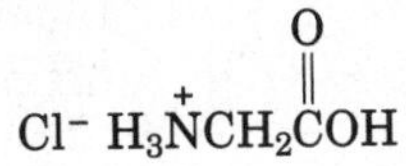

$$Cl^-\ H_3\overset{+}{N}CH_2\overset{O}{\overset{\|}{C}}OH$$

(a) Reaction with the first equivalent of NaOH removes the acidic –COOH proton:

$$Cl^- \; H_3\overset{+}{N}CH_2\overset{\overset{\displaystyle O}{\|}}{C}OH + NaOH \longrightarrow H_3\overset{+}{N}CH_2\overset{\overset{\displaystyle O}{\|}}{C}O^- + H_2O + NaCl$$

(b) Reaction with a second equivalent of NaOH removes the $-NH_3^+$ proton:

$$H_3\overset{+}{N}CH_2\overset{\overset{\displaystyle O}{\|}}{C}O^- + NaOH \longrightarrow H_2NCH_2\overset{\overset{\displaystyle O}{\|}}{C}O^- \; Na^+ + H_2O$$

PROBLEM 15.4 Draw phenylalanine in its zwitterionic form.

PROBLEM 15.5 Write structural formulas for these equations:
(a) Phenylalanine + 1 equiv NaOH $\longrightarrow$?
(b) Product of (a) + 1 equiv HCl $\longrightarrow$?
(c) Product of (a) + 2 equiv HCl $\longrightarrow$?

15.3 ISOELECTRIC POINTS

ISOELECTRIC POINT
The pH at which an amino acid exists primarily in its neutral zwitterionic form

In acid solution at low pH, an amino acid is protonated and exists primarily as a cation. In basic solution at high pH, an amino acid is deprotonated and exists primarily as an anion. Thus, at some intermediate pH, the amino acid must be exactly balanced between anionic and cationic forms and exist primarily as the neutral, dipolar zwitterion. This pH is called the **isoelectric point** of the amino acid.

$$H_3\overset{+}{N}\overset{\overset{\displaystyle R}{|}}{C}H\overset{\overset{\displaystyle O}{\|}}{C}OH \underset{H_3O^+}{\rightleftarrows} H_3\overset{+}{N}\overset{\overset{\displaystyle R}{|}}{C}H\overset{\overset{\displaystyle O}{\|}}{C}O^- \underset{^-OH}{\rightleftarrows} H_2N\overset{\overset{\displaystyle R}{|}}{C}H\overset{\overset{\displaystyle O}{\|}}{C}O^-$$

Low pH (protonated) → High pH (deprotonated)

Isoelectric point (neutral zwitterion)

The isoelectric point of an amino acid depends on its structure, with values for the 20 common amino acids given in Table 15.1. The 15 amino acids with neutral side chains have isoelectric points near neutrality, in the pH range 5.0–6.5. (These values aren't exactly at neutral pH 7 because carboxyl groups are stronger acids in aqueous solution than amino groups are bases.) The 2 amino acids with acidic side chains have isoelectric points at lower (more acidic) pH, which suppresses dissociation of the extra –COOH function, and the 3 amino acids with basic side chains have isoelectric points at higher (more basic) pH, which suppresses protonation of the extra amino function. For example, aspartic acid has its isoelectric point at pH = 2.77, and lysine has its isoelectric point at pH = 9.74.

ELECTROPHORESIS Technique for separating charged species by placing them in an electric field

We can take advantage of the differences in isoelectric points to separate a mixture of amino acids (or a mixture of proteins) into pure constituents. Using a technique known as **electrophoresis**, a solution of amino acids is placed near the center of a strip of paper or gel. The paper or gel is moistened with an aqueous buffer of a particular pH, and electrodes are connected to the ends of the strip. When an electric potential is applied, the amino acids with negative charges (those that are deprotonated because their isoelectric points are below the pH of the buffer) migrate slowly toward the positive electrode. Similarly, the amino acids with positive charges (those that are protonated because their isoelectric points are above the pH of the buffer) migrate toward the negative electrode.

Different amino acids migrate at different rates, depending on their isoelectric points and on the pH of the buffer. Thus, the different amino acids can be separated. Figure 15.1 illustrates this separation for a mixture of lysine (basic), glycine (neutral), and aspartic acid (acidic).

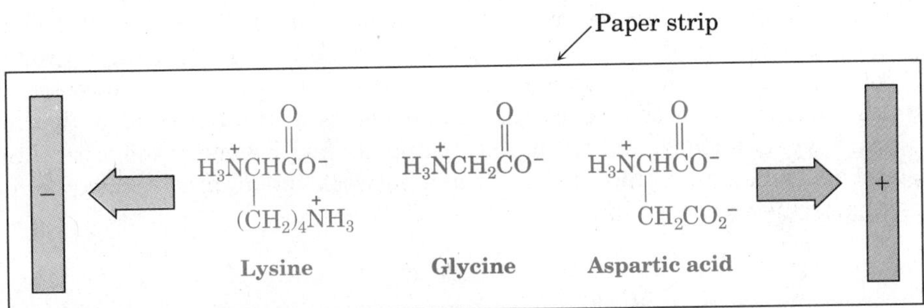

FIGURE 15.1 Separation of an amino acid mixture by electrophoresis. At pH 6.0, glycine molecules are primarily neutral and do not migrate; lysine molecules are largely protonated and migrate toward the negative electrode; aspartic acid molecules are largely deprotonated and migrate toward the positive electrode. (Lysine has its isoelectric point at pH = 9.7, glycine at pH = 6.0, and aspartic acid at pH = 2.8.)

PRACTICE PROBLEM 15.2 Draw structures of the predominant forms of glycine at pH 3.0, pH 6.0, and pH 9.0.

SOLUTION According to Table 15.1, the isoelectric point of glycine is 6.0. At a pH substantially lower than 6.0, glycine is protonated; at pH 6.0, glycine is zwitterionic; and at a pH substantially higher than 6.0, glycine is deprotonated.

$$\mathrm{H_3\overset{+}{N}CH_2\overset{O}{\overset{\|}{C}}OH} \qquad \mathrm{H_3\overset{+}{N}CH_2\overset{O}{\overset{\|}{C}}O^-} \qquad \mathrm{H_2NCH_2\overset{O}{\overset{\|}{C}}O^-}$$

At pH 3.0 | At pH 6.0 | At pH 9.0

PROBLEM 15.6

Draw the structure of the predominant form of each of these amino acids:
(a) Lysine at pH 2.0 **(b)** Aspartic acid at pH 6.0 **(c)** Lysine at pH 11.0
(d) Alanine at pH 4.0

PROBLEM 15.7

For the mixtures of amino acids indicated, predict the direction of migration of each component (toward the positive or negative electrode) and the relative rate of migration during electrophoresis.
(a) Valine, glutamic acid, and histidine at pH 7.6
(b) Glycine, phenylalanine, and serine at pH 5.7
(c) Glycine, phenylalanine, and serine at pH 6.0

15.4 PEPTIDES

PEPTIDE
Amino acid polymer

RESIDUE
Common term for an amino acid unit

Peptides are amino acid polymers in which the amino acid units, also called **residues**, are linked together by amide bonds. The amino group of one residue forms an amide bond with the carboxyl of a second residue, the amino group of the second residue forms an amide bond with the carboxyl of a third, and so on. For example, alanylserine is the dipeptide formed when an amide bond is formed between the alanine carboxyl and the serine amino group:

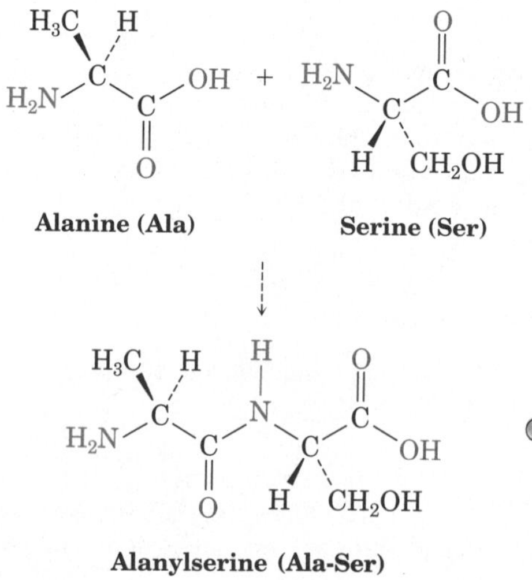

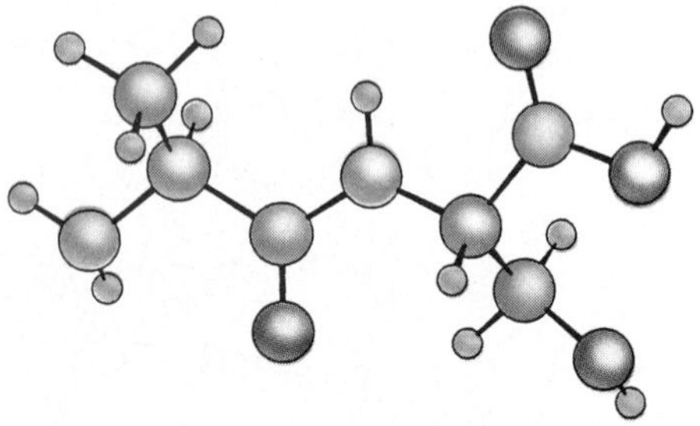

H_3C, H, H_2N—C—C(=O)—OH + H_2N—C(H)(CH_2OH)—C(=O)—OH

Alanine (Ala) **Serine (Ser)**

↓

H_2N—C(H_3C)(H)—C(=O)—N(H)—C(H)(CH_2OH)—C(=O)—OH

Alanylserine (Ala-Ser)

Note that two dipeptides can result from reaction between alanine and serine, depending on which carboxyl group reacts with which amino group. If the alanine amino group reacts with the serine carboxyl, serylalanine results:

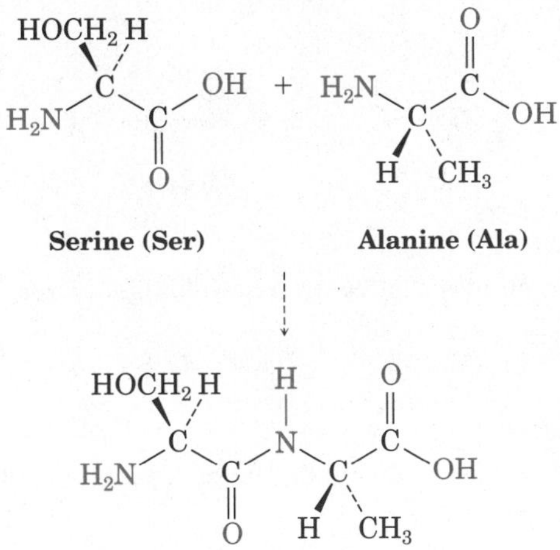

HOCH₂ H ... Serine (Ser) + Alanine (Ala) → Serylalanine (Ser-Ala)

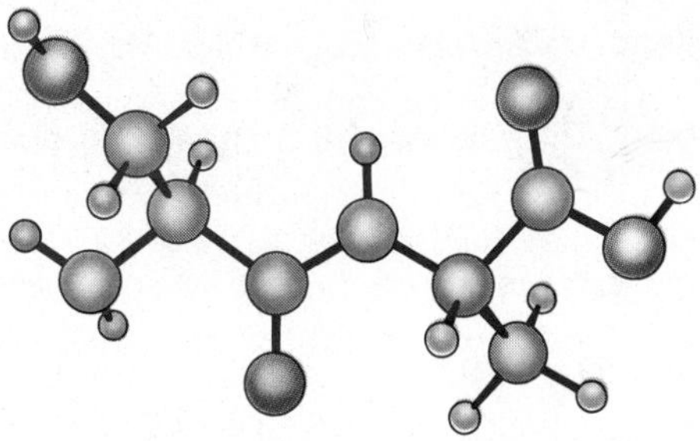

N-TERMINAL AMINO ACID
Amino acid with a free $-NH_2$ group at one end of a protein chain

C-TERMINAL AMINO ACID
Amino acid with a free $-COOH$ group at one end of a protein chain

By convention, peptides are always written with the **N-terminal amino acid** (the one with the free $-NH_2$ group) on the left, and the **C-terminal amino acid** (the one with the free $-COOH$ group) on the right. The name of the peptide is then indicated using the three-letter abbreviations listed in Table 15.1. Thus, serylalanine is abbreviated Ser-Ala, and alanylserine is abbreviated Ala-Ser.

The number of possible isomeric peptides increases rapidly as the number of amino acid units increases. There are six ways in which three amino acids can be joined, and more than 40,000 ways in which the eight amino acids in the hormone angiotensin II can be joined (Figure 15.2).

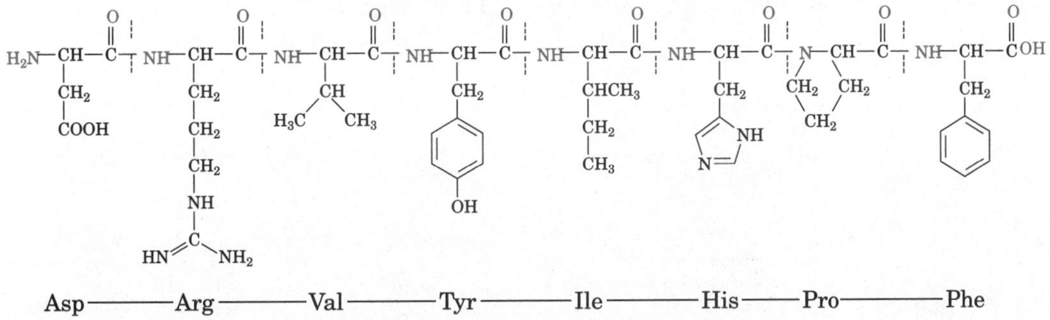

Asp—Arg—Val—Tyr—Ile—His—Pro—Phe

FIGURE 15.2 The structure of angiotensin II, a hormone in blood plasma that regulates blood pressure.

PRACTICE PROBLEM 15.3 Draw the structure of Ala-Val.

SOLUTION By convention, the N-terminal amino acid is written on the left and the C-terminal amino acid on the right. Thus, alanine is N-terminal, valine is C-terminal, and the amide bond is formed between the alanine –COOH and the valine –NH_2.

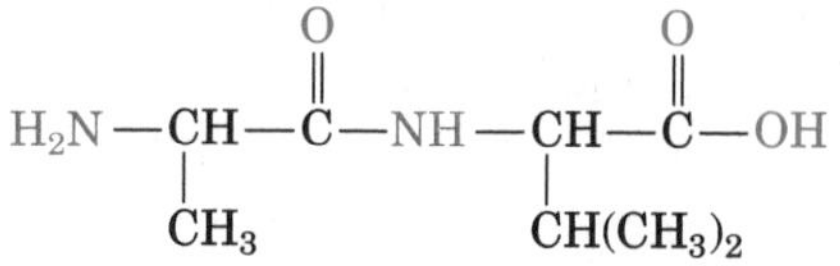

$$H_2N-\underset{CH_3}{\underset{|}{CH}}-\overset{O}{\overset{\|}{C}}-NH-\underset{CH(CH_3)_2}{\underset{|}{CH}}-\overset{O}{\overset{\|}{C}}-OH \quad \textbf{Ala-Val}$$

PRACTICE PROBLEM 15.4 Name the six tripeptides that contain methionine, lysine, and isoleucine.

SOLUTION

Met-Lys-Ile	Lys-Met-Ile	Ile-Met-Lys
Met-Ile-Lys	Lys-Ile-Met	Ile-Lys-Met

PROBLEM 15.8 Draw structures of the two dipeptides made from leucine and cysteine.

PROBLEM 15.9 Using the three-letter shorthand notations for each amino acid, name the six possible isomeric tripeptides that contain valine, tyrosine, and glycine.

PROBLEM 15.10 Draw the structure of Met-Pro-Val-Gly, and indicate where the amide bonds are.

15.5 COVALENT BONDING IN PEPTIDES

DISULFIDE LINK An S–S bond between two cysteine residues in a protein

In addition to the amide bonds that link amino acid residues in peptides, a second kind of covalent bonding occurs when a **disulfide linkage**, RS–SR, is formed between two cysteine residues. The linkage is sometimes indicated by writing CyS, with a capital "S" (for sulfur), and then drawing a line from one CyS to the other: $\overset{\frown}{\text{CyS CyS}}$. As we saw in Section 8.12, disulfides are formed from thiols (RSH) by mild oxidation and are converted back to thiols by mild reduction:

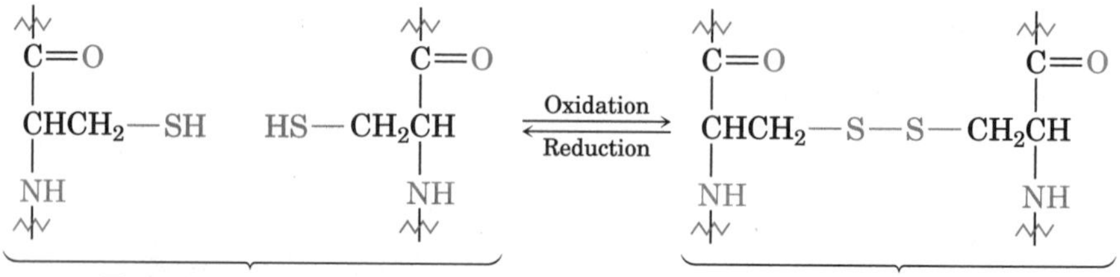

$$\underset{\textbf{Two cysteines (thiols)}}{\underset{NH}{\underset{|}{\overset{C=O}{\overset{|}{CHCH_2}}}}-SH \quad HS-\underset{NH}{\underset{|}{\overset{C=O}{\overset{|}{CH_2CH}}}}} \underset{\text{Reduction}}{\overset{\text{Oxidation}}{\rightleftharpoons}} \underset{\textbf{Disulfide}}{\underset{NH}{\underset{|}{\overset{C=O}{\overset{|}{CHCH_2}}}}-S-S-\underset{NH}{\underset{|}{\overset{C=O}{\overset{|}{CH_2CH}}}}}$$

Disulfide bonds between cysteine residues in two separate peptide chains link the otherwise separate chains together. A disulfide bond between cysteine residues in the same chain creates a loop in the chain.

Such is the case with the nonapeptide vasopressin, an antidiuretic hormone involved in controlling water balance in the body. Note that the C-terminal end of vasopressin occurs as the primary amide, $-CONH_2$, rather than as the free acid.

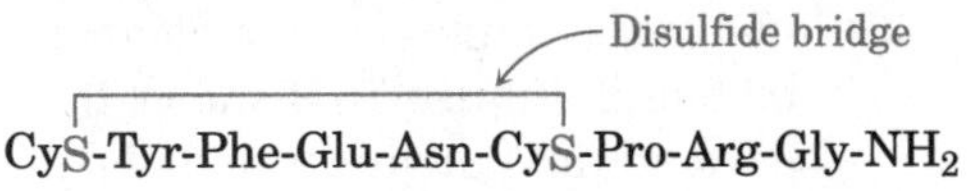

Disulfide bridge

CyS-Tyr-Phe-Glu-Asn-CyS-Pro-Arg-Gly-NH_2

Vasopressin

15.6 PEPTIDE STRUCTURE DETERMINATION: AMINO ACID ANALYSIS

Determining the structure of a peptide requires answering three questions: What amino acids are present? How much of each is present? In what sequence do the amino acids occur in the peptide chain? The answers to the first two questions are provided by an instrument called an *amino acid analyzer*.

An amino acid analyzer is an automated instrument based on techniques worked out in the 1950s by W. Stein and S. Moore at the Rockefeller University in New York. In preparation for analysis, the peptide is broken into its constituent amino acids by reducing all disulfide bonds and hydrolyzing all amide bonds with aqueous HCl. The resultant amino acid mixture is then analyzed by placing it at the top of a glass column filled with a special adsorbent material and pumping a series of aqueous buffers through the column. The various amino acids migrate down the column at different rates depending on their structures and are thus separated as they come out (*elute* from) the end of the column.

As each amino acid elutes from the end of the glass column, it mixes with a solution of *ninhydrin*, a reagent that forms a purple color when it reacts with α-amino acids. The purple color is detected by a spectrometer, which measures its intensity and charts it as a function of time.

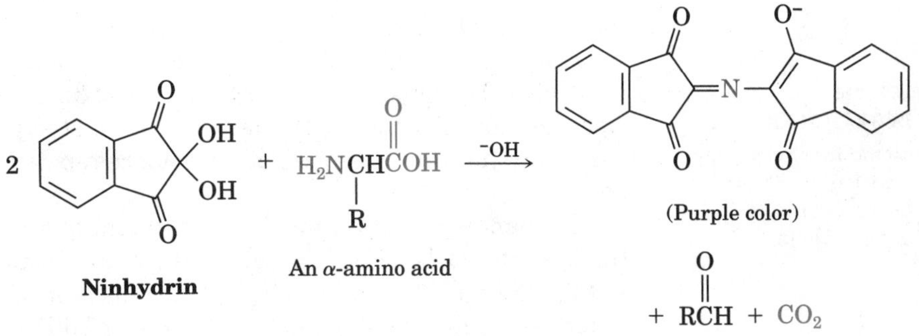

$$2\ \text{Ninhydrin} + H_2NCH(R)COOH \xrightarrow{^-OH} \text{(Purple color)} + RCHO + CO_2$$

Ninhydrin

An α-amino acid

(Purple color)

Since the time required for a given amino acid to elute from a standard column is reproducible, the identities of all amino acids in a peptide are determined simply by noting the various elution times. The amount of each amino acid in the sample is determined by measuring the intensity of the purple color resulting from its reaction with ninhydrin. Thus, the identity and percentage composition of each amino acid in a peptide can be found. Figure 15.3 shows the results of amino acid analysis of a standard equimolar mixture of 17 α-amino acids.

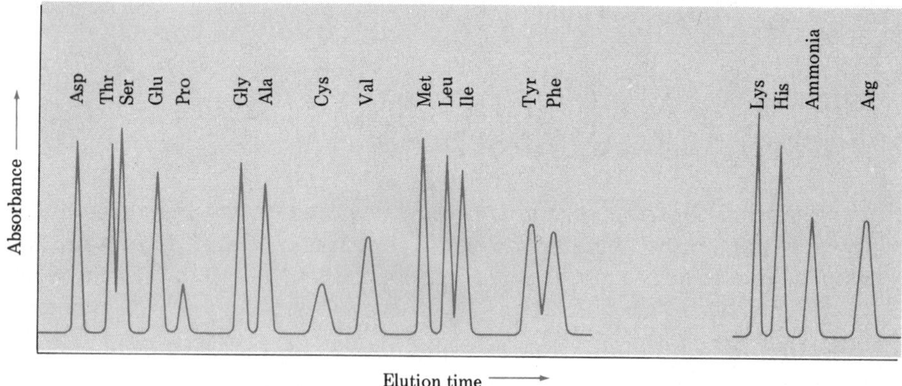

FIGURE 15.3 Amino acid analysis of an equimolar amino acid mixture.

PROBLEM 15.11 Write an equation for the reaction of valine with ninhydrin.

15.7 PEPTIDE SEQUENCING: THE EDMAN DEGRADATION

EDMAN DEGRADATION
Method for selectively cleaving the N-terminal amino acid from a peptide chain

After the identities and amounts of the amino acids in a peptide are known, the peptide is *sequenced* to find the order in which the amino acids are linked. The general idea of peptide sequencing is to cleave selectively one amino acid residue at a time from the end of the peptide chain (either C terminus or N terminus). That terminal amino acid is then separated and identified, and the cleavage reaction is repeated on the chain-shortened peptide until the entire peptide sequence is known. Most peptide sequencing is now done by **Edman degradation**, an efficient method of N-terminal analysis. Automated protein sequenators are available that allow a series of 20 or more repetitive sequencing steps to be carried out.

Edman degradation involves treatment of a peptide with phenyl isothiocyanate, $C_6H_5\text{–}N{=}C{=}S$, followed by mild acid hydrolysis, as shown in Figure 15.4. The first step attaches a marker to the $-NH_2$ group of the N-terminal amino acid, and the second step splits the N-terminal residue

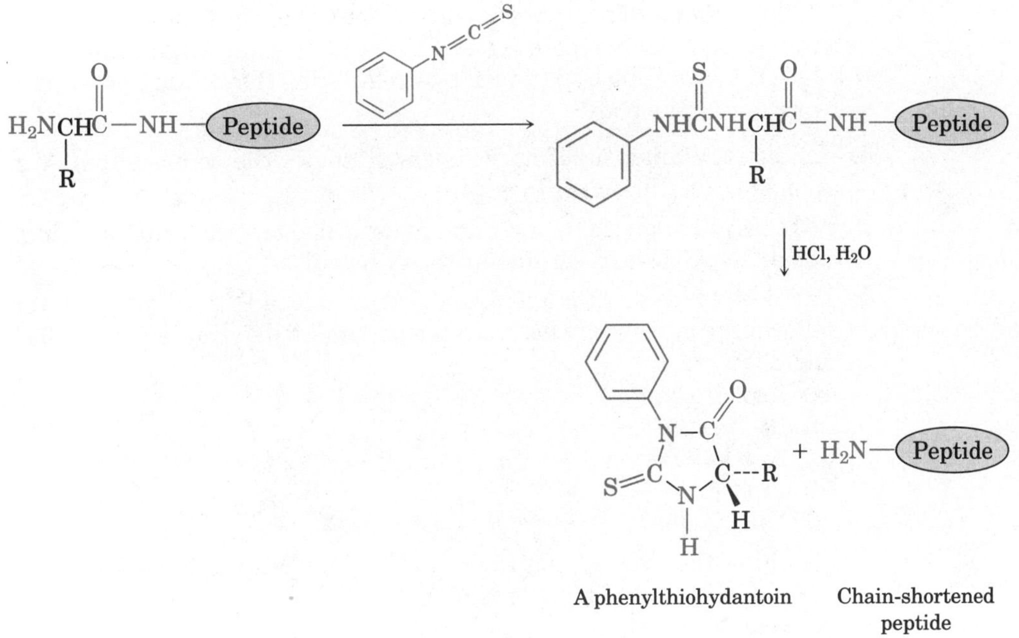

FIGURE 15.4 Edman degradation of a peptide chain.

from the chain, yielding a *phenylthiohydantoin* derivative plus the chain-shortened peptide. The phenylthiohydantoin is identified by comparison with known derivatives of the common amino acids, and the chain-shortened peptide is resubmitted to another round of Edman degradation.

Complete sequencing of large peptides and proteins by Edman degradation is impractical since the method is limited by buildup of unwanted by-products to about 20 cycles. Instead, a large peptide chain is first cleaved by partial hydrolysis into a number of smaller fragments, the sequence of each fragment is determined, and the individual pieces are then fitted together.

Partial hydrolysis of a peptide can be carried out either chemically with aqueous acid or enzymatically with enzymes such as trypsin and chymotrypsin. Acid hydrolysis is unselective and leads to a more or less random mixture of small fragments. Enzymic hydrolysis, however, is quite specific. Trypsin catalyzes hydrolysis only at the carboxyl side of the basic amino acids arginine and lysine; chymotrypsin cleaves only at the carboxyl side of the aryl-substituted amino acids phenylalanine, tyrosine, and tryptophan. For example:

Val-Phe-Leu-Met-Tyr-Pro-Gly-Trp-Cys-Glu-Asp-Ile-Lys-Ser-Arg-His

Chymotrypsin cleaves these bonds. Trypsin cleaves these bonds.

As an example of peptide sequencing, look at a hypothetical structure determination of angiotensin II (see Figure 15.2), a hormonal octapeptide involved in controlling hypertension by regulating the sodium–potassium salt balance in the body.

1. Amino acid analysis of angiotensin II shows the composition: Arg, Asp, His, Ile, Phe, Pro, Tyr, Val.
2. N-Terminal analysis by the Edman method shows that angiotensin II has an aspartic acid residue at the N terminus.
3. Partial hydrolysis of angiotensin II with dilute HCl might yield the following fragments, whose sequences can be determined by Edman degradation:
   (a) Asp-Arg-Val
   (b) Ile-His-Pro
   (c) Arg-Val-Tyr
   (d) Pro-Phe
   (e) Val-Tyr-Ile
4. Matching the overlapping regions of the various fragments provides the full sequence of angiotensin II:

```
(a) Asp-Arg-Val
(c)     Arg-Val-Tyr
(e)         Val-Tyr-Ile
(b)                 Ile-His-Pro
(d)                         Pro-Phe
    Asp-Arg-Val-Tyr-Ile-His-Pro-Phe
```

Angiotensin II

The structure of angiotensin II is relatively simple—the entire sequence can be done easily by a protein sequenator—but the methods and logic used here are the same as those used to solve far more complex structures. Indeed, single protein chains with more than 400 amino acids have been sequenced by these methods.

PRACTICE PROBLEM 15.5 A hexapeptide with the composition Arg, Gly, Leu, Pro_3 has proline at both C-terminal and N-terminal positions. What is the structure of the hexapeptide if partial hydrolysis gives Gly-Pro-Arg, Arg-Pro, and Pro-Leu-Gly?

SOLUTION Line up the overlapping fragments:

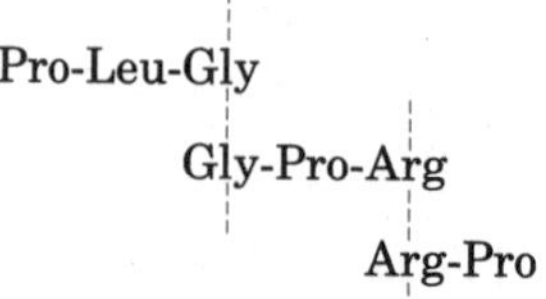

The final sequence is Pro-Leu-Gly-Pro-Arg-Pro.

PROBLEM 15.12 What fragments would result if angiotensin II were cleaved with trypsin? With chymotrypsin?

PROBLEM 15.13 Give the amino acid sequence of a hexapeptide containing Arg, Gly, Ile, Leu, Pro, and Val that produces these fragments on partial acid hydrolysis: Pro-Leu-Gly, Arg-Pro, Gly-Ile-Val.

PROBLEM 15.14 Propose two structures for a tripeptide that gives Leu, Ala, and Phe on hydrolysis but doesn't react with phenyl isothiocyanate.

15.8 PEPTIDE SYNTHESIS

After a peptide's structure has been determined, synthesis is often the next goal as a means of obtaining larger amounts of the peptide for biological evaluation. Although simple amides are usually formed by reaction between amines and acid chlorides (Section 10.7), peptide synthesis is much more complex because of the requirement for specificity. Many different amide links must be formed in a specific order rather than at random.

The solution to the specificity problem is *protection* (Section 9.11). We can force a reaction to take only the desired course by protecting all the amine and carboxylic acid functional groups except for those we want to react. If we wanted to couple alanine with leucine to synthesize Ala-Leu, for example, we could protect the amino group of alanine and the carboxyl group of leucine to render them unreactive. With only the alanine carboxyl and the leucine amine available, we then form the proper amide bond and remove the protecting groups:

$$\underset{\text{Alanine}}{H_2NCH(CH_3)\overset{O}{\overset{\|}{C}}OH} \xrightarrow[\text{—NH}_2]{\text{Protect}} \boxed{H_2N}\text{—}CH(CH_3)\overset{O}{\overset{\|}{C}}OH$$

$$\underset{\text{Leucine}}{H_2NCH(CH_2CH(CH_3)_2)\overset{O}{\overset{\|}{C}}OH} \xrightarrow[\text{—COOH}]{\text{Protect}} H_2NCH(CH_2CH(CH_3)_2)\text{—}\boxed{\overset{O}{\overset{\|}{C}}OH}$$

$$\xrightarrow[\text{2. Deprotect}]{\text{1. Form amide}} \underset{\text{Ala-Leu}}{H_2NCH(CH_3)\overset{O}{\overset{\|}{C}}\text{—}NHCH(CH_2CH(CH_3)_2)\overset{O}{\overset{\|}{C}}OH}$$

Carboxyl groups are often protected simply by converting them into methyl esters. Ester groups are easily made from carboxylic acids and are easily hydrolyzed by mild treatment with aqueous sodium hydroxide:

$$H_2NCH(CH_2CH(CH_3)_2)COOH \xrightarrow[HCl]{CH_3OH} H_2NCH(CH_2CH(CH_3)_2)COOCH_3 \xrightarrow[2.\ H_3O^+]{1.\ NaOH,\ H_2O} H_2NCH(CH_2CH(CH_3)_2)COOH$$

Leucine **Methyl leucinate** **Leucine**

Amino groups are often protected as their *tert*-butoxycarbonyl amide (BOC) derivatives. The BOC protecting group is easily introduced by reaction of the amino acid with di-*tert*-butyl dicarbonate and is removed by brief treatment with a strong acid such as trifluoroacetic acid, CF_3COOH.

$$H_2NCH(CH_3)COOH + (CH_3)_3COC(=O)OC(=O)OC(CH_3)_3 \xrightarrow{(CH_3CH_2)_3N} (CH_3)_3COC(=O)\text{—}NHCH(CH_3)COOH$$

Alanine **Di-*tert*-butyl dicarbonate** **BOC-Ala**

Formation of the peptide bond is accomplished by treating a mixture of the protected acid and amine components with dicyclohexylcarbodiimide (DCC). Although its mechanism of action is complex, DCC functions by first converting the acid into a reactive intermediate, which then undergoes further nucleophilic acyl substitution reaction with the amine.

$$RC(=O)OH + R'NH_2 \xrightarrow{DCC} RC(=O)NHR' + C_6H_{11}NHC(=O)NHC_6H_{11}$$

An acid An amine An amide **Dicyclohexylurea**

The five steps needed to synthesize Ala-Leu are summarized:

STEP 1 Protect the amino group of alanine as the BOC derivative:

$$\text{Ala} + (\text{BOC})_2\text{O} \longrightarrow \text{BOC-Ala}$$

STEP 2 Protect the carboxyl group of leucine as the methyl ester:

$$\text{Leu} + CH_3OH \longrightarrow \text{Leu-OCH}_3$$

STEP 3 Couple the two protected amino acids using DCC:

$$\text{BOC-Ala} + \text{Leu-OCH}_3 \xrightarrow{\text{DCC}} \text{BOC-Ala-Leu-OCH}_3$$

STEP 4 Remove the BOC protecting group by acid treatment:

$$\text{BOC-Ala-Leu-OCH}_3 \xrightarrow{\text{CF}_3\text{COOH}} \text{Ala-Leu-OCH}_3$$

STEP 5 Remove the methyl ester protecting group by base treatment:

$$\text{Ala-Leu-OCH}_3 \xrightarrow[\text{H}_2\text{O}]{\text{NaOH}} \text{Ala-Leu}$$

These steps can be repeated to add one amino acid at a time to a growing chain or to link two peptide chains together. Many remarkable achievements in peptide synthesis have been reported, including a complete synthesis of human insulin. Insulin, whose structure is shown in Figure 15.5, is composed of two chains totaling 51 amino acids and linked by two cysteine disulfide bridges. Its structure was determined by Frederick Sanger, who received the 1958 Nobel Prize for his work.

A chain (21 units):
Gly
Ile
Val
Glu
Gln-CyS-CyS-Thr-Ser-Ile-CyS-Ser-Leu-Tyr-Gln-Leu-Glu-Asn-Tyr-CyS-Asn

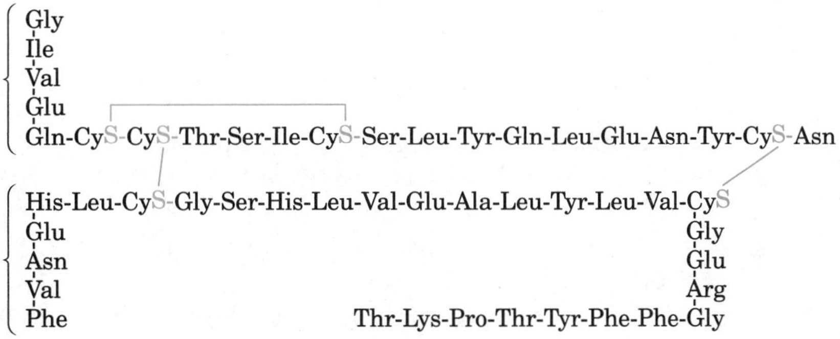

B chain (30 units):
His-Leu-CyS-Gly-Ser-His-Leu-Val-Glu-Ala-Leu-Tyr-Leu-Val-CyS
Glu ... Gly
Asn ... Glu
Val ... Arg
Phe ... Thr-Lys-Pro-Thr-Tyr-Phe-Phe-Gly

FIGURE 15.5 Structure of human insulin.

PRACTICE PROBLEM 15.6 Write equations for the reaction of methionine with:
(a) CH_3OH, HCl **(b)** Di-*tert*-butyl dicarbonate

SOLUTION

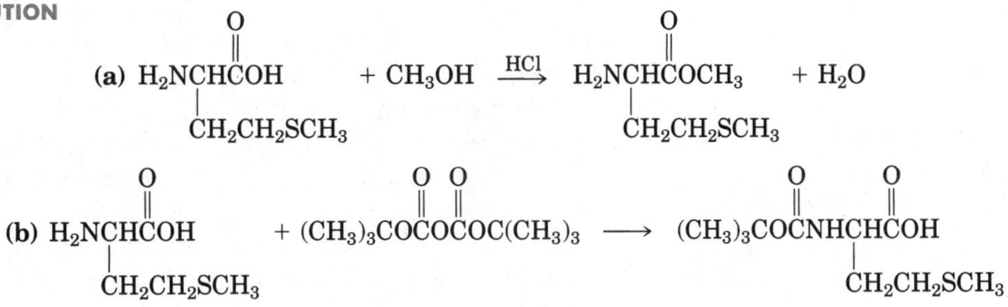

(a) $$\text{H}_2\text{NCH(CH}_2\text{CH}_2\text{SCH}_3)\text{COH(=O)} + \text{CH}_3\text{OH} \xrightarrow{\text{HCl}} \text{H}_2\text{NCH(CH}_2\text{CH}_2\text{SCH}_3)\text{C(=O)OCH}_3 + \text{H}_2\text{O}$$

(b) $$\text{H}_2\text{NCH(CH}_2\text{CH}_2\text{SCH}_3)\text{C(=O)OH} + (\text{CH}_3)_3\text{COC(=O)OC(=O)OC(CH}_3)_3 \longrightarrow (\text{CH}_3)_3\text{COC(=O)NHCH(CH}_2\text{CH}_2\text{SCH}_3)\text{C(=O)OH}$$

PROBLEM 15.15 Write the structures of the intermediates in the five-step synthesis of Leu-Ala from alanine and leucine.

PROBLEM 15.16 Show all the steps involved in the synthesis of the tripeptide Val-Phe-Gly.

15.9 CLASSIFICATION OF PROTEINS

SIMPLE PROTEIN
Protein composed entirely of amino acids

CONJUGATED PROTEIN
Protein composed of both an amino acid part and a non-amino acid part

Proteins are classified into two major types according to their composition. **Simple proteins**, such as blood serum albumin, are those that yield only amino acids and no other compounds on hydrolysis. **Conjugated proteins**, which are much more common than simple proteins, yield other compounds in addition to amino acids on hydrolysis. As shown in Table 15.2, conjugated proteins can be further classified according to the chemical nature of the non-amino acid portion.

TABLE 15.2 Some Conjugated Proteins

NAME	COMPOSITION
Glycoproteins	Proteins bonded to a carbohydrate; cell membranes have a glycoprotein coating.
Lipoproteins	Proteins bonded to fats and oils (lipids); these proteins transport cholesterol and other fats through the body.
Metalloproteins	Proteins bonded to a metal ion; the enzyme cytochrome oxidase, necessary for biological energy production, is an example.
Nucleoproteins	Proteins bonded to RNA (ribonucleic acid); these are found in cell ribosomes.
Phosphoproteins	Proteins bonded to a phosphate group; milk casein, which stores nutrients for growing embryos, is an example.

FIBROUS PROTEIN
Tough, insoluble protein used in nature for structural purposes

GLOBULAR PROTEIN
Spherical, water-soluble protein found primarily inside cells

Another way to classify proteins is as either *fibrous* or *globular*, according to their three-dimensional shape. **Fibrous proteins**, such as collagen and keratin, consist of polypeptide chains arranged side by side in long filaments. Because these proteins are tough and insoluble in water, they're used in nature for such structural materials as tendons, hoofs, horns, and muscles. **Globular proteins**, by contrast, are usually coiled into compact, nearly spherical shapes. These proteins are generally soluble in water and are mobile within cells. Most of the 2000 or so known enzymes, as well as hormonal and transport proteins, are globular. Table 15.3 lists some common examples of both fibrous and globular proteins.

TABLE 15.3 Some Common Fibrous and Globular Proteins

NAME	OCCURRENCE AND USE
FIBROUS PROTEINS (INSOLUBLE)	
Collagens	Found in animal hide, tendons, and other connective tissues
Elastins	Found in blood vessels, ligaments, and other tissues that must be able to stretch
Fibrinogen	Found in blood; necessary for blood clotting
Keratins	Found in skin, wool, feathers, hooves, silk, fingernails
Myosins	Found in muscle tissue
GLOBULAR PROTEINS (SOLUBLE)	
Hemoglobin	Protein involved in oxygen transport
Immunoglobulins	Proteins involved in immune response
Insulin	Regulatory hormone for controlling glucose metabolism
Ribonuclease	Enzyme controlling RNA synthesis

Yet a third way to classify proteins is according to function. As shown in Table 15.4, there is an extraordinary diversity to the biological roles of proteins.

TABLE 15.4 Some Biological Functions of Proteins

TYPE	FUNCTION AND EXAMPLE
Enzymes	Proteins such as chymotrypsin that act as biological catalysts
Hormones	Proteins such as insulin that regulate body processes
Protective proteins	Proteins such as antibodies that fight infection
Storage proteins	Proteins such as casein that store nutrients
Structural proteins	Proteins such as keratin, elastin, and collagen that form the structure of an organism
Transport proteins	Proteins such as hemoglobin that transport oxygen and other substances through the body

15.10 PROTEIN STRUCTURE

Proteins are so large that the word *structure* takes on a broader meaning when applied to such immense molecules than it does with other organic compounds. In fact, chemists speak of four different levels of structure when describing proteins. At its simplest, protein structure is the sequence in which amino acid residues are bound together. Called the **primary structure** of a protein, this is the most fundamental structural level.

PRIMARY STRUCTURE
Amino acid sequence of a protein

SECONDARY STRUCTURE
Specific way in which segments of a protein chain are oriented into a regular pattern

TERTIARY STRUCTURE
Specific way in which a protein molecule is oriented into an overall three-dimensional shape

QUATERNARY STRUCTURE
The way in which several protein molecules aggregate together to yield a larger structure

There is much more to protein structure than amino acid sequence, though. The chemical properties of a protein are also dependent on higher levels of structure—on exactly how the peptide backbone is folded to give the molecule a specific three-dimensional shape. Thus, the term **secondary structure** refers to the way in which *segments* of the peptide backbone are oriented into a regular pattern, **tertiary structure** refers to the way in which the *entire* protein molecule is coiled into an overall three-dimensional shape, and **quaternary structure** refers to the way in which several protein molecules come together to yield large aggregate structures. Let's look at three examples—α-keratin (fibrous), fibroin (fibrous), and myoglobin (globular)—to see how higher structure affects a protein's properties.

α-Keratin

α-Keratin is the fibrous structural protein found in wool, hair, nails, and feathers. Studies show that α-keratin is coiled into a right-handed helical secondary structure, as illustrated in Figure 15.6. This so-called **α-helix**

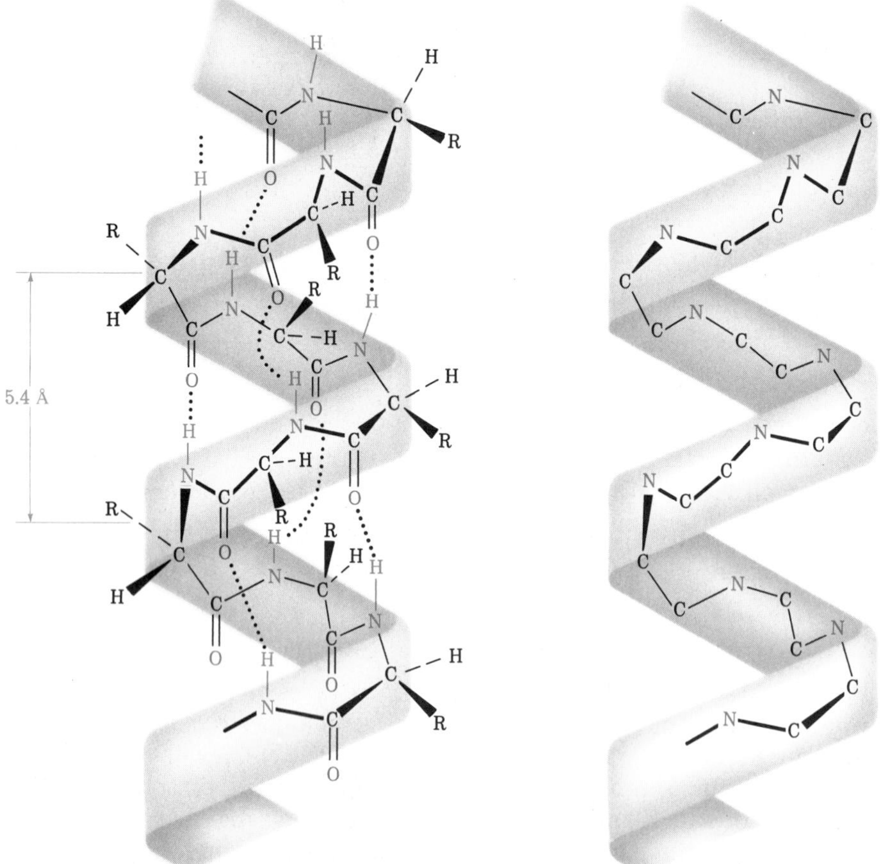

FIGURE 15.6 The helical secondary structure of α-keratin.

α-HELIX Common kind of secondary structure in which a protein chain coils into a spiral

is stabilized by hydrogen bonding between amide N–H groups and other amide C=O groups four residues away. Although the strength of a single hydrogen bond (about 5 kcal/mol) is only about 5% of the strength of a C–C or C–H covalent bond, the large number of hydrogen bonds made possible by helical winding imparts a great deal of stability to the α-helical structure. Each coil of the helix (the *repeat distance*) contains 3.6 amino acid residues, with a distance between coils of 5.4 Å.

Further evidence shows that the α-keratins of wool and hair also have a quaternary structure. The individual helical strands are themselves coiled about one another in stiff bundles to form a *superhelix* that accounts for the threadlike properties and strength of these proteins. Although α-keratin is the best example of an almost entirely helical protein, most globular proteins contain α-helical segments. Both hemoglobin and myoglobin, for example, contain many short helical sections in their chains.

Fibroin

β-PLEATED SHEET Common kind of secondary structure in which a protein chain folds back on itself so that two sections of the chain run parallel

Fibroin, the fibrous protein found in silk, has a secondary structure known as a **β-pleated sheet** in which polypeptide chains line up in a parallel arrangement held together by hydrogen bonds between chains (Figure 15.7). Although not as common as the α-helix, small β-pleated-sheet regions are often found in proteins where sections of peptide chains double back on themselves.

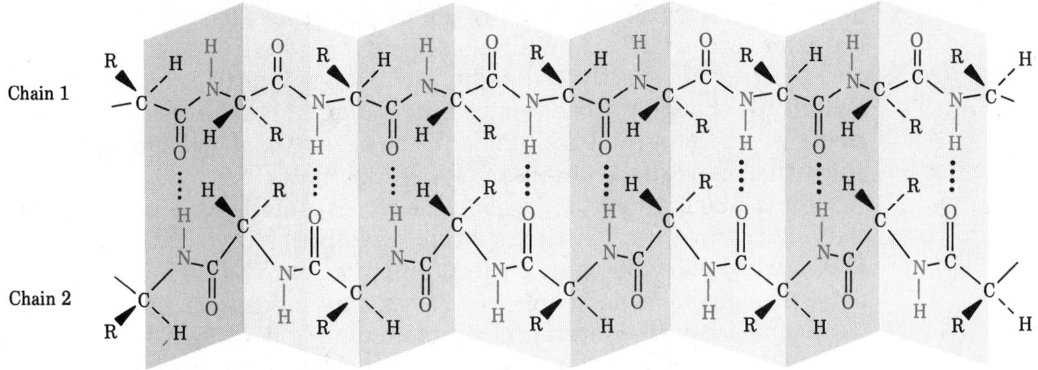

FIGURE 15.7 The β-pleated-sheet structure in silk fibroin.

Myoglobin

Myoglobin is a small globular protein containing 153 amino acid residues in a single chain. A relative of hemoglobin, myoglobin is found in the skeletal muscles of sea mammals, where it stores oxygen needed to sustain the animals during long dives. Myoglobin consists of eight straight segments, each of which adopts an α-helical secondary structure. These helical sections are connected by bends to form a compact, nearly spherical, tertiary structure (Figure 15.8).

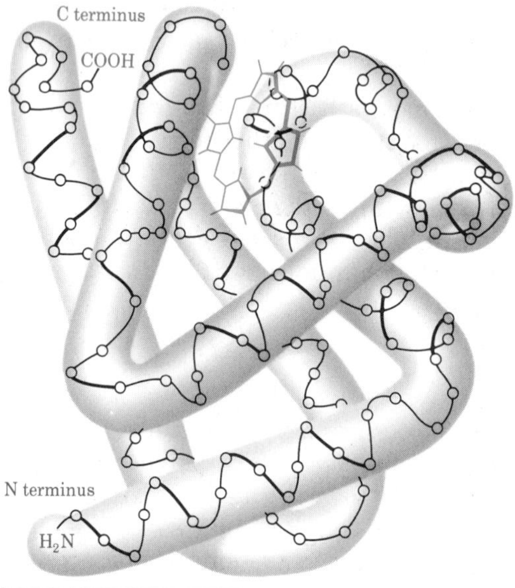

FIGURE 15.8 Secondary and tertiary structure of myoglobin.

Why does myoglobin adopt the shape it does? The forces that determine the tertiary structure of myoglobin and other globular proteins are the same forces that act on all molecules, regardless of size. By bending and twisting in exactly the right way, myoglobin achieves maximum stability. Although the bends appear irregular and the three-dimensional structure appears random, this isn't the case. All myoglobin molecules adopt this same shape because it's the most stable.

Particularly important among the forces stabilizing a protein's tertiary structure are the hydrophobic (water-repelling) interactions of hydrocarbon side chains on neutral amino acids. The amino acids with neutral, nonpolar side chains have a strong tendency to congregate on the hydrocarbon-like interior of a protein molecule, away from the aqueous medium. The acidic or basic amino acids with charged side chains, by contrast, tend to congregate on the exterior of the protein where they can be solvated by water.

Also important for stabilizing a protein's tertiary structure are the formation of disulfide bridges between cysteine residues, the formation of hydrogen bonds between nearby amino acids, and the formation of ionic attractions, called *salt bridges*, between positively and negatively charged sites on the protein. The various kinds of stabilizing forces are summarized in Figure 15.9.

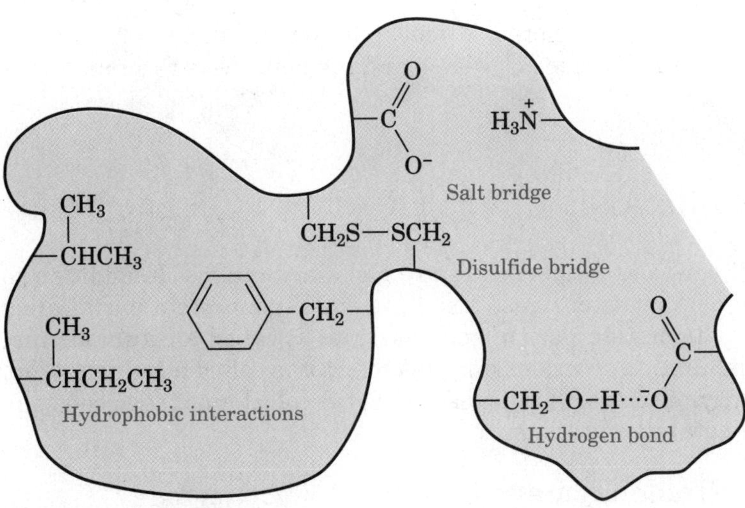

FIGURE 15.9 Interactions among amino acid side chains that stabilize a protein's tertiary structure.

15.11 ENZYMES

ENZYME
Large globular protein that acts as a catalyst for a specific biological reaction

Enzymes are large proteins that act as catalysts for biological reactions. Unlike many of the simple catalysts that chemists use in the laboratory, enzymes are usually specific in their action. Often, in fact, an enzyme can catalyze only a single reaction of a single compound, called the enzyme's *substrate*. For example, the enzyme amylase found in the human digestive tract catalyzes only the hydrolysis of starch to yield glucose; cellulose and other polysaccharides are untouched by amylase.

Different enzymes have different specificities. Some, such as amylase, are specific for a single substrate, but others operate on a range of substrates. Papain, for instance, a globular protein of 212 amino acids isolated from papaya fruit, catalyzes the hydrolysis of many kinds of peptide bonds. In fact, it's this ability to hydrolyze peptide bonds that makes papain useful as a meat tenderizer and a contact-lens cleaner.

$$\text{+NH}\underset{\underset{\text{R}}{|}}{\text{CH}}\overset{\overset{\text{O}}{\|}}{\text{C}}\text{—NH}\underset{\underset{\text{R}'}{|}}{\text{CH}}\overset{\overset{\text{O}}{\|}}{\text{C}}\text{—NH}\underset{\underset{\text{R}''}{|}}{\text{CH}}\overset{\overset{\text{O}}{\|}}{\text{C}}\text{+} \xrightarrow[\text{H}_2\text{O}]{\text{Papain}} \text{+NH}\underset{\underset{\text{R}}{|}}{\text{CH}}\overset{\overset{\text{O}}{\|}}{\text{C}}\text{OH} + \text{H}_2\text{N}\underset{\underset{\text{R}'}{|}}{\text{CH}}\overset{\overset{\text{O}}{\|}}{\text{C}}\text{—NH}\underset{\underset{\text{R}''}{|}}{\text{CH}}\overset{\overset{\text{O}}{\|}}{\text{C}}\text{+}$$

Like all catalysts, enzymes don't affect the equilibrium constant of a reaction and can't bring about chemical changes that are otherwise unfavorable. Enzymes act only to lower the activation energy, thereby making the reaction take place faster or at lower temperatures. Starch and water, for example, react very slowly in the absence of a catalyst because the activation energy is too high. When amylase is present, though, the energy barrier is lowered, and the hydrolysis reaction occurs rapidly.

15.12 STRUCTURE AND CLASSIFICATION OF ENZYMES

COFACTOR Small nonprotein part of an enzyme necessary for biological activity

APOENZYME Protein part of an enzyme that needs a cofactor for biological activity

HOLOENZYME Combination of apoenzyme and cofactor

COENZYME Small organic molecule that acts as an enzyme cofactor

VITAMIN Small organic molecule that must be obtained in the diet and is required for proper growth

All the 2000 or so known enzymes are globular proteins. In addition to the protein part, most enzymes also have small nonprotein parts called **cofactors**. The protein part in such enzymes is called an **apoenzyme**, and the combination of apoenzyme plus cofactor is called a **holoenzyme**. Only holoenzymes have biological activity; the cofactor or the apoenzyme alone can't catalyze reactions.

Holoenzyme = Cofactor + Apoenzyme

Cofactors can be either inorganic ions such as Zn^{2+} or small organic molecules, called **coenzymes**. The requirement of many enzymes for inorganic cofactors is the main reason for our dietary need of trace minerals. Iron, zinc, copper, manganese, and numerous other metal ions are all essential minerals that act as enzyme cofactors, though the exact biological role isn't known in all cases.

A variety of organic molecules act as coenzymes. Many, though not all, coenzymes are **vitamins**, small organic molecules that must be obtained in the diet and are required in trace amounts for proper growth. Table 15.5 lists the 13 known vitamins required in the human diet and their enzyme functions.

Enzymes are grouped into six main classes according to the kind of reaction they catalyze (Table 15.6). *Hydrolases* catalyze the hydrolysis of substrates; *isomerases* catalyze the isomerization of substrates; *ligases* catalyze the bonding together of two substrates with participation of adenosine triphosphate (ATP); *lyases* catalyze the breaking away of a small molecule such as H_2O from a substrate or the reverse reaction; *oxidoreductases* catalyze oxidations and reductions of substrate molecules; and *transferases* catalyze the transfer of a group from one substrate to another.

Although some enzymes, like papain and trypsin, have uninformative common names, the systematic name of an enzyme has two parts, ending with *-ase*. The first part identifies the enzyme's substrate, and the second part identifies its class. For example, *hexose kinase* is an enzyme that catalyzes the transfer of a phosphate group from adenosine triphosphate (ATP) to glucose.

PROBLEM 15.17 To what classes do these enzymes belong?

(a) Pyruvate decarboxylase **(b)** Chymotrypsin

(c) Alcohol dehydrogenase

TABLE 15.5 Vitamins and Their Enzyme Functions

VITAMIN	ENZYME FUNCTION	DEFICIENCY SYMPTOM
WATER-SOLUBLE VITAMINS		
Ascorbic acid (vitamin C)	Hydrolases	Bleeding gums, bruising
Thiamin (vitamin B_1)	Reductases	Fatigue, depression
Riboflavin (vitamin B_2)	Reductases	Cracked lips, scaly skin
Pyridoxine (vitamin B_6)	Transaminases	Anemia, irritability
Niacin	Reductases	Dermatitis, dementia
Folic acid (vitamin M)	Methyltransferases	Megaloblastic anemia
Vitamin B_{12}	Isomerases	Megaloblastic anemia, neurodegeneration
Pantothenic acid	Acyltransferases	Weight loss, irritability
Biotin (vitamin H)	Carboxylases	Dermatitis, anorexia, depression
FAT-SOLUBLE VITAMINS		
Vitamin A	Visual system	Night blindness, dry skin
Vitamin D	Calcium metabolism	Rickets, osteomalacia
Vitamin E	Antioxidant	Hemolysis of red blood cells
Vitamin K	Blood clotting	Hemorrhage, delayed blood clotting

TABLE 15.6 Classification of Enzymes

MAIN CLASS	SOME SUBCLASSES	TYPE OF REACTION CATALYZED
Hydrolases	Lipases	Hydrolysis of an ester group
	Nucleases	Hydrolysis of a phosphate group
	Proteases	Hydrolysis of an amide group
Isomerases	Epimerases	Isomerization of stereogenic center
Ligases	Carboxylases	Addition of CO_2
	Synthetases	Formation of new bond
Lyases	Decarboxylases	Loss of CO_2
	Dehydrases	Loss of H_2O
Oxidoreductases	Dehydrogenases	Introduction of double bond by removal of H_2
	Oxidases	Oxidation
	Reductases	Reduction
Transferases	Kinases	Transfer of a phosphate group
	Transaminases	Transfer of an amino group

INTERLUDE

PROTEIN AND NUTRITION

Dietary protein is needed by everyone, from weightlifters to infants. Children need large amounts of protein for proper growth, and adults need protein to replace what's lost each day by the body's normal biochemical reactions. Dietary protein is necessary because our bodies can synthesize only 10 of the 20 common amino acids from simple precursor molecules; the other 10 amino acids must be obtained from food by digestion of edible proteins. Table 15.7 shows the estimated essential amino acid requirements of an infant and an adult.

TABLE 15.7 Estimated Essential Amino Acid Requirements

	DAILY REQUIREMENT (mg/kg body weight)	
AMINO ACID	Infant	Adult
Arginine	?	?
Histidine	33	?
Isoleucine	83	12
Leucine	35	16
Lysine	99	12
Methionine + Cysteine	49	10
Phenylalanine + Tyrosine	141	16
Threonine	68	8
Tryptophan	21	3
Valine	92	14

Not all foods provide sufficient amounts of the 10 essential amino acids to meet our minimum daily needs. Most meat and dairy products are satisfactory, but many vegetable sources such as wheat and corn are *incomplete*; that is, many vegetable proteins contain too little of one or more essential amino acids to sustain the growth of laboratory animals. Wheat is low in lysine, for example, and corn is low in both lysine and tryptophan.

Using an incomplete food as the sole source of protein can cause nutritional deficiencies, particularly in children. Vegetarians must therefore be careful to adopt a varied diet that provides proteins from several sources. Legumes and nuts, for example, are useful for overcoming the deficiencies of wheat and grains. Some of the limiting amino acids found in various foods are listed in Table 15.8.

TABLE 15.8 Limiting Amino Acids in Some Foods

FOOD	LIMITING AMINO ACID
Wheat, grains	Lysine, threonine
Peas, beans, legumes	Methionine, tryptophan
Nuts, seeds	Lysine
Leafy green vegetables	Methionine

SUMMARY AND KEY WORDS

Proteins are large biomolecules consisting of **α-amino acid residues** linked together by amide, or **peptide**, bonds. Twenty amino acids are commonly found in proteins: All are α-amino acids, and all except glycine have stereochemistry similar to that of L sugars.

Determining the structure of a large polypeptide or protein requires several steps. The identity and amount of each amino acid present in a peptide can be determined by *amino acid analysis*. The peptide is first hydrolyzed to its constituent α-amino acids, which are then separated and identified. Next, the peptide is *sequenced*. **Edman degradation** by treatment with phenyl isothiocyanate cleaves off one residue from the N terminus of the peptide and forms an easily identifiable derivative of that residue. A series of Edman degradations can sequence the peptide chains up to 20 residues in length.

Peptide synthesis involves the use of protecting groups. An N-protected amino acid with a free carboxyl group is coupled using DCC to an O-protected amino acid with a free amino group. Amide formation occurs, the protecting groups are removed, and the sequence is repeated. Amines are usually protected as their *tert*-butoxycarbonyl (BOC) derivatives; acids are usually protected as esters.

Proteins are classified as either **globular** or **fibrous**, depending on their **secondary** and **tertiary structures**. Fibrous proteins such as α-keratin are tough and water-insoluble; globular proteins such as myoglobin are water-soluble and mobile within cells. Most of the 2000 or so known enzymes are globular proteins.

Enzymes are globular proteins that act as biological catalysts. Like all catalysts, enzymes speed up the rate of a reaction without themselves being changed. They are classified into six groups according to the kind of reaction they catalyze: *oxidoreductases* catalyze oxidations and reductions; *transferases* catalyze transfers of groups; *hydrolases* catalyze hydrolysis; *isomerases* catalyze isomerizations; *lyases* catalyze bond breakages; and *ligases* catalyze bond formations.

In addition to their protein part, many enzymes contain **cofactors**, which can be either metal ions or small organic molecules. If the cofactor is an organic molecule, it is called a **coenzyme**. The combination of protein (**apoenzyme**) plus coenzyme is called a **holoenzyme**. Often, the coenzyme is a **vitamin**, a small molecule that must be obtained in the diet and is required in trace amounts for proper growth.

ADDITIONAL PROBLEMS

15.18 What does the prefix "α" mean when referring to an α-amino acid?

15.19 What amino acids do these abbreviations stand for?
(a) Ser **(b)** Thr **(c)** Pro **(d)** Phe **(e)** Glu

15.20 What kinds of molecules are found in the following conjugated proteins in addition to the protein part?
(a) Nucleoproteins (b) Glycoproteins (c) Lipoproteins

15.21 Why is cysteine such an important amino acid for determining the tertiary structure of a protein?

15.22 The *endorphins* are a group of naturally occurring compounds in the brain that act to control pain. The active part of an endorphin is a pentapeptide called an *enkephalin*, which has the structure Tyr-Gly-Gly-Phe-Met. Draw the structure of this enkephalin.

15.23 What kinds of reactions do these classes of enzymes catalyze?
(a) Hydrolases (b) Lyases (c) Transferases

15.24 What kind of reaction does each of the following enzymes catalyze?
(a) A protease (b) A kinase (c) A carboxylase

15.25 Although only *S* amino acids occur in proteins, several *R* amino acids are found elsewhere in nature. For example, (*R*)-serine is found in earthworms and (*R*)-alanine is found in insect larvae. Draw Fischer projections of (*R*)-serine and (*R*)-alanine.

15.26 Draw a Fischer projection of (*S*)-proline, the only secondary amino acid.

15.27 Define these terms:
(a) Amphoteric (b) Isoelectric point (c) Peptide (d) N terminus (e) C terminus
(f) Zwitterion

15.28 Using the three-letter code names for each amino acid, write the structures of the peptides containing the following amino acids:
(a) Val, Leu, Ser (b) Ser, Leu_2, Pro

15.29 Draw these amino acids in their zwitterionic forms:
(a) Serine (b) Tyrosine (c) Threonine

15.30 Draw structures of the predominant forms of lysine and aspartic acid at pH 3.0 and pH 9.7.

15.31 At what pH would you carry out an electrophoresis experiment if you wanted to separate a mixture of histidine, serine, and glutamic acid? Explain.

15.32 Which of the following amino acids are more likely to be found on the outside of a globular protein, and which on the inside? Explain.
(a) Valine (b) Aspartic acid (c) Isoleucine (d) Lysine

15.33 Predict the product of the reaction of valine with these reagents:
(a) CH_3CH_2OH, H^+ (b) NaOH, H_2O (c) Di-*tert*-butyl dicarbonate

15.34 Write out full structures for these peptides, and indicate the positions of the amide bonds:
(a) Val-Phe-Cys (b) Glu-Pro-Ile-Leu

15.35 The amino acid threonine, (2*S*,3*R*)-2-amino-3-hydroxybutanoic acid, has two stereogenic centers and a stereochemistry similar to that of the four-carbon sugar D-threose. Draw a Fischer projection of threonine.

15.36 Draw the Fischer projection of a diastereomer of threonine (see Problem 15.35).

15.37 The amino acid analysis data in Figure 15.3 indicate that proline is not easily detected by reaction with ninhydrin. Suggest a reason.

15.38 Cytochrome *c*, an enzyme found in the cells of all aerobic organisms, plays a role in respiration. Elemental analysis of cytochrome *c* reveals it to contain 0.43% iron. What is the minimum molecular weight of this enzyme?

15.39 Draw the structure of the phenylthiohydantoin product you would expect to obtain from Edman degradation of these peptides:
(a) Val-Leu-Gly **(b)** Ala-Pro-Phe

15.40 Arginine, which contains a *guanidino* group in its side chain, is the most basic of the 20 common amino acids. How can you account for this basicity? (*Hint:* Use resonance structures to see how the protonated guanidino group is stabilized.)

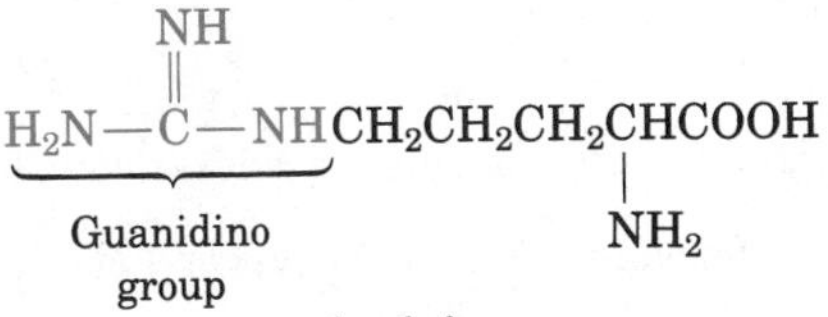

$$\underbrace{H_2N{-}\overset{\overset{\displaystyle NH}{\|}}{C}{-}NH}_{\text{Guanidino group}}CH_2CH_2CH_2\underset{\underset{\displaystyle NH_2}{|}}{C}HCOOH$$

Arginine

15.41 Show the steps involved in a synthesis of Phe-Ala-Val.

15.42 When unprotected α-amino acids are treated with dicyclohexylcarbodiimide (DCC), 2,5-diketopiperazines result. Explain.

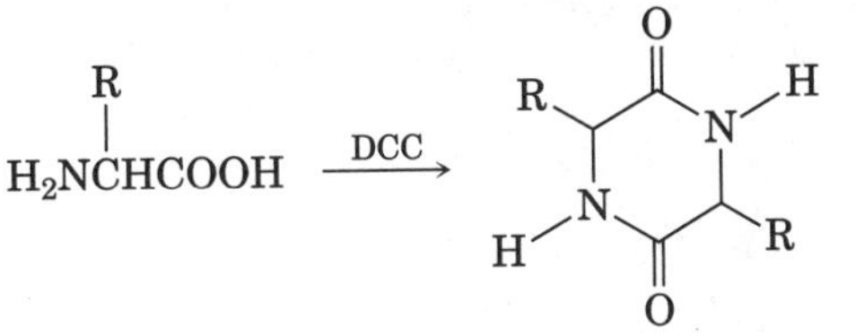

$$H_2N\overset{\overset{\displaystyle R}{|}}{C}HCOOH \xrightarrow{\text{DCC}}$$

A 2,5-diketopiperazine

15.43 Which amide bonds in the following polypeptide are cleaved by trypsin? By chymotrypsin?

Phe-Leu-Met-Lys-Tyr-Asp-Gly-Gly-Arg-Val-Ile-Pro-Tyr

15.44 Look up the structure of human insulin (Figure 15.5) and indicate where in each chain the molecule is cleaved by trypsin and by chymotrypsin.

15.45 A heptapeptide shows the composition Asp, Gly, Leu, Phe, Pro_2, Val on amino acid analysis. Edman degradation shows glycine to be the N-terminal group. Acidic hydrolysis gives the following fragments:

Val-Pro-Leu Gly Gly-Asp-Phe-Pro Phe-Pro-Val

Propose a structure for the starting heptapeptide.

15.46 Give the amino acid sequence of hexapeptides that produce these fragments on partial acid hydrolysis:
(a) Arg, Gly, Ile, Leu, Pro, Val gives Pro-Leu-Gly, Arg-Pro, Gly-Ile-Val
(b) Asp, Leu, Met, Trp, Val_2 gives Val-Leu, Val-Met-Trp, Trp-Asp-Val

15.47 How can you account for the fact that proline is never encountered in a protein α-helix? The α-helical segments of myoglobin and other proteins stop when a proline residue is encountered in the chain.

15.48 Draw as many resonance forms as you can for the purple anion obtained by reaction of ninhydrin with an amino acid:

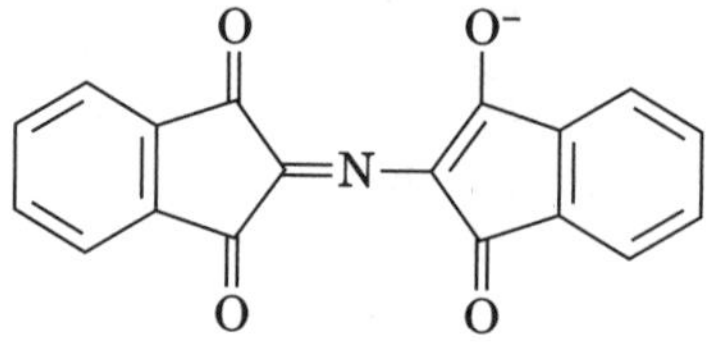

15.49 A nonapeptide gives the following fragments when cleaved by chymotrypsin and by trypsin:

Trypsin cleavage: Val-Val-Pro-Tyr-Leu-Arg and Ser-Ile-Arg

Chymotrypsin cleavage: Leu-Arg and Ser-Ile-Arg-Val-Val-Pro-Tyr

What is the structure of the nonapeptide?

15.50 Oxytocin, a nonapeptide hormone secreted by the pituitary gland, stimulates uterine contraction and lactation during childbirth. Its sequence was determined from the following evidence:

1. Oxytocin is a cyclic peptide containing a disulfide bridge between two cysteine residues.
2. When the disulfide bridge is reduced, oxytocin has the constitution Asn, Cys_2, Gln, Gly, Ile, Leu, Pro, Tyr.
3. Partial hydrolysis of reduced oxytocin yields seven fragments:

Asp-Cys	Ile-Glu	Cys-Tyr	Cys-Pro-Leu
Leu-Gly	Tyr-Ile-Glu	Glu-Asp-Cys	

4. Gly is the C-terminal group.
5. Both Glu and Asp are present as their side-chain amides (Gln and Asn) rather than as free side-chain acids.

On the basis of this evidence, what is the amino acid sequence of reduced oxytocin? What is the structure of oxytocin?

15.51 *Aspartame*, a nonnutritive sweetener marketed under the trade name NutraSweet, is the methyl ester of a simple dipeptide, Asp-Phe-OCH_3.

(a) Draw the full structure of aspartame.
(b) The isoelectric point of aspartame is 5.9. Draw the principal structure present in aqueous solution at this pH.
(c) Draw the principal form of aspartame present at physiological pH 7.6.
(d) Show the products of hydrolysis on treatment of aspartame with H_3O^+.

16 BIOMOLECULES: LIPIDS AND NUCLEIC ACIDS

CHAPTER

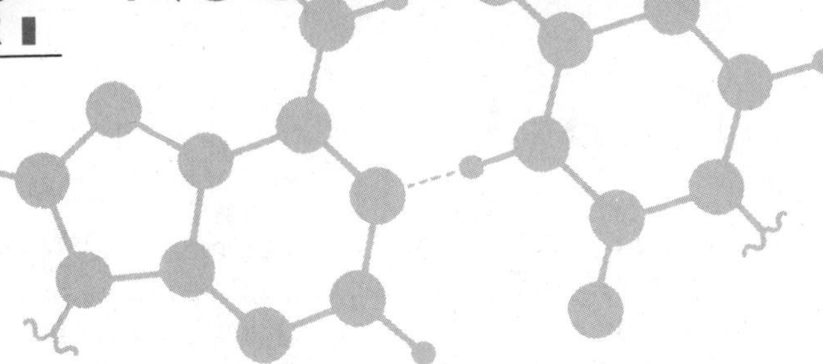

In the previous two chapters, we've discussed the organic chemistry of carbohydrates and proteins, two of the four major classes of biomolecules. Let's now look at the two remaining classes: *lipids* and *nucleic acids*. Though chemically quite different from one another, all four classes are essential for life.

16.1 LIPIDS

LIPID
Naturally occurring substance that can be isolated from plants or animals by extraction with a nonpolar organic solvent

Lipids are naturally occurring organic substances that can be isolated from cells and tissues by extraction with nonpolar organic solvents. Since they usually have large hydrocarbon portions in their structures, lipids are insoluble in water but soluble in organic solvents. Note that this definition differs from those used for carbohydrates and proteins in that lipids are defined by a physical property (solubility) rather than by their structure.

Lipids are further classified into two general types: those like fats and waxes, which contain ester linkages and can be hydrolyzed; and those like cholesterol and other steroids, which don't have ester linkages and can't be hydrolyzed.

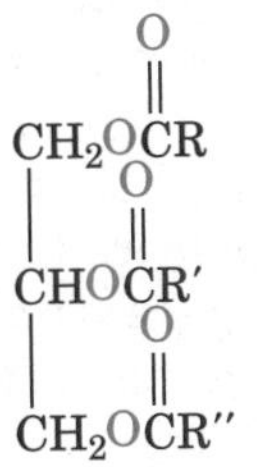

$$\begin{array}{l} CH_2OC(=O)R \\ | \\ CHOC(=O)R' \\ | \\ CH_2OC(=O)R'' \end{array}$$

Animal fat, an ester
(R, R′, R″ = $C_{11} - C_{19}$ chains)

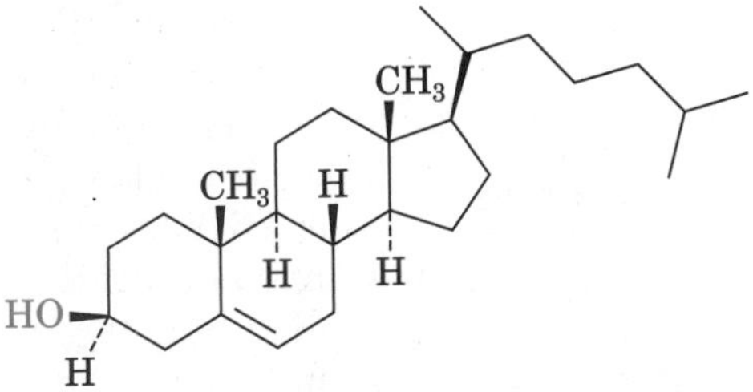

CH_3 CH_3 H H H HO H

Cholesterol

PROBLEM 16.1 Beeswax contains, among other things, a lipid with the structure $CH_3(CH_2)_{20}COO(CH_2)_{27}CH_3$. What products would you obtain by reaction of this lipid with aqueous NaOH followed by acidification?

16.2 FATS AND OILS

TRIACYLGLYCEROL
Triester of glycerol with three fatty acids; an animal fat or vegetable oil

Animal fats and vegetable oils are the most widely occurring lipids. Although they look different—animal fats such as butter and lard are solids, whereas vegetable oils such as corn oil and peanut oil are liquids—their structures are closely related. Chemically, fats and oil are **triacylglycerols** (also called *triglycerides*), triesters of glycerol with three long-chain carboxylic acids. Hydrolysis of a fat or oil with aqueous sodium hydroxide yields glycerol and three fatty acids:

$$\begin{array}{l} CH_2O-\overset{O}{\overset{\|}{C}}-R \\ | \\ CHO-\overset{O}{\overset{\|}{C}}-R' \\ | \\ CH_2O-\overset{O}{\overset{\|}{C}}-R'' \end{array} \xrightarrow[2.\ H_3O^+]{1.\ ^-OH} \begin{array}{l} CH_2OH \\ | \\ CHOH \\ | \\ CH_2OH \end{array} + \begin{array}{l} RCOOH \\ R'COOH \\ R''COOH \end{array}$$

A fat — **Glycerol** — Fatty acids

The fatty acids obtained by hydrolysis of triacylglycerols are generally unbranched and contain an even number of carbon atoms between 12 and 20. If double bonds are present, they usually have Z (cis) geometry. The three fatty acids of a specific molecule need not be the same, and a fat or oil from a given source is likely to be a complex mixture of many different triacylglycerols. Table 16.1 lists some of the commonly occurring fatty acids, and Table 16.2 lists the approximate composition of fats and oils from various sources.

POLYUNSATURATED FATTY ACID (PUFA)
Fatty acid with more than one double bond in its chain

About 40 different fatty acids occur naturally. Palmitic acid (C_{16}) and stearic acid (C_{18}) are the most abundant saturated fatty acids; oleic and linoleic acids (both C_{18}) are the most abundant unsaturated ones. Oleic acid is monounsaturated since it has only one double bond, whereas linoleic, linolenic, and arachidonic acids are **polyunsaturated fatty acids**, or **PUFA's**, because they have more than one double bond. Linoleic and linolenic acids occur naturally in cream, and are essential in the human diet; infants grow poorly and develop skin lesions if fed a diet of nonfat milk for prolonged periods.

TABLE 16.1 Structures of Some Common Fatty Acids

NAME	CARBONS	STRUCTURE	MELTING POINT (°C)
SATURATED			
Lauric	12	$CH_3(CH_2)_{10}COOH$	44
Myristic	14	$CH_3(CH_2)_{12}COOH$	58
Palmitic	16	$CH_3(CH_2)_{14}COOH$	63
Stearic	18	$CH_3(CH_2)_{16}COOH$	70
Arachidic	20	$CH_3(CH_2)_{18}COOH$	75
UNSATURATED			
Palmitoleic	16	$CH_3(CH_2)_5CH{=}CH(CH_2)_7COOH$ (cis)	32
Oleic	18	$CH_3(CH_2)_7CH{=}CH(CH_2)_7COOH$ (cis)	4
Ricinoleic	18	$CH_3(CH_2)_5CH(OH)CH_2CH{=}CH(CH_2)_7COOH$ (cis)	5
Linoleic	18	$CH_3(CH_2)_4CH{=}CHCH_2CH{=}CH(CH_2)_7COOH$ (cis,cis)	−5
Arachidonic	20	$CH_3(CH_2)_4(CH{=}CHCH_2)_4CH_2CH_2COOH$ (all cis)	−50

TABLE 16.2 Approximate Fatty Acid Composition of Some Common Fats and Oils

	SATURATED FATTY ACIDS (%)				UNSATURATED FATTY ACIDS (%)		
SOURCE	C_{12} LAURIC	C_{14} MYRISTIC	C_{16} PALMITIC	C_{18} STEARIC	C_{18} OLEIC	C_{18} RICINOLEIC	C_{18} LINOLEIC
ANIMAL FAT							
Lard	—	1	25	15	50	—	6
Butter	2	10	25	10	25	—	5
Human fat	1	3	25	8	46	—	10
Whale blubber	—	8	12	3	35	—	10
VEGETABLE OIL							
Coconut	50	18	8	2	6	—	1
Corn	—	1	10	4	35	—	45
Olive	—	1	5	5	80	—	7
Peanut	—	—	7	5	60	—	20
Linseed	—	—	5	3	20	—	20
Castor bean	—	—	—	1	8	85	4

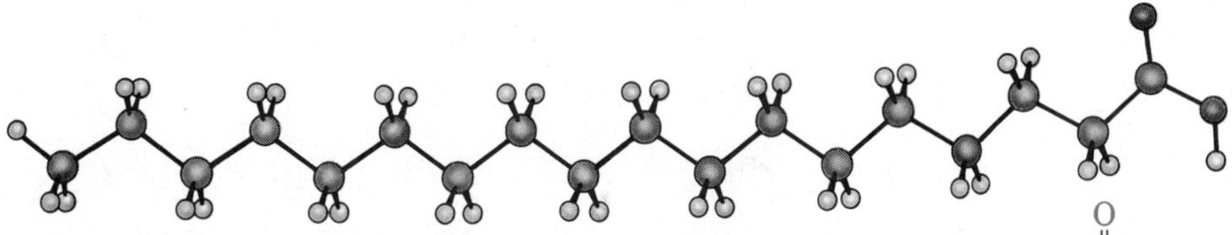

$$CH_3CH_2CH_2CH_2CH_2CH_2CH_2CH_2CH_2CH_2CH_2CH_2CH_2CH_2CH_2CH_2CH_2\overset{\overset{O}{\|}}{C}OH$$

Stearic acid, a saturated acid

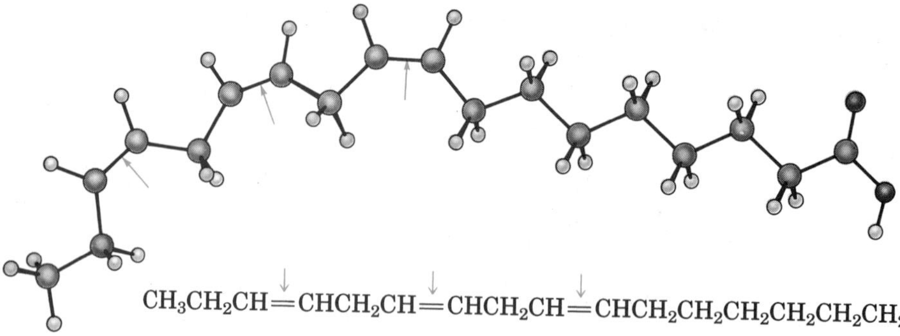

$$CH_3CH_2CH{=}CHCH_2CH{=}CHCH_2CH{=}CHCH_2CH_2CH_2CH_2CH_2CH_2CH_2\overset{O}{\overset{\|}{C}}OH$$

Linolenic acid, a polyunsaturated fatty acid (PUFA)

The data in Table 16.1 show that unsaturated fatty acids generally have lower melting points than their saturated counterparts, a trend that's also true for triacylglycerols. Since vegetable oils generally have a higher proportion of unsaturated to saturated fatty acids than animal fats do (Table 16.2), they have lower melting points. This behavior is due to the fact that saturated fats have a uniform shape that allows them to pack together easily in a crystal. In vegetable oils, however, the carbon–carbon double bonds in unsaturated fatty acids such as linolenic acid introduce bends and kinks into the hydrocarbon chains, making crystal formation difficult. The more double bonds there are, the harder it is for the molecule to crystallize, and the lower the melting point of the oil.

The carbon–carbon double bonds in vegetable oils can be reduced by catalytic hydrogenation (Section 4.6) to produce saturated solid or semi-solid fats. Margarine and solid cooking fats such as Crisco are produced by hydrogenating soybean, peanut, or cottonseed oil until the preferred consistency is obtained.

PRACTICE PROBLEM 16.1 Draw the structure of glyceryl tripalmitate, a typical fat molecule.

SOLUTION Glyceryl tripalmitate is the triester of glycerol with three molecules of palmitic acid, $CH_3(CH_2)_{14}COOH$

$$\begin{array}{l} CH_2O\overset{O}{\overset{\|}{C}}CH_2CH_2CH_2CH_2CH_2CH_2CH_2CH_2CH_2CH_2CH_2CH_2CH_2CH_2CH_3 \\ | \\ CHO\overset{O}{\overset{\|}{C}}CH_2CH_2CH_2CH_2CH_2CH_2CH_2CH_2CH_2CH_2CH_2CH_2CH_2CH_2CH_3 \\ | \\ CH_2O\overset{O}{\overset{\|}{C}}CH_2CH_2CH_2CH_2CH_2CH_2CH_2CH_2CH_2CH_2CH_2CH_2CH_2CH_2CH_3 \end{array}$$

Glyceryl tripalmitate

PROBLEM 16.2 Draw structures of the following compounds. Which would you expect to have the higher melting point?
(a) Glyceryl trioleate **(b)** Glyceryl monooleate distearate

PROBLEM 16.3 Fats and oils can be either optically active or optically inactive, depending on their structures. Draw the structure of an optically active fat that gives 2 equiv palmitic acid and 1 equiv stearic acid on hydrolysis. Draw the structure of an optically inactive fat that gives the same products on hydrolysis.

16.3 SOAPS

Soap has been known since at least 600 BC, when the Phoenicians prepared a curdy material by boiling goat fat with extracts of wood ash. The cleansing properties of soap weren't generally recognized, however, and the use of soap didn't become widespread until the eighteenth century. Chemically, soap is a mixture of the sodium or potassium salts of long-chain fatty acids produced by hydrolysis (*saponification*) of animal fat with alkali:

$$\underset{\substack{\text{A fat} \\ (R = C_{11}\text{–}C_{19}\text{ aliphatic chains})}}{\begin{array}{l} CH_2OC(=O)R \\ | \\ CHOC(=O)R \\ | \\ CH_2OC(=O)R \end{array}} \xrightarrow[H_2O]{NaOH} \underset{\text{Soap}}{3\ RC(=O)O^-\ Na^+} + \underset{\textbf{Glycerol}}{\begin{array}{l} CH_2OH \\ | \\ CHOH \\ | \\ CH_2OH \end{array}}$$

Crude soap curds contain glycerol and excess alkali as well as soap but can be purified by boiling with water and adding NaCl to precipitate the pure sodium carboxylate salts. The smooth soap that results is dried, perfumed, and pressed into bars. Dyes are added for colored soaps, antiseptics are added for medicated soaps, pumice is added for scouring soaps, and air is blown in for soaps that float.

HYDROPHILIC
Attracted to water (and repelled by hydrocarbons)

HYDROPHOBIC
Repelled by water (and attracted to hydrocarbons)

Soaps act as cleansers because the two ends of a soap molecule are so different. The carboxylate end of the long-chain molecule is ionic and therefore **hydrophilic** (water-loving); it tries to dissolve in water. The long aliphatic chain portion of the molecule, however, is **hydrophobic** (water-fearing); it tries to avoid water and dissolve in grease. The net effect of these two opposing tendencies is that soaps are attracted to both grease and water.

MICELLE
Spherical cluster formed by aggregation of soap molecules in water

When soaps are dispersed in water, the long hydrocarbon tails cluster together into a hydrophobic ball, while the ionic heads on the surface of the cluster stick out into the water layer. These spherical clusters, called **micelles**, are shown schematically in Figure 16.1. Grease and oil droplets are solubilized in water when they become coated by the hydrophobic nonpolar tails of soap molecules in the center of micelles. Once solubilized, the grease and dirt can be rinsed away.

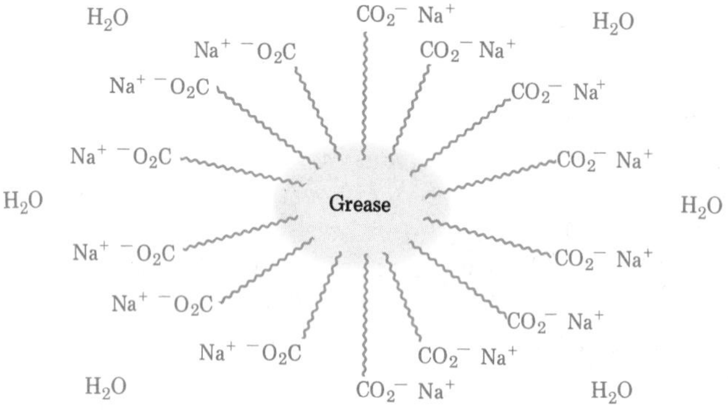

FIGURE 16.1 A soap micelle solubilizing a grease particle in water.

Soaps make life much more pleasant than it would otherwise be, but they also have drawbacks. In hard water, which contains metal ions, soluble sodium carboxylates are converted into insoluble calcium and magnesium salts, leaving the familiar ring of scum around bathtubs and the gray tinge on clothes. Chemists have circumvented these problems by synthesizing a class of synthetic detergents based on salts of long-chain alkylbenzenesulfonic acids. The principle of synthetic detergents is identical to that of soaps: The alkylbenzene end of the molecule is attracted to grease, but the sulfonate salt end is ionic and is attracted to water. Unlike soaps, though, sulfonate detergents don't form insoluble metal salts in hard water and don't leave an unpleasant scum.

A synthetic detergent

PROBLEM 16.4 Draw the structure of magnesium oleate, one of the components of bathtub scum.

PROBLEM 16.5 Formulate the saponification reaction of glyceryl monopalmitate dioleate with aqueous NaOH.

16.4 PHOSPHOLIPIDS

PHOSPHOLIPID
Lipid that contains an ester link to phosphoric acid, H_3PO_4

PHOSPHOGLYCERIDE
Phospholipid in which glycerol has ester links to two fatty acids and to phosphoric acid

Just as waxes, fats, and oils are esters of carboxylic acids, **phospholipids** are esters of phosphoric acid, H_3PO_4. There are two main kinds of phospholipids: *phosphoglycerides* and *sphingolipids*.

Phosphoglycerides are closely related to fats and oils in that they contain a glycerol backbone linked by ester bonds to two fatty acids and one phosphoric acid. Although the fatty acid residues can be any of the C_{12}–C_{20} units normally present in fats, the acyl group at C1 is usually saturated, and that at C2 is usually unsaturated. The phosphate group at C3 is also bound by a separate ester link to an amino alcohol such as choline, $HOCH_2CH_2\overset{+}{N}(CH_3)_3$, or ethanolamine, $HOCH_2CH_2NH_2$.

The most important phosphoglycerides are the *lecithins* and the *cephalins*. Note that these compounds are chiral and that they have the L, or *R*, configuration at C2.

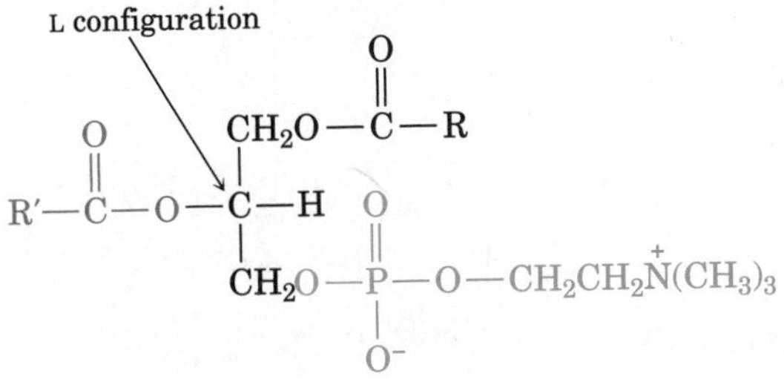

L configuration

$$R'-\overset{O}{\overset{\|}{C}}-O-\underset{CH_2O-\overset{O}{\overset{\|}{P}}(O^-)-O-CH_2CH_2\overset{+}{N}(CH_3)_3}{\overset{CH_2O-\overset{O}{\overset{\|}{C}}-R}{C}}-H$$

Phosphatidylcholine, a lecithin

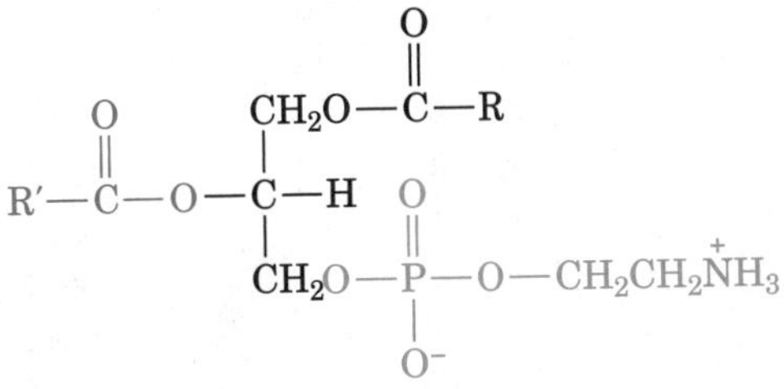

$$R'-\overset{O}{\overset{\|}{C}}-O-\underset{CH_2O-\overset{O}{\overset{\|}{P}}(O^-)-O-CH_2CH_2\overset{+}{N}H_3}{\overset{CH_2O-\overset{O}{\overset{\|}{C}}-R}{C}}-H$$

Phosphatidylethanolamine, a cephalin

LIPID BILAYER
Aggregation of phospholipids that makes up cell walls

Found widely in plant and animal tissues, phosphoglycerides are the major lipid component of cell membranes (approximately 40%). Like soaps, phosphoglycerides have a long, nonpolar hydrocarbon tail bound to a polar ionic head (the phosphate group). Cell membranes are composed mostly of phosphoglycerides oriented into a **lipid bilayer** about 50 Å thick. As shown in Figure 16.2, the hydrophobic tails aggregate in the center of the bilayer in much the same way that soap tails aggregate into the center of a micelle (Figure 16.1). The bilayer thus forms an effective barrier to the passage of ions and other components into and out of the cell.

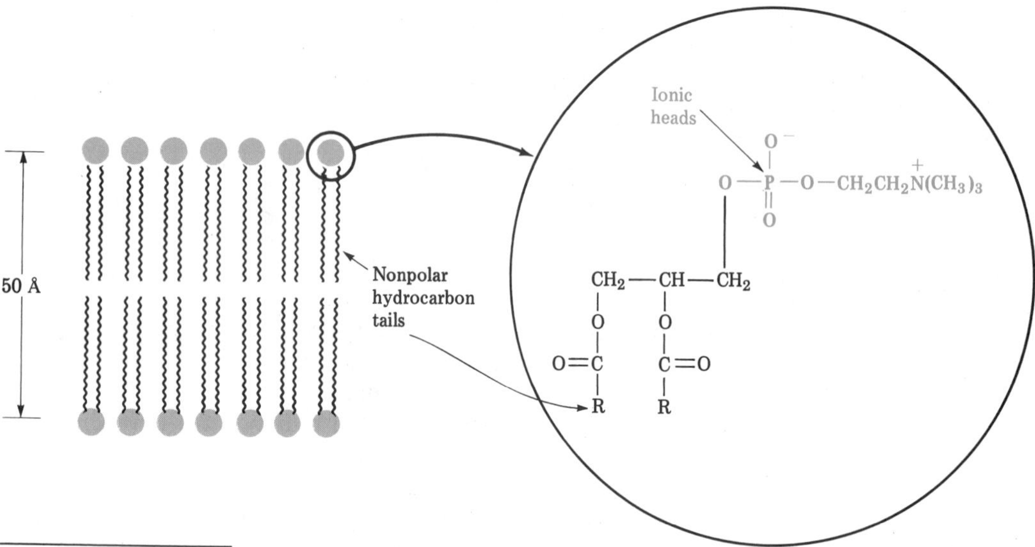

FIGURE 16.2 Aggregation of phosphoglycerides into the lipid bilayer that composes cell membranes.

SPHINGOLIPID
Phospholipid based on a sphingosine backbone rather than on glycerol

The second major group of phospholipids is comprised of the **sphingolipids**. These substances, which have *sphingosine* or a related dihydroxyamine as their backbones, are constituents of plant and animal cell membranes. They are particularly abundant in brain and nerve tissue, where *sphingomyelins* are a major constituent of the coating around nerve fibers.

$$\begin{array}{l} CH_2OH \\ | \\ CHNH_2 \\ | \\ CHOH \\ | \\ CH{=}CH(CH_2)_{12}CH_3 \end{array}$$

Sphingosine

$$\begin{array}{l} \qquad\qquad\; O \\ \qquad\qquad\; \| \\ CH_2O-P-OCH_2CH_2\overset{+}{N}(CH_3)_3 \\ | \qquad\quad\;\; | \\ | \qquad\quad\; O^- \\ CHNHCO(CH_2)_{16-24}CH_3 \\ | \\ CHOH \\ | \\ CH{=}CH(CH_2)_{12}CH_3 \end{array}$$

Sphingomyelin, a sphingolipid

16.5 STEROIDS

STEROID
Lipid whose structure is based on a characteristic tetracyclic ring system

In addition to fats and phospholipids, the lipid extracts of plants and animals also contain *steroids*. A **steroid** is an organic molecule whose structure is based on the tetracyclic ring system shown on page 481. The four rings are designated A, B, C, and D, beginning at the lower left, and the carbon atoms are numbered beginning in the A ring. The three six-membered rings (A, B, and C) adopt minimum-energy chair conformations, but are constrained by their rigid conformations from undergoing the usual cyclohexane ring-flips (Section 2.11).

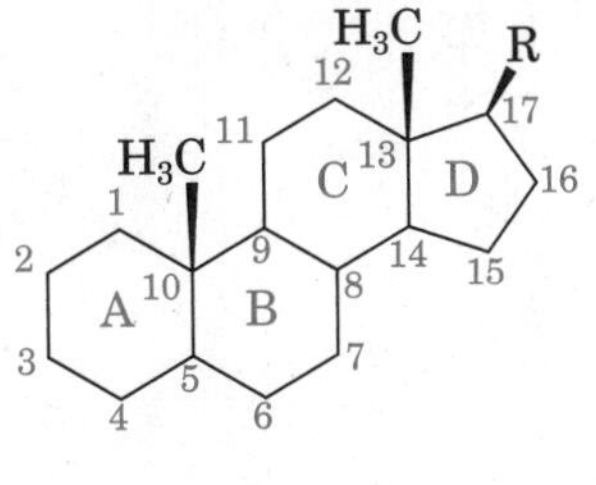

A steroid
(R = various side chains)

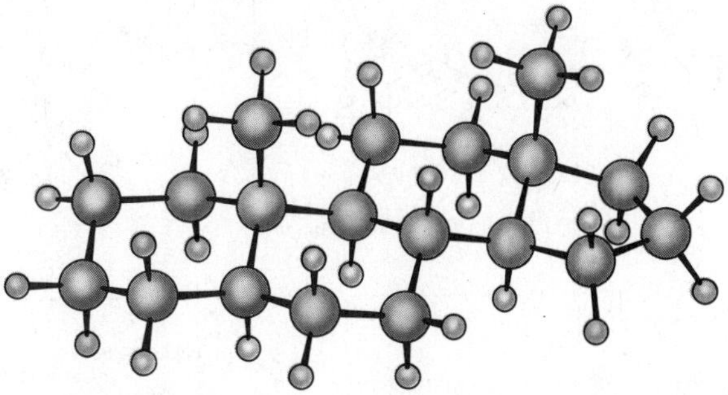

HORMONE
Chemical messenger secreted by a specific gland and carried through the bloodstream to affect a target tissue

In humans, most steroids function as **hormones**, chemical messengers that are secreted by glands and carried through the bloodstream to target tissues. There are two main classes of steroid hormones: the *sex hormones*, which control maturation and reproduction, and the *adrenocortical hormones*, which regulate a variety of metabolic processes.

Sex Hormones

ANDROGEN
Male steroid sex hormone

Testosterone and *androsterone* are the two most important male sex hormones, or **androgens**. Androgens are responsible for the development of male secondary sex characteristics during puberty and for promoting tissue and muscle growth. Both are synthesized in the testes from cholesterol.

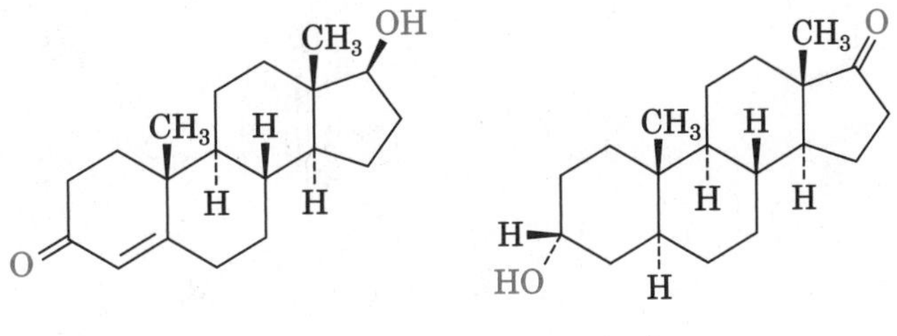

Testosterone **Androsterone**

Androgens

ESTROGEN
Female steroid sex hormone

Estrone and *estradiol* are the two most important female sex hormones, or **estrogens**. Synthesized in the ovaries from testosterone, estrogenic hormones are responsible for the development of female secondary sex characteristics and for regulation of the menstrual cycle. Note that both have a benzene-like aromatic A ring. In addition, another kind of sex hormone called a *progestin* is essential for preparing the uterus for implantation of a fertilized ovum during pregnancy. *Progesterone* is the most important progestin.

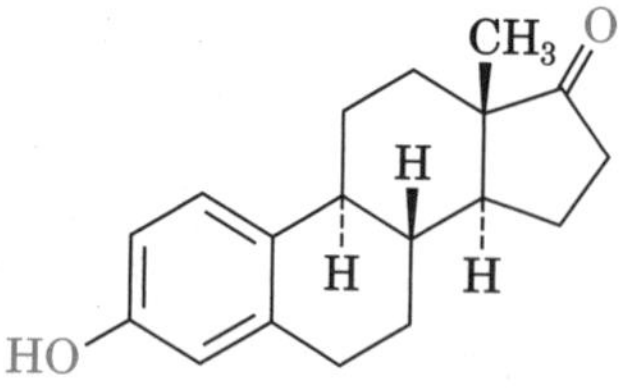

Estrone

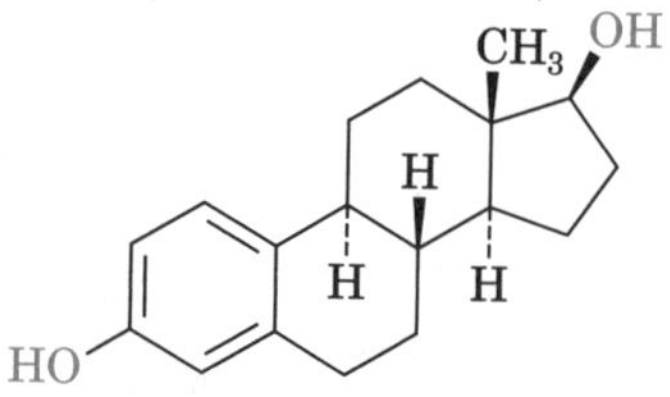

Estradiol

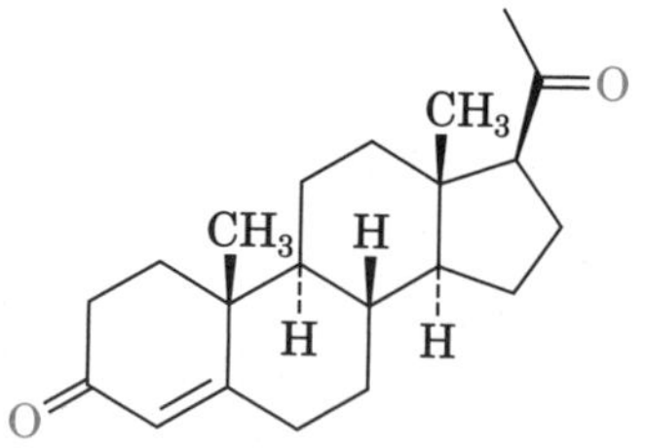

Progesterone
(a progestin)

Estrogens

Adrenocortical Hormones

Adrenocortical steroids are secreted by the adrenal glands, small organs located near the upper end of each kidney. There are two types of adrenocortical steroids, called *mineralocorticoids* and *glucocorticoids*. Mineralocorticoids such as *aldosterone* control tissue swelling by regulating cellular salt balance between Na^+ and K^+. Glucocorticoids such as *hydrocortisone* are involved in the regulation of glucose metabolism and in the control of inflammation. Glucocorticoid ointments are widely used to bring down the swelling from exposure to poison oak or poison ivy.

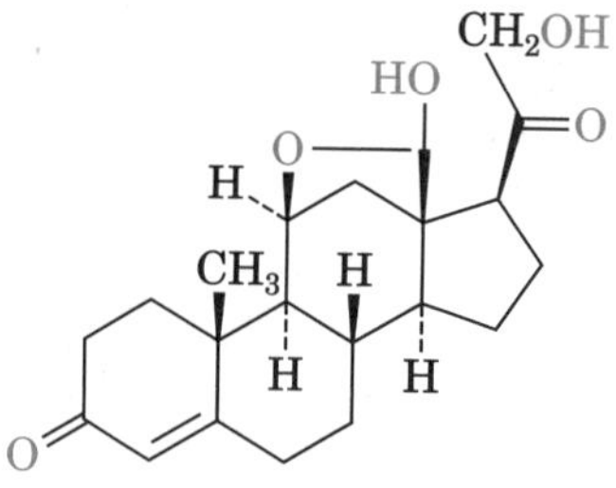

Aldosterone
(a mineralocorticoid)

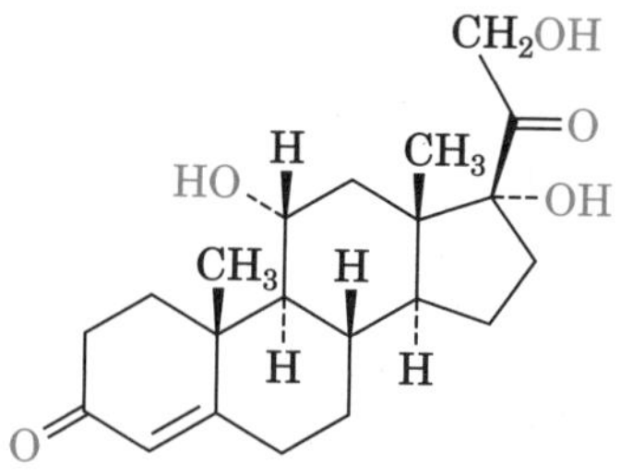

Hydrocortisone
(a glucocorticoid)

Synthetic Steroids

In addition to the many hundreds of steroids isolated from plants and animals, thousands more have been synthesized in pharmaceutical laboratories in the search for new drugs. The idea is to start with a natural hormone, carry out a chemical modification of the structure, and then see what biological properties the modified steroid has.

Among the best-known synthetic steroids are the oral contraceptive and anabolic agents. Most birth-control pills are a mixture of two compounds, a synthetic estrogen such as *ethynylestradiol* and a synthetic progestin such as *norethindrone*. Anabolic steroids such as *stanozolol*, detected in several athletes during the 1988 Olympics, are synthetic androgens that mimic the tissue-building effects of natural testosterone.

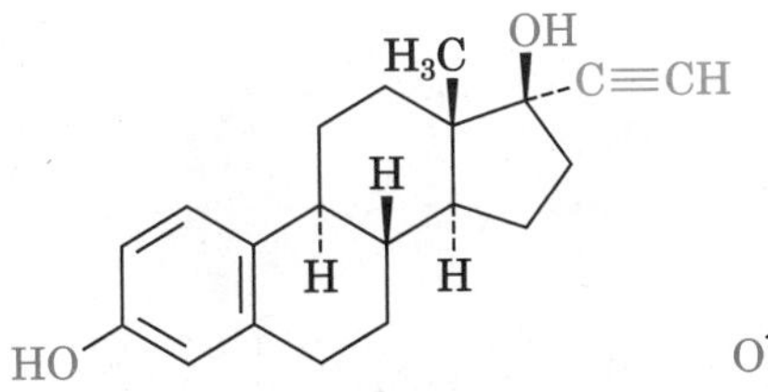

Ethynylestradiol
(a synthetic estrogen)

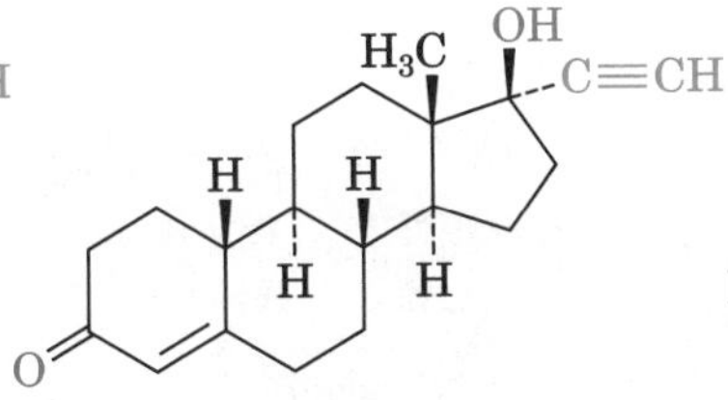

Norethindrone
(a synthetic progestin)

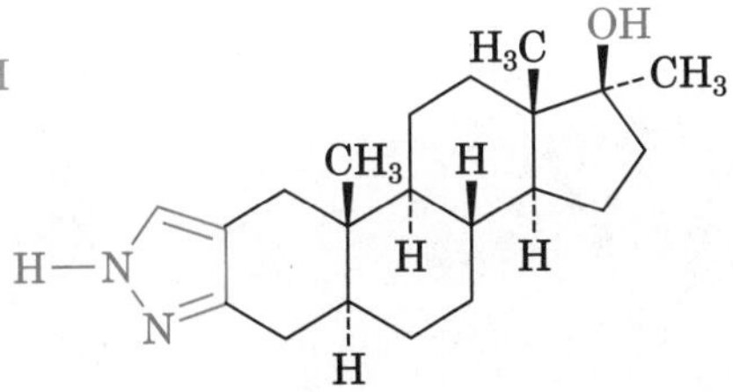

Stanozolol
(an anabolic agent)

PROBLEM 16.6 Look at the structure of cholesterol shown on the first page of this chapter and tell whether the hydroxyl group is axial or equatorial.

PROBLEM 16.7 Look at the structure of progesterone and identify all the functional groups in the molecule.

PROBLEM 16.8 Look at the structures of estradiol and ethynylestradiol and point out the differences. What common structural feature do they share that makes both estrogens?

16.6 NUCLEIC ACIDS

NUCLEOTIDE
Building block for the construction of nucleic acids, consisting of phosphoric acid, a pentose sugar, and a heterocyclic amine base

NUCLEOSIDE
The hydrolysis product of a nucleotide, consisting of a pentose sugar bonded to a heterocyclic amine base

The **nucleic acids, deoxyribonucleic acid (DNA)** and **ribonucleic acid (RNA)**, are the chemical carriers of a cell's genetic information. Coded in a cell's DNA is all the information that determines the nature of the cell, controls cell growth and division, and directs biosynthesis of the enzymes and other proteins required for all cellular functions.

Just as proteins are polymers made up of amino acid units, nucleic acids are polymers made up of individual building blocks called **nucleotides** linked together to form a long chain. Each nucleotide is composed of a **nucleoside** plus phosphoric acid, H_3PO_4, and each nucleoside is composed of a simple aldopentose sugar plus a heterocyclic amine base (Section 12.7).

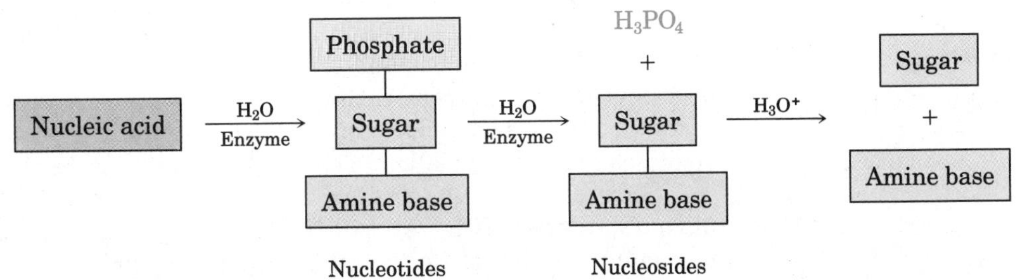

The sugar component in RNA is ribose, and the sugar in DNA is 2-deoxyribose (*2-deoxy* means that oxygen is missing from C2 of ribose).

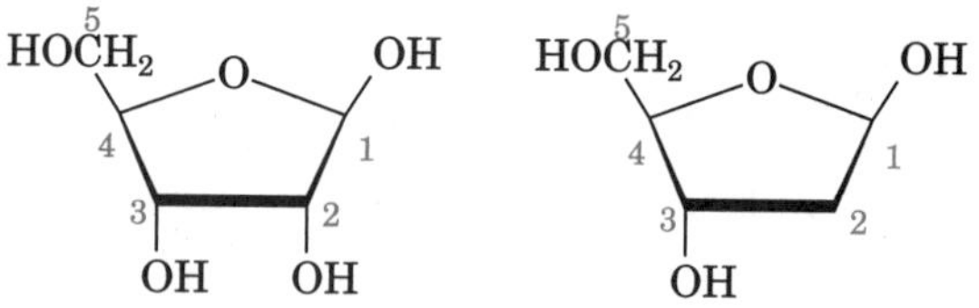

Ribose **2-Deoxyribose**

There are four different heterocyclic amine bases in DNA. Two are substituted *purines* (adenine and guanine), and two are substituted *pyrimidines* (cytosine and thymine). Adenine, guanine, and cytosine also occur in RNA, but thymine is replaced in RNA by a different pyrimidine base called uracil.

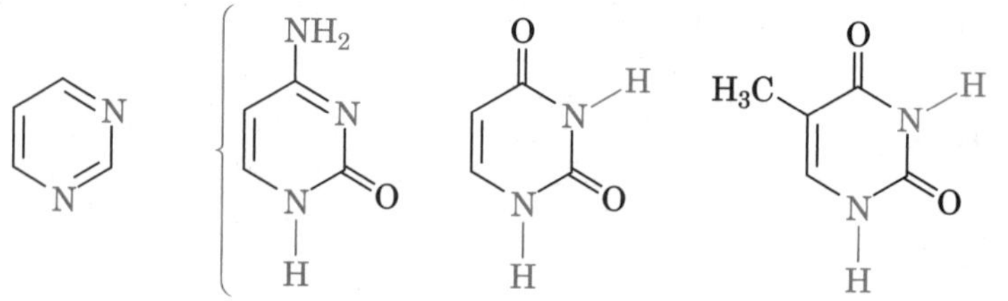

Pyrimidine **Cytosine** **Uracil (RNA)** **Thymine (DNA)**

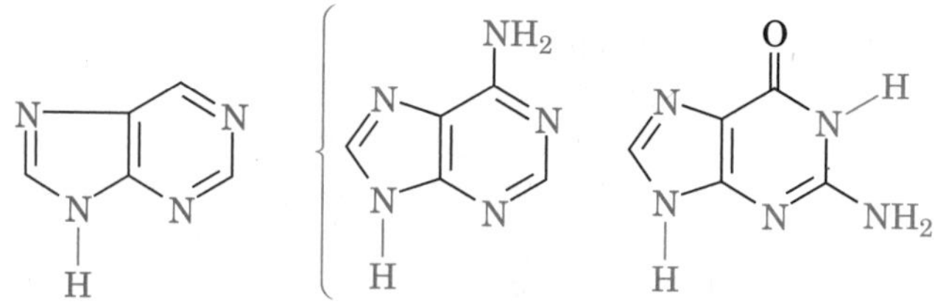

Purine **Adenine** **Guanine**

In both DNA and RNA, the heterocyclic amine base is bonded to C1′ of the sugar, and the phosphoric acid is bonded by a phosphate ester linkage to the C5′ sugar position. Thus, nucleosides and nucleotides have the general structure shown in Figure 16.3. (In discussions of RNA and DNA, numbers with a prime refer to positions on the sugar component of a nucleotide; numbers without a prime refer to positions on the heterocyclic amine base.) The complete structures of all four deoxyribonucleotides and all four ribonucleotides are shown in Figure 16.4, p. 486.

Though chemically similar, DNA and RNA are different in size and have different roles in the cell. Molecules of DNA are enormous. With molecular weights of up to 150 billion and lengths of up to 12 cm, they are found mostly in the nucleus of the cell. Molecules of RNA, by contrast, are much smaller (as low as 35,000 mol wt) and are found mostly outside the cell nucleus. We'll consider the two kinds of nucleic acids separately, beginning with DNA.

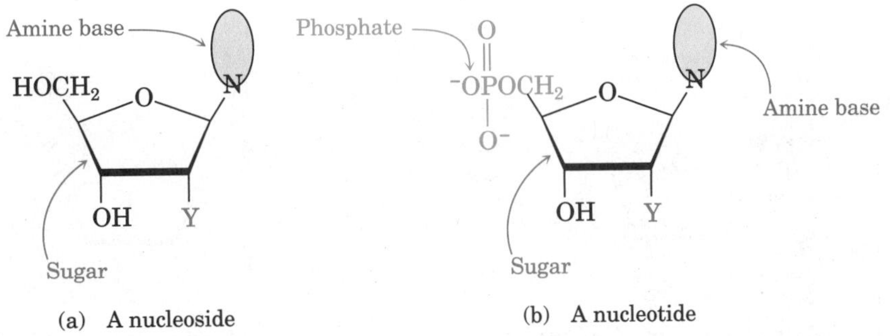

FIGURE 16.3 General structures of (a) a nucleoside and (b) a nucleotide. When Y = H, the sugar is deoxyribose; when Y = OH, the sugar is ribose.

16.7 STRUCTURE OF DNA

3′ END
End of a nucleic acid chain that has a free sugar hydroxyl group

5′ END
End of a nucleic acid chain that has a phosphoric acid unit

Nucleotides join together in DNA by forming a phosphate ester bond between the 5′-phosphate component of one nucleotide and the 3′-hydroxyl on the sugar component of another nucleotide. One end of the nucleic acid polymer has a free hydroxyl at C3′ (called the **3′ end**), and the other end has a phosphoric acid residue at C5′ (the **5′ end**).

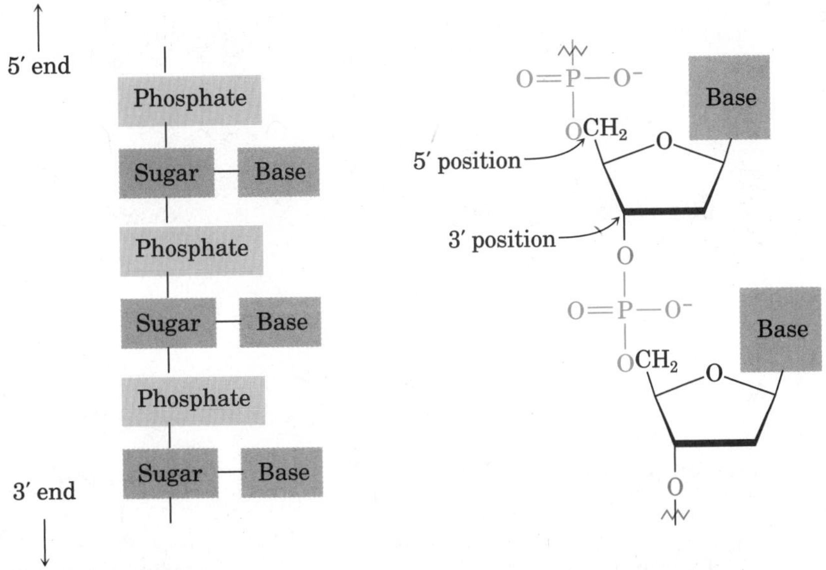

Just as the structure of a protein depends on the sequence in which individual amino acids are connected, the structure of a nucleic acid depends on the sequence of individual nucleotides. To carry the analogy

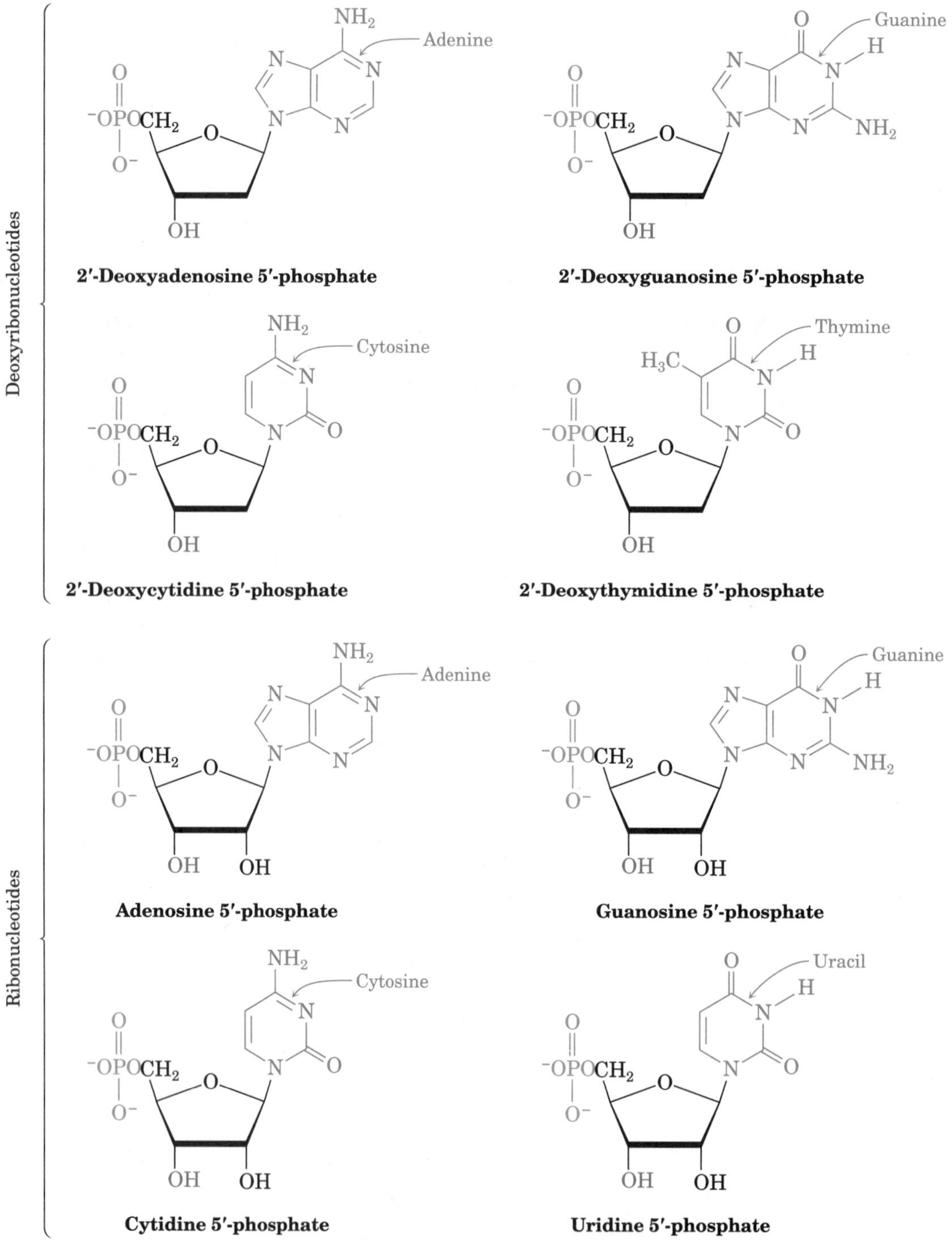

Deoxyribonucleotides

Adenine

2′-Deoxyadenosine 5′-phosphate

Guanine

2′-Deoxyguanosine 5′-phosphate

Cytosine

2′-Deoxycytidine 5′-phosphate

Thymine

2′-Deoxythymidine 5′-phosphate

Ribonucleotides

Adenine

Adenosine 5′-phosphate

Guanine

Guanosine 5′-phosphate

Cytosine

Cytidine 5′-phosphate

Uracil

Uridine 5′-phosphate

FIGURE 16.4 Structures of the four deoxyribonucleotides and the four ribonucleotides.

further, just as a protein has a polyamide backbone with different side chains attached to it, a nucleic acid has an alternating sugar–phosphate backbone with different amine base side chains attached at regular intervals (Figure 16.5).

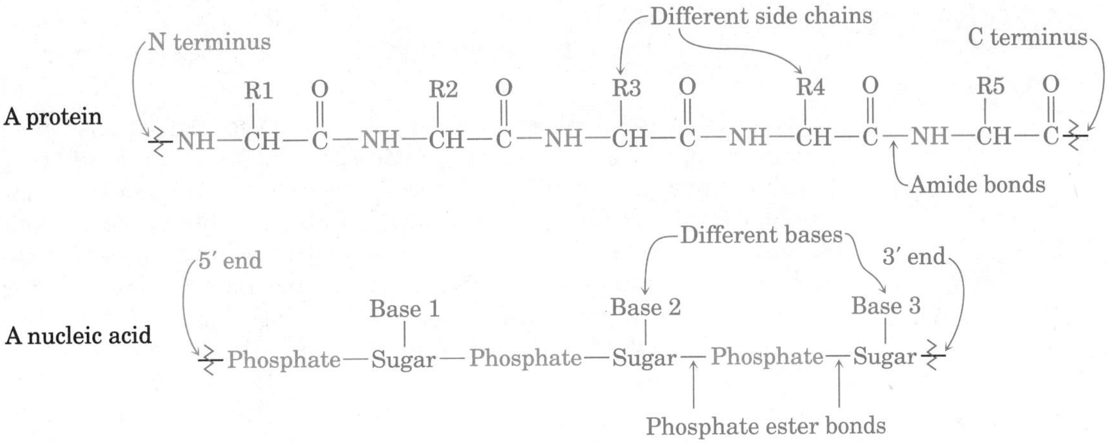

A protein

N terminus — Different side chains — C terminus

R1, R2, R3, R4, R5

–NH—CH—C(=O)—NH—CH—C(=O)—NH—CH—C(=O)—NH—CH—C(=O)—NH—CH—C(=O)–

Amide bonds

A nucleic acid

5′ end — Different bases — 3′ end

Base 1, Base 2, Base 3

–Phosphate—Sugar—Phosphate—Sugar—Phosphate—Sugar–

Phosphate ester bonds

FIGURE 16.5

The sequence of nucleotides is described by starting at the 5′ end and identifying the bases in order of occurrence. Rather than write the full name of each nucleotide, though, it's easier to use abbreviations: A for adenosine, T for thymine, G for guanosine, and C for cytidine. Thus, a typical sequence might be written as T-A-G-G-C-T.

PRACTICE PROBLEM 16.2 Draw the full structure of the DNA dinucleotide C-T.

SOLUTION

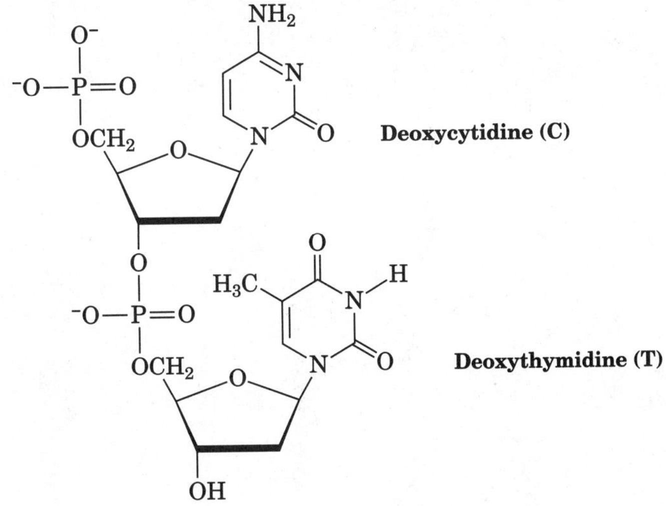

O^-, $^-O—P=O$, OCH_2, NH_2, N, O

Deoxycytidine (C)

O, $^-O—P=O$, OCH_2, H_3C, N, H, O, OH

Deoxythymidine (T)

PROBLEM 16.9 Draw the full structure of the DNA dinucleotide A-G.

PROBLEM 16.10 Draw the full structure of the RNA dinucleotide U-A.

16.8 BASE PAIRING IN DNA: THE WATSON–CRICK MODEL

Samples of DNA isolated from different tissues of the same species have the same proportions of heterocyclic bases, but samples from different species can have greatly different proportions of bases. Human DNA, for example, contains about 30% each of adenine and thymine and about 20% each of guanine and cytosine. The bacterium *Clostridium perfringens*, however, contains about 37% each of adenine and thymine and only 13% each of guanine and cytosine. Note that in both examples, the bases occur in pairs. Adenine and thymine are usually present in equal amounts, as are guanine and cytosine. Why should this be?

In 1953, James Watson and Francis Crick made their now classic proposal for the secondary structure of DNA. According to the Watson–Crick model, DNA consists of two polynucleotide strands coiled around each other in a *double helix*. The two strands run in opposite directions and are held together by hydrogen bonds between specific pairs of bases. Guanine (G) and cytosine (C) form strong hydrogen bonds to each other but not to A or T. Similarly, adenine (A) and thymine (T) form strong hydrogen bonds to each other but not to G or C.

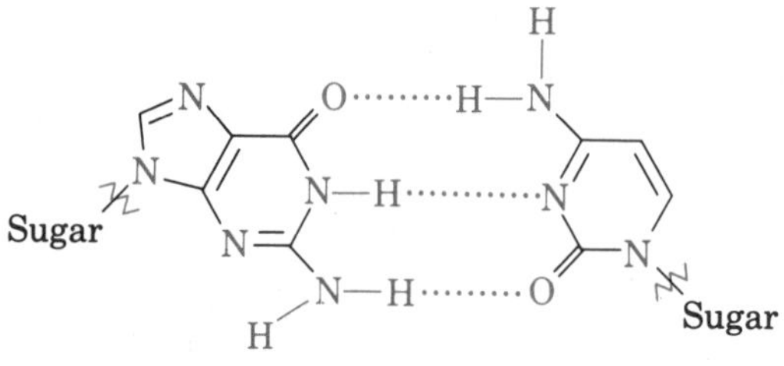

Sugar
Sugar

(Guanine) G : : : : : : C (Cytosine)

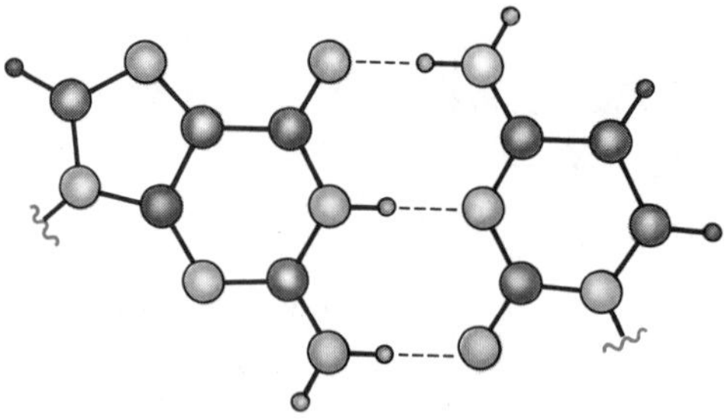

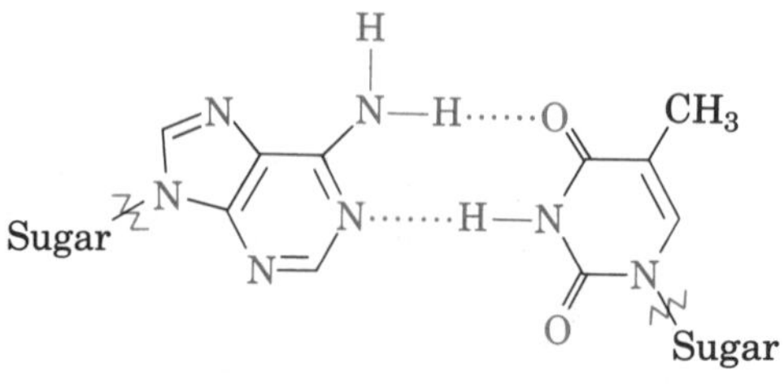

Sugar
Sugar

(Adenine) A : : : : : : T (Thymine)

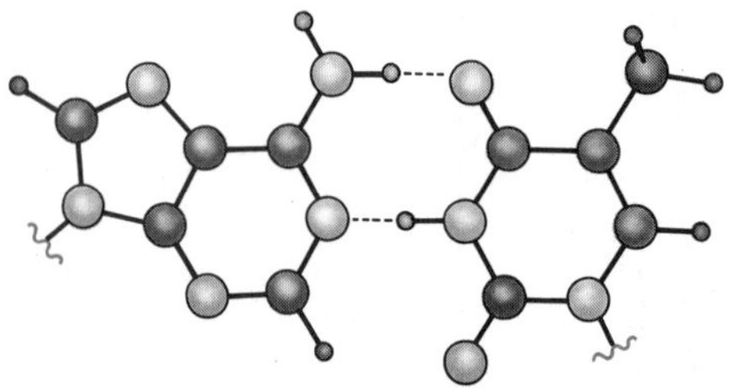

The two strands of the DNA double helix are not identical; rather, they're complementary. Whenever a G base occurs in one strand, a C base occurs opposite it in the other strand. When an A base occurs in one strand, a T base occurs in the other strand. This complementary pairing of bases explains why A and T, and G and C, are always found in equal amounts. Figure 16.6 illustrates this base pairing, showing how the two complementary strands coil into the double helix. Measurements show that the DNA double helix is 20 Å wide, that there are 10 base pairs in each full turn, and that each turn is 34 Å in height.

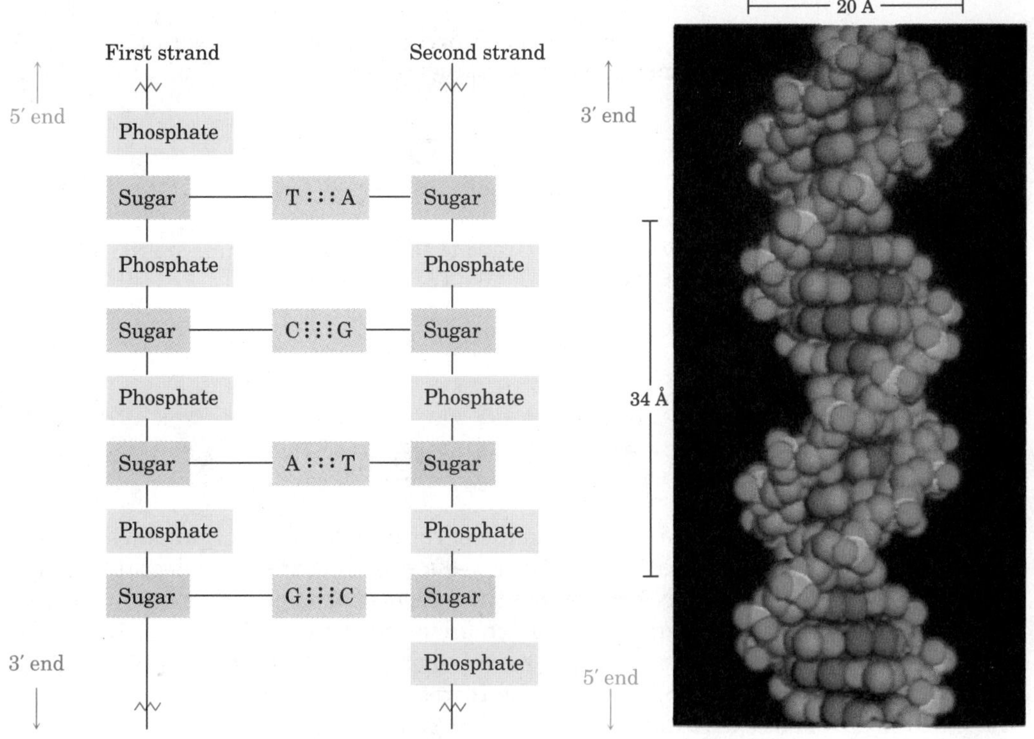

FIGURE 16.6 Complementarity of base pairing in the DNA double helix. The sugar–phosphate backbone of DNA runs along the outside of the helix; the atoms of the amine bases lie inside the helix.

A helpful mnemonic device to remember the pairing of the four DNA bases is the simple phrase "pure silver taxi":

Pure	Silver	Taxi
Pur	Ag	TC
The purine bases,	A and G,	pair with T and C.

MAJOR GROOVE The large groove in the double helix of DNA

MINOR GROOVE The small groove in the double helix of DNA

Notice in Figure 16.6 that the two strands of the double helix coil in such a way that two kinds of "grooves" result, a **major groove** that's 12 Å wide and a **minor groove** that's 6 Å wide. The major groove is slightly deeper than the minor groove, and both are lined by potential hydrogen bond donors and acceptors. Thus, a variety of molecules are able to *intercalate*, or fit into one of the grooves between the strands. A large number of cancer-causing and cancer-preventing agents are thought to function by interacting with DNA in this way.

PRACTICE PROBLEM 16.3 What sequence of bases on one strand of DNA is complementary to the sequence T-A-T-G-C-A-T on another strand?

SOLUTION Remembering that A and G (silver) form complementary pairs with T and C (taxi) respectively, go through the given sequence replacing A by T, G by C, T by A, and C by G:

Original: T-A-T-G-C-A-T
Complement: A-T-A-C-G-T-A

PROBLEM 16.11 What sequence of bases on one strand of DNA is complementary to the following sequence on another strand?

G-G-C-T-A-A-T-C-C-G-T

16.9 NUCLEIC ACIDS AND HEREDITY

A DNA molecule is the chemical repository of an organism's genetic information, which is stored as a sequence of deoxyribonucleotides strung together in the DNA chain. For the information to be preserved and passed on to future generations, a mechanism must exist for copying DNA. For the information to be used, a mechanism must exist for decoding the DNA message and implementing the instructions it contains.

The *central dogma of molecular genetics*, as Crick called it, says that the function of DNA is to store information and pass it on to RNA. The function of RNA, in turn, is to read, decode, and use the information received from DNA to make proteins. Each of the thousands of individual genes on a chromosome contains the instructions necessary to make a specific protein needed for a specific biological purpose. By decoding the right genes at the right time, an organism uses genetic information to synthesize the thousands of proteins necessary for smooth functioning.

Three fundamental processes take place in the transfer of genetic information:

$$\text{Replication} \circlearrowright \text{DNA} \xrightarrow{\text{Transcription}} \text{RNA} \xrightarrow{\text{Translation}} \text{Proteins}$$

1. **Replication** is the process by which identical copies of DNA are made so that genetic information can be preserved and handed down to offspring.
2. **Transcription** is the process by which the genetic messages contained in DNA are read and carried out of the nucleus to parts of the cell called ribosomes where protein synthesis occurs.
3. **Translation** is the process by which the genetic messages are decoded and used to build proteins.

16.10 REPLICATION OF DNA

Replication of DNA is an enzyme-catalyzed process that begins by a partial unwinding of the double helix. As the DNA strands separate and bases are exposed, new nucleotides line up on each strand in an exactly complementary manner, A to T and C to G, and two new strands begin to grow. Each new strand is complementary to its old template strand, and two new identical DNA double helices are produced (Figure 16.7).

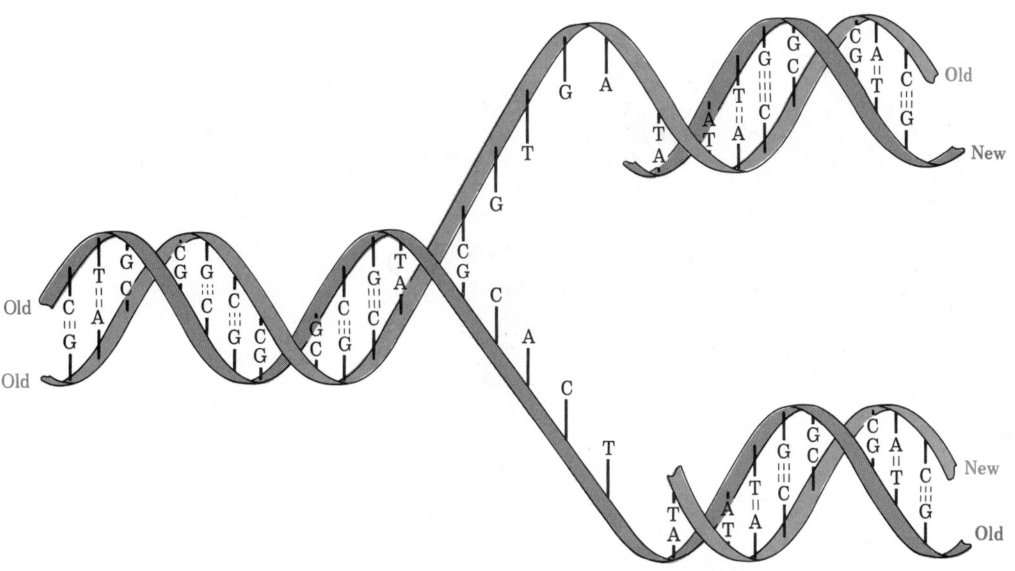

FIGURE 16.7 Schematic representation of DNA replication.

Crick probably described the process best when he used the analogy of the two DNA strands fitting together like a hand in a glove. The hand and glove separate, a new hand forms inside the glove, and a new glove forms around the hand. Two identical copies now exist where only one existed before.

The process by which the individual nucleotides are joined to create new DNA strands involves many steps and many different enzymes. Addition of new nucleotide units to the growing chain is catalyzed by the enzyme *DNA polymerase* and occurs by addition of a 5′-mononucleotide triphosphate to the free 3′-hydroxyl group of the growing chain, as indicated in Figure 16.8.

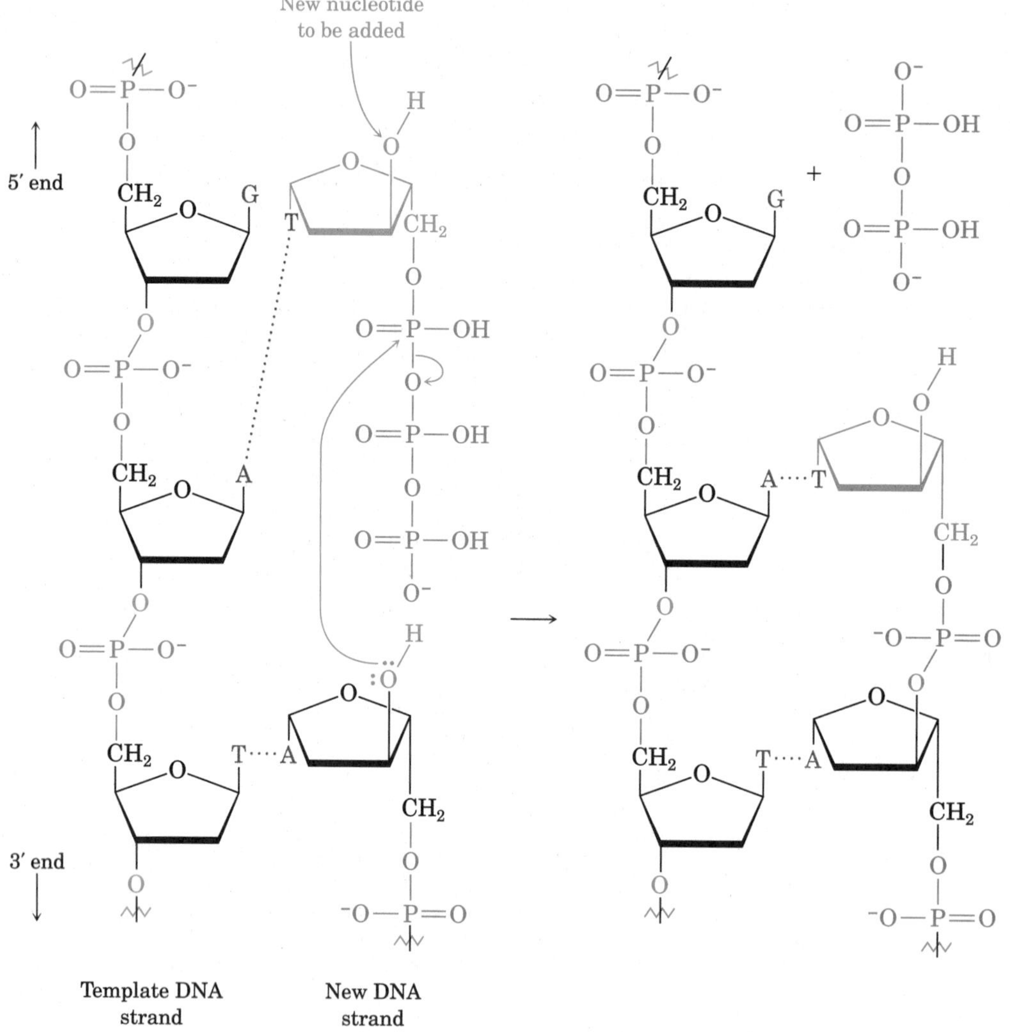

FIGURE 16.8 Addition of a new nucleotide to a growing DNA strand.

It's difficult to conceive of the magnitude of the replication process. The nucleus of a human cell contains 46 chromosomes (23 pairs), each of which consists of one very large DNA molecule. Each chromosome, in turn, is made up of several thousand DNA segments called *genes*, and the sum of all genes in a human cell (the *genome*) is estimated to be approximately three billion base pairs. Regardless of the size of these massive molecules, the base sequence is faithfully copied during replication, with an error occurring only about once each 10–100 billion bases.

16.11 STRUCTURE AND SYNTHESIS OF RNA: TRANSCRIPTION

RNA is structurally similar to DNA. Both are sugar–phosphate polymers and both have heterocyclic bases attached. The only differences are that RNA contains ribose rather than 2-deoxyribose and uracil rather than thymine. Uracil in RNA forms strong hydrogen bonds to its complementary base, adenine, just as thymine does in DNA.

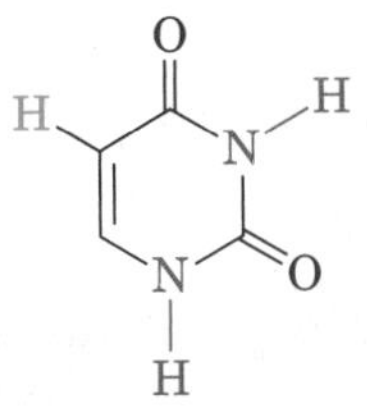

Uracil (in RNA) 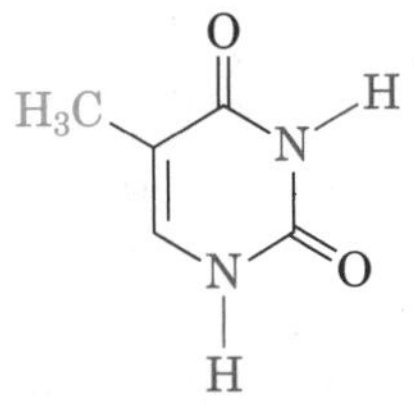 **Thymine (in DNA)**

There are three major kinds of ribonucleic acid, each of which serves a specific function. All three kinds of RNA are much smaller molecules than DNA, and all remain single-stranded rather double-stranded like DNA.

Messenger RNA (mRNA) carries genetic messages from DNA to *ribosomes*, small granular particles in the cell that act as "protein factories."

Ribosomal RNA (rRNA) is a structural component of ribosomes.

Transfer RNA (tRNA) transports specific amino acids to the ribosomes, where they are joined together to make proteins.

Molecules of RNA are synthesized in the nucleus of the cell by transcription of DNA. A small portion of the DNA double helix unwinds, and the bases are exposed. One of the two DNA strands acts as a template for complementary ribonucleotides to line up, and bond formation then occurs in the $5' \rightarrow 3'$ sense. Unlike DNA replication, though, the completed RNA molecule does not remain in a double helix with DNA but separates and migrates from the cell nucleus (Figure 16.9).

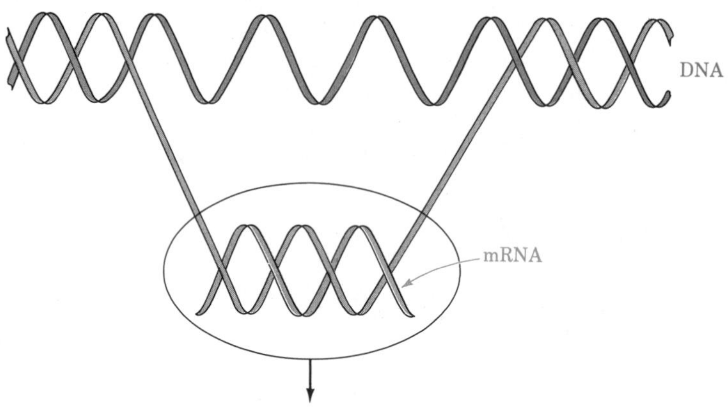

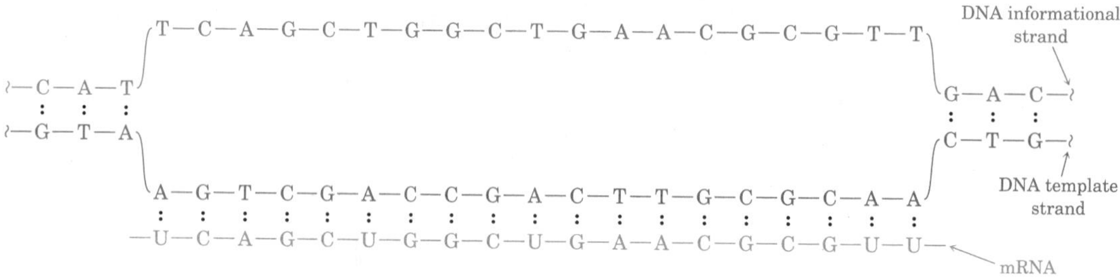

FIGURE 16.9 Synthesis of RNA using a DNA segment as template.

PRACTICE PROBLEM 16.4 What RNA base sequence is complementary to the following DNA base sequence?

T-A-A-G-C-C-G-T-G

SOLUTION Go through the sequence replacing A by U, G by C, T by A, and C by G:

Original DNA: T-A-A-G-C-C-G-T-G
Complementary RNA: A-U-U-C-G-G-C-A-C

PROBLEM 16.12 Show how uracil can form strong hydrogen bonds to adenine.

PROBLEM 16.13 What RNA base sequence is complementary to the following DNA base sequence?

G-A-T-T-A-C-C-G-T-A

PROBLEM 16.14 From what DNA base sequence was the following RNA sequence transcribed?

U-U-C-G-C-A-G-A-G-U

16.12 RNA AND PROTEIN BIOSYNTHESIS: TRANSLATION

Once the information in DNA has been transcribed into RNA, the information is used to synthesize proteins. The mechanics of protein biosynthesis are directed by messenger RNA and take place on *ribosomes*, small granular particles in the cytoplasm of a cell that consist of about 60% ribosomal RNA and 40% protein. The specific ribonucleotide sequence in mRNA acts like a long coded sentence to specify the order in which different amino acid residues are to be joined. Thus, each of the estimated 100,000 proteins in the human body is synthesized from a different mRNA that has been transcribed from a specific gene on DNA.

CODON
Sequence of three ribonucleotides on mRNA that codes for incorporation of a specific amino acid into a protein sequence

Each "word," or **codon**, along the mRNA chain consists of a series of three ribonucleotides that is specific for a given amino acid. For example, the series cytosine-uracil-guanine (C-U-G) on mRNA is a codon directing incorporation of the amino acid leucine into the growing protein. Similarly, guanine-adenine-uracil (G-A-U) codes for aspartic acid. Of the $4^3 = 64$ possible triads of the four bases in RNA, 61 code for specific amino acids (most amino acids are specified by more than one codon), and 3 of the 64 codons specify chain termination. Table 16.3 shows the meaning of each codon.

TABLE 16.3 Codon Assignments of Base Triads

FIRST BASE (5′ END)	SECOND BASE	THIRD BASE (3′ END)			
		U	C	A	G
U	U	Phe	Phe	Leu	Leu
	C	Ser	Ser	Ser	Ser
	A	Tyr	Tyr	Stop	Stop
	G	Cys	Cys	Stop	Trp
C	U	Leu	Leu	Leu	Leu
	C	Pro	Pro	Pro	Pro
	A	His	His	Gln	Gln
	G	Arg	Arg	Arg	Arg
A	U	Ile	Ile	Ile	Met
	C	Thr	Thr	Thr	Thr
	A	Asn	Asn	Lys	Lys
	G	Ser	Ser	Arg	Arg
G	U	Val	Val	Val	Val
	C	Ala	Ala	Ala	Ala
	A	Asp	Asp	Glu	Glu
	G	Gly	Gly	Gly	Gly

The code expressed in mRNA is read by transfer RNA (tRNA) in the process called translation. There are at least 60 different tRNA's, one for each of the codons in Table 16.3. Each specific tRNA acts as a carrier to bring a specific amino acid into place so that it can be transferred to the growing protein chain. A typical tRNA is roughly the shape of a cloverleaf,

ANTICODON
Sequence of three ribonucleotides on tRNA that is complementary to a codon on mRNA

as shown in Figure 16.10. It consists of about 70–100 ribonucleotides and is bonded to a specific amino acid by an ester linkage through the free 3′-hydroxyl on ribose at the 3′ end of the tRNA. Each tRNA also contains in its structure a segment called an **anticodon**, a sequence of three ribonucleotides complementary to the codon sequence. For example, the codon sequence C-U-G present on mRNA is "read" by a leucine-bearing tRNA having the complementary anticodon sequence G-A-C.

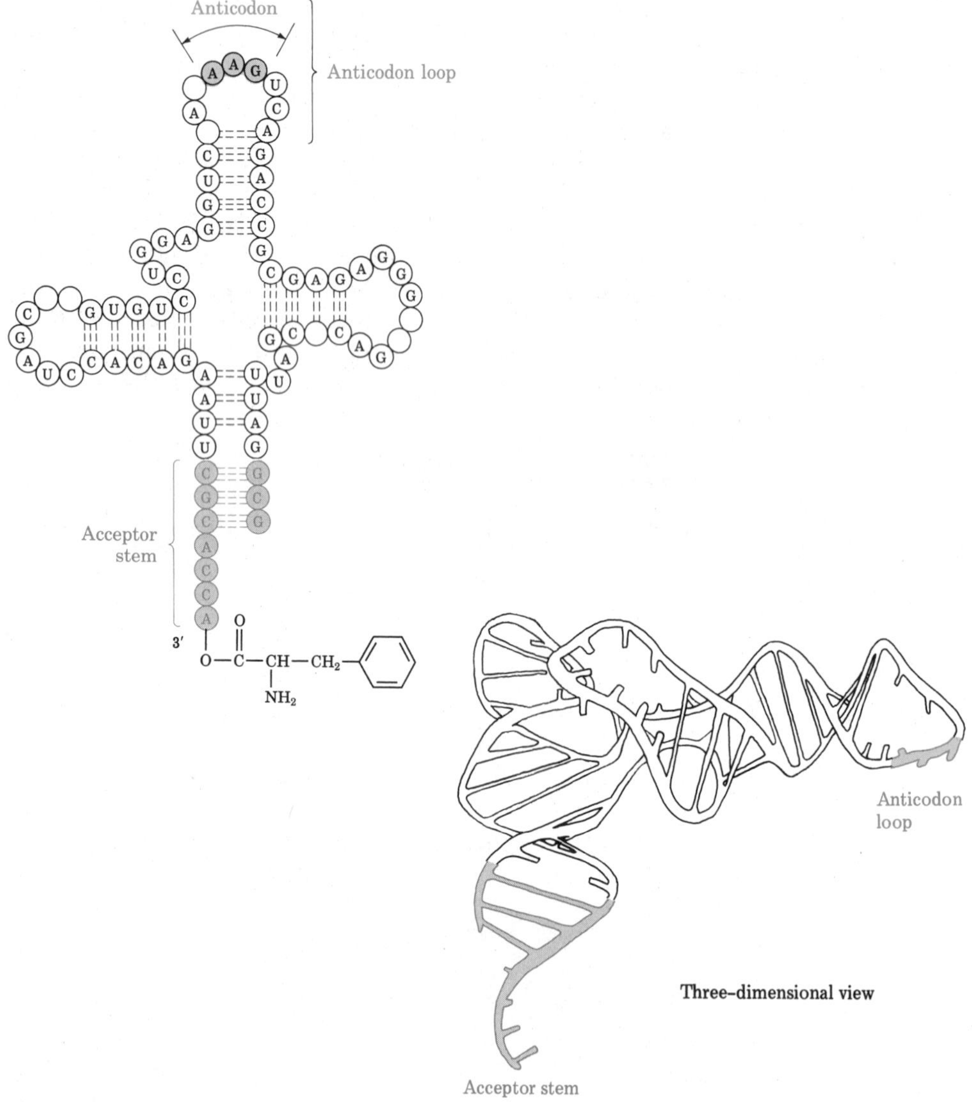

FIGURE 16.10 Structure of a tRNA molecule. The tRNA is a roughly cloverleaf-shaped molecule containing an anticodon triplet on one "leaf" and a covalently attached amino acid residue at its 3′ end. The example shown is a yeast tRNA that codes for phenylalanine. The nucleotides not specifically identified are chemically modified analogs of the four normal nucleotides.

As each successive codon on mRNA is read, different tRNA's bring the correct amino acids into position for enzyme-mediated transfer to the growing peptide. When synthesis of the proper protein is completed, a "stop" codon signals the end, and the protein is released from the ribosome. The entire process of protein biosynthesis is illustrated schematically in Figure 16.11.

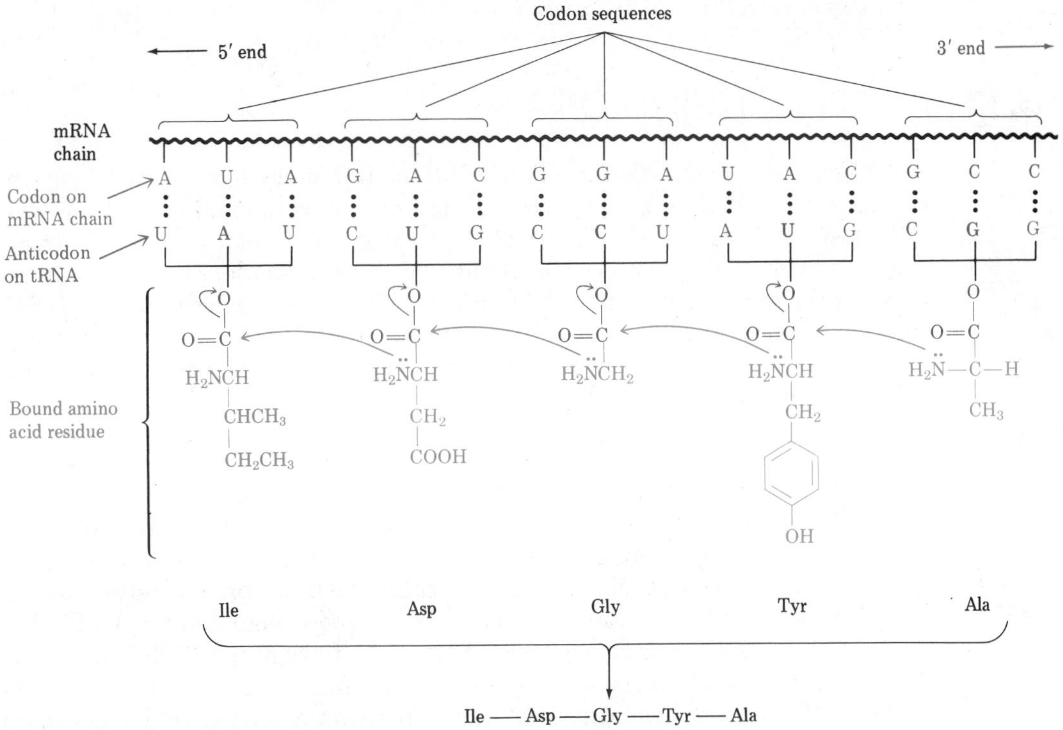

FIGURE 16.11 Schematic representation of protein biosynthesis. The mRNA containing codon base sequences is read by tRNA containing complementary anticodon base sequences. Transfer RNA assembles proper amino acids into position for incorporation into the peptide.

PRACTICE PROBLEM 16.5 Give a codon sequence for valine.

SOLUTION According to Table 16.3, there are four codons for valine: G-U-U, G-U-C, G-U-A, G-U-G.

PRACTICE PROBLEM 16.6 What amino acid sequence is coded by the mRNA base sequence AUC-GGU?

SOLUTION Table 16.3 indicates that AUC codes for isoleucine and GGU codes for glycine. Thus, AUC-GGU codes for Ile-Gly.

PROBLEM 16.15 List codon sequences for these amino acids:
(a) Ala **(b)** Phe **(c)** Leu **(d)** Tyr

PROBLEM 16.16 What amino acid sequence is coded by the following mRNA base sequence?

CUU-AUG-GCU-UGG-CCC-UAA

PROBLEM 16.17 What anticodon sequences of tRNA's are coded by the mRNA in Problem 16.16?

PROBLEM 16.18 What was the base sequence in the original DNA strand on which the mRNA sequence in Problem 16.16 was made?

16.13 SEQUENCING DNA

One of the greatest scientific revolutions in history is now under way in molecular biology as scientists are learning how to manipulate and harness the genetic machinery of organisms. None of the extraordinary advances of the past decade would have been possible, however, were it not for the discovery in 1977 of a method for sequencing immense DNA chains to find the messages therein. Much DNA sequencing is carried out by a remarkably efficient and powerful method developed by Allan Maxam and Walter Gilbert. There are five steps:

STEP 1 Since molecules of DNA are so enormous—some molecules of human DNA contain up to 250 million base pairs—the first problem in DNA sequencing is to find a method for cleaving the DNA chain at specific points to produce smaller, more manageable pieces. This problem has been solved by the use of enzymes called **restriction endonucleases**. Each different restriction enzyme, of which more than 200 are available, cleaves a DNA molecule between two nucleotides at a well-defined point along the chain where a specific base sequence occurs.

RESTRICTION ENDONUCLEASE
Enzyme that is able to cut a DNA strand at a specific base sequence in the chain

By cleavage of a large DNA molecule with a given restriction enzyme, many different and well-defined segments of manageable length (100–200 nucleotides) are produced. For example, the restriction enzyme *Alu I* cleaves the linkage between G and C in the four-base sequence AG-CT. If the original DNA molecule is cut with another restriction enzyme, other segments are produced whose sequences partially overlap those produced by the first enzyme. Sequencing of all the segments, followed by identification of the overlapping sections, then allows complete DNA structure determination.

STEP 2 After restriction enzymes have cleaved DNA into smaller pieces called *restriction fragments*, the various double-stranded fragments are isolated, and each is radioactively tagged by enzymatically incorporating a labeled ^{32}P phosphate group onto the 5′-hydroxyl of the terminal nucleotide. The fragments are then separated into two strands by heating, and the strands are isolated. Imagine, for example, that we now have a single-stranded DNA fragment with the following partial structure:

(5′ end) ^{32}P-G-A-T-C-A-G-C-G-A-T- - - (3′ end)

STEP 3 The labeled DNA strand is subjected to four parallel sets of chemical reactions under conditions that cause:

(a) Splitting of the DNA chain next to A
(b) Splitting of the DNA chain next to G
(c) Splitting of the DNA chain next to C
(d) Splitting of the DNA chain next to *both* T and C

Mild reaction conditions are used so that *only a few of the many possible splittings occur in each reaction.* Thus, the pieces shown in Table 16.4 would be produced.

TABLE 16.4 Splitting of a DNA Fragment Under Four Sets of Conditions

CLEAVAGE CONDITION	LABELED DNA PIECES PRODUCED
Original DNA fragment	^{32}P-G-A-T-C-A-G-C-G-A-T-
Next to A	^{32}P-G ^{32}P-G-A-T-C ^{32}P-G-A-T-C-A-G-C-G + Larger pieces
Next to G	^{32}P-G-A-T-C-A ^{32}P-G-A-T-C-A-G-C + Larger pieces
Next to C	^{32}P-G-A-T ^{32}P-G-A-T-C-A-G + Larger pieces
Next to C + T	^{32}P-G-A ^{32}P-G-A-T ^{32}P-G-A-T-C-A-G ^{32}P-G-A-T-C-A-G-C-G-A + Larger pieces

Cleavages next to A and G are accomplished by treatment of a restriction fragment with dimethyl sulfate [$(CH_3O)_2SO_2$]. Deoxyadenosine (A) is methylated at N3 (S_N2 reaction), and deoxyguanosine (G) is methylated at N7, but T and C aren't affected. Treatment of the methylated DNA with an aqueous solution of the secondary amine *piperidine* then brings about destruction of the methylated nucleotides and opening of the DNA chain at both the 3′ and 5′ positions next to the methylated bases. By working carefully, it's possible to find reaction conditions that are selective for cleavage either at A or at G.

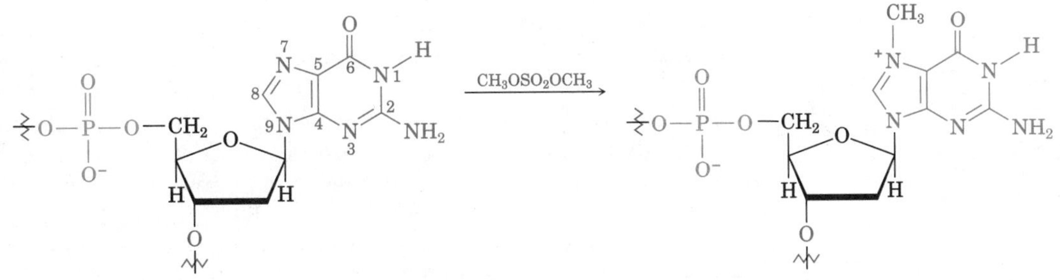

$CH_3OSO_2OCH_3$

Deoxyguanosine

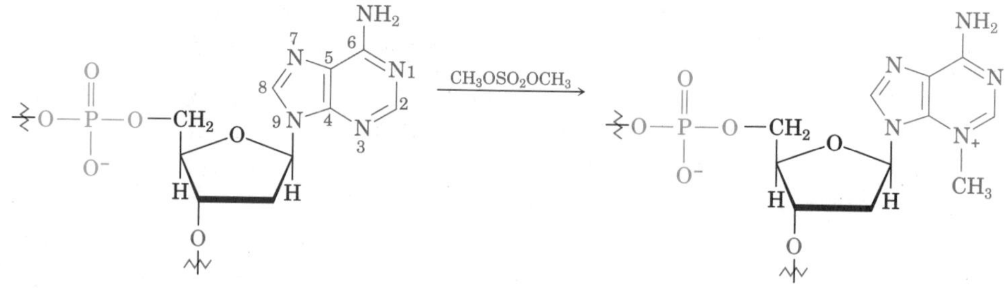

$$\text{Deoxyadenosine} \xrightarrow{CH_3OSO_2OCH_3} \text{N-methyldeoxyadenosine}$$

Deoxyadenosine

Breaking the DNA chain next to C and T is accomplished by treatment of DNA with hydrazine, H_2NNH_2, followed by heating with aqueous piperidine. Although conditions that are selective for cleavage next to T have not been found, selective cleavage next to C is accomplished by carrying out the hydrazine reaction in 5 M NaCl solution.

STEP 4 Product mixtures from the four cleavage reactions are separated by electrophoresis (Section 15.3). When a mixture is placed at one end of a strip of buffered polyacrylamide gel and a voltage is applied to the two ends of the strip, each DNA piece moves along the gel at a rate that depends on the number of negatively charged phosphate groups (that is, the number of nucleotides) it contains. Smaller fragments move rapidly, and larger pieces move more slowly. The technique is so sensitive that up to 600 DNA pieces differing in size by only one nucleotide can be separated.

Once separated, the location of each DNA fragment is detected by exposing the gel to a photographic plate. Each radioactive end piece containing a ^{32}P label appears as a dark band on the photographic plate, but nonradioactive pieces from the middle of the chain aren't seen. The gel electrophoresis pattern shown in Figure 16.12 would be obtained in our hypothetical example.

STEP 5 The DNA sequence is read directly from the gel. The band that appears farthest from the origin is the terminal mononucleotide (the smallest piece) and can't be identified. Since the terminal mononucleotide appears in the A column, though, it must have been produced by splitting *next to* an A. Thus, the *second* nucleotide in the DNA fragment is an A.

The second farthest band from the origin is a dinucleotide that appears only in the T + C column and is produced by splitting next to the third nucleotide, which must therefore be a T or C. Since this piece doesn't appear in the C column, though, the third nucleotide isn't a C and must therefore be a T. The third farthest band appears in both C and T + C columns, meaning that the fourth nucleotide is a C.

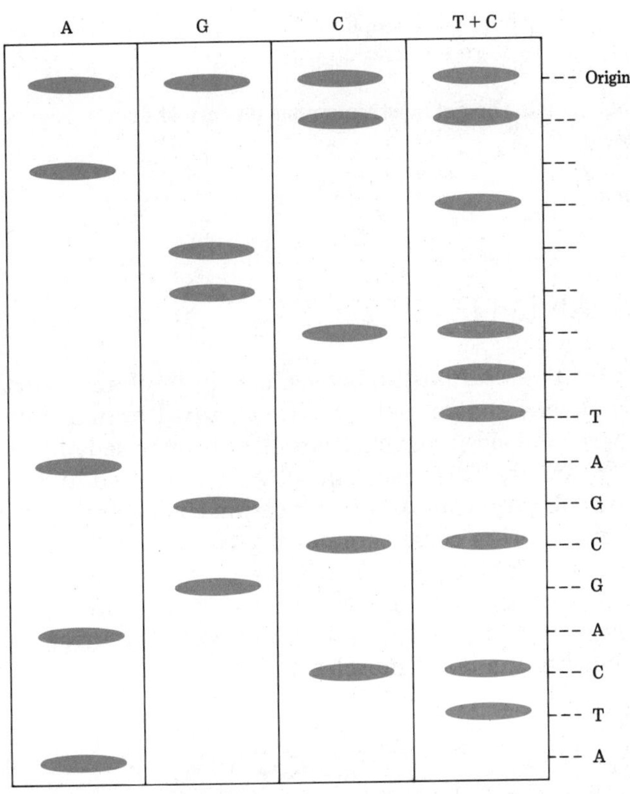

FIGURE 16.12 Representation of a gel electrophoresis pattern. The products of the four cleavage experiments are placed at the top of the gel, and a voltage is applied between top and bottom. Smaller products migrate along the gel at a faster rate and thus appear at the bottom. The DNA sequence is read from the radioactive spots.

Continuing in this manner, the entire sequence of the DNA fragment is read from the gel simply by noting in what column the successively larger labeled polynucleotide pieces appear. Once read, the entire sequence can be checked by determining the sequence of the complementary strand.

The Maxam–Gilbert method of DNA sequencing is so efficient that a trained person can sequence up to 2000 base pairs per day. DNA strands of up to 170,000 base pairs have already been sequenced, and work is now under way to sequence the entire human genome with 3,000,000,000 base pairs. At least 10 years and several billion dollars will be needed.

PROBLEM 16.19 Show the labeled products you would obtain if the following DNA segment were subjected to each of the four cleavage reactions:

^{32}P-A-A-C-A-T-G-G-C-G-C-T-T-A-T-G-A-C-G-A

PROBLEM 16.20 Sketch what you would expect the gel electrophoresis pattern to look like if the DNA segment in Problem 16.19 were sequenced.

PROBLEM 16.21 Finish assigning the sequence to the gel electrophoresis pattern shown in Figure 16.12.

INTERLUDE

CHOLESTEROL AND HEART DISEASE

What are the facts about the relationship between cholesterol and heart disease? It's well established that a diet rich in saturated animal fats often leads to an increase in blood serum cholesterol, at least in sedentary, overweight people. Conversely, a diet lower in saturated fats and higher in polyunsaturated fats (PUFA's) leads to a lower serum cholesterol level. Studies have shown that a serum cholesterol level greater than 300 mg/dL (a normal value is 150–240 mg/dL) is weakly correlated with an increased incidence of *atherosclerosis*, a form of heart disease in which cholesterol deposits build up on the inner walls of coronary arteries, blocking the flow of blood to the heart muscles.

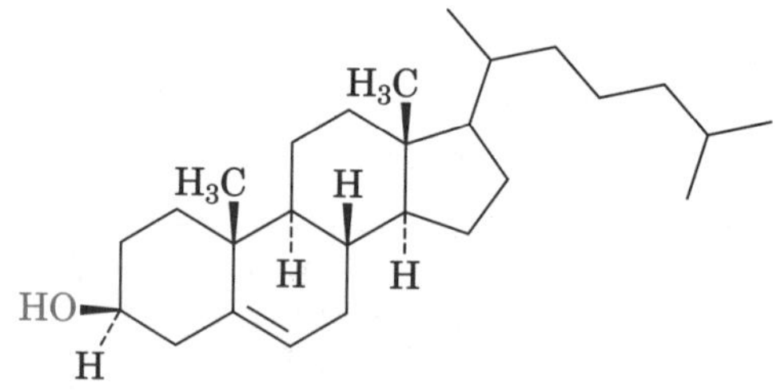

H_3C H_3C H H H HO H

Cholesterol

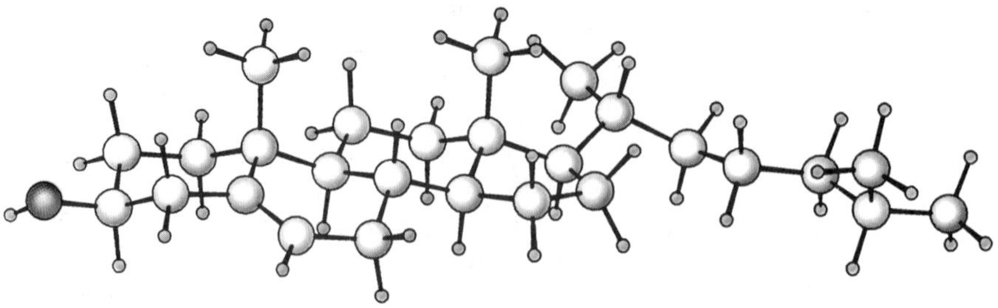

A better indication of a person's risk of heart disease comes from a measurement of blood lipoprotein levels. **Lipoproteins** are complex molecules with both lipid and protein parts that transport lipids through the body. They can be divided into four types according to density, as shown in Table 16.5. People with a high serum level of high-density lipoproteins (HDL's) seem to have a decreased risk of heart disease. As a rule of thumb, a person's risk drops about 25% for each increase of 5 mg/dL in HDL concentration. Normal values are about 45 mg/dL for men and 55 mg/dL for women, perhaps explaining why women are generally less susceptible than men to heart disease.

TABLE 16.5 Serum Lipoproteins

NAME	DENSITY (g/mL)	% LIPID	% PROTEIN
Chylomicrons	<0.94	98	2
VLDL's (very-low-density lipoproteins)	0.940–1.006	90	10
LDL's (low-density lipoproteins)	1.006–1.063	75	25
HDL's (high-density lipoproteins)	1.063–1.210	60	40

Chylomicrons and very-low-density lipoproteins (VLDL's) act primarily as carriers of triglycerides from the intestines to peripheral tissues, whereas LDL's and HDL's act as carriers of cholesterol to and from the liver. Present evidence suggests that LDL's transport cholesterol as its fatty acid ester *to* peripheral tissues, whereas HDL's remove cholesterol as its stearate ester *from* dying cells and transport it back to the liver. If LDL's deliver more cholesterol than is needed, and if insufficient HDL's are present to remove it, the excess is deposited in arteries. The higher the HDL level, the less the likelihood of deposits and the lower the risk of heart disease.

Not surprisingly, the most important factor in gaining high HDL levels is a generally healthy lifestyle. Obesity, smoking, and lack of exercise lead to low HDL levels, whereas regular exercise and a sensible diet lead to high HDL levels. Distance runners, in particular, have HDL levels nearly 50% higher than the general population.

SUMMARY AND KEY WORDS

Lipids and **nucleic acids**, along with carbohydrates and proteins, comprise the four major classes of biomolecules. Lipids are the naturally occurring materials that can be isolated from cells by extraction with organic solvents. Animal fats and vegetable oils are the most widely occurring lipids. Both fats and oils are **triacylglycerols**, triesters of glycerol with long-chain fatty acids. **Phosphoglycerides** such as lecithin and cephalin are closely related to fats. The glycerol backbone in these molecules is esterified to two fatty acids and to one phosphate ester. **Sphingolipids**, another major class of phospholipids, have an amino alcohol such as sphingosine for their backbone.

Steroids are plant and animal lipids with a characteristic tetracyclic carbon skeleton. Steroids occur widely in body tissue and have many different kinds of physiological activity. Among the more important kinds of steroids are the sex hormones (**androgens** and **estrogens**) and the **adrenocortical** hormones.

The nucleic acids, **DNA (deoxyribonucleic acid)** and **RNA (ribonucleic acid)**, are biological polymers that act as chemical carriers of an organism's genetic information. Enzyme-catalyzed hydrolysis yields **nucleotides**, which consist of a purine or pyrimidine heterocyclic amine base linked to C1′ of a simple pentose (ribose in RNA and 2-deoxyribose in DNA), with the sugar in turn linked through its C5′ hydroxyl to a phosphate group. Nucleotides are joined by ester links between the phosphate of one nucleotide and the 3′ hydroxyl on the sugar of another nucleotide.

Molecules of DNA consist of two polynucleotide strands held together by hydrogen bonds between heterocyclic bases on the different strands and coiled into a *double-helix* conformation. Adenine and thymine form hydrogen bonds to each other, as do cytosine and guanine. The two strands of DNA are complementary rather than identical.

Three main processes take place in deciphering the genetic information in DNA:

Replication of DNA is the process by which identical DNA copies are made and genetic information is preserved. This occurs when the DNA double helix unwinds, complementary deoxyribonucleotides line up in order, and two new DNA molecules are produced.

Transcription is the process by which RNA is produced to carry the genetic information from the nucleus to the ribosomes. This occurs when a segment of the DNA double helix unwinds, and complementary ribonucleotides line up to produce **messenger RNA (mRNA)**.

Translation is the process by which mRNA directs protein synthesis. Each mRNA has segments called **codons** along its chain. These codons are ribonucleotide triads that are recognized by small molecules of **transfer RNA (tRNA)**, which carry and then deliver the appropriate amino acids needed for protein synthesis.

Sequencing of DNA fragments is done by the Maxam-Gilbert method in which chemical reactions are carried out to cause specific cleavages of the DNA chain, followed by separation of the pieces by electrophoresis. The DNA sequence is read directly from the electrophoresis pattern.

ADDITIONAL PROBLEMS

16.22 Write representative structures for the following:
(a) A fat **(b)** A vegetable oil **(c)** A steroid

16.23 Write the structures of these molecules:
(a) Sodium stearate (b) Ethyl linoleate (c) Glyceryl palmitodioleate

16.24 Show the products you would expect to obtain from the reaction of glyceryl trioleate with the following:
(a) Excess Br_2 in CCl_4 (b) H_2/Pd (c) NaOH, H_2O (d) O_3, then Zn, CH_3COOH
(e) $LiAlH_4$, then H_3O^+

16.25 How would you convert oleic acid into these substances?
(a) Methyl oleate (b) Methyl stearate (c) Nonanal (d) Nonanedioic acid

16.26 Eleostearic acid, $C_{18}H_{30}O_2$, is a rare fatty acid found in tung oil. On ozonolysis followed by treatment with zinc, eleostearic acid yields one part pentanal, two parts glyoxal (OHC–CHO), and one part 9-oxononanoic acid [$OHC(CH_2)_7COOH$]. Propose a structure for eleostearic acid.

16.27 Stearolic acid, $C_{18}H_{32}O_2$, yields oleic acid on catalytic hydrogenation over the Lindlar catalyst. Propose a structure for stearolic acid.

16.28 Draw the products you would obtain from treatment of cholesterol with these reagents:
(a) Br_2 (b) H_2, Pd catalyst (c) O_3, followed by Zn

16.29 If the average molecular weight of soybean oil is 1500, how many grams of NaOH are needed to saponify 5.00 g of the oil?

16.30 Define these terms:
(a) Steroid (b) DNA (c) Base pair (d) Codon (e) Lipid (f) Transcription

16.31 The DNA from sea urchins contains about 32% A and about 18% G. What percentages of T and C would you expect in sea urchin DNA? Explain.

16.32 What DNA sequence is complementary to the following sequence?

G-A-A-G-T-T-C-A-T-G-C

16.33 Give codons for these amino acids:
(a) Ile (b) Asp (c) Thr

16.34 Draw the complete structure of the ribonucleotide codon UAC. For what amino acid does this sequence code?

16.35 Draw the complete structure of the deoxyribonucleotide sequence from which the mRNA codon in Problem 16.34 was transcribed.

16.36 What amino acids do the following ribonucleotide codons code for?
(a) AAU (b) GAG (c) UCC (d) CAU (e) ACC

16.37 From what DNA sequences were each of the mRNA codons in Problem 16.36 transcribed?

16.38 What anticodon sequences of tRNA's are coded by each of the codons in Problem 16.36?

16.39 If the gene sequence T-A-A-C-C-G-G-A-T on DNA were miscopied during replication and became T-G-A-C-C-G-G-A-T, what effect would the mutation have on the sequence of the protein produced?

16.40 Give an mRNA sequence that codes for synthesis of metenkephalin, a small peptide with morphine-like properties:

Tyr-Gly-Gly-Phe-Met

16.41 Give a DNA gene sequence that will code for metenkephalin (see Problem 16.40).

16.42 Human and horse insulin both have two polypeptide chains, with one chain containing 21 amino acids and the other containing 30 amino acids. How many nitrogen bases are present in the DNA to code for each chain?

16.43 Human and horse insulin (see Problem 16.42) differ in primary structure at two amino acids: at the ninth position in one chain (human has Ser and horse has Gly) and at the 30th position in the other chain (human has Thr and horse has Ala). How must the DNA differ?

16.44 What amino acid sequence is coded by the following mRNA sequence?

CUA-GAC-CGU-UCC-AAG-UGA

16.45 What anticodon sequences of tRNA's are coded by the mRNA in Problem 16.44? What was the base sequence in the original DNA strand on which this mRNA was made? What was the base sequence in the DNA strand *complementary* to that from which this mRNA was made?

16.46 Look up the structure of angiotensin II (Figure 15.2) and give an mRNA sequence that codes for its synthesis.

16.47 Diethylstilbestrol (DES) exhibits estradiol-like activity even though it is structurally unrelated to steroids. Once used widely as an additive in animal feed, DES has been implicated as a causative agent in several types of cancers. Look up the structure of estradiol (Section 16.5) and show how DES can be drawn so that it is sterically similar to estradiol

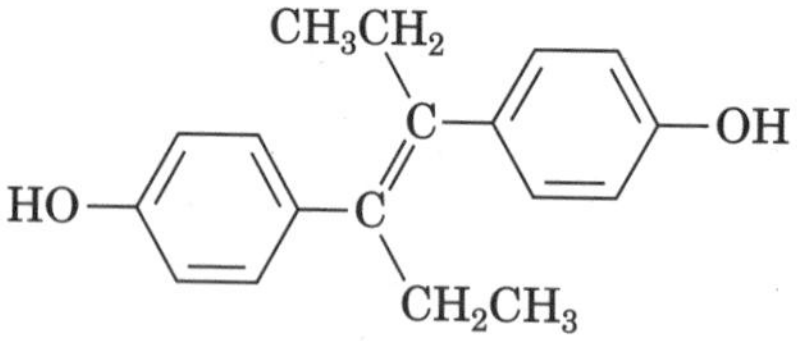

CH_3CH_2

HO—C$_6$H$_4$—C=C—C$_6$H$_4$—OH

CH_2CH_3

Diethylstilbestrol

16.48 How many stereogenic centers are present in estradiol (see Problem 16.47)? Label each of them.

16.49 What products would you obtain from reaction of estradiol (Problem 16.47) with these reagents?
(a) NaOH, then CH_3I **(b)** CH_3COCl, pyridine **(c)** Br_2 (1 equiv)

16.50 *Nandrolone* is an anabolic steroid sometimes taken by athletes to build muscle mass. Compare the structures of nandrolone and testosterone, and point out their structural similarities.

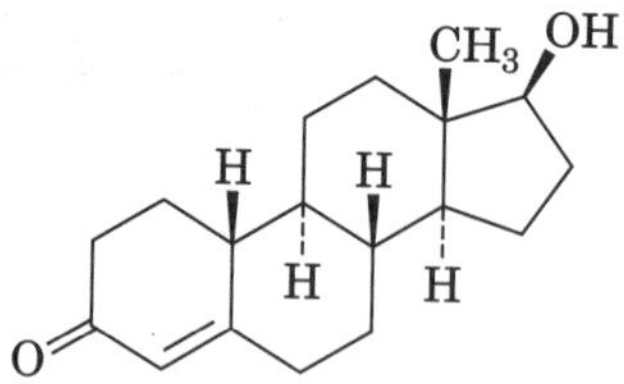

CH_3 OH

H H H H

O

Nandrolone
(an anabolic steriod)

17 CHAPTER

THE ORGANIC CHEMISTRY OF METABOLIC PATHWAYS

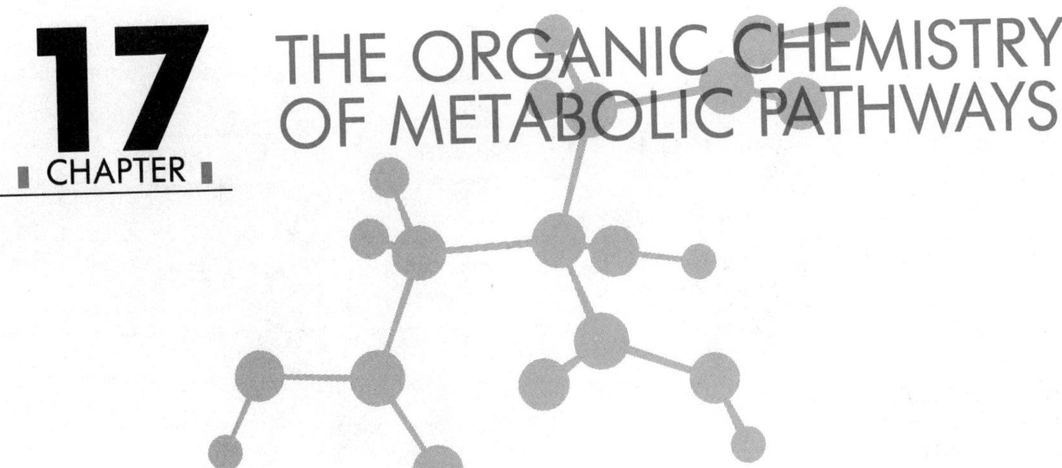

The organic chemical reactions that take place in even the smallest and simplest living organism are more complex than those carried out in any laboratory. Yet the reactions in living organisms, regardless of their complexity, follow the same rules of reactivity that we've developed in the preceding 16 chapters.

In this chapter, we'll look at some of the pathways by which organisms carry out their chemistry, focusing primarily on how they break down fats and carbohydrates. All the reactions in living organisms are catalyzed by enzymes (Sections 15.11 and 15.12), but our emphasis will be on the organic chemistry of the various pathways. We'll pay particular attention to recognizing the similarities between mechanisms of biological reactions and mechanisms of the analogous laboratory reactions.

17.1 AN OVERVIEW OF METABOLISM AND BIOCHEMICAL ENERGY

METABOLISM
Total of all reactions in living organisms

CATABOLISM
Metabolic reactions that break down large molecules

ANABOLISM
Metabolic reactions that synthesize larger molecules from smaller precursors

The many reactions that go on in the cells of living organisms are collectively called **metabolism**. Reactions that break down large food molecules are known as **catabolism**, while reactions that put smaller molecules together to synthesize larger biomolecules are known as **anabolism**. Catabolic reactions usually release energy, while anabolic reactions often absorb energy. Catabolism can be divided into the four stages shown in Figure 17.1.

The first catabolic stage, **digestion**, takes place in the stomach and small intestine when food is broken down by hydrolysis of ester, glycoside (acetal), and peptide (amide) bonds to yield fatty acids, simple sugars, and amino acids. These small molecules are further degraded in the second stage of catabolism to yield two-carbon acetyl groups attached by a

FOOD

Fats **Carbohydrates** **Proteins**

Fatty acids and glycerol

Glucose and other sugars

Amino acids

Stage 1
Bulk food is digested in the stomach and small intestine to yield small molecules.

β-Oxidation

Glycolysis

Stage 2
Small sugar, fatty acid, and amino acid molecules are degraded in cells to yield acetyl CoA.

$$CH_3-\overset{\overset{\displaystyle O}{\|}}{C}-SCoA$$ **Acetyl CoA**

Stage 3
Acetyl CoA is oxidized in the citric acid cycle to yield CO_2 and energy.

Citric acid cycle

CO_2

Energy

O_2

Stage 4
The energy produced in stage 3 is used by the respiratory chain to make ATP.

Respiratory chain

ATP + H_2O

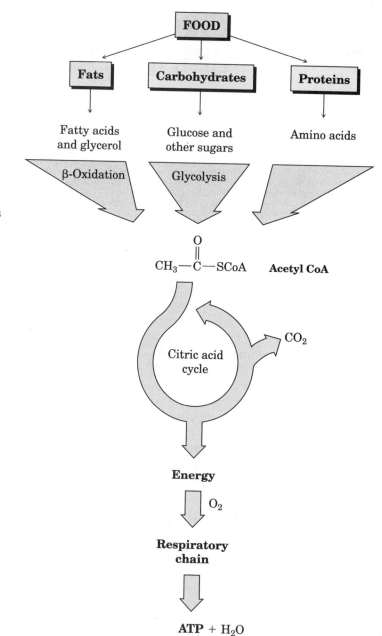

FIGURE 17.1 An overview of catabolic pathways for the degradation of food and the production of biochemical energy.

DIGESTION
First stage of catabolism: Food molecules are hydrolyzed to yield fatty acids, amino acids, and monosaccharides

thiol ester bond (Chapter 10 Interlude) to the large carrier molecule *coenzyme A*. The resultant compound, *acetyl coenzyme A (acetyl CoA)*, is an intermediate in the breakdown of all main classes of food molecules.

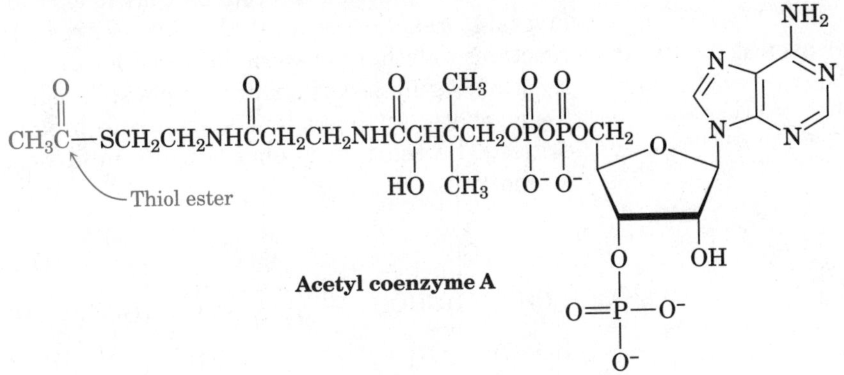

$$CH_3C(=O)-SCH_2CH_2NHC(=O)CH_2CH_2NHC(=O)CH(OH)C(CH_3)_2CH_2OP(=O)(O^-)OP(=O)(O^-)OCH_2-\text{(ribose–adenine, 3'-}OPO_3^{2-}\text{)}$$

Thiol ester

Acetyl coenzyme A

CITRIC ACID CYCLE
Third stage of catabolism: Acetyl groups are degraded to CO_2

RESPIRATORY CHAIN
Fourth stage of catabolism: ATP is synthesized

Acetyl groups are oxidized in the third stage of catabolism, the **citric acid cycle**, to yield carbon dioxide. This stage releases a large amount of energy that is used in the fourth stage, the **respiratory chain**, to make molecules of the nucleotide *adenosine triphosphate (ATP)*. ATP, the final result of food catabolism, has been called the "energy currency" of the cell. Catabolic reactions "pay off" in ATP by synthesizing it from *adenosine diphosphate (ADP)* plus phosphate ion, PO_4^{3-}. Anabolic reactions "spend" ATP by transferring a monophosphate group to other molecules, thereby regenerating ADP. The entire process of energy production thus revolves around the ATP $\rightleftharpoons$ ADP interconversion:

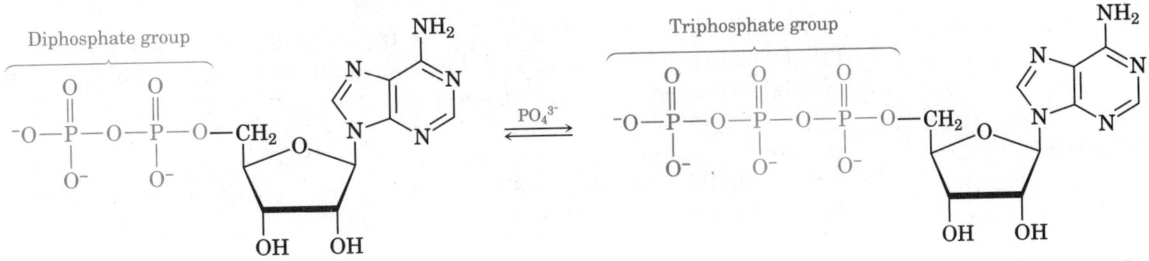

Diphosphate group

$$^-O-P(=O)(O^-)-O-P(=O)(O^-)-O-CH_2-\text{(ribose, OH, OH)–adenine}$$

Adenosine diphosphate (ADP)

$$\xrightleftharpoons{PO_4^{3-}}$$

Triphosphate group

$$^-O-P(=O)(O^-)-O-P(=O)(O^-)-O-P(=O)(O^-)-O-CH_2-\text{(ribose, OH, OH)–adenine}$$

Adenosine triphosphate (ATP)

PHOSPHORIC ACID ANHYDRIDE
Functional group containing the P–O–P linkage

Both ADP and ATP are **phosphoric acid anhydrides**, containing a $-P(=O)-O-P(=O)-$ linkage analogous to the $-C(=O)-O-C(=O)-$ linkage in carboxylic acid anhydrides. Like carboxylic anhydrides (Section 10.8), phosphoric anhydrides react with alcohols, transferring a phosphate group and breaking a P–O–P bond.

How does the body use ATP? Recall from Sections 3.9 and 3.10 that energy (actually, *free energy, G*) must be released in order for a chemical reaction to have a favorable equilibrium constant. This means that the free-energy change for the reaction (ΔG) must be negative. If ΔG is positive, then the reaction is unfavorable, and energy must be absorbed for the process to occur. What normally happens in order for an energetically unfavorable reaction to occur is that it is "coupled" to an energetically favorable reaction so that the *overall* free-energy change for the two reactions together is favorable. Take, for example, the **phosphorylation** reaction of glucose with hydrogen phosphate ion (HPO_4^{2-}) to yield glucose 6-phosphate, an important step in the breakdown of dietary carbohydrates. The reaction is energetically unfavorable by 3.3 kcal/mol (13.8 kJ/mol):

PHOSPHORYLATION Reaction that transfers a phosphate group from a phosphoric anhydride to an alcohol

$$\underset{\textbf{Glucose}}{HOCH_2\underset{HO}{CH}\underset{OH}{CH}\overset{OH}{CH}\underset{OH}{CH}\overset{O}{\overset{\|}{CH}}} \xrightarrow[-H_2O]{HPO_4^{2-}} \underset{\textbf{Glucose 6-phosphate}}{{}^-O\overset{O}{\overset{\|}{\underset{O^-}{P}}}OCH_2\underset{HO}{CH}\underset{OH}{CH}\overset{OH}{CH}\underset{OH}{CH}\overset{O}{\overset{\|}{CH}}} \quad \Delta G^\circ = +3.3\ \text{kcal}$$

When ATP is present, however, its reaction with water to yield ADP plus hydrogen phosphate ion is energetically favorable by about 7.3 kcal/mol (30.5 kJ/mol), making the *overall* process favorable by about 4.0 kcal/mol (16.7 kJ/mol). ATP thus "drives" the phosphorylation reaction of glucose:

$$\begin{array}{lrcll} & \text{Glucose} + HPO_4^{2-} & \longrightarrow & \text{Glucose 6-phosphate} + H_2O & \Delta G = +3.3\ \text{kcal/mol} \\ & \text{ATP} + H_2O & \longrightarrow & \text{ADP} + HPO_4^{2-} & \Delta G = -7.3\ \text{kcal/mol} \\ \hline \textit{Net:} & \text{Glucose} + \text{ATP} & \longrightarrow & \text{Glucose 6-phosphate} + \text{ADP} & \Delta G = -4.0\ \text{kcal/mol} \end{array}$$

It's this ability to drive otherwise unfavorable reactions that makes ATP so useful. In fact, most of the thousands of reactions going on in your body every minute are powered by energy from ATP.

PROBLEM 17.1 One of the steps in fat metabolism is the reaction of glycerol ($HOCH_2{-}CHOH{-}CH_2OH$) with ATP to yield glycerol 1-phosphate. Write the reaction, and draw the structure of glycerol 1-phosphate.

PROBLEM 17.2 The reaction of ATP with an alcohol to yield ADP and an alkyl phosphate is analogous to that of a carboxylic acid anhydride with an alcohol to yield a carboxylate ion and an ester (Section 10.8). Show the mechanism of the reaction of ATP with methanol to yield methyl phosphate, $CH_3OPO_3^{2-}$.

17.2 CATABOLISM OF FATS: β-OXIDATION PATHWAY

The catabolism of fats and oils (triacylglycerols) begins with their hydrolysis in the stomach and small intestine to yield glycerol plus fatty acids.

Glycerol is phosphorylated by reaction with ATP and then oxidized to yield glyceraldehyde 3-phosphate, which enters the carbohydrate catabolic pathway (Section 17.3):

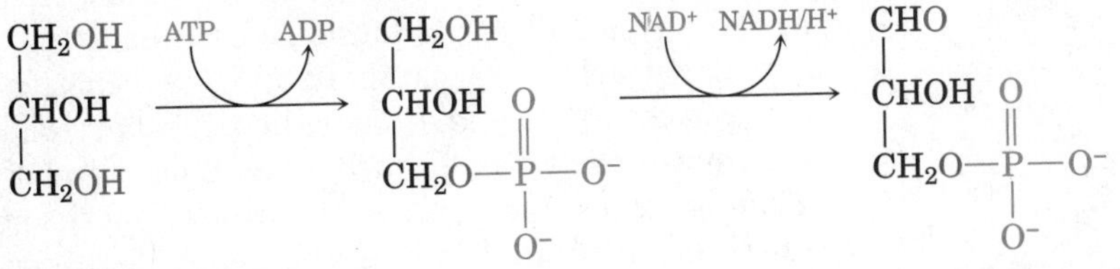

$$\underset{\textbf{Glycerol}}{HOCH_2CH(OH)CH_2OH} \xrightarrow{ATP \to ADP} \underset{\textbf{Glycerol monophosphate}}{HOCH_2CH(OH)CH_2O{-}PO_3^{2-}} \xrightarrow{NAD^+ \to NADH/H^+} \underset{\textbf{Glyceraldehyde 3-phosphate}}{OHC{-}CH(OH)CH_2O{-}PO_3^{2-}}$$

Note how the above reactions are written. It's common when writing biochemical transformations to show only the structure of the substrate, while abbreviating the structures of coenzymes (Section 15.12) and other reactants. The curved arrow intersecting the usual straight reaction arrow in the first step shows that ATP is also a reactant and that ADP is a product. The coenzyme *nicotinamide adenine dinucleotide (NAD^+)* is required in the second step, and *reduced nicotinamide adenine dinucleotide (NADH)* plus a proton are the products. We'll see shortly that NAD^+ is often involved as a biochemical oxidizing agent for converting alcohols to ketones.

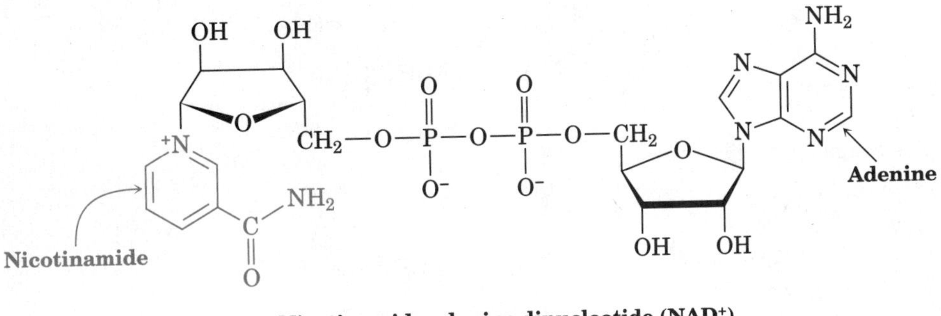

Nicotinamide

Adenine

Nicotinamide adenine dinucleotide (NAD^+)

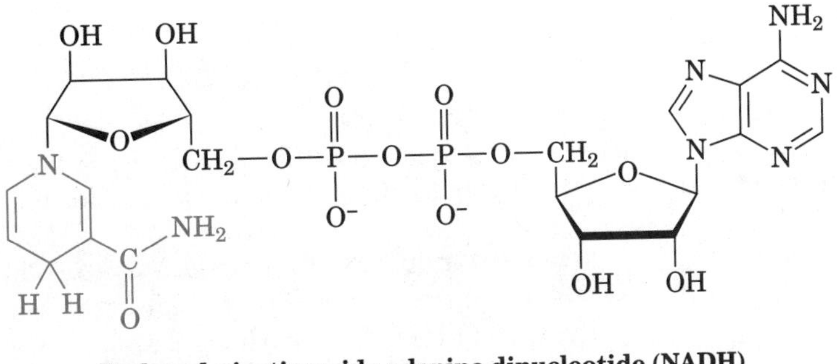

Reduced nicotinamide adenine dinucleotide (NADH)

Note also that glyceraldehyde 3-phosphate is written with its phosphate group dissociated; that is, $-OPO_3^{2-}$ rather than $-OPO_3H_2$. It's standard practice in writing biochemical structures to show carboxylic acids and phosphoric acids as their anions, since they exist in this form at the physiological pH found in organisms.

β-OXIDATION PATHWAY Series of four enzyme-catalyzed reactions that cleave two carbon atoms at a time from the end of a fatty acid chain

Fatty acids are catabolized by a repetitive four-step sequence of enzyme-catalyzed reactions called the *fatty acid spiral*, or **β-oxidation pathway**, shown in Figure 17.2. Each turn of the path results in the cleavage of a two-carbon acetyl group from the end of the fatty acid chain, until the entire molecule is ultimately degraded. As each acetyl group is produced, it enters the citric acid cycle and is further catabolized (Section 17.4).

$$RCH_2CH_2CH_2CH_2\overset{O}{\overset{\|}{C}}SCoA$$

Step 1
A double bond is introduced by enzyme-catalyzed removal of hydrogens from C2 and C3.

$$\xrightarrow{FAD \rightarrow FADH_2}$$

$$RCH_2CH_2CH{=}CH\overset{O}{\overset{\|}{C}}SCoA$$

Step 2
Water adds to the double bond to yield an alcohol.

$$\xrightarrow{H_2O}$$

$$RCH_2CH_2\overset{OH}{\overset{|}{C}}HCH_2\overset{O}{\overset{\|}{C}}SCoA$$

Step 3
The alcohol is oxidized to a ketone.

$$\xrightarrow{NAD^+ \rightarrow NADH/H^+}$$

$$RCH_2CH_2\overset{O}{\overset{\|}{C}}CH_2\overset{O}{\overset{\|}{C}}SCoA$$

Step 4
The bond between C2 and C3 is broken in a retro Claisen reaction to yield acetyl CoA and a chain-shortened fatty acid.

$$\xrightarrow{HSCoA}$$

$$RCH_2CH_2\overset{O}{\overset{\|}{C}}SCoA + CH_3\overset{O}{\overset{\|}{C}}SCoA$$

FIGURE 17.2 The reactions of the β-oxidation pathway, resulting in the cleavage of a two-carbon acetyl group from the end of the fatty acid chain.

STEP 1: INTRODUCTION OF A DOUBLE BOND The β-oxidation pathway begins when a fatty acid forms a thiol ester bond with acetyl CoA, and two hydrogen atoms from carbons 2 and 3 are removed by an *acyl CoA dehydrogenase* enzyme to yield an unsaturated acyl CoA. This kind of oxidation—the introduction of a conjugated double bond into a molecule—

occurs frequently in biochemical pathways and is usually carried out by the coenzyme *flavin adenine dinucleotide (FAD)*. Reduced $FADH_2$ is the by-product.

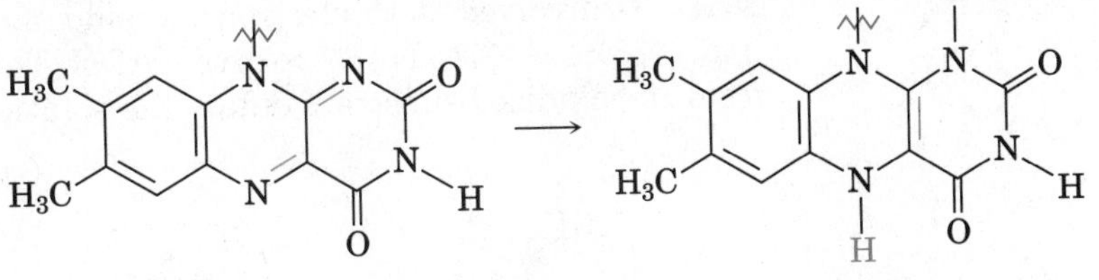

FAD → **$FADH_2$**

STEP 2: ADDITION OF WATER The unsaturated acyl CoA produced in step 1 reacts with water to yield a β-hydroxy acyl CoA in a process catalyzed by the enzyme *enoyl CoA hydratase*. Though we haven't specifically studied such processes in previous chapters, the reaction is closely analogous to the nucleophilic addition of water to a ketone to yield a hydrate (Section 9.8). In ketone hydration, water as nucleophile adds *directly* to a C=O group, yielding an alkoxide ion intermediate. In step 2 of the β-oxidation pathway, water as nucleophile adds to a double bond *conjugated with* a C=O group, yielding an enolate ion intermediate. In both reactions, the carbonyl oxygen atom withdraws electrons from the nearby carbon atom, making that carbon atom electrophilic.

Direct addition

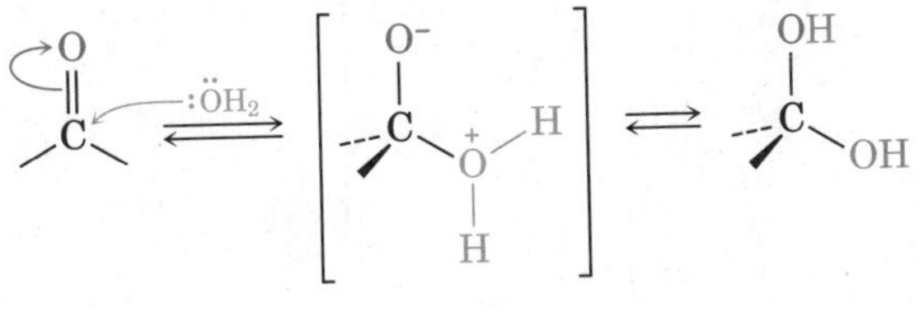

Conjugate addition

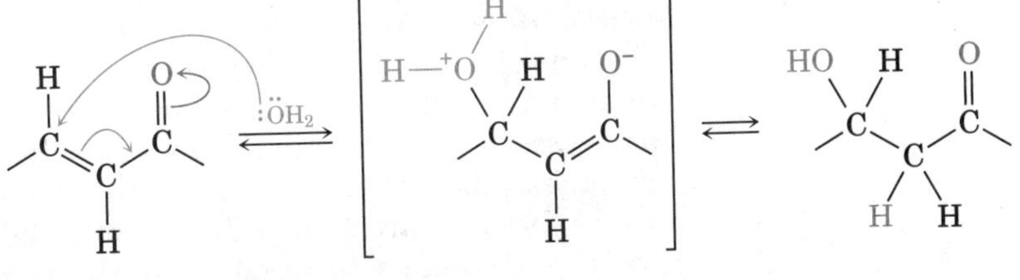

STEP 3: ALCOHOL OXIDATION The β-hydroxy acyl CoA from step 2 is oxidized to a β-keto acyl CoA in a reaction catalyzed by the enzyme *L-3-hydroxyacyl CoA dehydrogenase*. As in the oxidation of glycerol 1-phosphate to glyceraldehyde 3-phosphate mentioned earlier, alcohol oxidation in the β-oxidation pathway requires NAD^+ as a coenzyme and yields reduced $NADH/H^+$ as by-product.

It's useful when thinking about enzyme-catalyzed oxidation and reduction reactions to recognize that a hydrogen *atom* is equivalent to a hydrogen *ion*, H^+, plus an electron, e^-. Thus, for the two hydrogen atoms removed in the oxidation of an alcohol, 2 H atoms $= 2\ H^+ + 2\ e^-$. When NAD^+ is involved, both electrons accompany one H^+, in effect adding a hydride ion, $H:^-$, to NAD^+ to give NADH. The second hydrogen removed from the oxidized substrate enters the solution as H^+.

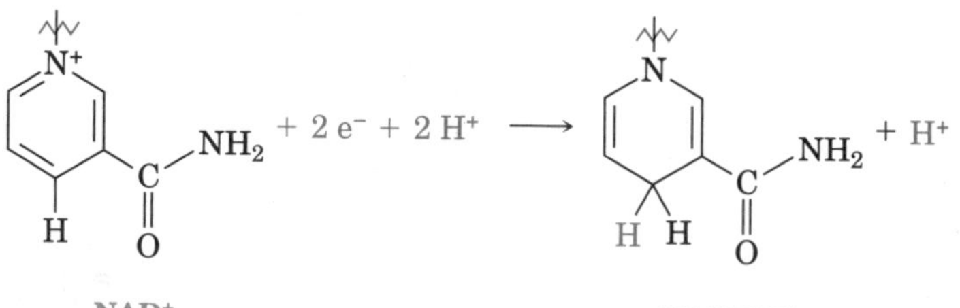

$NAD^+ + 2\ e^- + 2\ H^+ \longrightarrow NADH/H^+ + H^+$

NAD^+ NADH/H^+

The mechanism of oxidation of glycerol 1-phosphate has many analogies in the laboratory, and is similar in some respects to that of the hydration reaction in step 2. Thus, a hydride ion expelled from the alcohol acts as a nucleophile and adds to the $C{=}C{-}C{=}N^+$ part of NAD^+ in much the same way that water acts as a nucleophile and adds to the $C{=}C{-}C{=}O$ part of the unsaturated acyl CoA.

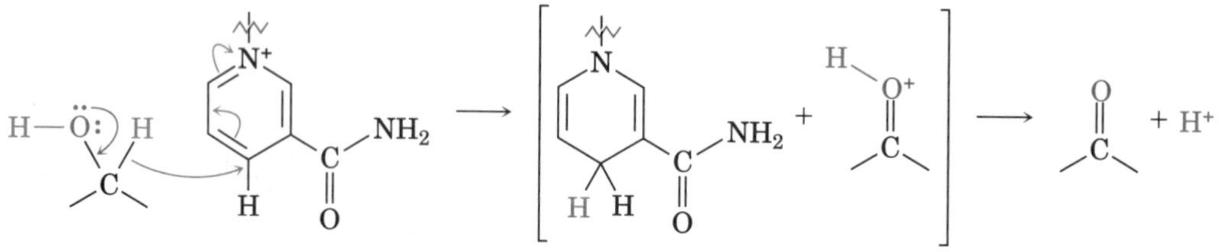

Alcohol NAD^+ NADH **Ketone**

STEP 4: CHAIN CLEAVAGE An acetyl group is split off from the acyl chain in the final step and is attached to a new coenzyme A molecule, leaving behind an acyl CoA that is two carbon atoms shorter. The reaction is catalyzed by the enzyme *β-ketothiolase* and is mechanistically the exact reverse of the Claisen condensation reaction discussed in Section 11.10. In the *forward* direction, a Claisen condensation joins two esters together to form a β-keto ester product. In the *reverse* direction, a retro Claisen reaction splits a β-keto ester (or β-keto thiol ester) apart to form two esters (or two thiol esters).

The reaction occurs by nucleophilic addition of coenzyme A to the keto group of the β-keto acyl CoA to yield an alkoxide ion intermediate, followed by cleavage of the C2–C3 bond with expulsion of an acetyl CoA enolate ion. Protonation of this ion gives acetyl CoA, and the chain-shortened acyl CoA enters another round of the pathway for further degradation.

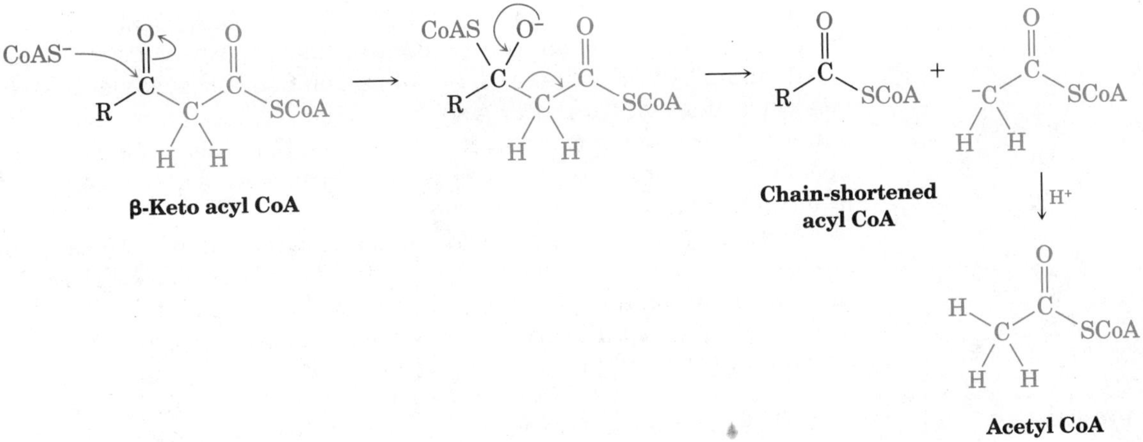

$$\text{CoAS}^- + \text{R}-\overset{\overset{\text{O}}{\|}}{\text{C}}-\text{CH}_2-\overset{\overset{\text{O}}{\|}}{\text{C}}-\text{SCoA} \longrightarrow \text{R}-\overset{\overset{\text{CoAS}\;\;\text{O}^-}{\backslash\;/}}{\text{C}}-\text{CH}_2-\overset{\overset{\text{O}}{\|}}{\text{C}}-\text{SCoA} \longrightarrow \text{R}-\overset{\overset{\text{O}}{\|}}{\text{C}}-\text{SCoA} + {}^-\text{CH}_2-\overset{\overset{\text{O}}{\|}}{\text{C}}-\text{SCoA}$$

β-Keto acyl CoA

Chain-shortened acyl CoA

$$\xrightarrow{H^+} \text{H}-\text{CH}_2-\overset{\overset{\text{O}}{\|}}{\text{C}}-\text{SCoA}$$

Acetyl CoA

Look at the catabolism of 14-carbon myristic acid shown in Figure 17.3 to see the overall results of the β-oxidation pathway. One turn of the path converts the 14-carbon myristyl CoA into the 12-carbon lauryl CoA plus acetyl CoA; a second turn converts lauryl CoA into the 10-carbon capryl CoA plus acetyl CoA; a third turn converts capryl CoA into the 8-carbon caprylyl CoA; and so on. Note that the last turn produces *two* molecules of acetyl CoA, since the precursor has 4 carbons.

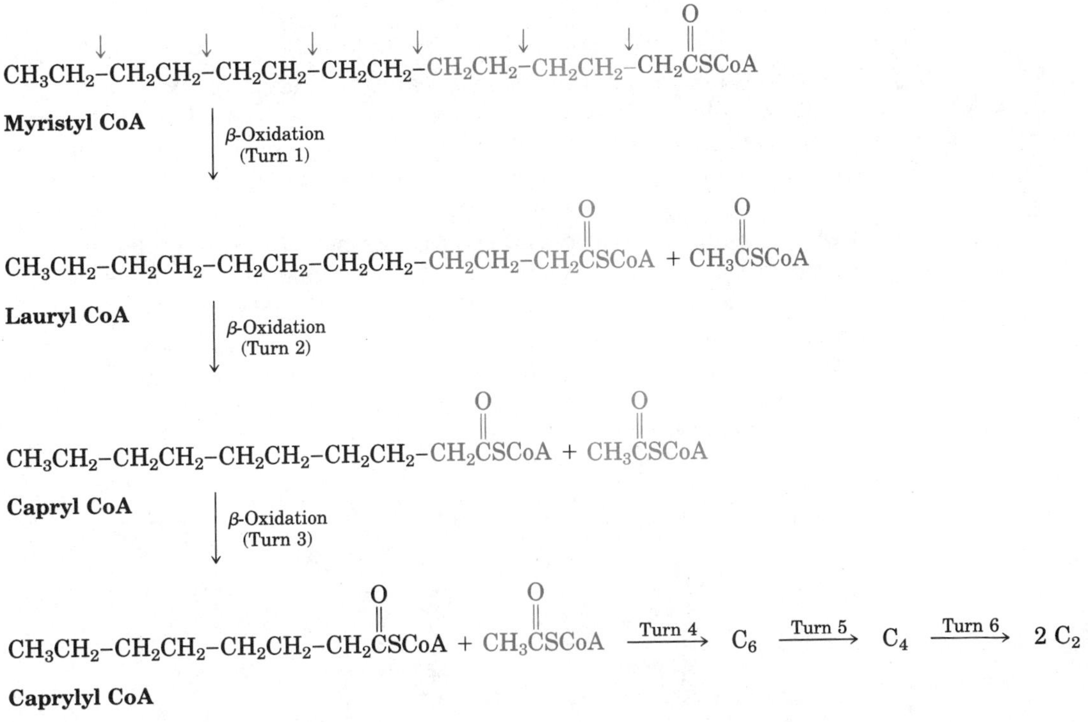

$$CH_3CH_2{\downarrow}CH_2CH_2{\downarrow}CH_2CH_2{\downarrow}CH_2CH_2{\downarrow}CH_2CH_2{\downarrow}CH_2CH_2{\downarrow}CH_2\overset{\overset{O}{\|}}{C}SCoA$$

Myristyl CoA

$$\Big\downarrow \text{β-Oxidation (Turn 1)}$$

$$CH_3CH_2-CH_2CH_2-CH_2CH_2-CH_2CH_2-CH_2CH_2-CH_2\overset{\overset{O}{\|}}{C}SCoA + CH_3\overset{\overset{O}{\|}}{C}SCoA$$

Lauryl CoA

$$\Big\downarrow \text{β-Oxidation (Turn 2)}$$

$$CH_3CH_2-CH_2CH_2-CH_2CH_2-CH_2CH_2-CH_2\overset{\overset{O}{\|}}{C}SCoA + CH_3\overset{\overset{O}{\|}}{C}SCoA$$

Capryl CoA

$$\Big\downarrow \text{β-Oxidation (Turn 3)}$$

$$CH_3CH_2-CH_2CH_2-CH_2CH_2-CH_2\overset{\overset{O}{\|}}{C}SCoA + CH_3\overset{\overset{O}{\|}}{C}SCoA \xrightarrow{\text{Turn 4}} C_6 \xrightarrow{\text{Turn 5}} C_4 \xrightarrow{\text{Turn 6}} 2\ C_2$$

Caprylyl CoA

FIGURE 17.3 Catabolism of 14-carbon myristic acid in the β-oxidation pathway yields seven molecules of acetyl CoA after six turns of the pathway.

You can predict how many molecules of acetyl CoA will be obtained from a given fatty acid simply by counting the number of carbon atoms and dividing by 2. For example, the 14-carbon myristic acid yields seven molecules of acetyl CoA after six turns of the β-oxidation pathway. The number of turns of the pathway is always 1 less than the number of acetyl CoA molecules produced, because the last turn cleaves a 4-carbon chain into two acetyl CoA's.

Most fatty acids have an even number of carbon atoms, so that none are left over after the β-oxidation pathway. Those fatty acids with an odd number of carbon atoms or with double bonds require additional steps for degradation, but all carbon atoms are ultimately released for further oxidation in the citric acid cycle (Section 17.4).

PROBLEM 17.3 Write the equations for the remaining turns of the β-oxidation pathway following those shown in Figure 17.3.

PROBLEM 17.4 How many molecules of acetyl CoA are produced by catabolism of these fatty acids, and how many turns of the β-oxidation pathway are needed?
(a) Palmitic acid, $CH_3(CH_2)_{14}COOH$ **(b)** Arachidic acid, $CH_3(CH_2)_{18}COOH$

17.3 CATABOLISM OF CARBOHYDRATES: GLYCOLYSIS

GLYCOLYSIS
Series of ten enzyme-catalyzed reactions that break down a glucose molecule into two pyruvate molecules

Glycolysis is a series of ten enzyme-catalyzed reactions that break down a glucose molecule into two pyruvate molecules, $CH_3COCO_2^-$. The steps of glycolysis, also called the *Embden–Meyerhoff pathway* after its discoverers, are summarized in Figure 17.4, pp. 518–519.

STEPS 1–3: PHOSPHORYLATION AND ISOMERIZATION Glucose produced in the digestion of dietary carbohydrates is first phosphorylated by reaction with ATP in a reaction catalyzed by the enzyme *hexokinase*. The glucose 6-phosphate that results is then isomerized by *phosphoglucose isomerase* to fructose 6-phosphate. As the open-chain structures in Figure 17.4 show, this isomerization reaction takes place by keto–enol tautomerism (Section 11.1), since both glucose and fructose share a common enol:

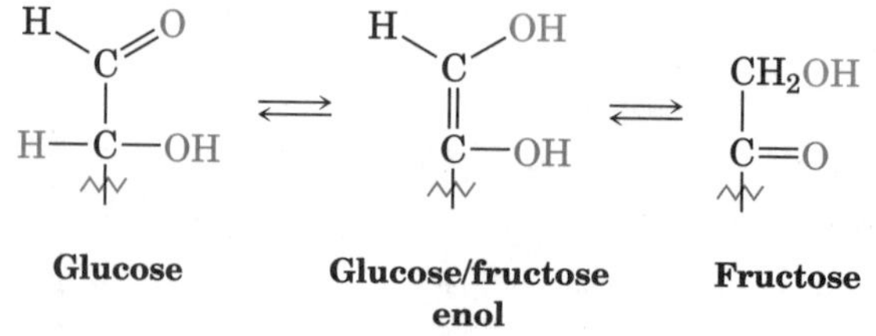

H–C(=O)–C(H)(OH)– ⇌ H(HO)C=C(OH)– ⇌ CH_2OH–C(=O)–

Glucose | Glucose/fructose enol | Fructose

Fructose 6-phosphate is then converted to fructose 1,6-diphosphate by *phosphofructokinase*-catalyzed reaction with ATP. The result is a molecule ready to be split into the two three-carbon intermediates that will ultimately become two molecules of pyruvate.

STEPS 4–5: CLEAVAGE AND ISOMERIZATION Fructose 1,6-diphosphate is cleaved in step 4 into two, three-carbon monophosphates, one an aldose and one a ketose. The bond between carbons 3 and 4 in fructose 1,6-diphosphate breaks, and a C=O group is formed. Mechanistically, this cleavage is the reverse of an aldol reaction (Section 11.8), and is carried out by an *aldolase* enzyme. A forward aldol reaction joins two ketones/aldehydes to give a β-hydroxy ketone or aldehyde; a retro aldol reaction cleaves a β-hydroxy ketone or aldehyde into two ketones/aldehydes:

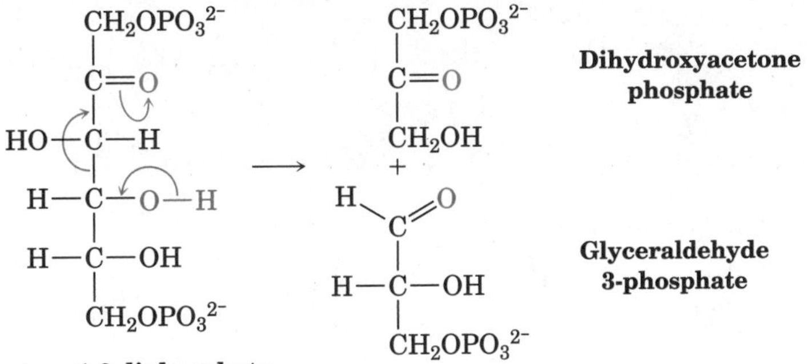

$CH_2OPO_3^{2-}$ | C=O | HO—C—H | H—C—O—H | H—C—OH | $CH_2OPO_3^{2-}$ ⟶ $CH_2OPO_3^{2-}$ | C=O | CH_2OH + H(C=O) | H—C—OH | $CH_2OPO_3^{2-}$

Fructose 1,6-diphosphate
(a β-hydroxy ketone)

Dihydroxyacetone phosphate

Glyceraldehyde 3-phosphate

Glyceraldehyde 3-phosphate continues on in the glycolysis pathway, but dihydroxyacetone phosphate must first be isomerized by the enzyme *triose phosphate isomerase*. As in the glucose-to-fructose conversion of step 2, the isomerization of dihydroxyacetone phosphate to glyceraldehyde 3-phosphate takes place by keto–enol tautomerization through a common enol. The net result of steps 4 and 5 is the production of *two* glyceraldehyde 3-phosphate molecules that each pass separately down the rest of the pathway. Thus, each of the remaining five steps of glycolysis takes place twice for every glucose molecule that enters at step 1.

STEPS 6–8: OXIDATION AND PHOSPHORYLATION Glyceraldehyde 3-phosphate is oxidized and phosphorylated by the coenzyme NAD^+ in the presence of the enzyme *glyceraldehyde 3-phosphate dehydrogenase* and phosphate ion, HPO_4^{2-} (abbreviated P_i). Transfer of a phosphate group from the carboxyl of 1,3-diphosphoglycerate to ADP then yields 3-phosphoglycerate, which is isomerized to 2-phosphoglycerate. The phosphorylation is catalyzed by *phosphoglycerate kinase*, and the isomerization is catalyzed by *phosphoglyceromutase*.

STEPS 9–10: DEHYDRATION AND DEPHOSPHORYLATION Like the β-hydroxy carbonyl compounds produced in aldol reactions (Section 11.9), 2-phosphoglycerate undergoes a ready dehydration by an E2 mechanism, yielding *phosphoenolpyruvate (PEP)*. The process is catalyzed by *enolase*.

Glucose

Step 1
Glucose is phosphorylated by reaction with ATP to yield glucose 6-phosphate.

ATP
ADP

Step 2
Glucose 6-phosphate is isomerized to fructose 6-phosphate.

Step 3
Fructose 6-phosphate is phosphorylated by reaction with ATP to yield fructose 1,6-diphosphate.

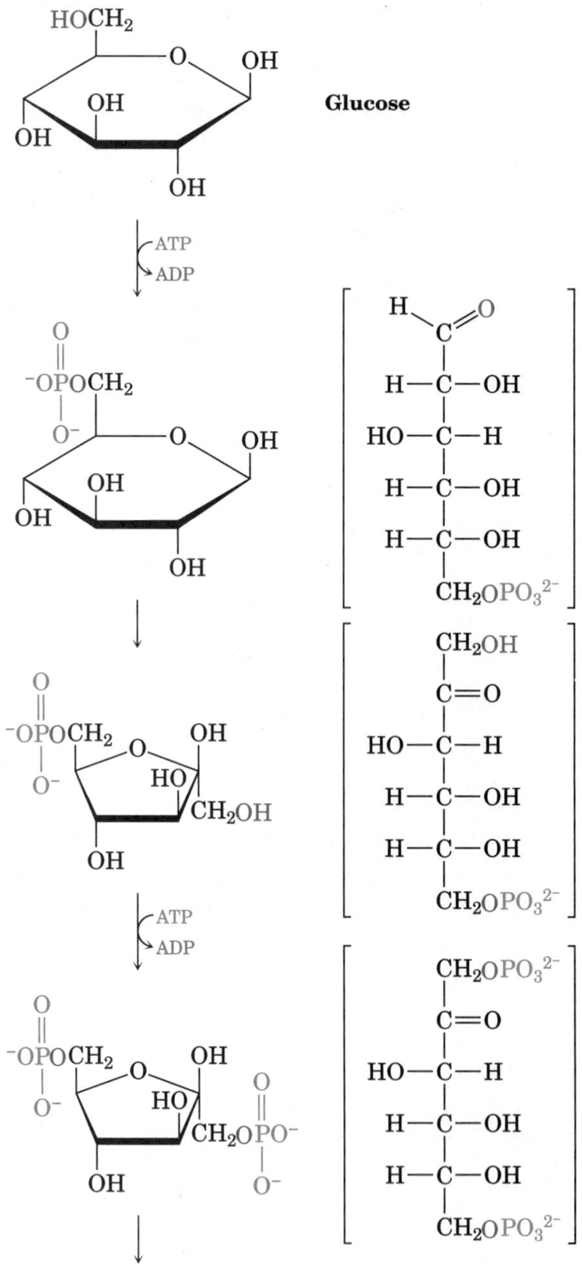

ATP
ADP

FIGURE 17.4 The ten-step glycolysis pathway for catabolizing glucose to pyruvate.

Step 4
Fructose 1,6-diphosphate is cleaved into two three-carbon pieces by the enzyme aldolase.

Step 5
Dihydroxyacetone phosphate, one of the products of step 4, is isomerized to glyceraldehyde 3-phosphate, the other product of step 4.

$$^{2-}O_3POCH_2CH(OH)CH(=O) \rightleftharpoons {}^{2-}O_3POCH_2C(=O)CH_2OH$$

Step 6
Glyceraldehyde 3-phosphate is oxidized to a carboxylic acid and then phosphorylated to yield 1,3-diphosphoglycerate.

NAD^+, P_i → $NADH/H^+$

$$^{2-}O_3POCH_2CH(OH)C(=O)OPO_3^{2-}$$

Step 7
A phosphate is transferred from the carboxyl group to ADP, resulting in synthesis of an ATP and yielding 3-phosphoglycerate.

ADP → ATP

$$^{2-}O_3POCH_2CH(OH)C(=O)O^-$$

Step 8
A phosphate group is transferred from the C3 hydroxyl to the C2 hydroxyl, giving 2-phosphoglycerate.

$$HOCH_2CH(OPO_3^{2-})C(=O)O^-$$

Step 9
Dehydration occurs, to yield phosphoenolpyruvate (PEP).

→ H_2O

$$H_2C{=}C(OPO_3^{2-}){-}C(=O)O^-$$

Step 10
A phosphate is transferred from PEP to ADP, yielding pyruvate and ATP.

ADP → ATP

$$CH_3C(=O){-}C(=O)O^-$$

$$\text{HOCH}_2\text{CH(OPO}_3^{2-}\text{)CO}_2^- \xrightarrow{\text{E2 reaction}} \text{H}_2\text{C=C(OPO}_3^{2-}\text{)CO}_2^- + \text{H}_2\text{O}$$

:Base

2-Phosphoglycerate
(a β-hydroxycarbonyl compound)

Phosphoenolpyruvate (PEP)

Transfer of the phosphate group to ADP then generates ATP and gives pyruvate, a reaction catalyzed by *pyruvate kinase.*

The net result of glycolysis can be summarized by the following equation:

$$\underset{\textbf{Glucose}}{C_6H_{12}O_6} + 2\ NAD^+ + 2\ P_i + 2\ ADP \longrightarrow \underset{\textbf{Pyruvate}}{2\ CH_3\overset{O}{\overset{\|}{C}}-\overset{O}{\overset{\|}{C}}O^-} + 2\ NADH + 2\ ATP + 2\ H_2O + 2\ H^+$$

Pyruvate produced in the catabolism of glucose can undergo several further transformations, depending on the conditions and on the organism. Most commonly, pyruvate is converted to acetyl CoA in an overall process that can be written as

$$\underset{\textbf{Pyruvate}}{CH_3\overset{O}{\overset{\|}{C}}-\overset{O}{\overset{\|}{C}}O^-} \xrightarrow[\text{HSCoA}]{\text{NAD}^+ \quad \text{NADH/H}^+} \underset{\textbf{Acetyl CoA}}{CH_3\overset{O}{\overset{\|}{C}}SCoA} + CO_2$$

The conversion of pyruvate to acetyl CoA is a complex, multistep sequence of reactions, requiring three different enzymes and four different coenzymes. All the individual steps are well understood and are well precedented by simple laboratory analogies, though their explanations are a bit outside the scope of this book.

PROBLEM 17.5 Identify the two pairs of steps in glycolysis in which ATP is produced.

PROBLEM 17.6 Propose a mechanism for the isomerization of dihydroxyacetone phosphate to glyceraldehyde 3-phosphate in step 5 of glycolysis.

17.4 THE CITRIC ACID CYCLE

The first two stages of catabolism result in the conversion of fats and carbohydrates into acetyl groups that are bonded through a thiol ester link to coenzyme A. These acetyl groups then enter the third stage, the

citric acid cycle, also called the *tricarboxylic acid (TCA) cycle* or *Krebs cycle* after Hans Krebs, who unraveled its complexities in 1937. The eight steps of the citric acid cycle and a brief description of each are given in Figure 17.5.

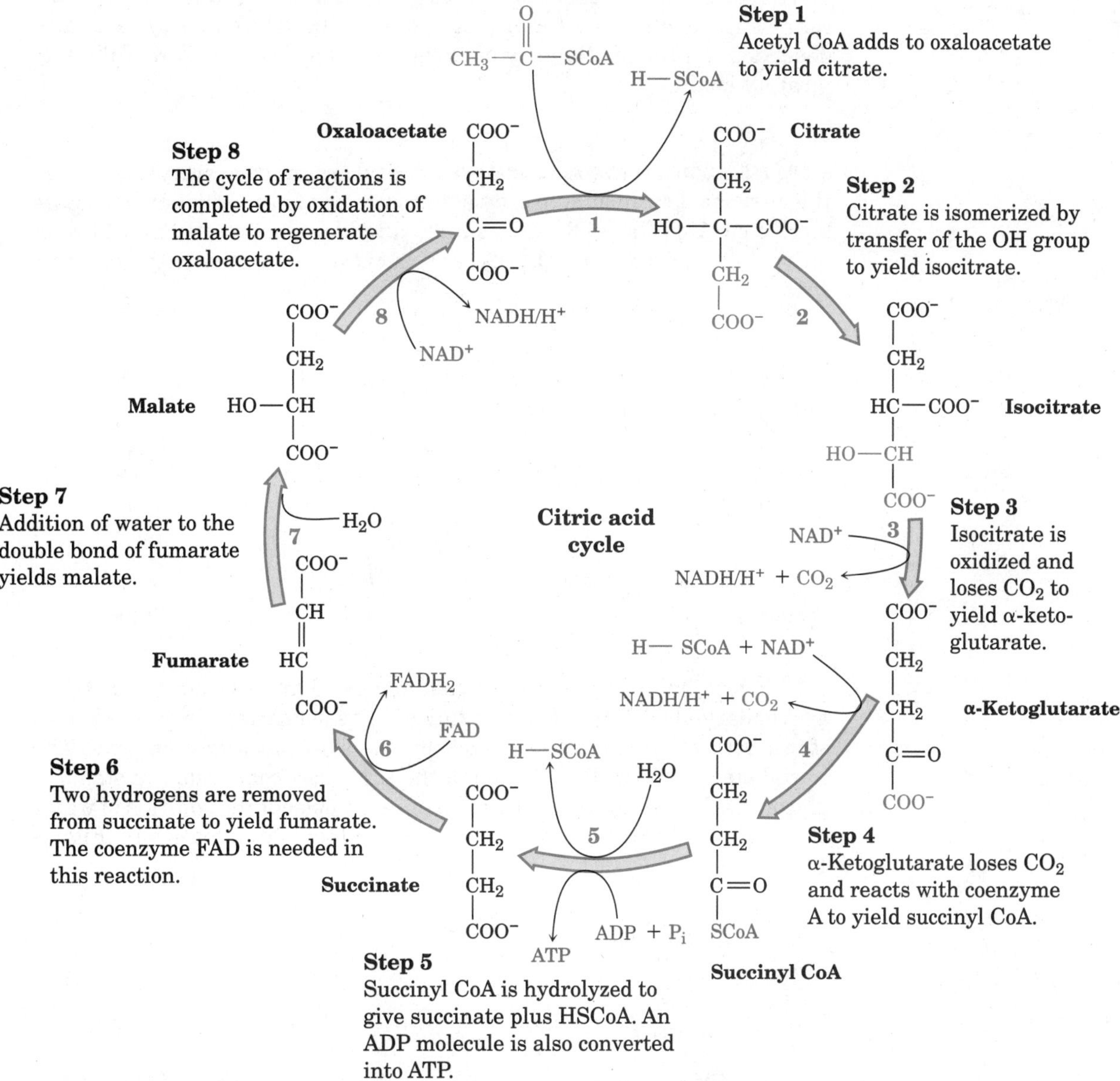

Step 1
Acetyl CoA adds to oxaloacetate to yield citrate.
Step 2
Citrate is isomerized by transfer of the OH group to yield isocitrate.
Step 3
Isocitrate is oxidized and loses CO_2 to yield α-ketoglutarate.
Step 4
α-Ketoglutarate loses CO_2 and reacts with coenzyme A to yield succinyl CoA.
Step 5
Succinyl CoA is hydrolyzed to give succinate plus HSCoA. An ADP molecule is also converted into ATP.
Step 6
Two hydrogens are removed from succinate to yield fumarate. The coenzyme FAD is needed in this reaction.
Step 7
Addition of water to the double bond of fumarate yields malate.
Step 8
The cycle of reactions is completed by oxidation of malate to regenerate oxaloacetate.
Citric acid cycle
Oxaloacetate
Citrate
Isocitrate
α-Ketoglutarate
Succinyl CoA
Succinate
Fumarate
Malate
H—SCoA
H_2O
NAD⁺
NADH/H⁺
NADH/H⁺ + CO_2
H— SCoA + NAD⁺
FAD
$FADH_2$
ADP + P_i
ATP

FIGURE 17.5 The citric acid cycle, an eight-step series of reactions that results in the conversion of an acetyl group into two molecules of CO_2 plus reduced coenzymes.

As its name implies, the citric acid cycle is a closed loop of reactions in which the product of the final step is a reactant in the first step. The intermediates are constantly regenerated and flow continuously through the cycle, which operates as long as the oxidizing coenzymes NAD^+ and FAD are available. To meet this condition, the reduced coenzymes NADH and $FADH_2$ must be reoxidized via the respiratory chain, which in turn relies on oxygen as the final electron acceptor. Thus, the cycle is also dependent on the availability of oxygen and on the operation of the respiratory chain.

STEPS 1–2: ADDITION TO OXALOACETATE Acetyl groups from acetyl CoA enter the citric acid cycle at step 1 by addition to the four-carbon oxaloacetate to give citrate. The addition is simply an aldol reaction (Section 11.8) of an enolate ion from acetyl CoA, and is catalyzed by the enzyme *citrate synthetase*.

$$\underset{\textbf{Oxaloacetate}}{O{=}C(CH_2COO^-)(COO^-)} + {}^-CH_2\overset{O}{\overset{\|}{C}}SCoA \longrightarrow \left[{}^-O{-}C(CH_2COO^-)(COO^-)(CH_2COO^-)\right] \longrightarrow \underset{\textbf{Citrate}}{HO{-}C(CH_2COO^-)(COO^-)(CH_2COO^-)}$$

Citrate, a tertiary alcohol, is then converted into its isomer, isocitrate, a secondary alcohol that can be oxidized. The isomerization occurs in two steps, both of which are catalyzed by the same *aconitase* enzyme. The initial step is an E2 dehydration of the same sort that occurs in step 9 of glycolysis (Figure 17.4). The second step is a nucleophilic addition of water of the same sort that occurs in step 2 of the β-oxidation pathway (Figure 17.2).

$$\underset{\textbf{Citrate}}{HO{-}C(CH_2COO^-)(COO^-)(CH_2COO^-)} \xrightarrow[\text{Aconitase}]{-\,H_2O} \underset{\textbf{Aconitate}}{{}^-OOCCH_2{-}C(COO^-){=}CH{-}COO^-} \xrightarrow[\text{Aconitase}]{+\,H_2O} \underset{\textbf{Isocitrate}}{{}^-OOCCH_2{-}CH(COO^-){-}CH(OH){-}COO^-}$$

STEPS 3–4: OXIDATIVE DECARBOXYLATIONS Isocitrate, a secondary alcohol, is oxidized by NAD^+ in step 3 to give a ketone intermediate, which loses CO_2 to give α-ketoglutarate. Catalyzed by the enzyme *isocitrate dehydrogenase*, the decarboxylation is a typical reaction of a carboxylic acid that has a second carbonyl group two atoms away (a β-keto acid). A similar kind of decarboxylation reaction occurs in the malonic ester synthesis (Section 11.6).

$$^{-}OOC{-}C({=}O){-}CH(CH_2COO^-){-}C({=}O){-}O^- \longrightarrow \left[\,^{-}OOC{-}C(O^-){=}CH{-}CH_2COO^-\,\right] \longrightarrow \,^{-}OOC{-}C({=}O){-}CH_2{-}CH_2COO^- + CO_2$$

α-Ketoglutarate

The transformation of α-ketoglutarate to succinyl CoA in step 4 is a multistep process analogous to the transformation of pyruvate to acetyl CoA that we saw in the previous section. Like the pyruvate conversion, the α-ketoglutarate conversion requires a number of different enzymes and coenzymes.

STEPS 5–6: DEHYDROGENATION OF SUCCINATE Succinyl CoA is hydrolyzed to succinate in step 5. The reaction is catalyzed by *succinyl CoA synthetase* and is coupled with phosphorylation of guanosine diphosphate (GDP) to give guanosine triphosphate (GTP). Succinate is then dehydrogenated by FAD and *succinate dehydrogenase* to give fumarate in a process analogous to that of the first step in the β-oxidation pathway.

STEPS 7–8: REGENERATION OF OXALOACETATE Catalyzed by the enzyme *fumarase*, nucleophilic addition of water to fumarate yields malate in a reaction similar to that of step 2 in the β-oxidation pathway. Oxidation with NAD^+ catalyzed by *malate dehydrogenase* then gives oxaloacetate, and the cycle has returned to its starting point, ready to revolve again. The net result of the citric acid cycle can be summarized as

$$\text{Acetyl CoA} + 3\,NAD^+ + FAD + ADP + P_i + 2\,H_2O \longrightarrow HSCoA + 3\,NADH + 3\,H^+ + FADH_2 + ATP + 2\,CO_2$$

PROBLEM 17.7 Which of the substances in the citric acid cycle are tricarboxylic acids (thus giving the cycle its alternate name)?

PROBLEM 17.8 Write mechanisms for step 2 of the citric acid cycle, the dehydration of citrate and the addition of water to aconitate.

17.5 CATABOLISM OF PROTEINS: TRANSAMINATION

The catabolism of proteins is much more complex than that of fats and carbohydrates, because each of the 20 amino acids is degraded through its own unique pathway. The general idea, however, is that the amino nitrogen atoms are removed, and the substances that remain are converted into compounds that enter the citric acid cycle.

TRANSAMINATION Reaction in which the $-NH_2$ group of an amine changes places with the keto group of an α-keto acid

Most amino acids lose their nitrogen atom by a **transamination** reaction in which the $-NH_2$ group of the amino acid changes places with the keto group of α-keto glutarate. The products are a new α-keto acid and glutamate:

$$\underset{\text{An amino acid}}{\overset{NH_3^+}{\overset{|}{R C}}HCOO^-} + \underset{\alpha\text{-Keto glutarate}}{{}^-OOCCH_2CH_2\overset{O}{\overset{\|}{C}}COO^-} \rightleftharpoons \underset{\text{An } \alpha\text{-keto acid}}{R\overset{O}{\overset{\|}{C}}COO^-} + \underset{\text{Glutamate}}{{}^-OOCCH_2CH_2\overset{NH_3^+}{\overset{|}{C}}HCOO^-}$$

Transaminations use pyridoxal phosphate, a derivative of vitamin B_6, as cofactor. As shown in Figure 17.6 for the reaction of alanine, the key step in transamination is nucleophilic addition of the amino acid $-NH_2$ group to the pyridoxal aldehyde group to yield an imine (Section 9.10). Loss of a proton from the α position then results in a bond rearrangement to give a new imine, which is hydrolyzed (the exact reverse of imine formation) to yield pyruvate and a nitrogen-containing derivative of pyridoxal phosphate. Pyruvate is converted into acetyl CoA (Section 17.3), which enters the citric acid cycle for further catabolism. The pyridoxal phosphate derivative transfers its nitrogen atom to α-keto glutarate by the exact reverse of the steps in Figure 17.6, thereby forming glutamate and regenerating pyridoxal phosphate for further use.

PROBLEM 17.9 Show the product from transamination of leucine.

INTERLUDE

BASAL METABOLISM

The minimum amount of energy an organism expends per unit of time to stay alive is called the organism's **basal metabolic rate (BMR)**. This rate is measured by monitoring respiration and finding the rate of oxygen consumption, which is proportional to the amount of energy used. Assuming an average dietary mix of fats, carbohydrates, and proteins, approximately 4.82 kcal are required for each liter of oxygen consumed.

The average basal metabolic rate for humans is about 65 kcal/hr, or 1600 kcal/day. Obviously, the rate varies for different people, depending on sex, age, weight, and physical condition. As a rule, the BMR is lower

$^{2-}O_3POCH_2$ H C O :NH_2 + CH_3CHCOO^- ^+N H OH CH_3

Pyridoxal phosphate

Nucleophilic attack of the amino acid on the pyridoxal phosphate carbonyl group gives an imine.

$^{2-}O_3POCH_2$ H H CH_3 C N C COO^- + H_2O ^+N H OH CH_3

Loss of a proton moves the double bonds and gives a second imine intermediate.

$^{2-}O_3POCH_2$ H CH_3 C N C COO^- N H OH CH_3

Hydrolysis of the imine then yields an α-keto acid along with a nitrogen-containing pyridoxal phosphate derivative.

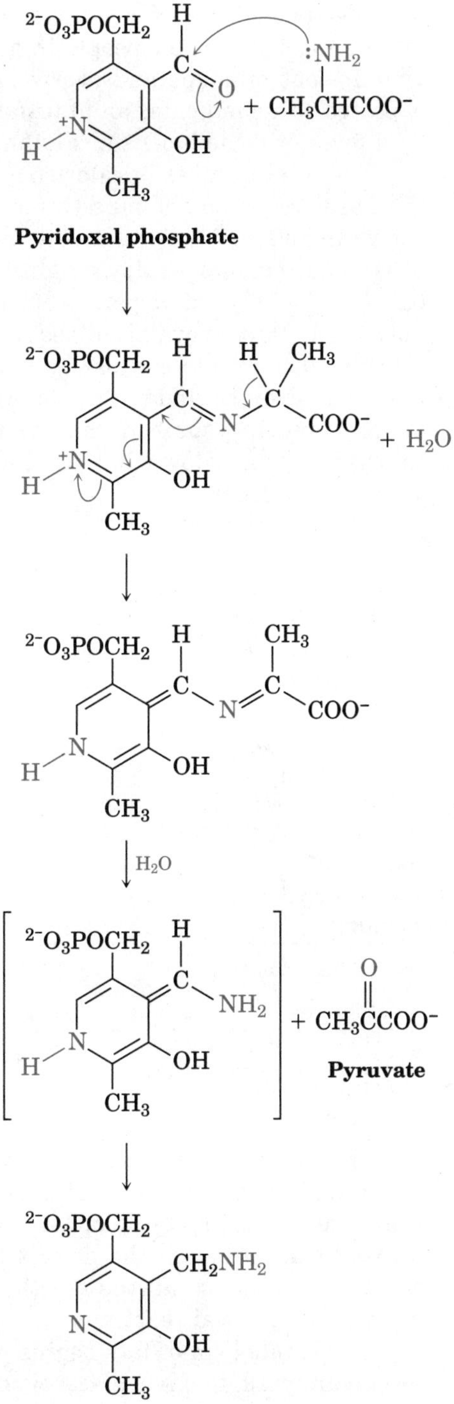

H_2O

$^{2-}O_3POCH_2$ H C NH_2 N H OH CH_3 + $CH_3\overset{O}{\overset{\|}{C}}COO^-$

Pyruvate

$^{2-}O_3POCH_2$ CH_2NH_2 N OH CH_3

FIGURE 17.6 Oxidative deamination of alanine requires the cofactor pyridoxal phosphate and yields pyruvate as product.

for older people than for younger people, is lower for females than for males, and is lower for people in good physical condition than for those who are out of shape and overweight. A BMR substantially above the expected value indicates an unusually rapid metabolism, perhaps caused by a fever or some biochemical abnormality.

The total number of calories a person needs each day is the sum of the basal requirement plus the energy used for physical activities, as shown in Table 17.1. A relatively inactive person needs about 30% above basal requirements per day, a lightly active person needs about 50% above basal, and a very active person such as an athlete or construction worker may need 100% above basal requirements. Some endurance athletes in ultradistance events can use as many as 10,000 calories per day above the basal level. Each day that your caloric intake is above what you use, fat is stored in your body and your weight rises. Each day that your caloric intake is below what you use, fat in your body is metabolized and your weight drops.

TABLE 17.1 Energy Cost of Various Activities[a]

ACTIVITY	ENERGY USED (kcal/min)
Sleeping	1.2
Sitting, reading	1.6
Standing still	1.8
Walking	3–6
Tennis	7–9
Basketball	9–10
Walking up stairs	10–18
Running	9–22

[a]For a 70 kg man.

SUMMARY AND KEY WORDS

Metabolism is the sum of all chemical reactions in the body. Reactions that break down large molecules into smaller fragments are called **catabolism**; reactions that build up large molecules from small pieces are called **anabolism**. Although the details of specific biochemical pathways are sometimes complex, all the reactions that occur follow the normal rules of organic chemical reactivity.

The catabolism of fats begins with **digestion**, in which ester bonds are hydrolyzed to give glycerol and fatty acids. The fatty acids are degraded in the four-step **β-oxidation pathway** by removal of two carbons at a time, yielding acetyl CoA. Catabolism of carbohydrates begins with the hydrolysis of glycoside bonds to give glucose, which is degraded in the ten-step **glycolysis** pathway. Pyruvate, the initial product of glycolysis, is then converted into acetyl CoA. The acetyl groups produced by degradation of fats and carbohydrates next enter the eight-step **citric acid cycle**, where they are further degraded into CO_2.

Protein catabolism is more complex than that of fats or carbohydrates, because each of the 20 different amino acids is degraded by its own unique pathway. In general, though, the amino nitrogen atoms are removed, and the substances that remain are converted into compounds that enter the citric acid cycle. Most amino acids lose their nitrogen atom by **transamination**, a reaction in which the $-NH_2$ group of the amino acid changes places with the keto group of an α-keto acid such as α-keto glutarate. The products are a new α-keto acid and glutamate.

The energy released in all catabolic pathways is used in the **respiratory chain** to make molecules of *adenosine triphosphate* (ATP). ATP, the final result of food catabolism, couples to and drives many otherwise unfavorable reactions.

ADDITIONAL PROBLEMS

17.10 What chemical events occur during the digestion of food?

17.11 What is the difference between digestion and metabolism?

17.12 What is the difference between anabolism and catabolism?

17.13 Draw the structure of adenosine *mono*phosphate (AMP), an intermediate in some biochemical pathways.

17.14 What general kind of reaction does ATP carry out?

17.15 What general kind of reaction does NAD^+ carry out?

17.16 What general kind of reaction does FAD carry out?

17.17 What substance is the starting point of the citric acid cycle, reacting with acetyl CoA in the first step and being regenerated in the last step?

17.18 Lactate, a product of glucose catabolism in oxygen-starved muscles, can be converted into pyruvate by oxidation. What coenzyme do you think is needed? Write the equation in the normal biochemical format using a curved arrow.

$$\begin{array}{c} \text{OH} \\ | \\ \text{CH}_3\text{CHCOO}^- \end{array} \quad \textbf{Lactate}$$

17.19 How many moles of acetyl CoA are produced by catabolism of these substances?
(a) 1.0 mol glucose **(b)** 1.0 mol palmitic acid ($C_{15}H_{31}CO_2H$) **(c)** 1.0 mol maltose

17.20 How many grams of acetyl CoA (mol wt = 809.6 amu) are produced by catabolism of these substances?
(a) 100.0 g glucose **(b)** 100.0 g palmitic acid **(c)** 100.0 g maltose

17.21 Which of the substances listed in Problem 17.20 is the most efficient precursor of acetyl CoA on a weight basis?

17.22 List the sequence of intermediates involved in the catabolism of glycerol from hydrolyzed fats to yield acetyl CoA.

17.23 Write the equation for the final step in the β-oxidation pathway of any fatty acid with an even number of carbon atoms.

17.24 Show the products of each of the following reactions:

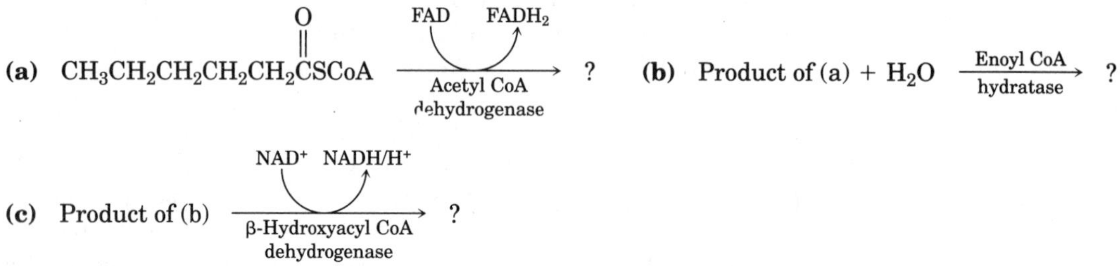

(a) $CH_3CH_2CH_2CH_2CH_2\overset{O}{\overset{\|}{C}}SCoA \xrightarrow[\text{Acetyl CoA dehydrogenase}]{\text{FAD} \rightarrow \text{FADH}_2} ?$

(b) Product of (a) + $H_2O \xrightarrow[\text{hydratase}]{\text{Enoyl CoA}} ?$

(c) Product of (b) $\xrightarrow[\beta\text{-Hydroxyacyl CoA dehydrogenase}]{\text{NAD}^+ \rightarrow \text{NADH/H}^+} ?$

17.25 What is the structure of the α-keto acid formed by transamination of each of these amino acids?
(a) Valine **(b)** Phenylalanine **(c)** Methionine

17.26 What enzyme cofactor is associated with transamination?

17.27 Fatty acids are synthesized in the body by the *lipogenesis* cycle that begins with acetyl CoA. The first step in lipogenesis is the condensation of two acetyl CoA molecules to yield acetoacetyl CoA, which undergoes three further enzyme-catalyzed steps, yielding butyryl CoA. Based on the kinds of reactions that occur in the β-oxidation pathway, what do you think the three further steps of lipogenesis are?

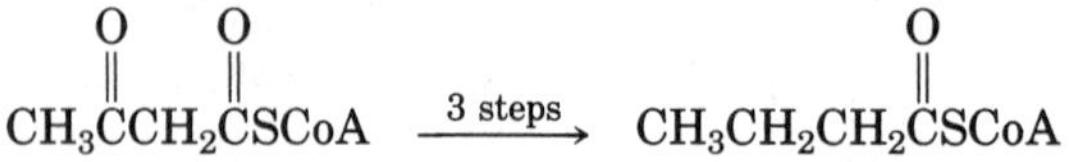

$$CH_3\overset{O}{\overset{\|}{C}}CH_2\overset{O}{\overset{\|}{C}}SCoA \xrightarrow{\text{3 steps}} CH_3CH_2CH_2\overset{O}{\overset{\|}{C}}SCoA$$

Acetoacetyl CoA **Butyryl CoA**

17.28 In the *pentose phosphate* pathway for degrading sugars, ribulose 5-phosphate is converted to ribose 5-phosphate. Propose a mechanism for the isomerization.

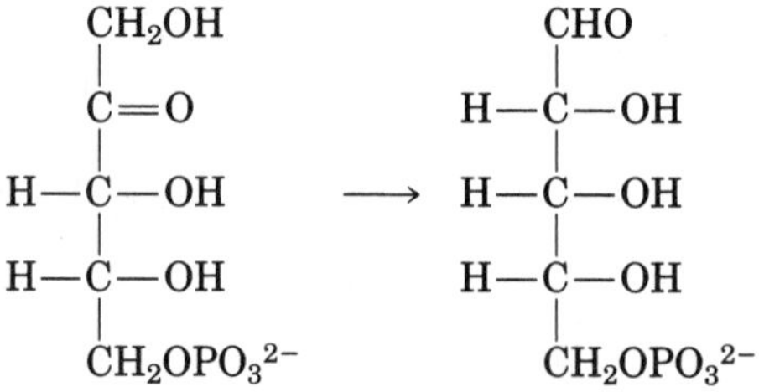

$$\begin{array}{c} CH_2OH \\ | \\ C{=}O \\ | \\ H{-}C{-}OH \\ | \\ H{-}C{-}OH \\ | \\ CH_2OPO_3^{2-} \end{array} \longrightarrow \begin{array}{c} CHO \\ | \\ H{-}C{-}OH \\ | \\ H{-}C{-}OH \\ | \\ H{-}C{-}OH \\ | \\ CH_2OPO_3^{2-} \end{array}$$

Ribulose 5-phosphate **Ribose 5-phosphate**

17.29 Another step in the pentose phosphate pathway for degrading sugars (see Problem 17.28) is the conversion of ribose 5-phosphate to glyceraldehyde 3-phosphate. What kind of organic process is occurring? Propose a mechanism for the conversion.

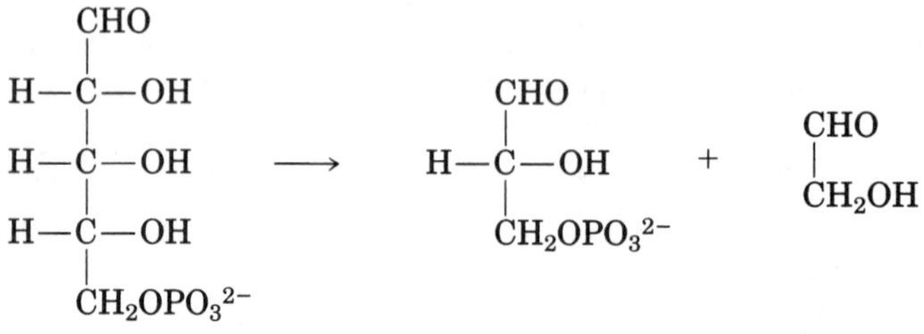

$$\begin{array}{c} CHO \\ | \\ H{-}C{-}OH \\ | \\ H{-}C{-}OH \\ | \\ H{-}C{-}OH \\ | \\ CH_2OPO_3^{2-} \end{array} \longrightarrow \begin{array}{c} CHO \\ | \\ H{-}C{-}OH \\ | \\ CH_2OPO_3^{2-} \end{array} + \begin{array}{c} CHO \\ | \\ CH_2OH \end{array}$$

Ribose 5-phosphate **Glyceraldehyde 3-phosphate**

17.30 One of the steps in the *gluconeogenesis* pathway for synthesizing glucose in the body is the reaction of pyruvate with CO_2 to yield oxaloacetate. Tell what kind of reaction is occurring and suggest a mechanism.

$$CO_2 + CH_3\overset{O}{\overset{\|}{C}}-\overset{O}{\overset{\|}{C}}O^- \longrightarrow {}^-O\overset{O}{\overset{\|}{C}}CH_2\overset{O}{\overset{\|}{C}}-\overset{O}{\overset{\|}{C}}O^-$$

Pyruvate **Oxaloacetate**

17.31 Another step in gluconeogenesis (see Problem 17.30) is the conversion of oxaloacetate to phosphoenolpyruvate by decarboxylation and phosphorylation. Propose a mechanism for the transformation.

$${}^-O\overset{O}{\overset{\|}{C}}CH_2\overset{O}{\overset{\|}{C}}-\overset{O}{\overset{\|}{C}}O^- \xrightarrow{ATP \to ADP} H_2C{=}\overset{{}^{2-}O_3PO}{\overset{|}{C}}-\overset{O}{\overset{\|}{C}}O^- + CO_2$$

Oxaloacetate **Phosphoenolpyruvate**

17.32 The primary fate of acetyl CoA under normal metabolic conditions is degradation in the citric acid cycle to yield CO_2. When the body is stressed by prolonged starvation, however, acetyl CoA is converted into compounds called *ketone bodies*, which can be used by the brain as a temporary fuel. The biochemical pathway for the synthesis of ketone bodies from acetyl CoA is shown:

$$2\ CH_3\overset{O}{\overset{\|}{C}}SCoA \xrightarrow{\nearrow HSCoA} CH_3\overset{O}{\overset{\|}{C}}CH_2\overset{O}{\overset{\|}{C}}SCoA \xrightarrow{H_2O \to ?} CH_3\overset{O}{\overset{\|}{C}}CH_2\overset{O}{\overset{\|}{C}}O^-$$

Acetyl CoA **Acetoacetyl CoA** **Acetoacetate**

Acetoacetate $\xrightarrow{?}$ $CH_3\overset{O}{\overset{\|}{C}}CH_3$ (**Acetone**)

Acetoacetate $\xrightarrow{? \to ?}$ $CH_3\overset{OH}{\overset{|}{C}}HCH_2\overset{O}{\overset{\|}{C}}O^-$ (**3-Hydroxybutyrate**)

Ketone bodies

Fill in the missing information represented by the four question marks.

17.33 The initial reaction in Problem 17.32, conversion of two molecules of acetyl CoA to one molecule of acetoacetyl CoA, is a Claisen reaction. Assuming that there is a base present, show the mechanism of the reaction.

APPENDIX

NOMENCLATURE OF POLYFUNCTIONAL ORGANIC COMPOUNDS

Judging from the number of incorrect names that appear in the chemical literature, it's probably safe to say that relatively few practicing organic chemists are fully conversant with the rules of organic nomenclature. Simple hydrocarbons and monofunctional compounds present few difficulties, because the basic rules for naming such compounds are logical and easy to understand. Problems are often encountered with polyfunctional compounds, however. Whereas most chemists could correctly identify hydrocarbon **1** as 3-ethyl-2,5-dimethylheptane, few could correctly identify polyfunctional compound **2**. Should we consider **2** as an ether? As an ethyl ester? As a ketone? As an alkene? It is, of course, all four, but it has only one correct name: ethyl 3-(4-methoxy-2-oxo-3-cyclohexenyl)propanoate.

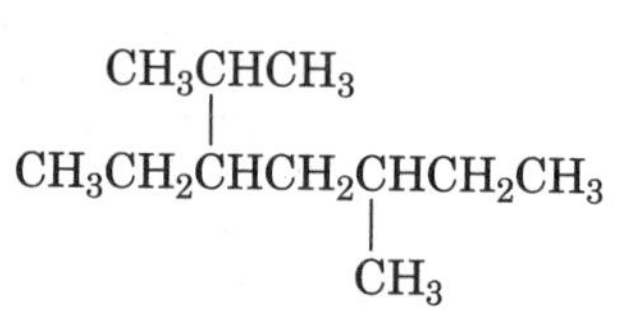

1. 3-Ethyl-2,5-dimethylheptane

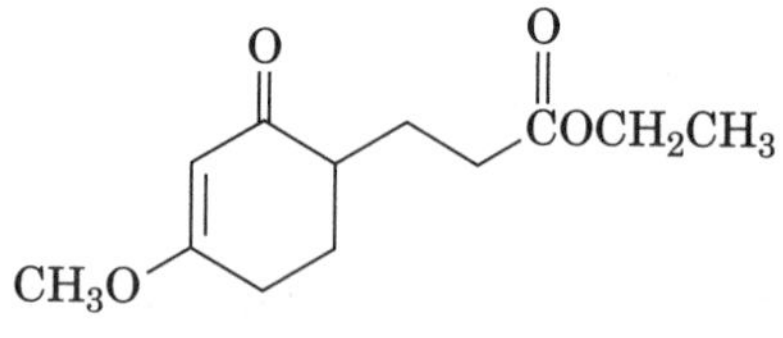

2. Ethyl 3-(4-methoxy-2-oxo-3-cyclohexenyl)propanoate

Naming polyfunctional organic compounds isn't really much harder than naming monofunctional ones. All that's required is a knowledge of nomenclature for monofunctional compounds and a set of additional rules. In the following discussion, it's assumed that you have a good command of the rules of nomenclature for monofunctional compounds that were given throughout the text as each new functional group was introduced. A list of where these rules can be found is given in Table A.1.

TABLE A.1 Where to Find Nomenclature Rules for Simple Functional Groups

FUNCTIONAL GROUP	TEXT SECTION	FUNCTIONAL GROUP	TEXT SECTION
Acid anhydrides	10.1	Amines	12.1
Acid halides	10.1	Aromatic compounds	5.4
Alcohols	8.1	Carboxylic acids	10.1
Aldehydes	9.3	Cycloalkanes	2.7
Alkanes	2.3	Esters	10.1
Alkenes	3.1	Ethers	8.1
Alkyl halides	7.1	Ketones	9.3
Alkynes	4.13	Nitriles	10.1
Amides	10.1		

The name of a polyfunctional organic molecule has four parts:

1. **Suffix**—the part that identifies the principal functional-group class to which the molecule belongs.
2. **Parent**—the part that identifies the size of the main chain or ring.
3. **Substituent prefixes**—parts that identify what substituents are located on the main chain or ring.
4. **Locants**—numbers that tell where substituents are located on the main chain or ring.

To arrive at the correct name for a complex molecule, you must identify the four name parts and then express them in the proper order and format. Let's look at the four parts.

The Suffix—Functional-Group Precedence

A polyfunctional organic molecule can contain many different kinds of functional groups, but for nomenclature purposes, we must choose just one suffix. It's not correct to use two suffixes. Thus, keto ester **3** must be named either as a ketone with an *-one* suffix or as an ester with an *-oate* suffix, but it can't be named as an *-onoate*. Similarly, amino alcohol **4** must be named either as an alcohol (*-ol*) or as an amine (*-amine*), but it can't properly be named as an *-olamine*. The only exception to this rule is in naming compounds that have double or triple bonds. For example, the unsaturated acid $H_2C{=}CHCH_2COOH$ is 3-butenoic acid, and the acetylenic alcohol $HC{\equiv}CCH_2CH_2CH_2CH_2OH$ is 5-hexyn-1-ol.

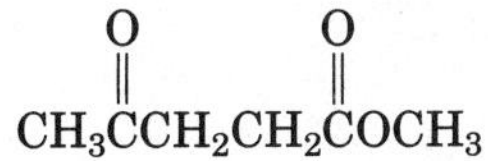

3. Named as an ester with a keto (oxo) substituent:
Methyl 4-oxopentanoate

$$\begin{array}{c}\text{OH}\\ |\\ CH_3CHCH_2CH_2CH_2NH_2\end{array}$$

4. Named as an alcohol with an amino substituent:
5-Amino-2-pentanol

How do we choose which suffix to use? Functional groups are divided into two classes, **principal groups** and **subordinate groups**, as shown in Table A.2. Principal groups are those that may be cited either as prefixes or as suffixes, while subordinate groups are those that may be cited only as prefixes. Within the principal groups, an order of precedence has been established. The proper suffix for a given compound is determined by identifying all the functional groups present and then choosing the principal group of highest priority. For example, Table A.2 indicates that keto ester **3** must be named as an ester rather than as a ketone, since an ester functional group is higher in priority than a ketone. Similarly, amino alcohol **4** must be named as an alcohol rather than as an amine. The correct name of **3** is methyl 4-oxopentanoate, and the correct name of **4** is 5-amino-2-pentanol. Further examples are shown below.

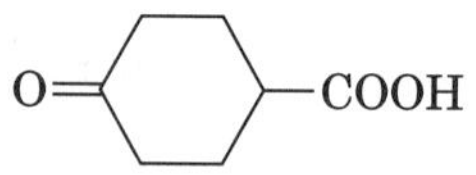

O=⟨cyclohexane⟩—COOH

5. Named as cyclohexanecarboxylic acid with an oxo substitutent:
4-Oxocyclohexanecarboxylic acid

$HOC(=O)—C(CH_3)_2—CH_2CH_2CH_2C(=O)Cl$

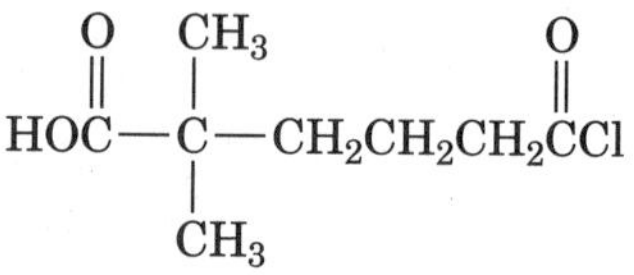

6. Named as a carboxylic acid with a chlorocarbonyl substituent:
5-Chlorocarbonyl-2,2-dimethylpentanoic acid

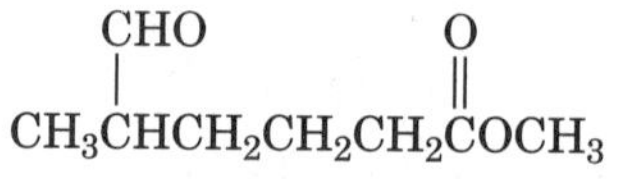

$CH_3CH(CHO)CH_2CH_2CH_2C(=O)OCH_3$

7. Named as an ester with an oxo substituent:
Methyl 5-methyl-6-oxohexanoate

The Parent—Selecting the Main Chain or Ring

The parent, or base, name of a polyfunctional organic compound is usually easy to identify. If the group of highest priority is part of an open chain, we simply select the longest chain that contains the largest number of principal functional groups. If the highest-priority group is attached to a ring, we use the name of that ring system as the parent. For example, compounds **8** and **9** are isomeric aldehydo acids, and both must be named as acids rather than as aldehydes according to Table A.2. The longest chain in compound **8** has seven carbons, and the substance is therefore named 6-methyl-7-oxoheptanoic acid. Compound **9** also has a chain of seven carbons, but the longest chain that contains both of the principal functional groups has only three carbons. The correct name of **9** is 3-oxo-2-pentylpropanoic acid.

$CH_3CH(CHO)CH_2CH_2CH_2CH_2C(=O)OH$

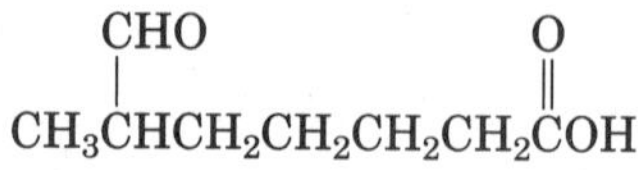

8. Named as a substituted heptanoic acid:
6-Methyl-7-oxoheptanoic acid

$CH_3CH_2CH_2CH_2CH_2CH(CHO)C(=O)OH$

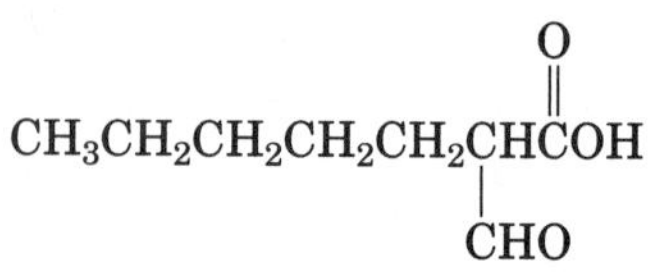

9. Named as a substituted propanoic acid:
3-Oxo-2-pentylpropanoic acid

TABLE A.2 Classification of Functional Groups for Purposes of Nomenclature[a]

FUNCTIONAL-GROUP CLASS	STRUCTURE	NAME WHEN USED AS SUFFIX	NAME WHEN USED AS PREFIX
PRINCIPAL GROUPS			
Carboxylic acids	—COOH	-oic acid -carboxylic acid	carboxy
Carboxylic anhydrides	$-\overset{O}{\overset{\Vert}{C}}-O-\overset{O}{\overset{\Vert}{C}}-$	-oic anhydride -carboxylic anhydride	
Carboxylic esters	—COOR	-oate -carboxylate	alkoxycarbonyl
Acid halides	—COCl	-oyl halide -carbonyl halide	halocarbonyl (haloformyl)
Amides	$-CONH_2$	-amide -carboxamide	amido
Nitriles	$-C\equiv N$	-nitrile -carbonitrile	cyano
Aldehydes	—CHO	-al -carbaldehyde	formyl
	=O		oxo (either aldehyde or ketone)
Ketones	=O	-one	oxo
Alcohols	—OH	-ol	hydroxy
Phenols	—OH	-ol	hydroxy
Thiols	—SH	-thiol	mercapto, sulfhydryl
Amines	$-NH_2$	-amine	amino
Imines	=NH	-imine	imino
Alkenes	C=C	-ene	
Alkynes	$C\equiv C$	-yne	
Alkanes	C—C	-ane	
SUBORDINATE GROUPS			
Ethers	—OR		alkoxy
Sulfides	—SR		alkylthio
Halides	—F, —Cl, —Br, —I		halo
Nitro	$-NO_2$		nitro
Azides	N=N=N		azido
Diazo	=N=N		diazo

[a]Principal functional groups are listed in order of decreasing priority, but the subordinate functional groups have no established priority order. Principal functional groups may be cited either as prefixes or as suffixes; subordinate functional groups may be cited only as prefixes.

Similar rules apply for compounds **10–13**, which contain rings. Compounds **10** and **11** are isomeric keto nitriles, and both must be named as nitriles according to Table A.2. Substance **10** is named as a benzonitrile since the –CN functional group is a substituent on the aromatic ring, but substance **11** is named as an acetonitrile since the –CN functional group is on an open chain. The correct names are 2-acetyl-4-methylbenzonitrile (**10**) and (2-acetyl-4-methylphenyl)acetonitrile (**11**). Compounds **12** and **13** are both keto acids and must be named as acids. The correct names are 3-(2-oxocyclohexyl)propanoic acid (**12**) and 2-(3-oxopropyl)cyclohexanecarboxylic acid (**13**).

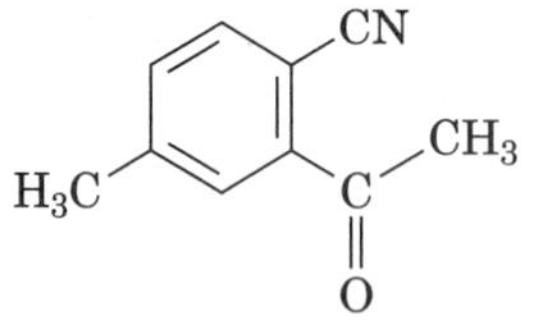

CN, CH_3, H_3C, C, O

10. Named as a substituted benzonitrile:
2-Acetyl-4-methylbenzonitrile

CH_2CN, CH_3, H_3C, C, O

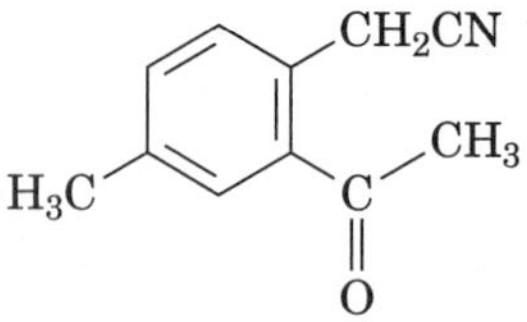

11. Named as a substituted acetonitrile:
(2-Acetyl-4-methylphenyl)acetonitrile

O, O, C, OH

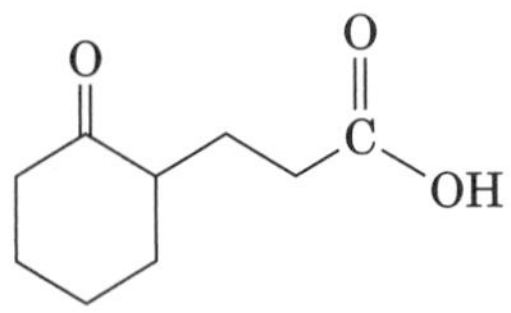

12. Named as a carboxylic acid:
3-(2-Oxocyclohexyl)propanoic acid

O, OH, C, O, C, H

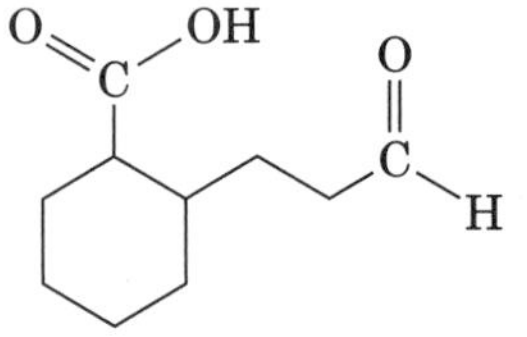

13. Named as a carboxylic acid:
2-(3-Oxopropyl)cyclohexanecarboxylic acid

The Prefixes and Locants

With the suffix and parent name established, the next step is to identify and number all substituents on the parent chain or ring. These substituents include all alkyl groups and all functional groups other than the one cited in the suffix. For example, compound **14** contains three different functional groups (carboxyl, keto, and double bond). Because the carboxyl group is highest in priority, and because the longest chain containing the functional groups is seven carbons long, **14** is a heptenoic acid. In addition, the main chain has an oxo (keto) substituent and three methyl groups. Numbering from the end nearer the highest-priority functional group, we find that **14** is 2,5,5-trimethyl-4-oxo-2-heptenoic acid. Note that the final *-e* of heptene is deleted in the word *heptenoic*. This deletion occurs when the name would have two adjacent vowels (thus, *heptenoic* has the final "e" deleted, but *heptenenitrile* retains the "e"). Look back at some of the other compounds we've named to see other examples of how prefixes and locants are assigned.

$$CH_3CH_2{-}C(CH_3)_2{-}C(=O)CH{=}C(CH_3)C(=O)OH$$

14. Named as a heptenoic acid:
2,5,5-Trimethyl-4-oxo-2-heptenoic acid

Writing the Name

Once the name parts have been established, the entire name is written out. Several additional rules apply:

1. **Order of prefixes:** When the substituents have been identified, the main chain has been numbered, and the proper multipliers such as *di-* and *tri-* have been assigned, the name is written with the substituents listed in alphabetical, rather than numerical, order. Multipliers such as *di-* and *tri-* are not used for alphabetization purposes, but the prefix *iso-* is used.

$$H_2NCH_2CH_2CH(CH_3)CH(OH)CH_3$$

15. 5-Amino-3-methyl-2-pentanol
(*NOT* 3-methyl-5-amino-2-pentanol)

2. **Use of hyphens; single- and multiple-word names:** The general rule is to determine whether the principal functional group is itself an element or compound. If it is, then the name is written as a single word; if it isn't, then the name is written as multiple words. For example, methylbenzene (one word) is correct because the parent—benzene—is itself a compound. Diethyl ether, however, is written as two words because the parent—ether—is a class name rather than a compound name. Some further examples are shown below:

$$H_3C{-}Mg{-}CH_3$$

16. Dimethylmagnesium
(one word, since magnesium is an element)

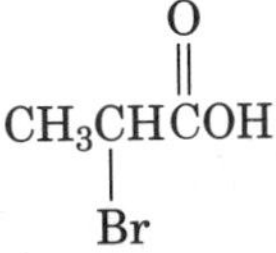

17. 2-Bromopropanoic acid
(two words, since "acid" is not a compound)

$$(H_3C)_2N{-}C_5H_4N \text{ (pyridine ring, N at position 4)}$$

18. 4-(Dimethylamino)pyridine
(one word, since pyridine is a compound)

$$C_5H_9{-}C(=O)OCH_3$$

19. Methyl cyclopentanecarboxylate

3. **Parentheses:** Parentheses are used to denote complex substituents when ambiguity would otherwise arise. For example, chloromethylbenzene has two substituents on a benzene ring, but (chloromethyl)benzene has only one complex substituent. Note that the expression in parentheses is not set off by hyphens from the rest of the name.

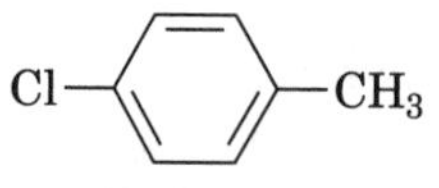

Cl—C_6H_4—CH_3

20. *p*-Chloromethylbenzene
(two substituents)

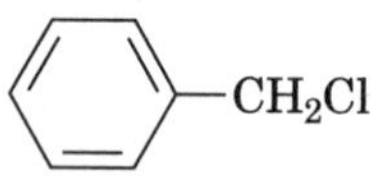

C_6H_5—CH_2Cl

21. (Chloromethyl)benzene
(one complex substituent)

$$\begin{array}{c} \text{O} \qquad\qquad\quad \text{O} \\ \| \qquad\qquad\quad\; \| \\ \text{HOCCHCH}_2\text{CH}_2\text{COH} \\ | \\ \text{CH}_3\text{CHCH}_2\text{CH}_3 \end{array}$$

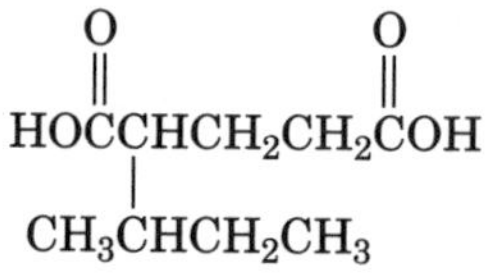

22. 2-(1-Methylpropyl)pentanedioic acid
(The 1-methylpropyl group is a complex substituent on C2 of the main chain.)

Additional Reading

Further explanations of the rules of organic nomenclature can be found in the following references:

1. J. H. Fletcher, O. C. Dermer, and R. B. Fox, "Nomenclature of Organic Compounds: Principles and Practice," Advances in Chemistry Series No. 126, American Chemical Society, Washington, D. C., 1974.
2. International Union of Pure and Applied Chemistry, "Nomenclature of Organic Chemistry, Sections A, B, C, D, E, F, and H," Pergamon Press, Oxford, 1979.
3. J. G. Traynham, "Organic Nomenclature: A Programmed Introduction," Prentice–Hall, Englewood Cliffs, NJ, 1985.

INDEX

The page references given in color refer to pages where terms are defined.